KIRK'S FIRE INVESTIGATION

University Centre at
Blackburn
College

Telephone: 01254 292165

Please return this book on or before the last date shown

KIRK'S FIRE INVESTIGATION
Seventh Edition

John D. DeHaan, Ph.D.
Fire-Ex Forensics, Inc.
Vallejo, California

David J. Icove, Ph.D., P.E.
The University of Tennessee
Knoxville, Tennessee

Pearson

Boston Columbus Indianapolis New York San Francisco Upper Saddle River Amsterdam
Cape Town Dubai London Madrid Milan Munich Paris Montreal Toronto Delhi
Mexico City Sao Paulo Sydney Hong Kong Seoul Singapore Taipei Tokyo

Library of Congress Control Number: 2011923252

Publisher: Julie Levin Alexander
Publisher's Assistant: Regina Bruno
Executive Editor: Marlene McHugh Pratt
Senior Acquisitions Editor: Stephen Smith
Development Editor: Pamela A. Powell
Associate Editor: Monica Moosang
Editorial Assistant: Samantha Sheehan
Director of Marketing: Dave Gesell
Marketing Manager: Brian Hoehl
Marketing Specialist: Michael Sirinides
Marketing Assistant: Crystal Gonzalez
Managing Production Editor: Patrick Walsh
Production Liaison: Julie Boddorf
Production Editor: Lisa S. Garboski, bookworks
Manufacturing Manager: Ilene Sanford
Manufacturing Buyer: Alan Fischer
Art Director: Christopher Weigand
Cover Designer: Karen Salzbach
Cover Photo: Jamie Novak, St. Paul Fire Department
Composition: Aptara®, Inc.
Printing and Binding: R.R. Donnelley/Willard
Cover Printer: Lehigh-Phoenix Color/Hagerstown

PEARSON

Education

Credits and acknowledgments borrowed from other sources and reproduced, with permission, in this textbook appear on appropriate pages within the text. Unless otherwise stated, all photos have been provided by the authors.

Note: A portion of this book was prepared by David J. Icove, a former employee of the Tennessee Valley Authority (TVA). The views expressed in this book are the views of Dr. Icove and his coauthor. They are not necessarily the views of TVA or the United States Government. Reference in this book to any product, process, or service by trade name, trademark, manufacturer, or otherwise does not necessarily constitute its endorsement or recommendation by TVA or the United States Government.

Library of Congress Cataloging-in-Publication Data
DeHaan, John D. (John David)
    Kirk's fire investigation / John D. DeHaan, David J. Icove.—7th ed.
        p. cm.
    Includes bibliographical references and index.
    ISBN-10: 0-13-508263-3
    ISBN-13: 978-0-13-508263-8
1. Fire investigation.   I. Kirk, Paul Leland, 1902–1970.   II. Icove, David J., 1949-III. Title.   IV. Title: Fire investigation.

    TH9180.K5 2012
    363.37'65—dc22   2011005408

Pearson® is a registered trademark of Pearson PLC.

10 9 8 7 6 5 4 3 2

ISBN 10:   0-13-508263-3
ISBN 13: 978-0-13-508263-8

www.pearsonhighered.com

For Mick Gardiner, in recognition of his unlimited dedication to improving fire investigation across the UK and the rest of Europe through better training.

To the memory of Monty McGill—friend, colleague, teacher, and investigator of great integrity, skill, and courage.

CONTENTS

Chapter 3 Fundamentals of Fire Behavior and Building Construction 32

Basic Combustion 33

Flaming Fire 34

Structure of Flames 36

Smoldering Fire 38

Explosive Combustion 41

Heat 42

Flame Plume 51

Sequence of a Room Fire 54

Flow of Hot Gases 63

Effects of Environmental Conditions 65

Elements of Building Construction 69

Chapter 4 Combustion Properties of Liquid and Gaseous Fuels 85

Chapter 5 Combustion Properties of Solid Fuels 123

Chapter 6 Sources of Ignition 167

Chapter 7 Structure Fires and Their Investigation 231

Chapter 8 Wildland Fires and Their Investigation 327

Chapter 10 Electrical Causes of Fires 401

Chapter 13 Chemical Fires and Hazardous Materials 539

Chapter 16 Arson as a Crime 655

Chapter 17 Other Investigative Topics 695

PREFACE

The last 25 years have seen a dramatic change in the standards of performance expected of fire and explosion investigators. Such changes have been brought about by Supreme Court decisions, a development of professional standards including several certification programs, and acceptance of published texts and guides such as *NFPA 921*. These years have also seen a dramatic improvement in the frequency and quality of interactions among fire investigators, fire scientists, and engineers involved in fire safety and fire protection. The intuitive extrapolation or interpolation of data to explain fire development or fire indicators had been standard practice among fire investigators, and it has been faulty far too often. The scientific method has finally been recognized as the core analytical process that leads to accurate and defensible conclusions in fire investigations. That method, however, requires reliable data and information, which often have been lacking in fire topics. The integration of the wealth of information, knowledge, and experience of fire engineers and of those scientists involved in the chemistry and physics of fire development into fire investigation has proceeded along many paths—personal, educational, and professional—and on an international basis. The ATF Fire Research Laboratory at Beltsville, Maryland; the Center for Fire Research at NIST; the Fire Research Station in the United Kingdom; and numerous private researchers have all contributed significantly to fire investigations for many years. Recently, Vyto Babrauskas published a virtual encyclopedia of information about the ignition and combustion of materials. The *Ignition Handbook* is a comprehensive summary of decades of fire research in an accessible, fully referenced source. This seventh edition of *Kirk's Fire Investigation* includes new material reflecting "new" knowledge from that remarkable book. Fire engineers are now involved directly in investigations and also teach investigators how to apply fire engineering principles.

The National Fire Protection Association's *NFPA 921: Guide for Fire and Explosion Investigations* has focused the attention of investigators and the legal profession on the scientific principles behind investigation. Coauthors David Icove and John DeHaan were active technical consultant members in the development of *NFPA 921* (DeHaan from 1991 to 1999, and Icove from 1990 to date), and this seventh edition of *Kirk's Fire Investigation* reflects a closer parallel between practices and information in both sources that can only enhance the accuracy and reliability of all investigations. Correlations are also offered to information in *NFPA 1033: Standard for Professional Qualifications for Fire Investigator*.

Eight years ago, a companion book for this text was released. *Forensic Fire Scene Reconstruction* was created to explore more of the engineering principles behind fire behavior and the mechanisms of production of many of the post-fire indicators that *Kirk's* describes. Taken together, these books provide a sound basis for the fire expert to use when evaluating fire scenes, preparing reports, or offering testimony. Today's investigators are being held to a higher standard of professional practice than ever before. It is no longer adequate to claim expertise based on years of experience alone. A professional must demonstrate that what he or she is doing follows the practices *and* the knowledge base of the relevant professional community. Such knowledge and practices are based on texts such as this one, *Fire Scene Reconstruction*, and *NFPA 921*, which are continuously peer reviewed and revised to reflect the most current knowledge. The revisions in this edition follow the same path as in previous editions but include many new photographs, published experimental data, and case examples. There is revised material on ignition and fire dynamics, supported by new references and color photos. This text reflects an international database as offered by fire and explosion investigators, scientists, and engineers from all over the world. It is offered in the hope it will augment the knowledge and improve the skills of investigators everywhere and help them find the right answers for the right reasons.

It is hard to believe that it has been 30 years since John DeHaan took over responsibility for the text that what was then *Fire Investigation* by Dr. Paul Kirk. So many years have passed that a whole new generation of fire investigators is now practicing, many of whom have asked, Why is it called *Kirk's Fire Investigation*? It is clear they are not aware of Professor Kirk's contribution to the discipline. Paul L. Kirk was a professor of biochemistry and criminalistics at the University of California at Berkeley, but it was his specialty of microchemistry that focused his attention on physical evidence and its analysis. Professor Kirk was part of the Manhattan Project (where separation and identification of trace quantities of particular chemicals was a critical step in the development of the atomic bomb). After the war, he focused on analytical chemistry as an adjunct to criminal investigation. He was in charge of the criminalistics program at Berkeley until his death in 1970 and launched the careers of many criminalists who now practice around the world. He wrote the landmark text *Crime Investigation* in 1953 and maintained a private criminalistics consulting practice. It was this practice that led to his involvement in fire and arson investigation, where he was consulted in a wide variety of fire and explosion cases. He published *Fire Investigation* in 1969 as the first textbook on fire investigation written by a scientist rather than a field investigator. It became a standard reference and was still in print some 11 years after his death. His concern with using science to solve the puzzles of fire and explosion presaged the current emphasis on using the scientific method to investigate fires by more than 30 years. It is clear that good, knowledgeable investigators have been using that approach for years, even if they were not aware of it.

In honor of Dr. Kirk's pioneering work in bringing science to fire investigation, his name is included in the title, and the spirit, of this text.

J. D. DeHaan, Ph.D., F-ABC, IAAI-CFI, FSSDip, CFEI-NAFI
D. J. Icove, Ph.D., P.E., CFEI

What's New

The Seventh Edition is one of the most adventurous editions over the last decade. John DeHaan has been joined by David Icove to produce the keystone textbook in the fire investigation field. The following highlights are changes to this edition that set it apart from previous ones.

- Completely updated chapters with learning objectives
- Reference tracking of the National Fire Academy–developed Fire and Emergency Services Higher Education (FESHE) curriculum
- New case examples and results of recent fire tests
- Substantial new artwork and photographs, many in color
- Updated bibliographic references and appendices, which can be found on the MyFireKit for this text

PEARSON
myfirekit

For additional review materials, appendices, and suggested readings, visit **www.bradybooks.com** and follow the **MyBradyKit** link to register for book-specific resources.

Register for **MyFireKit** by following directions on the **MyFireKit** student access card provided with this text. If there is no card, go to **www.bradybooks.com** and follow the **MyBradyKit** link to Buy Access from there.

As an added bonus, *Kirk's Fire Investigation,* Seventh Edition, features a MyFireKit, which provides a one-stop shop for online review materials, appendices, suggested readings, chapter support materials, and other resources.

You can prepare for class and exams with multiple-choice and matching questions, Web links, study aids, and more! To register for MyFireKit for this text, please visit www.bradybooks.com and follow the MyBradyKit link.

Scope of This Book

Kirk's Fire Investigation is divided into the following chapters.

Chapter 1, *"Introduction,"* describes the field of *fire investigation,* which is the formal process of determining the origin, cause, and development of a fire or explosion.

Chapter 2, *"The Elementary Chemistry of Combustion,"* describes how a fire is a chemical reaction that produces physical effects. This chapter also introduces some of the concepts and terms that the fire investigator might encounter when reviewing laboratory reports or meeting with experts.

Chapter 3, **"Fundamentals of Fire Behavior and Building Construction,"** describes fire as an exothermic oxidation reaction that proceeds at such a rate that it generates detectable heat and light. This chapter also describes the basic elements of building construction and materials to give the fire investigator an understanding of how they can affect fire growth and patterns.

Chapter 4, **"Combustion Properties of Liquid and Gaseous Fuel,"** discusses some of the concepts that are basic to understanding how gaseous and liquid fuels burn, the conditions and limitations that apply to the combustion of such fuels, and the conventional methods of expressing combustion properties in terms of laboratory tests.

Chapter 5, **"Combustion Properties of Solid Fuels,"** describes the more complex ignition and combustion of solid fuels, which usually depend on pyrolysis to create combustible gases and vapors.

Chapter 6, **"Sources of Ignition,"** describes a wide variety of heat sources and their thermal properties. This chapter also discusses the fundamental processes of heat transfer, heat release rate, fire propagation, and assessment that must be applied to each possible situation.

Chapter 7, **"Structure Fires and Their Investigation,"** describes the principles of fuels, ignition, and fire behavior with which investigators should be reasonably familiar before undertaking the probe of a fire. This chapter also discusses the necessity of having a clear understanding of the purposes and goals of the investigation and a rational, orderly plan for carrying it out to meet those purposes, as well as the value and limitations of post-fire indicators and the basic physical processes that create them.

Chapter 8, **"Grass and Wildland Fires and Their Investigation,"** describes how the investigator who understands fuels, fire behavior, and the effects of environmental conditions is in a better position to interpret the subtle and sometimes delicate signs of fire patterns in wildland fires and therefore is better able to identify the origin and cause, no matter what type of fire is involved.

Chapter 9, **"Automobile, Other Motor Vehicle, and Ship Fires,"** describes the fuels, ignition sources, and dynamics as encountered in vehicles with which vehicle fire investigators must be familiar to carry out correct investigations.

Chapter 10, **"Electrical Causes of Fires,"** describes the basics of electricity, its control, uses, and measurement of which the investigator must have a working knowledge to accurately assess its effects on fuels in the fire environment. The chapter also discusses the many ways in which electrical power can cause fires and diagnostic indicators of use to investigators.

Chapter 11, **"Clothing and Fabric Fires,"** describes how despite governmental regulations, fires in which fabrics are the first materials to be ignited are still a very common occurrence. The chapter discusses the nature of common fabrics and upholstery materials and their contributions to both ignition hazard and fuel load in current studies.

Chapter 12, "Explosions and Explosive Combustion," describes the range of explosion violence—from the diffused type of rolling, progressive flame resulting from the combustion of a rich mixture of flammable gases or vapors in air, to the violent, almost instantaneous detonation of condensed-phase explosives. The chapter also discusses the chemical structure and mechanisms of low and high explosives, as well as mechanical and electrical explosions.

Chapter 13, "Chemical Fires and Hazardous Materials," describes how to identify the common causes of fires involving chemicals and hazardous materials.

Chapter 14, "Laboratory Services," describes the role of laboratory services in fire and explosion investigation and the types of examinations that can be requested. These include not only fire debris analysis but the application of a wide range of physical, chemical, optical, and instrumental tools on a variety of substances.

Chapter 15, "Fire-Related Deaths and Injuries," identifies the common causes of (and factors contributing to) fire-related deaths and injuries and describes indicators and postmortem tests. Postmortem fire effects on the body are a primary focus.

Chapter 16, "Arson as a Crime," explains the basic concepts of arson as a crime along with its common motives, and describes common incendiary devices and mechanisms.

Chapter 17, "Other Investigative Topics," identifies specialized tools that are relevant and useful in fire investigation, including fire modeling. The chapter also discusses on-scene safety considerations for the fire investigator and the implications of the scientific method in analyzing fire events.

ACKNOWLEDGMENTS

Peer review is important for ensuring that a textbook is well balanced, useful, authoritative, and accurate. The following agencies, institutions, companies, and individuals provided invaluable support during the peer-review process of this edition.

Vyto Babrauskas, Ph.D.
Fire Science and Technology Inc.
Issaquah, WA

Nick Carey
Fire Investigation Group, London Fire Brigade,
London, UK

Detective Mike Dalton
Knox County Sheriff's Office/Fire Investigation Unit
Knoxville, TN

Gary Edwards
Fire Science Program Director/Instructor
Montana State University–Billings College of Technology
Billings, MT

Tom Goodrow
ATF National Academy (Retired)
West Chatham, MA

Gary S. Hodson, IAAI-CFI
Sgt., Provo UT Police Department (retired)
Adjunct Instructor, Utah Valley University
Special Investigator, Unified Investigations and Sciences
Provo, UT

Jeffrey Lee Huber
Professor of Fire Science
Lansing Community College
Lansing, MI

Judith Kuleta
Bellevue College
Bellevue, WA

John E. Malooly
President of Malooly & Associates, Inc.
Senior Special Agent (Retired), U.S. Department of Justice, Bureau of ATF
Chicago, IL

Matthew Marcarelli
Lieutenant, City of New Haven Fire Department
Adjunct Instructor, Connecticut Fire Academy

J. Ronald McCardle
Major, Florida Bureau of Fire & Arson Investigations (Retired)
Currently Instructing, Consulting, & Researching in Fire and Explosion Causation
Florida

C. W. Munson
Chemeketa Fire Technology
Salem, OR

J. Graham Rankin, PhD
Forensic Science Program
Marshall University
Huntington, WV

Steve Riggs
Public Agency Training Council
Indianapolis, IN

James P. Ryan
Fire Investigator, Arson Bureau
New York State
Department of Homeland Security and Emergency Services
Academy of Fire Science
Montour Falls, NY

Nathan Sivils
Director, Fire Science
Blinn College
Bryan, TX

Aaron S. Woolverton
Adjunct Professor
Austin Community College
Austin, TX

Rather than extend an already substantial list from previous editions, we focus on those persons who contributed the most to this Seventh Edition. As with any other evolutionary process, the result is the product of many generations. You know who you are from those earlier lists. Rest assured that your contributions are still greatly appreciated. We want to acknowledge the following individuals who reviewed and generously offered counsel and new material to improve this edition: Doug Wood; Morris Polich & Purdy, San Francisco, CA; Daniel Madrzykowski, P.E., National Institute of Standards and Technology, Gaithersburg, MD; Major Ron McCardle, Florida State Fire Marshal's Office (retired); Dr. Niamh Nic Daéid, University of Strathclyde; Kim R. Mniszewski, P.E., FX Engineering; Jeff Morrill, MorrFire Investigations; Gordon Damant, Sacramento, CA; Steve Riggs, Public Agency Training Council; Senior Special Agent Paul Steensland (retired), United States Forest Service, Susanville, CA; Luis Velazco, Luis Velazco Investigations, Ltd., St. Simons Is., GA; and Special Agent Dino Balos and Steven J. Avato, Bureau of Alcohol, Tobacco, Firearms and Explosives, Falls Church, VA. Your efforts were substantial, and your ideas were greatly welcomed. Our special appreciation is offered to Doug Wood and his staff for all the new material on spoliation and other legal issues.

Vyto Babrauskas, Mick Gardiner, Gary White, Steve Mackaig, Ron Parsons, Chris Korinek, Rick Korinek, Bob Svare, and Mark Svare all reviewed critical portions of text and contributed significantly to the accuracy of this book. In addition, Vyto Babrauskas and Dan Madrzykowski always generously shared ideas, insight, and information whenever we asked, for which we are deeply grateful.

A number of people generously shared their case histories, test results, and photographs for this edition, including Jim Albers, Santa Ana Fire Dept. (retired); David Barber, Goleta, CA; Steve Bauer; Dr. Roger Berrett; Lou Bilancia, Synnovation Engineering and HTRI Forensics; Calvin Bonenberger, Fire Marshal, Lafayette Hill Fire Department; Dr. Bernard R. Cuzzillo; Chris Bloom, CJB Consultants, Grants Pass, OR; Joe Bloom, Bloom

Fire Investigation, Grants Pass, OR; Helmut and Peter Brosz, Brosz & Assoc., Markham, Ontario, Canada; Paul Carolan, Toronto Fire Department, Canada; Randy Crim, Fire Marshal, Lake Jackson, TX; Donna Deaton, U.S. Forest Service (retired); Andrew Derrick; Denise DeMars, Streich DeMars, Inc.; Jack Deans; Det. Richard Edwards, Los Angeles County Sheriff's Department (retired); Ryan B. Fields, Orca Fire Investigation, Medford, OR; Capt./Inv. Bruce Fusselman, Phoenix Fire Department, Phoenix, AZ; Nick Carey, John Galvin, and Paul Spencer, London Fire Brigade, London, UK; Edward Garrison, Fire/Explosion Investigator, Raleigh, NC; Thomas Goodrow, Fire/Explosives Technical Specialist, ATFE (retired); Tony Grissim, Leica GeoSystems; Don Perkins, Curt Hawk, and Diane Spinner, Fire Cause Analysis; Gerald Haynes, Forensic Fire Analysis, LLC, Fredericksburg, VA; Dr. Robin Holleyhead; *Science &Justice;* Chief Kurt Hubele, Richland Fire Department, WA; John Jerome; Capt. Thomas Kinkaid, Knoxville, Fire Department; Chris W. Korinek, P.E., and Richard E. Korinek, P.E., Synergy Technologies LLC; Ken Legat, Christchurch, New Zealand; SA/CFI Michael A. Marquardt, ATFE, Grand Rapids, MI; Vic Massenkoff, Contra Costa County Fire Department; Marion Matthews, U.S. Forest Service; Lamont "Monty" McGill, retired Fire Investigator and Bomb Technician (deceased); Wayne Moorehead, Forensic Consultant; Jamie Novak and Cameron Novak, Novak Fire Investigations and St. Paul Fire Department; Dr. Said Nurbakhsh, California Bureau of Home Furnishings, North Highlands, CA; Chief Mike Oakes and Tony Hudson, Clallam County Fire Investigation Team, Port Angeles, WA; Keith Parker, Marin County Fire Department, Woodacre, CA; David W. Powell, SYTEK Consultants, East Syracuse, NY; Steve Riggs and Tim Yandell, Public Agency Training Council; Susan Sherwin, Scottsdale, AZ; Stuart Sklar, Fabian, Sklar & King, P.C., Farmington Hills, MI; Robert Toth, Iris Investigations; Inv. Jeff Weber, San Jose Fire Department, CA; and Capt. Sandra Wesson, Little Rock Fire Department, AR. Our special thanks to Det. Michael Dalton and Inv. Greg Lampkin, Knox County Sheriff's Office, Knoxville, TN, for *all* their great photos.

We also want to offer our special acknowledgment to Jamie Novak, who seems to be able to burn (and blow up) more buildings than anyone else. Jamie generously opened his awesome collection of photos, and many were selected for this book. We all know that fire and explosion investigation is largely a visual endeavor, and such photos are vital to the usefulness of an investigation text. The Bureau of Home Furnishings and the Fire Research Station (Garston, UK) have repeatedly provided us with opportunities to do special tests and share the results.

Firsthand observations of fire behavior and patterns are a critical element in qualifications for every fire expert. Organizations such as Gardiner Associates (UK), the European Working Group on Fires & Explosions, ATF (Glynco, GA), Florida State Fire Marshal, and various chapters of the IAAI are to be commended for the special efforts they have made in providing "live burn" training and research. The efforts of Dan Madrzykowski and his team from NIST at CCAI conferences were greatly appreciated for the careful science they demonstrated. The Training Committee of the CCAI has done an exemplary job of providing extraordinary opportunities for firsthand fire observations. Its motto is "Build it and they will come—to burn it and learn." The Forensic Fire Death Investigation Courses provided recently by the San Luis Obispo Fire Investigation Strike Team provided unique opportunities to observe the effects of fire on human cadavers. These results will benefit investigators through the illustrations presented here.

Our very patient editors, Monica Moosang, Pam Powell, and Barbara Liguori, saw this text through all manner of crises. John DeHaan's office manager, Shirley Runyan, performed exemplary work in myriad roles—text, proofreading, graphics, and especially in unraveling the mysteries of electronic communication. Our deepest thanks to everyone.

Finally, we want to acknowledge the personal inspiration we gain from working with fire investigators such as Monty McGill, Jamie Novak, Jack Malooly, Ross Brogan, Jeff Campbell, Nick Carey, Randy Crim, Bob Toth, Wayne Miller, Jim Allen, Mike Dalton, Jim Munday, and Mike Marquardt. Their dedication to finding the right answer through scientific analysis is an example for all of us.

ABOUT THE AUTHORS

This textbook is coauthored by two of the most experienced fire scientists in the United States. Their combined talents total more than 80 years of experience in the fields of investigation, fire behavior, fire protection engineering, criminalistics, fire science, and crime scene reconstruction.

An internationally recognized forensic scientist, Dr. DeHaan is the senior author of *Kirk's Fire Investigation* and coauthor of *Forensic Fire Scene Reconstruction,* the two leading textbooks in the field of fire and arson investigation. He is also a former principal member of the NFPA 921 Technical Committee on Fire Investigations.

John D. DeHaan, PhD, FABC, CFI–IAAI, CFEI-NAFI, FFSS, FSSDip

Dr. DeHaan has been a criminalist for more than 40 years and has gained considerable expertise in the analysis of fire and explosion evidence as well as shoe prints, and in instrumental analysis and crime scene reconstruction. He has been employed as a criminalist by the Alameda County Sheriff's Office, the U.S. Treasury Department, and the California Department of Justice.

His research into forensic fire scene reconstruction is based on firsthand fire experiments on fire behavior involving more than 500 observed full-scale structure, and 100 vehicle, fires under controlled conditions, as well as laboratory-scale studies. Dr. DeHaan has testified as an expert witness in civil and criminal trials across the United States and overseas. He is currently the president of Fire-Ex Forensics Inc. and consults on civil and criminal fire and explosion cases across the United States and Canada and overseas.

Dr. DeHaan graduated from the University of Illinois–Chicago Circle in 1969 with a BS degree in Physics and a minor in Criminalistics. He was awarded a PhD in Pure and Applied Chemistry (Forensic Science) by Strathclyde University in Glasgow, Scotland, in 1995. Dr. DeHaan is a Fellow of the American Board of Criminalistics (Fire Debris), a Fellow of the Forensic Science Society (UK), and holds Diplomas in Fire Investigation from the Forensic Science Society and the Institution of Fire Engineers, and a Certified Fire Investigator certification from the International Association of Arson Investigators. He is also a Certified Fire and Explosion Investigator in the National Association of Fire Investigators.

An internationally recognized forensic fire engineering expert with more than 40 years of experience, Dr. Icove is coauthor of *Kirk's Fire Investigation* and *Forensic Fire Scene Reconstruction,* the two leading textbooks in the field of fire and arson investigation. He is also coauthor of *Combating Arson-for-Profit,* the leading textbook on the crime of economic arson. Since 1992 he has served as a principal member of the NFPA 921 Technical Committee on Fire Investigations. As a retired career federal law enforcement agent, Dr. Icove served as a criminal investigator on the federal, state, and local levels. He is a Certified Fire and Explosion Investigator (CFEI).

David J. Icove, PhD, PE, CFEI

He retired in 2005 as an Inspector in the Criminal Investigations Division of the U.S. Tennessee Valley Authority (TVA) Police, Knoxville, Tennessee, where he was assigned for the last 2 years to the Federal Bureau of Investigation (FBI) Joint Terrorism Task Force (JTTF). In addition to conducting major case investigations, Dr. Icove oversaw the development of advanced fire investigation training and technology programs in cooperation with various agencies, including the U.S. Fire Administration of the Federal Emergency Management Agency (FEMA).

Before transferring to the U.S. TVA Police in 1993, he served 9 years as a program manager in the elite Behavioral Science and Criminal Profiling Units at the FBI, Quantico, Virginia. At the FBI, he implemented and became the first supervisor of the Arson and

Bombing Investigative Support (ABIS) Program, staffed by FBI and ATF criminal profilers. Prior to his work at the FBI, Dr. Icove served as a criminal investigator at arson bureaus of the Knoxville Police Department, the Ohio State Fire Marshal's Office, and the Tennessee State Fire Marshal's Office.

His expertise in forensic fire scene reconstruction is based on a blend of on-scene experience, conduction of fire tests and experiments, and participation in prison interviews of convicted arsonists and bombers. He has testified as an expert witness in civil and criminal trials, as well as before U.S. congressional committees seeking guidance on key arson investigation and legislative initiatives.

Dr. Icove holds BS and MS degrees in Electrical Engineering and a PhD in Engineering Science and Mechanics from the University of Tennessee. He also holds a BS degree in Fire Protection Engineering from the University of Maryland–College Park. He is presently a Research Professor in the Department of Electrical Engineering and Computer Science at the University of Tennessee, Knoxville; serves on the faculty of the University of Maryland's Professional Master of Engineering in Fire Protection; and is a Registered Professional Engineer in Alabama, Arkansas, Louisiana, Tennessee, Texas, Virginia, Maryland, and Pennsylvania.

NFPA 1033 Matrix

NFPA 1033 identifies the professional level of job performance requirements for fire investigators. This standard specifies the minimum job performance requirements for service as a fire investigator in both the private and public sectors. Job performance requirements (JPRs) for each duty are the tasks an investigator must be able to perform to successfully carry out that duty and are summarized in the following matrix:

	Professional Levels of Job Performance for Fire Investigators as Cited in *NFPA 1033*, 2009 Edition
General Requirements for a Fire Investigator	**4.1.2** Employ all elements of the scientific method as the operating analytical process **4.1.3** Complete site safety assessments on all scenes **4.1.4** Maintain necessary liaison with other interested professionals and entities **4.1.5** Adhere to all applicable legal and regulatory requirements **4.1.6** Understand the organization and operation of the investigative team and incident management system
Scene Examination	**4.2.1** Secure the fire ground **4.2.3** Conduct an interior survey **4.2.4** Interpret fire patterns **4.2.5** Interpret and analyze fire patterns **4.2.6** Examine and remove fire debris **4.2.7** Reconstruct the area of origin **4.2.8** Inspect the performance of building systems **4.2.9** Discriminate the effects of explosions from other types of damage
Documenting the Scene	**4.3.1** Diagram the scene **4.3.2** Photographically document the scene **4.3.3** Construct investigative notes
Evidence Collection/ Preservation	**4.4.1** Utilize proper procedures for managing victims and fatalities **4.4.2** Locate, collect, and package evidence **4.4.3** Select evidence for analysis **4.4.4** Maintain a chain of custody **4.4.5** Dispose of evidence
Interview	**4.5.1** Develop an interview plan **4.5.2** Conduct interviews **4.5.3** Evaluate interview information
Post-Incident Investigation	**4.6.1** Gather reports and records **4.6.2** Evaluate the investigative file **4.6.3** Coordinate expert resources **4.6.4** Establish evidence as to motive and/or opportunity **4.6.5** Formulate an opinion concerning origin, cause, or responsibility for the fire
Presentations	**4.7.1** Prepare a written report **4.7.2** Express investigative findings verbally **4.7.3** Testify during legal proceedings **4.7.4** Conduct public informational presentations

Sources: NFPA 1033, 2009 ed.; S. Sklar presentation, 2008.

Fire and Emergency Services Higher Education (FESHE) Grid

The National Fire Academy-developed Fire and Emergency Services Higher Education (FESHE) curriculum serves as a national training program guideline which is a requirement for many fire service organizations and training programs. The following grid outlines the needs for Fire Investigator I and II levels and where specific content can be located in this text:

Course Requirements	1	2	3	4	5	6	7	8	9	10	11	12	13	14	15	16	17	App E	App H
Identify and explain the responsibilities of the fire department from a firefighter's perspective when responding to the scene of a fire, including the possibility of incendiary devices often encountered. (FI-I)							X	X											
Define criminal law and explain the constitutional amendments (4th, 5th, 6th, 8th, and 14th) as they apply to fire investigations. (FI-I)																X	X	X	
Analyze the precedents set by constitutional law case studies that have affected fire investigations. (FI-I)																X	X		
Define and explain the common terms used in fire investigations	X	X	X	X	X	X	X	X	X	X	X	X	X	X	X	X	X		
Describe the basic elements of fire dynamics and how they affect cause determination including fire behavior, characteristics of fuels and methods of heat transfer. (FI-I)			X	X	X						X								
Analyze the relationship of building construction on fire investigations including types of constructions, construction and finish materials. (FI-I)			X																
Evaluate fire protection systems and building services and discuss how their installation affects the ignition of fires in buildings. (FI-I)							X	X											
Discuss the basic principles of electricity. (FI-I)										X									
Explain the role of the fire investigator in recognizing health and safety concerns including potential hazardous materials awareness. (FI-I)												X					X		
Describe fire scene investigations and the process of conducting fire investigations using the scientific method. (FI-I)	X							X											
Explain how an investigator determines the point of origin in a room. (FI-I)							X	X	X	X	X								
Identify the types of fire causes and differentiate between accidental and incendiary causes. (FI-I)							X	X											
Describe and explain the basic procedures used for investigating vehicle fires. (FI-I)									X										
Identify the characteristics of arson and common motives for the firesetter. (FI-I)															X				
Identify and analyze the causes involved in line of duty firefighter deaths related to structural and wildland firefighting, training and research and the reduction of emergency risks and accidents. (FI-I)								X					X		X				
Explain the rule of law as it pertains to arrest, search and seizure procedures and their application to fire investigations. (FI-II)																X	X	X	
Recognize and interpret fire scenes common to various types of fires. (FI-II)							X	X	X	X		X	X		X	X			
Describe the chemistry of combustion and the relationship of atoms, elements, compounds, and organic compounds on fire. (FI-II)		X		X	X							X	X						
Explain the nature and behavior of fire including the effects of heat. (FI-II)			X	X	X	X	X												

Course Requirements	1	2	3	4	5	6	7	8	9	10	11	12	13	14	15	16	17	App E	App H
Explain and identify the combustion properties of liquids, gases, and solid fuels. FI-II)		X		X	X							X	X						
Identify and explain electrical causes of fires. (FI-II)									X	X									
List and explain the procedures for lifting fingerprints, evidence collection and preservation. (FI-II)														X	X		X		
List and identify the make-up and use of incendiary devices, explosives, and bombs. (FI-II)												X				X			
List the procedures for documenting fire scenes, including sketching, photography, and report writing. (FI-II)							X					X				X			X
Analyze fire-related deaths and injuries and describe methods of documentation. (FI-II)															X				X
Identify the techniques for interviewing and questioning suspects and subjects. (FI-II)							X									X			X
Explain the role of the fire investigator in courtroom proceedings including courtroom demeanor and testifying. (FI-II)																X			
Identify and list the sources and technology available for fire investigations. (FI-II)	X	X	X	X	X	X	X	X	X	X	X	X	X	X	X	X	X		

Courtesy of Jamie Novak, Novak Fire
Investigations and St. Paul Fire Dept.

KEY TERMS

accidental fire, *p. 2*

arson, *p. 4*

combustion, *p. 10*

explosion, *p. 2*

fire, *p. 2*

fire behavior, *p. 10*

incendiary fire, *p. 3*

suspicious, *p. 7*

OBJECTIVES

After reading this chapter, you should be able to:

- Describe the role of fire investigators in their determination of the origin, cause, and development of a fire or explosion.
- List the four major peer-reviewed publications in the field of fire investigation.
- Identify the common job performance requirements for fire investigators.
- Explain the problem of fire loss statistics in the United States and in the United Kingdom, particularly in the estimation of incendiary fires.
- Demonstrate an understanding of the role of the scientific method as it applies to fire and explosion investigations.
- Discuss the levels of confidence as they apply to expert opinions in fire and explosion investigations.

Fire Investigation

This textbook focuses on the field of fire investigation, which is the formal process of determining the origin, cause, and development of a fire or explosion.[1] Often called *origin and cause* (O&C) investigation, the probe is launched after a **fire** or **explosion** is extinguished and investigators then strive to determine what circumstances caused or contributed to the incident. Due to the complex nature of the event, where fire often deforms or distorts the evidence, fire investigation is among the most difficult forensic sciences to practice. Not knowing whether the fire was intentionally set, investigators must undertake to prove or disprove any allegations or suspicions and to piece together the pre- and post-fire events. If the investigation indicates a fire was deliberately set, reasonable **accidental fire** causes must be evaluated and eliminated.

Fire investigators must have broad and up-to-date knowledge of topics such as fire science, fire chemistry, thermodynamics, thermometry, fire dynamics, explosion dynamics, computer fire modeling, forensic sciences, fire analysis, scientific methodology, photography, hazardous materials, building construction, gas and electricity utility systems, human behavior (related to evacuation, victims, firesetting, witnesses, and information), forensic lab analysis, failure analysis, criminal and civil law, and analytical tools.

The latest edition of *Kirk's Fire Investigation* is considered the leading worldwide authoritative textbook in the field. The textbook is used both for training and as the basis for certification in many jurisdictions. A majority of the overall methodology of fire investigation relies upon *Kirk's* as well as on the following three peer-reviewed sources:

- The second edition of *Forensic Fire Scene Reconstruction*, published in 2008, which is also considered a leading forensic textbook, serves as a "bridge textbook" that ties the concepts of fire engineering analysis to fire investigations.
- The National Fire Protection Association's *NFPA 921: Guide for Fire and Explosion Investigations*, approved by the NFPA Standards Council on December 14, 2010, and issued on January 3, 2011 (NFPA, 2011). The 2011 edition of *NFPA 921* supersedes all previous issues and is considered by practitioners and judicial authorities the standard of care for conducting fire and explosion investigations.
- The National Fire Protection Association's *NFPA 1033: Standard for Professional Qualifications for Fire Investigator*, approved by the NFPA Standards Council on May 30, 2008, with an effective date of July 18, 2008. The 2009 edition of *NFPA 1033* supersedes all previous issues and was approved as an American National Standard on July 18, 2008.

NFPA 1033 identifies the professional level of job performance requirements for fire investigators.[2] This standard specifies the minimum job performance requirements for

fire ■ Rapid oxidation with the evolution of heat and light; uncontrolled combustion.

explosion ■ The sudden conversion of potential energy (chemical or mechanical) into kinetic energy with the production of heat, gases, and mechanical pressure.

accidental fire ■ A fire occurring without design or intent.

TABLE 1-1	Professional Levels of Job Performance for Fire Investigators as Cited in *NFPA 1033*, 2009 Edition
General Requirements for a Fire Investigator	**4.1.1** Employ all elements of the scientific method as the operating analytical process **4.1.2** Complete site safety assessments on all scenes **4.1.3** Maintain necessary liaison with other interested professionals and entities **4.1.4** Adhere to all applicable legal and regulatory requirements **4.1.5** Understand the organization and operation of the investigative team and incident management system
Scene Examination	**4.2.1** Secure the fire ground **4.2.2** Conduct an interior survey **4.2.3** Interpret fire patterns **4.2.4** Interpret and analyze fire patterns **4.2.5** Examine and remove fire debris **4.2.6** Reconstruct the area of origin **4.2.7** Inspect the performance of building systems **4.2.8** Discriminate the effects of explosions from other types of damage
Documenting the Scene	**4.3.1** Diagram the scene **4.3.2** Photographically document the scene **4.3.3** Construct investigative notes
Evidence Collection/Preservation	**4.4.1** Utilize proper procedures for managing victims and fatalities **4.4.2** Locate, collect, and package evidence **4.4.3** Select evidence for analysis **4.4.4** Maintain a chain of custody **4.4.5** Dispose of evidence
Interview	**4.5.1** Develop an interview plan **4.5.2** Conduct interviews **4.5.3** Evaluate interview information
Post-Incident Investigation	**4.6.1** Gather reports and records **4.6.2** Evaluate the investigative file **4.6.3** Coordinate expert resources **4.6.4** Establish evidence as to motive and/or opportunity **4.6.5** Formulate an opinion concerning origin, cause, or responsibility for the fire
Presentations	**4.7.1** Prepare a written report **4.7.2** Express investigative findings verbally **4.7.3** Testify during legal proceedings **4.7.4** Conduct public informational presentations

Sources: NFPA 1033, 2009 Edition; S. Sklar, personal communications, 2008.

service as a fire investigator in both the private and public sectors. Job performance requirements for each duty are the tasks an investigator must be able to perform to successfully carry out that duty. These are summarized in Table 1-1.

Not only must fire investigators refer to and rely on these guides and standards, as well as on expert treatises, they must conduct their investigations using a systematic and scientifically guided approach. In cases of **incendiary fires** (i.e., those set intentionally to destroy), in addition to determining the origin and cause of the fire, the investigator has the added responsibility of identifying the person who set the fire and providing proof to a trier of fact. Incendiary fires represent a significant percentage of all fires and, because they are set specifically with the intent of destroying property, an inordinately high percentage of dollar losses. Finally, a majority of the fire investigators who testify as expert witnesses usually have certification in the field by one or more internationally recognized professional organizations.

incendiary fire ■ A deliberately set fire.

The Fire Problem

Internationally, fire continues to be the most costly of all public safety problems today, as it has been for the past several decades. The losses in human lives and injuries due to fires and explosions continue to occur. Fire-caused property losses are far in excess of those caused by all classes of crime and rival those produced by hurricanes and earthquakes. The major issue in identifying and quantifying the fire problem is the failure to obtain accurate incident data. This problem is not limited to any single country.

FIRE STATISTICS IN THE UNITED STATES

For example, in the United States, fire incident data are collected by the U.S. Fire Administration and the National Fire Information Council using the National Fire Incident Reporting System (NFIRS) on a voluntary basis. Shrinking state budgets have often resulted in the downgrading of the priorities of fire incident data collection programs. Uniform Crime Reporting (UCR) statistics are amassed through the Federal Bureau of Investigation's (FBI's) National Incident Based Reporting System (NIBRS), but **arson** incident data are collected only when a fire or police investigator actually submits a report into NIBRS. Subsequently, arson cases go unreported when fire departments fail to pass along and fill out a separate police report on intentionally set fires.

arson ■ The intentional setting of a fire with intent to damage or defraud.

Although the National Fire Protection Association (NFPA) collects data on fire incidents, its estimates are based on data the association receives from fire departments that respond to its National Fire Experience Survey.[3] NFPA estimated that public fire departments in the United States responded to a total of 1,451,500 fires in 2008. Of these, an estimated 515,000 were structure fires, which represented a decrease of 2.9 percent from the number in 2007. The total monetary property loss for structure fires was $12.4 billion, an increase of 16.2 percent over the previous year's total. The average overall loss per structure fire was $24,002, an increase of 19.7 percent from the 2007 average. Residential fires accounted for about 78.2 percent of all structure fires (estimated at 403,000), with 291,000 occurring in one- and two-family dwellings and accounting for 56.5 percent of all structure fires. In addition, in 2008 there were an estimated 236,000 vehicle fires and 594,000 "outside" fires.

No accounting can accurately reflect the loss in lives and injuries that results from fires. In 1987, the NFPA estimated that 5,810 civilians died as a result of fires, and 28,295 suffered injuries.[4] By 1999, these losses were reduced to 3,570 deaths in structure fires (and 470 in vehicle fires) and 21,895 injured.[5] In 2003, estimated losses were 3,925 deaths (an increase of 16 percent over 2002 losses) in structure fires and 475 in vehicle fires, with 18,125 injuries.[6]

NFPA estimates on civilian fire deaths totaled 3,320 and 16,705 civilian fire injuries in 2008, a decrease from 2007 totals of 3.2 and 5.5 percent, respectively. An estimated 2,780 civilians perished and 13,560 were injured in residential fires. Of these, 390 died and 3,975 were injured in apartment fires, and 2,365 died and 9,185 were injured in one- and two-family dwelling fires.[7] Thus, even though the overall estimated number of fires has remained steady since 1999, injury and death rates are dropping.

During the period 1994–2007, an average of 100 firefighters died on duty each year (not including the 343 firefighters who died at the World Trade Center in 2001).[8] There were an estimated 36,880–44,210 fireground injuries reported to U.S. firefighters per year between 2003 and 2007, reflecting a fluctuating rate of 22–27 fireground injuries per 1,000 fires per year since 1987.[9] The National Safety Council reports that accidental fire-related burn injuries are the sixth leading cause of all accidental deaths.[10]

In addition to these direct losses are the incalculable indirect fiscal costs. The actual dollar property losses previously listed probably represent only 10 percent of the total costs. There are inestimable losses due to lost business income, unemployment, reduced property values, and reduced tax bases as a result of nearly every major fire. In a 2005 *Fire*

Protection Engineering article, Frazier estimated that the total cost of fire in the United States is on the order of $130–$250 billion per year.[11] Personnel and equipment costs are incurred in the suppression of every fire, historic structures are lost forever, and wildland fires destroy watersheds, timber, and wildlife, which cannot be replaced at any cost.

It should be noted that virtually all U.S. fire "statistics" are only educated guesses, in the authors' opinion. The Consumer Product Safety Commission (CPSC) once estimated that fewer than 5 percent of structure fires are even reported to authorities. Most of them were presumably too minor and resulted in no major injuries or property losses. Many state fire marshals do not forward their own statistics to NFIRS (sometimes because of staffing shortages). California, which represents more than 10 percent of the U.S. population, has not been represented in NFIRS analyses for several years. Further, without adequate training for the person checking off the boxes in NFIRS reporting forms, those responses are often no more than guesses based on first impressions and on-the-fly assessments made by incident commanders.

FIRE STATISTICS IN THE UNITED KINGDOM

The United States is not alone in the problem of fire loss. The Office of Fire and Rescue Services, Communities and Local Government (CLG) in the United Kingdom (UK) collects, maintains, and analyzes UK fire statistics. In May 2006, the CLG assumed responsibility for the fire and rescue service. Previously, responsibility had been held by different UK government departments: the Home Office; the Department for Transport, Local Government (DTLG); and the Office of the Deputy Prime Minister (ODPM).

In the United Kingdom, a total of 385,000 fires were reported in 2007, of which 77,000 (23.5 percent) were in buildings. In 2008, the number of building fires fell by 8 percent to 77,000; and the number of dwelling fires fell by 6 percent, to 51,000. The total number of fires in buildings other than dwellings (commercial premises, schools, etc.) also fell 12 percent, to 231,000. Outdoor fires totaled 186,000, and there were 44,000 road vehicle fires.[12]

In 2003 the ODPM estimated that the average costs (property damage, health care, and lost employment) were £24,280 (approximately $47,000) per domestic fire, £58,100 (approximately $107,000) per commercial fire, and £4,670 ($8,600) per vehicle fire. The ODPM estimated that £1.1 billion ($2.03 billion) in losses were incurred from material damage and business interruption claims in the United Kingdom in 2003, but another £6.6 billion ($12.2 billion) in costs fell on society in the form of suppression costs, cleanup, environmental damage, and disruption to households.[13]

In 2007, there were 443 fire-related deaths in the United Kingdom, a 2 percent increase from the 2006 figure. The UK study noted that a majority of fire-related deaths (greater than three quarters) occur in dwelling fires, with more fire-related deaths per 1,000 fires than in any other location.[14]

Findings from the 2004/05 Survey of English Housing on all outbreaks of fire experienced by households in England suggested that the fire and rescue service attended approximately one fifth of all domestic fires. The explanation for this was that many of the fires recorded in that survey were minor and could be put out by someone in the home or bystanders; therefore, the fire and rescue service was not called.[15]

ROLE OF THE FIRE INVESTIGATOR IN ACCURATELY REPORTING THE CAUSES OF FIRES

What role does the fire investigator play in this tragedy? By accurately and efficiently identifying the causes of fires, whether accidental or incendiary, investigators can make a real and substantial contribution to reducing these terrible losses. Think of the number of people who would have died had the flammability of clothing and bedding not been identified in the 1960s or had the problems with instant-on televisions not been tracked down in the 1970s, or with coffeemakers and portable heaters in the 1980s. What of the scores of potential victims in several large cities in the 1990s had the serial arsonists there not been identified and incarcerated? Saving lives and preventing injuries are the real reasons fire investigators spend long hours searching a fire scene or doing laboratory analysis and fire research.

The role thorough fire investigation plays in the regulatory and code-making process is also significant. Much of the progress in reducing the frequency of fires and fire deaths is due to the advances made in improving life safety and fire safety codes. Although these codes are written on the basis of the best predictive models, the only way to be sure they have the desired effect and do not create other, unanticipated risks is to document the effects of fire in code-compliant buildings and transport. In the early 1900s builders were proud of the fire-resistant buildings they had built of stone, concrete, steel, and glass. It was not until major tragedies like the Triangle Shirtwaist fire in 1911 that people realized that building contents and not just the structures were major factors in fires, and having adequate means of escape was critical to survival.[16] Manual alarm and firefighting systems were also found to be woefully inadequate.

The Detection of Incendiary Fires

Few other investigations are as daunting as a fire scene. Because a fire scene has to be investigated before it can be established whether the crime of arson has been committed, *every* fire scene must be considered a *possible* crime scene (from the standpoint of scene security, processing, and evidence collection) until clear proof is found that the fire was accidental. As a crime scene, the typical fire scene is the antithesis of the ideal. There has been wholesale destruction of the physical material and substance of the scene, first by the fire itself and second by fire suppression and overhaul by the firefighters. Naturally, this destruction affects evidence of both accidental and incendiary causes. There has been contamination of the scene by vehicles, personnel, equipment, hoses, and large quantities of water. Scene security and control of unauthorized persons are difficult because of the attraction fire offers to the public and the press and may be further complicated by the large area involved in many fire scenes. Often, the scene investigation is carried out days or weeks after the incident, after weathering and vandals may have further compromised whatever evidence remains. Considering these negative factors, it is not surprising that police and fire personnel have often let fire investigation fall through the cracks of responsibility. As a result, it is surprising that fire causes are ever accurately determined and civil and criminal cases successfully brought to trial.[17]

REPORTING ARSON AS A CRIME

The *Uniform Crime Reports (UCR),* published by the FBI of the U.S. Department of Justice, provides enlightening statistics on arson in the United States. Statistics for 2007, published in 2008, are used in this chapter. Annually, the FBI publishes *Crime in the United States*, which catalogs the entire crime picture of the United States. Note that a 2006 study of statistics for previous years indicates that there is a fairly constant pattern in arson statistics for past years.[18] *UCR* reported in 2008 that the number of arson offenses decreased 3.6 percent compared with those reported in 2007.[19]

The FBI's UCR Program defines arson as

> any willful or malicious burning or attempting to burn, with or without intent to defraud,
> a dwelling house, public building, motor vehicle or aircraft, personal property of another, etc.

The FBI collects data only on the fires that through investigation were determined to have been willfully set. Fires are not included in the data collection whose cause is labeled as "suspicious" or are "unknown or undetermined."

The FBI points out that arson rates are calculated based on data received from all law enforcement agencies that provide the UCR Program with data for 12 complete months. Unlike other national reporting systems, UCR Program data collection does not include any *estimates* for arson, because the degree of reporting arson offenses varies from agency to agency.

According to the 2008 *UCR*, 14,011 law enforcement agencies (providing 1–12 months of arson data) reported 62,807 arsons. Of those agencies, 13,980 provided expanded offense data on about 56,972 arsons. These data show that arsons involving structures (residential, storage, public, etc.) accounted for 24,750 fires, or 43.4 percent of the total number of arson offenses. Mobile property was involved in 28.9 percent of arsons, and other types of property (such as crops, timber, and fences) accounted for 27.7 percent of reported arsons.[20] See Table 1-2 for a detailed breakdown of the 2008 statistics.

The 2008 FBI statistics showed that the average dollar loss per incident due to arson was $16,015. Arsons of industrial/manufacturing structures resulted in the highest average dollar losses (an average of $212,388 per arson incident). Regarding arson rates, the 2008 *UCR* data showed that nationwide, the rate of arson was 24.1 offenses for every 100,000 inhabitants in the United States.

In contrast, the NFPA statistics for 2008 estimated that arson caused 30,500 structure fires and 17,500 vehicle fires, a decrease of 6.2 percent from 2007 totals. NFPA noted in its 2008 report that in Version 5.0 of the NFIRS the classification of arson was changed to *intentionally set* from the previous *incendiary* category. NFIRS no longer carries a **suspicious** fire category in its reporting system.[21]

suspicious ■ Fire cause has not been determined, but there are indications that the fire was deliberately set and all accidental fire causes have been eliminated.

TABLE 1-2	Arson in the United States by Type of Property, 2008						
PROPERTY CLASSIFICATION	NUMBER OF ARSON OFFENSES	PERCENT DISTRIBUTION[a]	PERCENT NOT IN USE	AVERAGE DAMAGE	TOTAL CLEARANCES	PERCENT OF ARSONS CLEARED[b]	PERCENT OF CLEARANCES UNDER 18
Total	56,972	100.0		$16,015	10,277	18.0	37.4
Total structure:	24,750	43.4	19.0	29,701	5,520	22.3	35.2
Single occupancy residential	11,435	20.1	21.4	28,788	2,389	20.9	25.7
Other residential	4,062	7.1	13.6	29,343	983	24.2	25.0
Storage	1,629	2.9	20.7	18,021	298	18.3	48.7
Industrial/ manufacturing	251	0.4	22.3	212,388	48	19.1	41.7
Other commercial	2,331	4.1	15.7	55,531	450	19.3	28.0
Community/public	2,667	4.7	16.2	15,799	857	32.1	65.3
Other structure	2,375	4.2	21.4	13,668	495	20.8	47.1
Total mobile:	16,454	28.9		8,766	1,461	8.9	20.3
Motor vehicles	15,572	27.3		8,186	1,314	8.4	18.7
Other mobile	882	1.5		18,991	147	16.7	34.7
Other	15,768	27.7		2,099	3,296	20.9	48.7

Source: Crime in the United States: Uniform Crime Reports (Washington, DC: U.S. Dept of Justice, Federal Bureau of Investigation, 2008).

Note: 13,980 agencies; 2008 estimated population 250,243,947.

[a]Because of rounding, the percentages may not add to 100.0.

[b]Includes arsons cleared by arrest or exceptional means.

The NFPA statistics for 2008 estimated that intentionally set fires in structure fires resulted in an estimated 315 civilian deaths and $866 million in property losses, an increase of 18.2 percent in property losses over those in 2007. Deliberate vehicle fires caused $139 million in property damage, a decrease of 4.1 percent from 2007 data.[22] These estimates are almost certainly very low, since many U.S. jurisdictions do not report arson incidents (structural or vehicular) (in the authors' experience). Many investigators think that the total of intentionally set structure fires is closer to 40 percent.[23] These statistics may become further compromised as departments devote larger portions of their resources to Emergency Medical Services (EMS) operations and less to fire investigation.

This is a startling contrast to the United Kingdom experience. The UK authorities (CLG) estimate that some 31 percent of structure fires in 2003 were arson caused and 26 percent in 2004/05 after an intense national campaign,[24] and that deliberately set fires in structures and vehicles took 117 lives, caused 3,200 casualties, and cost the economy there some £2.8 billion ($5.2 billion) (including costs in anticipation of arson fires).[25]

In the United Kingdom, the CLG report notes that "deliberately set" fires include those where deliberate ignition was identified or was "merely suspected" and recorded by the fire and rescue service as "doubtful." "Accidental" fires included those for which the listed cause was "not known" or "unspecified." The number of deliberate primary fires (buildings, vehicles, outdoor structures, and any involving casualties, as defined by CLG) fell for the third consecutive year, from 79,700 in 2005 to 72,600 (9 percent) in 2007. Of the 21,400 deliberate fires in buildings recorded in 2007, more than half (59 percent) occurred in buildings other than dwellings. Of these 12,600 deliberate fires in other buildings, around a third occurred in private garages or sheds. The number of "accidental fires" decreased by 6 percent, to 82,000, in 2007 from 2005 (in all major categories). Arson-caused road vehicle fires accounted for 56 percent of all deliberate fires and 68 percent of all vehicle fires in the United Kingdom in 2007.[26]

PROBLEMS ASSOCIATED WITH ESTIMATING INCENDIARY FIRES

The U.S. estimates are certainly on the conservative side, since many fires are never properly investigated because of lack of time, or are misidentified inadvertently as accidental fires due to lack of experience, or intentionally to avoid the complications that arise from identifying a fire as a criminal act. These complications can range from administrative, such as requiring more investigative time or involving another public agency, to political, such as generating negative publicity about the community. The sharp contrast between the UK estimates and the NFPA estimates that only 6.1 percent of the reported structure fires were incendiary demonstrates a systematic problem with NFPA's data gathering and statistical processing. Recent estimates (2003–2007) represent a decrease of 15.7 percent from 2002's already low percentages (traditionally around 15 percent of all fires).

One significant reason for this decrease is that the NFIRS changed its categories in 2003, eliminating the outmoded "suspicious" causation category. Only fires that were specifically determined to be intentionally set are now reported in the "incendiary category." In the past, estimates of half of "suspicious" and "undetermined" fires were added to the "incendiary" total. Fire investigators across the United States suspected that even the "old" percentage (15 percent) was far below the reality of actual investigations. In response to a nonscientific random survey by DeHaan, many experienced public-sector investigators estimate that 35–40 percent of all the structure fires they see (and an even higher percentage of vehicle fires) are deliberately set. Lack of evidence, lack of time and resources, and limited cooperation of other public agencies make the selection of "undetermined" or even "accidental" causation the expedient solution. These problems have recently been exacerbated by budgeting problems at all levels of state and local governments, which have forced severely curtailing or even ceasing fire investigation services in both fire and police agencies. Consequently, the decision-making process has reverted to fire company officers who may or may not be trained or prepared to identify evidence supporting an incendiary cause.

In 2000, the National Institute of Justice published *Fire and Arson Scene Evidence: A Guide for Public Safety Personnel* for first responders in police, fire, and medical services.[27] It was hoped that by familiarizing these professionals with the basics of fire-related investigation and physical evidence a higher percentage of them could contribute toward successful prosecutions. With the cutbacks in public services experienced across the United States, it is not known whether this guide has been successfully employed. From the small percentages of fires actually reported as intentional via NFIRS, it would appear not.

The situation is made even more critical by the governmental agencies that find themselves unable to cope with the onslaught of criminal cases of homicide, rape, assault, and drug-related activity and decide to ignore arson as a crime. This practice is conveniently expedited by categorizing arson as a property crime so that it can be allotted reduced priority and less tactical support than the more urgent crimes against persons. It must be recognized that arson is a crime against people. It is aimed against people where they live, where they work, and where they do business or go to church. Fire is just as much a weapon as a gun, a knife, or an axe. When aimed at persons, fires may be set directly on them or, more often, set deliberately to trap the victims in a structure or vehicle. Every set fire is a crime against the personnel of fire and police agencies who are obligated to respond to and enter fires in lifesaving efforts.

Fire is a most horrendous weapon. Successful fire investigation, culminating in an accurate determination of cause, is the only way to minimize these terrible losses. Accidental fires can be prevented by the identification and elimination of hazardous products and processes or careless practices, as well as the development and enforcement of better building codes and fire safety practices. As the recent British experience with vehicle arsons has shown, incendiary fires can be reduced by recognizing them, identifying those responsible, and prosecuting them.[28] Sadly, many cities, large and small, in the United States have been forced to eliminate or severely reduce their investigation units, leaving line officers to do what they can "on the fly."

With such massive losses involved, it is not surprising that fire losses are one of the most common subjects of civil litigation. Yet, for all its economic and legal importance, fire remains one of the most difficult areas of investigation. It is challenging because the destruction of the fire and, sometimes, its suppression create difficulties in reaching firm and unassailable conclusions as to the fire's cause. It is the investigator's responsibility not only to identify the origin and cause of the fire but to be prepared to defend those conclusions in a rational, logical manner supported by scientifically valid data. In the absence of a rational reconstruction or scientific data, conclusions cannot be substantiated, and successful prosecution for arson is virtually impossible. Civil cases, even with their reduced requisite of proof, are made more difficult to decide fairly without such scientific analysis.

From the standpoint of public safety and economic considerations, it is important that investigative methods for *accurately* determining the causes of fires be understood and applied to the fullest extent possible. There should be no presumption that a fire is either accidental or incendiary. It is vital that the investigator not prejudge any fire as to cause without evaluating all the collected evidence. If an investigator decides that a fire is arson before collecting any data, then only data supporting that premise are likely to be recognized and collected. As Sherlock Holmes pointed out, "It is a capital mistake to theorize before one has data. One begins to twist facts to suit theories instead of theories to suit facts."[29] The same caution is necessary for those who decide that all fires are accidental, no matter what their true cause may be. Although fire has been known since prehistoric times, it is still poorly understood by many people, including some firefighters, fire investigators, and others whose services depend on such understanding.

Scientifically Based Fire Investigation

There is a natural tendency for everyone to be impressed by the magnitude and violence of the fire itself and the degree of destruction that results from a large fire. In the case of insurance investigators, this tendency is aggravated by the fact that they are expected to assess this damage and possibly pay a claim for it. For the accurate investigation of fires, this preoccupation with the magnitude and results of the fire is unfortunate. Cause, not extent, should be the investigator's concern, and too much attention to extent will obscure and complicate the search for the origin and cause.

COMPREHENSIVE METHODOLOGIES FOR FIRE INVESTIGATION

Because of the difficulty and complexity of a complete and accurate fire investigation, and preconceptions, there is a special need for every investigator to develop a comprehensive analytical approach to the task. This analytical approach recognizes that to be a successful fire investigator, a person must understand and master numerous facets of fires, fuels, people, and investigative procedures. The investigator must really understand how a fire burns, what factors control its behavior, and that all fires do not necessarily behave in a precisely predictable fashion. Such deviations from typical or expected **fire behavior** must be correlated with the fire conditions and their causes established. These causes usually involve the nature of the fuel involved or the physical circumstances and environment of the fire.

fire behavior ■ The manner in which a fuel ignites, flame develops, and fire spreads. Unusual fire behavior may reveal the presence of added fuel or accelerants.

One of the most important, yet often neglected, areas of fire investigation is an understanding of the fundamental properties of the fuels involved. These properties, which include the chemistry, density, thermal conductivity, and heat capacity of the fuel, determine the ignitability and flame spread characteristics of the fuel. These, in turn, control the nature of ignition as well as events that follow. These properties are determined primarily by the chemistry of the fuel, and the investigator should have a thorough knowledge of the relevant chemistry and what happens in the reaction. Different ignition sources, whether glowing, flaming, or electric arc, will affect fuels differently, and so the investigator must rely on scientific data about the interaction of such sources with the first fuel ignited to help establish whether a particular source is even competent to accomplish ignition and what time factors may be involved. The focus tends to be simply on identifying fuels in fire scenes, but ventilation factors are equally important (as without air, there won't be much of a fire to investigate). Investigators need to document and evaluate the sizes of rooms, windows, and doors, and changes in their condition during the fire. Ventilation factors must be considered as carefully as fuel factors. The fundamental requirement for a fire is the presence of fuel (in suitable form), oxygen, and heat (in suitable quantities), all linked by a self-sustaining chemical reaction. These factors must be taken into account by investigators when evaluating every stage of a fire's development from ignition to extinguishment. The physical principles of heat transfer and effects on surfaces must also be understood, since they are fundamental to fire spread and to the production of fire patterns on which investigators rely after the fire.

Any attorney who is involved in fire cases, whether criminal or civil, should be generally familiar with the same areas of chemistry. An attorney who has no understanding of what happens in chemical reactions and the results of those reactions will be seriously handicapped when questioning both lay and technical witnesses. Highly significant facts of the fire, which may be crucial in causing an erroneous decision in court, may not be recognized. Such knowledge can also influence the decision whether to try a case or settle it on the best available terms. Thus, it is advantageous for both the investigator and the attorney involved to have a basic understanding of the simple chemistry of fuels, fire, and **combustion** and the physical laws that control the production of post-fire indicators.

combustion ■ Oxidation that generates detectable heat and light.

Because most fires investigated involve either residential or industrial structures, there is a tendency to think of fires *only* in those terms. This error becomes apparent when a forest or wildland fire is under investigation. Although wildland fires involve the same basic combustion of fuel in air, the dynamics of fire spread and the analysis and interpretation of diagnostic signs of fire behavior are quite different from those for structural fires. Factors such as weather, topography, and fuel density must be assessed in such fires. Today we commonly face the problems of investigating fires in the urban/wildland interface, where houses and trees coexist. Wherever the fire is of a highly local but very specialized nature, such as a fire in an industrial facility, the investigator must carefully analyze what is found and seek technical assistance if the issues are even marginally in doubt. Such fires are increasingly common when dealing with new products or complex processes of modern technology with which even experienced investigators have no familiarity.

Explosions, closely akin to fires and sometimes accompanying them, also must be understood by investigator and attorney alike. Explosions, like fires, can be accidental or intentional in cause. The difference between a fire and an explosion may be so small as to allow confusion and may almost be a matter of definition. The interrelation and distinction between the two types of chemical events must be clearly understood so that the combined event is properly interpreted when it occurs. The explosion that accompanies a fire is usually quite different from that produced by a bomb, commercial explosive, or a premixed fuel–air mixture, although all involve the rapid exothermic oxidation of a fuel or starting material. That distinction is sometimes lost both in the investigative and legal phases of subsequent actions. The investigation of a bombing may be very different in character from that of a fire-related explosion, but both rely on systematic collection and analysis of data and careful formulation of accurate conclusions. These distinctions and comparisons are developed in greater detail in Chapter 12.

THE SCIENTIFIC APPROACH TO FIRE INVESTIGATION

The scientific approach to fire investigation to be discussed here extends not only to the review of the entire case but also to the examination of the scene, where patient, thorough, and systematic evaluation of the scene and its contents will usually pay off in information critical to the case. It must be appreciated that the destruction of some scenes is so complete due to prolonged burning that virtually nothing can be ascertained reliably from the ashes regarding the origin, cause, and development of the fire. There can be virtually no reliable interpretation of burn patterns to establish direction or duration of fire exposure. Even such scenes can yield information about the contents of the structure, which can reveal fraud. There is a curious tendency for some investigators to show an inherent bias as to cause. These investigators appear to believe that all fires have a single cause, for example, electrical appliances or smoking materials; they then expend a vigorous effort to prove that the fire had this cause, rather than spend the time seeking the true cause. Prejudgment of fire causation is as dangerous as the prejudgment of any other unexplained event or crime. It is unfortunate that this appears to be an occupational issue within many fields of investigation, but perhaps it is more noticeable in fire investigation than in other areas. Adhering to a systematic approach (which is not to say it is dogmatic) such as the one described here will minimize the unfortunate effects of such prejudicial tendencies.

There is also an inclination to extrapolate observations about fuels and combustion in small-scale fires to full-blown incidents without knowing whether they are, in fact, appropriate or accurate. It is here that the data from fire engineering and fire dynamics studies can best be applied, but only if the user is fully aware of the conditions and limitations of the information.

Applying the Scientific Method

In recent years, the term *scientific method of fire investigation* has been more widely introduced, sometimes to the confusion of fire investigators.[30] The concept does not mean that a fire investigator has to be a scientist, or to analyze things in a laboratory, or to replicate a fire scene in a field test or computer model. It means to apply the scientific method of logical inquiry to the investigation. It is an approach most good fire investigators already apply; it simply means the investigation follows a series of logical steps.

The universally accepted methodology cited by *NFPA 921* in conducting investigations into the origin and causes of fires centers on the use of the *scientific method*, which embraces sound scientific and fire protection engineering principles. *NFPA 921* defines the scientific method as

> the systematic pursuit of knowledge involving the recognition and formulation of a problem, the collection of data through observation and experiment, and the formulation and testing of a hypothesis.[31]

The basic concepts of the scientific method are simply to *observe*, *hypothesize*, *test*, and *conclude*. The scientific method is considered the best approach for conducting fire scene origin and cause determinations, analysis, and reconstruction. The process of the scientific method is that it continuously refines and explores various working hypotheses until arriving at a final expert conclusion or opinion, as illustrated in Figure 1-1.

STEPS IN THE SCIENTIFIC METHOD

As applied to fire investigation and reconstruction, the scientific method is a logical and iterative seven-step systematic process. Properly applied, this systematic process addresses how to correctly formulate, test, and validate a hypothesis.

As illustrated in Figure 1-1, the following is a summary of this seven-step process:

1. *Recognize the Need.* A paramount responsibility of first responders is to protect a scene until a full investigation can be started. After initial notification, the investigator should proceed to the scene as soon as possible and determine the resources needed to thoroughly conduct the investigation. It is necessary not only to determine the origin and cause of the event but to determine whether future fires, explosions, or loss of life can be prevented through new designs, codes, or enforcement strategies. In public-safety sectors, determination of the origin and cause of every fire is often a statutory requirement. In some cases, the fire investigator may be called on to establish responsibility for the fire.

2. *Define the Problem.* Devise a tentative investigative plan to preserve and protect the scene, determine the cause and nature of the loss, conduct a needs assessment, formulate and implement a strategic plan, and prepare a report. This step is usually focused on the determination of the cause and nature of the loss but may also include determining who has the primary responsibility and authority to interview witnesses, protect evidence, and review preliminary findings and documents describing the loss.

3. *Collect Data.* Facts and information about the incident are collected through direct observations, measurements, photography, evidence collection, testing, experimentation, case histories, and witness interviews. All collected data should be subject to verification of how they were legally obtained, their chain of custody, and notation as to their reliability and authoritative nature. Data collection should include the documentation of the property; construction and occupancy; fuel loading; the processing and layering of the debris and evidence as found; examination of the fire (heat and smoke) patterns; char depths and calcination; and the application of electrical arc mapping surveys where relevant.

4. *Analyze the Data.* Using inductive reasoning (which involves formulating a conclusion based on a number of observations), analyze all the data collected. The investigator relies on his or her knowledge, training, and experience in evaluating the totality of the data. This subjective approach to the analysis may include knowledge of similar loss histories (observed or obtained from references), training in and understanding of fire dynamics, fire testing experience, and experimental data. Data

Step 1 - Recognize the Need

- Respond to loss, protect scene, assess resources
- Investigate to identify conditions or persons responsible, prevent future similar losses

Step 2 - Define the Problem

- Devise and implement a tentative investigative and strategic plan
- Determine primary responsibility and authority
- Determine cause and nature of fire

Step 3 - Collect Data

- Describe, observe, and document incident
- Photograph, sketch, and collect evidence
- Conduct tests and experiments
- Interview witnesses
- Assemble historical loss histories of similar incidents

Step 4 - Analyze the Data (Inductive Reasoning)

- Evaluate, where applicable, fire pattern damage, heat and flame vectoring, arc mapping, fire engineering analysis
- Test predictions with established scientific knowledge of fire dynamics, loss histories, fire tests, and experimental data
- Recommend changes in working hypothesis

Step 5 - Develop a Working Hypothesis

- Develop a tentative working hypothesis from data analysis
- Address causal or mathematical relationships
- Formulate and test alternative hypotheses with established scientific and fire engineering knowledge

Step 6 - Test the Working Hypothesis (Deductive Reasoning)

- Compare hypotheses to all known facts, incidence of prior loss histories, authoritative fire test data, sound published treatises, and experiments
- Recommend collection and analysis of additional data
- Solicit peer review to identify alternative hypotheses
- Modify working hypothesis
- Eliminate all other fire causes, if possible

Step 7 - Select the Final Hypothesis

"Expert Conclusion or Opinion"

- Make authoritative conclusion as to area and point of fire origin(s), and cause
- Identify competent ignition sources
- Establish level of confidence of final hypothesis
- Prepare expert report

FIGURE 1-1 A flowchart outlining the scientific method as it applies to fire scene investigation and reconstruction. Derived from *NFPA 921* [NFPA 2008 and 2011, pt. 4.3]. *Source: Icove, D. J., and DeHaan, J. D.* Forensic Fire Scene Reconstruction, *2nd ed., Pearson Brady Publications, Upper Saddle River, NJ, 2009.*

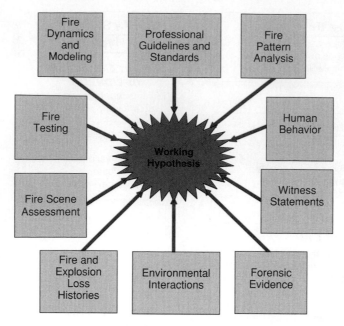

FIGURE 1-2 The working hypothesis relies on a wide spectrum of factors. *Source: Icove, D. J., and DeHaan, J. D. Forensic Fire Scene Reconstruction, 2nd ed., Pearson Brady Publications, Upper Saddle River, NJ, 2009.*

evaluated can include the pattern of fire damage, heat and flame vectoring, arc mapping, and fire engineering and modeling analysis.

5. ***Develop a Working Hypothesis.*** A *hypothesis* is defined as "a supposition or conjecture put forward to account for certain facts, and used as a basis for further investigation by which it may be proved or disproved."[32] Based on the data analysis, the investigator develops a tentative or working hypothesis to explain the fire's origin, cause, and development that is consistent with on-scene observations, physical evidence, and testimony from witnesses. The hypothesis may address a causal mechanism or a mathematical relation (e.g., plume flame height, impact of differing fuel loads, locations of competent ignition sources, room dimensions, impact of open or closed doors and windows). Figure 1-2 shows how working hypotheses take into account a wide spectrum of factors.

6. ***Test the Working Hypothesis.*** Deductive reasoning involves a conclusion based on previously known facts. Using deductive reasoning, compare the working hypothesis with all other known facts, the incidence of prior loss histories, relevant fire test data, authoritative published treatises, and experiments. Use the hypothesis testing to eliminate all other reasonable origins and causes for the fire or explosion, recommend the collection and analysis of additional data, seek new information from witnesses, and develop or modify the working hypothesis. Interactively repeat steps 4, 5, and 6 until there are no discrepancies with the working hypothesis. A critical feature of hypothesis testing is to create alternative hypotheses that also can be tested. If the alternatives are in opposition to the working hypothesis, their evaluation may reveal issues that need to be addressed. By testing all hypotheses rigorously against the data, those that cannot be conclusively eliminated should still be considered viable.

The scientific process as used in fire investigation is fundamentally different from that used in other fields, because in most fire cases, the solution or final hypothesis will not be tested by burning another house, vehicle, or wildland to see if the outcome is similar. The investigator must gather adequate reliable data and apply robust, reliable analyses (usually of basic indicators of fire movement, intensity, or duration) to develop a working (primary) hypothesis. Because these findings will often be "tested" in an adversarial legal proceeding, the fire investigator is expected to have devised and tested alternate explanations (hypotheses). The evaluation of all these hypotheses is sometimes called *abductive reasoning*. The final outcome of this process is the selection of the hypothesis that fits all available data and the rejection of all other reasonable hypotheses. When applied in medical evaluations, this process is called *differential diagnosis*. This process best describes most modern fire investigations. Even though it does not follow every sequential step of the pure "scientific method," it is a well-recognized analytical process.

It should be noted that this application of the scientific method is not the same as that used in experimental science. There, observation of an event causes the scientist to formulate a hypothesis about a possible explanation and then design experiments to gather data (information) with the specific intent of testing or rejecting that primary hypothesis. Only if the primary hypothesis has been rejected (or shown to be incorrect) will the experimenter return to the process and create another testable scientific hypothesis. The crucial test for scientific inquiries is to use the "proven" hypothesis to predict the outcome of further tests conducted by others (demonstrating its reproducibility and reliability). In fire/explosion investigations, it is very rarely possible to re-create an event to fully test the final hypothesis.

7. *Select Final Hypothesis (Conclusion or Opinion).* An *opinion* is defined as "a belief or judgment based upon facts and logic, but without absolute proof of its truth."[33] When the working hypothesis is thoroughly consistent with evidence and research, it becomes a final hypothesis and can be authoritatively presented as a conclusion or opinion of the investigation.

LEVELS OF CONFIDENCE

Upon selection of a final hypothesis, the confidence of an expert opinion is often determined. *Levels of confidence* may also vary with the profession of the person preparing the report. For example, *fire protection engineers* apply science and engineering principles to protect people, homes, workplaces, the economy, and the environment from the devastating effects of fires. Fire protection engineers also analyze how buildings are used, how fires start and grow, and how fires affect people and property. They commonly state in their written reports that their *expert opinion* is "to a reasonable degree of engineering certainty." Forensic opinions should be offered only at a very high level of certainty; that is, no other logical solutions offer the same level of agreement with the available data. In forensic origin and cause determination, the investigator applies scientific knowledge to re-create the path of spread of the fire, trace it back to its origin, and, there, establish the cause of the fire. Such opinions are often expressed "to a reasonable degree of scientific certainty."

When various hypotheses about the cause of the fire or explosion are tested, if only one hypothesis fits the available data and others are conclusively eliminated, the opinion can be expressed to a reasonable degree of scientific certainty (within the limitations of the available data). All scientific conclusions are subject to continual retesting or reevaluation if new reliable data are presented. For example, early observational "data" led the ancients to believe Earth was at the center of the universe. Later data showed that belief wrong and suggested a heliocentric universe. Modern data, of course, show our entire solar system to be circulating in an enormous galaxy moving across the universe.

If two (or more) hypotheses about the origin or cause of a fire or explosion exist and neither can be demonstrated to be false, the degree of certainty or confidence is reduced to "possible" or "suspected," and the conclusion should be "undetermined." The posing of alternative hypotheses is in the U.S. adversarial judicial system a fair way of testing the strength or certainty of expert conclusions.

An expert opinion should be expected to withstand the challenges of a reasonable examination by peer review or through cross-examination in the courtroom, according to *NFPA 921*.[34] Curiously, *NFPA 921* currently discusses only two levels of confidence with respect to opinions—probable and possible. There is no mention of a "conclusive" opinion! A "conclusive" opinion can be offered when all the available data fit the final hypothesis and all the reasonable alternatives can be tested and have been conclusively accepted or rejected. Obviously, if there is not a conclusive opinion, but the weight of available data suggests one alternative is vastly more likely than the others, a "probable" opinion can be offered. If two or more hypotheses are equally likely, then a "possible" opinion is justified.

The collection and interpretation of information, and the decision-making process, will be addressed in a later chapter of this book, but the steps presented here are what every professional fire investigator follows even when he or she does not stop and enumerate them.

LEGAL OPINIONS REGARDING SCIENCE IN INVESTIGATION

Some of the most significant changes in fire investigation in the last 15 years have been the result of legal interpretations of the *Daubert, Kumho Tire, Joiner,* and *Benfield* decisions (which will be discussed in some detail in Chapters 16 and 17). Today's fire investigator must demonstrate to the court that he or she has the training and qualifications to offer an opinion about the fire, has data (information) to support that conclusion, has followed an appropriate protocol in the investigation and decision-making process, and has followed, in general, the scientific method of inquiry. A basic protocol is outlined in the NIJ *Guide* described earlier. More detailed protocols are discussed in *Forensic Fire Scene Reconstruction*[35] and *NFPA 921: Guide for Fire and Explosion Investigations.*[36] Protocols will be discussed in Chapters 7 and 8. Note that a protocol is not a cookbook with all the steps laid out in mandatory order but an outline of a systematic, comprehensive approach to scene investigation.

The complexities encountered in fire investigation are sometimes overwhelming to the investigator, but patience and adherence to fundamental scientific principles of combustion, fire behavior, and heat transfer will usually result in a reasonable and defensible diagnosis of the fire if there are sufficient reliable data. These elementary principles of combustion, fuels, and heat effects, as well as the equally important principles of scientific investigation, are the concerns of this book.

Review Questions

1. Give three examples of direct costs associated with fire losses. Give three examples of indirect costs associated with fire losses.
2. What is the fundamental organizational problem that interferes with the complete investigation of fires in most communities?
3. What points of logic support the idea that arson should be considered a crime against people rather than a property crime?
4. What are some of the negative factors to overcome when investigating a fire scene?
5. List the steps an investigator might follow when using the scientific method of fire investigation.
6. Name some reasons why the *reported* number of incendiary fires is considerably less than the actual number that occur.

References

1. *NFPA 921: Guidelines for Fire and Explosion Investigation* (Quincy, MA: National Fire Protection Association, 2008), pt. 3.5.59; (2011), pt. 3.3.62.
2. *NFPA 1033: Standard for Professional Qualifications for Fire Investigator* (Quincy, MA: National Fire Protection Association, 2009).
3. M. J. Karter Jr., "Fire Loss in the United States, 2008" (Quincy, MA: National Fire Protection Association Report, Fire Analysis and Research Division, August 2009). (See also *NFPA Journal*, September–October 2009).
4. M. J. Karter Jr., "U.S. Fire Loss in 1987," *Fire Journal* (September 1988).
5. M. J. Karter Jr., "1999 U.S. Fire Loss," *NFPA Journal* (September–October 2000).
6. M. J. Karter Jr., "Fire Loss in the United States, 2003" (Quincy, MA: National Fire Protection Association Report, Fire Analysis and Research Division).
7. Karter Jr., "Fire Loss in the United States, 2008," 9.
8. R. F. Fahy, P. R. LeBlanc, and J. L. Molis, "Firefighter Fatalities in the United States in 2008," *NFPA Journal* (July–August 2009): 60–67.
9. M. J. Karter Jr. and J. L. Molis, "Firefighter Injuries for 2007," *NFPA Journal* (November–December 2008): 46–54.
10. National Safety Council, "Report on Injuries in America" (NSC: Itasca, IL, 2008).
11. P. Frazier, "Total Cost of Fire in the United States," *Fire Protection Engineering* 26 (April 2005): 32–36.
12. P. McEvoy and J. Gamble, "Summary Fire Statistics, United Kingdom, 2007" (Garston, Watford, Hertfordshire: Communities and Local Government, Fire Statistics and Social Research Branch, Building Research Establishment (August 2009).
13. Office of the Deputy Prime Minister, "Economic Cost of Fires in England and Wales—2003" (London: ODPM, 2005).
14. McEvoy and Gamble, "Summary Fire Statistics."
15. Office of the Deputy Prime Minister, *Statistical Bulletin*, "Fires in the Home: Findings from the 2004/2005 Survey of English Housing" (London: ODPM, 2007).
16. D. Von Drehle, *Triangle: The Fire That Changed America* (New York: Atlantic Monthly Press, 2003).
17. J. D. DeHaan, "Challenge of Fire Investigation" in *Proceedings Interflam 2001* (London: Interscience Communications, 2001).
18. *Crime in the United States: Uniform Crime Reports, 2006* (Washington, DC: U.S. Department of Justice, Federal Bureau of Investigation).
19. *Crime in the United States: Uniform Crime Reports, 2008* (Washington, DC: U.S. Department of Justice, Federal Bureau of Investigation).
20. Ibid.
21. Karter, "Fire Loss in the United States, 2008," 14.
22. Idem.
23. J. D. DeHaan and D. J. Icove, *Forensic Fire Scene Reconstruction*, 2nd ed. (Upper Saddle River, NJ: Pearson/Brady, 2009), 234.
24. "Communities and Local Government Fire Statistics, United Kingdom, 2007," (Garston, Watford, Hertfordshire: CLG, Fire Statistics and Social Research Branch, Building Research Establishment, March 2008).
25. ODPM, 2005.
26. McEvoy and Gamble, "Summary Fire Statistics."
27. National Institute of Justice, *Fire and Arson Scene Evidence: A Guide for Public Safety Personnel* (Washington, DC: U.S. Department of Justice), 2000.

28. ODPM, 2005.
29. A. C. Doyle, "A Scandal in Bohemia" in *The Complete Sherlock Holmes* (Garden City, NY: Doubleday), 1970.
30. S. J. Avato, "Scientific Methodology," *Fire and Arson Investigator* (July 2000): 35–36.
31. *NFPA 921*, pt. 3.3.139; 2011, pt. 3.3.62.
32. *ASTM E1138-89: Terminology of Technical Aspects of Product Liability Litigation* (West Conshohocken, PA: American Society for Testing and Materials Committee E30.40 on Technical Aspects of Products Liability Litigation (withdrawn 1995), 1989).
33. Ibid.
34. *NFPA 921*, 2011, pt. 18.6
35. Icove and DeHaan, *Forensic Fire Scene Reconstruction*.
36. NFPA 921, 2011.

CHAPTER

2

The Elementary Chemistry of Combustion

Courtesy of Jamie Novak, Novak Fire Investigations and St. Paul Fire Dept.

KEY TERMS

absolute temperature, *p. 29*

aliphatic, *p. 24*

aromatic, *p. 24*

atom, *p. 20*

Btu, *p. 27*

combustible, *p. 28*

diatomic, *p. 20*

exothermic, *p. 21*

flame, *p. 21*

flaming combustion, *p. 28*

flammable (same as inflammable), *p. 22*

heat of combustion, *p. 27*

hydrocarbon, *p. 23*

inorganic, *p. 23*

joule, *p. 27*

naphthenics, *p. 26*

normal hydrocarbons, *p. 24*

organic, *p. 23*

oxidation, *p. 21*

paraffinic, *p. 24*

pyrolysis, *p. 28*

saturated, *p. 24*

vapor, *p. 21*

volatile, *p. 26*

OBJECTIVES

After reading this chapter, you should be able to:

■ Understand what is a chemical reaction and how it is involved in fire.

■ Describe the basis of how elements can combine with other atoms to form molecules by chemical interaction.

■ Explain simple oxidation reactions involving carbon, carbon monoxide, or sulfur with oxygen.

■ List several carbon-containing organic compounds.

■ Describe four items formed during the thermal decomposition of wood (pyrolysis).

■ List and describe three physical states of fuel.

F ire is a chemical reaction that produces physical effects. As a result, the fire investigator should have some familiarity with the simpler chemical and physical properties and processes involved. Because fire consists of a number of chemical reactions occurring simultaneously, it is important to understand first what a chemical reaction is and how it is involved in a fire. This chapter also introduces some of the concepts and terms that the fire investigator might encounter when reviewing laboratory reports or meeting with experts.

Elements, Atoms, and Compounds

Under normal conditions, all matter is composed of elements or combinations of elements called *compounds*. An *element* is a substance that cannot be decomposed into simpler substances by chemical or physical treatment. Examples of elements are iron, carbon, oxygen, chlorine, sulfur, hydrogen, and chromium. There are 92 naturally occurring elements, along with a number of unstable, synthetic elements that are never encountered outside the laboratory. In descriptions of chemical reactions, each element is represented by a symbol, usually the first letters of an element's English or Latin name. For example, hydrogen is represented as H, oxygen as O, carbon as C, lead (plumbum) as Pb, and iron (ferrum) as Fe. See Appendix A for a list of the elements and their physical properties.

atom ▪ Smallest unit of an element that still retains its properties.

All elements are composed of very small units called **atoms.** All the atoms of a single element are the same as each other in terms of their size, weight,* and chemical properties, but are different from those of every other element.

Although the atom is not the indivisible block of matter that it was once thought to be, in a chemical reaction like fire, it does not break up into its component particles. Therefore, the atom is the smallest unit of an element that takes part in a chemical reaction.

The atoms of nearly every element can combine with other atoms to form molecules by chemical interaction. A *molecule* may be defined as the smallest unit of an element or compound that retains the chemical characteristics of the original substance. For example, the atoms of hydrogen, the simplest and lightest element, like those of most other elements, do not exist for very long as single or free atoms. They prefer to be in combination with other atoms. If the hydrogen is pure, two atoms of hydrogen will combine with each other

diatomic ▪ Molecules consisting of two atoms of an element.

to form a relatively stable **diatomic** (having two atoms) molecule of hydrogen gas represented by the chemical formula H_2. The letter H represents the hydrogen atoms, with the subscript 2 meaning that there are two hydrogen atoms in that molecule.

To understand these chemical combinations, it should be understood that every atom consists of a central core of subatomic particles called protons (positively charged) and neutrons (neutral or uncharged) in various numbers, surrounded by a cloud of negatively charged electrons circulating about the nucleus like planets around our sun. It is these

Isotopes of an element have different atomic weights and radioactive properties, however.

electrons and the ease with which they are shared with other atoms that determine the chemical reactivity of the element. This chemical reactivity tells us how easily the atom will break the bonds or ties that hold it in its molecule and how easily it will form bonds with other elements. The electrons surrounding the nucleus of each atom are what determine how that atomic element will prefer to combine with other atoms in particular ratios. For example, at normal temperatures oxygen is a diatomic gas represented by the formula O_2, which constitutes about 21 percent of the air. Occasionally, however, it will combine as a triatomic molecule, O_3. This compound is called *ozone* and is produced most often by a high-voltage electrical arc in an oxygen environment. The ozone molecule is not very stable and will readily give up its "extra" atom to form a stable oxygen molecule, O_2. The free atomic oxygen produced then very quickly reacts with other atoms in the vicinity to form oxides or with another free oxygen atom to form additional O_2 molecules. Oxygen is the most critical component of all ordinary fires because it is essential for common combustion. Eliminating it will extinguish nearly all fires.

THE OXIDATION REACTION

Fire is not only a number of simultaneous chemical reactions, it is, overall, a series of oxidative reactions. This means that atoms in the fuel are being *oxidized*, that is, combining with the oxygen of the air. There are many kinds of chemical reactions, but the ones of most interest in a **flame** are oxidations.

It is interesting to see what happens in a very simple **oxidation**, common to most fires. When hydrogen, an excellent fuel, is oxidized, two diatomic molecules of hydrogen combine with one diatomic molecule of oxygen to form two molecules of water. Simply, the chemical formula for that reaction can be written as

$$2H_2 + O_2 \rightarrow 2H_2O$$

Because water is a more stable compound than the gases that form it, the reaction occurs with great vigor and the output of much heat and is called an **exothermic** (heat-producing) reaction. If the gases are mixed before being ignited, a violent explosion will result upon ignition. When they are combined in a flame, a very hot flame is produced.

Because hydrogen is found in almost all fuels, even in the complex molecules that make up wood, plastic, or oil, the burning of virtually any common fuel results in the production of water **vapor** in large quantities. This water vapor is sometimes seen condensing on the cold glass windows of a burning structure. The conversion of chemically bound hydrogen in the fuel into water vapor during the fire also invariably leads to the production of much heat, although less than what is produced when pure hydrogen is burned. The net energy content of complex fuels is less than that of pure hydrogen or pure carbon because some of the energy is consumed in breaking the chemical bonds that hold the oxidizable atoms in their molecular structure.

Compounds may be represented by their numerical or empirical formulas, such as H_2O, CO_2, and O_2, or by structural or molecular formulas in which lines are used to represent the sharing of electrons by the atoms of a molecule. In this way water, H_2O, is represented as

$$H \diagdown \overset{O}{} \diagup H$$

and carbon dioxide, CO_2, is

$$O = C = O$$

The lines represent the forces or bonds between the atoms of elements making up the molecule. A single line indicates that one pair of electrons is shared, two lines indicate two pairs shared, three lines mean three pairs, and so on. In this way, the chemist shows not only how many atoms are in a molecule but in what order they are arranged and the nature of the chemical bonds between them, giving a great deal of information in a small space.

flame ■ A luminous cloud of burning gases.

oxidation ■ Combination of an element with oxygen; chemical conversion involving a loss of electrons.

exothermic ■ Generating or giving off heat during a chemical reaction.

vapor ■ The gaseous phase of a liquid or solid that can be returned to that phase by the application of pressure.

CARBON COMPOUNDS

Another element, carbon, represented chemically as C, is of very special interest in connection with fires because it, like hydrogen, is almost invariably a major component of fuels. In fact, it is the element around which most of the **flammable** compounds are built.

Some materials, such as diamonds and graphite, are pure carbon, but in crystalline forms that are very difficult to burn. When diamond does burn, it produces the same products (CO_2 and CO) and energy as any other form of carbon. When graphite burns, it is consumed so slowly that high-temperature crucibles are often made from it. Charcoal and coke are both industrial products consisting primarily of carbon, but quite impure carbon. These materials do not ignite easily. However, when afire, they produce considerable heat and are consumed at a slow rate. The general chemical equation ordinarily applied to the oxidation of carbon is

$$C(solid) + O_2(gas) \rightarrow CO_2(gas)$$

Carbon dioxide (CO_2) is always produced in fires involving carbonaceous fuels. It is an end product of nearly all combustions, including those that occur in the animal body. Exhaled breath is markedly enriched by carbon dioxide formed by oxidation of foodstuffs in the tissues. This is an oxidation like that which occurs in a fire, although the foodstuffs are not carbon alone but compounds of carbon with other elements. As a practical matter, in all fires there is also another reaction, which may be quite secondary or may become primary depending on the supply of oxygen. It is

$$2C(solid) + O_2 \rightarrow 2CO$$

Carbon monoxide (CO), the product of this reaction, is the commonly known gas that produces an asphyxiating effect (which will be dealt with in some detail in Chapter 15). Although CO invariably accompanies combustion reactions involving any carbonaceous fuel, it will not reach dangerous proportions in a properly adjusted gas appliance. Its concentration is very dependent on fire conditions: very low in freely burning flames but very high in oxygen-deficient fires or smoldering fires.

The three reactions given (those producing H_2O, CO_2, and CO) merely indicate the basis of the final combustion products and do not define the complex mechanisms by which the products are actually formed; however, they constitute the three most basic reactions of the fire. Water and carbon dioxide are the major products of nearly all fires, with carbon monoxide in somewhat lower concentrations in the effluent gases. Some fuels, notably those of coal or petroleum origin, are composed almost entirely of carbon and hydrogen and have only small amounts of other elements. A possible exception is sulfur, which is an impurity in most raw fuels. It also is oxidized, to form sulfur dioxide, according to the simple reaction

$$S(solid) + O_2 \rightarrow SO_2$$

Sulfur dioxide (SO_2) is the very sharp smelling gas often noted around metal smelters and other industrial installations.

OTHER ELEMENTS

Numerous other elements are found in the types of fuels most commonly encountered by fire investigators. Nitrogen does not burn in the sense of generating an exothermic reaction. Its role in a fire is complex, and yet it is inconsequential as a fuel. The nitrogen in air (~70%) absorbs heat and expands proportionally. Nitrogen may be encountered when it is part of a molecule such as a nitrate, which delivers extra oxygen to a reaction. The role of such oxidizers will be discussed in later chapters dealing with explosives and hazardous materials.

flammable (same as inflammable) ■ A combustible material that ignites easily, burns intensely, or has a rapid rate of flame spread.

Other elements present in wood include sodium, silicon, aluminum, calcium, and magnesium. These are the constituents (in the form of their respective oxides) that form the white or gray ash that remains after wood is burned. They do not contribute significantly to the combustion itself, with certain exceptions that will be mentioned later in this text.

Organic Compounds

Most of the compounds we have discussed up to this point fall into the general class of molecules called the **inorganics**. They are the combinations of sulfur, lead, chlorine, iron, and, in fact, all elements except carbon. The compounds based on carbon are so numerous and so vital to life processes that they are considered a branch of chemistry themselves. The chemistry of carbon-containing compounds is called **organic** chemistry, and most of the important fuels involved in structure and wildland fires are organic compounds. For the fire investigator, hydrocarbons and carbohydrates are the most significant organic compounds and are discussed next.

inorganic ■ Containing elements other than carbon, oxygen, nitrogen, and hydrogen.

organic ■ Compounds based on carbon.

HYDROCARBONS

The compounds that form good fuels are legion, but they fall into relatively few classes. Preeminent as a class are the **hydrocarbons**, compounds composed solely of carbon and hydrogen. The simplest of these is methane (CH_4), which is the chief component of natural gas. Written structurally,

hydrocarbon ■ Chemical compound containing only hydrogen and carbon.

$$H-\overset{\overset{\displaystyle H}{|}}{\underset{\underset{\displaystyle H}{|}}{C}}-H$$

the formula indicates the important fact that carbon has four *valences* or combining points on each atom. Hydrogen has only one, which means that four hydrogen atoms can combine with a single atom of carbon. The complete combustion of methane yields CO_2 and H_2O as final combustion products, but even this seemingly simple reaction goes through numerous intermediate reactions. The combustion of methane in oxygen is thought to involve about 100 such elementary reactions.[1] These reactions involve the loss of hydrogen atoms and then various recombinations that produce ethene and acetylene within the flame as well as unstable molecular species such as —OH, —CH_2O, and —CHO.

These unstable species are called *free radicals*. The more complex the molecule, the more pathways the intermediate reactions can follow and the greater the variety of free radicals that are produced. In hydrocarbon combustion, the —CHO and —OH free radicals (as well as CO) combine with O and H to form H_2O and CO_2. Free radicals can exist only at relatively high temperatures, but as they cool they can condense and form the pyrolysis products such as the brown "varnish" or oily residues that are seen adhering to surfaces after a fire. (These processes are explored in later sections of this text.)

In addition, carbon is somewhat unique in that its atoms have a very strong capacity to combine with one another to form chains, rings, and other complex structures. This capacity to form chains might be illustrated by the compound butane, which is packaged as a liquid petroleum gas, whose formula is C_4H_{10}, and is written structurally as

$$H-\overset{\overset{\displaystyle H}{|}}{\underset{\underset{\displaystyle H}{|}}{C}}-\overset{\overset{\displaystyle H}{|}}{\underset{\underset{\displaystyle H}{|}}{C}}-\overset{\overset{\displaystyle H}{|}}{\underset{\underset{\displaystyle H}{|}}{C}}-\overset{\overset{\displaystyle H}{|}}{\underset{\underset{\displaystyle H}{|}}{C}}-H$$

These straight-chain compounds can be extended almost indefinitely. They are called **normal hydrocarbons** or *n*-alkanes. Such chains can be branched to produce more complex structures without including elements other than carbon and hydrogen. Because there are many possibilities for the location of such branching, a large variety of compounds may exist that have the same empirical formula but different structures. For example, the compound isobutane has the same empirical formula as butane but the structure

Here the prefix *iso-* indicates a branched structure. Both chain length and point of branching can be varied almost indefinitely, so that there are an enormous number of possible compounds. All these compounds are called **aliphatic** or **paraffinic** (meaning they have the maximum number of hydrogen atoms attached to carbon atoms with no double bonds) and are named according to the number of carbon atoms in the longest carbon chain in the molecule. See Appendix B for some examples of the nomenclature for these compounds.

Another important class of hydrocarbons has a six-membered ring of carbon atoms as part of its structure. The simplest such compound is benzene (C_6H_6):

Note that there is an alternating sequence of double and single bonds between the carbons in the benzene ring. Because these ring structures impart a characteristic smell to many of their compounds, they are called **aromatics**. One or more of the hydrogen atoms of a ring compound can be replaced with any number of chemical structures, some of which can be very large and complex. Again, the simplest of the substituted benzene-ring or aromatic hydrocarbons is toluene (C_7H_8), with the structural formula

Just because the aliphatic compounds listed thus far are all **saturated**, or lacking double bonds, does not mean that all aliphatic compounds are in this category. For example, the gas eth*e*ne (not eth*a*ne, which is saturated) is C_2H_4, written structurally as while propene (C_3H_6) is

$$H-\overset{\overset{\displaystyle H}{|}}{\underset{\underset{\displaystyle H}{|}}{C}}-\overset{\overset{\displaystyle H}{|}}{C}=C\overset{\displaystyle H}{\underset{\displaystyle H}{\big<}}$$

Another category of unsaturated hydrocarbons involves triple bonds between adjacent carbons. Such compounds are usually designated by the suffix -*yne* as in pentyne (or 1-pentyne to designate the carbon atom to which the triple bond attaches):

$$H-C\equiv C-\overset{\overset{\displaystyle H}{|}}{\underset{\underset{\displaystyle H}{|}}{C}}-\overset{\overset{\displaystyle H}{|}}{\underset{\underset{\displaystyle H}{|}}{C}}-\overset{\overset{\displaystyle H}{|}}{\underset{\underset{\displaystyle H}{|}}{C}}-H$$

The most common triple-bond hydrocarbon encountered in fire investigation is ethyne (C_2H_2), or, as it is more commonly known, acetylene:

$$H-C\equiv C-H$$

Another class of hydrocarbons commonly encountered in fuels has rings consisting of five, six, or seven carbons but without the combination of single and double bonds that characterize the aromatics. These *cycloparaffins,* as they are called, appear as in cyclohexane (C_6H_{12}):

See Appendix B for additional hydrocarbon nomenclature and structure.

PETROLEUM PRODUCTS

All hydrocarbons are good fuels, but very few are ever used commercially as *pure* compounds. In fact, it is very expensive to isolate any single compound from petroleum or coal tar mixtures in a pure state. Once isolated, the pure compounds might not have the desired physical or chemical properties and would have to be blended with another compound in any event. Virtually all commercial fuels associated with fires are mixtures of relatively large numbers of individual compounds but with sufficient similarity in chemical structure that their combustion behavior may be quite similar.

Petroleum Distillates

Petroleum products are first separated from one another by heating raw petroleum and collecting the vapors as they distill off at various temperatures. Such products are labeled *petroleum distillates.* Each of these fractions or *cuts* will contain a mixture of all the compounds that boil off between two set temperatures. For example, petroleum ether is a cut of distillates that boil off at temperatures between 35°C and 60°C (95°F–140°F), while kerosene contains those that boil between 150°C and 300°C (300°F–572°F).[2] Most of the compounds found in these two fractions are aliphatic straight-chain or branched alkanes, and most are saturated—that is, with no double bonds. Some aromatics will also be present.

Cycloparaffins and aromatics have better characteristics as motor fuels, and their proportions are enhanced through chemical processing of the crude oil by such processes as cracking and reforming. They are then blended with various distillate cuts into fuels such as gasoline with the desired combustion properties. Gasoline is not a true petroleum distillate but rather a blended product that typically covers the boiling point range from 40°C to 190°C (100°F–375°F). Because all these common petroleum products are complex mixtures of hydrocarbons, not all their components evaporate or burn at the same rate. Their evaporation starts with the lightest and most **volatile** compounds and generally progresses to heavier compounds as the temperature rises (petroleum products that have undergone such progressive partial evaporation are sometimes called "*weathered*." For a discussion on the forensic impact of weathered petroleum distillates commonly found in fire scene debris, see Chapter 14 and the work by Stauffer, Dolan, and Newman[3]). During combustion, the process has the same progression but occurs at a much faster rate, as radiation from the flames causes the temperature of the liquid pool of fuel to rise. As a result, the residues of partially burned petroleum products found in fire debris have different chemical and physical properties (vapor pressure, flash point, specific gravity, viscosity, etc.) from those of the original fuel. At the same time, the flammability properties of the fresh or unburned fuel are determined in large part by the lighter, more easily vaporized components of the mixture. Winter-blend gasolines, for instance, may contain 6 to 10 percent by weight methyl butane and 5 to 6 percent *n*-pentane, while summer-blend gasolines may contain 4 to 8 percent methyl butane and 4 to 5 percent *n*-pentane to maintain the same bulk properties of vapor pressure and flash point at very different temperatures.[4] The physical and chemical properties of many of these liquid fuels will be discussed in more detail in Chapter 4.

Hydrocarbons with carbon backbone (chain) lengths greater than C_{24} are waxes at room temperatures. In general, the longer the chain length, the higher the melting and boiling points of the compound will be. Mixtures of various high boiling point hydrocarbons will be encountered as petrolatum (liquid or petroleum jelly) or paraffin wax. The heaviest (longest chain) hydrocarbons are found in asphalt.

Nondistillates

A number of new petroleum products are not true petroleum distillates but are used in a variety of consumer and industrial products that will be encountered in fire debris. One class of compounds consists of blends of isoparaffinic (branched hydrocarbon chain) compounds, which have good solvent properties but little of the odor commonly associated with petroleum solvents. They are used in insecticide solvents, hand cleaners, lamp oils, and some lighter fuels. Another class of petroleum products, called **naphthenics**, has had the aromatic and paraffinic components stripped out (by catalytic reactions or absorption/elution methods). These products consist largely of cyclohexane and cycloheptane with various substitutions. They have found very wide use in solvents and torch fuels. Others are blends of aromatics in various volatility ranges used in insecticides and adhesives for their specific solvent properties. There are also products that consist entirely of *n*-alkanes (sometimes just one or two, such as *n*-dodecane and *n*-tridecane) intended for use in liquid fuel candles. All these product types constitute possible volatile fuels but are synthesized products and not true petroleum distillates. Their identification and characterization will be discussed in more detail in Chapter 14.

CARBOHYDRATES

Perhaps the most important type of organic compound concerned in the study of combustion processes is in a different group, known as *carbohydrates*. These make up the bulk of wood, which is the most common fuel of structural fires. They differ from the hydrocarbons in very significant ways. Here the molecules are very large and complex. More significantly, they contain a relatively high content of oxygen; that is, they are already partially oxidized. The process of burning wood is simply a completion of the oxidation that started in the natural synthesis of the fuel itself.

volatile ■ A liquid having a low boiling point; one that is readily evaporated into the vapor state.

naphthenics ■ Class of hydrocarbons based on cycloalkanes.

Carbohydrates are so named because their chemical formulas include the elements carbon, hydrogen, and oxygen in multiples of the simple formula

$$-CH_2O-$$

which is a combination of a carbon with water (hydrate). Actually, the simplest carbohydrate has the formula

$$C_6H_{12}O_6$$

which is the empirical formula for several sugars, of which the most common are glucose (or dextrose), the sugar found in the blood; and fructose, which is obtained from grapes and other fruit. Cellulose, the major constituent of wood, is made up of many units of glucose attached to one another in chains. [This process takes place when each glucose molecule loses a water molecule (the reaction is represented by the formula $C_6H_{12}O_6 \rightarrow C_6H_{10}O_5 + H_2O$), and the molecules link up together at the open "holes" left in the structure.] Because cellulose is simply an infinite replication of glucose units (a natural polymer), the main reaction that occurs on burning cellulose is the same as that for burning glucose:

Glucose: $C_6H_{12}O_6 + 6O_2 \rightarrow 6CO_2 + 6H_2O$

Cellulose: $C_6H_{10}O_5 + 6O_2 \rightarrow 6CO_2 + 5H_2O$

Ordinarily, as with other fuels, not all the carbon is oxidized to carbon dioxide. In most fires, less oxygen is available and some CO (carbon monoxide) is produced instead of CO_2. Note that the hydrogen originally present is not as effective a fuel as it is in hydrocarbons, since it has already been partially oxidized in forming the original molecule. This partial oxidation accounts in part for the fact that wood fires do not readily achieve the high heat output of many other fuels.

Fats are a form of carbohydrates and can be significant fuels in some fires. Whether they are vegetable oils (such as cotton, corn, or linseed) or animal fats, they are carbohydrates (called *fatty acids*) with the characteristic molecular structure that they are straight-chain or branched hydrocarbons at one end and —COOH groups at the other, and are, in turn, substituted onto a glycerol backbone, as shown here:

heat of combustion ▪ The quantity of heat released from a fuel during combustion, measured in kilojoules per gram or Btus per pound.

Glycerol (Glycerin)   Linoleic acid ($C_{18}H_{32}O_2$)

Because of their largely carbon–hydrogen makeup, fats can burn readily and yield significant amounts of heat.

The amount of heat that any fuel can produce when burned is called its **heat of combustion**, and it is a measure of the number of **joules** or **Btus** (British thermal units) of heat released per kilogram (or pound) of fuel. The heat of combustion is a means of estimating how much total heat a given amount of fuel can produce; it is not a measure of the rate at which the heat can be released, or of the flame temperature a fuel will produce. (See Table 4-5 for some typical heats of combustion.) Heat of combustion is measured by *calorimetry* techniques. Heat of combustion is represented by the symbol ΔH_c, where Δ (delta) represents change, H represents heat (enthalpy), and the subscript c denotes combustion. Other constituents of wood, such as resins, not only have a higher heat of combustion

joule ▪ The SI unit of measurement of heat energy (1 J = 0.238 cal = 0.00095 Btu).

Btu ▪ British thermal unit. A standardized measure of heat, it is the heat energy required to raise the temperature of 1 pound of water 1 degree Fahrenheit.

than cellulose, but their chemistry produces a higher flame temperature. Therefore, resinous woods (such as pine or fir) are more likely to produce hotter fires than are nonresinous woods such as birch or balsa.

PYROLYSIS OF ORGANICS

Except for the simplest hydrocarbon fuels that merely need to evaporate to produce molecules simple enough to combine directly with oxygen in a flame, all fuels have molecules that have to be broken into small enough "pieces" to undergo combustion. The primary effect of heat on wood or other solid fuel is to decompose or pyrolyze it. The words *pyrolyze* or *pyrolysis* stem from the Greek words *pyro* (meaning fire) and *lysis* (meaning decompose or decay). Therefore, **pyrolysis** can be defined as the decomposition of a material into simpler compounds brought about by heat. Pyrolysis of wood, for example, yields

- burnable gases such as methane;
- volatile liquids such as methanol (methyl alcohol) in the form of vapors;
- **combustible** oils and resins, which may be as vapors of their original forms or pyrolyzed from more complex structures; and
- a great deal of water vapor, leaving behind a charred residue, which is primarily carbon or charcoal.

pyrolysis ■ The chemical decomposition of substances through the action of heat, in the absence of oxygen.

combustible ■ A material that will ignite and burn when sufficient heat is applied and when an appropriate oxidizer is present.

flaming combustion ■ Rapid oxidation of gases and vapors that generates detectable heat and light.

The gases and vapors generated diffuse into the surrounding air to burn via **flaming combustion**. The charcoal then burns with a considerably greater heat output on a weight-for-weight basis than does the original wood, since the noncontributing oxygen has been driven off, largely as water. The carbon combines directly with the oxygen in the air in contact with it to support glowing combustion. (From Table 4-5: the ΔH_c of wood is around 16 kJ/g, while that of charcoal is over 34 kJ/g.)

CONCLUSIONS ABOUT ORGANIC COMPOUNDS

It is apparent from the preceding considerations that fuels differ not only in their chemical nature but also in their mode of oxidation, the heat output of the reaction, and in other respects. They also have different characteristics with respect to the temperature at which they ignite, the quantities of oxygen used per unit of fuel, the rate at which they will oxidize, and other properties related to their physical state. Some of these characteristics will be discussed in Chapters 3, 4, and 5 of this text as they relate to the combustion of gaseous, liquid, and solid fuels.

State of the Fuel

Solids, liquids, and gases are the three physical states of matter of interest to the fire investigator. The state in which a fuel exists must be related to its other combustion properties, as we shall see in Chapters 3 and 4.

A *solid* is a substance whose molecules are held in a fixed, three-dimensional relationship to one another by molecular forces. Therefore, a solid has fixed volume and shape. Some solid fuels (like candle wax) melt to a liquid (which then evaporates to form a vapor). A few solids (naphthalene) evaporate directly to a vapor state (a process called *sublimation*). Many solids do not melt but decompose by pyrolysis.

A *liquid* has "looser" bonds that allow the molecules to move about but not escape. Thus, a liquid has no fixed three-dimensional shape but still retains a fixed volume. Gasoline, like every other liquid fuel, itself does not burn when in the liquid state. Such liquids are, however, easy to vaporize, and the vapor that is formed burns like any gas—by mixing with oxygen and combusting as a flame. Note that because the molecular bonds in liquids are typically not as strong as those in solids, less heat is required to break these bonds in liquids compared with those in solids.

The molecules in a *gas* have only very weak forces between them because the molecules are generally very far apart (depending on temperature and pressure). Thus, a gas will expand to fill any available volume and diffuse (more or less) freely in air. There are relatively few gaseous fuels at room temperature. However, many materials become gaseous at the temperatures that occur in a fire, and virtually all materials can be converted to gases by applying sufficiently high temperatures. Another hydrocarbon, paraffin wax, requires much higher temperatures first to melt, then to vaporize it to a form whose vapors will burn in the familiar form of a candle flame.

SIGNIFICANCE OF STATE OF FUEL

The classification of liquids, solids, and gases is far from absolute. They are to a large extent interchangeable, and the conversion from one state to another is a function of the temperatures and pressures that exist during a particular fire. The fire investigator must keep in mind that a change in physical state (such as from a bulk liquid to a finely divided aerosol when hydraulic fluid leaks under high pressure) can dramatically change the ignitability of the fuel and the size of the resulting fire. In general, the physical state of the fuel controls, in many ways, the ignitability of the fuel, its burning rate once ignited, and the quantity and nature of combustion products that result.

DIFFICULTY IN CLASSIFYING SOME HYDROCARBONS

Among the hydrocarbons are a number of materials that are difficult to classify as gases or liquids. The liquid petroleum gases, for instance, of which propane and butane are the prime examples, can exist briefly in both states at ordinary temperature and pressure. Because they are so very volatile, when exposed they readily vaporize, leaving no liquid residue. Propane, with a boiling point of −42°C, vaporizes much more rapidly than butane, which is heavier (longer carbon chain) and has a higher boiling point (−0.6°C).[5]

GASES AND THE IDEAL GAS LAW

Hydrogen, methane, ethane, propane, and acetylene are the most common gaseous fuels. Like other gases, these gases also obey (usually) the ideal gas law, which relates the amount of gas to its pressure, volume, and temperature.

The ideal gas law relationship is

$$pV = nRT$$

Where:

p = pressure
V = volume
n = number of *moles* of gas
T = temperature (**absolute temperature**, measured in Kelvin)*
R = universal gas constant (to make all the units come out right!)

absolute temperature ■
Temperature above
absolute zero (in K
or °R).

Increasing the temperature increases the volume if pressure is held constant or increases pressure if the volume is held constant. It is the heat-driven expansion of gases (or liquids) that gives rise to buoyancy. The same number of molecules occupying a larger volume produce a lower density.

If the amount of gas present is measured in moles (one mole of any material is equal in weight to its molecular weight and always contains the same number of molecules (Avogadro's number, or 6.023×10^{23}). One mole of oxygen (O_2) then weighs 32 g, and one mole of butane (C_4H_{10}) weighs 58 g. At standard temperature and pressure [0°C, 760 mmHg (1 atm)], one mole of gas will always occupy 22.4 L (liters).

*Temperature above absolute zero (0°C = 273 K). See Chapter 3, Table 3-2.

When vapors such as these are placed under pressure, they condense to form liquids. (By definition, a vapor can be condensed to a liquid by the application of pressure, where a gas is above its critical temperature and pressure and cannot be condensed by pressure alone.) Commercial-grade butane and propane [and the mixture often called liquefied petroleum gas (LPG)] are stored and transported in tanks, largely in liquid form. On release of the the pressure, as when the overlying gas in the tank is fed to a burner, more of the liquid vaporizes and maintains the same pressure (critical pressure) in the tank until all the fuel has evaporated.

SOLIDS

Solid fuels, discussed in greater detail in Chapter 5, as a rule do not burn as solids (there are a few exceptions). When a solid is burning, a portion of its surface may be smoldering, even glowing, yet a nonglowing solid fuel may be surrounded by flames, and we say that it is burning. This is not strictly true, for the flames are the result of gas–gas combustion, so it is the gaseous pyrolysis products that are mixing with the air to produce the flames and not the solid fuel. The smoldering phase generally follows when the thermal decomposition of the solid has slowed to the extent where some oxygen can reach the hot solid surface. (Very porous solid fuels may allow diffusion through the fuel as well.) If the surface is hot enough, a gas–solid interaction can take place that we observe as a glowing combustion. In a wood fire, the turbulence of the flames or the presence of a strong draft often makes it possible for glowing and flaming combustion to occur at the same time on the same piece of fuel.

Very finely divided solid fuels (such as wood dust or grain dust) can burn very quickly, even to the point of creating explosive forces, because the small volume of the individual particles permits rapid heating and decomposition into combustible products, and the large surface area permits rapid mixing with surrounding air.

LIQUIDS

A variety of liquid fuels exist. Examples include gasoline, pools of liquefied plastics, paint, organic heat-transfer fluids, and oils used in cooking and manufacturing processes. Of these liquid fuels, gasoline and pooled liquid plastics are of particular interest to the fire investigator.

Gasoline, being much more complex and having heavier molecules, has a higher boiling point (actually a range, as described earlier) and is mostly a liquid at ordinary temperatures and pressures. Thus, it is not necessary to seal it in a tank to retain it as a liquid. A small opening or vent to the exterior of the tank allows only a minor escape of the most volatile portions of the gasoline as vapors. When gasoline combusts in the cylinders of an automobile engine, most of it is actually vaporized in the carburetor (or by the action of injecting it at high pressure), and the remainder is vaporized in the hot cylinder, which means it undergoes a totally gaseous reaction of combustion. With a cold motor, in which some of the gasoline may not vaporize, the portion that remains liquid is not combusted and tends to escape past the pistons into the crankcase. If a gasoline engine is exposed to conditions below the flash point of the fuel, say −45°C (−50°F), it will not start until it is externally heated to the point where the fuel can produce an explosive mixture of vapor and air in the cylinders (or where mechanical dispersion by an injector system can create the necessary vapor pressure). Such interpretations may appear to be elementary, but if overlooked in the interpretation of a fire, they can lead to incorrect conclusions.

Plastics that melt and form pools of liquefied fuel as they burn will be discussed in more detail in Chapter 5.

Summary

Although fire is a complex phenomenon, it relies on only a handful of basic chemical reactions. The oxidation of carbon, hydrogen, and sulfur accounts for the largest volume of combustion products of common fuels. Investigators who understand that virtually all fire processes can be reduced to such simple reactions will be better prepared to examine the nature of the combustion reaction in more depth. Combustion—or, more precisely, the processes of oxidation, heat transfer, and fuel chemistry—will be treated in succeeding chapters.

Review Questions

1. The smallest unit of an element that takes part in a chemical reaction is a/an _____.
2. What natural element is necessary for normal combustion?
3. What three compounds are produced by the combustion of hydrocarbon fuels?
4. What is the difference between hydrocarbons and carbohydrates?
5. What is the difference between glucose and cellulose?
6. What is the *heat of combustion* of a fuel?
7. Name four *classes* of petroleum products.
8. What chemical structure is characterized by the designation *alkane*?
9. What chemical structure is characterized by the designation *aromatic*?
10. Define an *organic* compound.
11. According to the ideal gas law, if the temperature of a fixed amount of gas rises from 300 K (27°C) to 600 K (327°C), what will be its change in volume?

References

1. W. C. Gardiner, "The Chemistry of Flames," *Scientific American* 246, no. 2 (1982): 110–25.
2. *ASTM D3699: Specification for Kerosene* (West Conshohocken, PA: American Society for Testing and Materials, 2008).
3. E. Stauffer, J. A. Dolan, and R. Newman, *Fire Debris Analysis* (Burlington MA: Academic Press, 2008).
4. J. D. DeHaan, "The Reconstruction of Fires Involving Highly Flammable Hydrocarbon Liquids" (PhD dissertation, Strathclyde University, Glasgow, Scotland, 1995) (Ann Arbor, MI: University Microfilms, 1996), 11, 288–89.
5. R. H. Perry and D. Green, *Perry's Chemical Engineers' Handbook*, 7th ed. (New York: McGraw-Hill, 1997), table 3-1.

CHAPTER

3
Fundamentals of Fire Behavior and Building Construction

Courtesy of Jamie Novak, Novak Fire Investigations and St. Paul Fire Dept.

KEY TERMS

ambient, *p. 35*

backdraft, *p. 41*

buoyancy, *p. 35*

conduction, *p. 45*

convection, *p. 46*

entrainment, *p. 35*

fire-resistive, *p. 73*

firestorm, *p. 47*

flameover, *p. 56*

flame spread, *p. 40*

flammable liquid, *p. 36*

flashback, *p. 41*

flashover, *p. 49*

fuel load, *p. 54*

glowing combustion, *p. 38*

heat release rate, *p. 43*

heat transfer, *p. 45*

incipient, *p. 54*

overhaul, *p. 40*

plume, *p. 35*

radiation, *p. 47*

self-heating, *p. 40*

smoldering combustion, *p. 33*

stoichiometry, *p. 39*

vented, *p. 64*

ventilation, *p. 38*

watt, *p. 43*

OBJECTIVES

After reading this chapter, you should be able to:

- Describe the four conditions that must exist for a fire to occur.
- Recognize the details involved in the typical flaming combustion of organic fuels.
- Describe several states of combustion.
- Explain the three basic methods of heat transfer.
- Calculate the visible flame height given the heat release rate and effective diameter of a fire.
- Discuss the four phases of fire development.
- Explain the effects of environmental conditions on fire.
- Distinguish between Type III (ordinary) construction and Type V (wood frame) construction as seen in single-family residences in the United States.

32

Fire is an exothermic oxidation reaction that proceeds at such a rate that it generates detectable heat and light. As indicated in the previous chapter, two types of combustion (or fire) are (1) flaming and (2) smoldering. Flaming combustion, as defined in Chapter 2, is a gaseous combustion in which both the fuel and oxidizer are gases. **Smoldering combustion** involves the surface of a solid fuel with a gaseous oxidizer (usually the oxygen in air). Nearly all destructive fires are flaming combustion. The smoldering fire is not uncommon, either alone or in combination with flames. For example, the nonflaming fire in a mattress or a pile of sawdust is a good illustration of a smoldering fire (which can do extensive damage), as is the charcoal fire used for barbecues. Many destructive flaming fires began as small smoldering fires.

smoldering combustion
■ The direct combination of a solid fuel with atmospheric oxygen to generate heat in the absence of gaseous flames; see glowing combustion.

The differences between flaming and smoldering combustion are a result of the nature and condition of the fuel, as well as the availability of oxygen, all of which influence the rate at which heat is being produced. If the smoldering mattress or sawdust is stirred up, it may develop into a flaming fire because more fuel is brought into contact with oxygen. On the other hand, the glowing charcoal has little or no flame because the compounds that could be volatilized into gases (and thereby contribute to an open flame) were lost at the time the original fuel was charred or coked. The differences between a wood fire and a fire of charcoal made from wood are discussed later, in the section dealing with the properties of fuels.

Basic Combustion

For a fire to occur, the following conditions must exist:

- Combustible fuel must be present.
- An oxidizer (such as oxygen in air) must be available in sufficient quantity.
- Energy as some means of ignition (e.g., heat) must be applied.
- The fuel and oxidizer must interact in a self-sustaining chain reaction.

The first three elements listed have long been described as the *fire triangle*, but the fourth must also be present if the fire created is to be continuous (self-sustaining), thus creating what is called the *fire tetrahedron*[1] (see Figure 3-1). Although these requirements appear obvious, it is true, nevertheless, that statements have been made in court describing fires burning on bare concrete floors devoid of fuel. Absolute absence of any source of heat is sometimes claimed. In general, disregard of the simple requirements just listed is not at all uncommon. It must be remembered that removal of any one of the four elements results in extinguishment, and a careful analysis of all the factors is required to establish what happened, since mechanisms of ignition, combustion, and even sources of fuel may not be obvious at first glance. Some are transient and require careful analysis of possible mechanisms before their presence and effects can be reasoned out.

FIGURE 3-1 The fire tetrahedron: fuel, oxidizer, and heat interacting in a self-sustaining chemical reaction.

Experience with the common gaseous, liquid, and solid fuels is so widespread that misjudgments of their contributions to a fire would not be expected. However, there are many less common materials (e.g., plastics and metals) with which experience is less widespread. In addition, the mere presence of a suitable fuel in conjunction with the other conditions does not guarantee that a fire will result. The fuel must be present in a suitable physical state to be ignited. For example, common furnace oil is an excellent fuel but when spread on a concrete slab, it generally resists every effort to ignite it, even with a blowtorch. If that same oil is soaked into a cloth wick, it is easily ignited. Another example of the importance of the fuel's state is newspaper, which is so helpful in starting fires but can be used to extinguish some small fires. Proper interpretation of a fire situation requires knowledge of the properties of the fuel, its physical state, and the character and dynamics of the fire itself. Thus, the knowledge of availability and suitability of fuels in specific instances may require more than everyday experience.

Flaming Fire

A fire characterized by flame is the most common type. Here the flame actually *is* the fire—the production of gaseous reaction products with the evolution of heat and light. The color of the light emitted is determined in part by the elements in the reacting mixture. The color emitted by a hydrocarbon gas burning well mixed with air (e.g., a correctly adjusted oxyacetylene torch) is clear blue. The gaseous flame is made more visible when carbon and other solid or liquid by-products resulting from incomplete combustion are raised to incandescent temperatures and glow—red, orange, yellow, or white, depending on their temperature. The reactions of a *flaming fire* are, under ordinary circumstances, oxidative, with oxygen from the air as the oxidizer. An important fact to remember is that there cannot be a flaming fire unless a gas or vapor is burning. This holds true whether the fuel is a gas to begin with or a vapor evaporated from a liquid or distilled or driven off from a solid. The flame is a totally gaseous reaction. Liquid fuels can produce only flaming fires, since liquids themselves do not burn and do not pyrolyze to leave a solid char (except in unusual circumstances) to support smoldering.

Solid fuels often produce predominantly flaming fires, although they are frequently accompanied by glowing fire. Wood decomposes readily under the influence of heat to generate combustible gases; therefore, it tends to burn in flaming combustion, especially in the early stages of the fire. Later, the charcoal formed will continue to combust as a smolder fire. Coal, in contrast, has less volatile components and tends to give off proportionately

FIGURE 3-2 Typical flaming combustion of organic fuel showing decomposition region where volatized fuel decomposes to simpler species before combustion, and intermediate products are formed.

Buoyant smoke plume

Entrainment
(mixing, diffusing, and cooling)

Radiant heat

Flame plume

Radiant heat

Air   N₂

O₂

Decomposition region

Vaporized fuel

Pyrolysis zone

Fuel

O₂

N₂   Air

Entrainment
(supplying O₂)

smaller quantities of combustible gases than wood and a lesser amount of open flame. Figure 3-2 illustrates the dynamics of flaming combustion of a solid or liquid. Note that heat generated by the flame radiates onto the surface of the fuel, raising its temperature, evaporating it if it is a liquid, and pyrolyzing it if it is a solid. [The equilibrium surface temperature of a solid fuel is controlled by its chemistry. For most fuels it is on the order of 350°C to 500°C (660°F to 930°F)].[2] The molecules of vapor boiled out of the liquid or pyrolyzed out of the solid are dissociated or torn apart by the chemical conditions within the flame (sometimes called *reducing* conditions). They can then readily combine with oxygen coming from the outside. The **plume** of hot gases (and solid [soot] or liquid [aerosols] products from incomplete combustion) rises by **buoyancy** (being hotter and therefore less dense than the **ambient** air in the room). This movement draws surrounding air and its oxygen into the flame, entraining and mixing it with the fuel gases, thereby maintaining the reaction. **Entrainment** is the drawing of air (or other gases) into the buoyant plume.

If a flaming fire is confined to a closed room where the oxygen being used in the flames is not replaced, the oxygen concentration in the surrounding air will fall from its normal level of about 20.9 percent. The rate at which the concentration of oxygen drops depends on the size of the fire, the size of the room, and the amount of fresh air that can leak into the room from any small openings. Generally speaking, when the oxygen concentration falls below about 15 percent in the vicinity of the flame, the combustion rate of ordinary combustibles begins to decrease. Eventually, the concentration reaches a limiting oxygen concentration that will no longer support flaming combustion, and the flames will die out.

plume ■ The convective column of hot gases generated by a flame.

buoyancy ■ Tendency or ability to rise or float in air or liquid as a result of a difference in density.

ambient ■ Surrounding conditions.

entrainment ■ The mixing of two or more fluids as a result of laminar flow or movement.

The oxygen limit that will actually extinguish the flames is dependent on the nature of the fuels involved and on the temperature of the combustion gases—the higher the temperatures produced, the lower the oxygen concentration can drop and still support flames. For instance, oxygen concentrations between 5 and 8 percent have been observed in room tests with **flammable liquids** where the ceiling temperatures were on the order of 900°C to 1,000°C; and in post-flashover room fires, oxygen concentrations between 0 and 5 percent have been observed when temperatures in the hot gas layer were over 1,000°C.[3]

Structure of Flames

Most of the flames we encounter in fire investigation are *diffusion flames*. This means that the gases or vapors that are supporting the flame diffuse outward (or upward) from the surface of the fuel and the oxygen for combustion diffuses toward the fuel from the surrounding air. If the concentration of fuel and the concentration of oxygen are plotted as a function of distance from the fuel surface, a plot like Figure 3-3c results. At some distance,

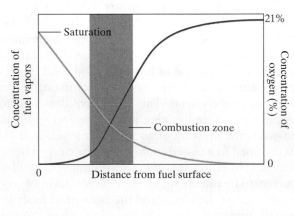

FIGURE 3-3 (a) Diffusion of fuel vapor away from vertical fuel surface as oxygen diffuses toward fuel surface. (b) Diffusion of fuel vapor upward from horizontal fuel surface. (c) Plot of concentration of O_2 and fuel vapor as a function of the distance from the fuel surface.

the concentrations of fuel vapors and oxygen are correct, the mixture can be ignited, and a flame can be maintained in that zone as long as the rate of fuel generation is maintained.[4] (The heat from the flame radiating onto the fuel surface helps maintain that supply from liquid or solid fuels.) The rate at which fuel is delivered to the flame is dependent on the total heat flux reaching the fuel surface and the latent heat of vaporization (for liquids) and latent heat of gasification (for solids). In combustion explosions, the fuel vapors and air are mixed together and then ignited in what is called a *premixed flame*. In gas appliances, this mixing is created by the gas jet. If properly adjusted, the gas is fed into the combustion zone at a rate and at a concentration at which nearly complete (and controlled!) combustion occurs in the desired location.

A small flame such as that of a candle is called a *laminar flame*, because the zones of high fuel concentration, dissociation (reducing), mixing, and combustion are well defined in layers surrounding the candle wick. The airflow around the flame is such that an envelope of very high temperatures is maintained around the source of fuel. In a candle, this high-temperature zone allows the combustion of the soot created by the pyrolysis (described in Chapter 2) of the wax to take place at very high temperatures (with the soot particles radiating visible light [incandescence] in the yellow-white colors of very high temperature surfaces). By the time the soot escapes the high-temperature zone it has been (almost) completely consumed and turned into CO_2 (as in Figure 3-4). If the flame becomes too large, the smooth flow of the air cannot be maintained, or if the envelope is disturbed by a sudden draft, this, in effect, tears open the high-temperature zone, allowing some of the soot and other intermediate products to escape unburned. The presence of the ragged tips of flames like these are a sign that the combustion has

(a)

(b)

FIGURE 3-4 (a) Typical laminar flame of candle showing blue luminosity at base. *Courtesy of John D. DeHaan.* (b) Temperature distribution in laminar candle flame showing a zone of very high temperature. *Source: From* Fire, *H. Rossotti, St. Anne's College, Oxford University, United Kingdom.*

FIGURE 3-5 Gasoline pool fire on plywood produces a typical turbulent flame plume.
Courtesy of Jamie Novak, Novak Fire Investigations and St. Paul Fire Dept.

become *turbulent* (as in Figure 3-5). Turbulent combustion dominates nearly all fires, causes the flicker that we associate with most fires, and in some cases can be extremely turbulent.

Smoldering Fire

A smoldering fire is characterized by the absence of flame but the presence of very hot materials on the surface of which combustion is proceeding. If the temperature of that surface is high enough [500°C (900°F) or higher], a visible glow or incandescence can be seen. As we shall see later, the color of the incandescent glow on the surface is related to its temperature, as is also true of the airborne particles in the flame. Some visible colors and their related temperatures are illustrated in Table 3-1. (Note that these are not the temper colors visible on polished metal surfaces after heating and cooling.) "Glowing" and "smoldering" are often used interchangeably [although Babrauskas describes **glowing combustion** as a non-self-sustaining condition (usually as a result of forced **ventilation**), while smoldering is self-sustaining solid gas combustion].[5]

glowing combustion ■ The rapid oxidation of a solid fuel directly with atmospheric oxygen creating light and heat in the absence of flames.

ventilation ■ A technique for opening a burning building to allow the escape of heated gases and smoke to prevent explosive concentrations (smoke explosions or backdrafts) and to allow the advancement of hose lines into the structure.

TABLE 3-1	Visual Color Temperatures of Incandescent Hot Objects	
COLOR	APPROXIMATE TEMPERATURE (°C)	APPROXIMATE TEMPERATURE (°F)
Dark red (first visible glow)	500–600	930–1,100
Dull red	600–800	1,110–1,470
Bright cherry red	800–1,000	1,470–1,830
Orange	1,000–1,200	1,830–2,200
Bright yellow	1,200–1,400	2,200–2,550
White	1,400–1,600	2,550–2,910

Source: Data taken from C. F. Turner and J. W. McCreery, *The Chemistry of Fire and Hazardous Materials* (Boston: Allyn and Bacon, 1981), 90. See also D. D. Drysdale, *An Introduction to Fire Dynamics,* 2nd ed. (Chichester, UK: Wiley, 1999), 53.

A charcoal or coal fire is a good example of smoldering combustion. In the past it was common to see a blacksmith heating iron objects in a fire made with coal through which a forced-air draft passed. These fires showed very little flame, but the glowing fire induced by the air draft was intensely hot, more so than most flaming fires. Similar situations are sometimes seen in the deep piles of hot coals left on collapse of large wood-frame structures. When there is sufficient draft through the coals and a layer of insulative ash to retain the heat, the temperatures reached can melt copper, steel, silver, and even cast iron, producing unusual artifacts. The duration of such high temperatures in glowing coals can be considerably longer than the high temperatures reached during the open-flame stages of the fire, making it possible to melt even massive solid objects.

Because the smoldering fire involves two phases, a solid and a gas, the ratios of oxidizer and fuel are not of direct concern, although the chemistry of the reaction still determines the total consumption of each. The proportions of fuel, oxygen, and final products are called the **stoichiometry** of the reaction. Perfectly balanced chemical reactions, which result in the complete reaction of all starting materials with no waste, are called *stoichiometric* or *ideal reactions*. Although glowing/smoldering combustion is often considered the end (decay) stage of a fire, it can occur at any time during a fire depending on the ventilation conditions. Figure 3-6 shows glowing combustion on the surface of a burning wood timber where the incoming draft is so strong it is pushing the vapors generated from the wood aside and forcing very rapid glowing combustion of the surfaces. The vapors burn as flames to the sides and rear of the timber, where they can mix with air.

Smoldering combustion can continue even in an oxygen-deficient atmosphere (below 5 percent), in part due to the very high temperatures at the oxidizing surface. In the combustion of ordinary solid fuels, a transition from flaming to smoldering may be seen as the oxygen level drops in an enclosure, which can revert back to open flames if a fresh supply of oxygen is introduced. In this instance, unlike in a flaming fire, the fuel and oxidizer

stoichiometry ■ Balance of chemical reactants and products.

FIGURE 3-6 Glowing combustion on timber with a very strong draft pushing combustion gases away from the burning surface. *Courtesy of Lamont "Monty" McGill, Fire/Explosion Investigator.*

are not premixed or even mixed at the moment of combustion but react as solid surfaces over which the oxidizing gas passes. Inasmuch as fuels are generally organic materials that are often reduced to charcoal (or coke) as the volatile decomposition products are distilled out and burned, most fires reach a smoldering state in their later phases. It must be remembered that the smoldering material here is not the original fuel but a derived fuel resulting from the earlier portions of the fire.

An important feature of the glowing or smoldering fire—especially when there is a forced draft, as in the blacksmith's forge—is that the limited flame that accompanies it results chiefly from the burning of carbon monoxide (CO) to carbon dioxide (CO_2). The carbon monoxide is generated by incomplete combustion of carbon due to insufficient oxygen and is further oxidized with a hot, but rather small, generally bluish, flame.

In the investigation of most fires, the *smoldering phase* generally represents either the incipient stage or the last part of the fire process. As an incipient fire, smoldering may be very prolonged and may do a great deal of damage before flames occur. The investigator and indeed the firefighter may be much more concerned with flaming fire and its effects in the earlier stages of the fire. Much of the post-fire cleanup or **overhaul** by firefighters at fire scenes, however, is directed at seeking out smoldering spots and ensuring that they do not rekindle into a destructive fire later. The investigator is primarily concerned with the initial small fire, whatever its nature, that grew into the large one.

Some exceptions to these observations must be considered. A fair proportion of flaming fires follow a period of smoldering. When the fire is ignited in a suitable organic fuel or in an environment in which ventilation is very limited, it does not generate the heat release rate necessary to burst into flames. This slow combustion may proceed for an appreciable time before the flaming fire erupts and yet do a great deal of damage. The rate of spread of a smoldering fire may be on the order of 0.0001 m/s (0.2 to 0.25 in./min) as compared with the 0.001 m/s to 0.01 m/s (0.04 to 0.4 in./s) typical of **flame spread** across solid fuels.[6] The conditions for transition from smolder to flame are not well understood at this time and have not been successfully modeled.

Smoldering fires may be produced by the ignition of latex rubber or natural fiber (cotton) mattresses or upholstery padding in residences; hay or grain piles in barns; or coal dust, sawdust, or other finely divided cellulosics in industrial premises. Fires that appear to start at times when there is no obvious source or reason for ignition are often attributed to such prolonged ignition mechanisms. Proof of such mechanisms requires careful consideration of all the following factors:

- initial source of ignition (there must be some source of heat to trigger the process);
- evidence of prolonged smoldering (adequate fuel mass);
- nature of the fuel (its ability to form a rigid porous char that will allow air (oxygen) to percolate into the mass of fuel); and
- cause (or sequence of events) that caused the fire to reach the open-flame state, that is, "burst into flame."

Even if the fire never reaches flaming combustion, considerable destruction may result from the slow, long-term burning of closely packed fuels such as paper, grain, cloth, or upholstery stuffing. The extensive production of carbon monoxide typical of smoldering fires makes them serious life safety risks. The smoke and heat generated by even modest smoldering fires can seriously damage other contents of a room or building.

Even in the absence of open flame, a smoldering fire is still a fire because it can accomplish the destruction of the fuel by the thermal processes of pyrolysis and combustion. All smoldering fires produce high concentrations of CO and visible smoke, since such combustion is never complete. The mechanisms of **self-heating**, by which a mass of fuel can reach its ignition conditions in the absence of any outside ignition source will be considered in Chapter 6. A destructive smoldering fire need not be accidental. Some incendiary fires are started in a suitable fuel but in a restricted space where the oxygen

overhaul ▪ The fire-fighting operation of eliminating hidden flames, glowing embers, or sparks that may rekindle the fire, usually accompanied by the removal of structural contents.

flame spread ▪ The rate at which flames extend across the surface of a material (usually under specific conditions).

The extensive production of carbon monoxide typical of smoldering fires makes them serious life safety risks.

self-heating ▪ An exothermic chemical or biological process that can generate enough heat to become an ignition source.

supply is deficient and flames are suppressed. Enough air may be available to sustain a smoldering fire until the fuel is exhausted or more oxygen becomes available.

Another consequence of smoldering results from the discarding of previously flaming, but supposedly extinguished, materials that are still combusting. Discarded matches that are not specially treated will continue to smolder for some time after the flame is extinguished and can initiate a fire in a suitable fuel. The smoldering cigarette certainly falls into this category (see later section). Hot coals, spilled from fireplaces or removed from barbecues or other similar sources, will start larger fires if they contact suitable fuel. These will be discussed in Chapter 6 on sources of ignition.

Explosive Combustion

Explosive combustion may not be recognized as a fire. It requires consideration here because this type of combustion does accompany fire, often as an initiating factor and sometimes during a conventional fire when favorable circumstances develop. Because this type of combustion is considered in a later chapter, it will be discussed only briefly at this point.

Explosive combustion can occur when vapors, dusts, or gases, premixed with an appropriate amount of air, are ignited. Under these circumstances, the combustion that results is not different from that which happens in the burning of these materials, except that the premixing allows the entire combustion to occur in a very short period of time. Thus, all the heat generation, creation of combustion products, and the expansion of those products—which normally would require an appreciable time—become an almost instantaneous event that is recognized as an explosion by its mechanical effects. The event may be very forceful and produce great damage, including the blowing apart of an entire building, or it may be so small as to produce only an audible pop.

Sometimes, the first phase of a fire is an explosion, as flammable gases or vapors have accumulated from some source, mixed with appropriate amounts of air, and become ignited. Such explosions are sometimes followed by flaming fire, especially if there is residual unburned fuel after the initial combustion. As we shall see later, the brief duration of such events precludes ignition of all but the thinnest, lightest fuels to create a following fire. In such instances, it is the cause of the initial explosion that is the concern of the investigator, since it is known that flammable gases were generated by some source and that they were later ignited. It remains for the investigator to identify the source of the fuel vapor, a suitable ignition source, and the conditions that brought them together.

When a vapor explosion occurs during a fire, it is usually an indication that a new source of fuel was made available locally and in some quantity. The most common source of this material is a sealed can, bottle, or drum of an ignitable liquid that breaks or bursts because the heat of the fire has expanded the contents beyond the strength of the container. Being heated above its normal boiling point, this fuel vaporizes immediately upon release and forms an explosive mixture, which is then ignited when an ignition source such as a flame is present. From the investigative standpoint, this event has two values. First, it proves that the container was sealed initially and was therefore not involved in starting the fire. Second, it proves that an extensive fire and much heat surrounded the container before it burst, and this situation was the result of fire progression from elsewhere, an unusual accident, or the work of an arsonist.

These facts can be of importance in the interpretation of many fires. Another cause of an explosion during a fire is a smoke explosion, also called a **backdraft** or **flashback**. In energetic fires in confined spaces, a great deal of fuel may be pyrolyzed into combustible gases and vapors that cannot burn because there is insufficient oxygen. If fresh air is suddenly admitted to the compartment (by the failure of a window or the opening of a door) and mixed with this hot, fuel-rich mixture, a destructive combustion explosion can occur. The dynamics of explosive combustion of gases and vapors are discussed in Chapters 4 and 12. Explosions involving detonating (high) explosives are not combustion

backdraft ▪ A deflagrative explosion of gases and smoke from an established fire that has depleted the oxygen content of a structure, most often initiated by introducing oxygen through ventilation or structural failure.

flashback ▪ The ignition of a gas or vapor from an ignition source back to a fuel source (often seen with flammable liquids).

based, as they do not involve the combination of a fuel with atmospheric or chemical oxygen. They occur when the explosive material itself dissociates (comes apart) and then recombines with an extremely high rate of reaction and heat release rate. These chemical reactions will be discussed in more detail in Chapter 12.

Heat

Aside from the significance of the chemical reactions that produce a fire, the most fundamental and important property of the fire is heat, which is a measure of energy. Heat initially starts a fire, and the fire produces heat. As will be discussed in the chapter on ignition, every method by which a fire may be ignited involves the application or generation of heat, and nothing but heat is ever necessary to start the fire *when the fuel and its environment are suitable.*

However, most of the destructiveness of a fire is the direct result of the heat generated. Heat produces the damage to the structure, intensifies the fire, is the means by which the fire spreads and enlarges, and provides the greatest barrier to the extinguishment of the fire. When the heat can be properly studied and understood in connection with a fire, the sequence and cause of that fire will generally be very clear. There are several special considerations of heat, as applied to a fire investigation, that require understanding:

1. Heat as it applies to igniting the fire (This topic is developed further in Chapter 6 on ignition.)
2. Heat as it applies to the increase of the rate of chemical reactions (including fires) as they develop
3. The transfer of heat as the factor controlling the spread of fire when the additional conditions of available fuel and oxygen are met
4. The effects of heat on materials that survive the fire to bear indicators of the fire's intensity, duration, or direction of spread

HEAT AND THE RATE OF REACTION

The rate of all chemical reactions is dependent on temperature. With rare exceptions, all reactions have a higher rate at greater temperatures. The Q_{10} *value* is the increase in rate of the reaction that results from raising the temperature 10°C (18°F). For most oxidation reactions, this value is two or more. Therefore, it can be generalized that the rate of a combustion reaction usually doubles with every increase of 10°C. The great importance of temperature stems, at least in part, from the fact that the fire generates much heat and raises the temperature of the reacting components, thus increasing the rate of reaction. This in turn generates more heat, thus again increasing the rate of reaction. This is the key to the fourth element of the fire tetrahedron—the *chain reaction*. If it were not for the diminishing availability of fuel and oxygen, combined with loss of heat to the larger surroundings, that brings this chain reaction under control, every fire would become a violent and rapid holocaust. Because these factors do act to control fire, the most immediate significance of the concept of rate of reaction is in connection with those processes that lead to spontaneous combustion. This subject is treated separately in a later section; however, it can be briefly considered in the following light.

Assume that a system of fuel and oxygen is properly contained, insulated, and proportioned, and a very limited exothermic reaction is proceeding. This might be a slow oxidation, bacterial activity, or some other type of chemical transformation in which heat is generated. If the heat cannot escape as fast as it is generated, the temperature will rise. As it rises it speeds up the very reaction that is generating the heat, thus leading to an increasing rate, which results in greater and greater quantities of heat. This heat will finally be sufficient to raise the temperature of the system to the ignition temperature of at least part of the fuel. Naturally, when this temperature is reached, the fuel ignites to a free-burning fire, and we say that it resulted from spontaneous combustion. If the heat *can* escape (due to convective, radiative, or conductive losses), the reaction rate does not increase as rapidly, and the reaction may continue at a reduced rate until the temperature stabilizes at some point and will not increase further.

HEAT AND TEMPERATURE

Heat is energy in a kinetic form, or energy of molecular motion. Except at absolute zero ($-273°C$, $-460°F$, 0 K) all matter contains heat because its molecules are in motion. Temperature is merely an expression of the relative amount of this energy that a body possesses. Heat or temperature is described using four units of measure: Celsius (°C), Kelvin (K), Fahrenheit (°F), or Rankine (°R). The conversions are as follows:

$$T_C = (T_F - 32)\, 5/9$$
$$T_K = (T_F - 32)\, 5/9 + 273$$
$$T_F = T_C\, (9/5) + 32$$
$$T_R = T_F + 460$$

Table 3-2 presents selected temperatures in those four units of measure, as well as conversions.

The total amount of heat contained in a body is determined by the mass of that body and by its *thermal capacity*. The thermal or *heat capacity* of material is a measure of how much heat must be added to a given mass of material to increase its temperature. The value is given in joules/kilogram kelvin (J/kg · K) (since it is dependent on the temperature of the mass as well). Thermal capacity is usually represented by the symbol c_p. The thermal capacities of some materials are shown in Table 3-3 (see also Appendix C). The fundamental properties of a material that influence its *ignitability* are thermal capacity, thermal conductivity, and density. These properties will be explored in later sections of this text. For the purposes of fire investigation, it must be remembered that heat is a measure of energy, while temperature is merely a means of comparing the heat content of two objects.

HEAT RELEASE RATE

The *amount* of heat produced by a fire or any heat source is measured in calories or, more commonly, in British thermal units (Btu) or joules (or kilojoules, kJ) (1 Btu = 252 cal = 1,055 J = 1.055 kJ). The *rate* at which that heat is being produced by the heat source is measured most conveniently in joules per second, Btu per second, or calories per second. Because 1 J/s = 1 **watt**, the rate is measured in watts (W) or Btu/s (0.95 Btu/s = 1000 W = 1 kilowatt, or 1 kW). This value, the **heat release rate,** is usually represented by $\dot{Q}$ (the dot meaning per unit time). $\dot{Q}$ is the quantity that characterizes the size of a fire. Another way of looking at

watt ■ Unit of heat release; 1 watt = 1 joule/second; unit of power or work (in electrical circuits, equivalent to voltage multiplied by amperes).

heat release rate ■ The rate at which heat is generated by a source, usually measured in watts, joules per second, or Btu per second.

TABLE 3-2	Temperatures Measured in Different Units		
K	**°C**	**°F**	**R**
773	500	932	1392
373	100	212	672
298	25	77	537
273	0	32	492
255	−18	0	460
233	−40	−40	420
0	−273	−460	0
Kelvin	Celsius	Fahrenheit	Rankine

TABLE 3-3	Heat Capacities and Thermal Conductivities of Some Common Materials		
MATERIAL	**HEAT CAPACITY** (c_P) (@20°C) (J/kg · K)	**THERMAL CONDUCTIVITY** (k) (W/m · K)	**THERMAL INERTIA** $(k\rho c)^a$ $(W^2/m^4K^2)/s$
Copper	380	387	1.3×10^9
Aluminum	900*	273*	—
Steel (mild)	460	45.8	1.6×10^8
Glass (window)	840	0.76	1.7×10^7
Brick (common)	840	0.69	9.3×10^5
Gypsum	840	0.48	5.8×10^5
Wood (oak)	2,850	0.17	3.2×10^5
Polyurethane foam	1,400	0.034	9.5×10^2
Cotton	1,300*	0.06*	—
Air	1,040	0.026	—
Water	4,180[†]	0.60	2.5×10^6

Sources: D. D. Drysdale, *An Introduction to Fire Dynamics*, 2nd ed. (Chichester, UK: Wiley,1999), 33.

*R. H. Perry and D. Green, *Perry's Chemical Engineers' Handbook* (New York: McGraw-Hill, 1984); *SFPE Handbook of Fire Prevention Engineering*, 2nd ed. (Quincy, MA: NFPA, 1995), table B-7.

[†]*SFPE Handbook of Fire Protection Engineering*, table B-4, p. A-30.

[a]ρ = density of the material in kg/m^3

the heat release rate of a fire is to consider the amount of fuel being consumed in it per second and multiplying that mass burning rate by the heat of combustion. In a real fire, there is a certain inefficiency due to incomplete combustion, so the relationship is not perfect. The heat release rate is a most useful way of comparing fires or predicting their behavior and their influence on other fuels nearby. The heat release rates of some typical fires are shown in Table 3-4.

TABLE 3-4	Heat Release Rates of Some Typical Fires
Smoldering cigarette	5 W
Wooden kitchen match or cigarette lighter	50 W
Candle	50–80 W
Office wastebasket with paper	50–150 kW
Small chair (some padding)	150–250 kW
Armchair (modern)	350–750 kW (typical)—up to 1.2 MW
Recliner (synthetic padding and covering)	500–1,000 kW (1 MW)
Sofa (synthetic padding and covering)	1–3 MW
Pool of gasoline (2 qt, on concrete)	1 MW
Christmas tree (dry, 6–7 ft)	1–2 MW (typical)—up to 5 MW
Living or bedroom (fully involved)	3–10 MW[a]

Sources: J. Quintiere, *Principles of Fire Behavior* (Albany, NY: Delmar, 1998), 114–22; J. Krasny, W. Parker, and V. Babrauskas, *Fire Behavior of Upholstered Furniture and Mattresses* (Park Ridge, NJ: Noyes, 2001).

[a]Depending on ventilation.

HEAT TRANSFER AND HEAT FLUX

An integral part of every fire is the transfer of heat, both to the fuel (which is critical to continuity of the fire) and away from regions of combustion, whatever their type. The principles of **heat transfer** are relatively simple and are vital to the understanding of the fire itself. The *rate* at which heat is falling on a surface (or passing through an area) is called the *heat flux* and is measured in watts per square centimeter (W/cm^2) or kilowatts per square meter (kW/m^2) (1 W/cm^2 = 10 kW/m^2). The heat flux is represented by the symbol $\dot{q}''$ (the double prime being notation for "per unit area").

Heat is transferred in three fundamental ways: (1) *conduction*, (2) *convection*, and (3) *radiation*. All these processes play a part in fires; however, the relative importance of each will vary with the intensity and size of the fire, as well as with the shape and content of the environmental system that is burning. Each will be considered separately.

heat transfer ▪ Spread of thermal energy by convection, conduction, or radiation.

Conduction

Conduction is the transfer of heat energy through a material by contact between its moving molecules. Because actual contact of the vibrating molecules is necessary, conduction is limited, for the most part, to localized action. Its effects are most noticeable in solid materials where molecular contact is at its highest and where convection (large-scale, physical movement of the molecules) does not occur. Heat always travels from hot areas of a solid to cold ones by conduction. This can be illustrated readily by heating one end of a metal rod and observing the temperature at the other. Heat traveling through the rod will cause a rise in temperature at the other end, but with considerable lag. The amount of heat flowing through the rod is proportional to the time, the cross-sectional area, and the difference in temperature between the ends, and is inversely proportional to the length.

conduction ▪ Process of transferring heat through a material or between materials by direct physical contact.

The rate at which heat is transmitted through a material by conduction is measured as the thermal conductivity (k), in units of W/m · K (when the heat impinging on a surface is measured in W/m^2, distances or thicknesses in meters, and temperatures in °C or K). Scientists and engineers find it convenient to express such relationships in the form of an equation, because it shows the way in which factors interrelate and allows for prediction of results. The relationship for conductivity is given by

$$\dot{q} = kA(T_2 - T_1)/l \text{ or by rearranging, } k = \dot{q}l/A(T_2 - T_1)$$

Where:

$\dot{q}$ = the rate at which heat is being conducted (in kJ/s or kW)
k = thermal conductivity of the material
A = the area through which the heat is being conducted
$T_2 - T_1$ = the difference in temperature between the "hot" and "cold" zones
l = the length (or distance) through which the heat is being conducted[7]

Plugging different values into this equation shows that heat is conducted faster ($\dot{q}$ gets larger) if T_2 is much larger than T_1, if the area is larger, or if the distance through which the conduction is taking place is smaller. The thermal conductivity of a material (k), is usually specified at 0°C or 20°C, since it varies with temperature. It also varies dramatically among materials. The thermal conductivities of some common materials were shown in Table 3-3 (see also Appendix C).

An illustration of the importance of thermal conduction might involve the relative behavior of a piece of metal—for example, a length of wire or a galvanized metal sheet—as compared with the behavior of a piece of wood in the same fire. The thermal conductivity of metals is high, so that if the metal piece is heated, the heat rapidly spreads to unheated areas. If the temperature in these portions rises above the ignition temperature of any fuel material in contact with the metal, this fuel may be ignited at a distance from the initial source of heat. A piece of wood, in contrast, being a poor conductor of heat, may be intensely heated and even burn at one location for a long interval. The heat, however, does not spread through it well, and the wood will not be expected to ignite at a distance from the point at which the heat is applied. One side of a simple board will show deep charring in a fire, and the opposite side, unless exposed to flame from another source, remains uncharred and apparently normal wood.

The importance of the thermal conductivity of the material through which the heat is being conducted in the development and consequence of fire is considerable. Not only is conductivity involved in the ignition of materials, in which the heat must flow from the heat source to the fuel and to some degree into it before a fire will start, but it also has a marked effect on the degree of fire damage, as we shall see in later sections. The thermal conductivity of easily ignitable fuels such as wood, plastic foam, or paper is very low, which means that heat applied to a surface tends to accumulate there, raising the local temperature, possibly to the fuel's ignition temperature. The conductivity of metals is very high, meaning that heat applied is quickly dissipated into the bulk of the metal, and the local temperature does not approach ignition temperatures as readily. Copper conducts heat almost 2,000 times more efficiently than does wood. After a fire, thermal damage to the insulation on a copper wire may be seen to extend for some distance from the actual point of heat application or the location of a heat-producing electrical fault. Wood may be damaged very heavily on only one side, and even then only in the area of direct application of heat. Other areas not directly exposed to fire may be undamaged. Poor thermal conductors such as felt, plastic foam, and the like are used as insulation. They also tend to be poor conductors of electricity as well as heat. The high thermal conductivity of some metals means that it is more difficult to melt them by application of localized heat. For example, the melting point of copper (1,082°C; 1,981°F) is often exceeded in the combustion zone of a wax candle, but only the finest gauge copper wire is likely to be melted in a candle flame because larger cross sections permit the heat to be conducted away so quickly that the temperature of the bulk metal never exceeds the melting temperature. The high thermal conductivity of metals contributes to their rapid loss of heat when taken out of a flame. A hot metal particle may contain a great deal of heat due to its high thermal capacity but loses its heat to its surroundings much more quickly than will wood of the same mass.

Convection

convection ■ Process of transferring heat by movement of a fluid. In convective flow, the warm fluid becomes less dense than the surrounding fluid and rises, inducing a circulation.

Convection, as a means of transferring heat, is extremely important in fires. It can be defined as the distribution of heat by means of a circulating medium or the transfer of heat to or from a moving medium. As such, it will occur in gases and liquids but obviously not within solids. Convection is, however, responsible for transferring heat from solids to liquids or gases (and vice versa) by convective heat transfer. In most fires it is driven by differences in density caused by temperature variations (buoyant flows). Such buoyant flows are usually referred to as *convective* or convection flows.

When water is heated, its temperature can rapidly be raised despite its very low heat conductivity. The temperature rises rapidly because the heated water on the bottom of a container expands, becomes less dense, and is displaced by the heavier cold water, which is heated in turn. Another very familiar example is the heating of buildings by furnaces or heaters. Here the fire is small and controlled, but the heated air coming from it is often circulated exclusively by convection. Forced-draft furnaces have a fan or blower to supplement this convection.

In fires the moving masses of hot materials are the gaseous products of combustion, along with surrounding air, which is also heated. These expand and become lighter and move upward at a rapid rate. This buoyancy is the form of heat transfer that accounts for most heat movement in a normal fire. It also largely determines the fundamental properties of fire with respect to its movement, spread, and ultimately its pattern.

Convective heat transfer is represented by the equation

$$\dot{q} = h(T_2 - T_1)A \qquad \text{(W)}$$

where, once again, $\dot{q}$ represents the rate at which the heat moves, A is the area through which it is being transferred, T_2 and T_1 are the temperatures of the two materials, and h is a value that represents the efficiency of the heat exchange between those two materials. Because convective heat transfer varies with different combinations of materials, surface textures, speeds of movement, and so on, it is a value that one looks up in a reference. For a buoyant flame plume in air, h is between 5 and 10 W/m^2 · °C. Application of the preceding equation, with a typical flame temperature of 800°C, would give a convective heat transfer to a nearby surface on the order of 4 to 8 kW/m^2.[8] This is considerably less than the radiative heat transfer of 40 or 50 kW/m^2 one would expect from the same flame at near contact.

In very large fires, especially outdoors, the upward movement of hot gases may be so great as to contribute to the formation of what is called a **firestorm**. In such cases, drafts are created that can suck lighter fuels at ground level into the fire, and the buoyant current lifts masses of burning gases and solid debris hundreds of feet into the air, debris that falls as burning brands among other fuels downwind. The burning gases can form a separate mass of flames called a *fireball* or *firewhirl*, which can ignite fuels that come into contact with it. The intensity of the fire produces radiative heat of such magnitude that fuels some distance away are heated to their ignition temperatures and burst into flames. The firestorm is the mechanism responsible for the terrible damage in infamous fires such as those in Dresden, Germany, and Tokyo, Japan, (both as a result of air raids during World War II), and in Chicago, Illinois, and Peshtigo, Wisconsin (accidental fires, one urban, the other wildland in origin, both in October 1871).

Radiation

Transfer of heat by **radiation** is less commonly appreciated than transfer by conduction or convection, and yet it plays the most critical role in fire growth and spread, particularly in larger fires. Radiation aids fire spread across a surface, promotes ignition of other fuels, and may produce burn patterns that survive the fire. All objects having a temperature above absolute zero (0 K, −273°C, −460°F) radiate energy in the form of electromagnetic energy. Radiated heat energy can be transferred to another body without any contact or circulating medium. As we shall see later, radiation plays a critical role in the spread of a fire later in its development. At low temperatures, this radiation is solely in the form of infrared emission. At temperatures above 500°C (900°F), some of the radiation is in the visible portion of the spectrum, and the color perceived depends on the temperature of the body, as described in Table 3-1. Infrared radiation behaves exactly like visible light except that it is not detectable by the unaided eye. The infrared radiation from the sun is the primary source of heat on Earth. As soon as the sun is visible, the heat it radiates can be felt, just as an observer feels the radiated warmth of an open fire. Such warmth is felt because it is absorbed by the body, and the absorption of heat is greater than the radiation of heat by the body under these circumstances. In the same way, a person in a cold environment is chilled because his or her body radiates heat faster than it receives heat from the environment. In surroundings where everything is at the same temperature, all objects are both radiating and absorbing radiant heat at the same rate. The result is no change in temperature. If the temperature is measured in K (°C + 273), the intensity of the total emitted radiation (E) is related to the temperature (in K) raised to the fourth power (T^4), as in the Stefan-Boltzmann relationship:

$$E = \epsilon \sigma T^4 \qquad (\text{W/m}^2)$$

where ϵ is the emissivity of the source and σ is the Stefan-Boltzmann constant.[9] This means that a modest increase in the temperature of a surface, say, doubling it from a room temperature of 300 K (27°C; 81°F) to 600 K (327°C; 621°F), increases the intensity of radiant heat emitted by that surface by a factor of 16 (2 to the 4th power = 16). The effects of radiant heat are determined by its magnitude and duration. Table 3-5 gives some examples

TABLE 3-5 | **Effects of Thermal Radiation on Thick Solids (Wood, Plastic, Human Tissue) in Still Air @ 20°C**

SOURCE	RADIANT HEAT FLUX ($\dot{q}''$) (kW/m²)	EQUILIBRIUM SURFACE TEMPERATURE*	OBSERVED EFFECT
Direct summer sun	1	40°C (100°F)	None
Distance from fireplace	2–4	45°C (120°F)	Pain after 30 s
Proximity to fireplace	4–6	54°C (130°F)	Pain after 8–10 s 2nd degree burns to skin in 20–30 s
Near proximity to fireplace	10	100°C–150°C (200°F–300°F)	Scorching of some materials Melting of some thermoplastics
Face of fireplace	20	150°C–250°C (300°F–500°F)	Some cellulosics and synthetics ignite in <60 s
Inside of fireplace	30	300°C–400°C (600 °F–800°F)	Autoignition of many fuels in 0–30 s (wood in > 60 s)
Adjacent to flames	50	400°C (800°F)	Autoignition of nearly all materials in <5 s
Post-flashover	120–150	>500°C (>800°F)	Rapid combustion

*Estimated from various sources including *ASTM E1321-97: Standard Test Method for Determining Material Ignition and Flame Spread Properties.*

of the effects of thermal radiation. These effects are useful in estimating the intensity of heat exposure on surfaces after a fire as an aid to reconstructing the positions of heat sources and "target" (receiving) surfaces.

The heat flux is not the only factor that determines the transfer of radiant heat. There is also a dependency on the view factor; that is, does the receiving surface sit at right angles to the path of the radiation, or is it exposed uniformly to radiation from the source? The intensity of the radiation being received from a point source of heat falls off with the square of distance.[10] In other words, if an object is 1 m (3 ft) away from a small wastebasket fire, directly facing it, and is receiving 6 kW/m² radiant heat, and the object is moved twice as far away (to 2 m; 6 ft), the radiant heat intensity (heat flux) drops to 1/4 of 6 kW/m², or 1.5 kW/m². The heat flux on a surface some distance r away from a small, localized fire is represented by

$$\dot{q}'' = X_{rad}\, \dot{Q}/4\pi r^2$$

where $\dot{Q}$ is the total heat release rate of the wastebasket fire, and X_{rad} is the fraction of heat being released as radiant heat (for a trash fire, that might be 0.4).[11] One can see that as r gets larger, $\dot{q}''$ falls off very rapidly (the inverse-square relationship).

If an object is very close to a large fire or hot surface, the fire can no longer be represented as a single point source, and the summed contributions from various parts of the large source overwhelm the inverse-square law. The relationship becomes

$$\dot{q}''_1 = \epsilon\sigma T_2^4 F_{12}$$

where F_{12} represents the view factor between the source (at temperature T_2) and the receiving or target surface (surface 1), and $\dot{q}''_1$ is the heat flux falling on that target surface.[12] Calculation of the view factor F_{12} requires an estimation of dimensions of the radiant surface and distance of separation, as in Figure 3-7. The value is calculated as the ratio a/c, and the appropriate curve is selected in Figure 3-8. The y-value is calculated from ratio b/c. The intersection of curve x with the vertical line corresponding to y yields the factor F_{12} on the vertical axis.[13]

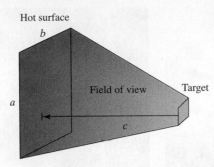

FIGURE 3-7 A small target facing a large heat source (*a* by *b* in size) a distance *c* away is subject to radiant heating from a wide angle of view

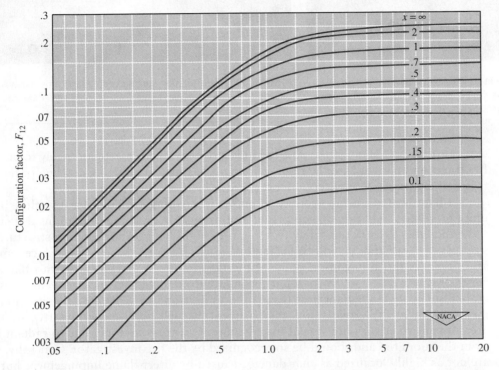

FIGURE 3-8 The configuration factor is taken from this nomograph, where $x = a/c$ and $y = b/c$. The closest curve for *x* is selected. Where that curve intersects with the *y* value determines F_{12}. *From Hamilton and Morgan. NACA Tech. Note 2836, December 1952.*

The calculation of the view factor for such exposures is much more complex, but notice that distance alone is no longer a factor. For instance, if the hot gas layer in a room ignites, the entire ceiling layer represents a radiant heat source. If we measure the heat flux on an object 1.2 m (4 ft) below the ceiling layer and the heat flux on an object 2.4 m (8 ft) below the ceiling layer, the heat fluxes will be very similar. That is one reason why radiant heat from a large heat source seems to bring so much of a room's contents to ignition at about the same time. These factors will control how much radiant heat is being received on the target surface. The "edge" or "corner" effect is one reason why the targets in the center of a room ignite long before the carpet at the edges as a room approaches **flashover**.

The effect of the heat flux is also subject to the nature of the exposed surfaces and especially to the degree to which they reflect heat or infrared rays. Heat (or visible light) falling on a surface may be absorbed or reflected. Objects that radiate heat readily also absorb it readily. In general, dark-colored objects both absorb and radiate heat better than light-colored ones, but some light-colored materials absorb infrared (heat) energy very effectively. However, polished surfaces, such as chrome plate, resist both absorption and emission. The same mirror that reflects visible light also reflects the infrared or heat rays in the same manner, but the glass from which it is made absorbs the infrared, and its

flashover ▪ The final stage of the process of fire growth; when all exposed surfaces of combustible fuels within a compartment are ignited, the room is said to have undergone flashover.

FIGURE 3-9A Massive fire in condominium building igniting building across the street by radiant heat (and fire balls). *Courtesy of Calvin Bonenberger, Fire Marshall, Lafayette Hill, PA, FD.*

FIGURE 3-9B Structure fire igniting vehicle (and ground cover) by radiant heat. *Courtesy of Greg Lampkin, Knox County, TN, Fire Investigation Unit.*

temperature rises. Many of the post-fire indicators on which investigators rely to indicate fire travel and intensity are the result of radiant heating, so an understanding of the factors that contribute to it is very important.

Radiant heat is a major factor that makes firefighting difficult, because close proximity to a large fire heats up equipment and firefighters, making operations difficult if not impossible to carry out safely. When a fire is burning in a portion of a structure, all surfaces that face the fire will be heated by radiant heat, and when this surface temperature reaches the autoignition point of the material itself, it will burst into flame. In very large fires, it is not uncommon for adjacent buildings and nearby vehicles to be ignited solely by the radiant heat (as in Figures 3-9a and b). This radiative ignition is a major factor in fire growth in a compartment, as will be discussed shortly. Charred areas inconsistent with fire spread from item to item are commonly due to radiant heat from the room fire and should be so recognized by the fire investigator. Generally, they are not as highly localized as char damage caused by direct flame impingement but can exhibit "shadow" effects where there are intervening solid objects to block some of the radiation. When there are no intervening or overlying solids, or prolonged post-flashover burning, radiation char patterns tend to be uniform in degree of charring. They can complicate the search for a fire origin because radiant heat can produce charring at low levels in a room even when the fire generating the heat is not as low in the room (see Figure 3-10). However, the ignition of floor coverings by radiant heating and their burning can result in char damage to the floor that is not as uniform as direct radiant heat char (and can be very irregular). These effects will be discussed in more detail in Chapter 7.

DIRECT FLAME IMPINGEMENT

In a real fire, flames are spread not only by the three basic mechanisms of heat transfer described previously but also by direct flame impingement, and its role cannot be ignored. On a microscopic scale, direct flame contact or *impingement* is a combination of both convective and radiative mechanisms. If the plume of hot gases from the established flame rises by buoyant flow into contact with more fuel, heat is transferred from the plume by convective heat transfer and radiation until the new fuel pyrolyzes and generates gaseous fuels that are ignited by the flames of the plume. The flames then spread upward and,

FIGURE 3-10A As fire extends across the ceiling, heat radiating downward from the ceiling layer chars and ignites exposed surfaces; areas under chair and table are protected until later in the fire. When flashover is completed, all fuels in the room will be on fire, and the undersides of tables, chairs, and shelves will be exposed to flames from floor-level burning.

FIGURE 3-10B A structure fire demonstrating ignition by radiation. Intense localized fire on the floor and railing of the porch was ignited by radiation coming from fire plume exiting the top portion of the door. *Courtesy of John D. DeHaan.*

more slowly, outward. Because fresh fuel is continually being added to the reaction with little loss of heat, the fire can spread quickly through the fuel presented. As more fire is created, the faster fuel is involved and the temperature of the room or enclosure rises, thereby increasing the rate of oxidation of other fuels, and the fire spreads at a faster and faster rate as long as the supply of oxygen is maintained.

Flame Plume

A fire burning in the middle of a room or out in the open creates a rising column of hot gases (plume) by the buoyant flow just described. The temperature in a simple fire is at a maximum in the well-mixed combustion zone just above the fuel package. As the gases (and solids and liquid aerosols mixed therein) rise they lose their heat, some by mixing with room air (entrainment) and convective transfer to the surrounding air, and some by radiant

FIGURE 3-11
Temperature distribution in buoyant flame and smoke plume along vertical centerline.

Temperature distribution in buoyant flame and smoke plume

loss (particularly from soot particles and aerosol droplets of pyrolysis products from incomplete combustion). As a result, if we measured the temperature of the flames, we would see a predictable pattern to the temperature as a function of height above the fuel package. Temperatures decrease from 700°C to 900°C (1,500°F to 1,800°F) just above the fuel to about 500°C (930°F) at the tip of the flame, where it becomes very turbulent and intermittent.[14] As the gases continue to rise they continue to lose heat, and the temperature continues to decrease in a predictable pattern until the gases have lost so much heat that they are no longer buoyant. (Remember that most of the combustion products—CO_2, SO_2—are heavier than air at the same temperature; CO is nearly the same density; only water vapor (H_2O) is lighter.) A temperature plot of flame temperature versus height would look something like Figure 3-11. The relationship between the flame height and the heat release rate of the fire is quite predictable. One form of the relationship for a symmetrical pool fire burning with no nearby walls or ceiling is the Heskestad equation:

$$Z_f = 0.23\dot{Q}^{2/5} - 1.02D$$

where Z_f is the visible flame height in meters, $\dot{Q}$ is the heat release rate in kilowatts, and D is the diameter of the pool (in meters).[15]

Entrainment draws fresh air into the base of the fire to keep the reaction going and diffuses and cools the rising gases in the plume, as in Figures 3-12a and b. If the fire is against a noncombustible wall, the entrainment of cooling air is reduced by 50 percent, as in Figure 3-12c. If fire gases have more time to rise before cooling to the point they are no longer incandescent, the visible flame plume is taller. If the fire is in a corner, the entrained air is reduced even more (as in Figure 3-12d), and the flame plume is even taller. Radiant heat from nearby walls further increases $\dot{Q}$. This situation is dramatically illustrated in Figure 3-13. These fires can be represented by the following equation developed by Alpert and Ward:

$$Z_f = 0.17(k\dot{Q})^{2/5}$$

where k is called the wall factor. The *wall factor* is 1 for no wall, 2 for a single wall, and 4 for a corner.[16] These relationships allow us to predict how big a fire will look to an observer (or what kind of effect its gases will have on materials that come into contact with it at various heights) given the heat release rate of the fire. Due to the turbulence of flames, the height of the flame tip will oscillate up and down, and the tip will sometimes separate from the main body of the flame. The human eye tends to average these variations, and the height is normally fairly close to the predicted Z_f.

FIGURE 3-12A Side view of entrainment—unrestricted fire.

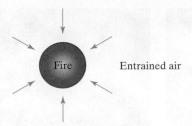

FIGURE 3-12B Top view of entrainment around periphery of unrestricted fire.

FIGURE 3-12C Top view of reduced entrainment of wall fire.

FIGURE 3-12D Top view of corner fire—restricted entrainment.

FIGURES 3-13A & B Fires in similar vertical cushions—free-standing (away from walls). Maximum flame height: about 0.7 m (30 in.) above cushion. *Courtesy of Jamie Novak, Novak Investigations and St. Paul Fire Dept.*

FIGURE 3-13C Maximum flame height against noncombustible wall (about 1.5 m). *Courtesy of Jamie Novak, Novak Investigations and St. Paul Fire Dept.*

FIGURE 3-13D Maximum flame height in corner (about 3 m). *Courtesy of Jamie Novak, Novak Investigations and St. Paul Fire Dept.*

Sequence of a Room Fire

fuel load ▪ All combustibles in a fire area, whether part of the structure, finish, or furnishings.

incipient ▪ Beginning stages of a fire.

A generalized room fire in a room with a *normal* **fuel load** goes through fairly predictable stages of development in its lifetime. Each stage has characteristic features, and the effects on fuels vary in each. The fire investigator must appreciate what these effects are to properly interpret the post-fire scene. The length of time each stage takes will vary (dramatically in some cases) with the circumstances of ignition, fuel, and ventilation, but the sequence from **incipient** or beginning stage to growth to decay stages will remain essentially the same (see Figure 3-14).

BEGINNING OR INCIPIENT STAGE

The beginning or incipient is the first stage of fire development after the moment of ignition. The flames are localized in the first fuel ignited, and a free-burning open flame is typical. The room has a normal oxygen content (about 21 percent) throughout, and the overall temperature has not yet begun to rise. There is a plume of hot gases rising from the flame. The contents of this plume will depend on the fuels being burned but usually include soot (carbon and other solids), water vapor, carbon dioxide, carbon monoxide, sulfur dioxide, and traces of other toxic gases (as well as heated nitrogen entrained from

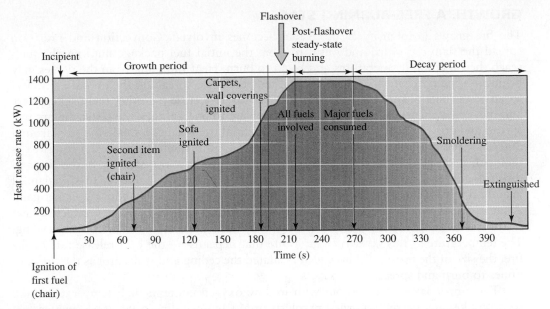

FIGURE 3-14 Total heat releases rate versus time for a fire in a typical small furnished room. As each item ignites it contributes to the growth of the fire. In this case, the critical heat release rate (and hot gas temperature) is reached and the room goes into flashover, followed by steady-state burning, and then a decay period as the fuel is exhausted. Finally, the fire dies down to smoldering embers before being extinguished.

the surrounding air). Convection or buoyant flow carries these products and heat to the upper part of the room and draws oxygen in at the bottom of the flames to sustain combustion. If there is a solid fuel above the flame, convection and direct flame contact spread flames upward and outward, producing the typical vertical pattern of char (with the flammability/flame spread of the fuel determining the nature of the pattern). If there is no fuel above the flames, the gases in the plume rise, losing heat to the room air by radiation and convective (often turbulent) mixing. An incipient fire has such a low heat release rate (HRR) that overall temperatures may not be affected measurably (and sometimes not even the temperature of the upper "smoke" layer) (see Figures 3-14 and 3-15).

FIGURE 3-15 The heat-release curve of a lighter-ignited fire in a bedroom. The incipient fire was less than 100 kW for more than 720 seconds (12 minutes). The growth phase involved the bed mattress and box spring (mattress platform). At ~1100 seconds (18 minutes) the fire spread to the face of a nearby wooden dresser. Rapid fire growth on the dresser caused ignition of the smoke layer to trigger flashover. Fire was quickly extinguished by hose spray. Peak HRR was 3.9 MW. *Courtesy of Bureau of Electronics, Appliance Repair, and Home Furnishings, North Highlands, California.*

GROWTH, A FREE-BURNING STAGE

The fire grows in intensity as more fuel becomes involved. Convection and radiation spread the flames upward and outward from the initial fuel package until nearby fuels reach their ignition temperatures and begin to burn. Heat radiating from the flames may begin to bring other fuels to their ignition temperatures, spreading the fire laterally. The speed of this lateral spread is highly dependent on how close the other fuel packages are and their physical arrangement. The closer they are to one another, the faster the fire will spread and grow in heat release rate. Until the initial fire reaches a moderate size (around 200 kW), radiant heat transfer to target surfaces even a small distance (0.5 m) away is minimal, and most fire spread is driven by convective heat transfer above the fire and direct flame contact. [Radiant heat transfer to a vertical surface facing the fire is balanced against convective transfer *from* that surface to the room air, thus reducing its net gain. A surface above the flame (or in contact with it) will not have the same convective heat losses so will get hotter faster.] Hot gases generated by the fire form a layer at the ceiling. The temperature of this layer rises (at a rate that depends on the heat release rate of the fire, the size of the room, and how well insulated the ceiling and walls are) as the fire continues to burn and spread.

This upper layer of the room with its low oxygen content; high temperature; and soot, smoke, and partially burned pyrolysis products (including toxic gases such as carbon monoxide, hydrogen cyanide, hydrogen chloride, acrolein, and the incompletely reacted chemical species called free radicals) is now very different from that near the floor. The floor layer is still relatively cool and contains fresh air (with an oxygen concentration close to normal) that is being mixed into the flames below the layer of reaction products. Survival is still possible in the room if one stays in the cooler lower layer and does not breathe the gases higher up. If ventilation to the room is limited, the flames will produce more incompletely burned materials including pyrolysis products, carbon monoxide, and solid fuels, which can be seen as soot and smoke. The level of this fuel-rich layer grows lower and lower in the room (deepens), and its temperature increases as the fire continues and the flames grow. If the layer deepens to reach the level of the fire in the room, its reduced oxygen content will reduce the combustion rate, possibly to the point of extinguishing any flaming fire. The mass of combustible gases, aerosols, soot, and smoke accumulates, often until one or more of the fuels present in the layer reaches its ignition temperature whether from accumulated heat from the flaming fire or direct contact with open flames. The layer of airborne "fuel" then catches fire, and the flames may extend throughout the room at ceiling level. This is the process often called **flameover** or *rollover*. The flame front in the roiling cloud of fuel and air can reach speeds of 3 to 5 m/s (10 to 15 ft/s). With flaming combustion proceeding in that layer, the temperature of the ceiling layer increases even faster. The rapid increase in temperature of the hot gas layer causes expansion of the layer, filling more of the room. In observed tests, flameover in a residential living room caused flames to fill the compartment nearly to floor level in 7.5 seconds after ignition, prompting firefighters to exit for their safety.[17]

FIRE GROWTH TO FLASHOVER

Even in the absence of flaming ignition of the hot gas layer (flameover), the radiant heat from the hot gas layer at the ceiling is striking all other materials in view in the room with an intensity (heat flux) that is dependent on the temperature of the layer. When this layer reaches a critical temperature of approximately 600°C (1,100°F), it is generating approximately 20 kW/m² (2 W/cm²) radiant heat at floor level in a normally proportioned room. This is sufficient to bring most other fuels, such as furniture and floor coverings in the room, to their ignition temperatures. At this time, all fuels in the room become involved in flaming combustion, and the room is said to have undergone flashover.[18] Once this occurs, all exposed fuels in the room are involved, temperatures throughout the room reach their maxima, and survival for more than a few seconds is impossible. Once the floor or floor covering ignites, the flames cause high temperatures near floor level, which meet and often exceed the high temperatures that existed only in the ceiling layer prior to flashover. Noncombustible floors, of course, will not contribute flames or heat, so the floor temperatures will remain lower than in the rest of the room. The enormous turbulence caused by the generalized flames fills the room with hot gases as the two-layer pattern breaks down. The rapid extension of open-flame fire across the hot gas layer of a room is often the event firefighters refer to as flashover because it means that complete ignition is imminent. Flashover to the fire scientist refers to the process of extension, radiation, and ignition of all fuels in a room or compartment. See Figures 3-16a–i for a graphic representation of the process. It will be discussed in greater detail in a later section.

As each fuel package in a compartment ignites, it contributes a quantity of heat to the room at a rate influenced by what fuel is present, how it is arranged, and even how it is ignited. If we plot the heat release rate (or average temperature of the hot gas

FIGURE 3-16A Room fire at 5 minutes is still limited to paper in wastebasket; HHR < 100 kW. *Courtesy of BRE (Incorporating the Fire Research Station), Garston, United Kingdom, by permission.*

FIGURE 3-16B Fire at 6 minutes has spread to papers and the fabric on the sofa by direct flame impingement; HRR < 200 kW. *Courtesy of BRE (Incorporating the Fire Research Station), Garston, United Kingdom, by permission.*

FIGURE 3-16C Fire at 7 minutes has spread to draperies by direct flame contact and radiant heating from sofa. HRR increasing very rapidly as vertical flame spread dominates; HRR = 1,000 kW (1MW). *Courtesy of BRE (Incorporating the Fire Research Station), Garston, United Kingdom, by permission.*

FIGURE 3-16D Fire at 7.5 minutes. The draperies are fully involved. Flames in the smoke layer are not sustained; HRR = 2,000 kW (2 MW). *Courtesy of BRE (Incorporating the Fire Research Station), Garston, United Kingdom, by permission.*

FIGURE 3-16E Fire at 9 minutes. Radiant heat ignition of tops of furniture, and the flames in the smoke layer become continuous (rollover); HRR = 2.5 MW. *Courtesy of BRE (Incorporating the Fire Research Station), Garston, United Kingdom, by permission.*

FIGURE 3-16F Fire at flashover transition (10.75 minutes). All contents are on fire, and there is a large plume exterior to room; HRR = 5.2 MW. *Courtesy of BRE (Incorporating the Fire Research Station), Garston, United Kingdom, by permission.*

FIGURE 3-16G Post-flashover fire at 11.75 minutes. The larger furniture is fueling the majority of fire; HRR = 2.5 MW. *Courtesy of BRE (Incorporating the Fire Research Station), Garston, United Kingdom, by permission.*

FIGURE 3-16H Decaying fire at 13 minutes. Massive furniture (sofa on right and cabinet with folded draperies on left) is still on fire; HRR = 1.4 MW. *Courtesy of BRE (Incorporating the Fire Research Station), Garston, United Kingdom, by permission.*

FIGURE 3-16I Decayed fire at 15 minutes; HRR = 900 kW. *Courtesy of BRE (Incorporating the Fire Research Station), Garston, United Kingdom, by permission.*

FIGURE 3-17 Heat-release curve for living room fire in Figure 3-16. *Courtesy of BRE (Incorporating the Fire Research Station), Garston, United Kingdom by permission.*

layer) for a typical room fire as a function of time, we might see something like Figure 3-17. As each item (chair, sofa, carpet, etc.) ignites, it contributes to the hot gas layer, which is continuously being depleted by conductive losses through the ceiling and walls, radiative and convective losses, and mass flows out doors, windows, and vents. When enough heat is released quickly enough into a room, the room can go to flashover. It is obvious that the higher the heat release rate (HRR) of significant fuel packages, the better the chances that their additive nature can make the development of flashover possible. If we examine various fuel packages using a furniture calorimeter (see Chapter 11), we can estimate their respective contributions. Organizations such as the Building and Fire Research Laboratory at the National Institute for Standards and Technology (NIST) and the California Bureau of Home Furnishings have conducted hundreds of tests to examine the HRR of various items. Depending on their covering and padding, upholstered chairs have a maximum HRR of 150 to 700 kW, sofas: 250 to 3,000 kW, and mattresses: 150 to 2,800 kW.[19] Other fuel packages can play a big part: Christmas trees: 800 to 1,600 kW,[20] wood shelves with 100 videocassettes: 800 kW,[21] a pool of 1.9 liters (L) of camp fuel: 900 to 1,000 kW.[22] If one large (700 kW) or several small (150 to 250 kW each) packages reach their maxima at or near the same time, the average-dimensioned room can go to flashover. A complication is that *external* radiant heat can increase the HRR of a typical package, so an encompassing fire involving multiple fuel packages can grow at a faster rate than a simple additive process would indicate.

The development of a well-ventilated, nonaccelerated fire in a furnished room is shown in Figure 3-16. The room was set up with a modern (flame-retardant) sofa (meeting Home Office Statutory Instrument 1324-1988 or BS 5852) with one wall open and venting beneath a large calorimeter. The fire was set in papers in a wastebasket just in front of the sofa. Figure 3-16a shows the incipient fire, still confined to the wastebasket after 5.0 minutes of burning. One minute later, the front of the sofa has ignited and the fire is growing (Figure 3-16b). At 7.0 minutes (Figure 3-16c), the draperies have ignited, and within 30 seconds, the draperies have spread the fire across the rear of the room (Figure 3-16d). By 9.0 minutes (Figure 3-16e), the smoke layer is beginning to ignite (rollover), and most of the fuel packages in the room have ignited by radiant heat. By 10.75 minutes, the fire has reached flashover, in which all the fuels in the room are on fire, and a large plume of waste gases is venting out of the room and igniting outside the room (Figure 3-16f). At this point (post-flashover), the maximum heat release rate of 5.2 MW was recorded by the calorimeter. By 11.75 minutes, the majority of the fuels have been burned, and the fire is going into a decay phase (Figure 3.16g). At 13.0 minutes, many of the fuel packages have burned out except the remains of the sofa in the right rear corner and the wood credenza on the left (which contained a number of folded draperies) (Figure 3-16h). The fire continued to burn for some 25 minutes before extinguishment. The heat release curve of this test, Figure 3-17, shows the various stages described. Due to the prolonged growth phase (~10 min), the post-flashover steady-state burning was very brief, as much of the fuel had already been consumed, but the decay phase is well defined. The increase at 17 to 20 minutes is due to the delayed combustion of a wood cabinet filled with combustible draperies.

It should be noted that the flashover transition triggers several significant effects in the room, which may or may not be visible to an observer. Starting with a hot gas layer temperature that exceeds 600°C (1,100°F), the temperatures throughout the room spike quickly, especially at floor level (as floor coverings ignite). The overall oxygen content in the room drops dramatically as the ventilation limit is reached and exceeded. This causes a dramatic increase in the concentration of CO (and also of HCN somewhat later). The radiant heat flux increases on target surfaces in the lower portions of the room. This triggers ignition (or at least off-gassing) of exposed fuels. Their subsequent ignition to flames depends on their individual thermal inertias.[23] When monitoring compartment fires, it is not uncommon to observe these events at slightly different times. The "time" of flashover, then is not a single exact point but is largely dependent on what indicator one is relying on. Window glass tends to shatter as a room fire approaches flashover, and the rustle and popping of a limited room fire becomes a violent rumble due to the turbulence of the post-flashover combustion. These events are audible to bystanders.

The approximate HRR needed to trigger flashover can be calculated if the dimensions of walls, ceiling, windows and doors, and insulative properties are known. The simple calculations in FPETool or in the *NRC Fire Dynamics Tools* spreadsheets can be used.[24] The *Fire Dynamics Tools* spreadsheets include the following three calculations:

Babrauskas:

$$\dot{Q}_{FO} = 750 A_O \, (h_O)^{0.5}$$

Thomas correlation:

$$\dot{Q}_{FO} = 7.8 \, A_T \; + \; 378 \, A_O \, (h_O)^{0.5}$$

McCaffrey/Quintiere/Harkleroad (MQH):

$$\dot{Q}_{FO} = 610 \left[h_K \, A_T \, A_O \, (h_O)^{0.5} \right]^{0.5}$$

Where:

$\dot{Q}_{FO}$ = heat release rate needed to include flashover
A_T = total surface areas of walls, floor, and ceiling (m²) (not including areas of windows or doors)
A_O = area of opening (m²)
h_O = height of opening (m)
h_K = convective heat transfer coefficient (for specific wall covering)

For a room 3 m wide × 4 m long × 2.4 m high (10 ft × 13 ft × 8 ft) with a single door 0.8 m (32 in.) wide × 2 m (78 in.) high, these estimates are

Babrauskas $\dot{Q}_{FO} = 1,697 \text{ kW}$
Thomas $\dot{Q}_{FO} = 1,292 \text{ kW}$
MQH $\dot{Q}_{FO} = 2,437 \text{ kW}$

There is a spread (1,300–2,400 kW) of estimated HRR needed for flashover, since each formula is based on different experimental data and assumptions about proportions (room size versus vent size) and thermal properties, so if the HRR calculated or estimated from the fuel packages present in a room falls anywhere in the 1,300–2,400 kW range, the possibility that that room will be driven to full involvement has to be considered.

Any fire is a chemical balancing act among fuel, air, heat, and the rate at which the fuel is being consumed. If the ventilation is limited, the fire will progress at a slower rate, the temperature rise will be more gradual, the smoke production will be enhanced, and the ignition of the airborne fuel (smoke layer) may be delayed until sufficient air is available for its open-flame ignition to occur. This ignition may not happen within the room at all; the fuel-rich smoke will then be seen igniting as it leaves the door and window openings to mix with the oxygen-rich air outside. This combustion can be sustained if a large enough fire is burning within to maintain a flow of smoke and gases. If the fuel does not burn quickly enough or does not produce enough heat when it does burn, heat losses, through convection and conduction, will prevent the fire from going to full involvement. One interesting result of observing room fires ignited with a significant (one liter or more) amount of a flammable liquid is that if the liquid is distributed over a sufficient number of ignitable substrates, the radiant heat or direct flame contact from the liquid-fueled fire can cause ignition of all the exposed ordinary fuels in the room within seconds of ignition without the formation of the hot gas layer. Under these conditions, full room involvement has been achieved without the hot gas layer/radiant heat flashover process.[25]

A well-insulated room is more likely to retain enough heat to develop a fully developed fire than is one with highly conductive walls and ceilings. Not all room fires reach flashover. If the ventilation openings are too small, of course, there will not be enough ventilation to support a large enough fire to bring about flashover. If vent openings are too large, the fire has to be very energetic to overcome the losses and form the hot gas layer that is critical to flashover. If the hot gas layer never reaches its critical temperature, there will not be enough radiant heat to involve the remainder of the room. Flashover is not a specific moment or event but a transition from a fire that is growing by involving one fuel package after another to a fire in which all the exposed fuel in the compartment is on fire.

POST-FLASHOVER STAGE

Once the fire has reached post-flashover burning, all involved fuels will continue to burn as quickly as oxygen is made available to them until all available fuels are nearly consumed. It is not a threshold that can be crossed again and again. A *post-flashover fire* in a typical room is usually a ventilation-controlled fire. The heat release rate of the fire is no longer controlled by how much of the exposed fuel is involved, because it is all on fire. The size of the fire from that point on may be limited by the amount of air that can enter the compartment. This is usually determined by the buoyant flows of hot gases and air into and out of the room. The area, height, and arrangement of ventilation openings into the room determine that flow, as noted in the relationship

$$\dot{Q}_{max} = \Delta H_{air} \cdot \dot{m}_{max} = 3{,}000(0.5A_o(h_o)^{0.5}) = 1{,}500A_o(h_o)^{0.5}$$

Where:

$\Delta H_{air} = 3{,}000$ kJ/kg

$\dot{m}_{max}$ = maximum buoyant flow
A_O = area of opening m^2
h_O = height (m)

A single entry door (0.8 m wide and 2 m tall) can allow enough air in (and enough waste gases out) to support a nearly 3 MW fire in a room.[26] Windows positioned high up in a wall will be largely filled with buoyant waste gases (including products of incomplete combustion) escaping from a room and will not be as effective in allowing fresh air in as windows placed lower down in the walls would be. In a post-flashover fire, the locations of the most energetic combustion (and accompanying thermal effects) are no longer controlled by where the fuel packages are but by where the best ventilation is located. Recent tests by Carman have demonstrated that post-flashover rooms can exhibit marked ventilation effects dependent on where fresh air rich in oxygen can enter a room or where the fire is burning in a continuously oxygen-deficient area. Entrained flows of air through a doorway can produce "clean burn" fire patterns on walls opposite the door, mimicking patterns from localized fuel packages.[27] Such effects can also produce burn patterns across carpets from doorways or windows. Areas where oxygen can enter the room and mix with the combustion gases will support the most intense combustion. The mixing is extremely turbulent to the point at which it is not a steady, torchlike flame but moves about in a chaotic fashion.[28] This flame behavior can produce irregular heat effects on nearby surfaces. If a post-flashover room fire is suddenly deprived of oxygen (such as by structural collapse), flaming combustion will rapidly decrease to return later when more air is available, but it cannot return through flashover. Some of these effects will be illustrated in Chapter 7.

DECAY STAGE

Eventually, the available fuel is exhausted, and open-flame burning gradually becomes less and less prevalent. (During the fire, if the oxygen concentration drops much below 16 percent, open-flame combustion will decrease even in the presence of unburned fuel and may even cease completely at oxygen concentrations below 5 percent.) As an adequately ventilated fire dies down the amount of fuel vapors that remain to be pyrolyzed out of the solid fuels decreases, and the amount of flaming combustion dies back. As a result, smoldering combustion becomes predominant, which can continue as long as there is suitable fuel. This *decay stage* is often called the "smoldering" phase of a room fire. High temperatures may continue as long as there is suitable fuel, depending on the ventilation and insulation present. Some products continue to pyrolyze to form carbon monoxide and other gaseous fuels as well as toxic gases, solid soot, liquid aerosols, and other fuels in the form of smoke.

If the room is not adequately ventilated, these combustion products can build up to form an ignitable vapor mixture. When a supply of fresh air is introduced in the presence of an ignition source, such vapor accumulations can deflagrate into a secondary fire, sometimes called a *backdraft, flashback,* or *smoke explosion.* This ignition can proceed with explosive speed, and although the pressures produced are low compared with fuel vapor–air deflagrations, they can produce pressures on the order of 0.5 to 1.5 psi (3 to 10 kPa, 30 to 100 mbar) or higher.[29] This pressure is enough to do structural damage and endanger lives. Although such smoke explosions are rare, they can occur, especially when smoky fuels such as latex or urethane foam rubber are present. An excellent study by Fleischmann, which explored the dynamics of these events, showed that even under controlled conditions, ignition of the layer could not be assured, since it involved having the right fuel-air mixture in contact with a suitable flaming ignition source.[30] One well-documented case involved a relatively small but smoky fire that filled an apartment with a hot, fuel-rich smoke. When fire crews entered the apartment, the smoke leaving the apartment ignited in the stairwell, trapping three firefighters.[31] The use of positive-pressure ventilation (PPV) in firefighting may introduce enough fresh air to a hot, fuel-rich environment to trigger such events. Testing by NIST has shown that PPV used prior to suppression causes higher temperatures and heat fluxes in the fire room and in areas "downstream" of the vent from that room.[32]

This generalized sequence does not occur in every fire. The entire fire may be ignited by a smoldering process, which then has to grow to the point at which it has a high enough heat release rate to initiate and sustain flaming combustion. Thus, smoldering should not be considered only to be the final end product of a fire. Glowing combustion can also take place at the same time as flaming combustion. On wood, for instance, if the pyrolysis and volatilization occur at such a rate that oxygen can reach the hot solid char surface, glowing combustion can take place. This situation is noted most often in strong draft conditions where the impressed airflow can push aside the outflowing gases, or late in a fire in which the rate is dropping due to exhaustion of the fuel. An example of such a fire was shown in Figure 3-6.

Flow of Hot Gases

Buoyant flow is the dominant mechanism by which fire gases move. Normal air (and most smoke) has a density of 1.2 kg/m^3 at 20°C (293 K). Its density (ρ) then varies with its absolute temperature (from the ideal gas law, introduced in Chapter 2):

$$\rho_{T1}/\rho_{T2} = T_2 \text{ (K)}/T_1 \text{ (K)}$$

Thus, the density of gas at flame temperature (800°C; 1073 K) is

$$\rho_{T2} = 1.2 \text{ } (293/1073)$$

$$\rho_{T2} = 0.33 \text{ kg/m}^3$$

Flame gases then will rise rapidly due to their much-decreased density. As the gases cool, of course, they lose some of that buoyancy in proportion to their cooling. The horizontal flow of a gas (or other fluid) due to gravity-driven forces is called *advective flow*. As a hot gas flows along a ceiling it loses its heat to the underside of the ceiling by convective transfer. Thus, its temperature (and its convective heat transfer rate) is at a maximum where it first contacts the cold ceiling. The horizontal, buoyancy-driven advective flow along the underside of a ceiling is called the *ceiling jet*. When that jet hits a vertical surface (like a wall), it piles up against it like water waves pile up against a breakwater (sometimes called *momentum flow* or "*mushrooming*"). If it flows under a door header or window soffit, it leaves a weak turbulent flow on the downstream side of the soffit (or other barrier), as shown in Figure 3-18. If the flow takes it outside, it forms a door or window plume, which then continues its vertical rise driven by its buoyancy. As the plume rises vertically, entrainment can push it back against the vertical wall above the opening, especially if there is no balcony or other horizontal projection to push it

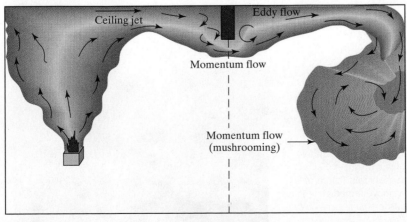

FIGURE 3-18 Buoyant flow against a vertical barrier causes downward momentum flow (mushrooming). Eddy flow on other side produces a protected area.

FIGURE 3-19
Entrainment of window
plume against vertical
wall is reduced by a
balcony or "eyebrow"
above the opening.

High heat
flux

Entrainment

Flames

Window plume

Airflow in

Flames

Window plume

Airflow in

away from the wall (see Figure 3-19). Such plumes, if they are flaming, can quickly cause failure of the spandrel panels or windows directly above the opening through which they **vented**, as shown in Figure 3-20. This is the mechanism by which many high-rise fires inflict damage on multiple stories. Fires such as the Caracas Parque Central (2004) (16 stories involved) or Madrid's Windsor building (2005) (12 stories involved) are examples.[33] A similar fire venting from the Cook County Building in Chicago (October 2003) (Figure 3-21) did not spread upward from its 12th floor of origin because the windows of each floor were set back from the face of the building by one foot or so. That inset protected the windows on floors above. The total heat flux against a wall or window above such an entrained window plume from a post-flashover fire can be as high as 80

vented ■ The fire has
extended outside the
structure or compart-
ment by destroying the
windows or burning an
opening in the roof or
walls.

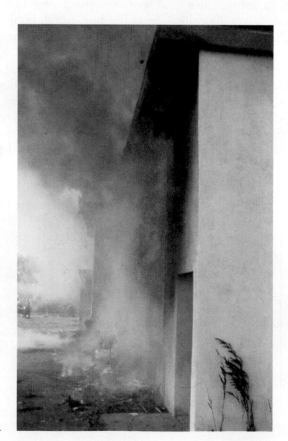

FIGURE 3-20
Entrainment of window
plume on sheer wall
causes rapid failure of
window immediately
above. *Courtesy of John D.
DeHaan*

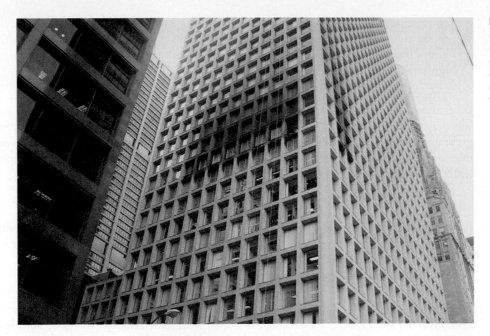

to 120 kW/m^2, well above the 20 kW/m^2 needed to crack ordinary window glass or the 35 kW/m^2 needed to cause the window to fall out or the 20 to 40 kW/m^2 needed to ignite most combustible materials.[34]

Effects of Environmental Conditions

The investigator of fires will usually be concerned with the role played by such environmental conditions as the outdoor temperature or humidity that preceded the fire. These factors need to be taken into account at times and should be documented. It is well known that when an area has become excessively hot and dry, there is greatly increased hazard of forest and grass fires, and structural fires may be more common or of greater severity. Wetness, an environmental condition subject to rapid alteration in a local region (e.g., water from fire hoses), is also a real factor in the rate of burning, the danger of ignition, and the intensity of the resulting fire. Although these factors may be secondary to the investigation, they merit careful consideration by the investigator, since they sometimes can affect the ignition, rate of fire spread, and even the post-fire burn indicators.

TEMPERATURE

All other things being equal, fires are more likely to become severe on a hot rather than on a cold day. However, some of the most destructive structure fires have occurred in weather when the water from fire hoses froze on the firefighters' ladders. The reason for the greater fire hazard on a warm day is easily explainable.

1. Heat dehydrates; therefore, fuel in a heated environment is dryer (has a lower moisture content) than in an unheated one. Dry materials ignite more readily than damp ones, although the heat from the fire itself once established is usually sufficient to dry out most materials ahead of the flames when there is no renewal of the wet condition.
2. As developed elsewhere in this chapter, the rate of a combustion reaction (such as fire) is approximately doubled with every rise of 10°C. In a fully developed fire, this factor in the initial environment is negligible because the heat from the fire itself is

so much greater than an ambient temperature. However, the warmer the environment, the faster the reaction in the extreme initial (incipient) portion of the fire, when the fire's own heat is still not so overwhelming. In other words, it is easier to ignite a fire with hot than with cold fuel.

The variation of temperature in normal interior living environments has little influence on most fires simply because the variations are not great enough, nor the influence direct enough, to make any appreciable difference. The pre-fire temperature may have great influence on self-heating processes, the vapor pressure of flammable liquids, and even the distribution (mixing) of vapors. For these reasons, the pre-fire temperatures (interior and exterior) should be documented by a thorough investigator.

HUMIDITY

Firefighters tend to be greatly concerned with humidity because they note a greater incidence of fires on dry days than on wet days. Again, this observation refers far more to exterior fires than to interior ones. When it is hot and dry for protracted periods, it is well known that wildland fires become an extreme menace; because the fuel moisture increases markedly on cool, humid days, there is far less hazard. The tendency to apply the same qualitative reasoning to interior structural fires is often a misguided one, although not always. The question as to the actual importance of humidity is not a simple one to develop, because it involves several secondary factors.

Humidity refers to the amount of water vapor in the air. At saturation, this value is determined by the air temperature. The greater the temperature, the greater the amount of water that can be held by the air without condensing into fog, dew, or rain. Thus, it is not so much the absolute humidity that is relevant as the relative humidity, that is, the percentage of the saturation value that is actually present. A relative humidity of 50 percent means only that the air contains half of the saturation value for that temperature. However, if the temperature rises without any change in moisture content, the relative humidity drops to a lower percentage, without changing the total moisture. For practical purposes, it is always the relative humidity that is quoted and of importance because this is the best measure of the moistness of the air at any given temperature.

The effects of humidity on fire stem from several basic factors. When porous objects are exposed for sufficient time to low relative humidity, they dry out (lose internal moisture) and are therefore more combustible than when less dry (see next section on fuel moisture). However, the difference in fuel moisture content is not so great as is generally believed; the moisture content of many materials, such as paper, varies within surprisingly small limits with ordinary changes in humidity. The relationship between humidity and fuel moisture content is shown in Figure 3-22 for fine cellulosic materials. Fuels outside are also exposed to wind, which causes lower fuel moisture content than the temperature

FIGURE 3-22
Approximate relationship between moisture content of fine cellulosic fuels and relative humidity at room temperature (25°C).
Courtesy of Vytenis Babrauskas, Fire Science and Technology.

and exterior air humidity alone would dictate. Perhaps the greatest effect from this standpoint is on living vegetation, a major fuel of forest fires. Plants not only dry out as would a sheet of paper, but, much more important, they transpire moisture from pores in the leaf structure. Despite the closing of these stomata in dry weather, a leaf may lose more moisture proportionately than a sheet of paper when exposed to drying conditions. The process is enhanced by wind, as the moving air forces faster moisture loss. The difference is clear when it is noted that dry periods correlate with increased wildland fires more than with increases in primary structural fires. For ignition to occur, the small amount of energy of some ignition sources has to be transferred to the fuel to trigger a self-sustaining reaction. If some of this energy goes into evaporating moisture from the first fuel, there may not be enough left to promote ignition. Storage and watering conditions of cut Christmas trees have been shown to be critical to their relative risk of ignition.[35] One published set of tests showed how difficult it was to ignite Christmas trees using modest ignition sources (such as matches) when the tree had been thoroughly hydrated.[36]

Humidity can also play a role in self-heating processes, for example, as in the fuel moisture content of agricultural materials, where too little will preclude the initial growth (and heat generation) of microorganisms. An excess may play a part in promoting self-heating in charcoal and other materials. These mechanisms will be discussed in more detail in Chapter 6.

Another effect that is rarely given its proper significance is the relation of humidity to small electrical arcs resulting from static discharge and from small sparks produced by striking materials such as a steel object with a rock. Static electricity tends to accumulate excessively in dry periods. This is not due to the insulating value of dry air (since moist air actually has a higher dielectric constant than dry air). Rather, the charge dissipates by leakage via surface moisture when the relative humidity is high, so it rarely reaches the potential necessary to produce a visible arc. Mechanical sparks tend to be more vigorous in dry atmospheres, since dry air conducts heat more poorly than moist; therefore, the particle retains its heat longer. The spark can also more readily ignite dry fuel materials than moist ones that it may contact.

Most important to understand is that once the fire is actually burning, the humidity alone has little or no effect on it. The effects are virtually confined to the primary ignition, not to the magnitude of the ensuing blaze. The reason is rather obvious, namely, that the heat generated by the fire is so great that all humidity considerations of the fire environment are totally overshadowed. It is also true that when fuel in a humid environment is exposed for a considerable period of time to localized heat, it is dried by that heat, regardless of the ambient humidity.

WETNESS OF FUEL (FUEL MOISTURE CONTENT)

Fuel moisture content is not the same as humidity, which relates to wetness of air. The former is of far greater importance than humidity when the degree of wetness exceeds that which is in equilibrium with the air (controlled partially by humidity). An object that is exposed to air reaches equilibrium (fuel moisture) with it and is stated to be "dry" as long as excess moisture in the form of rain, dew, or water from an artificial source is not applied to it. Although all "dry" fuel materials burn readily, the excess moisture of wet fuel must first be evaporated before the fuel can be raised to its ignition temperature. Because 4.4 kJ is required to evaporate one gram of free water, it is apparent why so much extra heat must be applied to wet fuel before it can ignite.

Wetness of fuel must not be confused with humidity, because they are totally different even though related in some aspects. Dew or even heavy fog tends to moisten and accumulate on surfaces and makes them hard to ignite. Fuel that is wet internally (high fuel moisture) from absorption of moisture is even more difficult to burn than fuel moistened on the surface only. The principle is identical, however, in that the water boils out at a temperature far below the ignition point and keeps the fuel temperature low only until there is no more free water to boil. Because of the excess water content of a water-logged piece of wood, there is little comparison between it and a piece of wood that has stood in humid air until it is in equilibrium with it, that is, is "dry." Preliminary studies indicate that measuring the moisture content of wood or char after a fire may give an indication of whether prolonged heating occurred prior to its ignition (such as in the formation of reactive charcoal) or whether fuel moisture conditions were suitable for ignition by low-energy ignition sources such as cigarettes or sparks from "hot work."[37] The prefire humidity in the area of ignition of grass and other vegetation should be documented, since it may greatly affect ignitability.

Tests by the California Bureau of Home Furnishings on dry Christmas trees produced maximum heat releases on the order of 800 kW, but a similar tree thoroughly wetted before ignition had a peak heat release rate of only about 200 kW.[38] Tests published by NIST on tests of Scotch pine Christmas trees verified these results.[39] This effect of fuel moisture is true whether the fuel is a solid fuel with water present or complex organic materials such as meat or fat, which incorporate water as part of their structures. If sufficient heat can be applied to drive off this water, meat or fat can certainly burn (as any hapless barbecuer can attest). The effect of excess water is the basis for fighting fires with water. If water can be added faster than the fire can boil it away, the fire eventually will be extinguished because the temperature is lowered. The steam generated by putting water on a large fire in a room may be sufficient to displace enough air in the vicinity of the flames to further reduce the combustion rate. Some exceptions exist. Some fuels, such as magnesium metal fires, cannot usually be fought with water because water on hot masses of magnesium generates the excellent fuel hydrogen gas, and its combustion increases the size of the fire spectacularly.

WIND

Wind is usually considered to be a factor only in fire spread in wildland fires, but a strong wind can affect fire development in a building, particularly after windows begin to fail. Open doors can admit enough wind to promote more horizontal spread than would be expected from normal dynamics, and additional movement can be induced as windows break by thermal shock. In some cases, the induced wind through the structure can overwhelm normal spread. In one documented fire test in a large wood-frame building, the area of origin was actually less damaged than were other areas where the fire was driven and then enhanced by a strong wind blowing through window and door openings.[40] Recent tests by NIST have demonstrated that even moderate external winds entering a high-rise building through an open window cause much higher temperatures and heat fluxes in the fire room and quickly spread hot gases and smoke into adjacent hallways.[41]

As mentioned previously, wind causes drying of living grasses and forest products beyond what would result from high temperatures alone, so wind conditions in the hours and days prior to a wildland fire may be critical factors in ignition. The careful investigator should establish the wind, temperature, and other weather conditions that existed at the time of the fire.

OXYGEN CONTENT

Although most fires occur in normal oxygen concentrations (20.9 percent in air at sea level), oxygen concentrations may play a role in both ignition and fire growth. Oxygen concentrations below 15 percent in air will retard ignition and fire growth. Concentrations below 10 percent will make flaming fire growth impossible (except in post-flashover fires), but smoldering fires will propagate even at oxygen concentrations below 3 percent. Elevated oxygen concentrations will dramatically enhance ignitability of fuels (even by weak ignition sources) and cause very rapid flame spread. Fuels that are

FIGURE 3-23 Plastic O₂ cannula burning with bright flame. *Courtesy of Jamie Novak, Novak Investigations and St. Paul Fire Dept.*

normally difficult to ignite, such as rubber insulation or petroleum lubricants, are very susceptible to ignition in oxygen environments. Fires in surgical or dental facilities may be very severe in areas (such as wall cavities) where plastic oxygen lines fail due to fire exposure. Ignition of oxygen-saturated surgical drapery, dressings, clothing, bedding, and even hair during surgeries has been reportedly caused by small electrical arcs in instruments such as cauterizers. The transition from smoldering to flaming combustion is dramatically enhanced by excess oxygen. Recently, smokers using supplemental (medical) oxygen have discovered that a dropped cigarette can readily ignite clothing or bedding, sometimes with tragic consequences (Figure 3-23).

Elements of Building Construction

It is very important for an investigator to be familiar with buildings—how they are constructed, what materials were used in their construction, and how they react to fire. The manner and materials of construction can have an enormous effect on manner and rate of fire spread as well as on the nature of post-fire patterns of damage.

STRUCTURAL SHELL

The outermost portion of a building is often called the *structural shell*. The structural shell is the outside or supporting portion of a structure, whether it is residential, business, or industrial in use. It includes especially the outer walls and supports and occasionally the columns and steel skeletons at the center of the building, which are sometimes used in modern construction as supporting structures. The materials, arrangement, and interconnections of these structural elements can have major effects on the spread of fire and the resulting collapse of burning buildings. The most comprehensive source of useful information in these disciplines is Francis Brannigan's *Building Construction for the Fire Service*, and the reader is referred there for a detailed review.[42] Structural shells can be considered to have one of the basic constructions seen in Figure 3-24 and discussed next.

Masonry

In this category are concrete (cast and block), tile, brick, and similar materials, even adobe. Shells constructed of masonry are clearly not flammable and will not directly contribute to a fire. To assume that they are not subject to fire loss is a mistake, because

FIGURE 3-24 Typical ordinary, wood frame, and heavy-timber construction, and commonly used terms.

within such shells there are ordinarily walls, floors, and other structures of wood, combustible furnishings, and other combustible materials.

Masonry walls are found both in heavy-timber (industrial) structures and in ordinary construction of many residential and light commercial structures. They can form the structural shell, supporting the ends of beams and girders, as well as the load-bearing walls within the structure. Masonry walls can be two interconnected wythes or thicknesses, with a hollow space in between, or the space may be filled with insulation, mortar, cement, or rubble, or even be a single thickness. Hollow masonry walls offer an opportunity for flames and hot gases to travel upward within a building.

A fire that develops within such a masonry shell may actually be more intense than if the walls were directly involved because of the insulation effect and chimney action that may be present. Some of the worst conflagrations have involved this type of structure. The large thermal mass of masonry walls, floor, and ceiling holds in the heat produced by fire within. The higher temperatures can cause more complete combustion of the contents and inhibit fire suppression activities. Furthermore, higher and sustained temperatures in nonreinforced masonry can also lead to the premature unexpected failure of exterior supporting walls. The remaining rubble stays hot for considerable periods due to this "oven effect."

Wood

In some portions of the world, wooden shells for residences are the rule, and almost everywhere at least some residential buildings are basically of wood. Business and industrial buildings are less commonly constructed in this manner at present, but many older buildings still in existence were so constructed. These may be referred to as "fire traps" and often they are, but the reasons relate more to the interior construction, finish, and content than to the shell, which is rarely the point of fire initiation, regardless of what material is used for its construction. Wood stud walls are particularly dangerous from a fire spread standpoint when they are used in *balloon frame* form. In these multistory structures, the wall studs form a continuous cavity from floor to attic. Balloon frame construction has not been used in the United States since the 1940s. Platform construction minimizes fire spread, as each stud frame is no more than one story, and there is separation between floor, wall, and ceiling cavities, as was seen in Figure 3-24.

A special case of wood structural shells is that in which the exterior coating is some form of plaster such as stucco. The presence of such material on the outside of the exterior walls has little influence on the fire hazard, except that such a building is less likely to ignite from an external source such as a brush fire or a fire in an adjoining building. However, interior fires, which make up the vast majority of all structural fires, will be quite independent of the existence of a stucco coating. The stucco and wire mesh coverings on wood-framed structures may remain standing after the structural wood has burned away. Manufactured or modular homes often have wood-framed exterior shells. They will be discussed in more detail in a later chapter.

Metal framing and studs are being used as a substitute for wood framing. Top plates, sole plates, sills, and studs are made of formed galvanized sheet steel. They interlock to form direct replacements for wood framing. Sheathing (inside and out) is then affixed using adhesives or screws. The performance of these wall panels in fire exposures appears to meet or exceed those of traditional wood-framed walls.

Glass and Metal

Some modern buildings have shells that are virtually all steel with panels of aluminum and glass hung between the vertical members to form the walls. Although the steel and aluminum panels themselves have no intrinsic fire hazard, they are sometimes covered with synthetic coatings of plastic sheet or insulating materials that can contribute greatly to the fuel load of the building.

The large glass panels are subject to rapid failure on direct exposure to flames from fires in adjacent areas or lower floors. With their failure, the fire can ignite draperies or furnishings immediately within, and the fire leapfrogs upward through the building. Such failures are thought to be responsible for fire spread in tragic fires in high-rise buildings like those in Madrid (2004), in the First Interstate Bank Building (Los Angeles, 1988), and in the One Meridian Plaza (Philadelphia, 1990).[43] The flexible plastic or rubber seals that retain the metal and glass panels in such buildings can also fail on fire exposure, allowing flames to penetrate the wall even if the panel itself remains in place. Similar failure of sealants between floor decks and exterior walls can allow fire to move upward between floors even when the windows remain intact and in place.

Sandwich Panels

A relatively recent innovation, especially in industrial and commercial structures, is the use of sandwich panels of sheet steel or aluminum glued to both sides of a rigid insulating material such as rigid polyurethane or polyisocyanurate foam or expanded polystyrene, or onto a built-up frame with mineral wool. The panels can be fitted to a metal external framework or interlocked together to form walls, corners, and ceilings. They offer easily maintained surfaces, good thermal insulative capacity, and ease of construction and portability, all at relatively low cost. They are often encountered in food processing or refrigerated storage facilities (see Chapter 5). Panels may also be faced with fiberglass or rigid plastic, gypsum wallboard, plywood, or chipboard. Depending on the type of foam used, they can be subject to very rapid collapse if exposed to fire.[44] (Figure 3-25.)

FIRE RESISTANCE RATINGS

The NFPA currently uses the following classes of construction per *NFPA 220, Types of Building Construction.*[45]

Type I (Formerly "Fire Resistive")

Only noncombustible materials are permitted for the structural elements of the building (materials must meet relevant NFPA fire resistance specifications. Structural materials include walls (interior and exterior), partitions, columns, floors, and roofs. The use of some combustible materials is permitted for nonstructural elements such as roof coverings, some insulation materials, and limited amounts of wood for interior finish and flooring.

FIGURE 3-25 Sandwich Panel Building. Walls are glued laminates of gypsum wallboard, expanded polystyrene foam, and chipboard. Floor and ceiling are reinforced with stamped steel channels. *Courtesy of Jamie Novak, Novak Investigations and St. Paul Fire Dept.*

Type II: (Formerly "Non-combustible"

Structural elements must be noncombustible or limited combustible (as permitted by the local code) that are protected to have some degree of fire resistance (1 hour or 2 hours) for exterior walls. Other components may be completely unprotected per the local code.

Type III (Formerly "Ordinary Construction" or "Exterior-Protected Combustible")

All or part of interior structural elements may be of combustible materials (or other materials as permitted by local code). Exterior walls must be of noncombustible materials. All concealed spaces must be appropriately fire-stopped. They can be further classified as Type III (211)-Protected, where floors and structural elements have a 1-hour fire resistance rating, or Type III (200)-Unprotected, where there are no specified fire resistance ratings for floors or structural elements.

Type IV (Formerly "Heavy Timber")

All structural members (columns, beams, arches, floors, and roofs) are of unprotected wood (solid or laminated) with large cross-sectional areas. No concealed spaces are permitted. Noncombustible or limited combustible materials (minimum fire rating of 2 hours) can be used in both interior and exterior load-bearing walls. Brick and stone are the most common wall materials. "Mill" construction often denotes floor supports designed to fail during an extreme fire, leaving the supporting walls or posts intact.

Type V (Formerly "Wood Frame")

Structural members are entirely of wood or other permitted materials. Exterior walls may or may not be required to be **fire resistive** based on the horizontal separation distance to the adjoining structure. Type V construction is divided into two subclasses: Type V (111), where there is a 1-hour fire resistance throughout including the exterior walls, and Type V(000), which has no fire resistance ratings except exterior walls when the horizontal separation distance is less than 10 ft.

fire-resistive ■ A structure or assembly of materials built to provide a predetermined degree of fire resistance as defined in building or fire prevention codes (calling for 1-, 2-, or 4-hour fire resistance).

INTERNAL STRUCTURE

As used in this chapter, internal structure refers to floors, carpets and pads, ceilings, walls and interior finish, and doors.

Floors

In multistory buildings especially, the material of which the floors are constructed is often critical to fire development. Wood floors—however they may be finished—are likely to be breached from beneath in any major fire, spreading the fire upward from floor to floor and limiting the chances for isolating it before it grows too large. Concrete floors, frequently used between stories of industrial and commercial buildings, tend to prevent the spread of flames from story to story, thereby making containment of the fire much easier. However, under sustained direct fire exposure, concrete floors can buckle and crack, allowing fire to penetrate. Such decks can be lightweight concrete poured onto corrugated steel panels or precast panels with internal steel reinforcement to provide strength and fire resistance. Combustible forms of wood or plastic foam may also be left in place around the concrete, accelerating its failure. The lowest floor of a structure will usually be breached only when the fire begins on the floor itself (or in some open space beneath it), when burning debris falls onto it from above, or when post-flashover radiant heat is maintained for long enough.

Some residences built on steep slopes may have an elevated lower floor, allowing exposure to exterior grass or trash fires. A thorough external examination of a burned building will reveal such exposure problems. Even on flat land, the ground floor is sometimes raised sufficiently to allow a fire to develop beneath it. The ventilation provided by the footings or posts under such buildings (usually old frame dwellings) produce perfect conditions for complete involvement of all wooden structural members. Such buildings can be destroyed virtually completely, leaving only charred stubs of the posts and the remains of a noncombustible roof covering. Gas may escape under the floor, and trash or grass fires that are too close to the building may also get under it, providing accidental hazards.

The material and construction of every floor can have a great influence on the rate of fire spread and can contribute to the overall fuel content of the structure and therefore must be considered whenever the overall fire pattern is being assessed. Floor decks comprised of plywood or chipboard over manufactured wood trusses have been shown to fail in as little as 10 minutes of direct flame contact.[46] Manufactured floor joists that use plywood or oriented strand board (OSB) can fail in less time than will joists of solid wood. These considerations are important when considering the time of development and spread in a building. Fire penetration rates of wood floors are discussed in Chapter 5. Babrauskas has compiled extensive references on fire-related failures.[47] Floor coverings vary greatly in their fire resistance. Tile, terrazzo, and heavy linoleum are highly resistant, while carpet, bare wood floors, and some vinyl floor coverings can burn more easily from above or beneath.

Carpets and Pads

Although not part of the structure itself, carpet and its padding are often overlooked as potential contributors to the fuel load. As illustrated in Chapter 5, carpet is made of a fiber pile woven or glued to a backing. The pile may be of a natural fiber like wool or cotton or, more commonly, a synthetic like nylon, polyester, or acrylic fiber (or a blend of two or more different types). The backing was traditionally made of a natural fiber like jute, but today it is much more common to find a synthetic fiber, like polypropylene, which is characterized by its white, flat, ribbonlike appearance. These carpets may have a rubber or urethane-foam backing affixed or may be installed over a separate fiber, butyl latex rubber, or urethane foam pad. A plastic-film moisture barrier may be integral with the pad or the carpet backing. These changes affect the spread of fire and the contributions to the fuel load and heat release rate of the furnishings. Wool and cotton carpets require high heat fluxes to sustain their ignition and tend to smolder rather than burn with open flame. Synthetic carpet piles tend to ignite more easily than natural fibers (and some, not intended for interior use, will ignite very readily) and tend to support open-flaming combustion. As a room fire approaches flashover conditions carpets will ignite and add to the heat release rate of the contents. When carpets are made with a jute backing, the jute tends to char but remains largely in place, offering some protection to the padding underneath. When a polypropylene woven backing is used, being a thermoplastic [melting point 165°–186°C (330°F–365°F)], it melts when heated. This allows the entire carpet to melt and shrink, exposing the underlying pad to the radiant heat and promoting its rapid involvement in the fire. As both synthetic carpet and pad melt and burn in a post-flashover fire they tend to form molten puddles of burning plastic, creating localized and irregular patterns of floor damage surrounding areas of undamaged floor. Such patterns can be mistaken for the pool/halo patterns formed by burning ignitable liquids, as shown in Figure 3-26.[48] The localized flames from such floor coverings can be sustained long enough to melt and oxidize aluminum floor moldings. It is strongly suggested that a comparison sample be taken of the unburned carpet and pad in any room suspected of being an origin. This comparison sample can be tested to establish the type of floor covering used (content, depth, construction), which may influence its contribution to the fire as well as reveal the volatile residues it can contribute to the debris.

Ceilings

In multistory buildings, the same considerations for floors also apply to ceilings, because the ceiling of one floor is often the lower side of the floor of the story above. One very important difference exists, however. Because hot gases rise and fires burn upward preferentially, the ceiling is always exposed to heat when any substantial fire exists in a room.

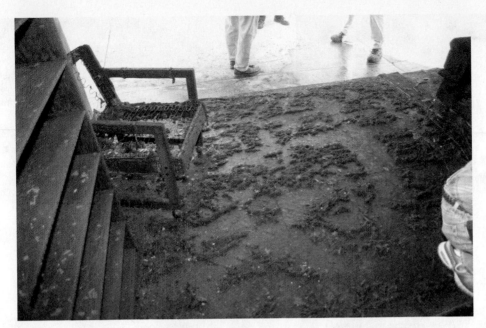

FIGURE 3-26 Brief post-flashover fire caused the synthetic carpet to melt and burn in a highly irregular pattern. *Courtesy of Jamie Novak, Novak Investigations and St. Paul Fire Dept.*

If the ceiling is noncombustible, as with gypsum board, the fire is inhibited from burning through to the story above. As a typical accidental fire develops in a room the buoyant hot gas layer induces more damage to the upper one third of the room prior to flashover. A combustible ceiling is exposed continuously to the hottest gases from the fire and is quickly breached. In room fire tests conducted by the authors, the time to penetrate ceilings has been observed as follows: 1/2-in. plywood—5 to 6 minutes, mineral fiber suspended (T-bar) ceiling—5 to 10 minutes, wood lath and plaster—10 to 15 minutes, 1/2-in. drywall—15 to 20 minutes of direct flame contact.

The situation with floors is different in that the buoyant flow takes the hot gases upward and the radiant heat from the flame plume heats more fuel above the original fire, so the fire burns upward, away from the floor, with the hot gases gathering in the upper portions of the room. Because the floor is exposed to less heat, it is much less likely to be breached during the *growth* of a fire involving ordinary combustible fuels, except where falldown of burning debris occurs. A fire accelerated with an ignitable liquid, however, may generate pockets of burning fuel that may char and even penetrate the floor (around joints and seams). In post-flashover room fires where the floor covering is ignited by the radiant heat, the heat fluxes and temperatures at floor level meet (and often exceed) the high temperatures that previously existed only near the ceiling. At this point damage to the floor becomes much more severe. In cases where the post-flashover burning has not occurred or has not been prolonged, then the fire damage to the floor is normally much less than seen in the upper portions of the room. Falldown of combustible ceiling coverings can cause both falldown fire damage and protected areas, as shown in Figure 3-27. In an accidental fire, until flashover occurs, involvement of fuels at floor level tends to occur late in the fire (except sometimes in the vicinity of the origin), where an accelerated incendiary fire *may* produce floor-to-ceiling damage (full-room involvement) early in the fire. Prolonged post-flashover burning may obliterate this distinction. When the ceiling in a room is made of combustible material like wood, the critical heat flux needed to ignite it is often reached early in the fire development. Early critical heat flux results in quick failure, and the fire can spread upward very rapidly. In a test conducted by one of the authors, the flame plume from a wood-and-cardboard bonfire reached the lath-and-plaster ceiling of a room, causing rapid heating of the wood lath. The ceiling failed within 10 minutes of ignition of the fire, allowing venting of the

FIGURES 3-27A & B Two views of post-flashover room fire. Note rectangular areas of floor damage and adjacent protected areas from falldown of low-density cellulose ceiling tiles. Furring strips and nails on ceiling (and remaining tiles) help identify ceiling lining that aided fire spread. *Courtesy of Jamie Novak, Novak Investigations and St. Paul Fire Dept.*

hot gas layer into the attic above. When this caused failure of the roof, it prevented the development of full flashover conditions in the room, since without the hot gas layer there was inadequate radiant heat to support post-flashover burning. At the other extreme, if the ceiling is wood with a noncombustible layer above it (such as the original plaster ceiling), the fire development in the room is extremely rapid, and time to flashover is shortened significantly, as the hot gases cannot escape the room. The presence of combustible ceilings *or* walls makes development to flashover much more likely and can dramatically reduce (often by 50 percent) the time needed for a room fire to progress to flashover.[49] Figure 3-28 shows a very combustible cellulose ceiling tile, once very common.

As we saw earlier in this chapter, a fire of 650 kW heat output (a level attainable by full flaming combustion of most large upholstered furniture items) has the potential for generating a flame plume that will reach the ceiling of a normal residential room from the floor (2.4 m; 8 ft), even if located away from the walls. Such flame contact will produce a scorching of the ceiling material and may char the covering away completely if the flames last long enough. Smaller fires (lower HRR) near walls or in corners may produce a ceiling-high plume due to the corner effect.

FIGURE 3-28 Low-density cellulose ceiling tiles may be found in older structures. They can support fire at ceiling level and prolonged smoldering fire as falldown at floor level. *Courtesy of John D. DeHaan.*

Walls and Interior Finish

The surfaces of walls are especially subject to attack by fire, both because of their vertical orientation and because of their large surface. Next to ceilings, walls constitute the most vulnerable portions of the main structure of any building. With the exception of masonry walls (and those walls with steel substructure), it is normal for the main structural components, like wall studs, to be of wood and therefore to be combustible. These structural components are usually covered and are relatively immune to attack for a limited time if the covering is noncombustible. Interior wall surfaces were at one time made of plaster, which is itself noncombustible. The support may be of metal wire or wooden lath. Even though the plaster will not burn, it will conduct heat and therefore will not protect underlying combustible lath and studding of the wall for as long as gypsum board, in more common use today, even when intact. Even a small opening, however, will allow flames to enter the wall cavity and play directly on the lath, quickly igniting it. Such walls are often papered, which again adds to the problem of fire resistance. A single layer of wallpaper is not especially hazardous; it quickly chars away without adding a lot of fuel to the fire. When a wall has been repeatedly papered without removing the old paper, the multilayer surface that results creates an increased hazard. As long as the paper is glued tightly to the plaster, heat applied to its face is conducted into the plaster. When the paper lifts away from the plaster, it ignites easily. Both the paper and the adhesive with which it is attached are combustible, and thickening the layer increases the amount of fuel available to the fire. Paint used similarly has generally less effect, although very thick layers of paint have been shown to contribute briefly to the intensity of a fire involving them and ignite "falldown" spot fires.[50] (See Figure 5-24.)

The use of various types of wallboard as a substitute for plaster has been a more recent development of wide application. Early wallboards were of cellulose materials and were generally quite combustible. The introduction of gypsum board or sheetrock is possibly the most important development in improving fire resistance in housing and commercial buildings. The terms *gypsum board, drywall, sheetrock, gyp rock, and plasterboard* are often used interchangeably to denote a rigid panel of calcium sulfate (plaster) coated on both sides with a heavy paper bonded to the plaster. Wallboards come in various thicknesses, and some may come pre-coated with a vinyl coating rather than raw paper. Some earlier forms may not be paper covered on both sides and may come in nonstandard thicknesses. Fire-rated gypsum board is thicker than standard board and contains glass or mineral fibers to offer additional mechanical strength to the board even after fire exposure. Even fire-rated gypsum board has a finite heat capacity and can conduct heat to underlying combustible members, eventually allowing them to char. As plaster (calcium sulfate dihydrate) is heated, it gives up its water of hydration and loses its mechanical strength by the time it reaches ~200°C (400°F). Water-resistant wallboards are made with a cement instead of plaster and are used in bathrooms and enclosures where moisture/water exposure is likely. One older variant is called *button board*, in which the gypsum board is pre-perforated, and a coating of cement-based plaster is applied over it.

Such products significantly reduce the speed of fire spread and can help contain even a large fire in a compartment due to the dehydration process and thermal mass of the calcium sulfate. The paper coating is bound tightly to the gypsum, and heat applied is conducted away into the bulk material, allowing it to resist rapid combustion. The fire properties of gypsum wallboard have been extensively examined by NRC-Canada and others.[51] As the room approaches flashover the paper across the entire exposed wall surface may ignite, adding briefly but substantially to the heat release rate of the fire. The vinyl coatings used on wood and gypsum walls can also add to the fuel load in the room. The flame speed ratings of such wall coverings are determined by tests such as ASTM E84 and E1321 and can be used to evaluate the relative risks of fire spread. There is also a new generation of tests that more closely simulate real-world performance by fitting the material to be tested to both the walls and ceiling of a corner-configuration life-size test cell (ISO 9705 and UBC Standard 26-3).[52]

Wood surfacing, which was originally standard in nearly all primitive construction, was replaced by plaster and other materials but has again become popular because of the aesthetic appeal of the many manufactured plywood and veneer boards that are currently available. These materials are relatively inexpensive, highly decorative, readily applied, and durable. They can also raise the fire hazard by a very significant amount in even more ways than solid wood walls, as described earlier. If ignited, they tend to burn rapidly because of their thinness. They allow the fire to penetrate into the interior of the wall, where it can spread on both sides of the board. And some plywoods have been known to delaminate, with separation of the very thin layers from which they are constructed. When delamination occurs, the thin layers exposed to hot gases from both sides ignite very rapidly, causing very intense fires. No matter how plywoods ignite, they always add to the fuel load of the room, with commensurate effects on fire behavior and intensity. It is not likely that materials as useful as these will fall into disuse, but it can be recommended that they be secured over a layer of gypsum board, which at least prevents burning from both sides and keeps the fire restrained, both in total intensity and from entering the interior of walls.[53] Figures 3-29a–g illustrate the influence of wall-covering material on vertical fire spread, where the same initial fire is ignited against gypsum board, marine plywood, and thin-veneer paneling. Laminated wall materials can delaminate during a fire, and thin combustible layers can curl to the floor, producing fire damage to floors and walls, as in Figure 3-30.

It should also be noted that nonstandard materials can be encountered as wall or even ceiling coverings that can have very fast flame-spread characteristics. These nonstandard materials include carpeting, unprotected urethane or polystyrene foam, or the kraft-paper covering of fiberglass insulation. These materials may burn away so completely, even in a short-duration fire, that they are not readily detectable by post-fire inspection alone. If there is any doubt, the investigator should establish what the wall and ceiling linings were from either interviews or pre-fire photographs. Careful examination of the walls behind baseboards or metal receptacle plates may reveal the presence of combustible coverings. The structural and covering materials of walls, floors, and ceilings should be documented for each room involved in the fire (particularly the room of origin) on forms such as that in Appendix J.

FIGURE 3-29A Small (150 kW) trash fire ignited at the base of a gypsum board wall does not spread. Post-fire damage is limited to surface scorching of paper coating (unpainted). *Courtesy of Susan Sherwin, Scottsdale, AZ.*

FIGURE 3-29B A small (150 kW) trash fire at the base of a ¾-in. marine plywood wall penetrates into the wall and sustains some vertical spread before self-extinguishing. *Courtesy of Susan Sherwin, Scottsdale, AZ.*

FIGURE 3-29C As it self-extinguishes fire leaves behind an inverted V pattern typical of limited-fuel fires. *Courtesy of Susan Sherwin, Scottsdale, AZ.*

FIGURE 3-29D A similar fire ignited against wood-veneer paneling. *Courtesy of Susan Sherwin, Scottsdale, AZ.*

FIGURE 3-29E Within 2 minutes, the fire in Figure 3-29d travels rapidly upward. *Courtesy of Susan Sherwin, Scottsdale, AZ.*

FIGURE 3-29F Flames reach the ceiling with nearly vertical spread from ignition. Notice the V pattern at the ceiling. *Courtesy of Susan Sherwin, Scottsdale, AZ.*

FIGURE 3-29G The burn pattern is nearly vertical, spreading into a V only where the plume encountered the ceiling. *Courtesy of Susan Sherwin, Scottsdale, AZ.*

FIGURE 3-30 Wood paneling on walls contributed to fire spread from ignition in left rear corner. Note paneling curling from ceiling downward (especially along seams). Falldown of burning, curled remnants contributed to ignition of chair and sofa on right. Had this fire burned longer, the exposed paneling would have burned away completely, leaving only irregular heat/smoke patterns on walls and remains behind baseboards or switchplates. The investigator must establish whether such paneling was present. *Courtesy of Investigator Greg Lampkin, Knox County, TN, Fire Investigation Unit.*

FIGURE 3-31A Kitchen fire test—underventilated. Fire in cabinets. *Courtesy of Jamie Novak, Novak Investigations and St. Paul Fire Dept.*

FIGURE 3-31B One door has failed—intense fire near door—near flashover. *Courtesy of Jamie Novak, Novak Investigations and St. Paul Fire Dept.*

FIGURE 3-31C Both doors have failed. Very intense fire at doors—post-flashover. *Courtesy of Jamie Novak, Novak Investigations and St. Paul Fire Dept.*

Doors

Doors provide critical separation of interior spaces, so when they fail, fire behavior can change. See Chapter 5 for times of door failure. The nature (construction and materials) and starting position of all doors in the fire's path should be documented by physical evidence. Changes in their position (from recovery or firefighting) should be documented via interviews. Figures 3-31a–c show the effect of door position on fire behavior and patterns.

CHAPTER REVIEW

Summary

It should now be apparent that fire is a complex process subject to many conditions that influence its growth, spread, and development. These conditions include the physical state of the fuel, its mixture with oxygen, and the transmission of heat throughout the fire environment. The physical state of the fuel (solid, liquid, or gas) has such an important effect on the combustion reaction that the different states will be examined separately in upcoming chapters. The behavior of a fire follows certain patterns (depending on fuel load, ventilation, and the like). These fairly predictable patterns may already be apparent to the astute reader, but they will be brought into clearer focus in later chapters, which deal with the interpretation of the diagnostic signs left after the fire.

Similarly, building construction offers a complex array of configurations and materials. Understanding the role of the structural shell, fire resistance ratings, and internal structure is particularly important for the fire investigator to understand fire spread.

Review Questions

1. What conditions must exist for a fire to occur?
2. Define the two basic types of combustion, and give an example of each.
3. What needs to be burning to produce a flaming fire?
4. Give a brief description of the four stages of a typical fire development—incipient, growth, free-burning, decay.
5. Define *flashover* as the term applies to fire development in a room.
6. Describe a diffusion flame.
7. What is the difference between *heat* and *temperature?*
8. Define *heat release rate.*
9. Describe the three basic means of heat transfer—conduction, convection, and radiation.
10. What is a ventilation limit?
11. What roles can concealed spaces play in allowing fire spread in some buildings?

References

1. Society of Fire Protection Engineers, *Handbook of Fire Protection Engineering,* 4th ed. (Quincy, MA: SFPE, 2008), 4–158, fig. 4–6.4.
2. D. D. Drysdale, *An Introduction to Fire Dynamics,* 2nd ed. (Chichester, UK: Wiley, 1998), 173.
3. J. D. DeHaan, "The Dynamics of Flash Fires Involving Flammable Hydrocarbon Liquids," *American Journal of Forensic Medicine and Pathology* 17, no. 1 (1996), 24–31.
4. J. G. Quintiere, *Principles of Fire Behavior* (Albany, NY: Delmar, 1998), 33–36.
5. V. Babrauskas, *Ignition Handbook* (Issaquah, WA: Fire Science Publishers, 2003).
6. J. G. Quintiere, "Surface Flame Spread" in *SFPE Handbook of Fire Protection Engineering,* 4th ed. (Quincy, MA: NFPA, 2008), 2–235.
7. Quintiere, *Principles of Fire Behavior,* 49.
8. Ibid., 53–54.
9. Drysdale, *An Introduction to Fire Dynamics,* 60.
10. Quintiere, *Principles of Fire Behavior,* 58.
11. Idid.
12. Idid., 56–57.
13. Idid.
14. Drysdale, *An Introduction to Fire Dynamics,* 126–33.
15. Ibid., 132.
16. R. J. Alpert and E. J. Ward, "Evaluation of Unsprinklered Fire Hazards," *Fire Safety Journal* 2, no. 2 (1984): 127–43.
17. J. Albers, "Flashover in Santa Ana," California Fire Service (November 1995), 18–21.
18. Drysdale, *An Introduction to Fire Dynamics,* 292–96.

19. V. Babrauskas and J. Krasny, "Fire Behavior of Upholstered Furniture," NBS Monograph 173 (Gaithersburg, MD: U.S. Department of Commerce, National Bureau of Standards, 1985); G. H. Damant and S. Nurbakhsh, "Heat Release Testing of Furniture and Bedding Products," Fifth Fire Safety Conference and Exhibition, Orlando, FL, February 1993.

20. G. H. Damant and S. Nurbakhsh, "Christmas Trees—What Happens When They Ignite?" *Fire and Materials* 18 (1994).

21. Kent Fire Brigade (United Kingdom), "Video Tapes Fed Fatal Fire," *Fire* (June 1994).

22. DeHaan, "The Dynamics of Flash Fires," 24–31.

23. D.W. Belles, "Illustration of Flashover in a Furnished Room," *NFPA Journal* 79, no 2 (1985).

24. J. R. Lawson and J. G. Quintiere, "Slide Rule Estimates of Fire Growth," NBSIR 85-3196 (Gaithersburg, MD: National Bureau of Standards, 1985); see also H. E. Nelson, "FPETool: Fire Protection Engineering Tools for Hazard Estimation," NISTIR 4380 (Gaithersburg, MD: Building and Fire Research Laboratory, NIST, October 1990); N. Iqbal and M. H. Salley, *Fire Dynamics Tools* (Washington, DC: U.S. Nuclear Regulatory Commission, November 2004).

25. J. D. DeHaan, "Compartment Test Fires: 1996–98," *CAC News* (Spring 2001).

26. Quintiere, *Principles of Fire Behavior,* 188.

27. S. Carman, "Progressive Burn Pattern Development in Post-Flashover Fires" in *Proceedings Fire and Materials 2009* (London: Interscience Communications, 2009).

28. DeHaan, "Compartment Test Fires."

29. R. J. Harris, *Investigation and Control of Gas Explosions* (New York: E & FN Spon, 1983), 126.

30. C. M. Fleischmann, P. J. Pagni, and R. B. Williamson, "Exploratory Backdraft Experiments," *Fire Technology* (Fourth Quarter, 1993).

31. R. W. Bukowski, "Modeling a Backdraft: The Fire at 62 Watts Street," *NFPA Journal* (November–December 1995).

32. S. Kerber and W. D. Walton, *Effect of Positive Pressure Ventilation on a Room Fire,* NISTIR 7213 (Gaithersburg, MD: U.S. Fire Administration, National Institute of Standards and Technology, March 2005).

33. J. A. Moncada, "Fire Unchecked," *NFPA Journal* (March–April 2005), 47–49; "Fire Destroys Madrid High Rise," *Fire Prevention and Fire Engineers Journal* (March 2005), 5.

34. Babrauskas, *Ignition Handbook,* 527–33.

35. C. B. Wood, "Pining for Water," *NFPA Journal* 93, no. 6 (November–December 1999); Damant and Nurbakhsh, "Christmas Trees."

36. M. Bishop and D. DeMars, "Accidental Ignition of a Christmas Tree? Very Unlikely!" *Fire and Arson Investigator* 50, no.1 (October 1999).

37. J. Bloom, "The Use of a Wood Moisture Detector in Structural Fire Investigation," Bloom Fire Investigations, 1998.

38. Damant and Nurbakhsh, "Christmas Trees."

39. J. Stroup, *Scotch Pine Christmas Tree Tests,* FR4010 (Gaithersburg, MD: NIST, December 1999); V. Babrauskas, G. A. Chastagner, and E. Stauss, "Flammability of Cut Christmas Trees" in *International Association of Arson Investigators Conference,* Atlantic City, NJ, 2001.

40. J. Novak, St. Paul Fire Department, personal communication, August 1995.

41. S. Kerber and D. Madrzykowski, *Evaluating Positive-Pressure Ventilation in High-Rise Structures,* NISTIR 7468 (Gaithersburg, MD: BFRL, November 2007).

42. F. L. Brannigan, *Building Construction for the Fire Service,* 3rd ed. (Quincy, MA: NFPA, 1992).

43. Fire Destroys Madrid High Rise," 5; G. P. Gary, "Report on the Investigation of the First Interstate Bank Building, Los Angeles, CA." Paper presented at the Kansas City Arson Seminar, Kansas City, MO, May 1988; D. L. Ziegler, "The One Meridian Plaza Fire: A Team Response," *Fire and Arson Investigator* (September 1993).

44. S. Colwell and D. Smith, "Comparative Fire Tests on Sandwich Panels—Summary Report," *Fire Engineers Journal* 60, no. 205 (March 2000); M. Harris, "Testing Your Metal," *Fire Engineering Journal/Fire Prevention* (June 2005): 27–28.

45. *NFPA 220, Types of Building Construction* (Quincy, MA: NFPA, 2005).

46. Illinois Fire Service Institute Floor Panel Tests, 1998; A. R. Earls, "It's Not Lightweight Construction; It's What Happens When Lightweight Construction Meets Fire," *NFPA Journal* (July–August 2009): 38–45.

47. V. Babrauskas, *Ignition Handbook,* 942–65; V. Babrauskas, "Charring Rate of Wood as a Tool for Fire Investigators" in *Proceedings Interflam 2004* (London: Interscience Communications, 2004), vol. 2, 1155–70.

48. R. J. Powell, "Testimony Tested by Fire," *Fire and Arson Investigator*: (September 1992), 42–51; J. J Lentini, "The Lime Street Fire: Another Perspective," *Fire and Arson Investigator* (September 1992): 52–54; J. D DeHaan, "Fire: Fatal Intensity," *Fire and Arson Investigator* (September 1992): 55–59; J. Albers, "Pour Pattern or Product of Combustion?" *California Fire/Arson Investigator* (July 2002): 5–10.

49. B. Karlsson and J. G. Quintiere, *Enclosure Fire Dynamics* (Boca Raton, FL: CRC Press, 2000).

50. J. M. McGraw and F. W. Mowrer, "Flammability of Painted Gypsum Wallboard Subjected to Fire Heat Fluxes" in *Proceedings Interflam 1999* (London: Interscience Communications, 1999), vol. 2, 1325–30.

51. S. M. Cramer et al., "Mechanical Properties of Gypsum Board at Elevated Temperatures" in *Proceedings Fire and Materials 2003* (London: Interscience Communications, 2003), 33–42.

52. R. W. Bukowski, "Prediction of the Structural Fire Performance of Buildings" in *Proceedings Fire and Materials 2003* (London: Interscience Communications, 2003), 1–8.

53. N. Benichou, M. A. Sutton, and V. R. Kodur, "Fire Resistance Performance of Lightweight Framed Wall Assemblies" in *Proceedings Fire and Materials 2003* (London: Interscience Communications, 2003), 9–20.

CHAPTER

4

Combustion Properties
of Liquid and Gaseous Fuels

*Courtesy of Jamie Novak, Novak
Investigations and St. Paul Fire Dept.*

KEY TERMS

accelerant, *p. 106*

adiabatic, *p. 104*

autoignition temperature, *p. 94*

BLEVE, *p. 120*

boiling point, *p. 98*

combustible liquid, *p. 94*

flame point, *p. 94*

flammable liquid, *p. 94*

flammability limits, *p. 88*

flash point, *p. 87*

ignitable liquids, *p. 112*

ignition energy, *p. 97*

ignition temperature, *p. 94*

odorant, *p. 113*

olefinic, *p. 107*

vapor density, *p. 99*

OBJECTIVES

After reading this chapter, you should be able to:

- List and describe the three types of fuels.
- Describe the physical properties of fuels.
- Explain how the flammability limits of a vapor/air mixture are affected as a function of initial temperature.
- Explain the relationships of the terms *flash point, flame point/fire point, ignition temperature,* and *ignition energy.*
- Define the terms *boiling point* and *vapor density.*
- Describe how the *heat of combustion* (defined in Chapter 2) of a fuel is of importance to the fire investigator.
- Explain the difference between *hydrocarbon* and *nonhydrocarbon fuels.*
- Describe the behavior of liquid pools on porous and nonporous surfaces.
- Describe the distribution and effects of radiant heat from a burning pool of an ignitable liquid.
- Discuss several failure modes of fuel gas lines.

85

Now that we have examined the fundamental chemical and physical properties of fire and fuels, we are ready to investigate the combustion mechanisms of particular fuels in greater detail. Unfortunately, there is no subject about which even experienced fire investigators are so likely to err. Most fire investigators are not chemists, but some chemical knowledge is essential to the proper interpretation of fires, as we have seen fires to be strictly chemical reactions. To compound the investigator's confusion about fuels and the tests used to evaluate them, there is a whole vocabulary of terms that apply to tests devised in the laboratory to define the properties of fuels of investigative significance. It is the purpose of this chapter to discuss some of the concepts that are basic to understanding how gaseous and liquid fuels burn, the conditions and limitations that apply to the combustion of such fuels, and the conventional methods of expressing combustion properties in terms of laboratory tests.

Types of Fuel

Combustion properties have different significance, or at times perhaps no significance, when applied to various types of fuel. Thus, the concept of flash point, discussed later, is essential information when studying liquids, is rarely ever used for solids, and would have no meaning if applied to gases. The relationship between the type of fuel and the type of property is clarified later.

GASES

Flammable materials found in the gaseous form will burn whenever mixed with the proper amount of air and properly ignited. Flash point has no significance with such materials, nor has boiling point. However, the flammable range of mixture with air, the vapor density, and ignition energy and temperature are important properties.

LIQUIDS AND THEIR VAPORS

Vapors from liquids are the materials that directly support the flame over a liquid fuel. For the liquid/vapor combination, vapor pressure, flash point, and, to a lesser degree, boiling point are all important considerations. Ignition temperature, range of flammable concentrations of vapor/air mixtures, and vapor densities are critical properties of the vapors themselves. Liquids *themselves* do not undergo combustion, except in exotic cases; instead, it is the vapors emanating from liquids that can ignite and burn as flames.

SOLIDS

Solids burn by direct combination of oxygen with their surface (smoldering combustion), as volatilized materials that have melted and vaporized, or as complex fuels that pyrolyze to form combustible gases and vapors and leave a noncombustible solid residue. Reactive metals such as magnesium, sodium, potassium, or phosphorus, and carbonaceous fuels such as charcoal, burn only at their surface, as a glowing fire. If the vapors being pyrolyzed from common solid fuels are disregarded, most of the properties just mentioned do not apply to solids. Nonpyrolyzable solids have no flash point in the strict sense, no vapor density, and no single ignition temperature. They have a flammable mixture range only when they are finely divided and stirred into air as a combustible dust suspension. The physical properties that control their flammability are such factors as density, conductivity, and thermal capacity. Their porosity and melting point will also influence their performance as fuels under some conditions. The chemical properties of a solid fuel will determine the nature of volatile pyrolysis products and the rate at which they are generated. The physical and chemical properties of solid fuels and their ignitability will be examined in greater detail in Chapter 5.

Physical Properties of Fuels

Physical properties of fuels of particular interest to the fire investigator include vapor pressure, flammability or explosive limits, flash point, flame point/fire point, ignition temperature, ignition energy, boiling points, vapor density, and heat of combustion.

VAPOR PRESSURE

Any liquid exposed to air will evaporate as molecules escape its surface. If this process occurs in a sealed system, an equilibrium state will be reached in which the vapor reaches saturation, and there is no further evaporation. The partial atmospheric pressure exerted by the vapor at this stage is called the *saturated vapor pressure*, which is most often measured in millimeters of mercury (mmHg) (atmospheric pressure is 760 mmHg) or kilopascals (kPa), where 1 atmosphere (atm) is 101.32 kPa. The vapor pressure is a direct measure of the volatility of a liquid and is determined by its molecular weight, its chemical structure, and the temperature (see Figure 4-1). The higher the temperature, the more liquid will evaporate and the higher the vapor pressure. Table 4-1 displays several pressure equivalents.

As seen in Figure 4-2, the relationship between temperature and vapor pressure is not linear but is a logarithmic function with controlling constants that are different for each liquid. For a given class of compound (alkane, alcohol, branched alkane), the lower the molecular weight, the higher the vapor pressure. The vapor pressure of an aromatic or that of a normal alkane is lower than that of a branched alkane of the same molecular weight at the same temperature, so a branched (iso) alkane will have a lower **flash point** than an alkane of the same molecular weight. For example, benzene (molecular weight 78): flash point, −11°C; hexane (MW = 82): flash point, −22°C; isohexane (MW = 82): flash point, −29°C. The vapor pressure is a fundamental physical property that has a great influence on the flammability of liquid fuel. When the vapor pressure of a liquid reaches 760 mmHg, the liquid is at its boiling point, so the temperature at which that

FIGURE 4-1 Vapor pressure is the pressure created by a liquid (or solid) evaporating to saturation in a sealed vessel.

flash point ■
Temperature at which an ignitable vapor is first produced by a liquid fuel.

TABLE 4-1	Pressure Equivalents				
kPa	mmHg	psi	atm	bar	in. w.c.
689.2	5170	100	6.80	7	—
101.32	760	14.7	1	1.013 (1013 mbar)	406.7
6.89	51.7	1	0.068	0.07 (70 mbar)	27.67
0.69	5.2	0.1	0.007	0.007 (7 mbar)	2.77

FIGURE 4-2 Calculated saturation vapor pressure (in mmHg) as a function of temperature (°C) for three common hydrocarbons: *n*-pentane, *n*-hexane, and *n*-octane. Upper dashed line is atmospheric pressure (760 mmHg). Middle (long-dashed) line corresponds to the upper explosive limit (UEL) (8% in air, 61 mmHg) for *n*-alkanes. The lowest dashed line corresponds to the lower explosive limit. Note that at all temperatures above 5°C, pentane and hexane are above their UEL at saturation.

occurs is recorded as its boiling temperature. When the temperature of a liquid fuel is such that its vapor pressure reaches the percentage of 760 mmHg that is equivalent to the lower flammability limit of the fuel (see next section), the fuel is said to be at its flash point, as discussed later in this section. The vapor pressure is also dependent on the contour of the surface of the liquid, being higher for a convex surface than for a flat surface at the same temperature. This partly explains why an aerosol of small droplets or a liquid distributed on the fibers of a porous wick is often easier to ignite than a pool of the same liquid. Also playing a critical role in such ignitions is that the larger total surface area and small size of the aerosol droplets allows them to absorb heat more quickly from their surroundings. A thin film of liquid on a fiber of a wick can absorb heat very quickly from applied heat flux as well as from the surrounding air and supporting fiber. The increased temperature causes it to evaporate more quickly. Being a thin film, it has no bulk liquid to "cool" it by convective heat transfer. The capillary action of the porous wick quickly replaces the evaporated fuel to sustain the evaporation or combustion.

FLAMMABILITY (EXPLOSIVE) LIMITS

flammability limits ■ The lower and upper concentrations of an air/gas or air/vapor mixture in which combustion or deflagration will be supported.

Mixtures of flammable gases or vapors with air will combust only when they are within particular ranges of concentration. When a gas is present at a concentration below its *lower explosive limit* (LEL, or lower **flammability limit**—LFL), it is considered too lean a mixture to burn and cannot be ignited. At concentrations above its *upper explosive limit* (UEL), the fuel/air mixture is too rich to burn and will not ignite. The explosive range between these two limits is illustrated in Figure 4-3 and Tables 4-2 and 4-3. Some materials such as hydrogen, carbon monoxide, and carbon disulfide have very wide ranges within which flame will propagate to create an explosion. However, most fuels, such as petroleum distillate hydrocarbons, have quite narrow explosive ranges.

Fuel/air mixtures are often considered to be in either closed or open systems, as discussed next.

Closed Systems

Within a *closed* system, if the mixture will not explode, it will not ignite. For this reason, the explosive range and flammability range of a gas or vapor may be thought of as one and the same. The explosive ranges (and vapor pressures) of possible fuels at a fire scene must be carefully considered, for they can play a significant part in determining accidental or incendiary origin and in ruling out some hypothetical situations. As can be seen

100% by volume →   ← 0% by volume

No combustion possible →
due to too little oxygen
(mixture too strong)

Increasing fuel content →

Limited combustion possible →
around ignition sources

Upper explosion limit (UEL) →

Explosion occurs →
as soon as
ignition is applied

Lower explosion limit (LEL) →

No explosion →
due to too little gas
(mixture too weak)

0% by volume →   ← 100% by volume

Gas content   Air content

Increasing air content ↓

FIGURE 4-3 Schematic of flammable gas/air mixtures above and below explosion limits. *Based on Brannigan, F. L., R. G. Bright, and N. H. Jason, eds., Fire Investigation Handbook, NBS Handbook 134. Washington, DC: U.S. Government Printing Office, 1980, p. 139.*

from Figure 4-4, flammability limits are temperature dependent: the higher the initial temperature, the wider the range. Technically, the explosive limits and flammability limits are determined under different conditions—constant pressure versus constant volume—and are therefore very nearly the same.

Open Systems

In an *open* system, other variables control the ignitability of the fuel/air mixture. If a mixture of vapor and air is initially too lean for an explosion, efforts to ignite it may raise the temperature and increase the amount of material volatilized sufficiently to cause ignition. More important, probably, is the too-rich mixture that will not ignite. The application of a brief arc will not be expected to cause a fire; however, prolonged application of a flame may add so much additional heat that the flammability limit is affected or may produce

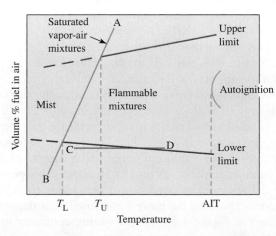

FIGURE 4-4 The flammability limits of a vapor/air mixture as a function of initial temperature at a constant pressure, showing the higher the temperature, the wider the explosion range. *From Zabetakis, M. G., "Flammability Characteristics of Combustible Gases and Vapors." U.S. Bureau of Mines, U.S. Government printing Office, 1965, p. 3.*

TABLE 4-2	Flammability (Explosive) Limits and Ignition Temperatures of Common Gases				
	EXPLOSIVE LIMITS (IN AIR)		IGNITION TEMPERATURE (MINIMUM)		MINIMUM IGNITION ENERGY
Fuel	Lower	Upper	°C	°F	mJ
Natural gas	4.5	15	482–632	900–1,170	0.25
Propane (commercial)	2.15	9.6	493–604	920–1,120	0.25
Butane (commercial)	1.9	8.5	482–538	900–1,000	0.25
Acetylene	2.5	81[a]	305	581	0.02
Hydrogen	4	75	500	932	0.01
Ammonia (NH_3)	16	25	651	1,204	—
Carbon monoxide	12.5	74	609	1,128	—
Ethylene	2.7	36	490	914	0.07
Ethylene oxide	3	100	429	804	0.06

Sources: Fire Protection Handbook, 19th ed. (Quincy, MA: National Fire Protection Association, 2003), table 8-9.2; and *SFPE Handbook of Fire Protection Engineering,* 4th ed. (Quincy, MA: Society of Fire Protection Engineers and the National Fire Protection Association, 2008), table 3-18.1.
[a]Higher concentrations (up to 100) may detonate.

so much turbulence as to mix in additional air and thus start a fire. Because of the initial overly rich mixture, this resultant fire is expected to develop into what is known as a *rolling fire* of some intensity. It must be emphasized that the reason fire is ignited in both cases is that the system was altered by introduction of heat from the ignition source. Such an alteration is not possible in a closed system and will not be effective in producing a fire any more than in producing an explosion. Thus, on a warm day, when the gasoline/air mixture inside a tank of gasoline is richer in gasoline vapor than the upper explosive limit, an arc could safely be produced inside the tank with no ignition. At *very* low temperatures, when fewer vapors are produced and therefore may form a mixture with the air that is within the explosive range, an explosion may result (see Table 4-2).

Vapor pressure considerations play a critical role in evaluating incidents where the flame external to a partially full container of flammable liquid is thought to have caused the vapor inside the container to explode by combustion. As we saw in Figure 4-1, the vapor pressure of pentane (the major, most volatile liquid component in automotive gasoline) is, at saturation, well above its UEL (8 percent) for any reasonable temperature (above −20°C; −4°F).[1] This means that if there is a significant quantity of liquid in the container and a reasonable time has elapsed, the concentration of vapors inside the container will be far too rich to support combustion. A flame can be sustained at the mouth of the vessel if there is enough vapor emanating to mix with the surrounding air, but no explosion will occur. (Heating of a closed vessel containing liquid can cause a mechanical explosion if the heating is sufficient to create a vapor pressure that exceeds the mechanical strength of the vessel. See Chapter 12 for a discussion of such mechanical explosions.)

To test the theoretical (calculated) concept, empirical testing of plastic fuel containers filled with gasoline was conducted by Novak and by DeHaan in cooperation with the New Zealand Police Service.[2] In each case, gasoline vapors emanating from the filler spout were ignited readily. There was no explosion, but there was a small clear flame about 12 cm (5 in.) high in each test, as shown in Figure 4-5a. As the plastic container melted, the opening increased in area (Figures 4-5b and c), and heat was absorbed by the gasoline (sometimes to boiling). The flame plume got larger, but there was no propagation into the can to cause an explosion. Eventually, the entire top of the can melted or burned away, exposing the entire horizontal surface of the gasoline, as seen in Figures 4-5d and e.

The maximum size of the fire is controlled by the area of the pool exposed multiplied by an experimentally determined kilowatts per area factor. For gasoline, that factor is around 1,800 to 2,000 kW/m^2. For a typical 20-L (5-gal) container 38 cm × 38 cm (15 in. × 15 in.) in size, this means a maximum fire of about 400 kW. Estimates of the HRR of these tests based on plume height revealed a maximum fire of ~150 kW (seen in Figure 4-5c), the reduction being due to the "lip" of plastic around the pool, which reduced entrainment efficiency. Even when the test was conducted with a flat-style 10-L container half full, lying on its side with no cap (Figure 4-6a), there was no explosion. In that test the upper side of the reclining can melted away over a period of minutes (Figures 4-6b–d). Eventually, the plastic containers failed as the gasoline overflowed from expansion, and a very large pool fire resulted, but there was no explosion (Figure 4-6e). The melted remains of such containers (Figures 4-5e–g and 4-6f) often contain pockets of unburned fuel and bear critical manufacturing data.

For liquid fuels with lower vapor pressures, propagation or ignition can occur within a fuel container. This is now recognized as a problem with ethanol-based motor fuels, where ignition inside fuel storage tanks is possible at normal temperatures.[3] Guidry reported an incident in which denatured (ethyl) alcohol (whose vapor pressure curve is much lower than that of pentane and has a much broader explosive range) caused an explosion within a container.[4] Under the ambient temperature and use conditions, an ignitable vapor concentration existed inside the container. When a flame ignited the stream of liquid pouring from the container, the flame propagated into the container. This caused an explosion inside the container that split the container and spewed burning alcohol over the person holding it, inducing serious burns.

In cases involving the home storage of gasoline, burn injuries can occur when the canister is tipped or knocked over, spilling its contents and dispersing vapors within the area of the container. In a case study by Kennedy and Knapp of 25 burned children less than 6 years old, the research showed that in each case vapors from overturned or spilled gasoline were ignited by a pilot light from either a water heater or dryer. The study noted that children at this age are totally unaware of the hazards gasoline introduces into their environment. Other problems cited in the study are that older children and adults who may misuse gasoline as an inhalant, solvent, or fire starter thus also may place themselves in harm's way by unwittingly exposing the vapors to a competent ignition source.[5]

The tragic explosion of flight TWA 800 off the coast of New York in July 1996 was caused by a very unusual set of conditions when a nearly empty jet fuel tank was heated (by ambient summertime conditions and the operation of onboard air conditioners). The fuel was then exposed to a reduced atmospheric pressure as the aircraft gained altitude (increasing the fuel's partial vapor pressure). At 4,000 m (13,000 ft) an accidental electrical ignition source within the tank ignited the vapors as they reached their LEL. The resulting internal fuel tank explosion caused massive damage to the fuselage and brought the plane down.[6] Automotive fuel tank explosions in movies are created by using high explosives. In real life, automotive fuel tanks very rarely explode, but when a tank does, it is usually as a result of a collision that ruptures it and allows all the liquid to drain out; the concentration of vapor in the tank can then fall below the UFL, and the vapor may explode if ignited.

FLASH POINT

Characteristics of Flash Point
The term *flash point* is basic to the description of many common liquid fuels, since it can be used to assess the fire hazards of ignitable liquids under various conditions (*ignitable*

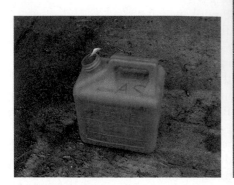

FIGURE 4-5A Vapors of one gallon of gasoline in a plastic container ignited at open spout. *Courtesy of Jamie Novak, Novak Investigations and St. Paul Fire Dept.*

FIGURE 4-5B Burning vapors at 10 minutes. *Courtesy of Jamie Novak, Novak Investigations and St. Paul Fire Dept.*

FIGURE 4-5C Gasoline boiling as vapors burn at 20 minutes. *Courtesy of Jamie Novak, Novak Investigations and St. Paul Fire Dept.*

FIGURE 4-5D Melted remains of plastic can with burning gasoline. *Courtesy of Jamie Novak, Novak Investigations and St. Paul Fire Dept.*

FIGURE 4-5E Melted gasoline container after fire (top). *Courtesy of Jamie Novak, Novak Investigations and St. Paul Fire Dept.*

FIGURE 4-5F Underside of melted container bears molded information. *Courtesy of Jamie Novak, Novak Investigations and St. Paul Fire Dept.*

FIGURE 4-5G Container cut open showing pockets that often trap residual liquid fuel. *Courtesy of Jamie Novak, Novak Investigations and St. Paul Fire Dept.*

FIGURE 4-5 Ignition test of 3.8-L (1-gal) plastic container filled with gasoline, with open filler spout.

FIGURE 4-6A Ignition—no explosion

FIGURE 4-6B One minute—quiet flame at mouth of filler.

FIGURE 4-6C Five minutes—filler neck is melting, flames growing longer.

FIGURE 4-6D Ten minutes—side of can melting, exposing more gasoline to air.

FIGURE 4-6E After 14 minutes, can melts in a pool fire.

FIGURE 4-6F Melted container after fire.

FIGURE 4-6 Plastic 10-L gas can lying on its side, half full of gasoline, with no cap. *All Courtesy of New Zealand Police Service, Canterbury CIB.*

liquids being those that fall under the common classifications of flammable or combustible liquids). The flash point of a material is the lowest temperature at which it produces a flammable vapor. At its flash point temperature, the vapor pressure of the fuel (as a percentage of 760 mmHg) is equal to that fuel's lower limit of flammability. For example, the flash point of *n*-decane is 46°C (115°F). Its vapor pressure at that temperature is 6 mmHg, which is 0.78 percent of 760 mmHg. This value is very close to the measured lower explosive limit of 0.75 percent. A liquid fuel must be able to generate a vapor in sufficient quantity to reach that lower limit in air before it can burn. This does not mean that the vapor will ignite spontaneously at this temperature but only that it can be ignited by a flame, a small arc, or other local source of heat. For example, the flash point of low-octane automotive gasoline is about −43°C (−45°F). This flash point value means only that gasoline produces a vapor that can be ignited at any temperature above about $-43°C$, not that gasoline will spontaneously ignite at such an extremely low temperature.

Determining Flash Point

The flash point is determined by placing a small sample of the fuel in the cup of a testing apparatus and heating or cooling it to the lowest temperature at which an arc or small pilot flame will cause a little flash to occur over the surface of the liquid. No fire results from such a test other than the little flash of flame, which immediately extinguishes itself. Several recognized flash point testers are used today, including the Pensky-Martens, Cleveland, and Tagliabue (Tag) testers. Each device has a particular temperature range and fuel type for which it is most accurate. The Tag closed-cup tester, for example, is the accepted method (described in *ASTM D56*) for testing nonviscous liquids with a flash point below 93°C (200°F).[7] The Pensky-Martens tester (closed-cup) is best suited for liquids with a flash point above 93°C (200°F) and is described in *ASTM D93*.[8] The open-cup tests are used for evaluating hazards of ignitable liquids where they are encountered in open-air situations—during transportation or spills, for instance. The Tag open-cup method is described in *ASTM D1310*,[9] and the Cleveland open-cup method is described in *ASTM D92*.[10] Most tables of flash point data for fire safety purposes list the closed-cup flash point of most materials. As a general rule, the open-cup flash points for the same materials would be a few degrees (<10 percent) higher than the closed-cup values. The flash point tests used in the industry are described in detail in the NFPA *Fire Protection Handbook* and will not be repeated here. The same source contains extensive tables of flash point data for all types of flammable liquids. Some representative flash points are listed in Table 4-3 for general reference.

NFPA[11] classifies ignitable liquids according to their reported flash points:

- Class I: Flashpoint below 38°C (100°F) (**flammable liquids**)
- Class II: Flashpoint between 38°C and 60°C (100°F–140°F) (**combustible liquids**)
- Class III: Flashpoint above 60°C (140°F) (combustible liquids)

FLAME POINT/FIRE POINT

The **flame point** or *fire point* is the lowest temperature at which a liquid produces a vapor that can sustain a continuous flame (rather than the instantaneous flash of the flash point). It is usually just a few degrees above the rated flash point. Although it is not often reported, the flame point is probably a more realistic assessment of a fuel's contribution to a fire environment than the flash point. The fire points of a few common fuels are shown in Table 4-3. The exact values of flash point and flame point temperatures of tested fuels depend on the following factors:[12]

- the size of the ignition source,
- how long it is held over the source each time,
- the rate of heating of the liquid, and
- the degree of air movement over the fuel.

Note that Glassman and Dryer (see Table 4-3) reported minimum fire points for alcohols to be equal to their minimum flash points and observed a significant difference between values obtained using flame igniters and those using electrical arc ignition, apparently as a result of the infrared absorption characteristics of alcohol fuels.

IGNITION TEMPERATURE

The **ignition temperature** is important to all considerations of a fire. Sometimes called the **autoignition temperature** (AIT) or spontaneous ignition temperature (SIT), it is the temperature at which a fuel will ignite on its own without any additional source of ignition. For purposes of comparing liquid and gaseous fuels, ignition temperature can be

flammable liquid ■ A liquid having a flash point below 38°C (100°F).

combustible liquid ■ A liquid having a flash point at or above 38°C (100°F).

flame point ■ Temperature at which a flame is sustained by evaporation or pyrolysis of a fuel.

ignition temperature ■ (same as **autoignition**) The minimum temperature to which a substance must be heated in air to ignite independently of the heating source (i.e., in the absence of any other ignition source; non-piloted ignition).

autoignition temperature ■ The temperature at which a material will ignite in the absence of any external pilot source of heat; spontaneous ignition temperature.

TABLE 4-3 Flash Points and Flame (Fire) Points of Some Ignitable Liquids

Fuel	FLASH POINT (CLOSED CUP)*		FLASH POINT (OPEN CUP)		FIRE POINT	
	°C	°F	°C	°F	°C	°F
Gasoline (auto, low octane)	−43	−45	—	—	—	—
Gasoline (100 octane)	−38	−38				
Petroleum naphtha[a]	−29	−20				
JP-4 (jet aviation fuel)	−23 to −1	−10 to −30				
Acetone	−20	−4				
Petroleum ether	<−18	<0				
Benzene	−11	12				
Toluene	4	40				
Methanol	11	52	1(13.5)[†]	34(56)	1(13.5)	34(56)
Ethanol	12	54				
n-Octane	13	56				
Turpentine (gum)	35	95				
Fuel oil (kerosene)	38[b]	100				
Mineral spirits	40	104				
n-Decane	46	115	52[†]	125	61.5[†]	143
Fuel oil #2 (diesel)	52 (min)	126				
Fuel oil (unspecified)	—	—	133[‡]	271	164[‡]	327
Jet A	43–66	110–150				
p-Xylene	25[§]	77[§]	29[§]	84[§]	—	—
JP-5 (Jet aviation fuel)	66	151				

Sources: *Fire Protection Guide to Hazardous Materials* (Quincy, MA: National Fire Protection Association, 2001), except as noted.

[†]Calculated minimum limit for open-cup tests with flame igniter. Higher values in parentheses are for spark source ignition. *Source:* I. Glassman and F. L. Dryer, "Flame Spreading across Liquid Fuels," *Fire Safety Journal* 3 (1980–1981): 123–38.

[‡]*SFPE Handbook of Fire Protection Engineering*, 4th ed. (Quincy, MA: NFPA, 2008), table 2-8.5.

[§]ScienceLab.com, MSDS p-xylene, 2008.

[a]Generic name for miscellaneous light petroleum fractions used in such consumer products as cigarette lighter fuels and fuels for camping stoves and lanterns.

[b]The flash point of kerosene is set by law in many jurisdictions and may be higher in compliance with local laws.

measured by injecting a small quantity of fuel into a hot-air environment (usually a closed container) at various temperatures. In the standard test, *ASTM E659*, the test chamber is an electrically heated 1-L glass flask (sometimes called a Setchkin apparatus).[13] When the container is at or above the fuel's ignition temperature, the fuel will burst into flame upon injection. For example, low-octane gasoline, with a flash point of −43°C (−45°F), requires a minimum temperature of about 280°C (536°F) to catch fire (100 octane gasoline has an AIT of 456°C). This is the minimum temperature that must be reached by the match, spark, lighter, or other igniting instrument in order for that material to burn. With one or two exceptions, such as catalytic action, the temperature of at least a portion of the fuel must be raised to its ignition temperature before fire can result from any fuel. Ignition temperatures of common materials are generally so high as to rule out spontaneous combustion, except for a very small category of materials and only under very special conditions. One exception, for instance, is carbon disulfide (CS_2), which is encountered as an industrial cleaning solvent. Carbon disulfide has various reported AITs of 90°C to 120°C (194° to 240°F). In general, for a given series of hydrocarbons, the longer the carbon backbone, the lower the AIT of the material. Note the sequence of AITs for *n*-alkanes in Table 4-4, where *n*-pentane has an AIT of 260°C, while *n*-octane and higher alkanes have an AIT of around 208°C.

Catalytic oxidation of some fuels can occur at temperatures much lower than the AIT, but these reactions are very rare and occur under special conditions. For example, platinum in a finely divided state may serve to allow (catalyze) combustion of certain flammable gases in air at low initial temperatures. In a fuel cell or a hand warmer this may be important, but it is very rare in general fires. For solids, in which pyrolytic decomposition must precede ignition, there is another value called the piloted ignition temperature, which will be discussed in Chapter 5.

TABLE 4-4	Minimum Autoignition Temperatures, Flammable Ranges, and Specific Gravities of Some Common Ignitable Liquids				
FUEL	**IGNITION TEMPERATURE** (°C)	(°F)	**MINIMUM IGNITION ENERGY (mJ)**	**FLAMMABLE RANGE (% IN AIR @ 20°C)**	**SPECIFIC GRAVITY**
Acetone	**465**	**869**	**1.15**	**2.6–12.8**	**0.8**
Benzene	498	928	0.2	1.4–7.1	0.9
Diethyl ether	160	320	—	1.9–36.0	0.7
Ethanol (100 percent)	363	685	—	—	0.8
Ethylene glycol	398	748	—	—	1.1
Fuel oil #1 (kerosene)	210	410	—	—	
Fuel oil #2	257	495	—	—	
Gasoline (low octane)	280	536	—	1.4–7.6	0.8
Gasoline (100 octane)	456	853	—	1.5–7.6	0.8
Jet fuel (JP-6)	230	446	—		
Linseed oil (boiled)	206	403	—	—	
Methanol	464	867	0.14	6.7–36.0	0.8
n-Pentane	260	500	0.22	1.5–7.8	0.6
n-Hexane	225	437	0.24	1.2–7.5	0.7
n-Heptane	204	399	0.24		0.7
n-Octane	206	403	—	1.0–7.0	0.7
n-Decane	210	410	—	—	0.7
Petroleum ether	288	550	—	1.1–5.9	0.6
Pinene (alpha)	255	491	—	—	
Turpentine (spirits)	253	488	—	—	<1

Sources: Data taken from *Fire Protection Guide to Hazardous Materials* (Quincy, MA: National Fire Protection Association, 2001); C. F. Turner and J. W. McCreery, *The Chemistry of Fire and Hazardous Materials* (Boston: Allyn and Bacon, 1981).

Virtually all fires originate where there is a localized high temperature in a region in which an appropriate fuel/air mixture occurs. The local region may be very small, as in the case of an electrically induced arc, a mechanical spark, or a minute flame. The important concept is that at this very small point in space, a temperature in excess of the ignition temperature occurs in the presence of appropriate fuel and air (or oxygen), and this energy is transferred to an adequate volume of ignitable fuel. Such circumstances are not unusual, but from the investigative standpoint, it must always be accepted that these circumstances constitute a *minimum* requirement for *any* fire whatever to result. That tiny initial flame then propagates through adjoining mixtures in the flammable range. Many times, no source of such relatively high temperatures appears to be present, and the origin of a fire seems very mysterious. In such instances it should be remembered that it is actually not difficult to produce a source of energy in a very small area. Fires can be kindled by rubbing two sticks together, by the flint-and-steel method (which strikes a very small but hot spark), and by other similar means. In the start of a fire, it is only the heat content of the source (as reflected in part by its temperature) that counts—not the area over which the temperature holds. A nail in a shoe heel striking a rock may ignite a fire, not because the shoe is hot but only because an almost infinitesimally small region (the spark) reaches a very high temperature. Except in industrial fires, where pressurized vessels and delivery lines make for special cases, it is rare indeed for any large amount of fuel to be heated past its ignition temperature. If even a small portion of it is so heated, and the temperature is above the ignition point, a fire can result. This is why so many fires occur under conditions that do not seem to be hazardous. As long as the fuel gas or vapor concentration is within its flammability range in the vicinity of this small but intensely energetic source, there *can* be ignition. The fundamental consideration is not the temperature of an ignition source compared with the ignition temperature of the fuel but the *transfer* of heat energy from the source to the fuel. An ignition source may be present that is at a temperature that is technically higher than the AIT listed for that fuel, but there still may not be ignition if enough energy is not transferred to the fuel.

It should be noted that the AIT is a *minimum* environmental temperature measured by a particular *laboratory* technique (*ASTM E659* being the most common) and is dependent on the size, shape, and even surface material and texture of the confinement vessel used in the test apparatus. The real-world ignition sources that a fuel might encounter will usually have different size, shape, and confinement and so will almost always have to be at a temperature considerably higher than the listed AIT. AIT data are more for reference and comparison than for a specific determination. Some *minimum* ignition temperatures of common fuels are listed in Tables 4-2 and 4-3. Because the minimum ignition temperature is dependent on the method used for its determination, that value should be used primarily for comparisons between fuels. Testing has revealed that "hot surface" ignition of gasoline in open air, for instance, requires that the temperature of the *surface* be at least 200°C (360°F) *above* the listed autoignition temperature [i.e., 500°C−700°C (930°F −1,300°F)]. It is thought that the hot surface produces rapid evaporation, which minimizes contact time.[14] The material and texture of the hot surface will also affect minimum ignition temperatures. The "grade" of gasoline is also critical: the higher the grade or octane rating, the higher the AIT. (See the discussion in Chapter 9.)

IGNITION ENERGY

Every fuel gas or vapor has a minimum **ignition energy**. This is the amount of energy that must be transferred to the fuel to trigger the first oxidation. For most fuels, this is a very small amount of energy; for hydrocarbon fuels, the minimum ignition energy is on the order of 0.25 millijoule (mJ) (see Table 4-2). This amount is dependent on the concentration of the vapor and is often at its minimum when the fuel vapor is at its stoichiometric or ideal concentration (as shown in Figure 4-7). For a vapor/air mixture to ignite, there

ignition energy ■ The quantity of energy that must be transferred into a fuel/oxidizer combination to trigger a self-sustaining combustion.

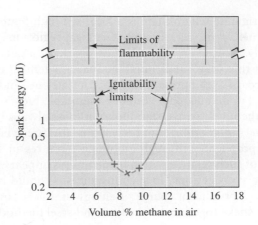

FIGURE 4-7 Ignitability curve and flammability limits for methane/air mixtures at atmospheric pressure and 20°C. *From Zabetakis, M. G., "Flammability Characteristics of Combustible Gases and Vapors," Washington, DC: U.S. Bureau of Mines, U.S. Government Printing Office, 1965, p. 2.*

are four required conditions. First, an ignition source must have adequate energy (above the minimum ignition energy for that fuel); second, the fuel must be present at a concentration within its flammability range; third, there must be contact between the source and the fuel while it is in that range; and finally, the contact must be of sufficient duration for enough energy to be transferred from the source to the fuel. Unless all four of these conditions are met, there will not be an ignition.

Several of the flammability properties are interrelated. For instance, the minimum ignition energy of a fuel is dependent on fuel concentration, as can be seen in Figure 4-7. This means that a very energetic source can ignite mixtures closer to the flammability limits for that fuel, while a very weak source may be competent only when the mixture is near its stoichiometric mixture. In addition, the flammability limits themselves are dependent on starting, or initial, temperature, as was shown in Figure 4-4: the warmer the starting (pre-fire) mixture, the wider the limits.

Although gaseous-phase explosions are treated in depth in Chapter 12, there are several observations about fire behavior that are related to concentration to be discussed here. The speed with which a flame front progresses through a fuel/air mixture is dependent on concentration. The flame speed reaches a maximum when the mixture is near its stoichiometric or ideal mixture. As a result, overpressures (caused by rapidly expanding gases) are also at or near their maxima. A premixed hydrocarbon fuel/air mixture near its stoichiometric mixture theoretically can produce pressures up to 120 psi (8 bar) of pressure under ideal conditions.[15] A lean mixture propagates at a slower speed and produces less heat and, therefore, lower pressures. Because all the fuel is consumed in the reaction, there is no subsequent fire or flame. In contrast, a rich mixture results in a less vigorous explosion with less mechanical damage due to lower pressures and slower speeds. The excess fuel produces smoke or soot and a following flaming fire. Thus, the heat effects and blackening are often greater and the blast damage less when the mixture is rich than when it is lean. The explosive limits of some of the more hazardous liquids and gases are discussed in Chapter 13.

BOILING POINTS

boiling point ■ The (pressure-dependent) temperature at which a liquid changes to its gas phase.

Boiling points are sometimes important in fire investigations, but they are of secondary significance compared with other properties discussed in this chapter. Most liquids may be expected to be heated to their **boiling point** in a fire. However, the very important distinction between a liquid fuel that is reasonably pure and will therefore have a reasonably definite boiling point and one that is a mixture of many components of variable boiling points must be grasped, since boiling point ranges are used to characterize many fuels encountered in fires (See a later section of this chapter for examples of boiling point ranges.)

A pure, single-component liquid fuel will have a definite boiling point at which an entire sample may be distilled without a change in temperature of the vapors. The general public has available to it relatively few pure liquid fuels compared with the much larger number of mixtures that have a boiling *range* rather than a boiling *point*. Such mixtures range from gasoline, with more than 200 individual compounds, to a large number of manufactured mixtures for special purposes such as cleaning or industrial solvent action. When heated, such mixtures allow the distillation of their most volatile constituents first, followed in turn by constituents of decreasing volatility (a process often called *weathering,* which will be discussed more fully in Chapter 14). In general, it is the most volatile constituent of a mixture that is of most significance as it relates to the fire hazard. When a liquid fuel that has a boiling point range (e.g., a mixture such as gasoline) burns, the more volatile species evaporate first and fastest from the surface layer. The bulk liquid beneath the surface will retain the lighter species for some time, since they will have to migrate by diffusion (aided by convective flows) through the bulk liquid to reach the surface before they can evaporate.[16] A consequence of such fractional distillation is that the residue of a petroleum distillate may have a lower autoignition temperature than the bulk liquid, possibly posing a risk of delayed ignition.[17]

In initiating a fire, materials of high boiling point will rarely be above their flash point at ambient temperatures, and only the more volatile (low boiling point) will be expected to show special hazard. As a rule, the boiling point and the flash point tend to parallel each other, so the low-boiling material will usually have a low flash point as well. The presence of a mixture does not greatly diminish the fire hazard of the most volatile component in that mixture. A mixture of gasoline and diesel fuel, for instance, exhibits a flash point indistinguishable from that of gasoline alone until the gasoline represents less than about 5 percent of the mixture.

VAPOR DENSITY

The fundamental property of a vapor that predicts its behavior when released in air is its **vapor density**. The vapor density, that is, the density of a vapor relative to that of air, may be simply calculated by dividing the molecular weight of the vapor by the mean molecular weight of air, which is approximately 29. This relationship is represented by the equation

> **vapor density** ■ The ratio of the weight of a given volume of gas or vapor to that of an equal volume of air.

Vapor density (gas or vapor) = Molecular wt of gas/Molecular wt of air

For example, the lightest gas, hydrogen, has a molecular weight of only 2; therefore, its vapor density (V.D.) is

$$V.D. = 2/29 = 0.07$$

Methane, the simplest hydrocarbon fuel gas found in swamp gas and natural gas, has a molecular weight of 16. Its vapor density, then, is roughly 16/29, or 0.55. Ethane, the other major component of natural gas, has a molecular weight of 30, making it almost exactly the same density as air, and giving it a vapor density of 1.03. The vapor densities of some common fuels are shown in Table 4-5.

Gases or vapors with a vapor density greater than 1.0 are heavier than air and tend to settle through the air into which they are released until they encounter an obstruction, such as a floor, after which they tend to spread outward at this level, much in the same manner as if they were liquids. In contrast, vapors lighter than air tend to rise through the air until an obstruction, such as a ceiling, is encountered, after which they spread at the high level. The tendency to spread is less absolute than with liquids because the difference in density as compared with air is much less than compared with any liquid, and air currents produce mixing. Some mixing with the air is inevitable, and the nearer the vapor density is to that of air, the greater the mixing of fuel and air. This effect of greater mixing is due to the process known as *diffusion.* All gases, regardless of their molecular weight or vapor density, tend to diffuse into each other when the opportunity exists.

TABLE 4-5	Vapor Densities of Common Fuel Gases and Vapors
FUEL	**VAPOR DENSITY (AT STANDARD TEMPERATURE AND PRESSURE)**
Hydrogen	0.07
Methane	0.55
Acetylene	0.90
Carbon monoxide	0.97
Ethane	1.03
Propane	1.51
Butane	1.93
Acetone	2.00
Pentane[a]	2.50
Hexane[b]	3.00

[a]Lightest liquid component of gasoline.
[b]Lightest major component of camping fuels.

Differences in diffusion rates of gases depend on their vapor density: when a fuel vapor is much heavier than air (V.D. >>1.0), its diffusion rate will be much less than when its density is much closer to that of the air into which it is diffusing, that is, when its vapor density is close to 1.0. The diffusion of a vapor is a very slow process. In a closed container, diffusion of a vapor will eventually produce a uniform distribution after some hours. In an open system such as a room, diffusion of vapors from a volatile liquid does not produce a uniform distribution but rather a gradient of concentration from very high (saturation vapor pressure) to very low, as represented in Figure 4-8.

Methane, as noted, is lighter than air and thus tends to rise in a room (as long as its temperature is the same as that of the room air). Its rate of diffusion is greater than that for heavier molecules, and so it tends to mix with air and form a gradient of concentration, as illustrated in Figure 4-8a. The same is true for natural gas, which is a mixture of methane and ethane and other light gases, so its vapor density will be between 0.55 and 1.0, depending on the relative proportions of the components. Ethane is rarely encountered by itself, but because its vapor density is close to 1, it will diffuse readily in air and neither rise nor settle in air, there being no gravitational separation. (It should be noted that diffusion acts to keep vapors diffused and mixed in air and *once mixed in air,* vapors will not "settle out" to re-form layers of high concentration.) The same is true for acetylene (C_2H_2) with a molecular weight of 26 and therefore a vapor density of 0.9, as well as ethene (C_2H_4) and carbon monoxide (CO), each with a molecular weight of 28 and therefore a vapor density of 28/29, or 0.97. All other commonly encountered fuel gases and vapors have a vapor density significantly greater than 1 and will tend to sink when released in air. For example, propane, the next heavier hydrocarbon, (V.D. = 1.51) will mix readily with air, but it will *tend* to sink as it does so. [It should be noted that a flammable propane/air *mixture* (2 − 9.6 percent) will have a vapor density very close to that of air.] Butane (V.D. = 1.93) sinks more readily and tends not to diffuse as quickly. All higher hydrocarbons, characteristic of most petroleum products, will of course have slower diffusion rates (diffusivity) along with their greater vapor density. Gasoline vapors, for instance, contain a mixture of all the most volatile components of the liquid blend. This mixture, being considerably heavier than air, will sink rapidly to the floor or ground surface. Gasoline vapors will therefore pour into low regions or down drains and will carry their intrinsic hazard to the lowest possible region (see Figures 4-8b and c for a graphic representation).

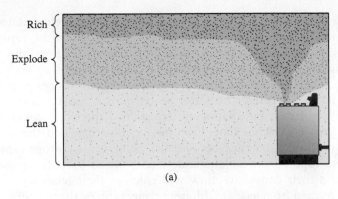

Rich {
Explode {
Lean {

(a)

FIGURE 4-8 Typical mixtures of natural gas or flammable fluid vapors in closed rooms. (a) Natural gas (methane) in a room. (b) Gasoline in a room. (c) Gasoline (or butane) in a room with various elevations.

Lean {
Explode {
Rich {

Liquid

(b)

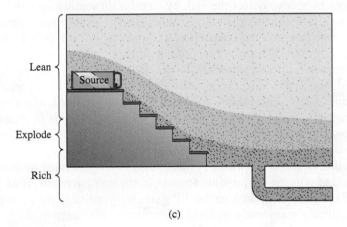

Lean {
Explode {
Rich {

Source

(c)

It should be remembered that from a fire hazard standpoint, the vapor density and diffusion rate of the lightest (most volatile) component of a mixture is what will determine the spread and ignitability of vapors, not the properties of the bulk liquid. For instance, gasoline has a range of components, some of which, like *n*-decane, have a very low vapor pressure at room temperature and a correspondingly high flash point. It is the isobutane and *n*-pentane content of automotive gasoline that generates the bulk of the vapors at ordinary room temperatures and produces the significant ignition risk. Gasoline's heavier components such as the xylenes do not contribute significantly to the vapors until the gasoline has been evaporating for many minutes, by which time the lighter components will have already been ignited.[18] In general, the vapors of fresh automotive gasoline will behave exactly like those of *n*-pentane in their movement.

It is apparent that various concentration layers will exist (with a gradient of intermediate concentrations between the heaviest and lightest layers). Some of these vapor layers will be within their explosive (flammable) limits, while the mixture in other layers will be either too rich or too lean to be ignited. These distributions represent release into a sealed room with still air in which there is no physical activity and in which the vapors are at the same temperatures. In assessment of real-world distribution of vapors, the effects of temperature and air currents must not be overlooked. Mechanical activity such as a fan, furnace, operating machinery, or even someone walking will stir vapors into mixtures throughout a room. The draft created by a water heater burner, furnace burner, or open window may be enough to redistribute vapors and may draw them into contact with an open-flame ignition source at the same time. Even butane if warmed sufficiently, will rise to the top of a room much like methane because the vapor density drops rapidly with increased temperatures. Vapors from a flammable liquid cooled by evaporative cooling will be even denser, and their convective flow will enhance their tendency to settle.[19] When a gas mixture is formed from gases at different temperatures, the temperature alone may control its movement initially until the temperature differences are equalized. The alert investigator must be aware of all these influences (including timer- or thermostat-controlled equipment that may cycle on with no one in attendance) that can distribute and circulate vapors and possibly provide an ignition source as well. In the United States, gas-fired water heaters have been required since 2005 to be designed so as to make it very difficult for their burners to ignite gasoline vapors. One design places the entire heater in an 18-in.-deep "bucket" to keep vapors out of the burner chamber. The other uses a flash suppressor on the combustion chamber to prevent propagation outward. These designs will be described in more detail in Chapter 6.

The concentration gradient established by a heavier-than-air gas (such as LP gas) leaking into *still* air has been shown to be very steep.[20] This means that a person standing in the room may not be able to detect the leak by its odorant while a layer below knee height in the same room is within its explosive range. Recent experiments have shown that thermal currents, even in a small space such as a camper or RV, result in almost complete mixing of propane into air, with no detectable gradient.[21]

One interesting property of dense vapors (V.D. >2.5) is their propensity to flow horizontally, like a viscous liquid. Pentane vapor released from a pool at room temperature will diffuse upward only a few centimeters, and then the mass of vapor will slump sideways and spread outward along the surface at a rate of about 0.05 m/s in still air. This so-called advective flow can spread vapors along a surface much more quickly than diffusion, but this flow is easily overcome by even a modest draft.[22]

Careful consideration of the distribution of fuel gases and vapors must be made when assessing the likelihood of ignition by possible sources in the compartment. If an ignition source like a candle flame is high up in a room, LP gases or gasoline vapors are going to require a long time to diffuse sufficiently to reach their flammable range in the vicinity of the candle flame unless there is some external means of circulation like a fan or strong draft, or the vapors are released under pressure (jet). Similarly, a kerosene heater burning normally at floor level in a room is unlikely to ignite natural gas leaking into the room until the gas can fill the room from the top down in the absence of mechanical circulation. The mere presence of an ignition source and a fuel source in the same room is not a guarantee that there will be ignition. The assessment of possible ignition sources for such vapor/air mixtures should take into account all these contributing factors, including the heights of, and protective enclosures around, appliance burners.

It was once thought that the distribution of blast effects from a vapor/air deflagration was directly linked to the preignition distribution of vapors. However, experiments by one of the authors have demonstrated that even when a highly localized (floor) layer of *n*-hexane vapors is ignited, the pressures are equalized throughout a moderate-sized compartment within 5 milliseconds (ms).[23] This means that the walls of the average compartment will be exposed to the same pressure at the same time from within and will therefore fail where it is weakest structurally. It is frequently noted in explosions in wood-frame structures that it is the bottom of the wall that moves rather than the top, despite whether the

FIGURE 4-9 Pocketed burning of natural gas accumulated under this wooden floor produced significant charring of joists and floor. *Courtesy of Jamie Novak, Novak Investigations and St. Paul Fire Dept.*

deflagration was fueled by natural gas or LP gas. This result can mean that the bottoms of the walls were not as well anchored to the foundation or slab as the tops of the walls were to the structure above. The bottoms of wood-frame walls are also much more likely to be weakened by termites or dry rot. When a fire follows an explosion, however, if the turbulence of the explosion has not completely dissipated the layer of fuel vapor, the fire will tend to burn at the interface between the richest layer of fuel and the air (as discussed previously). Thus, a fire following a hexane (or gasoline) vapor explosion will tend to burn atop or within a remaining layer of dense vapor (where the fuel-rich layers will burn longest). The distribution of *fire* damage low down in the room after a vapor ignition then indicates a heavier-than-air vapor has been involved. The reverse will occasionally occur on the underside of a natural gas layer trapped or pocketed on the underside of a floor or roof structure, as in Figure 4-9. In either case, the location and distribution of postexplosion *fire* damage may give important clues as to the vapor density of the fuel responsible.

The volume of vapor that can be generated by the evaporation of a known volume of a volatile liquid may be important in evaluating the ignition potential of accidental or even intentional spills. If total evaporation can be assumed, the volume of vapor generated (at standard temperature and pressure) by 1 U.S. gallon (3.8 L) of liquid can be calculated from the empirical relationship

Vapor (in cubic feet, ft^3) = 111 × Specific gravity of liquid* ÷ Vapor density of liquid

For example, 1 U.S. gallon of diethyl ether, with a specific gravity of 0.7 and a vapor density of 2.55, will produce

$$111 \times 0.7/2.55 = 30.4 \text{ ft}^3 \text{ of diethyl ether vapor}$$

Because diethyl ether has a flammability or explosive range of 2 to 36 percent in air, 30.4 ft^3 of vapor, if uniformly distributed, will produce an explosive mixture in a room of 84 to 1,520 ft^3 in volume. In metric units the equivalent relationship for 1 L of liquid (0.001 m^3) is

Vapor (in cubic meters, m^3) = 0.85 × Specific gravity ÷ Vapor density

*See Table 4-4 for specific gravity values.

For commercial propane, the liquid/vapor expansion rate is on the order of 272; that is, 1 L of liquid propane will produce about 272 L of vapor at standard temperature and pressure. Commercial butane has an expansion factor of approximately 234.[24] The expansion rate for LP gas itself will vary with the composition of the LP gas, since the rates for each component are different.

HEAT OF COMBUSTION

Another property of fuels that is often of great importance to the investigator is the *heat of combustion* of the fuels involved, as introduced in Chapter 2. If the approximate amounts of fuels in a room can be estimated by weight, the intensity of burning observed can be compared with the possibilities of various combinations of fuels. As we saw in Chapter 3, the rate at which heat is released is more critical to reconstructing the events of a fire than the total amount of heat, but it is sometimes useful to be able to recognize when the fire damage to a room simply does not tally with the amount of heat that should have been produced by the fuels known to be present. The causes for such differences can then be explored. The scientific method question is: Was there so much damage that it cannot be accounted for by assessment of fuel loads normally present? Was there reduced fuel because stock had been removed prior to the fire? Was there reduced ventilation available, or was there enhanced combustion due to oxygen enrichment? Was the lack of damage due to the incomplete or inefficient combustion of the available fuels? Some of the most common fuels, such as wood, have such variable properties as to give only approximations of the actual output in a specific instance. Table 4-6 summarizes the heat of combustion of a few of the most important fuels from the standpoint of the fire investigation.

adiabatic ▪ Conditions of equilibrium of temperature and pressure.

The heat of combustion is used to calculate a theoretical **adiabatic** flame temperature for each fuel. These calculated values are sometimes erroneously used to estimate the potential effect the flames will have on surrounding materials. Work by Henderson and Lightsey indicates that the actual measured flame temperatures produced by such fuels when burned in air are much lower than previously thought.[25] The measured temperatures they reported are included in Table 4-6. The roles of the flame temperatures and heat of combustion in fire investigation will be explored in later chapters. It should be noted that the real-world maximum open-flame temperatures of nearly all "normal" fuels burning in turbulent diffusion flames is about 900°C (1,700°F), and this figure is used in most calculations. Some fuels (such as styrene, crude oil, and urethane foam) that burn with very smoky flames, so that the heat produced cannot be readily lost (low emissivity), produce very high temperatures in the flames. Fuels that burn very cleanly with no soot or pyrolysis products (alcohols) also have a low emissivity and high flame temperatures.

Hydrocarbon Fuels

By far the most important fuels from the standpoint of fire investigation are the hydrocarbons, which range from the light gas methane to much heavier compounds, including oils and asphalts. (For a discussion of the chemistry of hydrocarbons, see Chapter 2.) Because of the importance of hydrocarbons to fire investigation, some of their specific properties will be discussed in order of their molecular weight or chemical classification.

NATURAL GAS

Natural gas, consisting chiefly of methane, is of obvious significance. Natural gas from different geological formations shows considerable variation in composition and may contain some noncombustible gases. It is typically 70 to 90 percent methane, 10 to 20 percent ethane, and 1 to 3 percent nitrogen, with traces of other gases.[26] For most purposes, the investigator may consider its properties to be the same as those of methane, without introducing serious error.

FUEL	HEAT OF COMBUSTION (MJ/kg) NET		FLAME TEMPERATURE (°C) ADIABATIC	ACTUAL
Acetylene	49.9[a] (55.8MJ/m^3)	48.2[b]	2,325	
Butane	49.5[b] (112.4 MJ/m^3)[c]	45.7	1,895	
Carbon monoxide	(11.7 MJ/m^3)[c]	10.1		
Charcoal	33.7	34.3	2,200[d]	1,390[d]
Coleman fuel				770[e]
Coal (anthracite)	30.9–34.8			
Cotton	16			
Ethane	47.5 (60.5 MJ/m^3)[c]	47.4	1,895	
Ethyl alcohol	29.7	26.8		840[e]
Fuel oil #1 (kerosene)	46.1			
Hexane	48.3 (164.4 MJ/m^3)[c]	43.8	1,948/2,200[e]	
Hydrogen	130.8 (12.1 MJ/m^3)[c]			
Gasoline	43.7			810[e]
Methane	55.5 (34 MJ/m^3)[c]	50	1,875[e]	
Propane	50.4 (86.4 MJ/m^3)[c]	46.4	1,925[e]	970[e]
Polyethylene	43.2			
Polypropylene	43.2			
Polystyrene	40			
Polyurethane	23			
Wood	20	16–20	1,590[f]	~600–1,000[f]

[a]*Fire Protection Handbook,* 19th ed. Quincy(MA: National Fire Protection Association, 2003), table A.1.

[b]*SFPE Handbook of Fire Protection Engineering,* 4th ed. (Quincy, MA: Society of Fire Protection Engineers and the National Fire Protection Association, 2008), tables 1-5.3 and 1-5.6.

[c]R. J. Harris, *The Investigation and Control of Gas Explosions* (New York: E and FN Spon, 1983), 6–7.

[d]R. W. Henderson and G. R Lightsey, "Theoretical Combustion Temperature of Wood Charcoal," *National Fire and Arson Report* 3 (1985): 7.

[e]R. W. Henderson and G. R Lightsey, "The Effective Flame Temperatures of Flammable Liquids," *Fire and Arson Investigator* 35 (December 1984): 8.

[f]R. W. Henderson and G. R Lightsey, "Theoretical vs. Observed Flame Temperatures during Combustion of Wood Products," *Fire and Arson Investigator* 36 (December 1985).

LIQUEFIED PETROLEUM GAS

Liquefied petroleum gas (LPG) is commonly used in rural areas where natural gas is not available. It is a mixture of propane and *n*-butane, with small quantities of ethane, ethylene, propylene, isobutane, and butylene.[27] Depending on its source and its intended use, it can be quite variable in its composition, ranging from almost pure propane to a heavier mixture largely composed of butane. For LPG sold in the United States, the investigator can consider LPG the same as propane in its physical and combustion properties without serious error (in some other countries, LPG can comprise primarily butane). If butane is *known* to have been involved (from lighters, torches, or aerosol containers), it is best to use its reference values for calculations.

FIGURE 4-10 Boiling point ranges of straight petroleum distillates.

Boiling point ranges (°C) of straight petroleum distillates

PETROLEUM

Petroleum in its crude state is a thick oil varying in color from light brown to black. It contains a very large number of compounds and different types of compounds, which are separated to a considerable extent in the manufacture of petroleum fuels. Products obtained from the distillation of petroleum include gases, petroleum ether, straight-run gasoline, kerosene, medium distillates including paint thinner–type mixtures and solvents, diesel fuels, heavy lubricating oils, petroleum jelly, paraffin, waxes, asphalt, and coke. The distillation curve of petroleum products is represented in Figure 4-10.

GASOLINE

Gasoline, widely used in vehicles and appliances and perhaps the most important fuel of petroleum origin, is a mixture of volatile, low-boiling, and midrange hydrocarbons. It contains hydrocarbon compounds with boiling points between approximately 32°C (90°F) and 205°C (400°F). The average molecular weight of gasoline is sometimes taken to be close to that of *n*-octane, about 114. The petroleum distillate described as "straight-run gasoline" contains the raw hydrocarbons that distill off between 40°C and 200°C (100°F to 395°F) and is used as camping fuel and sometimes called "white gas" or naphtha. [The range of compounds included is often measured by the "carbon number" or molecular length of *n*-alkanes included. In this case, this distillate "cut" includes C_4 (butane) to C_{12} (dodecane).] Modern automotive gasoline contains more than 200 hydrocarbons in a complex mixture (which is not just the straight cut or fraction but a blend of many compounds, especially aromatics) whose relative component concentrations vary little from refinery to refinery. This means that it is not generally possible to identify one oil company's product from another based on a comparison of the basic hydrocarbon "profiles" of various gasolines, Recently, forensic chemists have noticed that component ratios are changing and substitutions are being made, so laboratory identification of gasoline as an **accelerant** is no longer as straightforward as it once was. Before gasoline is sold as a motor fuel, however, a variety of additives are added by the refinery. These additives often include compounds to improve the burning characteristics of the fuel. Dyes were once added to distinguish leaded from unleaded motor fuels. Today, nearly all automotive gasolines are unleaded, and dyes are rarely seen except to identify lower-taxed fuels for agricultural use.

Oxygenates, especially ethanol and methanol, are very common in automotive gasolines today. In many areas they are completely replacing MTBE (methyl *t*-butyl ether), once commonly used as an oxygenate additive. The additive packages used by one manufacturer are usually different from those used by others and offer some means of distinguishing fresh, unburned, and uncontaminated gasolines. Unfortunately, due to the industry practice of exchanging raw petroleum feedstocks, additive packages, and even finished gasoline depending on market and supply conditions, it is not possible in most cases to identify a particular retail brand of gasoline from a sample of the fuel as it is delivered to the retail customer. However, recent advances in gas chromatography and data interpretation are making it possible to compare unburned fuels against suspected sources with considerable specificity. See Chapter 14 for a further discussion of identification of such products.

accelerant ▪ A fuel (usually a flammable liquid) that is used to initiate or increase the intensity of speed of spread of a fire.

KEROSENE AND OTHER DISTILLATES

Kerosene is defined by *ASTM D3699* as having a boiling point range from 175°C to 300°C (350°F to 572°F) with a minimum flash point of 38°C (100°F).[28] Kerosene and the distillate fuels of higher boiling ranges were considered useful only for illumination until the diesel motor became a common device. Although diesel fuel is similar to kerosene, it spans a wider range of less-volatile components. The modern jet engine and other turbine-driven engines further increased the use and economic significance of kerosene-like fuels. The classification scheme used for laboratory characterization of ignitable liquids (those liquids considered either flammable or combustible) includes an intermediate range called *medium petroleum distillates* (having a boiling point range between about 125°C and 215°C (250°F to 400°F). This range includes carbon numbers from C_9 to C_{17}. Many household and commercial products fall into this class—paint thinners, mineral spirits, some charcoal starters, and even some insect sprays. This classification scheme will be discussed in more detail in Chapter 14.

Kerosene and similar petroleum distillates have long been of significance in the setting of deliberate fires. Their lower volatility presents a lesser hazard to the user than does gasoline. The liquid evaporates more slowly, so that less haste is required in its ignition, and there is much less danger of explosion. Kerosene-class compounds have very low autoignition temperatures, on the order of 210°C (410°F) (see Table 4-4). Interestingly, despite their lower AITs, their high flash points make them less likely to aid the spread of fire (see Table 4-3). Their compositions involve several series of higher-molecular-weight hydrocarbons. Kerosene (fuel oil #1, Jet Fuel A) generally contains paraffinic and **olefinic** hydrocarbons in the C_{10} to C_{16} range. (This range denotes hydrocarbons having 10 to 16 carbon atoms linked together; see Appendix B.) These have boiling points between 175°C and 260°C (350°F to 500°F) and therefore fall in the fraction that follows (with some overlap) gasoline in the simple distillation of crude oil.

olefinic ■ Hydrocarbons containing double carbon—carbon bonds (denoted C=C); nonsaturated hydrocarbons; alkenes.

DIESEL FUEL

Diesel fuel contains predominantly paraffinic hydrocarbons with low sulfur content and additives to improve combustion. It covers the boiling point range from 190°C to 340°C (375°F to 650°F), which includes carbon numbers from C_{10} to C_{23}. Fuel oil, or domestic heating oil, is the next-heavier product, representing a distillation range of approximately 200°C to 350°C (400°F to 675°F). (Diesel fuel and domestic heating oil may both sometimes be described as fuel oil #2, depending on local usage.) The product for home use is carefully controlled to minimize sulfur and water content, which could corrode heating units having minimal service. Those fuels intended for industrial service have higher sulfur and ash content and cover a range of products and applications. Diesel fuels have a higher autoignition temperature (250°C or higher) and a low vapor pressure. In a fire, unless present on a porous wick or as an aerosol, diesel fuels only reluctantly support fire. Dyes may be added to denote diesel fuel intended for farm use, which is often taxed at a lower rate than that for road use.

LUBRICATING OILS

Lubricating oils include a tremendous variety of hydrocarbons with a wide range of viscosities and thermal and frictional characteristics to suit every lubrication need. Although lubricating oils are not readily combustible at ordinary temperatures, at elevated temperatures they can add to the fuel load of an established fire. Release of an oil under pressure creates an aerosol of fine droplets that are more easily ignited than the bulk liquid. It should be kept in mind that used motor oil (which is often stored in a casual fashion) often contains a significant percentage of gasoline residues. These contaminated oils can have a much lower flash point than the pure oil and can be the source of accidental fire.

SPECIALTY PETROLEUM PRODUCTS

Specialty petroleum products are not true petroleum distillates but are used in such a wide variety of consumer products that they will show up in fire debris even if they are not involved directly in the fire's ignition. These products are manufactured or blended for specific chemical and physical properties. Isoparaffinic blends (see Chapter 2) are made in various boiling point ranges and may be found in copier toners, cigarette lighter fuels, charcoal starters, lamp oils, and even in some waterless hand cleaners.[29] Aromatic blends, sometimes called LPAs for low-paraffinic aromatics, are blends of aromatics and naphthenic hydrocarbons intended for use in adhesives and some insecticides. They can be found as background volatiles in household furnishings and are chemically similar to the aromatic fractions found in automotive gasolines.[30] There are also single-component products such as limonene (used as a cleaner) and *n*-tridecane (fuel for liquid candles). They are all combustible liquids and are rapidly replacing many traditional petroleum distillates in consumer products. Some are being encountered as arson accelerants.

Nonhydrocarbon Liquid Fuels

Many materials other than hydrocarbons have considerable significance to the fire investigator because they can be used as accelerants and because they can play a part in the ignition and spread of accidental fires. These nonhydrocarbon liquid fuels include alcohols, solvents, and similar materials, as well as alternative or biofuels, as discussed next. The chemistry of nonhydrocarbons was discussed in Chapter 2.

ALCOHOLS, SOLVENTS, AND SIMILAR NONHYDROCARBONS

Methyl alcohol was formerly widely used in spirit (mimeograph) duplicating machines, as an alternative motor fuel, and as a thinner for shellacs and finishes. Ethyl alcohol, especially in reagent (190 proof) or high-proof beverage form (85 or higher proof), can also be used to start fires even though its heat output is not very high. Isopropyl (rubbing) alcohol is flammable at the concentrations packaged for retail use. All the simpler alcohols can provide fuel for the incendiary fire and accidental fire as well and are difficult to detect in post-fire debris because of their extreme volatility and water solubility. *Ketones* such as acetone and methyl ethyl ketone are widely used as solvents for paints and finishes, as are ethyl acetate and related compounds. Although more and more paints are water based, thus requiring no volatile solvents, many wood finishes employ turpentine as thinner or linseed oil (or related organic oil) as part of the finish. A tremendous variety of proprietary mixtures used as mop cleaners, dust attractants, insecticide solvents, copier toners, glue solvents, and plastic resin solvents are encountered at fire scenes. They require mention because of a predisposition to think of liquid fuels only in terms of petroleum distillates. Such thinking can be totally erroneous in many fires. Industrial and commercial fires are often fed and intensified to extremes by special-purpose cleaning agents and solvents. The general properties of many of these compounds have already been discussed. Because of the very large number of specific compounds and the variety of mixtures in which they may be found, no attempt will be made here to examine each of them. Special considerations of some of the more hazardous of these fuels will be made in Chapter 13.

ALTERNATIVE FUELS OR BIOFUELS

Currently, there are motor fuels in wide use that are not petroleum derivatives. Alternative fuels or biofuels are based on vegetable materials—either processed food oil or ethanol produced by fermentation of corn or other vegetable materials. These fuels can be pure biofuels or, more commonly, a blend of biofuel and petroleum product.[31]

Biodiesel

Biodiesel (sometimes denoted as Bl00) is produced by chemically treating vegetable oils (corn, peanut, safflower, canola, etc.) or, more rarely, animal fats to yield a mixture of four fatty acid methyl esters. (The ratios of the few components depend on what oil or fat was the starting point.) There are presently more than 770 commercial outlets for B100 biodiesel in the United States and countless informal producers. Biodiesel can be used as a direct substitute for diesel fuel in vehicles with minor mechanical modifications. A 20/80 blend of biodiesel and petroleum diesel (denoted as B20) is also in widespread use, since it requires no mechanical modifications for use in most motor vehicles. Its performance as a fuel in a motor or in an open fire will be dominated by its diesel fuel constituents. Biodiesel fuels may contain residues of methanol and water from the production process. Because the molecular weights of biodiesel components are similar to C_{18} to C_{23} alkanes, the flash points are going to be similar (>100°C). The presence of significant quantities of residual methanol will obviously greatly reduce the flashpoint.

Ethanol

Pure ethanol is not suitable as a motor fuel, so the most common variety is E85 (85 percent ethanol and 15 percent gasoline). This product is already being sold across the United States for use in "Flex Fuel" vehicles. Motor gasolines in the United States have contained up to 10 percent ethanol as an oxygenated additive to reduce carbon monoxide content. As a motor fuel, this product must meet the same flash point, vapor pressure, and ignition temperature standards as gasoline fuel. The effect on hazard, then, is expected to be minimal. However, its detection after a fire with water extinguishment presents a problem to be discussed in Chapter 14.

Combustion of Liquid Fuels

It must be remembered that when an ignitable liquid is poured or spilled on a surface, only its vapors (rather than the liquid itself) actually combust. The behavior of the liquid and its vapors is generally predictable. On a nonporous surface (linoleum, vinyl, painted concrete), the viscosity and surface tension of the liquid will determine the depth of the freestanding pool that results. The depth, of course, determines how large an area the pool will cover for a given quantity of fuel. Viscous liquids like kerosene or diesel fuel form a deeper pool (~1 mm) that will not cover as large an area as a nonviscous fuel like methanol or gasoline (whose freestanding pool may be only 0.1 to 0.5 mm deep).[32]

On other surfaces, the depth of the pool will be determined by the nature of the surface, as shown in Figure 4-11. A semiporous surface like raw wood or concrete will allow some penetration, 2 or 3 mm or so, with a proportional reduction in the area of the pool. Carpets or other porous surfaces allow very deep penetration, so the resulting pools can be very small. The prefire evaporation rate from such porous surfaces is considerably faster than the rate from a freestanding pool (of the same diameter) of the same liquid at the same temperature due to the capillary drive of the substrate (the wick effect).[33] As described previously, prior to ignition the vapors generated by a pool will rise by diffusion

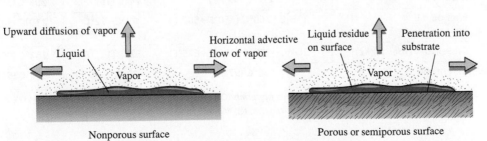

FIGURE 4-11 Liquid pools on nonporous and porous surfaces with vertical and horizontal movement of vapors.

Upward diffusion of vapor

Liquid

Vapor

Nonporous surface

Horizontal advective flow of vapor

Liquid residue on surface

Penetration into substrate

Vapor

Porous or semiporous surface

FIGURE 4-12 Distribution and effects of radiant heat from burning pool of ignitable liquid.

Central radiant heat into pool

Radiant heat at edges absorbed by substrate

Protected by overlying pool

Substrate

Localized scorching or other heat effects

and spread sideways by advective flow, as in Figure 4-11. Upon ignition, the flames are established everywhere the vapors have diffused into the surrounding air in a flammable mixture. The radiant heat from flames near the center of the pool is in part absorbed by the liquid remaining in the pool, and the radiant heat from flames at the edges is absorbed in part by the adjoining floor surface that is not protected by the liquid. This radiant heat may be enough to damage or even ignite the floor exterior to the pool, as in Figure 4-12. Meanwhile, the heat being absorbed by the liquid is being distributed through the liquid by convective circulation, and the temperature of the entire pool begins to rise. The liquid does not really cool the surface beneath it, but it can protect it because the temperature, obviously, cannot exceed the boiling point of the liquid. For a low-boiling-point fuel like methyl alcohol, the temperature of the floor under the pool will not exceed about 65°C (150°F) (the boiling point of methyl alcohol), and so the floor under the pool escapes all damage until the alcohol evaporates. With gasoline, there is progressive distillation and consumption of the components from light to heavy until only the heaviest remain (controlled by the diffusion mechanism through the bulk liquid described earlier). The bulk liquid temperature is limited only by the boiling point of the heaviest constituents, such as n-dodecane with a boiling point of 216°C (421°F). As the lighter components boil off, the

TABLE 4-7	Properties of Petroleum Products		
PETROLEUM DISTILLATE	**BOILING POINT/RANGE**	**FLASH POINT**	**AUTOIGNITION TEMPERATURE**
Ethanol	78.5°C (173°F)*	13°C (55°F)	363°C (685°F)
Toluene	110.6°C (231°F)*	4.4°C (40°F)	
Gasoline (low octane)	32°C–190°C (90°F–375°F)	−43°C (−45°F)	257°C (495°F)
Methanol	64.7°C (148.5°F)*		
Medium petroleum distillate	125°C–215°C (250°F–400°F)	13°C (55°F)	220°C (428°F)
Mineral spirits	Varies	40°C (104°F)	245°C (473°F)
VM & P Naphtha (regular)		−2°C (28°F)	232°C (450°F)
Kerosene (C_{10}–C_{16})	175°C–300°C (350°F–500°F)	38°C (100°F)	210°C (410°F)
n-Dodecane	216°C (421°F)*	74°C (165°F)	203°C (397°F)
Fuel oil #1	175°C–260°C (350°F–500°F)	43°C–72°C (110°F–162°F)	210°C (410°F)
Diesel fuel (fuel oil #2)	200°C–350°C (400°F–675°F)	52°C (125°F)	257°C (494°F)
n-Pentane	36°C (97°F)	−40°C (−40°F)	260°C (500°F)

Sources: Fire Protection Guide to Hazardous Materials (Quincy, MA: National Fire Protection Association, 2001).
*Merck & Co., *The Merck Index,* 11th ed. (Rahway, NJ: Merck, 1989).

liquid can reach temperatures such that some floor materials may be scorched. When heavy petroleum products like diesel fuel are burning, the temperatures of the liquid can be as high as 340°C (650°F), which is above the ignition temperatures of solid fuels like some woods. (See boiling points in Table 4-7.) (As discussed earlier, partial evaporation of petroleum distillates changes the ratio of components, since lighter ones evaporate faster than heavier ones.) This mechanism explains why some fuels produce halo-type burns, and others scorch and char the floor more extensively. Because they are so shallow, free-standing pools of most ignitable liquids burn away quite rapidly (the rate being dependent on the nature of the surface and the size of the pool, with very small pools (<0.1 m; 4 in. in diameter) burning away faster than larger pools. When there are cracks or crevices, localized burning can take place for longer periods of time, inducing localized charring of the floor at the joints and seams, as illustrated in Figure 4-13a. If extra ventilation can then

(a)

(b)

Radiant heat

Draft

FIGURE 4-13A Pocketed burning from liquid in cracks or seams in a floor may be enhanced by any upward draft and radiant heat feedback within cracks. *Courtesy of Lamont "Monty" McGill, Fire/Explosion Investigator.*

FIGURE 4-13B Radiant heat striking a wood plank floor can initiate charring of the exposed edges, causing a beveled appearance. Combustion between the planks can be enhanced by draft from below. This can be very difficult, if not impossible, to differentiate from (a).

FIGURE 4-13C Gasoline/diesel mix burning in seams of wood floor. Fuel on flat surfaces has already burned off.*Courtesy of Lamont "Monty" McGill, Fire/Explosion Investigator.*

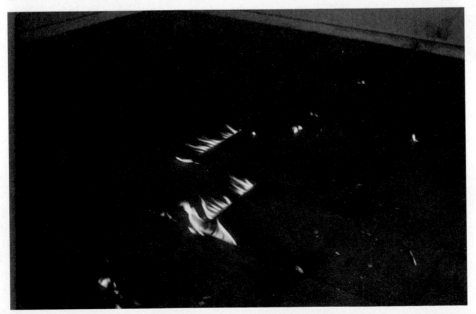

be drawn through these cracks, severe but localized floor damage can result, as shown in Figure 4-13c. Downward radiant heating can produce "beveled" edges on floorboards, as pictured in Figures 4-13b and 7-21. It can be very difficult, if not impossible, to differentiate between these effects.

Pyrolysis and Decomposition of Liquids

Although some unstable liquids decompose at fairly low temperatures, most fuel liquids reach their boiling point and boil away before being subject to pyrolytic decomposition. Even when liquids are exposed to an intense general fire, convection within a deep pool (as opposed to a thin film) helps mix the cooler and warmer liquids. The liquid at the surface boils off into the flame, with the heat absorbed from the flames being lost by evaporative cooling. This process maintains the temperature of the entire pool at or just below the boiling point of the liquid; the liquid pool protects the surface on which the pool is sitting. When dealing with viscous liquids such as asphalt, plastic resins, or some foods, it must be remembered that with impaired circulation, portions of the liquid in contact with an intense heat source can reach and exceed their pyrolysis temperatures and begin to char and decompose like a solid. The pyrolysis products then enter the flame in the same way as those driven off a solid.

In tank or pool fires involving heavy crude oils, the residue formed on the burning surface (by distillation of the lighter components) may become denser than the surrounding liquid. The hot residue can then settle into the tank and contact water in the bottom, producing a steam explosion that propels the burning fuel above outward with disastrous results. This situation can also occur with nonpetroleum liquids like cooking oils that can pyrolyze and form a true char.

Liquid fuels are categorized by their flash point (see Table 4-3.) Flammable liquids are all those that have a flash point below 100°F (37.8°C) (NFPA Classification: Class I), with Class IA and IB liquids having flash points below room temperature, 73°F (22.8°C). Combustible liquids are those with a flash point above 100°F (37.8°C) [NFPA Class II, flash point between 100°F and 140°F (37.8°C and 60°C), and Class III, flash point above 140°F (60°C)]. Class III combustible liquids are further divided into Class IIIA (flash point between 140°F and 200°F) and Class IIIB (flash point above 200°F (93°C)].[34] In the United Kingdom, highly flammable liquids are those that have a flash point below 32°C (90°F); flammable liquids are those with a flash point between 32°C and 60°C; and combustible liquids are those with a flash point above 60°C (140°F).[35] **Ignitable liquids** are a category of liquid fuels that includes both flammable and combustible liquids.

ignitable liquid ■ Classification for liquid fuels including both flammable and combustible classes.

Fuel Gas Sources

Among the many sources of fuel gases, of particular interest to the fire investigator are gas lines, natural gas, liquefied petroleum gas, containers, and appliances, as discussed next.

GAS LINES

Although gas lines do not normally represent a source of ignition, they do represent a source of readily ignitable fuel. Because of the properties of gas, outlined previously and in Chapter 12, it is clear that gas enclosed in a pipe does not normally pose any hazard. Unless mixed with appropriate quantities of air, the gas is not flammable. Only when the gas escapes (or air can enter) is there any danger associated with it, because then the gas can mix with air to form a combustible or explosive mixture. Such an escape is possible

FIGURE 4-14 Typical residential natural gas pressure regulator. *Courtesy of John D. DeHaan.*

because of (1) leakage due to inadequately sealed joints or corroded pipes, (2) mechanical fracture of lines from external causes, and (3) failure of lines because of excessive heat, which may melt essential sealing components of the line, valves, regulator, or gas meter. Each of these requires separate consideration.

NATURAL GAS

Fuel gas in the form of natural gas or LP gas is delivered to appliances via a system of tanks, regulators, pressurized delivery pipes, and flexible connectors. Natural gas in the United States is piped underground via transmission pipelines that may be pressurized to as high as 1,200 pounds per square inch (psi). Distribution pipelines (mains) operate at typical pressures of about 60 psi (but may be up to 150 psi). A line regulator (as shown in Figure 4-14) at each service location then drops the pressure to 0.14 to 0.36 psi, or 4 to 10 inches water column (w.c.) for delivery to residential appliances (1 psi = 27.67 inches w.c.).[36] Pressures in piping inside commercial buildings (for industrial furnaces) may be up to 5 psi (but may be higher under special circumstances).[37] The delivery piping may be steel, wrought iron (copper tubing if the gas is low sulfur), brass, coated aluminum alloy tubing, or stainless steel. Cast iron pipe is not to be used. Plastic pipe or tubing can be used only on outside underground installations. Natural gas has very little odor of its own, so an **odorant** mixture (usually, *t*-butyl mercaptan, thiophane, or another mercaptan) is added during pumping into the delivery pipeline.

odorant ■ Organic chemical (often a mercaptan) added to natural gas or LP gas prior to its distribution to make it detectable by a person with normal sense of smell at a concentration less than 20% of its LEL.

LP GAS

Characteristics and Uses

Although natural (or manufactured) gas is common in urban areas, many rural and isolated regions make use of LP (liquefied petroleum) gases. These are normally either propane, butane, or a mixture of the two with small quantities of other hydrocarbons. Propane has a boiling point of −42°C (−44°F) at atmospheric pressure and is kept in liquid form by pressure (its vapor pressure at the ambient temperature). Commercial propane contains propylene and traces of other gases and ethyl mercaptan as an odorant. Butane (pure *n*-butane) has a boiling point of −0.5°C (33°F) at atmospheric pressure, so it can briefly exist as a free liquid at very low temperatures. It, too, contains butylenes and other trace gases and an odorant such as ethyl mercaptan.[38] LP gas is supplied in fixed or portable containers of various sizes that are normally refilled by transferring liquid fuel under pressure from a delivery truck or facility. Most installations are designed so that only gas is vented to the appliance rather than the liquefied fuel (the most common exception being LPG tanks for forklifts and similar equipment, in which the tank is designed to deliver liquid fuel to the motor). Only gas is vented by having the delivery tube above the maximum fill level of the tank or cylinder, as shown in Figure 4-15. When the cylinder is filled to its normal maximum (80 percent of capacity), liquid propane spills out of the spill tube via the bleed valve, and the operator knows the tank is full. If the tank is improperly positioned or overfilled and then connected to an appliance, the liquid can overwhelm the delivery system and spill out via the main delivery valve.

The pressure within the vapor space of the tank is controlled by the temperature of the tank and its contents. This can range from 28 psi for propane at 0°F (−18°C) to 127 psi at 70°F (21°C) to 286 psi at 130°F (54°C). A pressure relief valve on the container is usually set to bleed off gas when pressures exceed 250 psi.[39] A one- or two-stage regulator reduces the tank pressure to a working pressure of 11 to 14 inches w.c. (0.4–0.5 psi) for introduction into a typical nonindustrial appliance.[40] Operating pressures for undiluted LP gas systems in buildings shall not exceed 20 psi according to *NFPA 54, National Fuel Gas Code*. Flexible rubber connectors (hoses) may be found on connections to portable cylinders such as for campers and barbecue grills. The same comments with regard to leaks and failures listed previously apply to LP gas delivery systems, with some additional conditions. The portability of tanks exposes them to weather and mechanical damage from handling (especially as they are refilled). The tanks are labeled to reflect their capacities and the dates of their manufacture and pressure (hydrostatic) testing. The necessity to disconnect and reconnect them repeatedly for filling makes them susceptible to damaged connectors that can cause leaks. Flexible connections can crack with age or misuse.

Because both propane and especially butane are heavier than air, their escape into the air is somewhat different from that of methane, which is a major constituent of natural gas. Being heavy, propane and butane vapors behave somewhat more like the fumes of

FIGURE 4-15 Cross sections of typical horizontal and vertical propane containers. *Courtesy of* Fire & Arson Investigator, *April 1999.*

Cutaway diagram of typical vertical propane container

Cutaway diagram of typical horizontal propane container for cooking or heating. (Tanks for vehicles are positioned to deliver liquid, not vapor.)

gasoline, which accumulate at low levels in enclosures, where they are less likely to be smelled but more likely to encounter ignition sources. When propane (with a vapor density of 1.5) is involved, its density is sufficiently close to that of air to mix more readily with air than will the heavier fuel gases. In its flammable range, a propane/air mixture is almost the same density as air itself, so it may not "settle" at all. Modest mechanical stirring in a room, or even the thermal stirring caused by sunlight heating one side of an enclosure, has been shown to mix propane quite uniformly in an enclosed space.[41] Release of gas under pressure causes turbulent mixing.

It is of interest that LP gases apparently give rise to fires (and asphyxiations) somewhat more frequently than does natural gas. The reasons are related to the nature of the fuels themselves and also to the method of installation and uses of the related appliances and related storage and distribution equipment. While natural gas is normally delivered from heavy iron pipe, permanently installed and under the specifications of local building ordinances, LP gases are more often delivered through semiflexible copper tubing installed in an uncontrolled manner by amateur installers, and from removable tanks. In addition, the supply of gas has to be replenished from tank trucks or with portable tanks rather than from a permanent installation. Thus, there is considerably increased chance of poor connections, broken tubing, leaking valves, and similar deficiencies. LP gas water heaters have also been installed to replace costly electric heaters in small closets where ventilation is inadequate when the door is closed. LP gas is delivered at considerably higher pressures and through differently sized orifices than is natural gas. Because of its higher heat of combustion and higher density, it also delivers 2.5 to 3 times as much energy per unit volume. Equipment fitted for one type of gas can seriously malfunction if used with the other. If malfunction of a gas appliance is suspected, the investigator must ensure that the correct type of regulator, burner orifice, and control valve(s) are all in place to match the type of gas used. See Chapter 15 for additional information on gas appliance failures.

Pressurized Containers

Pressurized containers of LP gas, propane, or butane are kept on hand in many types of occupancies to fuel everything from tools like handheld propane torches to camping stoves and barbecues (see Figure 4-15). These containers come in various sizes from 1-lb to 28-lb capacity (rated by weight of *water* equivalent) and are usually not hard to recognize even when fragmented by a BLEVE, which is defined later in the chapter. However, there is another source of LP gases that is not as readily identified. A variety of aerosol products, from lubricants, paints, and insecticides to air fresheners, hair sprays, and deodorants, are pressurized with dimethyl ether or mixtures of butane, isobutane, and propane instead of the inert fluorocarbons used for so many years.[42] These aerosol containers may hold as much as 120 ml of pressurized liquid fuel, which vaporizes immediately upon release. Such small quantities do not present an ignition risk when uniformly diluted throughout an average-size room. However, when confined to a cabinet or an appliance, they can form an explosive mixture on release within that space. Failures of single aerosol cans inside a vehicle will produce explosive concentrations that can do considerable damage. Such products are sometimes used to excess (in defiance of the manufacturer's advice), and entire apartments and houses have been damaged by the release and subsequent ignition of multiple cans, as shown in Figure 4-16. The solvents used in such products (kerosene for insecticides, alcohols for cosmetics) pose an additional risk of fire if released in the vicinity of an open flame.

Natural gas is also shipped and stored as a cryogenic liquid (CNG) with a boiling point of −160°C (−260°F). When released, CNG causes the formation of a dense vapor cloud denser than air, which poses a significant ignition risk as it spreads.[43]

Leaks

A gas line may show leakage from inadequate sealing of the joints during installation, from deterioration of joints, or from corrosion severe enough to penetrate the walls of the

FIGURE 4-16 Six cans of aerosol roach killer set off in the cabinets of this kitchen provided an ignitable concentration for the pilot light of the stove (left foreground) to ignite. The blast damaged the ceiling of the kitchen and adjacent living room and lifted the roof. Note the scorch mark on the inside of the cabinet door from the flame plume from the can. *Courtesy of Captain Investigator Bruce Fusselman, Phoenix Fire Department, Phoenix, AZ.*

line. Rigid metal lines can fracture if subjected to severe vibration, mechanical movement, or loading. Leaks may also occur from worn or defective valves, regulators, and fittings. In any of these events, it is expected that the odorant intentionally added to all natural gas will serve as a warning of escaping gas, because natural and some manufactured gases are inherently odorless. Leakage has also followed lightning strikes whose induced current caused perforations of stainless steel and copper gas lines. Goodson recently reported that corrugated stainless steel tubing (CSST) with a plastic covering intended for residential gas delivery systems has been shown to be perforated by the passage of induced current from lightning strikes, causing perforations 0.04 to 0.15 in. across. These perforations allowed gas leaks into attics and other enclosed spaces where ignitable concentrations could accumulate.[44] Sanderson has described multiple penetrations of polyethylene gas piping caused by discharge of static electricity built up on the interior of such piping by the passage of particulates (dust and rust) through the pipe. These very small diameter perforations would release gas at low rates that could accumulate in concealed spaces (such as the steel pipe riser near the meter).[45] Chemical corrosion can also cause multiple leaks in any metallic tubing, as shown in Figure 4-17.

Assuming that such a leak exists and that it is not noted, the sequence of events is as follows. Initially, the gas is too low in concentration to be ignited from any source. When the gas reaches its lower combustible limit it will have formed an explosive mixture throughout the area in which the gas concentration has reached this limit. This will

FIGURE 4-17A Corrosion of propane gas flex line caused sufficient leak to cause explosion. *Courtesy of Peter Brosz, Brosz & Associates Forensic Services Inc.; Professor Helmut Brosz, Institute of Forensic Electro-Pathology.*

FIGURE 4-17B Internal close-up of chemical corrosion. *Courtesy of Peter Brosz, Brosz & Associates Forensic Services Inc.; Professor Helmut Brosz, Institute of Forensic Electro-Pathology.*

FIGURE 4-17C Electric arc–induced perforation of CSST and ignition of gas. *Courtesy of Jamie Novak, Novak Investigations and St. Paul Fire Dept.*

normally occur only in a confined or partially confined space, but that space will be under great danger of an explosion—not a fire, initially at least. If there is a source of ignition, such as a pilot flame, or if a match is ignited by a smoker at this time, an explosion will occur. It can be very forceful, which is characteristic of a stoichiometric mixture, and without continuing flame. If the mixture is not ignited until it accumulates to form a rich mixture, it will generate less mechanical force but will sustain a flame after the explosion. If ignited right at the leak, where there may be some gas/air mixture within the combustible limits locally, a continuing fire is expected. This fire can play, in a blowtorch effect, on other fuels in a direct line and ignite them. The distance the flame will project depends on the pressure of the gas being fed to the line. Typical residential delivery pressures of 0.5 psi or less produce a horizontal jet of flame only 1 or 2 ft (0.3–0.6 m) in length. Industrial or service pressures are much higher, and the flame-jet length will be proportionately longer. It should be remembered that due to the odorants used in utility gas, most gas leaks are detected by residents long before combustible mixtures are produced.

Mechanical Fracture

Release of gas from a mechanically fractured line is expected rarely. However, it is probably far more common than is generally believed and occurs even in the absence of natural catastrophes such as earthquakes, landslides, and similar major causes of destruction. For example, a gas line was laid under a street over which there was heavy trucking. Failure to tamp the soil properly around the pipe allowed it to buckle under the recurrent loads until the line parted, filling the soil with gas, which eventually percolated through porous soil to accumulate inside a structure, which led to an explosion and fire. In another instance, a small gas stove was attached to the gas line with a copper tube and with a soldered connection. The stove was moved around somewhat, with recurrent flexure on the connection. Finally, the joint parted and resulted in both a fire and multiple asphyxiations by the combustion products. Misuse can also lead to leaks, as depicted in Figure 4-18. The corrugated flexible connector pipes used on many gas appliances are subject to corrosion by ammonia or sulfur-containing contaminants in the gases and can become dangerously leaky even when not moved. Many such lines today are coated to minimize such corrosion.

Gas (natural or LP) released from a buried pipe has been seen to migrate considerable distances along the loose earth or void space surrounding some gas lines (or through porous soils). This migration can result in gas accumulations inside buildings some

FIGURE 4-18 Improvised use of portable LP tanks for household furnace (note empty cylinders and wrenches). *Courtesy of Greg Lampkin, Knox County, TN, Fire Investigation Unit.*

distance from the actual leak. This process is enhanced if the overlying soil is sealed by pavement, ice, or even frozen soil, which minimizes upward losses. Such percolation through soil can also scrub the gas of its odorant, which means that a leak can exist without witnesses reporting the usual gas odors. Both natural gas and LP gas can also lose their odorants by chemical adsorption onto the walls of new pipe (steel or plastic) or new containers, or by oxidation by the red iron (oxide) rust inside old, disused or newly installed steel pipe.[46,47]

Improperly plumbed gas appliances that are connected using rubber tubing or even garden hose have been found responsible for leaking fuel that resulted in a fire or explosion, as illustrated in Figure 4-19. Such connections leak most often at joints or connections to metal fittings. The low delivery pressures in most residential structures rarely lead to failure by bursting, but such rubber products can degrade from age, sunlight, or exposure to chemicals and can crack or split as they become brittle.

Another type of mechanical failure is failure of pressure regulators in gas delivery systems. A failure due to corrosion, mechanical damage, or freezing in any of these regulators can result in massive overpressures being delivered to the appliances. These overpressures, in turn, result in flames exiting the appliances as the burners and vents are unable to accommodate the extra flames and exhaust volume.

As with leakage, release of gas under those circumstances can lead to fire when the gas is ignited directly as it first emerges from the fractured line. If the leaking gas mixes with enough air before it is ignited, there will be an explosion in the accumulated mixture. This explosion is almost always followed by a continuous flame at the source of the fuel gas.

Failure from Heat

In general, failure of a gas line from excessive heat will be limited to a general fire with which the gas is not causally related. Lines are frequently joined by low-melting alloys such as solder; if such a joint is heated by exterior flames or even hot gases, the joint may fail. In very hot structure fires, it is not uncommon to have brass or bronze fittings partially melt (see Figure 4-20). Because these fittings are regularly used in connection with gas lines, such failures will be found after the fire. Severe fire heating can cause differential expansion of brass connectors and steel/iron pipe fittings, sometimes resulting in "loose" connections after cooling.

Because there must already be a fire to melt the attachment in these situations, it is clear that the escaping gas will merely feed additional fuel to the fire locally and will produce

FIGURE 4-20A New LP tank fittings. *Courtesy of Greg Lampkin, Knox County, TN, Fire Investigation Unit.*

FIGURE 4-20B Post-fire. Partially melted valve and fittings from fire exposure may require X-ray examination to reveal whether they are open or closed. *Courtesy of Greg Lampkin, Knox County, TN, Fire Investigation Unit.*

a characteristic "blowtorch" type of burning in front of the opening. A melted connection is not to be considered as a cause of the fire but only as a result of and a contributing factor to the intensity of the fire. The gas cannot spread under these conditions to augment the flames except ahead of the line opening. Under no conceivable circumstances will gas escaping into a burning fire avoid burning locally to be consumed at a distance from the opening. The large blowtorch type of flame will have the effect of causing a burned-out area immediately in front of the opening or charring to adjacent surfaces, as pictured in Figure 4-21. Its effect on the remainder of the fire will be negligible in virtually every circumstance. At worst, the blowtorch type of flame may alter the fire pattern because of local intensity, which tends to deflect air currents to the hottest part of the fire.

Despite these rather obvious facts, it is quite common for claims to be made that escaping gas severely augmented the total fire throughout a building. This would necessitate the highly unlikely situation that the unburned gas passed through, around, or over burning areas to reach distant portions of the fire. Very unusual release conditions might provide mixtures that are too rich to ignite or gases that are moving too quickly to be ignited by a fire in the vicinity of the leak (until they are sufficiently diluted or slowed), but such movement would normally be expected to mix enough air and flames into the gas stream by turbulence that ignition would readily occur immediately.

Another dangerous situation occurs when ignitable liquids are packaged in sealed containers, such as solvents in metal cans or glass bottles. When these containers are

FIGURE 4-21 Plume damage (clean burn) on the back of the stove from the same gas plume. (The stove had been moved away from the wall prior to fire.) *Courtesy of Det. Richard Edwards (retired), Los Angeles County Sheriff's Department, Whittier, CA.*

BLEVE ■ Boiling-liquid, expanding-vapor explosion. A mechanical explosion caused by the heating of a liquid in a sealed vessel to a temperature far above its boiling point.

exposed to a fire, pressure built up by the expanding liquid prevents the liquid from evaporating. Because evaporation allows a liquid to maintain its temperature at or below its boiling point, the temperatures inside the containers exceed the boiling point of the liquid and may trigger pyrolysis of the liquid. The containers are then filled with a mixture of pressurized liquid, vapor, and (sometimes) gaseous pyrolysis products, which can add greatly to the fuel load of the fire when the containers erupt. These eruptions can occur with explosive force and suddenly and dramatically affect the course and intensity of the general fire. Failure of a fluid-filled container due to overpressure caused by extreme heating is often called a **BLEVE** (boiling- liquid, expanding-vapor explosion), even if the fluids and gases involved are not in themselves combustible. See Chapter 12 for further discussion of such mechanical explosions.

Summary

Flammable gases and liquids are common fuels in accidental structure fires and constitute a convenient source of fuel for a very high percentage of incendiary fires. Although such fuels can be generally classified and even compared on the basis of their physical and chemical properties, the investigator should be careful in using these properties blindly in assaying a fuel's contribution to ignition as well as to the fully developed fire. As we have seen, the fire environment is a complex one, and the intensity and duration of the heat produced may have dramatic effects on the physical and chemical properties of the fuels present. This is an example of the value of real-fire experience in observing flammable liquids and gases as they behave in actual fires.

Review Questions

1. What is saturation vapor pressure and how is it measured?
2. What is meant by the lower explosive limit?
3. What happens at the flash point of a liquid fuel?
4. What is the difference between flash point and autoignition temperature?
5. What is the equation used to determine the vapor density of a gas or vapor?
6. Which compounds have the widest flammability ranges?
7. Name five common petroleum-based fuels and describe each.
8. What are three differences between natural gas and LP gas?
9. At what pressures is natural gas delivered in distribution lines? Regulator to residential line? To appliance burner?
10. What is the difference between flammable, combustible, and ignitable liquids?

References

1. D. Drysdale, *An Introduction to Fire Dynamics,* 2nd ed. (Chichester, UK: Wiley, 1999).
2. J. D. DeHaan, "Exploding Gas Cans and Other Fire Myths," *CAC News* (Third Quarter 2004): 20–26.
3. D. Gardiner, M. Barden, and G. Pucher, "An Experimental and Modeling Study of the Flammability of Fuel Tank Vapors from High Ethanol Fuels," (Washington, DC: National Renewable Energy Laboratory, Subcontract Report NREL/SR-540-44040, Golden, Colorado, U.S. Department of Energy, October 2008).
4. D. Guidry, private communication, May 2005.
5. C. S. Kennedy and J. F. Knapp, "Childhood Burn Injuries Related to Gasoline Can Home Storage," *Pediatrics* 99 (March 1997): e3.
6. M. A. Dorsheim, "Fuel Ignites Differently in Aircraft, Lab Environments," *Aviation Week and Space Technology,* July 7, 1997, 61–63.
7. *ASTM D56-0: Standard Test Method for Flash Point by Tag Closed Tester* (West Conshohocken, PA: ASTM, 2005).
8. *ASTM D93-08: Test Methods for Flash Point by Pensky-Martens Closed Cup Tester* (West Conshohocken, PA: ASTM, 2008).
9. *ASTM D1310-01: Test Method for Flash Point and Fire Points of Liquids by Tag Open Cup Apparatus* (West Conshohocken, PA: ASTM, 2007)
10. *ASTM D92-05a: Standard Test Method for Flash and Fire Points by Cleveland Open Cup* (West Conshohocken, PA: ASTM, 2005).
11. R. Friedman, *Principles of Fire Protection Chemistry and Physics,* 3rd ed. (Quincy, MA: National Fire Protection Association, 1998), 44.
12. Friedman, *Principles of Fire Protection Chemistry and Physics,* 114.
13. *ASTM E659-78: Standard Test Method for Autoignition Temperatures of Liquid Chemicals* (Quincy, MA: ASTM, 2000).
14. American Petroleum Institute, "Ignition Risk of Hydrocarbon Vapors by Hot Surfaces in the Open Air," API Publication 2216 (January 1991).

15. R. J. Harris, *The Investigation and Control of Gas Explosions* (New York: E & FN Spon, 1983), 39.
16. J. D. DeHaan, "The Reconstruction of Fires Involving Highly Flammable Hydrocarbon Liquids" (PhD dissertation, Strathclyde University, Glasgow, Scotland, 1995; Ann Arbor, MI: University Microfilms, 1996).
17. P. Kennedy and A. Armstrong, "Fraction Vaporization of Ignitable Liquids: Flash Point and Ignitability Issues," paper presented at *Fire and Materials 2007*, 10th International Conference and Exhibition, Interscience Communications, San Francisco, January 29–31, 2007.
18. DeHaan, *The Reconstruction of Fires.*
19. Ibid.
20. R. L. Valentine and R. D. Moore, *The Transient Mixing of Propane in a Column of Stable Air to Produce a Flammable but Undetected Mixture* (New York: American Society of Mechanical Engineers, 1974).
21. L. Thatcher, RVFOG Results, personal communication, March 1999.
22. DeHaan, "The Reconstruction of Fires."
23. J. D. DeHaan et al., "Deflagrations Involving Heavier-than-Air Vapor/Air Mixtures," *Fire Safety Journal* 36 (2001): 693–710.
24. *NFPA 58: Standard for Storage and Handling of LP Gases* (Quincy, MA: NFPA, 2008).
25. R. W. Henderson and G. R. Lightsey, "The Effective Flame Temperatures of Flammable Liquids," *Fire and Arson Investigator* 35 (December 1984).
26. NFPA, *Fire Protection Handbook,* 19th ed. (Quincy, MA: NFPA, 2008), 8–120.
27. Ibid.
28. *ASTM D3699-08: Specification for Kerosene* (West Conshohocken, PA: ASTM, 2008).
29. Exxon Corp. *Isopars.* Houston, TX, 1990.
30. D. M. Gialamas, "Is It Gasoline or Insecticide?" *CAC News* (Spring 1994).
31. R. J. Kuk and M. V. Spagnola, "Extraction of Alternative Fuels from Fire Debris Samples," *J. Forensic Sciences* 53, no. 5 (September 2008): 1123–29; E. Stauffer and D. Byron, "Alternative Fuels in Fire Debris Analysis: Biodiesel Basics," *J. Forensic Sciences* 52, no. 2 (March 2007), 371–79.
32. DeHaan, "The Reconstruction of Fires."
33. J. D. DeHaan, "The Influence of Temperature, Pool Size, and Substrate on the Evaporation Rates of Flammable Liquids," paper presented at *Interflam,* Edinburgh, June 1999.
34. NFPA, *Fire Protection Handbook,* 3–44.
35. Drysdale, *An Introduction to Fire Dynamics,* 204.
36. *NFPA 921: Guide for Fire and Explosion Investigations* (Quincy, MA: NFPA, 2008 and 2011), pt. 9.3.3, p. 92.
37. *NFPA 54: National Fuel Gas Code* (Quincy MA: NFPA, 1999), 16.
38. R. Lapina, "Propane Fire & Explosion Investigation," *Fire and Arson Investigator* (April 2005): 46–49.
39. B. V. Ettling, "Venting of Propane from Overfilled Portable Containers," *Fire and Arson Investigator* (April 1999): 54–56.
40. *NFPA 921,* (2008 and 2011) pt. 9.4.3.2.
41. Thatcher, RVFOG Results.
42. J. D. DeHaan and W. Howard, "Combustion Explosions Involving Household Aerosol Products" in *Proceedings Fourth International Symposium of the Detection and Identification of Explosives,* Chantilly, VA, December 1995. See also *Fire and Arson Investigator* (September 1997): 25–27.
43. P. K. Raj, "Where in a LNG Vapor Cloud Is the Flammable Concentration Relative to the Visible Cloud Boundary?" *NFPA Journal* (March–June 2006): 68–70.
44. M. Goodson and M. Hergenrether, "Investigating the Causal Link between Lightning Strikes, CSST and Fire," *Fire and Arson Investigator* (October 2005): 28–31.
45. J. Sanderson, "Electrostatic Pinholing," *Fire Findings* 13, no. 4 (Fall 2005): 1–3.
46. Gas Odorization Manual, American Gas Association, (1996).
47. NFPA 921: Guide for Fire and Explosion Investigations (Quincy, MA: NFPA, 2011) Pt. 9.9.92, p. 104.

Courtesy of Jamie Novak, Novak Investigations and St. Paul Fire Dept.

KEY TERMS

cellulosic, *p. 125*

char, *p. 128*

flame resistant, *p. 156*

piloted ignition, *p. 128*

point of origin, *p. 138*

pyrolysis, *p. 125*

pyrophoric, *p. 132*

soot, *p. 143*

spontaneous ignition, *p. 132*

synthetic, *p. 143*

thermoplastic, *p. 145*

thermosetting (resin), *p. 125*

OBJECTIVES

After reading this chapter, you should be able to:

- Describe different modes by which fuel vapors are generated from a solid fuel.
- Define pyrolysis.
- Demonstrate a clear understanding of the combustion properties of wood, paper, plastics, paint, metals, and coal.
- Differentiate the ignition and combustion processes of thermoplastics from those of wood.
- Describe the reasons why wood does not have a fixed ignition temperature.
- Explain the importance of dust particle size and concentration in dust explosions.
- Explain the role of flame color and smoke production during fire development.

In addition to the liquids and gases examined in Chapter 4, there is a large group of very common fuels for which accurate and precise combustion data cannot generally be tabulated—solid fuels. Many of the typical combustion properties may have little or no meaning for this group. For this reason, it is not often possible to list such familiar properties as flash point or explosive range for the most common fuels of all: wood, plastic, paper, and the like. Despite this limitation, the properties of these fuels are of the greatest importance, and to some extent they can be described, either by numerical data or in general terms.

The ignition and combustion of solid fuels are more complex than that of liquid or gaseous fuels because solid ignition usually depends on pyrolyzing enough of the material into a combustible fuel and having those products mix with the correct amount of air to be ignited. The ignition of any solid fuel is a surface phenomenon, which means that the temperature of the surface is the critical factor, not the bulk temperature. The ignition of the typical solid depends on providing a high enough heat flux to bring about the necessary pyrolysis (as introduced in Chapter 2). The criterion of "high enough" depends, as we have seen, on the properties of the potential fuel, that is, its thermal conductivity, thermal capacity, and density, as well as its rate of absorption of heat. Ignition has been defined as the initiation of a self-sustaining combustion.[1] Because combustion can be either flaming or smoldering, when we examine ignition processes we have to be aware of the difference between those processes. As Babrauskas has pointed out, some ignition temperature data are the surface temperatures of fuels as they ignite into flames, and others are the temperature of an oven into which the object is placed that results (after some time elapses) in smoldering combustion.[2]

Reasonably pure cellulose is rare in nature, occurring in substances like cotton or linen fiber. Thus, an undyed cotton dress or linen shirt is fairly pure cellulose. A wooden beam is composed largely of cellulose, but in a different structural arrangement with many additional complex organic substances.

The chief reason that any numerical values at all can be attributed to the solid fuels is that when heated, most, if not all of them, undergo heat decomposition or pyrolysis with the production of simpler molecular species that do have definite and known properties. Even so, precise values are not often available, because in some instances the pyrolysis of solid materials has been inadequately studied and because the pyrolysis of a single solid material gives rise to a great many simpler products. The resulting complex mixture varies with the thermal and environmental conditions and has physical and chemical properties unlike those of any pure compound. The properties of solid fuels and pyrolysis, and the mechanism of their combustion, will be examined in this chapter.

Pyrolysis

As discussed earlier, the most significant part of all flaming fires is the flame itself, in which the combustion reaction takes place solely between gases. This remains true even though the fuel feeding the flame is a solid: wood, cloth, plastic, paper, or even coal. How then does a solid fuel maintain a gaseous reaction in the flames that burn around it? There are several routes by which this can occur, as shown in Figure 5-1. Some solid fuels can evaporate directly to vapors, by a process called *sublimation*. Naphthalene, a flammable solid formerly used as mothballs, and methenamine, used for ignition tests, sublime at room temperatures. Other solid fuels, such as candle waxes, melt and then evaporate. Thermoplastics melt, decompose into smaller molecular species, and then evaporate. Another class, such as polyurethane, decomposes when heated to form volatile liquid products, which evaporate. The final category is the very large class of products that when heated, decompose to yield both volatile products and a charred matrix. This category includes wood, paper, other **cellulosic** products, and most **thermosetting resins**. The key to most of these mechanisms is the phenomenon of pyrolysis, which occurs within the solid fuel as a result of being strongly heated.

In chemistry, **pyrolysis** is defined as the decomposition of a material brought on by heat in the absence of reactions with oxygen. In fires, pyrolysis of fuel also includes thermal decomposition that takes place in the presence of oxygen (since oxygen is nearly always present in the combustion zone). Oxidation by-products are therefore common in real-world combustion products. The phenomenon has been known for a very long time, but its mechanism is only now beginning to be understood, and then to only a limited degree. To understand what occurs during the pyrolysis of fuel, it is necessary to remember that all practical fuels are organic in nature; that is, they are complex compounds based on carbon.

Nearly all fuels of importance to the fire investigator are vegetable in origin or are derived by decomposition, bacterial action, or geologic processes from "living" material, both animal and vegetable. This is true of oil and coal as well as of natural gas. Wood, the most common fuel in ordinary fires, is the direct result of life processes in which very complex organic structures are synthesized by natural processes within living cells. To date, no one can write a complete structural formula for wood, although it is well known that its major constituent is cellulose, a very large molecule synthesized from many glucose (sugar) molecules in long chains of undetermined length (see page 27). In addition to cellulose, there are many other compounds in wood: hemicellulose and lignin are the most prevalent (up to about one quarter each), and there are various resins, pitches, oils, and other substances in various quantities. Certain softwoods, such as pine, have large quantities of volatile oils called *terpenes* and *oleoresins* (the source of commercial rosin), while most hardwoods contain little or no resin and only low concentrations of some of the terpenes.

Organic compounds (including the constituents of wood) when heated are subject to complex degradations to simpler compounds that are more volatile and therefore more

cellulosic ▪ Plant-based materials based on a natural polymer of sugars.

thermosetting (resin) ▪ Polymer that decomposes or degrades as it is heated rather than melts.

pyrolysis ▪ The chemical decomposition of substances through the action of heat, in the absence of oxygen.

FIGURE 5-1 Different modes by which fuel vapors are generated from a solid fuel. *Source:* Introduction to Fire Dynamics, 2nd ed., D. Drysdale. 1999. New York: John Wiley & Sons Limited. Reproduced with permission.

flammable than the original ones. It is these compounds that oxidize in the flame. If a sample is pyrolyzed by heating it electrically in an inert atmosphere, the decomposition products may be separated by gas chromatography, identified by mass spectrometry, and studied by a variety of tests. Such studies have been important in recent years in elucidating the mechanism of heat decomposition. Pyrolysis products from wood include water vapor, methane, methanol, and acrolein, for instance. For the purposes of fire investigation, it is not important to understand the exact mechanism of pyrolysis or all the intermediate products that result. However, it is important to understand the basic facts of heat decomposition because they are at the heart of understanding the basic nature of fire fed by solid fuel.

In the sense that they do not evaporate or sublime directly into fuel vapors, wood, paint, plastics, and coal do not burn. When they are heated, they decompose into smaller molecules with greater volatility and flammability, and, of course, into carbon, which is combustible directly. This is the process of pyrolysis, and it is fundamental to nearly all fires. It explains how fuel vapors generated by the solid are ignited to support the flames visible above the solid fuel surface. There is a zone just below the surface where pyrolysis occurs, and beneath (or behind) there is a region where no pyrolysis has yet occurred. The gaseous materials formed by the pyrolysis of the large, nonvolatile molecules of the solid fuel are the materials that burn in the flame. Without them there could be no flame. For instance, recent work by Stauffer has demonstrated that polyethylene pyrolyzes into a homologous series of alkanes, alkenes, and dienes; polystyrene forms aromatic products while burning; and polyvinyl chloride first releases noncombustible hydrogen chloride (HCl) gas and then subsequently yields burnable organic fragments.[3] Recent work by DeHaan has demonstrated that animal fats pyrolyze into homologous series of alkenes, aldehydes, and alkanes in ratios different from those produced by polyethylene.[4]

CROWN FIRES AND FIREBALLS

The pyrolysis process explains why fires may become "crown" fires in the burning of a forest. When foliage is heated to the point that it exudes a large quantity of volatile gases and vapors that are flammable, the vapor cloud can burst explosively into flames, and the entire bush or tree becomes almost instantaneously enveloped in fire. This is the phenomenon that gives rise to the comment that a tree or bush "exploded" into fire. The material composing it simply was heated to the point that it decomposed by pyrolysis to form a large quantity of volatile, flammable gases that when mixed with air, underwent rapid combustion, or deflagrated. Pyrolysis also explains the mechanism by which fires are set at a distance by radiant heat from a large fire nearby. The radiation is absorbed until the temperature produced starts pyrolytic action, and temperatures rise until the autoignition temperature of the pyrolysis products is reached. At this time, fire erupts all over the fuel exposed to the radiated heat.

An additional curious fire phenomenon that is explainable only on the basis of pyrolysis products is the *fireball,* or "firewhirl." It may be generated in a very intense fire and can travel through the air for appreciable distances while burning. The "ball" is formed by the emission of quantities of pyrolysis products far in excess of the air locally available to burn them. This means they accumulate in a limited region, burning primarily on the periphery. The strong updraft created by the fire carries the pyrolysis products upward into the air, where they continue to burn independently of the original fuel from which they were formed. In wildland fires, these products can drift out of the direct updraft and carry a mass of flaming gases into areas not yet on fire. It is possible for such fireballs to entrap personnel fighting the fire. Fatalities from fireballs are known. Anyone who has viewed a very intense fire will have observed this phenomenon, often in miniature, as masses of flame that detach from the fire and rise into the air, as shown in Figure 5-2.

NONPYROLYZING FUELS

Not all solid fuels have to undergo pyrolysis before they can combust. Reactive metals like sodium, potassium, phosphorus, and magnesium combust in air when oxygen combines directly on the exposed surface. The resulting heat vaporizes the fuel and produces hot gases and incandescent oxides (ash), but the fuel does not first pyrolyze into simpler compounds. A few "solid" fuels like asphalt and wax melt and vaporize when heat is applied, and the vapors support the flames. Carbon (as charcoal) can undergo combustion without flames (smoldering combustion) as a solid–gas interaction on its myriad surfaces as oxygen diffuses into it.

Combustion Properties of Wood

Far more wood is burned as fuel in structural and outdoor fires than any other solid material. Thus, its properties as a fuel, in regard to its behavior during combustion, are generally of greater importance to the fire investigator than those of any other solid combustible material.

COMPONENTS OF WOOD

The term *wood* is generic and covers a wide variety of materials, natural and man-made, whose chief component is vegetable in origin. The major constituent of wood is cellulose (~50 percent), while numerous other constituents are present: hemicellulose (~25 percent) and lignin (~25 percent), with resins, salts, and water making up variable percentages.[5]

The chemical makeup of cellulose was discussed in Chapter 2. Wood products, which are materials composed chiefly of cellulose, include all kinds of manufactured boards and panels, formed wood items such as furniture, and a host of paper and cardboard products manufactured, for the most part, from wood pulp.

Wood is obtained from many varieties of trees, some resinous, some not; some dense, others light. Woods vary greatly in their water content, volatile components, and other chemical properties. In addition to having diverse origins, wooden materials are also greatly altered by manufacturing processes that result in a variety of prepared sheet, board, and formed materials with all types of treatment and finishes.

IGNITION AND COMBUSTION OF WOOD

Wood, being a chemically and physically complex cellulosic fuel, introduces a number of variables. Each of the major constituents has different temperatures at which it undergoes pyrolysis: hemicellulose at 200°C to 260°C (400°F to 500°F), cellulose at 240°C to 350°C (460°F to 660°F), and lignin at 280°C to 500°C (536°F to 930°F).[6] The thermal conductivity of wood varies with orientation, as does its permeability to air, both being significantly higher in the direction of the grain than across it. This will affect the ignitability of a wood mass or surface, for some portions will be more easily ignited than others. Heat must penetrate the wood to trigger pyrolysis and charring, as illustrated in Figure 5-3. The production of volatile oils and resins is also faster along the grain than at right angles to it, adding further variation.

Wood discolors and chars relatively quickly at temperatures above 200°C to 250°C (400°F to 480°F), as shown in Figure 5-4, but prolonged heating at temperatures above 107°C (225°F) will have the same effect.[7] (There are limited data on long-term [days or weeks] exposure to lower temperatures, but it appears that destructive pyrolysis can occur very slowly at temperatures down to 85°C [175°F].) As wood chars, its absorptivity of incident heat flux becomes higher due to the darker surface and the lower thermal inertia ($k\rho c$) of the **char**, so its temperature begins to increase faster once charring begins. All this means that wood does not have a fixed ignition temperature; it varies with the rate and manner in which heat is applied to it. It is clear that on heating, a wood surface has one temperature at which it generates enough volatiles (which are evidenced by the visible smoke) that can be ignited by the application of a pilot (external) flame (**piloted ignition**), and another much higher temperature at which the vapors themselves will ignite without any external pilot flame (non-piloted or spontaneous ignition). Along with all the other variables, the nature of the heating (whether radiant or convective) will influence both piloted and spontaneous ignition temperatures. Other variables include moisture content and size thickness of the sample,

char ■ Carbonaceous remains of burned organic materials.

piloted ignition ■ Ignition aided by the presence of a separate external ignition source such as a flame or electric arc.

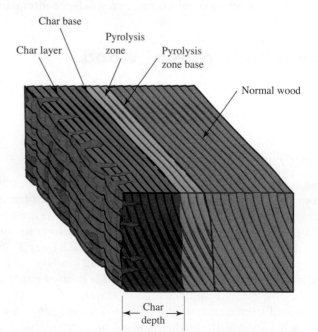

FIGURE 5-3 Normal combustion of wood resulting in progressive formation of char and pyrolysis zones. *Courtesy of the USDA Forest Service, Forest Products Laboratory, Madison, WI.*

Char base
Char layer
Pyrolysis zone
Pyrolysis zone base
Normal wood
Char depth

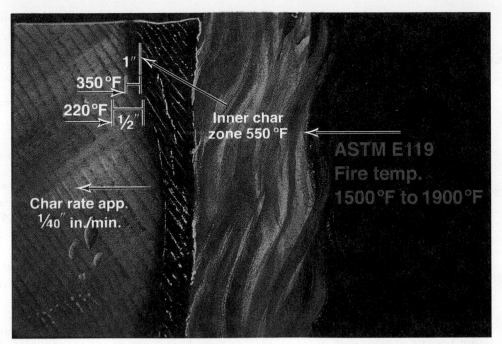

FIGURE 5-4 When Douglas fir and similar softwoods are exposed to a fire temperature of 700°C to 900°C (1500°F to 1900°F), the pyrolysis and accompanying char-ring occur at temperatures of approximately 177°C to 288°C (350°F to 550°F), as shown. The depth of the pyrolysis is dependent on the moisture content, density of the wood, and intensity of the approaching fire. Note that the char rate is only approximated even under these idealized laboratory test conditions. *Courtesy of the USDA Forest Service, Forest Products Laboratory, Madison, WI.*

orientation (vertical or horizontal), oxygen concentration, duration of heat exposure, and especially whether piloted or nonpiloted (autoignition) is being observed.

Due to its complexity, the ignition temperature of "wood" is not a single, clearly defined value for all woods and thermal conditions, as evidenced by the scattering of temperatures reported here. It is generally agreed that even with fresh (undecomposed) woods, the piloted ignition temperature will vary with the species of wood, the size and form of the sample itself, the ventilation, and the intensity, manner, and period of heating. The autoignition temperature is similarly variable. It might be thought to be dependent on the ignition temperatures of the pyrolysis products formed in the wood, but the major species—methane, methanol, and carbon monoxide—all have autoignition temperatures on the order of 450°C to 600°C (850°F to 1,100°F), which are much higher than some observed autoignition temperatures for wood. Clearly, other factors not yet identified are involved, perhaps the glowing combustion of surface char. The auto- or self-ignition temperature is closely related to the temperature at which the oxidation of the wood becomes sufficiently exothermic to raise the temperature of the surrounding fuel and thereby become self-sustaining. Various authors have reported autoignition temperatures to be on the order of 230°C to 260°C (440°F to 500°F). Browning reported that ignition temperatures could be as low as 228°C (442°F) for some woods, and temperatures ranging from 192°C (378°F) to 393°C (740°F) have been reported elsewhere in the literature, but it is not clear whether they were observing flaming ignition or self-sustaining smoldering combustion.[8] The reported variabilities reflect the effects of different test methods, availability of oxygen, and even different masses and geometries of the sample. Browning's tests were of small wood chips embedded with thermocouples in a heated chamber, and his criterion for ignition was the oven temperature at which the wood chip temperature exceeded that of the environment. Many of these earlier studies did not discriminate between smoldering and flaming combustion, so it is not surprising that reported results vary so much. Because ignition of wood is a surface phenomenon, the measurement of surface temperature at the location of ignition is the desired result. Clearly, the complexity of the ignition of wood makes measurement of relevant temperatures very difficult and subject to measurement error. Babrauskas has assembled the results of numerous studies and concluded that values around 250°C to 260°C (480°F) represent a defensible surface temperature for both piloted or nonpiloted ignition of fresh whole wood in a reasonably short period of time[9] [ignition near a minimum flux (immersion in oven) being smoldering combustion that may transition to flaming]. Piloted autoignition (surface) temperatures measured in tests where loose specimens were in open air being heated by a radiant heat source were typically 300°C to 400°C (570°F to 750°F). For smoldering ignition, surface temperatures under 300°C are typical. Higher temperatures favor gasification and thereby flaming combustion.[10]

Ignition Variables

Dry wood is more readily ignited than wood with a higher moisture content (for the reasons explored in Chapter 3). Continuous or even recurrent exposure to raised temperatures will dry wood, with the result that the ignition temperatures may be somewhat lowered. However, the work of McNaughton does not indicate a significantly increased hazard as the wood dries until the temperatures reach 275°C to 280°C (527°F to 536°F), by which point both paint and wood are thoroughly charred.[11]

One recent study of piloted ignition of Masson pine by radiant flux (in horizontal specimens 10 cm × 10 cm) revealed surface ignition temperatures of 301°C to 405°C (574°F to 761°F).[12] The higher the moisture content of the sample (0–30 percent MC), the higher the ignition temperature (301°C at 0 percent MC to 368°C at 30 percent MC, for example). Janssens reported that the ignition temperature is increased by 2°C for every 1 percent moisture increase.[13] Hemicellulose has the lowest ignition temperature and lignin the highest. Softwoods, with less hemicellulose and more lignin, have slightly higher autoignition temperatures (349°C–364°C; 660°F–687°F) than do hardwoods (300°C–311°C; 572°F–592°F) under the same test protocol.[14]

Some woods, like pine, will ignite into flames more readily and burn vigorously, because they contain resinous materials that easily produce volatile flammable vapors when heated. These add greatly to the limited ability of cellulose and lignin alone to support the fire. Ease of ignition of wood is correlated with the content of pitch and other components that readily decompose to generate combustible vapors. Some woods, like pitch pine and slash pine, contain so many combustible components that they may easily be ignited with a match flame and will burn furiously, as if soaked in kerosene.

Other than volatile resins found in many species of wood, the main fuel component is cellulose. Because the reader is probably familiar with the low heat output from such pure cellulose fuels as cotton or linen fabrics, it is readily appreciated that the cellulose (with its ΔH_c = 16 kJ/g) itself is not an efficient part of wood combustion. As we have seen, cellulose, as a derivative of glucose, is already partly oxidized, and therefore it has proportionately less fuel (per unit weight) to provide for oxidation in fire. Ease of ignition makes wood (ΔH_c ~ 16 kJ/g) or paper suitable for kindling a fire, but the total heat output is less on a weight-for-weight basis than that of fuels with a higher percentage of hydrogen such as coal or petroleum (ΔH_c = 46 kJ/g). Thus, fires that have only wood for fuel *may* be somewhat less intense than those that are fed by hydrocarbon liquids, which are often used as accelerants (because they can release more energy per second per unit of surface area, sometimes called *energy density*). Interestingly, charcoal has a ΔH_c of 34 kJ/g.

Decomposed Wood

The ignition temperatures of fresh, undecomposed wood as measured in the laboratory may be of less importance to the fire investigator, since fresh, new wood is not always involved in a structure fire. Of more interest is the effect of organic decomposition (decay) and thermal decomposition (pyrolysis) on the ignitability of wood. Angell reported that the ignition temperature of southern pine, measured at 205°C (400°F), when sound, dropped to 150°C (300°F) when the same wood decayed.[15] (These values seem far too low for autoignition and may represent values for piloted ignition or sustained smoldering, possibly as a result of self-heating.) The minimum radiant heat flux for ignition of wood has been measured at ~12 kW/m^2 for piloted ignition and 20 to 40 kW/m^2 for autoignition (depending on time of exposure and test protocol). It should be noted that these samples were small (0.5 in. × 2 in.) and exposed to convective, short-term heating. Ignition times were recorded as 427 seconds for the sound wood and 105 seconds for the rotted wood (which indicate that the samples were exposed to very high heat fluxes).[16] The relationship between applied heat flux and piloted ignition has been studied by Janssens. As we saw in a previous chapter, the higher the heat flux and the lower the thermal inertia, the faster ignition will take place. Because different woods have different densities and thermal conductivities, the ignition time will depend on the variety of wood involved. Moisture content will affect both, but since most interior wood is in the moisture content range of 9 to 12 percent, its effect can be ignored.

Regarding time to ignition, Babrauskas's data offered the empirical correlation

$$t_{ig} = 130\rho^{0.73}/(\dot{q}_{ex}'' - 11.0)^{1.82}$$

where ρ = density of the wood (kg/m^3), and $\dot{q}_{ex}''$ is the radiant heat flux to which the (thick) wood is exposed (over the range 20–40 kW/m^2).[17] The lower the density or the higher the radiant heat flux, the shorter the ignition time. There is a relationship between environmental (immersion) temperature and ignition time for wood. Very old data published by NFPA show that longleaf pine, for instance, does not produce piloted ignition at 157°C (315°F) even after exposure for 40 minutes, yet piloted ignition can occur after 14.3 minutes at 180°C (360°F), 11.8 minutes at 200°C (390°F), 6 minutes at 250°C (480°F), 2.3 minutes at 300°C (570°F), or 0.5 minute at 400°C (750°F).[18] (Very low reported ignition temperatures of 180°C–200°C, however, should be interpreted as coming from faulty measurements when only short-term heating has been performed.)

"LOW TEMPERATURE" IGNITION OF WOOD

Of more frequent interest to fire investigators are the effects of slow, prolonged heating of wood, which dehydrates and then decomposes the wood by pyrolytic action. The normal combustion of wood is a progressive process with zones of char and pyrolysis that progress inward from surfaces exposed to heat, as illustrated in Figure 5-4. At temperatures between 100° (212°F) and about 280°C (540°F) wood loses weight slowly as moisture and volatile oils are released gradually, and a large percentage (~40 percent) of the wood is turned to charcoal.[19] At temperatures above about 180°C (360°F), the pyrolysis of all three major solid constituents (i.e., cellulose, hemicellulose, and lignin) reaches its maximum rate, leaving a smaller percentage (10 to 20 percent by weight) as char. If the heat being accumulated by the char is retained, and there is an adequate supply of oxygen, the temperature of the mass can rise to the point at which combustion can take place (ignition temperatures of about 300°C; 570°F). The retention of heat, of course, depends on the amount of thermal insulation available and the amount of heat that is being lost to convective and conductive processes. If there is too much insulation, the supply of oxygen becomes inadequate to sustain combustion, although smoldering combustion can be sustained even at very low oxygen levels. It is for these reasons that the formation of charcoal by slow heating of wood can play a significant role in the initiation of some accidental smoldering fires[20] as depicted in Figure 5-5.

Failure of the insulation around metal chimneys or fireboxes can expose wooden building components to such temperatures, while the failure of the integrity of metal or masonry flues may allow flames or sparks to provide piloted ignition. Obviously, direct flame exposure from an opening in a flue or chimney will lead to relatively rapid ignition of exposed wood. There is also a relationship between incident heat flux, piloted ignition temperature, and time to ignition. Yudong and Drysdale demonstrated that the higher the heat flux (between 15 and 32 kW/m^2), the lower the observed ignition temperature and the shorter the time to ignition.[21]

Although whole wood itself may not ignite at "low" temperatures, it is well known that prolonged exposure to heat at temperatures below 120°C (248°F) over an extended

FIGURE 5-5 ■ Charring in ceiling from prolonged contact with electric radiant-heat ceiling panels. *Courtesy of Robert Toth, IRIS Investigations.*

pyrophoric ■ Capable of igniting on exposure to atmospheric oxygen at normal temperatures.

period of time can cause wood to degrade to charcoal by the distillation and pyrolysis process described.[22] Shaffer calculated that exposure to temperatures as low as 150°C (300°F) for long periods can decompose finely divided cellulosics to charcoal, which then may ignite.[23] This charcoal has been referred to as *pyrophoric carbon* or *pyrophoric charcoal,* alluding to the properties of activated (laboratory) charcoal to oxidize with room air even at modestly elevated temperatures. It has been argued that **pyrophoric** is a misnomer, since the U.S. Department of Transportation defines pyrophoric materials as those that can ignite spontaneously within 5 minutes of exposure to air. (The Chemical Manufacturers Association defines *pyrophoric* as a material that will ignite spontaneously in dry or moist air at a temperature of 54.4°C (130°F) or below.)[24] Bowes, in his comprehensive study of self-heating mechanisms, points out that *activated* charcoal is made by heating a carbonaceous fuel (often, coconut hulls for laboratory use) under reducing conditions (in the absence of all oxygen). Although this material is capable of self-heating when exposed to air, especially when warm, exposure to air over a long period of time simply permits oxidation of the carbon.[25] If a wooden surface is exposed to a radiant heat source when air is freely available, the pyrolysis products will distill away, but the activated form of charcoal will not be formed in bulk. The resulting char may not be susceptible to self-heating, and flames could not be supported because the required volatiles have already dissipated. Bowes's examination of the problem showed, however, that if a heated cylinder such as a flue or pipe runs through a massive wood member with minimal clearance, the conditions will be appropriate for the formation of a charcoal that can self-heat to smoldering com-

spontaneous ignition ■ A chemical or biological process that generates sufficient heat to ignite the reacting materials. *See* **self-heating**.

bustion if the temperatures are high enough **spontaneous ignition**. Bowes offered the opinion that the temperatures produced by ordinary steam pipes would not be adequate to create this condition but that superheated steam pipes or flues might. He commented, however, that if a suitably hot source (i.e., one with adequately high radiant heat output) was close to a wood surface overlaid with a noncombustible layer of sheet metal, tile, or similar barrier to oxygen, a reducing atmosphere could be produced, as shown in Figure 5-6. The failure of the barrier or some other change that would suddenly bring the heated charcoal into contact with fresh air could then result in flaming combustion. Bowes presented the thermodynamic argument that to accomplish this, the temperature of the source would have to be considerably hotter than steam pipes operating at normal pressures.

FIGURE 5-6 Fire in a wood floor under the hearth of an improperly installed fireplace. *Courtesy of Greg Lampkin, Knox County, TN, Fire Investigation Unit.*

Heaters and Flues

One study of the problem of ignition of wooden structures by heaters and flues has shed further light on this problem. Martin and Margot evaluated a series of fires in houses where a change had been made to high-efficiency wood stoves, sometimes 5 and 10 years prior to the fire.[26] They discovered that the affected wood surfaces had been faced with metal or tile, often with an intervening layer of solid insulation. With prolonged heating, even at temperatures below 120°C (250°F), the insulation would allow enough heat to be conducted into the wood to trigger degradation while at the same time excluding air from the surface. The lower the temperature was, of course, the longer the time required. If after a suitable mass of charcoal had been created and maintained at a high temperature by the surrounding mass of wood oxygen was admitted (by way of structural collapse, wood shrinkage, or movement of the responsible heater), the mass of charcoal could sustain self-heating combustion to the point of igniting flaming fire in adjoining fuels. Martin and Margot's analysis of the thermodynamics showed that temperatures had to be maintained for a very long period of time, that air circulation had to be minimal (to minimize heat losses as well as to exclude fresh air), and that the involved wooden members had to be fairly massive for this process to take place.

Degradation to Char

Recent work by Cuzzillo and Pagni has explored a number of factors associated with the heating (or "cooking" at temperatures well below normal ignition temperatures) of wood.[27] Their findings show that the degradation of wood to char over a long period of time decreases the thermal conductivity and dramatically increases the porosity of wood. Cracking of the char makes it even more accessible to atmospheric oxygen. (Fresh wood is much less permeable across its grain than it is along its grain, and it has higher thermal conductivity along its grain than across it as well.) When wood is heated at low temperatures, charcoal is formed (even in the presence of some air). If enough is produced to be a critical mass for self-heating, and the char is sufficiently porous to permit diffusion of oxygen through the mass, the heated material can be ignited as a self-sustaining, smoldering combustion. If conditions promote a sufficient increase in heat release rate, there can be a transition to flaming combustion in adjoining masses of wood that have not been degraded completely.

Cuzzillo's work showed that self-heating of the char formed by low-temperature "cooking" is a likely mechanism leading to runaway self-heating when the conditions are right. It has to be realized that runaway self-heating of such char does not necessarily lead to flaming combustion, because so much of the volatile content has been cooked out of the char. But a change in ventilation (such as shrinkage of wood timbers, failure of a metal or tile covering, or even a shift in the direction and intensity of ambient wind) could be the factor to push the combustion process into flames. The processes that prompt transition from smoldering to flaming are not well understood. Flaming may be triggered by better ventilation that causes a higher heat release rate (above a critical threshold), or the creation of hot spots in the char that cause piloted ignition. The role of the processes involved in long-term, low-temperature ignition of wood will be discussed in more detail in Chapter 6.

Areas in which oxygen-deprived "cooking" processes may take place include areas surrounding light fixtures and fireplaces. Flues, vents, and chimneys (see Figure 5-7) normally have adequate clearances built into them because they are recognized sources of heat, but untreated cellulose insulation (wood shavings, sawdust, or paper pulp) can inadvertently be brought into contact with such heat sources. Pipes carrying steam under pressure can reach temperatures of more than 150°C (300°F) and should be well ventilated with adequate clearances where they pass through wooden structural members. In a recent review, it was concluded that any heat source that can create surface temperatures on wood members in excess of 77°C (170°F) must be properly separated or insulated (as UL recommended years ago). In fact, that is a requirement of the City of New York Building Code.[28]

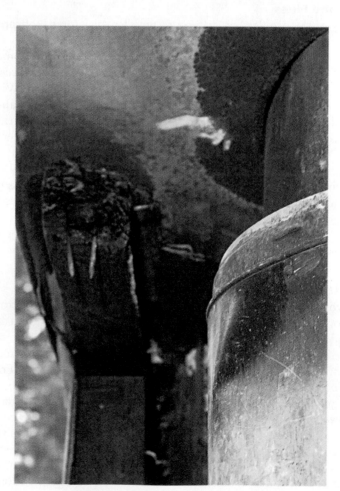

FIGURE 5-7 Displaced metal chimney sections allowed heat to ignite wood framing of chimney enclosure. *Courtesy of Lamont "Monty" McGill, Fire/Explosion Investigator.*

FIGURE 5-8 Massive floor timbers over 80 years old ignited by hot-water pipes. Residents smelled smoke for 2 weeks prior to detecting smoldering fire. *Courtesy of Thomas Goodrow, Fire/Explosives Technical Specialist, ATF (retired).*

The review of low-temperature ignition issues by Babrauskas revealed numerous well-documented cases where cellulosic materials (including sawdust, wood boards, and timber) were "ignited" by sources between 90°C and 200°C (196°F–390°F).[29] Charring of wood is known to take place at temperatures as low as 105° to 107°C (221°F–225°F), and that wood charred at such low temperatures exhibits similar exothermic oxidation as char produced at very high temperatures (>375°C; 700°F). It is also thought that moisture and cyclic heating and cooling play roles in making charred wood susceptible to self-heating. There are well-documented fire cases where *massive* wood timbers in a residence were ignited by prolonged contact with steam heating pipes whose surface temperatures never exceeded 120°C (250°F), as illustrated in Figure 5-8. As with other self-heating materials (as will be seen in Chapter 6), if there is a sufficient reactive mass, adequate oxygen, insulation, and time, such self-heating can reach runaway growth achieving smoldering that, in many cases, can transition to open flame.

In the 2004 *MagneTek* court decision, several scientific errors crept in and were "validated" by the published decision.[30] These included a claim that pyrolysis is just a theory that is not suitably supported by scientific data. Nothing could be further from the truth. Pyrolysis as described in this text has been known for over 200 years to be part of the combustion process, and thousands of papers support and prove its mechanisms. Somehow the court became convinced that wood has a fixed ignition temperature of 400°F (195°C) and that any ignition has to be caused by a source with at least that temperature. The court failed to note that wood is a very complex fuel that undergoes numerous chemical and physical changes as it is heated. The various forms created by these processes have different chemical and physical properties, some of which include an increased propensity to self-heat. Even though the term *pyrophoric* is misapplied to wood and charcoal (and that is presumably the term the court meant to discredit), the mechanisms of self-heating are well known. No, there is no tested and published "formula" that can accurately predict the relationship between heat (or temperature) exposure and time to ignition for such a complex and multivariate problem as "low-temperature" ignition of wood; however, that does not mean that scientifically defensible hypotheses cannot be offered and tested in fire investigation. Some solar water heaters (which are deliberately designed to capture the energy of the sun's heat radiation) can produce temperatures above 150°C (300°F) both in the heat-collecting element and in the pressurized transfer medium (if any). When the circulating pumps for these heaters fail, charring of underlying wood structures and roof fires have reportedly resulted.

Investigators tend to associate the time of discovery with the time of first ignition. This assumption may introduce serious errors to the fire analysis. Due to its slow output of heat and smoke, smoldering may proceed for an extended period without being noticed by passersby. When the combustion transitions to flame, it is almost certain to be discovered quickly.

Flame Temperatures

The actual flame temperatures produced by burning wood vary greatly, with such variables as oxygen content, forced-air draft, resin content, and degree of carbonization having significant effects. Using a typical net heat of combustion for wood of 18 MJ/kg (7755 Btu/lb) gives a calculated adiabatic (ideal condition) flame temperature of 1,590°C (2,900°F). The actual flame temperature of wood has been measured and reported as being some 500°C less (estimated to be 1,040°C, or 1,900°F).[31] It should be noted that the combustion of wood is so complex that a calculation of adiabatic flame temperature is of little relevance. It merely serves to illustrate the contrast between such calculated values and observed flame temperatures. It should also be remembered that measurements of turbulent flame temperatures produce time averages of rapidly fluctuating temperatures.

Char Rates

Once wood is ignited, the rate at which it will char is dependent on the heat flux to which the wood is exposed, its material, and its physical form. The standard exposure for wall and floor materials uses the E119 furnace, whose temperature and heat flux rise from near ambient to a maximum after 60 to 90 minutes. Under these test conditions the char rate for plywood ranged from 1.17 mm/min to 2.53 mm/min (0.05 to 0.1 in./min). For 1/2-in. (12.7-mm) plywood, burn-through times were 10 to 12 minutes. For 3/4-in. plywood (18–19 mm), burn-through times were 7.5 to 17.6 minutes. For boards (with tongue-and-groove edges or no gaps), burn-through times were 10.5 minutes for 19.8-mm pine, 14.17 minutes for 20.6-mm oak, and 24.3 minutes for 38-mm boards.[32] The presence of gaps dramatically decreased burn-through times, reducing them by half that of tightly fitted boards of the same thickness. Testing by the authors has shown that post-flashover burning of wood surfaces (such as plywood) can achieve 3 to 4 mm/min (0.12 to 0.16 in./min) char rates, particularly where ventilation is at a maximum (in doorways for example).[33] These rates are in accordance with results reported for "jet" fires against softwoods and for higher-temperature fires than the E119 protocol, in which for exposures to mild fluxes (~20 kW/m^2) the char rates are on the order of 0.6 to 1.1 mm/min (0.02 to 0.04 in./min).[34]

Butler tested the charring rate of one type of wood at a wide range of radiant heat fluxes and found an arithmetic relationship, as shown in Figure 5-9. Note that at a radiant heat flux of less than 10 kW/m^2, the char rate was zero, while at 20 kW/m^2 it was approximately 0.6 mm/min (0.02 in./min). At flame contact heat fluxes of 50 kW/m^2 the rate was 1.2 mm/min (0.05 in./min), and at 150 kW/m^2 the rate was on the order of 4.2 mm/min (0.17 in./min).[35] Babrauskas reported that other researchers reported similar correlation. With a correlation for mass loss rate, the best mathematical fit for surface char rate was $\beta = 0.028\ \dot{q}''_{\text{tot}}$ (where β, char rate, is in mm/min) for radiant flames below 200 kW/m^2.

FIGURE 5-9 Variation of charring rate of wood with radiant heat flux. Note the log/log scale and sharp decrease below 20 kW/m². *From Butler, C. P. "Notes on Charring Rates in Wood." Fire Research Note No. 896, 1971. Reproduced in Drysdale, D. An Introduction to Fire Dynamics. New York: John Wiley & Sons, 1985, p. 182.*

Doors often provide the only separation between compartments and provide delays to fire spread, so an appreciation of fire penetration times is important to timeline reconstructions. The critical feature is usually the tightness of fit, with loose or badly fitting doors providing only 3 to 5 minutes of resistance to flame spread. If the door is known to be a reasonably good fit, a solid-core door 44.5 mm (1-3/4 in.) thick can withstand 14 to 30 minutes of E119 exposure.[36] Shoub found "panel" doors [panels 9.5 mm (3/8 in.) thick] burned through in 5 to 6 minutes of E119 exposure (1.6 to 1.9 mm/min). Hollow-core doors with 4.8-mm (3/16-in.) wood panels burned through in 9 to 10 minutes.[37] Many hollow-core doors in low-cost housing or offices are coated fiberboard and have been seen to fail within 3 to 5 minutes of fire exposure in room fire tests by these authors. As noted previously, intensities (and exposure times) of real fires in modern rooms can be more severe than the E119 test protocol.[38]

Role of Adhesive

Another factor that may contribute to the influence of manufactured wood products is the combustion properties of the adhesive used. Often, the adhesive is less readily combustible than the wood layers; if it is more combustible, it may contribute what is effectively a fire accelerant. The resins used in oriented strand board (OSB) have the same ignition and combustion properties as the bulk solid timber, so burn-through times are reportedly similar.[39] In the evaluation of any fire involving these materials, special attention to the adhesive involved is essential. Some plastic binders can melt and support a flame like a large candle, leaving little residue after prolonged combustion.

CHARCOAL AND COKE

Although wood contains oxygen in its structure and could be considered partially oxidized, one of its main combustion products, charcoal, contains no significant oxygen and is therefore an entirely carbonaceous fuel. Charcoal is an excellent fuel, having a heat of combustion on the order of 34 MJ/kg. Charcoal is, of course, formed by pyrolyzing and destructively distilling the volatile materials out of the wood, leaving behind the nonvolatile constituents, chiefly carbon, as shown in Figures 5-3 and 5-4. Charcoal is correspondingly difficult to ignite and, lacking volatile materials, gives little flame. The small blue flames associated with its burning result from the formation of carbon monoxide (CO) gas, which burns rapidly close to the surface of the charcoal from which it was generated, to form carbon dioxide (CO_2). A charcoal fire is largely a smoldering fire, that is, one without flame, but intensely hot. Here the solid is reacting with gaseous oxygen to produce combustion on and slightly below the surface of the fuel.

Coke, formed similarly from coal or petroleum, displays burning characteristics much like those of charcoal and for the same reasons. It is not a cellulosic fuel, because coke (and coal) is almost entirely carbon and hydrocarbons.

WOOD PRODUCTS

Not all wood present in structures is in the form of structural lumber. There is an infinite variety of manufactured wood products in every building. The susceptibility of these products is not the same as for wood alone. These products deserve special mention because they greatly modify the burning characteristics of the wood from which they are derived, and they can add significantly to the fuel load and ease of ignition. As a general rule of heat transfer, the smaller the dimensions, the thinner the fuel, and the more edges presented to a heat source, the faster the fuel package will be ignited.

Plywood and Veneer Board

Plywood and veneer board are very common. Both are made from very thin sheets of wood (peeled from suitable logs) laminated to one another with layers of adhesive. The nature of the adhesive determines the suitability of the product for interior (dry conditions only), exterior (some weather exposure), or marine (extended exposure to water) use. Standard plywood is largely made from fir, with cheaper, imperfect sheets and pieces in the interior layers. Veneer board generally is made with a top layer of hardwood laminated to less expensive base layers.

point of origin ■ The specific location at which a fire was ignited.

The authors have investigated a number of fires in which plywood and veneer board were heavily involved. In no instance was the ignition of the fire traced to such materials, but the rapidity of buildup and spread of the fire was nearly always markedly increased. (See fire tests on vertical wood surfaces in Chapter 3.) In one instance of a fire of electrical origin, an extremely thin plywood finish at some distance from the origin had produced such severe local burning that the location was for a time suspected as the **point of origin**. This illustrates the necessity of checking on all possible types of causation and the behavior of involved fuels before adopting any one of them. Failure to do this has led to numerous errors on the part of some fire investigators. Plywood and veneer board curl as they delaminate as they burn. (See Figure 3-30.) This behavior can contribute to rapid downward flame spread and cause floor-level burns, sometimes on flooring some distance from the wall. (See Chapter 7 for an example.)

Particleboard or Chipboard

Particleboard or chipboard is made from small chips, sawdust, and waste from wood and paper mills bonded together under great pressure with a suitable adhesive. Particleboard is very dense but cheaper to make than plywood and is widely used in floors, cabinets, and furniture where appearance is not of importance. Particleboard has very limited strength under wet conditions and tends to swell and crumble when exposed to water for any length of time. Both plywood and particleboard are often finished with a thin vinyl covering that can be combustible. A product that is very widely used for exterior sheathing and manufactured structural components is called oriented strand board (OSB) or chipboard. These are large splinters, shards, and thin fragments of wood that are oriented so that their grain is roughly parallel and then glued together under great pressure and heat using formaldehyde resins. The resulting product is stronger than particleboard and cheaper than plywood because it can be made from waste from wood processing.

Burning Behavior of Thin-Cut Woods

With regard to the burning characteristics of wood products, it must be remembered that the finer or thinner wood is cut, the better it ignites, and the faster it burns, because the thin fuel is exposed to heat from both sides, and there is no underlying mass into which the excess heat can be conducted. As a result, the thin fuel reaches its surface ignition temperature more quickly and ignites. It then burns faster because of the large surface area exposed to heat and air. A laminated or glued-composite board will burn much like a solid-wood board of the same dimensions except for the influence of the adhesive or coating, which can be significant. If the adhesive softens under heat (or if it has been exposed to water or moist conditions), the layers will delaminate and open like the pages of a book. This will make the thin layers much easier to ignite, expose a greatly increased surface area to the flames, and cause comparatively fiercer burning. Thus, in this regard, the adhesive may be more important than other considerations. Plywood from the most common sources is not much different in its combustion properties from other wood of the same thickness. Some imported veneer boards are made with unsuitable adhesive and may greatly accelerate a fire involving them. Particleboard is generally more difficult to ignite because of its dense nature. Once ignited, it behaves in the same way as densely packed sawdust and is prone to smoldering combustion that is difficult to extinguish. Prolonged exposure to moisture causes many particleboard-type composite materials to disintegrate, with corresponding changes in their ignition and combustion properties.

Exterior Deck Materials

Several new products intended for exterior decks are being seen more frequently today. These consist of cellulose (wood flour, wood fiber, or ground paper) blended with a polyethylene or polypropylene plastic binder. They can vary from 41 to 65 percent wood fiber, with 31 to 50 percent plastic and ash (possibly added as a filler or opacifying agent).[40] They may be solid in cross section, hollow, or channeled (as seen in Figure 5-10). These products have been tested for ignitability by burning brand and for fire performance when ignited by an underdeck burner, or in a cone calorimeter. These composite materials have been found to exhibit increased heat release rate (i.e., a bigger fire) than the equivalent solid wood material.[41] They also exhibit a tendency to drip molten plastic as they burn (thus increasing the chances of spreading a fire down onto combustibles below) and to fail structurally rapidly and without warning (increasing risks of injury to firefighters).[42]

Other Cellulosic Building Materials

Other cellulosic building materials, not necessarily made from wood but alike in their combustion properties, include Celotex, Masonite, and similar boards made from compressed and bonded fibrous materials of a cellulosic nature. The low-density products (such as Celotex) were formerly used as acoustical tiles to cover ceilings and today are used, in large sheets, as insulation, floor underlay, or for other such purposes (see Figure 3-27). Tests show that some of these materials can support rapid flame spread across their surfaces and are subject to a smoldering fire along their edges or, occasionally, in their interiors. In a large surrounding fire, they burn well and contribute much fuel. Thus, it is common to see that in a room fire, all the acoustical tiles have fallen and burned on the floor. They fall readily because they are generally glued into place, and the heat softens the adhesive. If ceiling insulation or ceiling tiles are made of low-density cellulose (paper fibers), they can contribute significantly to the intensity of the fire, but only rarely are they involved in starting it. Once on fire, they are difficult to extinguish, since they tend to continue smoldering. The high-density products (Masonite) are considerably more ignition resistant.

Blown-in ceiling insulation can be shredded or macerated paper with a flame retardant added. A similar product is also made for application by being blown in while wet for use in walls, where the dried mass is more resistant to mechanical disturbance and fire penetration. Blown-in insulation can also be mineral fiber, which is noncombustible. Both types of products are typically gray in color, and careful examination (including laboratory analysis) is required to identify the product.[43] Loose blown-in cellulose insulation is a triple threat when it comes to accidental ignition for the following reasons:

- Its good insulative properties minimize heat loss from even small heat sources buried in it.
- Its loose texture means it can be moved or can fall into contact with heat sources.
- Its fire retardance can be compromised by moisture or mechanical separation, making it more susceptible to smoldering.
- Some brands of cellulose insulation have been formulated with fire retardants which can be highly corrosive to metals, although these are not common.

Fresh, proper flame-retardant treatment will, however, make the bulk material very resistant to both flaming and smoldering combustion. Samples of any suspected structural, decorative, or insulation materials should be retained for laboratory analysis.

Paper

Paper is one of the more interesting substances involved in fires, partly because of its frequent use in starting fires and partly because of its unusual combustion properties. Everyone who has ignited a fire with paper will realize that newspaper works well, while a picture magazine makes very poor kindling. The reasons are simple but not always understood.

The basic ingredient of all paper is cellulose, the same material as that in cotton and a major constituent of wood, both of which are used in the manufacture of paper. Although cellulose is readily combustible, not all the extraneous constituents of paper are. "Slick" papers have a very high content of clay, which is not combustible. Many writing papers also have a high clay content. The differences in the combustion properties of various papers are readily demonstrated by burning a piece of newspaper, which leaves a small and very light ash, followed by burning a piece of heavy, smooth, slick magazine paper, which leaves a very heavy ash. The ash in most instances is the clay, although many specialty papers also contain such constituents as titania (titanium dioxide), barium sulfate, and other noncombustibles.

Because paper is a cellulosic product like wood, its ignition properties are not easily measured. Graf studied the ignition temperatures of a variety of papers.[44] With few exceptions, he found the ignition temperature to be between 218°C and 246°C (425°F and 475°F). At 150°C (300°F) most papers appeared to be unchanged during the duration of his tests. Many papers tanned in the range up to 177°C (350°F), went from tan to brown by about 204°C (400°F), and if not ignited, went from brown to black at about 232°C (450°F). Such events are affected by the rate of heating, manner of ignition, nature of heat exposure, and amount of ventilation; therefore, a precise ignition temperature cannot be cited. Smith's data (as reported in the *Ignition Handbook*) indicate that most uncoated papers ignited at surface temperatures of 260°C to 290°C (500°F to 550°F) as a result of a validated test.[45] Other reported tests generally agreed. Long-term ignition was reported to occur in paper wrapped around a pipe maintained at 200°C (400°F).

Paper ignites easily because, like wood, it has a low *thermal inertia* [calculated from $k\rho c$, the product of the fuel's thermal conductivity (k), its density (ρ), and thermal capacity (c)], so its surface temperature goes up rapidly, and it is thermally thin, allowing rapid heat saturation of its entire thickness. The minimum ignition heat flux for various paper products has variously been reported from 20 to 35 kW/m^2.[46] If paper is exposed as a single sheet rather than compressed in a stack, its large surface area promotes rapid burning, and a high, but brief, heat release rate. When paper is stacked, it no longer has much exposed surface, and a pile of stacked papers is very difficult to burn, although it

FIGURE 5-10 Plastic
and cellulose composites
for building applications.
(a) Plank form for decks.
(b) Board form for fences.
Courtesy of John D. DeHaan.

(a)

(b)

will char on the exposed surface. Figure 5-11 illustrates the effect of thermal thickness. Ventilation is absent on the inside of such a stack, thus making paper a good insulator against penetration of heat from a surrounding fire. In fact, a small fire may be smothered by paper quite effectively, merely by covering the burning surface with a few layers of it. The air supply is thereby shut off, and the fire is quenched. This is true even of

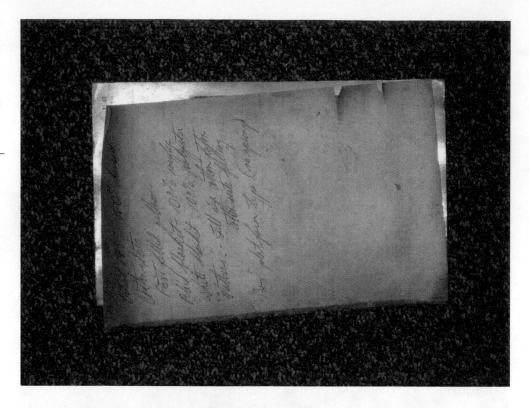

FIGURE 5-11 Loose paper exposed to radiant heat from a developing room fire demonstrates the effect of "thermal thickness"—scorched where it curled away and unaffected where in contact with book beneath.
Courtesy of John D. DeHaan.

newspaper, which is among the most combustible types of paper in existence. A stack of paper, however, can support smoldering fire within its mass. Single sheets of paper, either flat or crumpled, have proven very difficult, if not impossible, to ignite by contact with a glowing cigarette. Carbon paper, however, can be ignited with a cigarette and will support a smoldering fire. Once burning at a steady rate, cardboard cartons (flats) have been measured to yield a maximum mass flux of 14 g/m^2/s compared with that of wood cribs, 11 g/m^2/s.[47]

In the interpretation of a fire, it is not the presence of paper that is significant but its distribution in terms of exposed surface. Sometimes, the presence of a stack of papers is taken as the origin of a general fire. Although it is possible to burn a stack of folded newspapers to completion over an extended time, it is surprisingly difficult to do because of the limited surface area exposed to the air. In contrast, a pile of loosely crumpled sheets of the same paper is not only easy to ignite but quickly raises the temperature of nearby combustibles to their ignition points and thus initiates a fast-growing, very destructive fire. In our experiments, rooms with a small general fuel load (bed, dresser, carpet, and wooden doors) could be quickly brought to a fully involved stage of fire with a large pile of crumpled newspaper ignited with a match. The papers burned nearly to completion, leaving only small, unremarkable flakes of ash. Although fiberglass thermal insulation is not combustible, it is often faced with a kraft paper layer secured with an asphaltic binder. The thin paper, if exposed on walls or ceilings, can support extremely fast fire spread.

At the same time, it should be remembered that a flammable liquid poured onto a pile of papers is a common arson set. If the papers are closely packed, the accelerant burns at the surface of the papers, with the absorbent paper acting like a wick. The residue of unburned papers within such a pile is a good place to look for unburned traces of the liquid accelerant. Here the primary purpose of the paper is to retain the accelerant, not to contribute largely to the fire by its own combustion.

Plastics

Plastics are nearly universal in modern structures and vehicles and are found burned in nearly all fires. As a result, fire investigators need a working knowledge of the kinds of plastics, their applications, and their burning behaviors.

GENERAL CHARACTERISTICS

There are many kinds of plastics, with a variety of physical and chemical properties. Some, like Teflon (tetrafluoropolyethylene), are not readily combustible and will exhibit heat damage only at extremely high temperatures. Others, such as nitrocellulose, are readily ignitable and violently flammable. Most of the commonly used plastics range between these two extremes. Called **synthetics** as a class because they are not natural products but are manufactured in a factory (usually from petroleum feedstocks), most plastics are readily combustible. Some may burn to completion only in an intense surrounding fire, and others will readily burn almost to completion. Many laboratory tests of the ignition and combustion properties of plastics are used to identify them and classify them as to fire hazard. It is important to remember that nearly all plastics are made from long chains of hydrocarbons linked together in various ways. Thus, sufficient heat input will rupture the chemical bonds holding the chains together, and the plastic will pyrolyze to simpler, more volatile compounds. These pyrolysis products may be very toxic, as in the case of styrene monomer and cyanogens; extremely flammable, as in the case of carbon monoxide (CO) and the short-lived radicals CH and CH_2; or readily susceptible to further pyrolysis. In addition, some plastics and rubbers create pyrolysis products that can have serious consequences in the fire environment. Polyvinyl chloride plastic releases hydrochloric acid (HCl), and Teflon releases hydrogen fluoride (HF) and free fluorine (F_2) when it finally decomposes in a fire. Some rubbers generate hydrocyanic acid (HCN) during combustion. For a full description of pyrolysis products from plastics in fires, the reader is referred to other references.[48] In addition, thermoplastic polymers will melt and flow, sometimes dripping onto a source of flame to fuel it and sometimes collecting in a large pool of burning plastic, which burns fiercely and resists extinguishment. The behaviors of some of the more common plastics when exposed to a small flame in air are listed in Table 5-1. As with laboratory tests with other solid fuels, it must be remembered that the fire behavior of synthetics is largely dependent on their degree of molecular cross-linking, physical form (sheet, rod, fiber, etc.), test conditions, and presence of inorganic fillers and fire retardants, so it is best not to generalize about the fire behavior of a particular type of plastic.

Because the chemistry of plastics is similar to that of petroleum products, they tend to produce similar sooty flames when burning in air, and the open-flame temperatures produced by burning plastics are on the same order as those of common petroleum distillates. The flames of thermoplastics can be sustained by the formation of a pool of molten plastic, producing very intense fires. The flame temperatures produced by some plastics are quite high. Polyurethane is one example for which flame temperatures up to 1,300°C (2,400°F) have been measured over polyurethane mattresses during fire experiments.[49] The reported maximum mass burning fluxes for plastics are shown in Table 5-2 (for horizontal steady-state burning). Multiplied by the heat of combustion for each fuel, these mass fluxes can be used to approximate the HRR of horizontal fuel surfaces of various sizes. Note that these heat release rates are nearly the same as for ignitable liquids [for polypropylene, for example: 24 $g/m^2 \cdot s$ 44 kJ/g = 1 MW/m^2].

The combustion of many polymers contributes largely to the formation of the greasy or sticky dense **soot** found at many fire scenes and is responsible for the dense black smoke more frequently noted during the early stages of structure fires. The dense smoke many polymers produce is rich in flammable vapors, and the ignition of dense clouds of these vapors quickly involves an entire room upon flashover. Only polyethylene, polypropylene,

synthetic ■ Material that is human-made, usually referring to organic polymers.

soot ■ The carbon-based solid residue created by incomplete combustion of carbon-based fuels.

TABLE 5-1	Flaming Burning Characteristics of Small Samples of Common Polymers		
POLYMER	**FLAME COLOR**	**ODOR**	**OTHER FEATURES**
Nylon	Blue with yellow tip	Resembles burning grass	Tends to self-extinguish; clear melt
Polytetrafluoroethylene	Yellow	None	Chars very slowly; difficult to burn
Polyvinyl chloride	Yellow with green base	Acrid	Tends to self-extinguish; acidic fumes
Urea-formaldehyde	Pale yellow	Fishy formaldehyde	Very difficult to ignite
Butyl rubber	Yellow, smoky	Sweet	Burns
Nitrile rubber	Yellow, smoky	Sickly sweet	Burns
Polyurethane	Yellow, blue base	Acrid	Burns readily
Silicone rubber	Bright yellow-white	None	Burns with white fumes; white residue
Styrene-butadiene rubber	Yellow, very smoky	Styrene	Burns
Alkyd plastic	Yellow, smoky	Pungent	Burns readily
Polyester (contains styrene)	Yellow, very smoky	Styrene	Burns
Polyethylene	Yellow, blue base	Resembles burning candle wax	Burns readily; clear when molten
Polymethyl methacrylate	Yellow	Characteristic	Burns
Polypropylene	Yellow, blue base	Resembles burning candle wax	Burns readily; clear when molten
Polystyrene	Yellow, blue base, very smoky	Styrene	Burns readily

Source: K. J. Saunders, *The Identification of Plastics and Rubbers* (London: Chapman and Hall, 1966). See also *Fire Protection Handbook,* 19th ed. (Quincy, MA: NFPA, 2003), fig. 8.10.6.

and polyoxymethylene (Delrin) generally burn with a clear flame and little soot. Polyethylene and polypropylene also produce a candle-wax smell, since upon heating, they pyrolyze into much the same components as wax, with the resulting similar behavior.

Although the autoignition temperatures of plastics are generally on the order of 330°C to 600°C (650°F to 1,100°F), their universal applications in packaging, utensils,

TABLE 5-2	Maximum Mass Flux Rates (Steady-State Burning)
PLASTIC	**$g/m^2 \cdot s$**
Polystyrene (granular)	38
Acrylic plastic (PMHA, granular)	28
Polyethylene (granular)	26
Polypropylene (granular)	24
Rigid polyurethane foam	22–25
Flexible polyurethane foam	21–27
Polyvinyl chloride (granular)	16

Source: From J. Quintiere, *Principles of Fire Behavior* (Albany, NY: Delmar, 1998), 109.

TABLE 5-3	Thermal Properties of Polymeric Materials		

NATURAL POLYMERS	ΔH_c	THERMAL BEHAVIOR (DRYSDALE)	MELTING POINT (°C) (MERCK INDEX)[a]
Wool		Chars above 200°C	
Cellulose	16.1	Chars above 100°C	
Thermoplastic Polymers			
Polyethylene (HD)	46.5	Melts 130°C–135°C	(85–110)[a]°C
Polypropylene (isotactic)	46.0	Melts 186°C	(165)[a]°C
Polymethyl methacrylate	26.2	Melts 160°C	(250)°C
Polystyrene	41.6	Melts 240°C	(180)°C
Polyacrylonitrile		Melts 317°C	
Nylon 66	31.9	Melts 250°C–260°C	
Nylon 6		—	(225)°C
Polyvinyl chloride (PVC)	19.9	—	(Chars 200–300)°C
Thermosetting Polymers			
Polyesters			(Dec. 250)°C
Polyurethane foam	24.4	Decomposes 200°C–300°C: polyols, isocyanates	
Phenolic foams	17.9	Chars	
Polyisocyanurate foams	24.4	Chars	
Silicones	26.8[b]	Decomposes/oxidizes to SiO_2[b]	

Source: From D. Drysdale, *An Introduction to Fire Dynamics,* 2nd ed. (Chichester, UK: Wiley, 1999), 3.

[a]The melting point of many polymers is determined by the amount of cross-linking between the chemical structures and can vary considerably.
[b]Buch, RR, "Rates of Heat Release and Related Fire Parameters for Silicones", *Fire Safety Journal,* 17, 1991, 1–12.

furnishings, wall coverings, windows, walls, and even exterior structure panels provide many opportunities for them to come into contact with temperatures of this magnitude (see Table 5-3).[50]

BEHAVIOR OF PLASTICS

When considering plastics as fuels, it is helpful to classify them by their behavior when they are heated. Plastics that undergo reversible melting without appreciable chemical degradation are called **thermoplastics**. Plastics that do not melt but rather decompose and leave behind a solid char are called *thermosetting* plastics. There is a small group of materials that decompose into volatile materials that are not rigid and are sometimes called *elastomers*; these include polyurethane rubber. The chemical and physical behaviors of some of the common polymers are shown in Table 5-3.

Minimum autoignition temperatures of common plastics are shown in Table 5-4.

thermoplastic ■ Polymer that can undergo reversible melting without appreciable chemical change.

Polyvinyl Chloride

The behavior of a particular plastic in a general fire is the result of many factors, which all should be kept in mind when evaluating its potential contributions to the fire. Vinyl (polyvinyl chloride—PVC) may burn slowly when tested in solid form in a laboratory but has been observed to burn rapidly and contribute greatly to flame spread when present as a thin film on wall coverings (probably due to the adhesives and plasticizers used). PVC wire insulation softens and melts when heated by overheated wiring but chars if exposed to a flame.

TABLE 5-4	Autoignition Temperatures of Common Plastics		

PLASTIC	MINIMUM IGNITION TEMPERATURE		
	°C	°C	°F
Polyethylene	365*	488	910
Polyisocyanate		525	977
Polymethyl methacrylate	310*	467	872
Polypropylene	330*	498	928
Polystyrene	360*	573	1,063
Polytetrafluoroethylene		660	1,220
Polyurethane foam (flexible)		456–579	850–1,075
Polyurethane foam (rigid)		498–565	925–1,050
Polyvinyl chloride		507	944

Sources: NFPA, *Fire Protection Handbook*, 19th ed. (Quincy, MA: NFPA, 2003), Table A-6.
*V. Babrauskas, *Ignition Handbook* (Issaquah, WA: Fire Science Publishers, 2003), 244.

Nylon

One of the first synthetic fibers, nylon, is widely used in carpets. Most nylon-fueled flames tend to be self-extinguishing, but under intense radiant heat flux, nylon fiber carpet has been shown to burn with some enthusiasm.

Polyurethane and Polystyrene Foams

Unless it is properly treated with fire retardants, polyurethane foam is very flammable, as it is readily ignited with any open flame. It produces a hot, smoky flame that progressively decomposes more and more of the foam, with the "melt" joining the burning substance at the bottom of the "crater." Rigid (and flexible) polyurethane foam and polystyrene foam are used as thermal insulation in equipment and in structures and have contributed to the rapid spread of many destructive and often-fatal fires. The 2003 tragic nightclub fire in West Warwick, Rhode Island, killed 100 of the 400 patrons present when stage pyrotechnics ignited the unprotected urethane foam sound insulation covering the stage walls. The fire growth was captured on videotape and shows progression from localized flashover in the drummer's alcove into the main room (~10 m × 15 m) in less than 3 minutes as the fire spread across the walls with extreme rapidity.[51] Plastic foams in foam-core steel structural "sandwich" panels have become more commonly used in building commercial buildings.[52] Polystyrene foam–filled panels are especially susceptible to failure at joints and seams (see Figure 5-12). Others have been seen to delaminate, exposing the melting plastic foam to external flames.[53] Recent advances in this technology have improved the fire performance of some sandwich panels, with some being resistant to both ignition and melting.[54] Investigators should take a sample of any polymeric filler for analysis before reaching conclusions about sandwich panel contribution to fire spread.

Latex or Natural Rubber

Latex or natural rubber foam products have largely been replaced by polyurethane foams, but the former are still encountered in older furnishings and carpet pads. Although latex foam products sometimes physically resemble high-density urethane foams, their flammability properties are very different. Although urethane foams can be ignited by an open flame (Figure 5-13a) or intense radiant heat from an external source, most cannot be ignited by direct contact with a glowing cigarette (as in Figure 5-13b) unless they are covered with a susceptible fabric or are in contact with a sustained heat source (as in Figure 6-29).[55]

Unlike urethane foams, latex foams (Figure 5-14a) can be ignited by a small glowing ignition source (such as a cigarette) and once ignited will sustain a smolder and are capable of rekindling into a flaming fire many hours after suppression of flaming combustion, as illustrated in Figure 5-14b. Latex foams are also capable of self-heating to the point of spontaneous ignition if preheated to a modest level, whereas finished urethane foams do not self-heat. (The chemical reaction that creates the foamed elastomer, however, is exothermic, and incompletely cured blocks of foam and waste products have been known to self-heat to ignition.) Like urethane foams, latex foams can be ignited by a flaming source and will generate a dense, toxic smoke when burning. This smoke has been linked to "cold" smoke explosions.[56]

FIGURE 5-13A Polyurethane foam mattress topper burns readily after match flame ignition. *Courtesy of John Jerome and Jim Albers.*

FIGURE 5-13B A cigarette on polyurethane foam alone produces longitudinal char but no sustained combustion (glowing or flaming). If covered with heavy cotton fabric, the PU foam may sustain combustion. *Courtesy of John D. DeHaan.*

FIGURE 5-14A Typical latex foam with curing/ventilation holes. *Courtesy of John D. DeHaan.*

FIGURE 5-14B A close-up of the perforated latex foam from an upholstered sofa fire that rekindled owing to self-heating. Latex foam, unlike polyurethane, can self-heat if warmed. *Courtesy of John D. DeHaan.*

Thermoplastics

Thermoplastics are molten at the temperatures of the burning fuel surface, so they will run and drip as they burn, as pictured in Figure 5-15. If the droplets are on fire as they fall to floor level, they can ignite new flames along the vertical upholstered surfaces of sofas. Plastic skylights, lampshades, and light fixture diffusers have been known to be ignited by

FIGURE 5-15 Burning synthetic upholstery melts as it burns. Burning droplets ignite the carpet below and the lower portion of the vertical upholstery face. *Courtesy of Jamie Novak, Novak Investigations and St. Paul Fire Dept.*

FIGURE 5-16A Large polypropylene waste bins, if not flame retardant, are easily ignited to produce very large sustained pool fires, especially when empty. *Courtesy of Lamont "Monty" McGill, Fire/Explosion Investigator.*

FIGURE 5-16B Large polypropylene waste container ignited with crumpled paper becomes a very large sustained pool fire. *Courtesy of Lamont "Monty" McGill, Fire/Explosion Investigator.*

the ceiling temperatures in a general room fire. As these articles burn they soften and sag, finally falling, sometimes onto combustibles below, producing isolated, discrete points of origin. During post-fire examination, some investigators have been misled by these numerous separate fires, suspecting an incendiary cause. The investigator should be careful to attempt to correlate all unusual fire patterns with results from the fuel load before ascribing a criminal origin to a suspicious fire. Plastics are being used increasingly as a substitute for glass in windows, particularly in schools where vandalism is a problem. These plastics have been ignited by matches or lighters and gone on to ignite major structural fires. Large polypropylene waste bins, if not flame retardant, are easily ignited to produce very large, sustained pool fires, especially when empty, as seen in Figures 5-16a and b.

Plastic Building Materials

Several new building materials have been introduced for use as exterior decks, railings, and fences. These are basically extruded plastic (reinforced or nonreinforced) often made entirely from recycled polyethylene, PVC, or ABS,[57] as seen in Figure 5-17. These materials have been shown to be easily ignited with a match or other small open flame to produce a rapidly growing fire, with molten burning material dripping or running off as it burns. The result can be a large pool fire of the molten material or rapid ignition of other combustibles located below, as seen in Figure 5-16. Such rapid combustion also leads to loss of structural strength.[58]

SPECIAL CONSIDERATIONS FOR FIRE INVESTIGATORS

The potential of plastics as a source of fuel in both accidental and intentional fires should not be overlooked. Years ago, televisions, radios, and other small appliances were enclosed in wooden or metal housings, which would often contain the small flames of internal fires; today they may be made of synthetics that fuel the initial fire and may help spread it. Some but not all plastic appliance housings include fire retardants and will resist

FIGURE 5-17A Extruded plastic deck planks come in a variety of forms. *Courtesy of John D. DeHaan.*

FIGURE 5-17B Plastic decking ignited with a match. *Courtesy of John Jerome and Jim Albers, Barona Fire Dept.*

FIGURE 5-17C Dripping flaming plastic with rapidly growing flames above. *Courtesy of Jim Albers and John Jerome, Barona Fire Dept.*

FIGURE 5-17 Fire test of extruded plastic deck planks.

modest internal fire exposure. The prudent investigator should have such housings tested for fire retardants before ascribing an origin to such an appliance.

When the fire investigator is faced with reconstructing a fire, these considerations of fuel type are critical. The materials that make up the bulk of the fuels involved in most structure fires have changed dramatically over the last 30 or 40 years, affecting the way in which fires ignite and spread, the combustion products they create, the heat release rates and temperatures they produce, and the residues they leave behind.

Post–World War II Finishes and Furnishings

A typical house was finished in the 1940s or 1950s with plaster walls over wood lath or wire mesh, painted or covered with heavy paper or cloth wallpaper. Floors were bare wood or linoleum with wool carpets and rugs. Window coverings were heavy cotton (brocade) or wool fabrics (or steel-and-wood window blinds). Furniture was often bare wood. When it was upholstered, it was usually with cotton velour or wool mohair, or leather, over padding of horsehair or cellulosic (vegetable) fibers (cotton, coir, kapok, Spanish moss, etc.). Latex foam rubber was found in some cushions and pillows, and mattresses were stuffed with cotton or other vegetable fiber or even feathers. Rooms were usually poorly insulated, and glass windows were single glazed.

FIGURE 5-18
'Traditional' cotton mattress burned over a period of hours (unobserved). Heat release rate was never sufficient to ignite nearby combustibles except for tablecloth on nearest table. *Courtesy of Diane Spinner, Fire Cause Analysis.*

Such furnishings were all combustible, but how readily? Most of the materials just listed were susceptible to smoldering ignition sources such as a dropped cigarette, but, with a few exceptions, most would not be readily ignited (to a self-sustaining fire) by a short-lived flame source such as a common match. Ignition of all but latex foam and kapok took many seconds or even minutes of exposure, and all produced large quantities of smoke prior to ignition to flames. Once ignited, they would burn slowly, often reluctantly, producing small flames, preferring to smolder instead. Heat release rates would be very low (300–400 kW being typical for a cotton sofa), so spread to other fuels would require direct flame contact. See Figure 5-18 for an example. Although the localized flame temperatures would be "normal" [i.e., 800°C to 900°C (1,700° to 1,800°F)], because the heat release rates were so low (and room insulation so poor), overall hot gas layer temperatures would usually remain below the 600°C (1,200°F) range. Progression to flashover would therefore be rare and would take a long time to develop when it did occur. Combustion products included carbon monoxide (CO), soot, acrolein (from wood), and hydrogen cyanide (HCN) (from wool). Carpets and floors burned only with reluctance and only after a lot of radiant heat had been applied. Post-fire residues of liquid accelerants were relatively easy to identify based on odor and, later, by simple gas chromatography (GC), because the furnishings offered little in the way of interfering pyrolysis products.

Changing Contemporary Finishes and Furnishings

By the 1960s and 1970s, things were changing. Wall coverings tended to be paint or vinyl wallpapers over gypsum wallboard (Sheetrock; drywall), with its better durability under fire exposure. Floor coverings were often wall-to-wall carpet of nylon or polyester face yarns with jute fiber backings over a pad of butyl rubber or mixed fiber. Furnishings were often polyurethane (PU) foam or cotton padding covered by cotton or cotton/synthetic upholstery fabric (or vinyl). Window coverings were cotton or synthetic, or sometimes fiberglass fabrics. After 1972, mattresses were more commonly combinations of a cotton core with a polyurethane foam layer to improve resistance to smoldering sources (cigarettes). Thermal insulation in rooms became more common, and double glazing appeared.

As a result, fire behavior began to change. Synthetic fabrics and PU foam added smolder resistance to furnishings but made them more susceptible to ignition by short-lived open-flame sources. Fires became hotter, and heat release rates increased as more synthetics became involved. Fire growth was more rapid in this new style of furniture, and rooms could approach flashover more readily and in a shorter period of time. Numerous researchers have confirmed these results.[59] Carpets, however, and many other furnishings resisted ignition even by radiant heating. When carpets did ignite, the face yarns tended to burn off, but the heavy jute backings would burn slowly and remain in place to protect the rubber pad beneath. Vinyl products could resist ignition and slow the progress of the fire, but when they finally became involved, could contribute hydrogen chloride (HCl) vapors to the fire gases. Gas chromatography could discriminate flammable liquid residues from all the pyrolysis patterns, thanks to improved chromatography, because the peak patterns produced by these pyrolysis products were easily distinguished from the patterns from common flammable liquids.[60]

By the late 1980s, floor coverings still consisted of various synthetic face yarns, but the backing came to be made of polypropylene due to its lower cost and better resistance to odors and mildew (see Figures 5-19a and b). Polyurethane foam carpet padding (new or rebond) replaced latex rubber or fiber padding exclusively. Under severe radiant heating, face yarns could melt and even ignite, as could the backing, exposing the combustible PU foam underlay beneath. Such carpets would tend to resist localized flame and heat sources, but once alight could burn more completely than before and create much more heat and floor-level damage than ever before. Combustion products now included a wide range of aldehydes, ketones, and toxic intermediates known as free radicals.

Furnishings became almost exclusively polyurethane foam padding (rarely with flame retardants) with synthetic covering—very difficult to ignite with a smoldering cigarette but readily ignited by even a small flame. Once furnishings were ignited, flames could spread quickly and engulf all of a large chair or sofa in flames in less than 5 minutes.[61] Thermoplastic upholstery materials, with their low melting points, melt as they burn, producing molten,

FIGURE 5-19A Modern Carpet Pads. Upper left: Foam rubber "waffle." Lower left: High-density foam. Lower right: Rebond polyurethane. Upper right: "Waste" felt automotive. *Courtesy of John D. DeHaan.*

FIGURE 5-19B Carpet Backings. Top left: Traditional broadloom carpet with woven jute backing. Top right: Jute backing on wall-to-wall carpet. Lower left: Synthetic backing coated with sealant. Lower right: Polypropylene and other synthetic fiber backing. *Courtesy of John D. DeHaan.*

burning droplets of materials that fall to the base of the furniture and institute rapidly growing vertical-face fires on the sides of the furniture, as well as drop-down damage to floors and carpets beneath (see Figure 5-15). Ignition of a modern mattress and comforter is shown in Figure 5-20. Recent tests by DeHaan have shown that a modern queen-size mattress and comforter ignited by a single flame will result in flashover of a large 3.35 × 4.27 × 2.74-m (11 × 14 × 9-ft) bedroom in 5.5 minutes. Calorimetry tests have revealed that modern queen-size mattresses are capable of producing a heat release rate of over 3 MW.[62]. Since 2007 all mattresses and futons sold in the United States (even for residential use) are required to pass a stringent ignition test, *16 CFR 1633*. This test involves exposing the side of the mattress to a sustained 18 kW gas burner flame and mandates the maximum heat release rate and total heat released after removal of the burner.[63] This

(a)

(b)

FIGURE 5-20 Open-flame ignition by small gas burner causes rapidly growing fire on (a) common polyurethane mattress set and (b) polyester/cotton comforter. (Ignition with a match flame would add less than one minute to these development times.) *Courtesy of Bureau of Electronics & Appliance Repair, Home Furnishings.*

FIGURE 5-21 Traditional cotton padding (right) is easily ignited with a cigarette and forms a loose, powdery black ash. Polyester fiberfill is thermoplastic, not ignitable with a cigarette. It is easily ignited with a flame and burns readily, forming a black, hardened knob of molten plastic (arrow). *Courtesy of John D. DeHaan.*

requirement will significantly reduce the ignitability and size of any resulting fire originating from new mattresses. Sadly, this standard does not apply to foam mattress toppers or other bedding, and imported mattresses with forged approval labels have already been seen in the marketplace. This testing is described in more detail in Chapter 11.

Figure 5-21 illustrates the difference between traditional cotton padding and its modern replacement for pillow and cushion stuffing, polyester fiberfill. Polyester is a thermoplastic that ignites readily on contact with a flame and produces a fast-growing flaming fire.

Today, wall-to-wall carpets often include low-melting-point and readily ignitable polypropylene face yarns with polypropylene backing over polyurethane underlay (particularly in low- or medium-price-range installations). Except for "high-end" mattresses, for which latex foams are being offered once again (due to their better breathability and reduced moisture buildup), polyurethane foams are nearly universal in mattresses and furniture cushions. Such PU foams are very resistant to smolder ignition but are still easily ignited with a match. The acquisition of comparison samples of upholstery fabrics, padding (fillers), and floor coverings has become more important to the fire investigator than previously.

New regulations for mattresses do not imply a similar improvement for upholstered furniture. CPSC has recently proposed new regulations for upholstered furniture flammability, but these would do little more than enforce existing voluntary industry efforts to make furniture cigarette-resistant. Flaming-ignition behavior would not be regulated. Flaming-ignition behavior has been regulated in California for more than three decades by the California Bureau of Home Furnishings. However, these regulations have not resulted in consumer furniture that is difficult to ignite from flames, nor furniture with improved HRR characteristics. It is apparent, however, that the simple inclusion of a barrier layer of PVC, aluminized fabric, wool, or even fiberglass can dramatically improve all types of fire resistance at a very modest increase in cost.[64]

The polypropylene carpets are becoming a widespread problem. They will pass the methenamine tablet test (originally *DOC FF 1-70*, now known as *16 CFR 1630*) (meaning

FIGURE 5-22 Polypropylene carpet ignited with small pools of gasoline (at center of each panel) continued to burn at 0.5–1.0 m²/hr until self-extinguished.
Courtesy of John D. DeHaan.

that a dropped match or similar flame of ~50W HRR will not ignite the carpet to a self-sustaining fire) and are therefore legal for use in residences. Unfortunately, exposure to a slightly larger fire (such as a burning crumpled sheet of paper) with its higher heat flux can initiate a self-sustaining fire that spreads at about 1 m²/hr (10 ft²/hr) with very small flames 5 to 8 cm (2 to 3 in.) in height along the margins (i.e., a very low heat release rate).[65] If the fire starts in an unoccupied building, it can burn for hours, consuming entire rooms of carpet without running out of oxygen, due to its very low rate of combustion (see Figure 5-22). In addition, because of its low critical incident heat flux, such carpet is more readily ignitable by radiant heat even at some distance away from a large fire (such as a chair fire). As the face yarn burns, the polypropylene backing melts and shrinks (as shown in Figure 5-23), exposing the combustible PU foam underlay, which can now ignite and sustain the spread of fire through and across the carpet. The rate of spread is dependent on the interaction between the carpet and the pad and can be affected by such variables as whether the carpet is stretched tightly and securely fastened or lying loosely atop the pad with plenty of air between the two. Urethane foam carpet pads (particularly rebond) and all-synthetic carpets produced complex "stews" of pyrolysis products that can sometimes be quite difficult to distinguish from some non-distillate-type petroleum products. Comparison samples of unburned carpet and pad are very important evidence.

Today's furniture is markedly improved in its resistance to the most common type of accidental ignition: a dropped cigarette or other forms of glowing sources (electric heaters, glowing electrical connections, etc.). The significant disadvantage is little or no resistance to flaming sources. Once alight, such furnishings can be completely involved in 3 to 5 minutes and be reduced to a charred frame in 10 minutes while producing very high temperatures [in excess of 1,000°C (2,000°F) in the flame plume] and enormous heat release rates (2 to 3 MW is common for a sofa or recliner today).[66] Such performance virtually ensures that the average-size furnished room with an open door will be completely engulfed in a post-flashover fire within 3 to 10 minutes of first established flame. The intensity of such fires can obliterate traces of ignition sources, obscure fire patterns (if sustained for more than 5 minutes), induce the speed of spread and kind of damage once thought possible only for accelerated fires, and make the recovery and identification of possible accelerant traces very difficult. Gas chromatography/mass spectrometry is now needed to discriminate possible accelerant traces from pyrolysis products. Further changes for the future include **flame resistant** performance standards for bed covers and household furnishings. As these products appear fire growth will be affected. The reader is referred to Chapter 11 for more information.

The best the investigator can do is to be aware of the contributions such furnishings can make to the ignition and growth of a fire and to take whatever steps necessary to document the kind of furnishings present (new or old? cellulosic or synthetic fabrics and padding?) and, especially, to collect comparison samples of carpet, pad, upholstery, and padding whenever possible for later identification in the forensic lab (even partially burned samples are better than no samples at all). This may be the only way to accurately reconstruct the train of events of a fire.

flame resistant ■ Material or surface that does not maintain or propagate a flame once an outside source of flame has been removed.

Paint

Paint is a greatly misunderstood product, in part because of the tremendous diversity of surfacing coatings, all described as paint. Historically, paint consisted of a drying oil (such as linseed or tung oil) in which mineral-based pigments were suspended. The mixture was then thinned using turpentine or a petroleum distillate. The thinner evaporates shortly after application, and so it plays an important part in the flammability of the paint only while it is still wet. The drying oils and other resins present are combustible even after the paint has dried and constitute the major fuel contribution of many paints. The mineral powders once used universally as pigments were not combustible and so tended to retard the combustion of the paint itself. This is no longer the case. The cost and environmental hazards posed by most inorganic pigments have caused them to be replaced in large part by cheaper organic dyes and fillers, which will burn. The newer water-base paints are generally emulsions of latex, polyvinyl acetate, or acrylics in water when applied and form a plasticlike polymer coating when they dry. Their properties are then similar to the plastics on which they are based. Varnishes, shellacs, and lacquers tend to be more combustible, since they are often based on naturally occurring resins or on combustible plastic polymer equivalents.

Most paints and coatings will add slightly to the fuel present in a structure and can aid the spread of flames from one part of a room to another. In some cases the paint layers have been seen to soften and peel off the wall or ceiling as they ignite and thus aid the spread of fire downward if there is readily ignitable fuel below it, as pictured in Figure 5-24. This is especially true if there are many layers of paint. Then, even "noncombustible" surfaces of painted steel or gypsum board can contribute to the fire's spread.

There are also numerous coatings intended to protect the substrate from fire effects. These coatings, called *intumescents*, swell and bubble when heated, providing a noncombustible layer of thermal insulation that slows the incursion of heat into the substrate.[67] Such intumescents are widely used to protect steel structural members as well as combustible surfaces.

FIGURE 5-24 Test fire of gasoline and diesel caused ignition of paint on walls, which peeled off and fell as flaming sheets. *Courtesy of Jamie Novak, Novak Investigations and St. Paul Fire Dept.*

Whether a paint or other coating will aid or retard the flames of a small, growing fire cannot be generalized without testing the specific product, but the contribution of layers of paint to the fuel load of an ordinary structure cannot be overlooked by the careful investigator. Recent tests by McGraw and Mowrer using a cone calorimeter revealed that at applied heat fluxes of 50 kW/m^2, multiple layers of latex paint on gypsum wallboard did not support flame spread but doubled the heat release rate (from paper alone) to ~111 kW/m^2 (of wall area). At heat fluxes of 75 kW/m^2, fire spread was modestly accelerated, and the heat release rate rose to 134 kW/m^2 (of wall area). (At 25 kW/m^2 there was no ignition of wallboard—painted or unpainted.)[68]

Metals

Metals, as fuels, are so rarely of importance that they may well be overlooked. Actually, most metals can be burned, and many metals, when finely divided, are susceptible to direct combustion in air. Others can burn when they are in relatively massive form, but generally only in a very hot environment. Some metals are *pyrophoric*; that is, they burn spontaneously in air when finely divided. A metal like uranium, for example, is extremely hazardous when finely divided because of its tendency to oxidize in air, and the heat of the slow spontaneous oxidation may be sufficient to ignite a larger quantity of stored uranium powder. Iron powder is, to a considerable extent, pyrophoric, although without the special hazard that characterizes some other metals. Iron in the form of steel wool or fine powder will burn readily and requires only a modest ignition source (a match or electrical heating). Ignition temperatures are as low as 315°C (600°F) (for powder) and 377°C (710°F) (for fine steel wool). (See Figure 5-25.) Very fine iron powder is pyrophoric.[69] The important qualification of the dangers associated with metal combustion is the state of subdivision. The finer the metal powder, the more likely it is to ignite, but rarely without supplemental ignition. The combustion properties of some of the more seriously hazardous metals are treated in some depth in Chapter 13.

FIGURE 5-25 Steel-wool fire—rapid glowing combustion with emission sparks. *Courtesy of John D. DeHaan.*

The danger associated with burning of metals is confined almost exclusively to industrial plants, especially those in which the metals are handled in a finely divided state. Many disastrous fires of this type have been fed primarily by metal fuel. In general, such fires are so intensely hot and violent that they reduce the structure far more completely than is ever seen in a fire that is primarily fed by wood as fuel. Chemical laboratories or manufacturing facilities may contain enough reactive metals like sodium, potassium, or phosphorus to cause special fire hazards (see Chapter 13).

MAGNESIUM

Perhaps the most important metal that can serve as a fuel in structure fires is magnesium. Magnesium is not readily ignited unless it is rather finely divided—in the form of shavings, filings, or dust. In a large fire, massive magnesium castings—wheels, aircraft components, and the like—can burn very fiercely, producing an enormously hot fire (see Figures 5-26a and b). In a fire where large magnesium pieces have burned, it is more likely that the magnesium was ignited by a strong surrounding fire rather than the reverse. The reported ignition temperature of magnesium ranges from 520°C (968°F) (powder) to 623°C (1153°F) (solid). Magnesium/aluminum alloys with more than 10 percent magnesium have ignition temperatures in the 500°C to 600°C (930°F to 1,112°F) range.[70]

A matter that is of special concern with magnesium especially, and with some other metals that are rarely encountered, is that when water is sprayed on hot magnesium, large quantities of hydrogen gas will evolve and will either explode or increase the fire by mammoth proportions. In fighting a large magnesium fire, it is of the greatest importance that the firefighters use foams, carbon dioxide, or other essentially nonaqueous materials.

ALUMINUM

Aluminum is much more difficult to ignite than is magnesium because of its tendency to form a fine adherent film of oxide on the surface, which protects the metal that underlies the film. Thus, in structural fires, it is very common to find much aluminum that has melted but not burned. [Under special test conditions, ignition temperatures have been measured from 1,500°C to 1,750°C (2,822°F to 3,182°F)].[71] Residential fires are hot enough to exceed the melting point of aluminum (660°C; 1,220°F) in a majority of instances. This temperature can be reached during fire growth in the hot gas layer, directly in a flame plume, or in a post-flashover fire at floor level. Thin aluminum roofing

FIGURE 5-26A Magnesium alloy engine block from light aircraft produces a large sustained white flame. *Courtesy of David Barber, Goleta, CA.*

FIGURE 5-26B Magnesium extrusions burn with white-hot flames. *Courtesy of Lamont "Monty" McGill, Fire/Explosion Investigator.*

or sheathing may oxidize to white aluminum oxide ash, especially in commercial structures or mobile homes where aluminum sheathing is common. Even this material, however, is more likely just to melt or to blow away as ash in the updraft from the fire than to burn.

Coal

From the standpoint of its general use, coal ranks high as a fuel widely used in industrial power production and general heating purposes. It is of secondary consequence as a problem for the fire investigator. There is progressively less use of coal for heating structures than in the past, but it is of increasing industrial importance as a fuel because of new low-pollution combustion techniques. Both the mining and storage of coal are hazardous due to the combustibility of coal. For example, under proper conditions, both stored and unmined coal is subject to self-heating to the point of spontaneous ignition.[72] Moisture appears to be essential; large masses of coal and long time frames are typically needed, as well as both oxygen and moisture, to trigger self-heating.[73] The dimensions (mass) of piles of coal are critical to self-heating, with deep piles more susceptible (with content percentages dependent on type—anthracite versus bituminous).[74]

Coal is highly carbonaceous, being an organic fuel made up of approximately 80 to 90 percent carbon, 4 to 5 percent hydrogen, and 12 percent oxygen, nitrogen, sulfur, and other elements in (usually) trace quantities. The presence of these other elements suggests that coal contains a variety of compounds that can undergo pyrolysis. Coal differs from most other organic fuels in that the proportion of such pyrolyzable fuels is relatively small. The more readily pyrolyzable compounds that are present in a fuel, generally the easier it is to ignite. Coal requires considerable input of heat from an ignition source to initiate a self-sustaining, flaming combustion (in a bulk solid) because of the small quantities of pyrolyzable compounds. Coke is made by heating coal in the absence of oxygen to drive off light fractions and impurities to increase its heat content.

Explosions involving coal dust can occur in mines and in coal-handling equipment (such as conveyors) in mines, power plants, and railroad facilities. Smaller quantities and dry conditions are involved in such explosions. Reported autoignition temperatures range from 200°C to 600°C (400°F to 1120°F) depending on grade (lignite being the lowest).[75]

Dust Explosions

Any finely divided solid fuel is capable of undergoing explosive combustion if adequately mixed with air and suitably ignited. Such fuels as coal dust, metallic dusts, sawdust, flour, rice dust, cornstarch, milk powder, sugar, aspirin, plastic dust, dry copier toner, computer printer toners, and powdered sulfur have all been involved in such dust explosions.[76] Explosive limits (flammability limits) are defined in terms of units such as grams of fuel per cubic meter, which are difficult values to measure or estimate after the incident. Minimum explosive concentrations range from 30 g/m^3 for fuels like powdered aluminum, sulfur, or polypropylene; to 60 g/m^3 for coal dust; to 200 g/m^3 for PVC powder, sugar, or soy flour. The minimum concentration is also affected by particle size, with finer dusts having a lower minimum concentration.[77] Ignition energies are also particle-size dependent, with the finer dusts being the most easily ignited. As with vapor or gaseous fuels, the limits are sensitive to the energy of the ignition source—the stronger the source, the wider the explosive range. Flame speeds tend to be lower than for vapor/air explosions, and there tends to be more post-blast burning. Because there are usually layers of dust on adjacent surfaces that are stirred up as the blast wave passes, there is often a series of explosions, sometimes increasing in severity as the sequence progresses. If there is an inorganic oxidizer present as a dust suspension along with the fuel, the resulting explosion can be very energetic. One such mixture is aluminum powder and potassium perchlorate flash powder. If this mixture is suspended as a dust and ignited, the resulting deflagration can approach detonation velocities. This mixture has an extremely low ignition energy, and a major explosion can be triggered by even a small arc, friction, or even a heated surface.

Combustion Products of Solid Fuels

FLAME COLOR

The color of the flame, especially in the very early stages of a fire, can be highly significant but will rarely be remembered or recorded with accuracy, although it may be captured in videos or photos of the fire. It is clear that in most fires the flame color is not likely to have great significance, because most structural, as well as outdoor, fires involve only cellulosic material such as wood and some other organic construction materials such as asphalt, paint, and similar supplemental substances. For all these materials, the flame will be more or less yellow or orange in color, and only the amount and color of smoke that they give off will be related to the character of the fuel and the adequacy of the oxygen supply. It should also be noted that flame color, when not complicated by special elements that produce unusual colors, is related to the flame temperature, which is, in turn, generally related to the oxygen admixture with the fuel. Thus, the color of the flame is most often indicative of burning conditions as well as of the actual temperature of the flame.

However, unusual colors of flame are sometimes seen in fires. When this is the case, it would be poor judgment not to give them consideration. For example, the simpler alcohols burn with a relatively blue or purple flame. If such a flame was observed *immediately after the fire started*, it could well indicate the use of this type of material as an accelerant. It would also rule out hydrocarbon accelerants, all of which burn with a yellow flame. Natural gas (if not premixed), like liquid hydrocarbons, generally burns with a yellowish flame, tending to be blue in the periphery. If adequate air is mixed with it, as is done in a heating burner, the flame is colorless to blue. Hydrocarbon gases or vapors properly mixed with air burn with blue flames. Without the proper air mixture, they become yellow and smoky due to incomplete combustion.

One gas that is of special interest relative to flame color is carbon monoxide, which is extremely flammable. Most of it is produced by oxygen-deficient flames in which the combustion is not complete, and the gas escapes before it is consumed by further oxidation. Carbon monoxide (CO) also burns with a blue flame. In a confined space, if the amount of carbonaceous fuel involved greatly exceeds the amount of air available to combine with it, there will be strong generation of carbon monoxide. The CO may escape to a region richer in oxygen, where it may be reignited and ring the primary fire with blue flames. Such a situation will provide insight into the combustion of excess fuel with limited air in the initial fire area.

A good illustration of this property was reported by Kirk in an early edition of this text. In that case, the most intense part of a large hotel fire was in a subbasement and fed by hydrocarbons.[78] Firefighters noted blue flames coming from a sidewalk entrance to the basement while black smoke was emerging elsewhere. The blue flames were at a significant distance from the subbasement, and there was as yet little fire in the intervening space. The reasons for the existence of the bluish flame were not immediately perceived, and the only final explanation that fit the facts was a secondary burning of carbon monoxide produced in the basement fire because of limited ventilation.

Some other special flame colors may occasionally be of significance. Colored flames have long been generated for pyrotechnical purposes by adding certain metallic salts that generate the metals' spectral colors at flame temperature. Traces of sodium, present in most fuel, give a yellow flame, difficult to distinguish from the yellow produced in flames by glowing carbon particles. Some of the yellow that is so common in flames is partially due to sodium traces, but not specifically so. The significance of sodium in this connection is therefore minimal. Strontium salts give the bright red color used in many flares and decorative flame and explosion effects. Copper halides impart an intense green color to the flame, potassium salts give violet, and barium salts, a yellowish green. Brilliant white flames will accompany the burning of magnesium/alloy components (as shown in Figure 5-26). Many other variations of color will be characteristic either of the elements or of certain of their compounds.[79]

Clearly, these unusual colors of flame will only occasionally be of significance in the investigation of fires. Assume, for instance, that an intense fire burned in the interior of a motor vehicle. This fire might have been due to various types of fuel material and ignition sources, one of which could have been road flares. If a brilliant red color was noted in the early part of the fire, this information could assist in locating the early fuel as flares. Yellow flames could indicate more conventional fuels such as synthetic upholstery. The investigator who is aware of the potential value of flame color in the interpretation of fire will be alert for any color that is unusual or not explainable in the most obvious hypothesis, as this could lead to more accurate appraisal of the cause and sequence of many fires. The firefighters who were first on the scene should be interviewed to determine any unusual flame conditions on their arrival.

SMOKE PRODUCTION

Although the character of the smoke produced by a fire is variable, observation of the smoke may be helpful in obtaining an indication of the type of the fuel. When complete combustion occurs, most materials produce little or no visible smoke. Under these ideal conditions, all carbon is burned to carbon dioxide, a colorless gas. Incomplete combustion of carbon leads to carbon monoxide, also colorless and odorless but having very important toxic properties. Incomplete combustion of carbonaceous materials also produces various highly carbonaceous compounds that are opaque, such as soot, which is largely carbon.

Hydrogen, universally present in all organic materials, burns to water vapor, which is a major constituent of all combustion gases. It does not contribute to the smoke characteristics, although it may at times be recognized as steam. Other elements often present in organic materials, such as nitrogen, sulfur, and halogens, contribute gases to the combustion products. However, only under very special conditions do these alter the color of the smoke. They may contribute strongly to its odor when present in larger-than-ordinary amounts. Such effects will be encountered in some industrial fires but rarely in residential ones.

The color of smoke is almost entirely determined by the character and type of fuel and the availability of oxygen for complete combustion. Most materials, including all hydrocarbons, will never burn completely even in the presence of excess air, without premixing of the fuel with air. This is even true of natural gas when burning in large volumes.

As the hydrocarbon molecules become larger, such as those in petroleum oils, more air is required, and it becomes more difficult to completely mix the fuel with air. Thus, in an ordinary fire, materials such as petroleum distillates tend to produce a large quantity of dense black smoke. Therefore, the presence of much black smoke *may* indicate the burning of a highly carbonaceous material like the petroleum products sometimes used as accelerants. Dense black smoke, however, is usually more indicative of the amount of ventilation and the stage of development at which the fire is burning. For example, fire burning in a restricted space without adequate and complete ventilation will always form more smoke and darker smoke than when the same fire is well ventilated. The early stages of a developing structure fire may be dominated by the incomplete combustion and partial pyrolysis of fuels present. These products of incomplete combustion are often characterized by the dense clouds of dark smoke they form until flashover occurs, windows break, and the flammable vapors freely combust. At this point, the smoke becomes lighter colored and less opaque, sometimes almost transparent. If the initial fuel is cellulosic, the smoke will be light colored from the outset due to the release of water vapor and absence of soot.

Oxygenated organic materials (such as cellulosics) burning freely in the presence of air very often produce little or no colored smoke. Alcohols, for example, are in this category. If used as accelerants, they will not be detected by observing the smoke. Wood and most cellulosic building materials also fall into this category. They produce either a white or light gray smoke, except when the supply of air is greatly restricted. Even under these circumstances, they will not produce a heavy black smoke, such as follows the burning of heavier hydrocarbons. Asphaltlike materials, also common in building construction, and such materials as tar paper, some paints, sponge rubber upholstery, adhesives, sealing compounds, and most floor coverings, will generally produce a black smoke. These are more likely to burn late in a fire, so that the production of heavy black smoke late in a fire is not ordinarily an indication of the use of accelerant. Many plastics, especially polystyrenes and polyurethane, produce thick black smoke when burned in large quantities in air.

The optical density of smoke produced during oxygen-depletion calorimetry is measured by a standardized light-path method in the exhaust duct. The value obtained can be used to predict the visibility factors that affect tenability and escape paths for people in the fire room.[80] Once again, if comparison samples of suspected fuels are available, tests can be conducted to assess their contributions.

Smoke color, especially colors other than the neutral white, gray, or black, may give indications of special types of fuel being combusted if they are present in large amounts. A smoke that has an unusual hue, such as red-brown or yellow, should lead to an inquiry about the presence of materials other than conventional building materials and furnishings. This is another occasion when questioning the fire crew may reveal useful insight or even evidence.

Summary

Solid fuels account by far for most of the combustibles consumed in both structure and wildland fires. Because of pyrolysis and solid/air (smoldering) combustion, the behavior of solid fuels in fire is more complex than that of gases or liquids, which evaporate and then mix more freely with air. Those processes are, nonetheless, fairly predictable if all appropriate variables and vagaries are considered. For example, an appreciation of the mechanisms of heat transmission by radiation, convection, and flame impingement is important if the investigator is to understand the ignition and combustion of solids. Although the complete chemistry of pyrolysis is too complex to be fully examined here, its role in decomposing a solid fuel to simpler, oxidizable components must be understood. Only one element is now missing from our examination of the fire process—a source of ignition. No matter whether the fuel is solid, liquid, or gas, without ignition there will be no fire to investigate. Sources of ignition will be examined in the next chapter.

Review Questions

1. Describe the phenomenon of pyrolysis.
2. Describe four of the routes by which a solid can be converted to a vapor.
3. Describe what happens when wood burns.
4. What conditions usually accompany low-temperature ignition of wood?
5. Describe the difference between thermoplastic and thermosetting synthetic materials.
6. Why is latex foam different from synthetic foams in its combustion properties?
7. What behavior affects the combustion of thermoplastics like polypropylene, nylon, or polystyrene?
8. What metals burn readily in air?
9. What factors control the phenomenon of a dust explosion?
10. Describe the difference between cellulosic and synthetic materials. Which is more easily ignited by a glowing cigarette?

References

1. V. Babrauskas, *Ignition Handbook* (Issaquah, WA: Fire Science Publishers, 2003), 16.
2. Ibid., 9.
3. E. Stauffer, "Sources of Interference in Fire Debris Analysis" in *Fire Investigation*, ed. N. Nic Daéid (Boca Raton, FL: CRC Press, 2004).
4. J. D. DeHaan, D. Brien, and R. Large, "Volatile Organic Compounds from the Combustion of Human and Animal Tissue," *Science and Justice* 11, no. 4 (October–December 2004): 223–36.
5. D. D. Drysdale, *An Introduction to Fire Dynamics,* 2nd ed. (Chichester, UK: Wiley, 1999), 175.
6. Ibid.
7. Ibid., 179.
8. B. L. Browning, ed., *The Chemistry of Wood* (New York: Wiley-Interscience, 1951); C. R. Brown, "The Ignition Temperatures of Solid Materials," *NFPA Quarterly* 28, no. 2 (October 1934), quoted in *Fire Protection Handbook,* 17th ed. (Quincy, MA: NFPA, 1991), 3–26; H. O. Fleischer, *The Performance of Wood in Fire*, Report No. 2202 (Madison, WI: Forest Products Laboratory, 1960), quoted in *Fire Protection Handbook,* 17th ed. (Quincy, MA: NFPA, 1991), 3–26.
9. Babrauskas, *Ignition Handbook*, 944.
10. Ibid., 942–52.
11. G. C. McNaughton, *Ignition and Charring Temperatures of Wood* (U.S. Department of Agriculture, Forest Service, November 1944); see also Forest Products Laboratory Report No. 1464, *Wood Products* 50, no. 2 (1945): 21–22.
12. X. Z. Xiang and J. J. Cheng, "A Model for the Prediction of the Thermal Degradation and Ignition of Wood under Constant Variable Heat Flux" in *Proceedings Fire and Materials 2003*, San Francisco, CA (London: Interscience Communications): 71–79.
13. M. L. Janssens, cited in Babrauskas, *Ignition Handbook*, 946.
14. Xiang and Cheng, "A Model for the Prediction of the Thermal Degradation and Ignition of Wood."

15. H. W. Angell, *Ignition Temperature of Fireproofed Wood, Untreated Sound Wood, and Untreated Decayed Wood* (Forest Products Research Society, 1949).
16. Babrauskas, *Ignition Handbook*, 954.
17. Ibid., 947.
18. NFPA, *Fire Protection Handbook*, 3–26, Table 3-3B.
19. Drysdale, *An Introduction to Fire Dynamics*.
20. Babrauskas, *Ignition Handbook*, 943.
21. L. Yudong and D. D. Drysdale, "Measurement of the Ignition Temperature of Wood," First Asian Fire Safety Science Symposium, Hefei, Anhui Province, People's Republic of China, October 1992.
22. Drysdale, *An Introduction to Fire Dynamics*.
23. E. L. Shaffer, "Smoldering Initiation in Cellulosics under Prolonged Low-Level Heating," *Fire Technology* 23 (February 1980).
24. Chemical Manufacturers Association, cited in Babrauskas, *Ignition Handbook*, 18.
25. P. C. Bowes, *Self-Heating: Evaluating and Controlling the Hazards* (Amsterdam: Elsevier, 1984).
26. J. C. Martin and P. Margot, "Approche Thermodynamique de la Recherche des Causes des Incendies Inflammation du Bois I." *Kriminalistisk und forensische Wissenschaften* 82 (1994): 33–50.
27. B. R. Cuzzillo and P. J. Pagni, "The Myth of Pyrophoric Carbon," *Fire Safety Science—Proceedings of the Sixth International Symposium* (Bethesda, MD: IAFSS, 2000), 301–12; B. R. Cuzzillo and P. J. Pagni, "Low Temperature Wood Ignition, *Fire Findings* 7, no. 2 (Spring 1999).
28. V. Babrauskas, B. F. Gray, and M. L. Janssens, "Prudent Practices for the Design and Installation of Heat-Producing Devices near Wood Materials," *Fire and Materials* 31 (2007): 125–35.
29. V. Babrauskas, "Ignition of Wood: A Review of the State of the Art" in *Proceedings Interflam* 2001 (London: Interscience Communications), 71–88.
30. *Truck Insurance Exchange v. MagneTek, Inc.*, 360 F.3d 1206 (10th Cir. 2004).
31. G. R. Lightsey and R. W. Henderson, "Theoretical vs. Observed Flame Temperatures during Combustion of Wood Products," *Fire and Arson Investigator* 36 (December 1985).
32. V. Babrauskas, "Charring Rate of Wood as a Tool for Fire Investigations" in *Proceedings Interflam* 2004 (London: Interscience Communications), 1155–76.
33. J. D. DeHaan, "Compartment Fires 1996–98," *CAC News* (Spring 2001).
34. R. H. White and E. V. Nordheim, "Charring Rate of Wood for ASTM E 119 Exposure," *Fire Technology* 28 (February 1992): 5–30.
35. C. P. Butler, "Notes on Charring Rates in Wood," *Fire Research Note No FR 896* (Fire Research Station, Borehamwood, UK, 1971).
36. Babrauskas, "Charring Rate of Wood as a Tool for Fire Investigations."
37. Ibid.
38. J. Albers, "Straight Facts about the Standard Time-Temperature Curve," *American Fire Journal* (May 1998): 6–7.
39. Babrauskas, *Ignition Handbook*, 965.
40. Retrieved April 10, 2005, from www.cnr.berkeley.edu/forestry/structures/DeckbdMatls.htm; see also S. L. Quarles, L. G. Cool, and F. C. Beall, "Performance of Deck Board Materials under Simulated Wildfire Exposures," paper presented at Seventh International Conference on Woodfiber-Plastic Composites, Madison, WI, May 19–20, 2003.
41. N. M. Stark, R. H. White, and C. M. Clemons, "Heat Release Rate of Wood-Plastic Composites," *SAMPE Journal* 33, no. 5 (September–October 1997): 26–31.
42. Retrieved April 10, 2005, from www.cnr.berkeley.edu; see also Quarls, Cool, and Beall.
43. "Insulation Analysis," *Fire Findings* 8, no. 1 (Winter 2000).
44. S. H. Graf, "Ignition Temperatures of Various Papers, Woods, and Fabrics," *Oregon State College Bulletin* 26 (March 1949).
45. Babrauskas, *Ignition Handbook*, 896.
46. F. W. Mowrer, "Ignition Characteristics of Various Fire Indicators Subjected to Radiant Heat Fluxes" in *Proceedings Fire and Materials 2003* (London: Interscience Communications), 81–92.
47. J. G. Quintiere, *Principles of Fire Behavior* (Albany, NY: Delmar, 1998).
48. F. L. Fire, "Plastics and Fire Investigations," *Fire and Arson Investigator* 36 (December 1985); C. J. Hilado, *Flammability Handbook for Plastics*, 5th ed. (Lancaster, PA: Technomics, 1998); Stauffer, "Sources of Interference in Fire Debris Analysis."
49. G. Damant, California Department of Consumer Affairs, Bureau of Home Furnishings, personal communication, 1988.
50. Babrauskas, *Ignition Handbook*.
51. D. Madrzykowski et al., "Fire Spread through a Room with Polyurethane Foam Covered Walls" in *Proceedings Interflam 2004* (London: Interscience Communications), 1127–38.
52. G. Cooke, "When Are Sandwich Panels Safe in a Fire? Part 1," *Fire Engineers Journal* 58, no. 195 (July 1998); G. Cooke, "When Are Sandwich Panels Safe in a Fire? Part 2." *Fire Engineers Journal* 58, no. 196 (September 1998).
53. M. Shipp, K. Shaw, and P. Morgan, "Fire Behavior of Sandwich Panels Constructions" in *Proceedings Interflam 1999* (London: Interscience Communications), 98.
54. "Panel Perspectives," *Fire Engineering Journal and Fire Prevention* (September 2005): 44–46.
55. R. H. Holleyhead, "Ignition of Solid Materials and Furniture by Lighted Cigarettes," *Science and Justice*, 39, no. 2 (1999): 75–102.
56. W. D. Woolley and S. A. Ames, "The Explosion Risk of Stored Foam Rubber," Building Research Establishment Current Paper CP 36/75, 1975. (See also: "Fatal Mattress Store Fire at Chatham Dockyard," *Fire* 67 (1975), 388.)

57. Retrieved April 10, 2005, from www.cnr.berkeley.edu; see also Quarls, Cool, and Beall.
58. J. Albers, "Cover Story," *California Fire/Arson Investigator* (July 2008): 25.
59. J. F. Krasny, W. J. Parker, and V. Babrauskas, *Fire Behavior of Upholstered Furniture and Mattresses* (Norwich, NY: Noyes/William Andrew, 2001).
60. J. D. DeHaan and K. Bonarius, "Pyrolysis Products of Structure Fires," *Journal of Forensic Sciences* 28 (1988): 299-309.
61. J. Albers, "It's Time to Reduce Upholstered Furniture-Driven Flashover," *American Fire Journal* (July 1999): 6–8.
62. S. Nurbakhsh, California Bureau of Home Furnishings, personal communication, 2006.
63. CPSC, 16 *CFR 1633: Standard for the Flammability (Open Flame) of Mattress Sets. Federal Register* 70, no. 50 (March 15, 2006).
64. Holleyhead, "Ignition of Solid Materials and Furniture by Lighted Cigarettes."
65. J. D. DeHaan and S. Nurbakhsh, "The Combustion of Animal Carcasses and Its Implications for the Consumption of Human Bodies in Fires," *Science and Justice* (January 1999).
66. J. F Krasny, W. J., Parker, and V. Babrauskas, *Fire Behavior*; S. Sardqvist, *Initial Fires: RHR, Smoke Production and CO Generation from Single Items and Room Fire Tests* (Lund, Sweden: Lund University, April 1993).
67. P. E. Jackson, "Intumescent Materials," *Fire Engineers Journal* 58 (May 1998): 194.
68. J. M. McGraw and F. W. Mowrer, "Flammability of Painted Gypsum Wallboard Subjected to Fire Heat Fluxes." *Proceedings Interflam 1999* (London: Interscience Communications), 1320–30.
69. Babrauskas, *Ignition Handbook*, 870–74.
70. Ibid., 873.
71. Ibid., 871.
72. D. J. Hodges, *Colliery Guardian* 207, no. 678 (1963).
73. Babrauskas, *Ignition Handbook*, 719–24.
74. Ibid.
75. Ibid.
76. A. B. Spencer, "Dust: When a Nuisance Becomes Deadly," *NFPA Journal* (November–December 2008): 57–61.
77. R. G. Zalosh, "Explosion Protection." *SFPE Handbook on Fire Protection Engineering*, 2nd ed. (Boston: SFPE, 1995), 3–315.
78. P. L Kirk, *Fire Investigation* (New York: Wiley, 1969).
79. J. A. Conkling, *The Chemistry of Pyrotechnics and Explosives* (New York: Marcel Dekker, 1985), chap. 7.
80. *ASTM E 1354: Standard Test Method for Heat and Visible Smoke Release Rates for Materials and Products Using an Oxygen Consumption Calorimeter* (West Conshohocken, PA: ASTM, 1997).

6
Sources of Ignition

Courtesy of Jamie Novak, Novak Investigations and St. Paul Fire Dept.

KEY TERMS

arc, *p. 174*

piloted ignition, *p. 185*

resistance, *p. 169*

rollout, *p. 184*

self-ignition, *p. 208*

service, *p. 179*

short-circuit, *p. 179*

spark, *p. 174*

OBJECTIVES

After reading this chapter, you should be able to:

- Describe common ignition sources and their typical heat release rates.
- Recognize secondary sources of ignition.
- Identify the role of appliances in starting fires.
- Discuss the role of hot and burning fragments in kindling fires.
- Explain the role of smoking as a fire origin.

PEARSON
myfirekit™

For additional review materials, appendices, and suggested readings, visit **www.bradybooks.com** and follow the **MyBradyKit** link to register for book-specific resources.

Register for **MyFireKit** by following directions on the **MyFireKit** student access card provided with this text. If there is no card, go to **www.bradybooks.com** and follow the **MyFireKit** link to Buy Access from there.

Introduction to Ignition Sources

Ignition can be defined as the initiation of self-sustaining combustion in a fuel. The mechanisms of ignition were explored in detail in Chapters 4 and 5. As we have seen, the fundamental properties that influence an object to ignitability are its density, thermal capacity, and thermal conductivity. Taken together, these properties constitute the material's *thermal inertia*. Ignition requires transfer of enough energy quickly enough to overcome the thermal inertia and trigger sufficient combustion to become self-sufficient. If that occurs, the ignition source is said to be *competent* (for that fuel under those conditions). As we have seen from previous chapters, ignition involves bringing at least a part of the fuel to a characteristic temperature by means of conducted, convected, or radiant heat transfer until it can sustain combustion. In gaseous fuels, this process involves raising the temperature of only a small volume to ignition. In solids, it usually involves the surface of the fuel (the exception being self-heating, in which the heat is being generated within the bulk of the fuel). No matter what the nature of the fuel, without some source of ignition, that is, without some source of energy, there will be no fire. Without exception, the source is some type of hot object or mass, chemical reaction, flame, or electric current. Except for the self-heating case, the temperature of the source must exceed the ignition temperature of the fuel, and it must be able to transfer enough heat into a suitable mass of fuel before there can be ignition. In the case of self-heating, ignition will cause self-sustaining smoldering combustion, in which oxygen from the surrounding air diffuses into the surface of the charring fuel to create enough heat to advance the reaction, but smoldering combustion can, of course, be started equally well by an external heat source. Although ignition of smoldering combustion is sometimes of interest to the fire investigator, for the most part the onset of flaming combustion is the event of most concern. Flaming combustion of a solid material occurs when sufficient gases or vapors are generated from the pyrolyzing fuel to support flames.

As a general rule, the lower the energy of the ignition source, the closer the source and first fuel have to be for ignition to take place. The ignition source may be small and inconspicuous compared with the destructive fire that follows. There may be intermediate stages from initial ignition to full involvement of the first fuel package and then spread as other fuel packages in the room ignite, and this item-to-item growth must obey the same physical laws. Material that is intentionally heated by electricity, friction, chemical reaction, or a controlled flame may be dropped or blown into fuel that then ignites. These intermediate occurrences must always be taken into consideration in dealing with the source of ignition. The challenge to the fire investigator is to identify the fuel first ignited and then determine how the ignition source managed to transfer enough heat to that fuel for it to kindle into flame.

In this chapter we examine a wide variety of heat sources and their thermal properties. This text cannot discuss each source in combination with every possible fuel. The fundamental processes of heat transfer, heat release rate, and fire propagation must be applied to each possible situation. By applying the scientific method to ignition processes, the investigator must evaluate likely sources.

Primary Ignition Sources

The primary source of ignition of virtually every fire is heat. As we have seen, the heat can originate from a hot surface, a hot particle, a chemical reaction, or a distant radiant heat source, but the small open flame of a common match or lighter is among the most effective primary ignition source for fires. Without question, such sources are responsible for a large percentage of both accidental as well as deliberate fires. Other sources include mechanical sparks (incandescently hot or burning particles) or electric arcs, and the latter can range from very small to lightning bolt size. All these sources generate enough localized heat to ignite many fuels, although not all sources will ignite all fuels. Heat may also be generated by the passage of current through wires (particularly in an overload condition), through a high **resistance** connection or through an unintended conductive path. Heat is also generated by numerous chemical reactions of the exothermic type, some very slow, some very fast. To clarify the relationship of various ignition sources to various fuels, matches, lighters, torches, and candles will be considered briefly.

resistance ■ Opposition to the passage of an electrical current.

MATCHES

A very common means for initiating combustion, the match is specifically and exclusively designed for this purpose, and it is the most basic source of flame. Whether it directly initiates the destructive fire or merely sets a controlled fire that is later responsible for the larger fire is immaterial to the basic question of ignition but may be very material to the investigation of the larger fire.

The match is a stick that is treated to be readily combustible combined with a head that includes both fuel and oxidizer in a form that can be ignited by localized heat produced by friction. Matches are found in two general categories: the *strike-anywhere* match and the *safety* match. The strike-anywhere or kitchen match contains an oxidant such as potassium chlorate mixed with an oxidizable material such as sulfur or paraffin, a binder of glue or rosin, and an inert filler like silica. The tip, which is more easily ignitable by friction, contains a high percentage of both phosphorus sesquisulfide (P_4S_3) and ground glass. The head of the safety match contains an oxidizer and a fuel such as sulfur. It will ignite only when struck against the strip containing red phosphorus, glue, and an abrasive like ground glass. The stick of the match is often treated with a chemical to suppress afterglow. Paper matches are generally impregnated with paraffin to improve both water resistance and the burning characteristics of the head.

The heat output of a large kitchen match is on the order of 50 to 80 W (see Table 6-1), and the *average* flame temperature (once the incendiary tip has burned away) is on the order of 700°C to 900°C (1,300°F to 1,650°F). Because the match flame is laminar and involves wood (or cardboard) and wax, temperatures within the thin combustion zone near the outside of the plume are similar to those of a laminar candle flame and can be as high as 1,200°C to 1,400°C (2,200°F to 2,500°F).[1] Such energy and temperature are adequate for ignition of many fuels as long as that energy can be transferred into the fuel. Such a low heat output is not enough to ignite most fuels by radiant heat, so there is usually direct contact between the flame and fuel before ignition. This is true of ordinary combustible solids, but combustible gases or vapors can be drawn into the flame by entrainment flow, so there may appear to be ignition without direct contact.

U.S. and Foreign-Manufactured Matches

Minor variations in match formulation and manufacturing detail exist among U.S. manufacturers, although some matches are custom-made with special head or shaft designs for promotional purposes. Large differences in appearance (and sometimes in action) are expected only with matches of foreign origin. For instance, some foreign-made matches have been known to ignite by frictional contact between their heads.

Considerations for Fire Investigators

It is common to attempt to compare matches from a scene with suspected sources like partial matchbooks on the basis of their length, width, color, and the paper content of the cardboard stems. On some occasions, the torn bases of the paper matches can be physically compared, jigsaw fashion, to the stubs in the matchbook. Most U.S. book matches today, however, are pre-perforated, which makes them tear uniformly and thus reduces the chances for success in such comparisons. It is sometimes possible to trace the origin of partially consumed matches at a fire scene by conducting an elemental analysis on the unburned match heads.[2]

	TYPICAL OUTPUT (kW)	TYPICAL BURN TIME (s)[a]	MAXIMUM FLAME HEIGHT FLAME (mm)	MAXIMUM HEAT FLUX (kW/m²)
TABLE 6-1 Heat Release Rates and Burn Times of Common Ignition Sources				
Cigarette 1.1 g (not puffed, laid on solid surface) bone dry	0.005	1,200	—	42
Cigarette 1.1 g (not puffed, laid on solid surface) conditioned to 50% R.H.	0.005	1,200	—	35
Methenamine pill, 0.15 g	0.045	90		4
Candle (21 mm, wax)*	0.075	—	42	70[b]
Wood cribs, BS 5852 Part 2				
No. 4 crib, 8.5 g	1	190		15[c]
No. 5 crib, 17 g	1.9	200		17[c]
No. 6 crib, 60 g	2.6	190		20[c]
No. 7 crib, 126 g	6.4	350		25[c]
(Crumpled) brown lunch bag, 6 g	1.2	80		
(Crumpled) wax paper, 4.5 g (tight)	1.8	25		
Newspaper—folded double sheet, 22 g (bottom ignition)	4	100		
(Crumpled) wax paper, 4.5 g (loose)	5.3	20		
Newspaper —crumpled double sheet, 22 g (top ignition)	7.4	40		
Newspaper double sheet newspaper, 22 g (bottom ignition)	17	20		
Polyethylene wastebasket, 285 g, filled with 12 milk cartons (390 g)	50	200[d]	550	35[e]
Plastic trash bags, filled with cellulosic trash (1.2–14 kg)[f]	120–350	200[d]		
Small upholstered chair	150–250	—	—	—
Upholstered (modern foam) easy chair	350–750	—	—	—
Recliner (PU foam, synthetic upholstery)	500–1000	—	—	—
Gasoline pool on concrete (2 L) (1/m² area)	1000	30–60	—	—
Sofa	1000–3000	—	—	—

Sources: V. Babrauskas and J. Krasny, Fire Behavior of Upholstered Furniture, NBS Monograph 173 (Gaithersburg, MD: U.S. Department of Commerce, National Bureau of Standards, 1985).

*From S. E. Dillon and A. Hamins, "Ignition Propensity and Heat Flux Profiles of Candle Flames for Fire Investigation" in *Proceedings: Fire and Materials 2003* (London: Interscience Communications), 363–76.

[a]Time duration of significant flaming.

[b]Centerline—immediately above flame; <4 kW/m² outside.

[c]Measured from 25 mm away.

[d]Total burn time in excess of 1800 s.

[e]As measured on simulation burner.

[f]Results vary greatly with packing density.

LIGHTERS

Basically, there are two general types of lighters: the electric-element type found in vehicles, and liquid-fuel lighters. For most fire purposes, the electric lighter may be disregarded as a source of ignition because it depends on the battery of a vehicle to heat the wire heating element. It is not operative except with its electrical connections to a battery, which greatly restricts its use.

The basic liquid-fuel lighter ignites the flammable vapors from a fuel-soaked wick or liquid reservoir by means of a spark produced when a steel strikes a flint. Cigarette lighter "flints" are made of "misch metal," which includes a rare earth such as cesium, lanthanum, or neodymium whose sparking potential is very high.[3] Thus, a spark or hot particle ignites a fuel vapor under controlled conditions, leading to a small, predictable flame. Such lighters are a convenient substitute for a match and are, no doubt, responsible for many incendiary fires. They are rarely left behind and are therefore unlikely to be recovered as evidence. When lighters are recovered, the possible presence of latent fingerprints (especially on a removable fuel reservoir inside the case) and trace evidence should be considered. Some lighters ignite the vapors by catalytic oxidation or piezoelectric effect on a crystal, but since the flame produced is the same as from a "regular" flint lighter, their use as an ignition source remains functionally the same. The heat release from a wick-type lighter is on the same order as that of a large kitchen match (50–80 W), and maximum temperatures are similar to those reported for pool fires of hydrocarbon fuels, ≈1,000°C (1,830°F).[4]

Traditional lighters that the user refueled with a light liquid petroleum product have largely been supplanted by disposable lighters filled with butane. In the latter, the butane fuel compartment is under pressure, and the fuel is delivered through a metering valve to be ignited (in the manner described earlier) as a small, laminar jet flame. The adiabatic temperatures of such flames have been reported as 1,895°C (3,440°F), so real-world maximum temperatures might be expected on the order of 1,400°C (2,500°F). Research at the University of Maryland on cigarette lighters with port-type and Bunsen-type burner designs (adjusted to produce 75-W flames) reported temperatures reaching 2,022 K (1,749°C; 3,180°F) and peak heat fluxes as high as 169 kW/m^2 (25 mm from tip axially).[5]

Tests of a typical disposable butane lighter indicated a fuel release rate of 0.001 to 0.002 g/s and a flame less than 20 mm (0.8 in.) in height. Based on a ΔH_c of 46 kJ/g, this fuel consumption would produce an estimated 40 to 90 W.[6] Such lighters can explode if exposed to high temperatures and are subject to leakage when dropped or exposed to reduced atmospheric pressure such as in airplane cabins. The metering valve can sometimes be adjusted to release a fairly voluminous stream of flammable gas, and the resulting flame can be several inches high (with corresponding increase in heat release rate). Because these lighters are nonrefillable and intended to be disposable, they are more likely to be left at the point of ignition of a fire. Innumerable specialty lighters such as barbecue or fireplace lighters with an extended delivery tube are also available, as well as novelty lighters disguised as guns, pens, and toys.

TORCHES

Torches of a wide variety of types and fuels find uses in many industrial and residential applications. There are "traditional" blowtorches using manually pressurized "white gas" delivery. There are "drip torches" using kerosene or diesel fuel for setting controlled "back fires" in wildlands. Handheld or long-wand roofing torches use propane, butane or MAPP [methyl acetylene ($H_3C—C\equiv CH$), and propadiene ($H_2C=C=CH_2$)] as fuels. Acetylene/oxygen and acetylene/air handheld torches are very well known. Certain paint-stripping devices (heat guns) that are either propane-fueled or electrically powered can produce very high gas temperatures and high heat fluxes. Examples are shown in Figure 6-1.

FIGURE 6-1 Common hand torches. From left: propane, MAPP (with igniter), blow torch (white gas), butane microtorch. Foreground: acetylene/air hobby torch and oxy-acetylene welding torch.
Courtesy of John D. DeHaan.

Fire Findings reported that flame temperatures in a propane hand torch ranged from 332°C (629°F) at the tip to 1,333°C (2,071°F) at the flame center to 1,200°C–1,350°C (2,200°F−2,468°F) at the tip of the "inner flame" cone.[7] Any of these propane torches can cause accidental (and intentional) fires if they are applied to suitable fuels. Cellulosic materials are most susceptible to ignition by torch application especially if they are finely divided, since brief application of such a flame to a solid wood surface will cause only surface scorching and (possibly) brief glowing combustion that ceases as soon as the flame is withdrawn. Finely divided materials (dried leaves, needles, sawdusts) are very suscepti-ble to ignition, usually in two stages: smoldering followed by flame. More massive mate-rials are susceptible when the torch flame is played into narrow crevices or spaces, where radiant losses from surface ignition are minimized. In these cases, ignition can be caused within the cavity that may not be readily detectable or extinguishable from the outside. Such crevices can also retain sawdust and detritus from termite attack, making ignition even more likely. In hot torch work on roofs, torches are used to cause tar to adhere to roof structures or liquefy the asphaltic adhesive on the flexible membrane as it is laid. Under either condition, smoldering ignition triggered in crevices or in debris is then con-cealed under the hot tar or membrane. Such roof fires following several hours of con-cealed smoldering have been identified.

A special variety of small devices is a new generation of microtorches advertised as being intended for welding and brazing operations but that are primarily sold by smoke shops. These use butane as a fuel and premix it in a special torch tip that can produce extremely hot flames. (See Figure 6-1.) These microtorches are small enough to be concealed in the hand and can be ignited and left burning. Due to their ostensible application as welding devices, such microtorches normally lack any child-resistant features. The maxi-mum temperatures measured in the combustion zones of such a microtorch should be the same as for a full-size one (1,200°C–1,350°C; 2,200°F–2,470°F). Such torches are often associated with inhalation of drugs of abuse such as crack cocaine. See Chapter 14 for further discussion of their role in arson.

CANDLES

Once only a consideration of fire causation during power failures, candles have become extremely popular in many different forms and functions. The National Candle Association estimates that candles are used in 7 of 10 American homes and that candle sales increase by 15 percent each year. The USFA estimates that candles are responsible for more than 9,400 residential fires each year. The CPSC estimated an even higher number—12,800 fires causing 170 deaths and 1,200 injuries.[8] The NFPA estimated that in 2002 some 18,000 reported home fires were started by candles, causing 130 deaths, 1,350 injuries, and property losses of $333 million in the United States alone.[9] A traditional paraffin or beeswax candle produces about 50 to 80 W of heat with a laminar flame with an *average* flame temperature of 800°C to 900°C (1,470F–1,650°F), but like a match flame, a candle flame has an outer combustion zone that can have temperatures as high as 1,200°C to 1,400°C (2,200°F–2,550°F), as we saw in Chapter 2. Because of the candle's steady, prolonged flame and high temperature zone, it is possible to melt thin copper or even iron wire in a candle flame if it is held steadily in the very hot combustion zone, but the thermal conductivity of such metals conducts the heat away too quickly to permit the melting of larger-diameter wire.

The structure and dynamics of candle flames were discussed in detail in Chapter 3. Dillon and Hamins reported that a 21-mm (0.8-in.)-diameter candle had a mass loss rate of 0.09 to 0.11 g/min. With a net ΔH_c of 43 kJ/g, this produced a flame 40 mm (1.6 in.) high, with an HRR of ~65 to 75 kW. They measured the total heat flux along the vertical centerline and showed a maximum of about 70 kW/m² at the tip of the flame, dropping to ~40 kW/m² at 50 mm (2 in.) above the tip and ~20 kW/m² at 170 mm (6.7 in.) above the tip.[10] Just 13 mm (0.6 in.) from the centerline, the total heat flux at the level of the flame tip was about 27 kW/m², showing the significant amount of heat in the buoyant plume. These results confirm direct observations of the propensity for a candle flame to ignite fuels at some distance immediately above the flame but not target fuels at the side of the flame.

A well-designed candle melts at just the right rate to maintain a steady supply of wax for the wick to deliver to the flame: too little wax and the candle extinguishes for lack of fuel (as the wick burns up); too much wax and the wick gutters and drowns.[11] Ideally the wax burns completely to carbon dioxide (CO_2) and water. The combustion of carbon soot occurring in the outer high temperature zone is essentially complete; virtually no soot escapes the laminar flame. If the wick is too long and the flame becomes turbulent, soot can escape. If the wax is contaminated with oils or other organic materials, these may not burn completely, and a large quantity of soot can be released. If the wick is not symmetrically placed in the candle, the candle can fail and slump, exposing large lengths of wick. The heat release rate (and therefore the flame length) of a candle is determined by the amount of exposed wick. If charred matchsticks are dropped into a candle, they can act as supplemental wicks, dramatically increasing the size of the flame. The larger flame can impinge on adjacent fuels or heat the container in which the candle is placed (such as the metal cup of a small chafing dish or tea light candle), creating increased risk of accidental fires.[12] Candles cause fires when combustibles come into contact with the flame: loose material can move into the flame, the candle can tip over while lighted (especially with taper or column-style candles as opposed to pillar or votive styles), or the candle fails as the wick comes loose or otherwise becomes more exposed. This last situation causes a larger flame. Ignition can also occur when combustibles are included in the candle as decorations. Candles with aromatic oils are being used in large arrays for aromatherapy and then left to burn while the occupant goes to sleep. Large decorative candles with multiple wicks and burning times of several days mean prolonged risks of ignition of draperies, furniture, and clothing.[13]

Secondary Ignition Sources

For purposes of this chapter, *secondary ignition sources* include sparks and arcs, hot objects and surfaces, friction, radiant heat, and chemical reaction.

SPARKS/ARCS

spark ▪ Superheated, incandescent particle.

arc ▪ Flow of current across an air gap (or another nonconductor) between two conductors.

The definition of **spark** is ambiguous because the term can refer to one of two situations:

■ an electric **arc** of brief duration where electric current is discharging through air or another insulator
■ a tiny fragment of burning or glowing solid material moving through the air

The electric spark is not readily distinguished from the electric arc, except for duration. Babrauskas has suggested that an arc represents a stable electrical flow, whereas an electric spark represents an initial flow of electric current across a gap between conductors. The arc persists as a discharge for some considerable time interval, while an electric spark is thought of as being virtually instantaneous. It is therefore simpler to regard all such electrical phenomena as arcs of various durations and leave the term *spark* to represent a solid particle or molten droplet heated by some process to incandescence. The longer an arc persists, the more time it has to heat its immediate surroundings and to transfer heat to surrounding fuel. The energy that can be released from an electric arc can range from millijoules to millions of joules. Because the arc can persist anywhere from microseconds to hundreds of seconds, the total heat released can be within a very broad range—from a tiny, brief arc of static electricity on a household doorknob to a massive lightning strike. The capacity of arcs to ignite fuels, then, depends greatly on their duration, current flow, and the susceptibility (physical and chemical properties) of nearby fuels. A full discussion of electrical sources of ignition including arcs is found in Chapter 10.

The latter conception of a spark, that is, a tiny fragment of burning or glowing material moving through the air, has become the standard usage and will be used here. Sparks, then, can be produced by a sharp blow between two dissimilar materials such as steel and flint, by extreme frictional contact between two moving materials, by extreme heating and melting of conductors in electrical failures, or by the airborne debris of a solid-fuel fire. Such incandescently hot particles are capable of igniting fires in vapors and some solids and will be treated later in this chapter.

Mechanical sparks include burning brands or very hot embers, fragments of debris from an existing fire, droplets of metal from extremely hot combustion or electric arcing events, incandescent or burning metal from impact or mechanical friction, or soot/particulates from engines of many types. These will be considered as a special case of "hot objects" in the following section.

HOT OBJECTS/HOT SURFACES

Most hot objects are heated either by being in or close to a flame, by frictional heating, or by the flow of electric current through them. In electric appliances, ceramic or wire heating elements may be present that are designed to carry current and generate heat as a result. If combustible materials come into contact with these heating elements, they can cause fires by igniting and then spreading the fire to other fuels nearby. Such appliances, then, are not primary sources of ignition but lead to loss of control of a designed heat source. Malfunctions that allow a flow of electricity through conductive paths for which they were not intended or at rates that exceed the design of intended conductors can also lead to the production of hot surfaces. Hot objects play an important role in transmitting the heat of various chemical and physical processes to fuels, thereby spreading a fire. As we saw in Table 3-1, the visible light emitted by the surface of a hot object can give an indication of its temperature. The capacity of this object to ignite a fire depends not only on its initial temperature but also on its mass and thermal capacity, which together determine how much heat this mass contains.

The evaluation of a possible ignition of a fuel by hot object or surface is not a simple matter of noting the temperature of the surface or mass or comparing it with the listed AIT of the fuel. Ignition of any fuel will not occur unless enough heat is transferred into a sufficient mass of fuel to establish a persistent flame. The transfer of that heat from a surface depends on the nature and contour of the surface (even its roughness and cleanliness), the nature of contact, and whether the fuel can maintain the contact long enough. The AIT of a liquid or gaseous fuel, for example, varies with the test method used, the volume and shape of the test cell, and the material used for the test cell.[14] For example, clean, polished metal hot surfaces produce ignitions at lower temperatures than do similar metals that have an oxide layer. Very short contact or *residence times* do not allow enough heat to be transferred. A volatile liquid fuel dripped onto a flat, hot metal surface is likely to cool the immediate contact area by evaporation; the resulting vapors rise by convection away from the hot surface, thus reducing the residence time. The temperature of the surface, then, would have to be hundreds of degrees hotter than the AIT for that fuel for there to be ignition, and even then it would not be a guaranteed result.

Interestingly, in general for liquid fuels, the lower the flash point of the liquid, the higher the temperature of the hot surface necessary to achieve ignition.[15] Hot-surface ignition of liquid or gaseous fuels is achieved at lower temperatures if they are confined (in a tube or other enclosure). Other variables include surface type, droplet size, and airflows near the surface. When hot particles can penetrate or burrow into a solid fuel, residence time is no longer a factor, but the total amount of heat that the mass can carry is critical. If the thermal energy of the particle is too low (insufficient mass or density, inadequate thermal capacity), there may not be ignition even if the object is buried in the fuel and transfers all its heat. Hot objects that land on a rigid surface are even less likely to ignite a solid fuel due to the extra convective and radiative losses to the surrounding air.

It is tempting to measure the temperatures of hot surfaces as they occur in open air and then compare those temperatures against ignition temperatures of potential fuels (as has been reported in various recent publications, such as *Fire Findings*), but such temperatures are only a guide to the processes taking place (see later section in this chapter.) If the temperature on the surface of a glass lightbulb in the open is measured at, say, 200°C (400°F), that value reflects only the steady-state temperature of that surface achieved by the balance between the heat flux (input) striking the inside of that bulb from the hot filament and the heat losses from convective, radiative, and even conductive processes on the outside of the bulb. If the rate at which the heat is lost is changed by placing the bulb in contact with a surface or by burying it in an insulating material, the temperature of that surface will rise, often dramatically. The new temperature then poses a very different "source" for transfer of heat to a potential fuel. In short, burying even a low-wattage heat source in a deep layer of cellulose insulation poses an ignition risk even when the open-air temperature of that heat source would make it seem "safe."

As a reference point, however, knowledge of open-air temperatures of some hot surfaces is useful. The surface temperatures of halogen work-light lamps (bulbs) has been reported to be 593°C to 685°C (1,100°F to 1,266°F) (sufficient to ignite flammable vapor/air mixtures as well as solids), and the glass face of a 300-W halogen work light to range from 180°C to 215°C (357°F to 418°F).[16] The open-air surface temperatures of heating elements of electric range tops ranged from 370°C (700°F) for small burners to 590°C (1,100°F) for large burners in older appliances, and 500°C to 730°C (925°F to 1,350°F) for equivalent burners in new ranges.[17] Most modern electric ranges have a great deal of power in each cooking element, as well as high temperature. Goodson and Hardin reported wattage ratings for various electric range heating elements ranges from 1,200 to 2,500 W.[18]

As we have seen, with complex fuels such as cellulosics, the increase in temperature may bring about physical and chemical changes in the fuel that make it more susceptible to ignition than the original fuel. The heat transfer rate to the fuel surface is probably more useful than temperature as an indicator of ignitability. A table of minimum critical radiant heat fluxes is found in Appendix K. The chemical and physical nature of the fuel has to be considered, and the effects of radiant heat have to be tested as part of the hypothesis testing of the scientific method. The higher the radiant flux applied to a fuel, the lower the observed ignition temperature will usually be, dramatically decreasing the time to ignition. If the fuel is a thermoplastic, it may melt, sag away from the ignition source, or form a nonporous mass and become less likely to be ignited. Materials that char upon initial heating may become more susceptible with initial heating if the charring produces a porous residue that will support a smoldering combustion. Charring may also produce a pyrolyzed fuel that requires more energy to ignite.

FRICTION

As a source of ignition, friction is a special case of a "hot object." Friction between two moving surfaces generates heat (as in the disc brakes of an automobile, which can become extremely hot). As a source of fire, rubbing two sticks together is little used on the modern scene (except traditionally by Boy Scouts). Even they are more likely to supply a bow to spin a wooden point in a wooden depression because of the much greater speed and frictional heat that are generated. Wood is used, not primarily because it is a fuel, but because it is a very poor conductor of heat. It allows heat to be generated by friction faster than it is carried away and dissipated. With enough effort, the local heat may rise to the point of kindling a glow in fine tinder, which can then be aroused into a flame by fanning or blowing. Although this method of igniting fires has historical interest, it is of little or no practical consequence today.

However, friction in other contexts may be of great importance in igniting fires, especially in machinery. An overheated wheel bearing or "hot box" on older railroad cars—frictionally heated because of inadequate lubrication—may cause discharge of hot metal fragments or cause adjacent fuel material (e.g., cotton-waste oil wick) to catch on fire. Any bearing that does not have adequate lubrication can become hot through friction, and contact of the hot object with a readily ignited fuel can lead to a fire. This lack of lubrication is without a doubt one of the relatively common sources of fire where machinery is in use. Conveyor or power transmission belts can jam or be forced to run against frozen rollers and cause extreme frictional heating. Once ignited, the moving belt can spread fire very rapidly.

If frictional heat becomes extreme, one or both of the surfaces may disintegrate, producing a shower of incandescent particles. Such disintegration may be the end result of massive overheating or very localized surface heating where the temperature of the mass itself may not rise appreciably. The temperatures of hot particles generated by this process are limited by the melting temperatures of the material involved. For instance, a mechanical spark generated from a copper-nickel alloy surface may have a maximum temperature of 300°C (570°F) (and not be incandescent at all), while a fragment of tool steel may be at 1,400°C (2,550°F) and be bright sparkling white.[19] This situation can occur with any moving machinery. Bearings in conveyor systems, for instance, can clog with abrasive dust or seize from lack of lubrication and create large numbers of incandescent sparks for long periods of time. Bearings in automotive turbochargers can fail in seconds if lubrication is interrupted, due to their very high rotational speeds (up to 25,000 rpm).

Not only bearings give rise to dangerous temperatures from excessive friction. High-speed rotors, such as those found in airplane motors, that come into contact with the housing have been known to produce enough heat locally to melt the housing and ignite related hardware. Fragments of overheated brake shoes in malfunctioning train or truck brakes can be thrown from the moving vehicle for miles before being detected. Mufflers, tailpipes, tailgates, chains, and other equipment dragging on the road surface can generate substantial showers of incandescent metal particles with enough heat to ignite roadside fuel (dried grass, leaves, or litter). Machining operations (milling machines, lathes) can generate showers of hot particles that can ignite waste in or under the tool. Mechanical sparks (hot fragments) are discussed in more detail in a later section of this chapter.

RADIANT HEAT

Although the heat radiated from a fire may at times ignite other fires at a distance, such ignition is not primary but secondary to an already large, established fire. Radiant heat plays a great role in spreading most fires and was examined in Chapter 3. As a primary ignition source, radiant heat alone is much less frequently encountered. Radiant heat from fireplaces, stoves, and heaters has been seen to bring *nearby* cellulosics to their ignition temperatures and thus start fires. When such devices are permanent fixtures, their remains will usually still be detectable after the fire. Other sources may not be so easily identifiable. The investigator must remember that direct physical contact between the heat source and the first fuel is not always necessary when sufficient radiant heat can be brought to bear on the fuel surface. Here the reflective and absorptive qualities of the fuel are critical, as are its density and thermal conductivity. All that is required is for the target fuel to absorb more heat than it can dissipate and for that heat to raise the local (surface) temperature above the fuel's spontaneous ignition temperature. Rays from direct sunlight (at a typical heat flux of 1 kW/m^2) are not intense enough to ignite common fuels, but if they are concentrated or focused by a transparent object that is round in cross section (i.e., spherical or cylindrical) or by a concave reflective surface (such as a shaving mirror or the polished metal bottom of some aerosol cans), they can reach 10 to 20 kW/m^2 at the focal point of the light path. If a cellulosic fuel is located at or near that focal point, the fuel can be heated to its ignition temperature and catch fire. Two examples of ignition by sunlight focused by shaving mirrors are shown in Figures 6-2 and 6-3. This mechanism is often blamed for starting wildland fires, but experiments by DeHaan revealed that the conditions have to be just right: the object has to produce a sharply focused area of sunlight, and the dry fuel has to be located precisely at the focal point of that lens or mirror before ignition will take place. Empty glass bottles, broken glass, or flat reflectors do not have a measurable focal length and will *not* cause ignition by this means.

The maximum heat fluxes of some common small flaming ignition sources (at a distance of 25 mm; 1 in.) were shown in Table 6-1. Some tools and appliances, even when operating as designed, can act as nonflaming ignition sources. Radiant heat from floodlights

FIGURE 6-2 Sunlight focused on cloth target by shaving mirror can reach heat flux sufficient to ignite fabrics. *Courtesy of Jamie Novak, Novak Investigations and St. Paul Fire Dept.*

FIGURE 6-3A Shaving mirror (foreground) focused sunlight onto corner of box above it in parked car. *Courtesy of Scott R. Schuett, Lebanon, CT.*

FIGURE 6-3B Shaving mirror (on left) focused sunlight from window onto towels, which ignited on rack. *Courtesy of Curt Hawk, Fire Cause Analysis.*

(particularly quartz halogen), projection lamps, heat lamps, and flameless torches can be enough to ignite cellulosic fuels at close range.[20] A noncombustible covering (sheet metal, tile, asbestos, or plasterboard, for example) can allow heat to be transferred to a fuel surface beneath while impeding the loss of heat by convection, thus making it susceptible to ignition even when concealed. Hot surfaces associated with vehicles such as brake rotors, exhaust manifolds, and catalytic converters have the potential to ignite fuels by both radiant and conducted heat and will be discussed in Chapter 9.

Radiant heat is also linked to the ignition of wood even when the heat flux is not enough to create ignition temperatures directly. In installations of heaters, furnaces, boilers, and even chimneys and vent flues, radiant heat at low levels may induce degradation of the wood to charcoal. Charring of wood can occur when temperatures on the order of 105°C (230°F) or greater are maintained for very long periods of time. This mechanism was discussed in Chapter 5 and will be discussed in greater detail in a later section dealing with furnaces.

CHEMICAL REACTION

A number of chemical mixtures are capable of great heat generation and even formation of flame. They are only of occasional consequence in ordinary investigation of fires. Some accidental fires in warehouse-style home-supply stores have recently been linked to leaks or spills of corrosives (acids or bases) that came into contact with metals, or strong oxidizers (swimming pool chlorines) that came into contact with fuels such as automobile brake fluids and underwent an exothermic reaction. They play a role in a limited number of incendiary fires because of the lack of knowledge about them on the part of the arsonists. Such reactions are more likely to be encountered in industrial or clandestine drug lab fires than elsewhere, and they require the attention of the chemical specialist. Salvage and recycling operations may produce mixtures of chemicals that spontaneously combust on contact, producing dangerous, large fires.[21] Many chemical fires are due to ignition or decomposition of hazardous chemicals. The properties of the more common hazardous materials are examined in detail in Chapter 13.

There are, of course, biochemical reactions that generate heat—drying oils, decomposing cellulosic and organic materials, and the like—that can contribute to ignition. They are discussed later in this chapter under the heading "Spontaneous Combustion."

The Role of Services and Appliances as Ignition Sources

So many fires are started either by **service** facilities or appliances, or are attributed to these origins, that consideration of such possible origins is of great importance. When the origin of a fire is not determined, it is very convenient to decide that it was due to some malfunction of an appliance, a **short-circuit** in electrical wiring, or other similar cause. This decision is especially attractive because, in many fires, there is so much damage to wiring, or melting of appliances, that such an opinion can have some credibility. In addition, it is indisputable that malfunctions or failures of services and appliances do cause many fires. Thus, investigators may find themselves on the horns of a dilemma in the attempt to separate the real from the claimed origins of fires at such times. In modern building construction, electrical wiring is widely distributed as an integral part of nearly all parts of the structure, and gas lines and appliances are also very common. In all such structures, it is the rare fire indeed that does not engulf some such items to some extent at least, thus possibly focusing attention on them.

The investigator must remember that the mere presence of an electrical wire or a gas line does *not* constitute proof of responsibility for igniting a fire. If an electrical wire is not energized (that is, no voltage applied to it), it cannot provide energy to ignite any fuel. Even if it is energized, there has to be a mechanism, a reason, for it to create heat in an unintended manner. It is the investigator's responsibility to establish this mechanism and demonstrate how it can create enough heat to ignite the first fuel. Gas lines, even when charged or pressurized with gas, do not constitute an ignition source (except under the most exceptional circumstances when they can conduct misdirected electrical energy) but are merely a source of fuel.

service ■ The conductors and equipment for delivering electricity from the supply system to the equipment of the premises served.

short-circuit ■ Direct contact between a current-carrying conductor and another conductor.

GAS APPLIANCES AS IGNITION SOURCES

Gas appliances used for nonindustrial purposes are relatively limited to a variety of heating devices: stoves, ranges, room heaters, furnaces, clothes dryers, and water heaters. All these items have open flame burners that create heat. Thus, regardless of the exact type of heating device, the investigations, as well as all the possible hazards that attend the use of the device, are very much the same. The heat output rating of the appliance, whether the flame was correctly adjusted, how the flame could come into contact with combustibles, and whether the automatic control mechanisms are functioning, are the features that are of interest to the fire investigator.

One of the few ways in which the gas furnace or dryer in normal operation is likely to initiate a fire results from the simultaneous failure of the thermostat and the high-level or "high limit" control. The high-level device is a second thermostat or *thermal cutoff* (TCO) located in the circulating-air ducting of a furnace, for example. It is usually coupled in series with the regular thermostat and is adjusted to break the circuit (and close a valve on the gas delivery line) at a maximum temperature above which the furnace cannot be safely operated. This temperature varies but is usually above 90°C (200°F). A simultaneous failure of both thermostats allows the furnace to continue operating indefinitely with great production of heat and possibly a resulting fire.

The failure of pressure regulators (as mentioned in Chapter 4) can cause an appliance to become an ignition source. Although such regulators may or may not be part of the appliance, their failure can cause masses of flame to escape from the burner compartment of the stove, furnace, or water heater affected. These flames can then ignite nearby combustibles by radiant heat or by direct flame impingement. Even under normal operation, the burners create considerable heat. Some appliances may not have adequate standoff or may be installed improperly, thereby exposing combustible walls or floors to radiant or conducted heat (as illustrated in Figures 6-4a–c). Inadequate air supplies can result in

FIGURE 6-4A Fire in water heater closet as found (unit less than 1 year old). *Courtesy of Battalion Chief Kurt Hubele, Richland, (WA) FD.*

FIGURE 6-4B After cleanup—no floor left in closet. *Courtesy of Battalion Chief Kurt Hubele, Richland, (WA) FD.*

FIGURE 6-4C Exemplar heater in next apartment showed heater installed on plywood floor with no stand-off legs. *Courtesy of Battalion Chief Kurt Hubele, Richland (WA) FD.*

excessive soot formation in the combustion chamber. This hot soot can then induce smoldering ignition of nearby combustible surfaces.

The "standing" pilot flame that was once standard on gas water heaters (and furnaces, ranges, and ovens) and offered a small but continuous and potentially competent ignition source, has largely been replaced by electric igniters. These igniters may be in the form of a high-voltage arc, an electric spark created by capacitive discharge or piezoelectric effect, or by a glow bar (electrically heated hot surface). The investigator should examine all gas appliances carefully before concluding their status as an ignition source.[22]

Although gas appliances do start fires, they are probably less often the cause of a fire than is generally assumed. Gas is the chief fuel in many places, and its intrinsic hazards tend to encourage an abnormal fear of it as a source of fire. Gas can serve as a convenient "probable cause" when the "poor" investigator finds no other that is provable; this approach is *not* recommended. Occasionally, gas is released as a result of a failure within the electrical system of the appliance and then ignited as it is released.[23] The investigation of any fire involving fuel gases should include examination and documentation of the condition of the fuel supply, the ventilation available to the flame, electrical control system(s), and the system for removal of products of combustion (vents, flues, and chimneys).

A common residential gas water heater (as diagrammed in Figure 6-5) produces a fire of 24,000 to 36,000 Btu/hr (6.66 to 10 Btu/s, 7 to 10.5 kW). A steel baffle is inserted in the central flue to slow the exhaust gases and give them more time to transfer heat to the water. There is typically a draft hood at the top of the heater to entrain room air into the exhaust gases and cool them. The vent is usually single-walled metal except where it passes through walls or ceilings, where it is at least double-walled, as shown in Figure 6-6. The flue-gas temperature at the top of the tank for steady-state operation of a 30,000 Btu/hr residential water heater is on the order of 300°C to 315°C (569°F to 601°F) with

Typical Temperatures during Normal (Steady-State) Operation

Center of flue (top of tank)
 With baffle 300°C–315°C (569°F–600°F)
 Without baffle 565°C (1,050°F)
Exterior of vent connector 73°C (164°F)
Top of 4-ft vent stack (gas temp.) 183°C–205°C
 (362°F–400°F)
 Without baffle 400°C (750°F)
Water 60°C–80°C (140°F–175°F)
Top exterior of tank 36°C (96°F)
Underside of front 63°C (145°F)

FIGURE 6-5 Cross section of typical residential gas water heater and some typical temperatures developed. *Some data courtesy* Fire Findings. *Additional data from tests by author (unpublished).*

FIGURE 6-6 Typical double-wall heater vents. Exterior—galvanized steel. Inner—steel or aluminum. *Courtesy of John D. DeHaan.*

the baffle in place. If the baffle is left out, the gas temperature rises to 565°C (1,050°F). The draft hood allows entrainment of room air, reducing the gas temperature in the flue (with baffle) to around 200°C to 250°C (400°F to 450°F).[24]

On-Demand Heaters

Energy conservation issues are prompting the installation of "on-demand" or "tankless" water heaters instead of traditional 30- to 50-gal reservoir units. These units can be gas or electrically fueled. They are typically wall-mounted metal boxes approximately 0.9 m (36 in.) high, 0.5 m (20 in.) wide, and 0.3 m (12 in.) deep. They may be concealed in closets or service areas. They provide hot water only when they sense water flow. A gas-fueled burner provides 120,000–200,000 Btu/hr (31–57 kW) of heat to heat 4–8 gal of water per minute. Due to their very high heat outputs, on-demand units have to be properly installed and vented. Most gas units have external electrical connections for the ignition and control devices.[25]

Fire Cause Considerations

When a gas appliance starts a fire, the fire pattern itself will often, if not usually, reveal the fact. Associated with the pattern showing the origin to be behind, around, or over the gas appliance there must invariably be some malfunction, improper installation, or other detectable fault or defect. Only the combination of factors can be considered as proof of origin from a gas appliance. The reason for the double requirement is relatively obvious. The arsonist often will set the fire under or close to a gas appliance to make it appear that the appliance was responsible for the fire. In such cases, liquid flammables are likely to be poured around the appliance and the fire started so that the origin seems to be obviously associated with the appliance. An incorrect assessment can result if care is not exercised in the scene examination. In such an instance, such ignitable liquids are likely to penetrate below the appliance, and possibly below the floor, and produce the characteristic charring and damage in regions a gas fire could not reach. Natural gas fires are unlikely to burn much below their point of release, since the fuel is significantly lighter than the surrounding air. Fuel oil and the vapors from flammable liquids will settle downward. Propane released under pressure (i.e., from a delivery line or as a jet) mixes with air sufficiently that it will neither rise nor settle.

Common sense to the contrary, people insist on storing flammable liquids in furnace or water heater closets. Leaks or spills from stored containers create vapors that are quickly drawn into the combustion chamber of the appliance (by the convection of either

FIGURE 6-7 New FVIR water heater with closed controls and fuel supply. Note flash suppressor mesh in air intake. *Courtesy of Jim Albers, Santa Ana Fire Dept. (retired).*

the pilot flame or the main burner). If there is sufficient vapor mixed with the air, an explosion (usually followed by sustained fire of the liquid) will result.

Recent changes in water heater design have been aimed at reducing the chances of ignition of flammable liquid vapors (called FVIR—flammable vapor ignition resistant—designs) (see Figure 6-7). For many years, building codes have required gas water heaters to be at least 18 in. (45.7 cm) above floor level if they are placed in a garage (where accidental releases of flammable liquids are likely). Testing has revealed that unless the liquid is poured near the heater and the room tightly sealed (to prevent losses by advective flow under doors), this elevation precaution will prevent ignition. If spillage of flammable liquids is a possibility even inside homes, to prevent ignition one design places a standard water heater into a steel bucket (or "weir") with sides 18 in. high. Testing has revealed this design to be highly efficient in preventing ignitions.[26] Another new design employs a flash suppressor screen on the combustion chamber and sealed service, electrical, and plumbing entry ports. This design will prevent any ignition (of vapors or fugitive gas) starting in the combustion chamber from propagating into the room.[27]

Insulation

All approved gas appliances are surrounded by some type of insulation. In circulating heaters, this insulation consists merely of an outer wall separated from the wall of the firebox, through which circulates the cool air that is to be heated. Such a wall will not become significantly hotter than the ambient air that circulates behind it if air circulation is maintained both within the heat exchanger and outside the appliance. Presence of combustible trash against such an appliance wall will not normally lead to ignition of the trash. The same cannot be said of the wall of the combustion chamber, which may well be hot enough to ignite material against it. However, something would come into direct contact with it only under very unusual circumstances.

Shifting or even partial collapse of such air-space insulation may create hot spots on the exterior, as will settling of packed insulation. These hot spots then create a higher ignition risk. Improper clearances between vent stacks (flues) and wood framing can result in ignition. The absence of the draft hood or central baffle in a gas water heater will cause higher-than-design (normal) gas temperatures on the outside of the vent, which can trigger degradation and ignition of nearby wood.

Backfire or Rollout

A common claim that there has been a backfire from a furnace or other heating appliance carries somewhat more credibility than an ignition of combustibles against the heater. Nevertheless, a backfire is an extremely rare happening that can result only from gross maladjustment or malfunction within the appliance. To obtain such a "backfire," it is necessary for the gas to accumulate in some quantity in the combustion chamber before it is ignited. Such accumulation can occur as a result of overpressure in the delivery system, use of the wrong fuel delivery orifice, or blockage of the heat exchanger or exhaust vent.[28] On ignition, a small explosion occurs and flames may blow out of the open front of the appliance. If there is a suitable combustible fuel in the path of this small amount of flame, a fire could result. The combination of factors is so unlikely, however, that any claim of this type must be very carefully scrutinized before being accepted. With a pilot light, it is difficult to accumulate sufficient gas for any significant backfire, although maladjustment of some type may be effective. In any event, so gross a defect is not difficult to recognize. In the absence of a pilot light, the gas will not ignite until it diffuses or spreads to some other source of ignition, which in all probability will require enough gas to produce an explosion. In addition, escaped natural gas will often pass harmlessly up the vent pipe into a chimney. It is difficult to imagine any reasonably normal situation in which fires are caused by this type of situation. The common use of electric ignition systems on gas furnaces, stoves, and water heaters is replacing a continuous ignition source—the pilot light—with an intermittent source that does not function until demanded. This system could allow for the buildup of a volume of gas/air mixture with explosive potential instead of providing a continuous ignition source to burn off small quantities as they reach their lower flammable limit. With a fuel oil–fired furnace, the motor-driven pump can produce a sustained fire outside the furnace housing if the exhaust or heat exchanger is blocked. The failure of a fuel gas pressure regulator (either in the appliance or external in the delivery system) can cause continuous massive "overfueling" that results in the exit of burner flames via appliance service doors or vent openings (**rollout**)."

rollout ■ Exiting of burner flames in an appliance due to massive overfueling.

Direct Contact

Another manner in which a gas appliance can start a fire involves direct contact of a normal gas flame with flammable or combustible material. In general, this will occur only with open flames, such as gas range tops or broilers, although flammable materials can sometimes be drawn into a protected flame. The electric elements of stoves or broilers can also ignite solid fuels that come into direct contact, or the contents of skillets or pans.

Plugged Vent

It has been reported that if the vent of a gas water heater or furnace is plugged, the lack of a chimney can induce sufficient back pressure that the flames are forced out of the burner enclosure, possibly into contact with nearby combustibles.[29] Examination of such devices should include careful examination of the exhaust duct for the remains of bird nests, dead leaves, construction debris, and the like. Blockage of the vent stack alone should cause the hot gases to vent from the draft hood if it is properly installed. The investigator should also check to see that the lower flame shield (under the burner) has not been removed and that the service doors are in their proper locations. (Careless owners often neglect to replace the service doors after igniting the pilot light and should be interviewed to determine the last time the equipment was serviced or lighted.)

Open Flames

Open flames are most common on kitchen ranges, laboratory and industrial burners, and special types of heating appliances and are more hazardous as sources of fire. A common example is the ignition of cooking oils and greases in the kitchen, as well as towels, clothing, curtains, paper, and the like near open flames. Fires started by stoves and ovens are the leading cause of residential fires.[30] Autoignition temperatures of lard or vegetable cooking oils are quite high [over 350°C (662°F)], but enough boiling and splattering oils can splash onto the burner below to cause **piloted ignition** at lower temperatures. (The **piloted ignition temperature** is the minimum temperature at which a fuel will sustain a flame when exposed to a pilot source.)

Skillet fires can be very large (see Figure 10-40). Grease ignitions are especially common as a cause of fires in restaurants, where the amount of cooking is great, and time spent cleaning stove hoods may be minimal. Such fires may also spread upward in grease-coated exhaust ducts, where fans and vents carry the fire up with increasing magnitude. Most building codes require automatic fire protection of the cooking area and lower hood. Such systems can do little if the flame is established in the exhaust ducting above the hood.

The ignition temperatures and flash points of cooking oils are listed in Table 6-2. Note that the fire points are well above the "normal" cooking oil maximum temperature of 205°C (400°F).[31] This suggests that unattended cooking is a major contributing factor. Tests observed by one of the authors of 500 ml (20 fl oz) of corn oil in a skillet on a gas stove revealed that heavy white smoke with a marked lachrymatory property was generated after ~10 minutes, and autoignition occurred about 25 minutes after the start of the test. The flames were between 0.5 and 0.7 m (19 and 28 in.) high and caused ignition of the cabinets above and to the nearest side of the stove.[32] Note in Table 6-2 that the fire point, AIT, and hot-surface ignition temperatures of all cooking oils decrease with use.

Small gas appliances that are not part of a permanent installation can be displaced or upset, with occasional drastic results. Laboratory burners are especially subject to being tipped over. They can start fires by playing the flame on top of a wooden table, for example, or against some other combustible material. Similar considerations may hold for small portable gas heaters sometimes used in homes and similar areas. These heaters may be permanently installed or temporary and may not meet any standard of safety or building codes. The investigator is referred to *NFPA 54: National Fuel Gas Code* or *NFPA 58: Liquefied Petroleum Gas Code*, which give detailed instructions for

piloted ignition ■ Ignition aided by the presence of a separate external ignition source such as a flame or electric arc.

piloted ignition temperature ■ The minimum temperature at which a fuel will sustain a flame when exposed to a pilot source.

TABLE 6-2	Properties of Cooking Fats Measured at Fire Research Station							
FAT	**FLASH POINT (°C)**		**FIRE POINT (°C)**		**AIT (°C)**		**HOT-SURFACE IGNITION TEMP. (°C)**	
	VIRGIN	**AFTER 8 HEATING CYCLES**	**VIRGIN**	**AFTER 8 HEATING CYCLES**	**VIRGIN**	**AFTER 8 HEATING CYCLES**	**VIRGIN**	**AFTER 8 HEATING CYCLES**
Corn oil	254	227	NA	321	309	283	526	542
Drippings (bacon)	254	241	NA	331	348	276	553	537
Hydrogenated cooking fat	260	210	NA	331	355	273	568	554
Lard	249	218	NA	326	355	282	541	568
Olive oil	234	218	NA	316	340	280	562	543
Peanut oil	260	243	347	335	342	280	552	535

From *Ignition Handbook* by V. Babrauskas, © 2003, p. 886, Fire Science Publishers. Used by permission.

the proper installation and operation of all types of gas-burning appliances. *Fire Findings* reported the air venting from a 60,000 Btu/hr (17 kW) kerosene-fueled salamander portable heater had a temperature of 639°C (1,182°F). Due to turbulent mixing, the temperature of the maximum hot-air stream dropped to 238°C (460°F) at 0.3 m (1 ft) and 148°C (298°F) at 0.6 m (2 ft) from the heater.[33]

Altered Gas Nozzle

An unlikely, but possible, type of fire causation from a gas appliance involves some alteration in the gas nozzle in the venturi, possibly by partial or total unscrewing of the nozzle or installation of the incorrect fuel nozzle. If this should happen, much larger than normal quantities of gas would reach the flame, which, in turn, might become much bigger and spill from the top or front of the appliance. The authors have not witnessed such an accident, but it is conceivable, especially when tampering or ignorant efforts at repair or adjustment have occurred. Such fires would necessarily occur at the first use of the appliance after the opening was enlarged, which should again make their detection and proof of cause relatively simple. The service and repair history of the appliance may reveal recent problems or repairs. As discussed in Chapter 4, the venturi used must match the fuel gas and pressure regulator being used. Often, X-rays will reveal internal failures, blockages, or the open/closed position of fire-damaged valves and regulators.

PORTABLE ELECTRIC APPLIANCES

Heat-producing electric appliances can, of course, cause ignition of susceptible fuels even when operating normally. (Arcing, failures, and malfunctions will be discussed in Chapter 10.) Electric heaters, toasters, or toaster ovens with exposed Nichrome (resistance) heater wires offer an ignition source with surface temperatures on the order of 600°C (1150°F) or greater. Such hot surfaces are not able to ignite most flammable gases or vapors but can ignite most cellulosics (including foods) if they come into contact or near contact with them (see Figure 6-8).

FIGURE 6-8 Toaster pastries ignited by repeated operation of toaster cause flames 30 cm (~12 in.) high sufficient to ignite nearby combustibles. *Courtesy of Jamie Novak, Novak Investigations and St. Paul Fire Dept.*

The heater elements in clothes dryers are especially susceptible to such contact. Dryer lint is one of the most readily ignitable solids in the average residence. It is usually a mixture of fine cellulosic, synthetic, and natural fibers (wool and hair). Accumulations of lint on heater grids or on adjacent surfaces of venting (both inside the dryer and out) are one of the most common "first fuels" in dryer fires.

The radiant heat produced by most portable electric heaters is on the order of 10 kW/m^2 or less at the wire guard. This may be enough to ignite cellulosics or char elastomeric and thermosetting materials if contact is prolonged. Cellulosic fuels that are brought into contact with the heater element itself are likely to be ignited quickly. Today, most consumer-type portable electric heaters have "tilt" switches to shut power off if the unit is knocked over, in addition to thermostats and TCOs. Investigators must carefully examine such heaters to determine what, if any, safety devices are present and if they have failed or been tampered with.

KEROSENE HEATERS

The fuel in kerosene heaters may be delivered to a constant flame by vertical transport up a wick that is partially inserted in a fuel reservoir, in the same way as in a kerosene lamp. A more effective means of fuel delivery is a barometric delivery system in which the fuel is delivered to a shallow horizontal reservoir from a vertically oriented removable tank. The flow of fuel is controlled by a barometric valve in the tank. A partial vacuum above the fuel in the tank prevents the fuel from flowing into the reservoir until the level drops below that preset by the valve. Modern heaters have a shutoff mechanism that extinguishes the wick if the heater is tipped over or subjected to excessive vibration (movement). The main safety problem with kerosene heaters is that the capacity of the tank exceeds that of the reservoir, so that if there is a loss of the vacuum in the tank, the excess fuel can flood the reservoir and create a spill external to the heater, which then ignites with catastrophic result.[34] This loss of vacuum can occur as the result of a leak in the tank or its associated parts, or as the result of fueling the heater with a fuel such as gasoline or camping fuel with a higher vapor pressure than that of kerosene. In these cases, the vapor then replaces the partial vacuum that holds back the flow of fuel, resulting in a flood of fuel that overwhelms the reservoir. Although a newer anti-flare-up device is available, most older heaters encountered pose a serious fire risk from this type of failure.[35] Accidental spills during refilling can cause significant fires (especially if the burner is not extinguished). Any residual fuel left in such a heater should be collected and preserved for testing to establish its actual content.

STOVES AND HEATERS

A variety of liquid fuel portable heaters are also made for camping and boating use. These range from pressurized fuel delivery systems (Primus stoves) to denatured alcohol fuel burning on the surface of a noncombustible wick element. These heaters are subject to spillage of burning fuel if tilted or overturned. Although these heaters are not intended for use in enclosed spaces, they are often seen in tents and campers. Because such heaters may consist of a small circular burner pan and a perforated metal shield over it, they may not be recognized by an investigator as ignition sources (see Figure 6-9).

OIL STORAGE

External heating (fuel) oil storage tanks are common in both residential and commercial premises. Traditionally, these are welded steel tanks with rigid metal pipe connections. Those features, combined with the low vapor pressure (high flash point) of typical heating fuels, mean they do not constitute a high risk of ignition (either accidental or intentional). However, when such fuels are absorbed on porous materials that can act as a wick, their low ignition temperatures (~205°C; 400°F) pose a risk. If plumbing connections are

FIGURE 6-9 Alcohol-fueled units sold in camping stores can be used as tent heaters or cook stoves. *Courtesy of John D. DeHaan.*

exposed to an external fire, they can fail quickly. Some jurisdictions allow fuel storage tanks to be made of high-density polyethylene (owing to their corrosion resistance and lower costs). Testing has shown that such tanks (in 300- to 400-gal sizes) can be caused to fail by external fires. If 5 L(1.5 gal) of gasoline is poured over the outside, or a modest cardboard/newspaper fire is ignited underneath the tank, a catastrophic failure and release of the tank contents can result. Placement of a cloth wick into the mouth of the tank also causes failure, since it continues to burn within the tank (due to the low vapor pressure) until the tank melts. The melted plastic of the tank can support a large pool fire of molten polyethylene, so even an empty tank poses a fire risk.[36] Oil-fueled furnaces can cause fires by creating heavy soot if burners are maladjusted, or more commonly, leaks from fuel connections can support flames outside the combustion chamber, which then ignite nearby combustibles.

ELECTRICITY

Electricity represents a source of ignition in the form of arcs, electrically heated wires and heating elements, and even lightning. For a full discussion of all means of electrical fire causation, see Chapter 10.

The Role of Hot and Burning Fragments in Igniting Fires

Hot or burning particles, commonly termed *sparks*, are in a special category with regard to kindling of fires. As pointed out earlier in the chapter, the designation *spark* is an ambiguous one, since it can refer to incandescently hot metal from a mechanical or electrical source or to hot bits of debris from an established fire. Most fragments from an established fire are glowing, although they may at times be flaming. Bits of wood, paper, and other light organic fuels are most susceptible to the formation of glowing fragments, and larger pieces of paper may travel while still flaming. Bits of paper or tobacco may become detached during lighting of a cigarette or pipe. These particles naturally have the propensity to ignite other fuels with which they come in contact if they retain enough heat. It should be remembered that most such fragments will be carried upward by the draft from the fire that generates them, and air currents may carry them for significant distances to ignite other fuel.

WINDBLOWN SPARKS

The question frequently arises: How far can windblown sparks travel and still cause a secondary fire? Although no precise answer is possible, some factors can be stated. The ambient wind is highly important, because such sparks tend only to rise with the buoyant flow over the fire, drift a short distance, and fall nearby, unless propelled by wind. Another factor is the type of material that is burning. Very small fragments and thin materials such as paper will normally be totally consumed before they return to some combustible surface. Fragments of wood, excelsior, or corrugated cardboard will burn considerably longer and travel farther before igniting something else. For the latter, distances of about 12 m (40 ft) have been observed. Tests by DeHaan demonstrated that burning wood chips and straw could be carried up to 15 m (50 ft) from their source in a 32-km/hr (20-mph) wind to ignite cloth, straw, and wood shavings targets.[37] Light items will probably only rarely travel as much as 6 m (20 ft), except under significant wind conditions. Another factor is the height that they reach before being blown to one side. The higher they go, the more time they have to burn up totally or to cool off, and the longer their exposure to any air movement. In any event, distances in excess of 9 to 12 m (30 to 40 ft) are to be accepted only with the greatest caution. Documentation of the intensity and direction of wind at the time of the fire is critical to the evaluation of such mechanisms.

Throughout this discussion of spark travel, the role of wind has repeatedly been referred to as modifying the fire spread. No investigation involving the distance or effectiveness of such spark ignition is complete without ascertaining from meteorological records, or at least from eyewitnesses, the wind conditions at the time in question. It is evident that violent winds will produce very different effects from mild ones. However, the difference is not that which might be supposed. A strong wind will blow a spark farther than a weak wind but will also speed the rate of combustion. With a given size of burning fragment, the wind effect does not vary greatly with different wind velocities. The most significant effect of the strong wind, by far, is its ability to blow large fragments of material, which will naturally burn longer than small ones. A wad of burning newspaper on the ground will not ordinarily travel along the ground for any significant distance if propelled only by a breeze. In a strong wind, however, it may roll for a considerable distance before it is consumed. The relation of importance is the time necessary for it to burn as compared with the speed with which it is moving laterally with the wind. The consideration is far more relevant to materials at ground level than to those that emerge from chimneys, because very large burning objects are rarely carried up a chimney to any significant height. In very large fires, however, the updraft can loft large pieces of burning debris, boards, branches, cardboard, and rags high enough that they can be carried hundreds of yards by the prevailing wind, igniting remote fires. This has been observed to be a major means of fire spread in large fires in both urban and wildland settings, as depicted in Figures 6-10a and b.

In one recent major conflagration, burning wood fire brands (some as large as 2 × 4s) were lofted some 60 m (200 ft) above a multistory condo/retail complex under construction and were blown more than 680 m (0.4 mi) to ignite wooden shake-shingle roofs of two apartment complexes (see Figure 6-10a). At the time, the winds were 2–27 km/hr (13–17 mph).[38] See Figure 6-10b for an illustrative map.

There are several common origins for hot fragments, and since these are the primary source of the fire that may result, they will be considered individually.

FIGURE 6-10A Massive fire engulfing six-story residential/commercial complex (wood framed). *Courtesy of Don Perkins, Fire Cause Analysis.*

FIGURE 6-10B Map showing trajectory of burning brands 0.7 km (0.4 mi) from fire scene to ignite roof of apartment complex 0.8 km (0.4 mi) away. © 2009 Google—Map data—© Tele Atlas

FIREPLACES AND CHIMNEYS

Fireplaces are blamed for igniting many fires, some of which they undoubtedly do. Nevertheless, the mere fact that a possible fire origin is close to a fireplace is not to be considered as positive indication that the fire was initiated by the fireplace. Discussed next are six ways in which a fireplace can be instrumental in initiating a fire.

Emission of Sparks or Blazing Materials from the Front or Open Portion

Such sparks or blazing materials may ignite nearby combustible floor coverings or furnishings. Serious fires from this cause are rare because the top of a wooden floor is very difficult to ignite to a sustained combustion. Scorch marks are not uncommon and are annoying, but wood floors do not usually represent the first fuel ignited. Light fuels, such as paper, may readily be ignited, so prefire actions and conditions may be critical. Some rugs or carpets are sufficiently flammable to be ignited under these circumstances, and furniture may occasionally be ignited. Most building codes require a noncombustible hearth extending well to the front and sides of a fireplace opening. Burning fragments of Christmas wrappings apparently ignited a large pile of crumpled wrappings in front of the fireplace in the fire shown in Figures 6-11a and b. Thus, there is a hazard from this source, but it is readily interpreted when encountered (when prefire activities are described).

Intense Radiant Heat from a Fireplace Stove Filled with Inappropriate Fuels or Overfueled

Sufficient radiant heat can be generated by intense fires in the firebox to create heat fluxes of 10–20 kW/m^2 through the front opening. These can ignite ordinary combustibles in line of sight of the opening if they are within 1 m (3 ft). Inappropriate fuels include crumpled masses of paper, Christmas tree branches, plastics, foams, cardboard boxes, and similar commodities. Some fuels are intended for fireplace use (e.g., compressed-sawdust artificial logs) but may cause overheating if overfueled. The manufacturer of artificial logs typically includes package instructions requiring that not more than one log be put on the fire at a time. Overheating of the firebox is also liable to cause ignition of wood materials located behind the fireplace or inside the wall.

Emission of Sparks or Blazing Materials from a Chimney

Chimney fires, which are the greatest danger factor, arise because soot, bird or rodent nests, creosote, dust, cobwebs, and a variety of combustible materials sometimes are allowed to

FIGURE 6-11A Living room furniture burned when fragments of burning Christmas wrappings escaped from fireplace to ignite a pile of wrappings on the floor. *Courtesy of Derrick Fire Investigations.*

FIGURE 6-11B Heat and smoke layer enveloped the top of Christmas tree but only scorched it (oxygen-deprived smoke layer). *Courtesy of Derrick Fire Investigations.*

accumulate in the chimney, where they may be ignited by a spark or flame from the fire below and lead to a large blaze within the chimney itself. When this happens, large quantities of burning fragments will be expelled from the chimney top and greatly increase the danger of igniting a roof or other adjacent combustible structure. It is clear that this possibility is the most dangerous from the standpoint of starting large fires. It is also clear that a large proportion of them will be roof fires, although blazing residues (firebrands) or large glowing sparks (embers) may travel some distance and ignite fires elsewhere than on a roof. One consideration of the effect of wind on chimney action is also relevant. Wind blowing across a chimney opening exerts what is known as the *Bernoulli effect*, that is, a lowering of the pressure at the chimney mouth. The result is a stronger draft up the chimney, which can increase the size of fragment that can move up to the top and be ejected into the stronger wind stream. This is an additional reason for the investigator to be very careful considering wind velocity when there is a possibility that spark movement started a fire.

Emission of Sparks or Blazing Materials Resulting from Defects or Holes in a Chimney

Such sparks can ignite timbers or other construction material adjacent to the chimney. Fires kindled by this method are equally rare but may be very serious when they occur. Few chimneys become so defective by use or abuse as to pose any real danger of leaking sparks or burning materials through their walls. However, such defects can result from anything that causes serious deterioration of the house, such as earth settling or slippage, earthquakes, or extreme age and decrepitude. Some manufactured chimneys can separate at their joints due to corrosion or mechanical failure, as shown in Figure 6-12. A chimney fire may damage the chimney lining, so that gaps develop that allow hot gases to escape the flue and begin to char wooden structures. The next time there is a chimney fire, such gaps may also permit flames and sparks to contact combustible framing directly. Because of the characteristic pattern and behavior of a roof fire (to be discussed in Chapter 7), there will rarely be a serious problem in interpreting the cause of the fire. The recent history of use should be established by interviews with the occupants.

Overheating of Combustible Materials near or in Contact with a Fireplace or Chimney

Overheating of combustible materials by radiant heat from or direct contact with the hot surfaces of fireplaces or chimneys is probably less important in fires involving stone or

FIGURE 6-12 Chimney cap was dislodged and allowed hot gases to vent into wood-framed enclosure eventually igniting it. *Courtesy of Lamont "Monty" McGill, Fire/Explosion Investigator.*

masonry fireplaces than those with metal ones. Fires in the immediate vicinity of chimneys are not uncommon, and many people attribute them automatically to overheating of wood structural elements. The insulative capacity of intact stone or brick is such that these chimneys rarely transmit enough heat to damage the structure. Metal chimneys and fireboxes are much more common today. Loose insulation between the double walls of metal fireplaces can settle, allowing spaces with minimal insulation capacity. The outer walls can then reach temperatures far in excess of what they should, and these may be dangerously close to combustibles, as shown in Figure 6-13. Chimney enclosures today are often of wood construction. They are supposed to be lined with drywall, but this lining

FIGURE 6-13 A close-up of the fire area around a metal heater vent shows extensive charring of timbers due to inadequate clearance. *Courtesy of the Fire Investigation Unit, Metropolitan Police, London.*

192   **Chapter 6** Sources of Ignition

is not always installed. Manufacturers of metal fireplaces provide detailed installation instructions, and if these have been violated by the installer, a fire may well occur. Violations can include disobeying required clearance from combustibles, use of incorrect fluepipes, blocking of required air-circulation openings, or altering the manufactured unit. The commonly specified triple-wall flue pipes depend on proper flow of cooling air through the annular segments, and the design can be defeated by improper installation that blocks this flow of cooling air. It is occasionally found that a tradesperson "simplified" the installation of a fireplace by removing protuberant standoffs that are designed to enforce a certain distance of clearance. By the way, the term *zero clearance* is an inappropriate marketing catchphrase and should not be taken to mean that such a unit can be installed in any relationship to combustibles that the installer might wish. Fireplaces lacking an Underwriters Laboratories (UL) (or equivalent) listing obviously may be presumed to be intrinsically defective and unsafe.

Circumstances of recent use may be of great value in establishing the actual cause of chimney-related fires. Unusually large or hot fires (such as from burning a dry Christmas tree) or sustained fireplace use due to a very cold weather spell may trigger ignition where it had previously been avoided by "normal" use.

Similar considerations apply to furnace and water heater flue pipes where they pass through openings in wooden structures. It must be remembered that in such instances there is usually a chimneylike insulator structure surrounding the actual chimney or vent. Any fire that starts from any cause in the vicinity of such a penetration will be drawn up it, often localizing the fire in the region around the suspected chimney or vent. Such insulators or "thimbles" may be required around exhausts for heaters, stoves, and emergency generators (see Figure 6-14) For this and other reasons, such a postulated origin of fire should be scrutinized most carefully before accepting it. Considerations of the properties of purported first fuels are relevant.

Changes in equipment such as the installation of high-efficiency heaters or fireboxes (fireplace inserts) can change the airflow of an old chimney or dramatically increase the surface temperatures of fireboxes and flues, thus creating problems where none existed before. Such increases may be enough to overwhelm the designed insulation or induce more heating by conduction through brick or stone, resulting in degradation of wood structural members in contact. Gaps can also develop between the top front of the firebox and the facing brick or stone. Because the chimney and fireplace are not likely to be destroyed by the fire, evidence of gaps in its structure, added to the fire pattern, should

FIGURE 6-14 Sheet metal thimbles for passing exhaust vents through walls. *Courtesy of John D. DeHaan.*

reveal to the investigator its contribution to starting the fire. Some units require that firebox side edges be tightly sealed with noncombustible material. A gap there can allow hot air emanating from the front of the unit to get sucked into the chase behind the fireplace, causing overheating of wood materials there. This is another possible source of fire.

Embers and "Hot Ashes" Removed from Fireplaces and Barbecues

Embers from fireplaces, campfires, and barbecues have been identified as the cause of numerous "delayed ignition" fires remote from the fireplace, campfire, or grill. Due to low heat release rate (HRR) and slow combustion and the insulative properties of ashes and the charred wood itself, the embers are undetected when removed. Usually, the homeowner cleans out the ashes and places them in a combustible box, bag, or trash can, assuming they are "out." Wood or charcoal embers, insulated by ashes, can continue to smolder for 3 to 4 days under the right conditions and can result in ignition after being removed.[39]

LONG-TERM HEATING ("LOW-TEMPERATURE" IGNITION)

As we saw in Chapter 5, wood, especially in massive form, requires a considerable amount of heat to cause its ignition. For flaming combustion, there must be enough heat flux to pyrolyze volatiles from the wood and ignite those vapors. Temperatures around proper flues and chimneys in good repair are not expected to reach the 250°C (482°F) minimum ignition temperatures for fresh wood. However, it has been reported that exposure to much lower temperatures (below 105°C or 221°F) causes degradation of wood to charcoal.[40] This charcoal, also erroneously called "pyrophoric carbon," is sometimes found in wood around heat sources like furnaces and chimneys. (Although the term *pyrophoric* is technically misapplied here because it refers to materials that ignite spontaneously in air at moderate temperatures [54°C (130°F) or below], it is widely used in fire investigation in discussing wood that appears to ignite at temperatures far below its usual ignition temperatures.) These mechanisms were discussed in Chapter 5 (see Figures 5-4–5-6).

It is apparent to experienced fire investigators that fires have begun in confined spaces in wooden structural components where exposure to a source of some heat for long periods of time has caused far more charcoal to form than would be expected from the duration of the reported (or observed) fire. Our review of fire cases that present the strongest argument for the formation of self-heating charcoal shows that the circumstances are often as Bowes predicted: a heat source near wood that is protected by a barrier of metal or tile or in a very confined area (such as a pipe or flue snugly fitted through a wooden beam, as pictured in Figure 5-8, for a long period of time, followed by structural collapse or other change that allows a sudden inrush of fresh air, can precipitate a flaming fire.[41] Production of this charcoal is clearly a long-term phenomenon, often requiring many months, and more typically years, to reach combustible form and critical mass. Martin and Margot reported that temperatures as low as 105°C (221°F) were *capable* of degrading wood but that the lower the temperature, the longer the time required.[42]

The mechanism for this process is complex and has been the subject of some debate, especially concerning the applicability of a predictive thermodynamic technique called Frank-Kamenetskii after its creator.[43] There are several important features of such processes that have to be recognized:

- The conversion of fresh wood to a charred or "cooked" state can occur at low temperatures [well below 250°C (482°F) and possibly as low as 77°C (170°F)] over a long period of time.
- Fresh wood when heated produces gases and vapors that will support flaming combustion (but requires significant heat be applied to effect this pyrolysis and then ignition), whereas charcoal has much fewer volatile components and supports only very limited flaming combustion but will readily support glowing combustion.
- Fresh whole wood (sawn lumber) will not self-heat unless very hot, but charcoal will.

- Fresh wood has a much higher permeability to air along its growth axis than across it.
- Smoldering combustion can result from runaway self-heating.
- A mass of fuel undergoing smoldering combustion can transition to open flame if ventilation causes a high enough heat release rate (either external such as from a draft of wind or enhanced buoyant flow around or through the mass).

When all these factors are taken into consideration within the context of the previously described scenarios, we can see that there is a supportable hypothesis for these fires. First, heat is applied to a mass of wood that has, by its mass or covering, limited exposure to air (the work by Cuzzillo has shown the permeability of "cooked" wood to air to be a critical factor).[44] The wood degrades to a "cooked" form with the gradual release of the volatile fuel vapors insufficient to support flames (with the characteristic flat, shrunken surfaces with sharp-edged segments). The charred, "cooked" wood has a high permeability to air enhanced by the multiple cracks into the wood produced by the "cooking," a low thermal inertia, and an ability to self-heat, all very different from fresh wood. Continued application of heat to such a material can induce runaway self-heating, especially if air is admitted as the result of shrinkage or separation from sealing materials. This runaway self-heating is the critical step in the process. At this point, there is ignition (self-sustained smoldering combustion), but it may not be recognized from outside the cavity or concealed space where it is taking place. If enough air is admitted and the enhanced ventilation causes a high enough heat release rate, the smoldering combustion can transition to flaming combustion (with the rapid involvement of adjacent intact wood structural members). The flaming stage is often detected as the "fire." Figures 6-15a–c show some of the critical features of the process. The conditions have to be nearly perfect—temperatures typically between about 100°C and 200°C (212°F–392°F) with very limited ventilation, so that heat is not readily dispersed and oxygen generally excluded, and an adequate mass of wood. There are extensive data on the relationship between fuel mass, environmental temperature, and time to develop runaway self-heating. These conditions are sometimes found in accidental fires investigated, but the investigator must be careful of blaming "pyrophoric carbon" for any fire in the vicinity of a flue or hot water pipe simply because no other cause can be identified. As Bowes pointed out:

- The temperatures of heating surfaces have to be at high temperatures [although nowhere as high (200°C; 400°F) as his calculations indicated];
- The wood has to have substantial mass (plywood floors are more likely to simply char through than to ignite);
- The temperature of the hot surface has to be high enough to maintain heat flux on the wood to overcome losses due to conduction or convection;
- The time has to be long enough (weeks, months, or years, depending on the intensity of the applied heat); and
- There has to be exclusion (or limited circulation) of air in the vicinity until just prior to detection of the fire.[45]

One case involving all five of these elements was the result of a water heater that sat on a wooden platform "protected" by a steel sheet and an asbestos barrier. Some weeks after a new water heater was installed, the platform began to burn and the fire was detected. (A similar case was shown in Figure 6-4.) It should be remembered that smoldering combustion (in charcoal) requires only very low oxygen concentrations and that fresh air with its richer oxygen supply is required to sustain open flames. Because of the time required for the production of charcoal, low-temperature ignition of wood can generally be eliminated as a cause of fires in very new buildings. Interestingly, Babrauskas et al. reported that wood in contact with a surface with a temperature in excess of 77°C (170°F), even if periodic, should be considered a potential ignition scenario.[46] Both UL and the Building Code of the City of New York cite 170°F as the upper "safe" temperature for wood surfaces.[47]

FIGURE 6-15A An 89-mm cube of Douglas fir after being heated at 200°C (424°F). Solid-phase ignition (smoldering) occurred on the end-grain faces while the side-grain faces exhibited little damage. Note the greater shrinkage at the end-grain faces than at the center. Unlike isotropically porous materials where the highest self-heating temperatures are at the center, the highest temperatures in self-heating whole wood are near the end-grain faces. *Courtesy of Dr. Bernard R. Cuzzillo, Berkeley Engineering and Research, Berkeley, CA.*

FIGURE 6-15B An 89-mm wood cube heated at 200°C (424°F) and split open to show the hourglass heat pattern within. This indicates that the regions of maximum oxidation reaction are near the end-grain faces and along the axial centerline. *Courtesy of Dr. Bernard R. Cuzzillo, Berkeley Engineering and Research, Berkeley, CA.*

FIGURE 6-15C An 89-mm cube of Douglas fir that has been heated at 200°C after its end grain has been sealed with RTV silicone sealant, preventing the diffusion of air along its centerline. The exposed faces are sharply faceted into small, flat shrunken areas. The same pattern of fissuring extends uniformly into the center of the cube. This is similar to exposing the middle of a long piece of solid timber to a heat source mid-length. The fissuring allows air penetration into the wood block and could support runaway self-heating. *Courtesy of Dr. Bernard R. Cuzzillo, Berkeley Engineering and Research, Berkeley, CA.*

As with most other accidental ignition sources, careful investigation and evaluation must be carried out before concluding low-temperature ignition or the more general "overheated chimney" as a cause. In addition to there being much more char than expected, if the wood timber is cross sectioned, the char should also gradually change from complete char to partially pyrolyzed to undamaged as one moves away from the purported heat source. Unfortunately, the processes involved and their interaction are complex, and there are presently no data showing an exact relationship between temperature of exposure versus time to ignition. Babrauskas has recently published an informative review on the topic in his *Ignition Handbook*.[48]

TRASH BURNERS, INCINERATORS, AND BONFIRES

As sources of hot and burning fragments, trash burners, incinerators, and bonfires rate high.

Trash Burners and Incinerators

The types of material normally destroyed by burning in such devices are those most likely to form airborne burning fragments, and the necessity of openings in the incinerator to allow ventilation of the fire allows escape of burning fragments. Accidental fires particularly of dry grass or brush in the area are commonly associated with trash burners and incinerators. It is fortunate that fires started by trash burners will rarely require much investigation because of the obvious nature of the origin and the fact that the pattern will trace directly to the burner in most instances. The strength and direction of the wind and the ignitability of possible first fuels (including fuel moisture and nature of materials being burned) are all important considerations.

Bonfires

The bonfire poses a somewhat different problem than the trash burner, despite also being an exterior fire. To maintain a satisfactory bonfire, it is customary to use relatively massive wood instead of paper, packing, cardboard, and similar items of trash. The wood poses relatively much less hazard from rising fragments of burning material than does trash. However, there is no restraint or protection of the fire from winds, as is true of the trash burner. In a high wind, a bonfire becomes very dangerous from the standpoint of starting larger, uncontrolled fires. If it is adjacent to dry grass or leaves, it is always hazardous. Here also, the investigation is likely to be rather simple. An occasional difficulty is in distinguishing the remains of a small bonfire from those of a general fire that involved a concentration of fuel at a point that may leave excessive ash and charcoal, which appear similar to the remains of a bonfire. Careful examination will usually distinguish the two. In many instances the remains of bonfires contain, in addition to wood ash, the remains of food containers or other indication of human activity. The direction of the prevailing wind and the contour of the land are critical to correctly determining such origins. The considerations of ignition by such airborne brands include type of fuel being burned, size of fragments that can escape, and size of the fire (creating the upward buoyancy). The larger the fragments and the more massive they are, the longer they can burn. Flat brands such as fragments of cardboard or wood shingles have better aerodynamics than needle or spherical shapes and can remain airborne much longer and travel farther. See Chapter 8 on wildland fires.

HOT METALS

As a direct source of ignition of fires, metals are only of occasional importance. They can play a part as sparks of burning metal from grinding or cutting operations or as droplets of molten metals. Molten metal may carry enough heat to ignite susceptible fuels with which it comes in contact and is therefore capable of starting fires, as are burning fragments. Because a considerable amount of heat is required to melt the metal, such dangers are usually associated only with special operations in metal fabrication. Places in which such operations are carried out are more than normally susceptible to fires from this type of ignition. However, the dangers are often exaggerated, since many alloys, including types of solder, will not generally carry enough heat to ignite fuel material on which they fall. Babrauskas has published a significant review of such ignitions.[49]

Molten ferrous alloys, such as steel, are far more dangerous than solders because of their higher melting temperatures and often higher thermal capacities. Thus, welding and cutting of steel with an arc or a torch in the presence of susceptible organic fuels is a very dangerous practice. See *NFPA 51: Standard for Fire Prevention during Welding, Cutting, and Other Hot Work* for safety during those operations.

The state of subdivision of a solid fuel is critical in determining its susceptibility. It is, for example, almost impossible to ignite a large timber accidentally by the short, direct application of a torch flame or by the molten metal from it. In contrast, finely divided fuels, such as shredded packing materials, sawdust, shavings, or loose paper may very well be ignited under the same conditions. Directing a flame into a narrow crack or split in a piece of wood, however, may cause ignition of both surfaces. The close spacing causes reduced losses and enhanced feedback, resulting in ignition.

Molten metal can also be produced by the fusion caused by the heat of a short circuit or failure of a component in electrical equipment. Although ejection of such material from the equipment certainly presents a hazard, the greatest danger may be from direct ignition by the short circuit itself, but only if a current protection device does not interrupt the current. Thus, the metal may be considered a secondary and less probable source of ignition.

In considering hot or molten metal fragments as sources of ignition, the investigator must always keep in mind that most metals have a high density, and the hot fragment will fall rapidly, in contrast to bits of paper or wood, which tend to rise in the convective plume of the fire itself. Aluminum is of relatively low density. In flat form, such as a roof section, aluminum sheet may rise in a buoyant fire plume rather than fall, but this will not be true of solder, molten brass, iron, or copper. Aluminum, because of its low density, ease of oxidation, and low melting point, is very unlikely to play a role in spreading fire as airborne debris and in general is not a likely source of ignition, except in the case of high-voltage power lines. These are most commonly made from aluminum, and wildland fires can readily be ignited when globs of molten aluminum land on dry vegetation. Such molten masses are most commonly generated when two power lines unexpectedly make contact, resulting in "conductor clashing," because of extreme winds, failed insulators, failed supports, or excessively stretched conductors, for example.

Despite their higher melting temperature, copper particles are a much less effective ignition source than are aluminum particles. It should also be remembered that metals can burn, although most metals are extremely difficult to ignite except when very finely divided. Magnesium is the chief exception, either alone or alloyed with aluminum, being more easily ignited.

Occasionally, massive metal pieces heated by torch cutting or welding operations can ignite ordinary combustibles they contact. This is far more likely if the fuel is a char-forming material rather than a thermoplastic. This mechanism is extremely unlikely, however, if more than a few minutes have elapsed between the hot work and observed ignition (due to the high thermal conductivity of the metal). Investigation and thorough interviews will be needed to test such hypotheses. Fire watches after hot-work operations are typically only 30 minutes long, but some smoldering ignitions from hot-work operations have been known to take hours, instead of minutes, to manifest in a flaming fire.

MECHANICAL SPARKS

Iron, steel, and some other metals will spark when struck by a suitable object, as during grinding operations or when a portion of a vehicle strikes a stone-surfaced roadway. The actual mechanism is quite simple. A highly localized frictional contact rips a fragment of the ferrous alloy out with sufficient energy to heat it beyond its ignition temperature [on the order of 700°C to 800°C (1,300°F to 1,475°F)]. The particle ignites in the air and burns (at >1,600°C; 2,900°F), leaving a sometimes incandescent flake of carbon and iron oxide as an ash.[50] These sparks represent small but finite heat sources that can ignite some flammable vapors. If the "ash" flakes are hot enough and large enough to transfer sufficient heat to their surroundings, they can ignite readily combustible (finely divided) solid materials on which they fall. Such fires are seen along railroad tracks where mechanical sparks from overheated brake shoes can be dropped for many hundreds of yards. Witnesses may report fires associated with passing trains.

Many grass and brush fires along railroads are ignited by particles emitted from diesel locomotive exhaust or eductor tubes. These particles, consisting of carbon, calcium oxide, and the oily residues of fuel and lubricant, are emitted along with the exhaust. Under some conditions of engine operation they are emitted while still burning and can ignite dry grass along the track or even several feet from it.[51] Many states require spark arrestors on all engines operating in high fire-hazard areas, but they are sometimes overlooked or even intentionally eliminated during maintenance.

Sparks can also be created by cutting or grinding of iron or steel using a carborundum blade grinder. Incandescent particles have been observed up to 5 m (15 ft) from a handheld grinder being used on a steel reinforcing bar. Such sparks reportedly have a starting temperature of 1,600°C to 2,130°C (2,900°F to 3,866°F) and an energy on the order of 1 kJ. Impact of a steel tool on concrete (particularly quartz) or aluminum-coated or contaminated rusty steel can also create mechanical sparks capable of igniting flammable gases.[52]

The rail-grinding equipment used on railroads creates large quantities of hot sparks that can be projected some distance from the track. Because such equipment is not stationary, its operation is not controlled and may escape consideration, as shown in Figure 6-16. The exhaust systems of off-road motorcycles and trucks are also liable to emit hot carbon particles occasionally. Some vehicles may be required to have spark arrestors before being permitted to operate in forest areas. Catalytic converters on vehicles have been known to decompose and to spew chunks of their ceramic matrix at very high temperatures into grass and brush.

FIGURE 6-16 Sparks from this rail grinder were thought to be responsible for igniting a sawdust bin in a lumberyard. Typical water spraying would extinguish fires on the railroad right-of-way but would not necessarily affect hot sparks buried in sawdust adjacent to the track. *Courtesy of Joe and Chris Bloom, Bloom Fire Investigation, Grants Pass, OR.*

FIREARMS RESIDUES

Forest and grass fires, as well as structure fires, are occasionally ignited by the burning residues of firearms ammunition. Although modern smokeless powder is nearly completely consumed during the discharge of rifles and shotguns, some flakes escape while still burning from the muzzle of a short-barreled handgun. Although they are low in mass and thermal content, these flakes can ignite finely divided cotton and similar combustibles at very close range (less than 0.3 m; 1 ft).

Black powder, in contrast, produces large quantities of burning particles of powder and incandescent ash from both short- and long-barreled guns. Although generally limited to antique handguns, rifles, and cannons, the use of black powder for sporting purposes is undergoing a strong resurgence in popularity.[53] Because residues from such weapons can ignite dry fuels like paper, cloth, wood shavings, leaves, and the like even at distances of up to 1.3 m (4 ft), the possibility of accidental ignition by this means is one the investigator must consider in identifying the cause of exterior, and even some interior, fires.

Military Ammunition

More dangerous sources of ammunition fires are tracer and incendiary bullets, normally used only in military weapons. Tracer ammunition contains a mixture of strontium nitrate, magnesium powder, and an active oxidizer like calcium peroxide. Incendiary ammunition usually contains white phosphorus or a similar material with a low ignition temperature. Either type is a severe fire hazard because of the high-temperature flames produced in the fired projectile. Both will initiate destructive fires in dry leaves, brush, or needles (duff). Such ammunition has, on rare occasions, been used by arsonists to set remote fires that are inaccessible to fire crews. (See Figures 6-17a and b.)

Virtually all military ammunition is marked by characteristic color bands on each projectile, but some foreign military surplus ammunition brought into the United States has been cleaned and repackaged, precluding ready identification.[54] Hunters buying this recycled ammunition have inadvertently discharged these missiles during hunting and target practice, sometimes with serious consequences. With the characteristic paint removed, identification of such incendiary projectiles requires X-ray or visual laboratory examination. Signaling or rescue flares, some of which are available for even small-caliber handguns, present the same sort of risk because they, too, contain pyrotechnic or incendiary

FIGURE 6-17A Unfired tracer bullet with orange nose paint. *Courtesy of John D. DeHaan.*

FIGURE 6-17B Tracer ammunition emitting bright red flare in a "bench" test. *Courtesy of John D. DeHaan.*

mixtures that burn for many (up to 40) seconds. Some have a parachute to slow their descent and thereby prolong their burning time (and make them subject to winds aloft), while others are simply ballistic "meteors."[55]

Exploding Ammunition

Exploding ammunition, the projectile of which contains a percussion cap and a small powder charge, has at times been sold to the public and represents an occasional fire danger. Foreign armor-piercing rifle ammunition with a steel insert has been thought to be the cause of some wildland fires behind target ranges, because the steel is capable of striking mechanical sparks as it ricochets off bare rock. Exploding paper targets have also been implicated in accidental wildland fires.

Smoking as a Fire Origin

The widespread prevalence of smoking, as well as the general opinion as to its causation of fires, calls for some special comments because there are many misapprehensions relative to this matter. There is no question that many fires are started by smokers, both indoors and outdoors. Nor is there any question that the carelessness of the smoker is the primary reason for this high incidence. However, it is common to hear theories about how smoking caused a fire when the theory is totally inappropriate, and the alleged ignition would not have been successful under the circumstances.

Smoking involves both burning tobacco and a flame used to ignite it. The roles of matches and lighters as primary sources of ignition were discussed previously. Whether the flame is from a match or a lighter, this flame will inevitably start a fire if it comes in contact with suitable fuel. Thus, the discarding of a match into dry grass, before the match is extinguished, has been the cause of many exterior fires. Both match flames and lighters will cause explosions and fires with volatile flammables in the vicinity. Many persons have attempted to commit suicide by turning on the gas, but then they wish a last cigarette. Sometimes, the explosion that follows is their executioner, but more often it puts them in the hospital with serious burns. It is important to realize only the flame causes this type of incident.

CIGARETTES

In the authors' opinion, cigarettes as a source of ignition have been blamed in many more instances than warranted. It must be appreciated first that cigarettes vary considerably from brand to brand in their burning characteristics. Modern tobacco cigarettes contain a complex chemistry of additives to control the moisture content and burning rates of both the tobacco and the paper. U.S. cigarettes sometimes used to include a chemical additive to keep them burning even when not being "puffed." But in the last few years, cigarettes designed to self-extinguish have become common, instead (see further).

The temperature distribution within the burning tobacco and its ash is complex and difficult to measure. The most extensive work done in this area was by Baker, who reported that surface (solid) temperatures reached 850°C to 900°C (1,560°F to 1,650°F) at the edge of the glowing mass during the puff or draw using X-ray measurement techniques.[56] After the draw stopped, the hottest part of the mass shifted to the center, with temperatures of 775°C (1,427°F), and the temperature of the ash at the edge dropped to ≈300°C (572°F), as illustrated in Figure 6-18. These temperatures are in general agreement with test results reported by Beland and by the NFPA.[57] Variations in the types and cut of the tobacco used will affect the temperatures developed as well as the burning rates. This is especially true of hand-rolled cigarettes, whether of tobacco or other vegetable material, whose properties have not been studied in detail. In tests conducted by the California Bureau of Home Furnishings, the temperature characteristics of many common tobacco cigarette brands were measured. Maximum temperatures produced ranged from

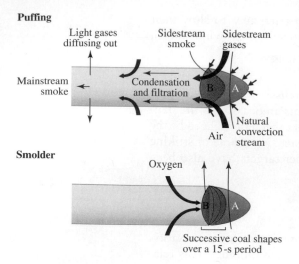

Puffing

Light gases diffusing out

Sidestream smoke

Sidestream gases

Mainstream smoke

Condensation and filtration

Air

Natural convection stream

Smolder

Oxygen

Successive coal shapes over a 15-s period

FIGURE 6-18A Airflow in a cigarette at rest and while being puffed. *Reprinted by permission from* Science & Justice.

Direction of gas flow

8 mm

100 400 500   500 400 100

750

775

300   600

700 300

°C

FIGURE 6-18B Temperature distribution in a burning cigarette at very start of puff. *Reprinted by permission from* Science & Justice.

400°C (760°F) for one variety to 780°C (1,440°F) for another. Temperatures rose by as much as 100°C as the burning reached the end of each cigarette. Some brands of cigarettes consistently self-extinguished when left unsmoked; others always burned to completion.[58] Although the rate at which a cigarette will burn depends on its orientation and the direction and intensity of any draft, the rate was also found to vary from brand to brand. Rates from 6 to 8 mm (1/4 to 1/3 in.) per minute and total burn times of 13 to 15 minutes for unaided horizontal burning were measured in a series of tests by the California Department of Forestry.[59]

The heat release rate for an average tobacco cigarette has been reported as ~5 W. Such a low level emphasizes the necessity for a fuel to be in direct contact with the smoldering tip of the cigarette before ignition can be ensured. A lighted cigarette will only occasionally set fire to dry fuel on which it rests uncovered, such as upholstery, dry grass, and the like. A linear scorch will result, but it will rarely cause ignition. If the fuel is loose enough that the burning cigarette can "burrow" into it as it burns, ignition is much more likely. When the same cigarette is covered by even a single layer of light cloth, the insulation prevents the radiation of the heat, and temperatures up to 100°C (212°F) higher can result. This temperature difference may be sufficient to allow ignition of the fuel. If the cigarette "burrows" down into loose cellulosic fuels or is pushed partially between two upholstered cushions, enough heat may build up to cause ignition if the arrangement conserves the heat without unduly restricting the air supply. On exposed polyurethane foam, which is often very susceptible to flame, heat from the cigarette melts the rubber away from it so that as the cigarette burns, it forms a trough in the foam (as in Figure 5-13b). In combination with some fabric coverings, however, polyurethane can support a smoldering fire.[60] Latex rubber foams can be ignited by a cigarette in contact and support glowing combustion that can transition to flame. Most modern upholstery foam rubbers are made to resist such ignition even when intentional. On average, a great many lighted cigarettes will have been improperly discarded for each fire that results from such a discard. (See Chapter 11 on fabric flammability.) One study showed that cigarettes discarded into crumpled paper trash transitioned to flames in only about 1 percent of tests.[61] One well-known fire investigator routinely demonstrates ignition of toilet paper rolls by placing a cigarette inside the roll (as sometimes occurs in public toilets). These often transition to a flaming fire within 20 minutes. He has been known to do this demonstration as an after-dinner entertainment (much to the dismay of his wife).[62] As discussed in Holleyhead's extensive review of cigarette ignition, many cellulosic products including toilet tissue, cardboard, paper towels, and office carbon paper are all susceptible if the contact and geometry are suitable.[63] This has to be taken as an indication that cigarettes can cause ignition of such fuels, and it takes only one ignition to start a fire!

It should be remembered that not all cigarettes have the same potential to ignite cellulosic fuels. *ASTM E2187* is a method for comparing the ignition potential of tobacco cigarettes. Recent efforts to reduce accidental fires caused by cigarettes have focused on redesigning cigarettes to make them self-extinguish if not being actively smoked (as do cigars) by changing the paper or the tobacco.

The majority of U.S. states have passed regulations mandating that cigarettes sold in those states be designed to self-extinguish when not being smoked.[64] This legislation is expected to have a significant impact on the frequency of cigarette-induced fires. These "Reduced Ignition Propensity" (R.I.P.) cigarettes are designed to self-extinguish within 5 minutes of being left unsmoked. This means there is insufficient contact time to ignite almost any susceptible fuel. In addition, Canada, Australia, and the European Union (EU) have enacted similar legislation. Early tests have revealed that although such cigarettes (employing bands of low-porosity paper) do not guarantee that all will self-extinguish and therefore will not prevent all fires, the frequency of ignition is reduced.

BEDDING AND FURNISHINGS

While changes in the construction materials of modern furnishings have made them less susceptible to ignition by cigarette, such fires still occur. Cigarettes and cigars, being sources of smoldering combustion, are much more likely to ignite finely divided cellulosic fuels like cotton padding than they are thermoplastics and foams that melt rather than char. Years ago, mattresses and upholstered furniture were made of cushions of cotton batting, kapok, or sisal (or latex foam rubber, which will also smolder), covered with natural fiber fabrics of linen or cotton. Thus, a dropped cigarette could burrow into fuel, which would in turn support a smoldering fire for an extended period of time. Such furniture is still frequently encountered.

Newer furniture is most often made with polyurethane foam cushions covered in thermoplastic fabric. Neither of these materials is likely to be ignited by a smoldering source such as a cigarette. The cushions are sometimes wrapped in cotton batting instead of a polyester fiber blanket, however, so the risk of cigarette ignition is not entirely absent. The synthetics, of course, are much more susceptible to ignition by an open flame and, once ignited, will support a much larger, more energetic fire than will the older furnishings.

Cotton sheets and blankets may similarly be ignited by cigarettes but only when they are presented as a wadded mass of fuel. Single flat layers of such fabrics will be penetrated by a smoldering cigarette but not ignited due to the heat lost to convection. Kapok, sometimes used as a pillow stuffing, is very susceptible to glowing ignition sources, but feathers are not (although both will support smoldering once established). Decorative pillows may be filled with anything and should not be overlooked as an intermediate fuel that is ignited by a cigarette and then grows into a flaming fire that involves a urethane mattress. The investigator must be aware of such complex interactions, must search the debris as well as interview the owner or tenant, and must obtain samples of all the likely first fuels, if possible.

Despite significant advances in improving the fire resistance of furnishings, smoking in bed remains a threat to life safety. The reasons are simple. Bedding often contains cotton, which is readily ignited by smoldering sources and is capable of sustaining a smoldering fire for an extended period. If the smoker should fall asleep, so that a lighted cigarette drops into wadded sheets, pillowcases, or onto a cotton-stuffed mattress or pad, a fire can readily result, and statistics indicate that such fires are common.[65] A special hazard of fires that involves upholstered furniture or bedding occurs when the drowsing

(a)

(b)

(c)

(d)

FIGURES 6-19A–D Cigarette ignition of a traditional arm chair taking 42 minutes from placement to flames. Note location of ignition and failure of other cigarettes on other surfaces. *Courtesy of Jamie Novak, Novak Investigations, and St. Paul Fire Dept.*

smoker is awakened by the heat but, being confused, adopts incorrect measures to control the situation. Such persons have been known to attempt to throw the burning bedding out the windows or carry it to the exterior and in the process have extended the fire over large areas of the building.

The time between ignition from a smoldering cigarette to open flaming combustion of upholstered furniture is highly variable. One test, depicted in Figure 6-19, shows a delay of 42 minutes between placement and flame ignition. The time depends on the combination of cellulosic and synthetic materials and the extent and location of contact with the cigarette. The National Bureau of Standards [now the National Institute of Standards and Technology (NIST)] reported that of six upholstered chairs in one test, three went to flaming ignition in 22 to 65 minutes, while the remaining three simply smoldered.[66] The California Bureau of Home Furnishings reported that of 15 upholstered chairs, 9 went to flaming combustion in times ranging from 60 to 306 minutes, 5 self-extinguished, and one was still smoldering when extinguished at 330 minutes.[67] Krasny,

Parker, and Babrauskas have published an extensive study of the variability of cigarette and flaming ignition of furniture, and the reader is referred there for more information.[68] In experiments observed by DeHaan, flaming combustion was achieved in older, cellulosic-filled upholstered furniture and bedding in times ranging from 22 minutes to over 2 hours from the placement of the burning cigarette. (If the fabric covering is torn, and the cigarette is placed against the cotton stuffing, the time to flame can be as little as 18 minutes). The correct combination of fabric and padding (stuffing) can reduce the ignitability of furnishings. Even cotton-padded furniture can be cigarette-ignition resistant if covered with a heavy fabric made of thermoplastic (synthetic) fiber.[69] If a piece of furniture appears to have been ignited by a cigarette, samples of covering and padding should be recovered for later examination.

An unpleasant accompaniment to bed and furniture fires is the danger of death by burns or smoke inhalation, even in the absence of flaming combustion. The fact that many persons who smoke in bed have also consumed excessive amounts of alcohol increases the total hazard greatly.

CIGARETTES AND FLAMMABLE LIQUIDS AND GASES

Another fallacy about cigarettes relates to the danger that they will readily ignite flammable liquids or gases. Repeated attempts to cause explosions by inserting a lighted cigarette into an explosive gasoline vapor–air mixture have resulted in failure, even with the ash knocked off and the cigarette being puffed strongly. Tests by the authors involving throwing or dropping lighted cigarettes into gasoline vapors above or alongside a spill on pavement have never resulted in an ignition (until a lighted match was dropped into the same vapor). This is not proof that such an explosion cannot happen, but any claim that it has occurred should be viewed with the greatest skepticism.

In his extensive review of the dynamics of tobacco cigarettes and flammable vapors, Holleyhead describes several factors revealed by the tests of Baker and others that make ignition of gasoline vapors by a glowing cigarette most unlikely.[70] Tests have shown the oxygen levels in the cigarette in the vicinity of the combustion zone to be very low and carbon dioxide levels to be very high, both factors that significantly reduce the chances of vapor ignition, as illustrated in Figure 6-20. Furthermore, the residence time of airborne vapors in the cigarette being puffed is so short that there is not enough time for any but the most reactive species to ignite. The spacing between the fuel elements in a commercially made cigarette is such that the quenching distance of all but the most reactive gases is not

FIGURE 6-20A Oxygen concentrations in a cigarette—at the start of a concentration in a cigarette at the start of a "puff." Note extremely low O_2 concentration throughout high-temperature zone preventing ignition of gases/vapors. *Reprinted by permission from* Science & Justice.

FIGURE 6-20B Carbon dioxide levels are very high whether the cigarette is being actively smoked or not. This will inhibit ignition of all but the most sensitive gases. *Reprinted by permission from* Science & Justice.

exceeded, suppressing any sustained ignition. It has been suggested by Kirk that the ash surrounding the burning tobacco may act in the same manner as the screen around a miner's safety lamp, acting to suppress any flame by quenching it between closely spaced elements. Finally, while the surface temperatures of the burning tobacco inside the cigarette tip are high, the presence of incombustible ash may reduce the efficiency of heat transfer to the fuel vapor.[71] These heat transfer considerations are validated by the published test data, which showed that only hydrogen, hydrogen sulfide, carbon disulfide, acetylene, ethylene oxide, phosphine, and diethyl ether are susceptible to cigarette ignition, while gasoline vapors and methane are not.[72] Recent tests by the ATF Fire Research Laboratory involved contact of burning tobacco cigarettes of different brands with gasoline vapors from a pool at room temperature. A total of 137 attempts were made using both smoldering (passive) and actively drawn (puffing) cigarettes with no ignitions observed. Further testing involved multiple applications of 157 burning cigarettes to or near a gasoline-saturated cotton blouse with no ignitions observed.[73] It must be noted that contaminants or irregularities in the filling or in the paper may occasionally promote a very brief, tiny flame. Such a flame would be capable of igniting a suitable fuel/air mixture. Although it is virtually impossible for a lighted cigarette or cigar to ignite most flammable vapors, this is not the case if either is actually lighted in such an atmosphere. That would ensure the ignition of a fire, explosion, or both. Recent efforts to reduce accidental fires caused by cigarettes have focused on redesigning cigarettes to make them self-extinguish if not being actively smoked (as do cigars) by changing the paper or the tobacco.

The damage to clothing or upholstery when it is scorched or a hole burned through it by a cigarette is annoying, but at least this damage will very rarely ignite the fire that burns the house. In any instance where smoking is claimed as the source of a fire, very careful attention must be given to assure that the claimed fuel is one that could be lighted by burning tobacco or, alternatively, the flame from a discarded burning match, which is actually a much more likely source. Consideration must also be given to time factors. Due to their low HRR, cigarettes require contact time to transfer enough heat to cause ignition of another fuel. This generally means at least 15 minutes have to elapse. (A slight draft, however, has been observed to shorten this time by 2–3 minutes.)[74] The authors have confirmed those ignition times by their own tests and that only incendiary materials (like gun cotton, gunpowder, or flash powder) can be ignited immediately on contact with a glowing cigarette. The use of medical oxygen by a smoker makes all materials saturated with oxygen much more ignitable by smoking materials. This includes clothing, bedding, facial hair, and the plastic mask and cannula (as seen in Figure 6-21).

PIPES AND CIGARS

Pipe and cigar smoking is not basically different in its fire potential from cigarettes. Cigars tend to go out when not puffed, but this does not prevent some fires from being ignited when cigars fall onto a suitable target. Discarded pipe embers, like lighted cigarette embers, can burn their way through crumpled papers in a wastebasket until the embers lodge where there is thicker fuel (and ventilation) and then ignite it. Pipe ashes will also scorch upholstery or clothing, actually burning holes in cloth in many instances. They rarely cause flaming fire except with sawdust, cotton packing, or similar readily ignitable materials. Hot pipe and cigar ashes are often physically larger than cigarette ashes and have more heat to transfer to fuels they contact but do not sustain combustion on their own for very long.

PLANTINGS

Finely divided cellulosic materials can be encountered in gardens and decorative plantings (of both real and artificial plants). Dry potting soils, peat moss, and Spanish moss have been shown to be readily ignitable by contact with a smoldering cigarette. Smokers often assume such plantings are in regular (mineral or garden) soil and are therefore noncombustible.[75]

FIGURE 6-21 Oxygen mask readily ignited by match flame burns with great intensity. *Courtesy of Jamie Novak, Novak Investigations, and St. Paul Fire Dept.*

Spontaneous Combustion (Self-Heating)

The claim is frequently made that a fire started from spontaneous combustion. This suggestion at first seems to violate the basic premise that there must be fuel, oxygen, and heat before there can be combustion and implies that in some mysterious manner fires simply set themselves on occasion. The term *spontaneous chemical causation* would be more accurate, for in such cases the mass of fuel, rather than igniting from one point or even one surface, ignites throughout its mass at very nearly the same time; and the heat has been accumulating, not from some exterior source, but from chemical processes occurring within the fuel mass. When such combustion occurs, it is a wonderful example of the balance of physical and chemical processes that occur in all fires but on a time scale that allows their individual roles to be better appreciated. Taken as a whole, however, their interactions make them hard for the fire investigator to understand. The single driving force is an exothermic reaction generating heat, and if that heat is not dissipated by normal mechanisms as fast as it is made, it can build up in a fuel mass and raise the temperature of the mass. Because reaction rate generally doubles with every increase in temperature of about 10°C (18°F), the reaction, by generating heat, tends to make itself go faster and faster with a constantly increasing heat production rate. When there is insufficient dissipation of the heat, the temperature may rise to the point of ignition of the entire reacting mass. The process is treated in more detail in the fire investigation literature.[76] Not all chemical reactions are sufficiently exothermic (self-heating) to cause such ignition, and not all exothermic reactions are sensitive enough to temperature rises to drive them to ignition. A list of commonly encountered materials, rated according to their susceptibility to self-heat, is published in the NFPA *Fire Protection Handbook*, but a few are discussed here by way of examples.

CHARACTERISTICS OF SELF-HEATING

The general rule that applies to self-heating is that the mass of material needed to accomplish ignition is inversely related to the reactivity of the material, and the time needed is directly related to the mass required (i.e., the lower the reactivity, the larger the necessary mass and the greater the time needed). Catalytic reactions (such as the setting of fiber-glass resin) require only a very small mass because the reaction is highly exothermic and can take place in a few seconds. Drying oils can trigger **self-ignition** in a few hours and require as little as 25 to 50 ml (1 to 2 fl oz). Activated charcoal can self-heat in masses of a few pounds and requires several hours to a few days. Laboratory studies have demonstrated that charcoal barbecue briquettes are unlikely to self-heat to ignition unless presented in large masses [more than 50 lb (20 kg)] or at high ambient temperatures (over 100°C).[77] However, these data must be understood as applying to pure and uncontaminated charcoal briquettes. Credible reports exist of fires that originated from small bags of charcoal briquettes where external ignition sources (e.g., cigarettes) were successfully excluded. These fires appear to have been due to briquettes contaminated by chemical substances that promoted combustion. Some charcoal briquette bag ignitions have taken place, however, after an unwise individual returned "hardly burned" briquettes to the bag.

Hays and grasses require significant masses (100 kg/220 lb or more) and days to weeks before ignition, even at moderate temperatures. Coal and bulk materials will not self-heat to ignition unless they are present in very large masses (many tons) and given weeks or even months. The higher the starting temperature, the faster the process can progress. Cotton laundry that has not been allowed to cool down properly from drying has been known to ignite after just a few hours. The presence of cooking oils that have not been removed by laundering may contribute to self-heating of laundry, as may residues of bleaches (oxidizers). It has been suggested that self-heating of residues of cooking oils in clothes dryer lint may lead to some fires in clothes dryers.[78] Tests have revealed that triglyceride oils such as corn, cotton, or linseed oils are not very soluble in water and will be retained in clothes after laundering (and may be detectable in the residual water in the washing machine pump if the clothes in the dryer have been fire damaged).[79] Ground-up or chipped plastics being stored for recycling have been involved in a number of apparent self-heating incidents. The example shown in Figure 6-22 was linked to cooking

self-ignition ■ *See* **autoignition**.

FIGURE 6-22 An opened box of chipped plastic shows deep charring in the center of the mass. Self-heating was linked to cooking oil contamination and grinding and storage in bulk in high ambient temperatures. *Courtesy of Edward Garrison, Fire/Explosion Investigator, Raleigh, NC.*

oil contamination in the plastic bottles that were ground up. The material was stored in large (4 × 4 × 4-ft; 1.2 × 1.2 × 1.2-m) boxes with plastic liners for extended periods of time. Two ignitions were discovered (in late July in very hot weather) a week apart. The first (shown here) was intercepted, but the second one led to a fire that destroyed the entire warehouse. The grinding or shredding process creates a great deal of frictional heat that can promote self-heating; also, metal or rock contaminants can create hot particles that can initiate ignition deep within the mass.

Some suspected sources of spontaneous heating can be tested by subjecting them to the Mackey test, which was developed to assess the hazards of oils used to improve the weaving qualities of cotton and linen fiber, or, preferably, by establishing their critical temperature for various mass sizes and shapes.[80] Some processes can be tested by differential scanning calorimetry (DSC) to determine the rate at which they yield heat. The likelihood that a process can be an ignition source can then be evaluated.[81] Because the environment has so much to do with self-heating (porosity, oxygen supply, thermal conductivity, heat transfer coefficient, etc.), simple chemical or thermal tests of materials may not reveal all their tendencies to self-heat. Although time-consuming, the most accurate analysis is to place various masses of the suspect material into ovens and expose them to slowly increasing environmental temperature. Internal temperature monitoring of the mass will reveal when runaway self-heating occurs. Detailed laboratory procedures and data analysis methods are described in the *Ignition Handbook*.[82] Brian Gray warned about simple reliance on small-scale tests in predicting self-heating in industrial situations where products have relatively high critical temperatures when tested in small, single packages but whose critical temperatures as stored "assemblies" are considerably lower. He cited two cases in which hot storage conditions of bulk product changed the thermodynamics in critical fashion. In one case, drums of calcium hypochlorate (whose generation of oxygen is an exothermic reaction) were stored in an exterior storage container and produced a runaway self-heating reaction. In the other case, small packages of instant noodles were originally packed at a packing temperature of 81°C (177°F), below the estimated critical temperature for the shipping assembly (in cartons on pallets) of 106°C (222.8°F). A change in manufacturing process caused the product to be packed at 113°C to 120°C (235°F–248°F), a change that caused a massive fire loss.[83]

In one well-documented case in New York, a very large warehouse fire was traced to a storage area where many cases of latex rubber surgical gloves had been stored for several years. Because the gloves were packaged in separate boxes in each case (as shown in Figure 6-23a), investigators could not believe that the latex had self-heated until they found cases on the periphery of the main fire whose contents displayed the entire range of self-heating damage from mildly discolored to deeply charred (see Figures 6-23b and c). The large masses of gloves had been stored in an unventilated area, and the high summer temperatures finally tipped the balance toward runaway self-heating.[84]

Powders (dusts) or dried aerosols (from catalytic-setting paints) have been seen to accumulate on ventilation filters or hot surfaces (such as lamps) and develop into self-heating masses, causing fires. These processes are often aided by airflow and elevated temperature of the substrate.[85]

SELF-HEATING OILS

The most commonly encountered consumer product that can self-heat to spontaneously ignite is a "drying oil" such as linseed (extracted from flax seeds), tung nut, fish, or soybean oil. As the name implies, such oils harden by oxidation of the double bonds of the fatty acids in the oils, particularly the linolenic acid component.[86] As they oxidize they polymerize to form a tough, natural plastic coating. This coating property is why these oils have been used in paints and finishes for centuries. The oxidation process relies on oxygen in the air and generates heat. The reaction rate depends, then, on the amount of

FIGURE 6-23A Self-heating of latex surgical gloves stored for years in a warehouse. *Courtesy of U.S. Department of Justice, Bureau of Alcohol, Tobacco and Firearms and Explosives.*

FIGURE 6-23B Showing gradations of damage in cartons. *Courtesy of U.S. Department of Justice, Bureau of Alcohol, Tobacco and Firearms and Explosives.*

FIGURE 6-23C Note limited damage to individual boxes. *Courtesy of U.S. Department of Justice, Bureau of Alcohol, Tobacco, Firearms and Explosives.*

drying oil that is presented to the air. A quantity of such oil in a can will not self-heat to a detectable degree, but if the oil is spread as a thin layer on a porous substrate like cotton cloth, there can be enough surface area to promote rapid oxidation and heating. If this heat is trapped by insulation in such a manner that oxygen can still diffuse into the mass, the reaction rate will rise.

Raw linseed oil will demonstrate this effect but often will not spontaneously ignite, because bulk raw linseed oil contains oleic and linoleic acids, which are not as reactive as linolenic acid. Raw linseed oil does not dry as fast and hard as boiled linseed oil, which is made by boiling the product to concentrate the amount of the less saturated, more reactive fatty acids or, more commonly, by adding a drying agent, most often a metallic oxide. The drying agent triggers rapid setting (and thereby more rapid heat generation). The effect is that a quantity of cotton rags smeared with boiled linseed oil can self-heat to the point of autoignition when kept at room temperature for a period of just a few hours (as pictured in Figures 6-24a and b). No containment or external insulation is necessary, for the reaction is enhanced by the availability of oxygen, and enough insulation is provided by the overlying rags. Such a reaction has been observed by DeHaan to occur in a pile of rags bearing an estimated 100 ml (4 fl oz) of boiled linseed oil, piled in the open against a concrete wall at an ambient temperature of 25°C (77°F) in just 5 hours.[87]

Howitt et al. reported that as little as 25 to 50 ml (1 to fl oz) is adequate if other factors are optimum.[88] Reactions monitored in one series of tests revealed temperatures in excess of 400°C (750°F) within the mass of rags, and these are confirmed by Howitt's observations.[89] Older literature is consistent with these estimates. Clearly, once the mass of rags bursts into open flame, there will be enough heat to ignite other ordinary fuels in the vicinity. This is the mechanism thought to be responsible for the fire in the Meridian Plaza high-rise office building, and has been identified as the heat source in numerous fires in furniture shops and home workshops where wood-finishing products containing boiled linseed oil were used and residues wiped up with cloth rags.[90] The key elements that must be present are presence of a drying oil (most commonly of vegetable or animal origin), a porous support medium that allows free access of oxygen to diffuse into the mass and that does not melt when the temperature rises, provision of an adequate oxygen supply, and time for the reaction to occur. The higher the ambient (or starting) temperature, the faster the reaction will proceed. If the ambient temperature is below about 10°C (50°F), linseed oil is unlikely to self-heat to flaming ignition. Many wood stains and varnishes contain both a petroleum solvent and a linseed oil base. The petroleum solvent plays no role in the self-heating process except to slow it down by evaporative cooling. If the product is sprayed and then wiped up, the solvent will largely have evaporated and will not serve as a delay mechanism.

Self-heating does not occur with hydrocarbon (lubricating) oils or animal lard (except at unusually high ambient temperatures) and does not *usually* occur with the cooking oils commonly found in household kitchens [e.g., peanut, corn, olive, safflower, rapeseed (canola) at relatively low ambient temperatures]. One exception occurs when such oils are absorbed into porous foods like flour or onto cotton towels and are then exposed to high starting temperatures. This mechanism is very likely when the oil has been used to deep fry fish. Then, the effluent will contain fish oils (that can self-heat), and porous bread or batter crumbs, all at high temperatures.[91] Rags used to wipe down hot cooking griddles with various oils and then thrown into waste (or laundry) bins while still hot can also self-heat to ignition. Self-heating fires have occurred when hydrocarbon oils leaked into porous insulation lagging covering pipes running at an elevated temperature, but such fires will occur only in an industrial environment.[92]

Gaw reported that stacks of cotton towels oiled with linseed oil, then laundered and washed and stored at 93°C (200°F) self-heated to 575°C (1067°F) and then ignited into flaming combustion. He noted that significant oil remained in the towels after laundering, that bleach had no effect on the amount of oil removed, and that residual oil was

FIGURE 6-24A Spontaneous ignition of cotton rags with linseed oil. Note burning from center of mass. *Courtesy of Jamie Novak, Novak Investigations and St. Paul Fire Dept.*

FIGURE 6-24B Linseed oil rags smoldering as they are removed. Notice discoloration. *Courtesy of John Galvin, London Fire Brigade.*

FIGURE 6-24C Moments later, rags ignite in flames as they are exposed to air. *Courtesy of John Galvin, London Fire Brigade.*

recovered from the clothes washer pump even after laundering oiled towels and then a full load of nonoiled towels.[93] A more extensive review of self-heating processes and the investigation of spontaneous ignition fires has been published by DeHaan.[94]

The reactivity of oils is measured by their so-called iodine number. While there are exceptions, generally, the higher the iodine number, the more susceptible the oil is to self-heating. For example, linseed oil has an iodine value of 173–201, tung oil: 150–176, soybean oil: 137–143, cottonseed oil: 108–110, corn oil: 111–130, and peanut oil: 85–105.[95] The NFPA has rated many common products (both natural and synthetic) with respect to their susceptibility to self-heating, and the reader is referred there for specifics.[96] Stauffer has carried out extensive work in the chemistry and forensic identification of these products and the reader is referred there for more detail.[97]

SELF-HEATING OF VEGETATION

Aside from occasional serious fires that occur in the storage of chemicals, the most common fires termed *spontaneous* are without doubt those that are ignited in hay, grasses, bagasse, or other vegetation residues. The phenomenon has been known since ancient times, and considerable experimental work has been performed in attempting to elucidate its mechanism.[98] A case of spontaneous fire in a hay mow was observed and recorded in detail by Ranke in 1873, who was himself a scientist and who performed further experiments in support of his theory of strong adsorption of oxygen by pyrophoric carbon. This was apparently the first detailed and well-documented description and served as the starting point for nearly all later work on the subject.[99]

There are several sources of heat that can contribute to self-heating of hays, grasses, and the like. First, there is the heat of respiration of microbial populations and living materials (such as freshly cut grass). This can raise the temperature of a fuel mass to over 70°C (160°F), at which temperatures the cells and microorganisms die. In order for most vegetable materials to undergo self-heating, there must be a favorable moisture content to allow a fermentative process to proceed. Unlike the fuels already described, which undergo chemical changes, self-heating of vegetable materials is often a biological process dependent on oxygen and living microorganisms, at least at first. If the hay is well cured (dried), it will not allow destructive fermentation; if it is too wet, not enough oxygen can diffuse into the mass, too much heat is conducted out, and no fire will occur. Partially cured (dried) hay is best for self-heating (moisture content between 12 and 21 percent). The activity of the microorganisms will heat the hay to the thermal death point of the organisms as a limiting temperature. This, too, is in the neighborhood of 70°C (160°F) and lower for many organisms. Regardless of insulation that exists in a large mass of the vegetable material, temperatures in this range fall far short of the necessary ignition point of the hay, which is on the order of 280°C (540°F) and above.

The biological temperature limit makes clear that microorganisms can raise the temperature only to a point at which some exothermic chemical process(es) can be initiated to raise the temperature much higher. This phase of the process has been the subject of much attention and has produced various theories, including the production of "pyrophoric carbon," pyrophoric iron, heat from enzyme action, and even auto-oxidation of the oils contained in seeds. It has been reported that much acid is generated in the early stages of the process,[100] and that this is accompanied by marked browning of the hay. This high acidity has been used as a means of determining whether spontaneous combustion has occurred as contrasted with external ignition. Perhaps the most widely accepted theory is that of Browne, who postulated the formation of unsaturated compounds by the action of microorganisms.[101] These could form peroxides with oxygen of the air, and later form hydroxy compounds along with heat. Others do not agree that acidic degradation products are formed.[102] Rathbaum's work explored the complex role of temperature, ventilation, and moisture content for hay and other grasses.[103] There are several possible chemical routes that do not rely on biological mechanisms. Gray has demonstrated that bagasse (the waste material from sugar cane extraction) can self-heat in the hot, dry climate of inland Australia even in the absence of any microorganisms.[104] Apparently, between 70°C and 170°C (160°F–340°F), chemical decomposition reactions supported by moisture dominate, and above 170°C (340°F) oxidation reactions dominate as long as oxygen is available.

Whatever the actual mechanism may be, it is certain that it consists of a number of steps, the most important being chemical in nature. The packing density of the hay or grass, the size of the stack or bale, and its moisture content when left are critical factors. Other facts that are of importance to the investigator concern the environment—rainfall prior to ignition and ambient temperature of storage. Spontaneous fires in stacked hay do not usually occur in less than 10 to 14 days after stacking and generally require 5 to 10 weeks [under ideal conditions of moisture and very high ambient temperature (35°C–40°C; 95°F–105°F), ignition could be observed in 6 to 7 days].[105] The fire that occurs is in the center of the stack and burns to the exterior, usually forming some type of chimney or flue to the exterior. A fire from an outside source (accidental or intentional) burns toward the inside.[106] Unburned hay in a stack in which spontaneous combustion has occurred will often be very dark in color (with a color gradation ranging from gold-brown to chocolate colored to blackened to char) and may have a higher acidity than normal hay. Odors described as caramel-like or tobacco-like have been described. If a burning stack is spread apart, hay that is not on fire may suddenly burst into flame when fresh oxygen is introduced if spontaneous combustion is the cause of the fire (which preheats the fuel mass). Depending on the geometry of the stack, some exterior fires may preheat the interior. This is not expected to occur when the hay has been fired exteriorly, but small interior fires have been found after deliberately ignited test fires.

One of the interesting artifacts of spontaneously ignited hay fires is the formation of the "hay clinker" (see Figure 6-25). This glassy, irregular mass, gray to green in color, can be found throughout the pile wherever temperatures were the highest and most prolonged but are likely to be concentrated near the center. The clinker is composed of the inorganic residues of silicon, sodium, and calcium from the plant stems as well as impurities baled with the hay (soil, dust) fused into a glassy mass by the heat developed in the fire. Hicks has done an in-depth study of hay clinkers that strongly indicates that the inorganic constituents of various grasses may be linked to their propensity to form such glasses.[107]

FIGURE 6-25A Hay clinker from haystack fire ignited by spontaneous ignition. Clinker is gray-green in color, foamy-glassy in appearance, and is a natural glass formed by the oxides of silicon, calcium, magnesium, and sodium present as trace elements in grasses and feedstocks. *Courtesy of John D. DeHaan.*

FIGURE 6-25B A much larger hay clinker from a haystack fire contains some carbonized stems, as well as the glassy residues of the inorganic oxides. *Courtesy of Ryan B. Fields, Orca Fire Investigation, Medford, OR.*

Recently, Tinsley, Whaley, and Icove performed a series of live fire experiments examining the formation of hay clinkers in externally ignited fires. The result of that study indicated that the formation of hay clinkers is dependent on a variety of factors, none of which reliably indicate the origin or cause of the fire.[108] Hay fires in pole barns with metal roofs may cause the formation of numerous large masses as opposed to a few smaller clinkers. Clinkers do not occur in all self-heating hay fires, so their absence cannot be taken as proof of incendiary cause, and their presence should not be taken as proof of spontaneous origin. A hay clinker may be mistaken for the residue of an incendiary device, but it is innocuous. Elemental analysis is essential. It must also be remembered that spontaneous combustion is an easy decision to reach when the cause of the fire is not known. It seems certain that many fires that have been attributed to spontaneous causes were actually set.

When other materials of vegetable origin such as fiberboard are involved, some input of external heat is necessary to trigger enough oxidation. Fiberboard, pressboard, and plywood are made by subjecting a wood product and a resin to high pressures at high temperatures. When these products are manufactured, if they are not stacked to allow excess heat to escape, residual heat remaining in the stacked mass may trigger self-heating. Sawdust piles and other materials of wood have also been thought to ignite spontaneously, but they are unlikely to do so unless there is a drying oil or similar external contaminant present. Sanding waste from freshly varnished wood has been linked to self-heating, triggered by residual heat from the sanding process. This tendency is aggravated if the wood has been recently coated with a catalytic polymer finish whose setting involves an exothermic reaction, not just simple evaporation of a solvent.

OTHER MATERIALS SUBJECT TO SELF-HEATING

In addition to the two best-known examples of systems subject to spontaneous ignition, oily rags and hay, a number of other fuel systems may ignite spontaneously. One of the more troublesome and common ones is coal, lying unmined in the ground or at times in large piles or storage bins above ground, since the wetting of coal can cause generation of heat.[109] As with hay, exposure to water and the accumulated moisture content has been found to be critical to self-heating.[110] Without doubt, the process itself is a slow exothermic chemical one in which heat is generated faster than it can escape in an environment that is well insulated. Other factors of importance are the type of coal (some grades of which ignite far more readily than others) and the coal size (finer dusts being more susceptible). Massive quantities are needed, and days or weeks are required before flaming ignition can be achieved.

IMPLICATIONS FOR THE FIRE INVESTIGATOR

There is a temptation to label many accidental fires as "spontaneous" because there is no identifiable ignition source or obvious human intervention, but this is not correct. If the materials and processes cannot be specifically characterized as susceptible to self-heating under the prevailing conditions, then the cause must be considered to be unknown. Spontaneous ignition (with very rare exceptions) does not occur instantaneously, and the time frame for development is linked to the chemistry and mass of the reactant. Flaming ignition is always preceded by smoke and odors that should be detectable by anyone in the vicinity for some time prior to flaming ignition.

Other Sources of Ignition

Although accidental fires are invariably ignited by a small flame, electric arc, burning fragments, or other hot objects, there are some sources of ignition that do not fall neatly into those classes. In this category are lightning, spontaneous combustion, electric lighting, discarded batteries, and animal interaction with sources of ignition.

LIGHTNING

Long before humans learned to start a fire, it is almost certain that their knowledge of fire came from fires that were ignited by bolts of lightning. Lightning is a tremendous electrical discharge of natural origin that is not fundamentally different from the electric arc, except in intensity. It is a massive discharge of the huge static electric charges built up in clouds of moisture or dust. The clouds and the earth act as the plates of a condenser or capacitor; the potential across them can reach up to 100 million volts. The discharge, when it finally occurs, usually consists of several strokes back and forth, each lasting a few millionths of a second and 40 to 80 thousandths of a second apart. The currents involved are typically around 20,000 amperes, although much higher values are occasionally reached.[111] Lightning strikes are usually accompanied by the physical destruction of any poor electrical conductor in their path. Windows can be blown out, wood is splintered and shattered, and electric appliances can be exploded (see Figures 6-26a–c). The air in the path of the main stroke can be heated to a temperature of 30,000°C (54,032°F) and is expanded at a supersonic speed as a result. The resulting pressure shock wave can damage nearby structures with explosive force. In a lightning flash, the heat generated by the flow of current through a struck solid object will often ignite it. This is especially true when the object is dry and readily ignited. The loss of buildings to lightning-induced fires is still large, although protective devices now available have markedly reduced this loss within the last 50 years or so.

Lightning and Trees

Trees are ignited only occasionally by lightning, the result being determined chiefly by their susceptibility to combustion. An old, dead, dry trunk, for example, is much more likely to catch fire than a living tree, which is both a better conductor and less flammable. In general, many bolts of lightning will strike for every fire that is initiated. In regions where thunderstorms are common, however, the risk is very great. The probability of ignition of ground fuels is controlled by their fuel moisture and the depth of the accumulated ground layer.[112] In most instances, such fires are readily traced. When a fire of undetermined origin occurs in a remote area of mountain or forest, the investigation as to whether it was caused by lightning may be difficult. Meteorological records will generally establish whether there was a storm in the questioned area at the right time. Some forestry services maintain a special lightning watch to monitor and record lightning strikes in wildland areas, and there are meteorological satellites that record lightning activity. A lightning detection service can be of help in confirming or eliminating lightning as an ignition source in the United States. Vaisala Inc. maintains a lightning strike service called StrikeNet, which uses a network of electronic detectors. Its reports claim accuracy of 500 m as to location and can be requested online for particular time frames and various distances from a street address or GPS location. A typical StrikeNet report is shown in Figure 8-19. Search of the area may uncover a tree or other object that was split or damaged by means other than the fire itself. If this finding corresponds to the origin of the fire as indicated by study of the pattern, it is very probable that lightning was the origin.

IMPLICATIONS FOR THE FIRE INVESTIGATOR

The investigator needs to be familiar with the effects of lightning when it strikes objects. The most common effects are splitting of wood and mechanical disruption of structures. Most of these effects arise from sudden volatilization of water inside the tree or structure by heat from the electric current, so that the effect is that of an internal explosion. The effects are extremely variable, being determined by both the intensity of the electrical discharge and the localized condition of the object that is struck. (Figure 8-18 shows the fused earth produced by a lightning strike in soil.) Lightning strikes can induce power surges in nearby power or phone lines (causing failure and sometimes ignition of appliances

FIGURE 6-26A Lightning strike on house roof. Fire damage to roof where struck. *Courtesy of Greg Lampkin, Knox County TN Fire Investigation Unit.*

FIGURE 6-26B Limited fire damage, down pipe dislodged. *Courtesy of Greg Lampkin, Knox County, TN, Fire Investigation Unit.*

FIGURE 6-26C Interior wall damage as lightning followed electrical cables. *Courtesy of Greg Lampkin, Knox County TN Fire Investigation Unit.*

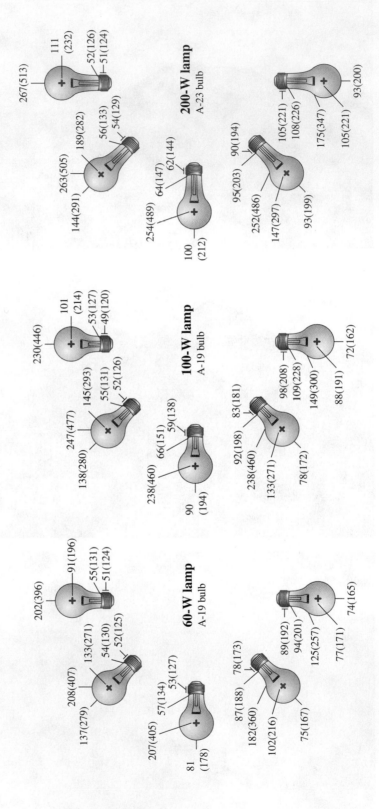

FIGURE 6-27 Surface temperatures of bulb and socket for typical incandescent lamps in open air (unlimited circulation) as a function of position. (Numbers in parentheses are temperatures in degrees Fahrenheit.) *Courtesy of GE Lighting.*

connected to them). Examination of appliances (including computers and telephones) throughout a structure or even in adjacent buildings is important in verifying the role of lightning (or lone power surges) in igniting a fire, as all appliances plugged in at the time are likely to be damaged.

IGNITION BY ELECTRIC LIGHTING

Incandescent lightbulbs can generate temperatures in open air at the surface of the glass bulb or envelope ranging from 74°C (165°F) to 267°C (513°F) depending on the wattage of the lamp and its position, as seen in Figure 6-27.[113] Such temperatures can scorch cellulosics or melt synthetics that come into contact but would be unlikely to result in flaming ignition. However, much higher temperatures can be produced if the bulb is enclosed in a housing or wrapped or buried in insulating material. Bulbs of low wattage normally produce a surface temperature inadequate to ignite common combustibles. As the wattage is increased (especially with bulbs in housings of some type that restrict ventilation), temperatures can build up to quite high values. Thus, potentially, a large incandescent light bulb (>50 W) can start a fire if it is in prolonged contact with suitable fuel. As was pointed out earlier in this chapter, the circumstances of contact between a heat source and a fuel control whether there is ignition. If a hot surface is placed in contact with a fuel long enough, the surface may provide insulation and reduce the heat losses from that surface, thereby increasing its temperature and its potential for ignition. (See Figure 6-28). Gruehn recorded smoldering ignition of a wood floor by prolonged contact of a 75-W lightbulb resting on it.[114] Even low-wattage lamps (<15W), if buried or wrapped in an insulating material, can generate sufficient temperatures to ignite paper, cloth, sawdust, grain, and the like. Such events are uncommon, not because of the lack of a high enough temperature, but rather, because something easily ignitable (such as a cellulosic material) must be in contact with the bulb, and this rarely occurs. The most likely areas for such ignitions are basements, storage rooms, and shops when packing material is allowed to be in contact with an operating bulb.

In tests conducted by DeHaan and Albers, a 60-W incandescent bulb buried in crumpled newspapers ignited them to flames in just 10 minutes 5 seconds. A 75-W bulb buried

FIGURE 6-28 A 75-W lamp igniting cellulose insulation. *Courtesy of Jamie Novak, Novak Investigations and St. Paul Fire Dept.*

in sawdust (uncontrolled fuel moisture) caused smoke in 6 minutes 13 seconds, but no flames were observed prior to suspension of the test at 28 minutes. There was melting and smoke generation of polyester fiberfill burying a 100-W bulb, but no ignition for 32 minutes. A 150-W decorative lamp had no effect on polyurethane foam resting atop it for some 30 minutes due to the large surface area and limited contact area but a 100-W lamp buried under polyurethane foam caused ignition after some two hours (see Figures 6-29a–c and 6-30).[115]

The breakage of any lit incandescent lamp, even low-voltage and low-wattage ones, introduces two additional sources of possible ignition: the filament and an arc. The operating temperature of a tungsten filament is on the order of 1,500°C (2,700°F). The filament begins to burn up as soon as air contacts it. The filament begins to cool as soon as the circuit is interrupted by failure of the filament, but it may retain more than enough

FIGURE 6-29A A 100-W lamp buried in polyurethane foam forces pyrolytic decomposition and char formation. *Courtesy of John Galvin, London Fire Brigade.*

FIGURE 6-29B Leading occasionally to flames. *Courtesy of John Galvin, London Fire Brigade.*

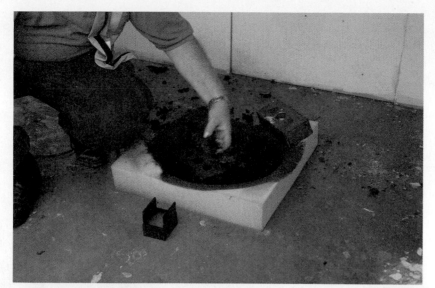

FIGURE 6-29C After flaming combustion, a rigid, porous, but very fragile char is formed around the heat source. *Courtesy of John Galvin, London Fire Brigade.*

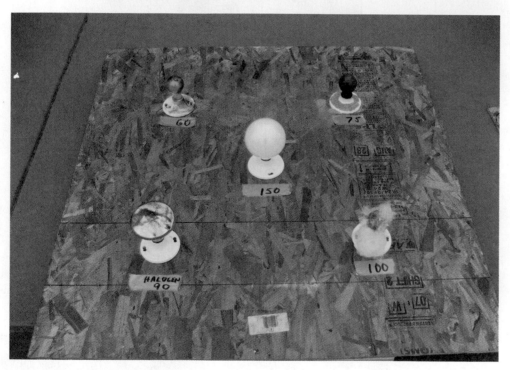

FIGURE 6-30 Test Panel for Lamp Ignition. The 60-W lamp ignited crumpled newspaper in ~10 minutes. The 75-W lamp caused smoldering of sawdust in 6 minutes, 13 seconds. The 100-W lamp caused melting and offgassing of polyester fiberfill. The 150-W lamp (center) failed to ignite a polyurethane foam pad. *Courtesy of John D. DeHaan.*

heat to ignite flammable vapors or solids that come into contact with it shortly after failure. In addition, the brief arc produced when the filament fails produces the same risk of ignition as any other electric arc if flammable gases or vapors are in the vicinity.

Fluorescent lights operate at much lower temperatures than incandescent lamps, with the exterior of the tube rarely exceeding 60°C to 80°C (140°F to 180°F).[116] Mechanisms involving failures of connectors for fluorescent lamps and power cords, and compact fluorescent lamps, will be discussed in Chapter 10 on electricity. Quartz halogen lamps contain a quantity of iodine or bromine vapor and a tungsten filament operating at very high temperatures. The surface temperature of the fused quartz envelope can be as high as 600°C to 900°C (1,100°F to 1,650°F), and the temperature of a nearby reflector can be proportionally high.[117] Even brief contact between a fuel and the lamp (or even the reflector) could result in flaming ignition (see Figure 6-31). Halogen lamps can also ignite

FIGURE 6-31 Halogen lamp ignites curtain valance after less than one minute of contact. *Courtesy of Jamie Novak, Novak Investigations and St. Paul Fire Dept.*

fuels by radiant heat and melt diffusers, shades, and fixtures. They have also been known to shatter while lit, scattering hot fragments of glass and metal onto combustibles.[118] An example of a high-wattage metal halide lamp whose element shattered, with the fragments penetrating both inner and outer enclosures is shown in Figure 6-32. The hot metal and quartz fragments were believed to have ignited cellulosic materials directly beneath. Lamps with an extra-shatter-resistant glass shroud are available but are rarely used because of their significant extra cost.

FIGURE 6-32 Metal halide lamp failure caused ignition. *Courtesy of Peter Brosz, Brosz & Associates Forensic Services Inc.; Professor Helmut Brosz, Institute of Forensic Electro-Pathology.*

IGNITION FROM DISCARDED BATTERIES

Most batteries for consumer devices (1.5 to 9 V) can support current discharges of up to 1 A when fresh. If metal can form a conductive path between the terminals, sufficient electric current heating can occur to ignite cellulosic materials in direct contact. Nine-volt batteries have both terminals at one end, so a coin or a piece of aluminum foil can act as a heating element if contact is made with both terminals. Lithium-ion battery failures can also cause fires.

ANIMAL INTERACTION WITH SOURCES OF IGNITION

The very mysterious origin of some fires may be clarified by considering the intervention of various animal species carrying out unnoticed activities of their own. It has been suggested that a rat may chew a match head and thereby ignite it. Considering the ingredients in common matches, it is unlikely that they would present much temptation; therefore, this explanation must be regarded with skepticism. Some evidence has been found that a rat or mouse carrying a strike-anywhere match in its mouth could strike it accidentally against a rough surface such as brick or concrete, lighting it in passing. The final effect will be the same as with any other match-ignited fire, but the combination of factors makes it most unlikely to occur.

Less consideration has been given to the possibility that both rodents and birds may carry lighted cigarettes or other burning objects, which then ignite combustibles, possibly even rooftops. Many rodents and birds have a strong instinct for acquiring and storing a variety of materials, either for nest building or other reasons. Paper objects are especially attractive, for example, to pack rats, whose nests have been found to be lined with facial tissues. The paper cigarette may be similarly attractive, even if still aglow, although we know of no proven instance of such an occurrence. (Admittedly, the ignition of a fire on a clear, dry day at the top of a tall palm tree occupied by birds notorious for their scavenging has no other rational explanation other than one has picked up a still-smoldering cigarette butt for its nest). Some birds are known to carry sticks, nest materials, and other objects and then drop them down the roof vents of heating appliances. Such a practice is clearly one that can produce a fire hazard. If the object was a lighted cigarette and it was dropped onto flammable material, a fire could result without human intent or carelessness, other than the discarding of an unextinguished cigarette. The effects of bird nests and rodent nests in chimneys and flues have been previously noted. Birds have also been known to build nests on top of or adjacent to high-wattage flood lamps or other intermittent heat sources. When the lamp or appliance is turned on, a delayed ignition can occur. DeHaan has ignited dried grass and leaf debris in a loose pile in front of a 250-W flood lamp. (In that case the lamp was mounted at ground level, and windblown leaves and grass had accumulated around it.)

Mice and rats have been known to chew the insulation off wires, either short-circuiting them directly or leaving the exposed conductors awaiting some other contact. A simple short across the exposed conductors will be very unlikely to ignite a fire rather than simply trip the circuit protection. If foreign materials (from saltwater to portions of the animal's anatomy) can sustain current flow, the heat generated can degrade the insulation and lead to ignition of the insulation and other fuels in the immediate vicinity. This mechanism probably represents the single most significant contribution to fire causation by animals. Post-fire signs may be minimal once the surrounding insulation has been destroyed, but unburned insulation has a gnawed appearance, as seen in Figure 6-33. (See also Figure 6-34.) Snails, slugs, birds, and other animals have been known to short out electrical equipment, thereby starting fires both inside structures and out. Dogs and cats have been known to knock over containers of flammable liquids, which are then ignited by remote ignition sources. They have also been known to knock candles, Christmas trees, and lamps (such as the heat lamps used to keep pets, poultry, and livestock warm) into contact with combustibles. Moths have been offered as a mechanism for starting spot

FIGURE 6-33 Rodent-gnawed wires in this RV refrigerator came close to being a fire cause. Fortunately, only the bare ground and neutral conductors had been exposed at this point. *Courtesy Joe Bloom, Bloom Fire Investigation.*

FIGURE 6-34 A grass fire started at the base of this power pole. The remains of the perpetrator can be seen as the tail and the foot on the left side of the transformer. Note "flash burn" on adjacent pole. *Courtesy David J. Icove, Ph.D., University of Tennessee.*

fires by catching fire in a candle flame and then flying some distance to ignite remote fuels. One author pointed out the absurdity of this explanation: "A moth on fire would be aerodynamically unstable and unable to fly very far even if the chitin from which the wings are made supported combustion."[119]

There appears to have been little or no study made of the habits of animal species in relation to fire hazard, since such studies would be very difficult to carry out. Even without more exact information, it is apparent that fires have at times been started by animals, and more attention to this possible source of fire hazard is clearly indicated. A good investigation would include consideration of such mechanisms and a search for evidence to refute or confirm the hypothesis.

Assessing Ignition Sources at the Fire Scene: The Ignition Matrix

Nearly every room (or vehicle) offers multiple possible ignition sources—not only from inherent energy sources such as fixed wiring, fixtures, outlets, heaters (gas or electric), or other "appliances"—but also from the variety of human activities that may bring more ignition sources into action. These additional ignition sources include smoking materials, use or misuse of electric tools or appliances and devices, extension cords and power strips, candles, and countless more. Today, more than ever before, fire investigators face the challenge of identifying all potential ignition sources in the area of origin and then eliminating them, down to one (based on reasonable data or methods). As discussed in Chapter 1, the scientific method of fire investigation expects that an investigator will reach a defensible conclusion by demonstrating not only how this source could have started the fire but also how all other reasonably possible sources have been excluded. If two or more ignition sources are deemed "possible" at the end of the investigation, the cause has to be declared to be "undetermined." Demonstrating these conclusions can be difficult, especially in a concise, yet comprehensive manner. Electrical engineer Lou Bilancia has developed an innovative way of systematically evaluating numerous ignition sources and also documenting how each was or was not competent to ignite a particular first fuel. The suggested ignition matrix is merely a grid on paper. Based on physical examination of the scene, interviews with occupants, or pre-fire photos or videos, the investigator identifies and lists all potential first fuels in the room in a vertical column (as shown in Figure 6-35). The investigator then lists all the possible ignition sources in a row across the top of the page. Each square, then, represents the interaction between an energy source and a first fuel.[120] Each combination is then evaluated on four primary values:

1. Is this ignition source competent to ignite this fuel? Yes or no.
2. Is this ignition source close enough to this fuel to be capable of igniting it? Yes or no.
3. Is there evidence of ignition? Yes or no.
4. Is there a pathway for a fire ignited in this first fuel to ignite the main fuel? Yes or no.

There are often comments or notes concerning each evaluation, such as the following:

- This source was not energized (appliance or power cord) or not in use (candle).
- "Too far away" or "Depends on duration being adequate," or "Unknown—must test exemplar."
- Physical or visual evidence of burning, ignition was witnessed or recorded in video.
- Plume developed from first fuel sufficient, or flashover produced ignition.

The matrix can then be color-coded to indicate which combinations are capable, which are excluded, and which need further data.

A typical bedroom, then, may have 15 or more potential first fuels: bed (mattress, box spring, frame), bedding (blankets, pillows, sheets, comforters), carpet and pad, dresser,

FIGURE 6-35 Ignition matrix—identified fuels along vertical axis; ignition sources along horizontal. *Courtesy of Lou Bilancia, Synnovation Engineering and HTRI Forensics. Copyright 2008–10. All rights reserved. Used with permission.*

Source / Fuel	Clock radio	Cell phone charger	Cigarette	Candle	Compact fluorescent lamp	Plug-in room refresher
Lace night table cover	1. yes (a) 2. yes close 3. no (d) 4. yes path	1. yes (a, c) 2. not close 3. 4. yes	1. 2. yes close 3. no evid. 4.	1. yes comp 2. yes 3. maybe 4. yes	1. yes (a, c) 2. yes 3. no (d) 4. yes gravity	1. yes (a) 2. not close 3. no (d) 4.
Sheets or covers	1. no (c) 2. not close 3. no (d) 4. yes gravity	1. yes (a, c) 2. yes close 3. no (d) 4. yes	1. 2. not close 3. no evid. 4.	1. yes comp 2. not close 3. 4.	1. yes (a, c) 2. no 3. no (d) 4. no	1. yes (a) 2. yes close 3. 4.
Curtains	1. no (c) 2. not close 3. no (d) 4. yes path	1. yes (a, c) 2. yes close 3. no (d) 4. yes	1. 2. not close 3. no evid. 4.	1. yes comp 2. not close 3. 4.	1. yes (a, c) 2. no 3. no (d) 4. yes shade	1. yes (a) 2. yes close 3. no (d) 4.
Plastic decorative flowers	1. no (c) 2. not close 3. no (d) 4.	1. yes (a, c) 2. not close 3. no (d) 4. no	1. not comp. 2. 3. no evid. 4.	1. yes comp 2. yes close 3. maybe 4. yes	1. no (a, c) 2. no 3. no (d) 4. yes shade	1. no (a, c) 2. not close 3. no (d) 4. no
Lamp shade	1. yes comp 2. not close 3. no (d) 4. yes path	1. yes (a, c) 2. not close 3. no (d) 4. no	1. 2. not close 3. no evid. 4.	1. yes comp 2. not close 3. indirect 4. yes path	1. no (a, c) 2. yes close 3. no (d) 4. yes	1. no (c) 2. not close 3. no (d) 4. no

1. Competent ignition source Y/N?
2. Proximity, ignition close to fuel Y/N?
3. Evidence of ignition Y/N?
4. Initial fuel path to fuel load Y/N?

Color legend
Red – Competent and close
Blue – Not competent
Yellow – Competent but ruled out

Codes
P – plume or flashover
W – witnessed
F – open flame
N – not energized

Notes
a. if the device failed
b. if fuel was cellulosic with a breeze
c. only with open flame
d. device was intact

TV cabinet, clothes, posters, chairs, draperies, books, bookshelves, and so forth. Wall and ceiling coverings must also be included if they are combustible. The bedroom may also have numerous potential ignition sources (based again on physical examination of the scene, pre-fire photos or videos, and interviews with recent occupants). These may include fixed wiring, outlets, switches, and fixtures; table lamps, clock, radio, TV, electric blanket, extension and power cords, cigarettes, candles, and matches. Although there may be many competent ignition-fuel combinations in such a room, many will be excluded based on physical separation alone. A candle on the dresser will not be capable, for instance, of directly igniting first fuels more than an inch or two away. (Movement of such ignition sources, however, must always be considered.) If the definable (and defensible) area of origin is small, then the lists of first fuels and ignition sources will be considerably shorter than if the whole room is being considered. The investigator must be careful, however, of excluding fuels or sources merely if they are "outside the area of origin" by a matter of a few inches.

This approach forces the investigator to consider a whole range of alternative hypotheses and consider each one on the basis of factors such as heat release rate, heat flux, separation distances, thermal inertia, and routes of fire spread. A completed matrix offers a concise demonstration that all potential ignition sources have been considered and all (but one, presumably) eliminated. Such a matrix is also more easily verified than the traditional item-by-item list, improving the investigator's own review process.

Summary

No matter what the nature of the fuel, its susceptibility to ignition, or the duration of the suspected ignition source, it is important to remember that an ignition source must be of sufficient energy and in contact with (or at least capable of transferring heat to) appropriate fuel long enough to raise the fuel to its ignition temperature to trigger the combustion of that fuel under the ambient conditions present. That combustion may be in the form of a self-sustaining smolder or open flames, and each requires particular conditions. As we have seen, flames, arcs, sparks, self-heating, and heated objects can all initiate fires under the right conditions, and all must be considered whenever a fire scene is being examined. It is not enough simply to have an ignition source with a given temperature. The cause of a fire is the source of energy, the first fuel ignited, and the circumstances under which those are brought together that resulted in ignition. The investigator must realize that the circumstances of ignition (and the contact that precedes it) may change the temperature of the source, the chemical properties of the fuel, or the physical state of the fuel in complex ways. All materials and reactions must follow certain physical and chemical laws, and fuels and oxidizers are no exceptions. The analytical fire investigator must ensure that a postulated sequence of events in a fire not only fits observed conditions but also follows these laws. Reliance on published data to confirm some ignition hypotheses may be enough if the conditions of the published tests are a reasonable match to the fire conditions. If they are not, properly designed and executed tests may be necessary to prove or disprove a particular hypothesis.

We have now completed an examination of all the basic elements of fire—chemistry of combustion, behavior of fuels, and the mechanisms of ignition. We are now in a position to better understand the behavior of fire in a structure or wildland and interpret the post-fire diagnostic signs in light of this knowledge.

Review Questions

1. List five different open-flame fires (ignition sources) and list their heat release rates.
2. Why are candles a more common accidental ignition source today?
3. What effect does "insulation" have on the temperature of an ignition source like a lightbulb?
4. Name three ways in which gas appliances can start accidental fires.
5. How can kerosene room heaters cause fires?
6. What fuel surface temperature is required to ignite fresh whole wood? What temperature is considered a potential ignition risk for wood exposed on a long-term basis?
7. Name three sources of frictional mechanical sparks.
8. Why will cigarettes not ignite gasoline vapors but will ignite acetylene or hydrogen?
9. Name three ways in which animals can cause fires.
10. List five conditions for spontaneous combustion to occur.

References

1. B. Karlsson and J. G. Quintiere, *Enclosure Fire Dynamics* (Boca Raton, FL: CRC Press, 2000), 13.
2. J. Andrasko, "Identification of Burnt Matches by Scanning Electron Microscopy," *Journal of Forensic Sciences* 24 (1978): 627–42.
3. V. Babrauskas, *Ignition Handbook* (Issaquah, WA: Fire Science Publishers, 2003), 509.
4. D. D. Drysdale, *An Introduction to Fire Dynamics*, 2nd ed. (Chichester, UK: Wiley, 1999), 165.
5. J. W. Williamson and A. W. Marshall, "Characterizing the Ignition Hazard from Cigarette Lighter Flames," *Fire Safety Journal* 40, no. 5 (July 2005): 491–92.
6. J. D. DeHaan, unpublished data, February 1996.

7. "Propane Torch Flame Exceeds 2400°F," *Fire Findings* 13, no. 2 (Spring 2005): 5.

8. S. E. Dillon and A. Hamins, "Ignition Propensity and Heat Flux Profiles of Candle Flames for Fire Investigation" in *Proceedings Fire and Materials 2003* (London: Interscience Communications): 363–76; M. Ahrens, *Home Candle Fires* (Quincy, MA: NFPA, 2002).

9. J. Nicholson, "When You Go Out—Blow Out," *NFPA Journal* (September–October 2005): 68–73.

10. Dillon and Hamins, "Ignition Propensity and Heat Flux Profiles."

11. M. Faraday, *The Chemical History of a Candle* (Atlanta: Cherokee, 1993).

12. D. Townsend, "Fires and Tea Light Candles," Fire *Engineers Journal* (January 2000): 10.

13. J. L. Sanderson, "Candle Fires," *Fire Findings* 5, no. 4, (Fall 1997); H. Singh, *Hazard Report on Candle-Related Incidents* (Washington, DC: U.S. CPSC, March 1998).

14. R. Holleyhead, "Ignition of Flammable Gases and Liquids by Cigarettes: A Review," *Science and Justice* 36, no. 4 (1996): 262–66.

15. Babrauskas, *Ignition Handbook*, 213.

16. S. Corwin, "Halogen Work Light Testing," *Fire Findings* 9, no. 2 (Spring 2001): 12–13.

17. "How Hot Does It Get?" *Fire Findings* 8, no. 2 (Spring 2000): 5.

18. M. Goodson and G. Hardin, "Electric Cooktop Fires," *Fire and Arson Investigator* (October 2003): 31–34.

19. *NFPA 921: Guide for Fire and Explosion Investigations* (Quincy, MA: NFPA, 2004), table 5.3.1.1, 19.

20. J. L. Sanderson. "Heat-Bulb Testing: Wood, Straw and Towels Respond Differently at Close Distances to Lamps," *Fire Findings* 8, no. 4 (Fall 2000); R. E. Lowe and J. A. Lowe, "Halogen Lamps II," *Fire and Arson Investigator* (April 2001): 39–42.

21. R. J. Hall, J. J. Wanko, and M. E. Nikityn, "Special Hazards Fire Investigation," *NFPA Journal* (November–December 2008): 68–73.

22. "Hot Surface Ignition," *Fire Findings* 13, no. 2 (Spring 2005): 6.

23. M. Goodson, D. Sneed, and M. Keller, "Electrically Induced Fuel Gas Fires," *Fire and Arson Investigator,* (July 1999): 10–12.

24. "How Hot Does It Get?" *Fire Findings* 7, no. 2 (Spring 1999); DeHaan, unpublished data.

25. "Instant Demand Water Heaters," *Fire Findings* 14, no. 1 (Winter 2006): 4.

26. J. M. Hoffman et al., "Effectiveness of Gas-Fired Water Heater Elevation in the Reduction of Ignition of Vapors from Flammable Liquid Spills," *Fire Technology* 39 (2003): 119–32.

27. J. L. Sanderson, "Manufacturers Vary Methods to Incorporate FVIR Technology," *Fire Findings* (Fall 2004): 1–4.

28. J. Bertoni, "The Essentials of Gas and Oil Fired Forced Warm Air Furnaces," *Fire and Arson Investigator* (December 1996): 15–18.

29. N. Dwyer, "Flame Rollout." *Fire Findings* 14, no. 1 (Winter 2006): 12-13.

30. J. Nicholson, "Watch What You Heat," *NFPA Journal* (September–October 2006), 67–73.

31. Babrauskas, *Ignition Handbook,* 886

32. J. D. DeHaan, CCAI tests (unpublished data), July 2002.

33. "How Hot Does It Get?" *Fire Findings* 11, no. 2 (Spring 2003).

34. J. J. Lentini, "Vapor Pressures, Flash Points, and the Case against Kerosene Heaters," *Fire and Arson Investigator* 40, no. 3 (March 1990).

35. R. W. Henderson and G. R. Lightsey, "An Anti-Flareup Device for Barometric Kerosene Heaters," *Fire and Arson Investigator* 45, no. 2 (December 1994): 8.

36. D. McAuley, "The Combustion Hazard of Plastic Domestic Heating Oil Tanks and Their Contents," *Science and Justice* 43, no. 3 (July–September 2003): 145–58.

37. J. D. DeHaan, unpublished data, July 2003.

38. USFA, Santana Row Development Fire, USFA-TR-153 (Emmitsburg, MD: FEMA, January 2004).

39. J. A. Bohn, "Woodstove Ashes Staying Hot for 110 Hours," personal communication, Longmont Fire Department, February 2003.

40. B. A. Schwartz, "Pyrophoric Carbon Fires," *Fire and Arson Investigator* (December 1996): 7–9; B. R. Cuzzillo and P. J. Pagni, "Low Temperature Wood Ignition," *Fire Findings* 7, no. 2 (Spring 1999): 7–10.

41. P. C. Bowes, *Self-Heating: Evaluating and Controlling the Hazards* (Amsterdam: Elsevier, 1984).

42. J. C. Martin and P. Margot, "Approche Thermodynamique de la Recherche des Causes des Incendies. Inflammation du Bois I." *Kriminalistik und Forensische Wissenschaften* 82 (1994): 33–50.

43. Drysdale, *An Introduction to Fire Dynamics;* V. Babrauskas, "Pyrophoric Carbon: The Jury Is Still Out," *Fire and Arson Investigator* (January 2001): 12–14.

44. B. R. Cuzzillo and P. J. Pagni, "The Myth of Pyrophoric Carbon" in *Proceedings Sixth International Symposium on Fire Safety Science,* Poitiers, France, 1999, 301–12.

45. Bowes, *Self-Heating.*

46. V. Babrauskas, B. F. Gray, and M. L. Janssens, "Prudent Practices for the Design and Installation of Heat-Producing Devices near Wood Materials," *Fire and Materials* 31 (2007): 125–35.

47. Building Code of the City of New York, Sec. [C26-1400.6] 27-792 and Sec. [C26-1409.1] 27-809. See also *UL127: Factory Built Fireplaces* (Northbrook, IL: Underwriters Laboratories).

48. Babrauskas, *Ignition Handbook.*

49. Ibid., 507–16.

50. Ibid., 509.

51. F. D. Maxwell and C. L. Mohler, *Exhaust Particle Ignition Characteristics* (Department of Statistics, University of California Riverside, February 1973).

52. Babrauskas, *Ignition Handbook,* 508–16.

53. V. Babrauskas, "Risk of Ignition of Forest Fires from Black Powder or Muzzle-Loading Firearms," Report for U.S. Forest Service (San Dimas T & D Center, 2005).

54. E. W. Few, "Ammunition Identification Guide," *Arson Investigator,* California Conference of Arson Investigators, Spring 1978, 24–29.

55. "Skyblazer Best in Low-Cost Aerial Pyrotechnics," *Powerboat Reports* (September 1994).

56. Holleyhead, "Ignition of Flammable Gases and Liquids by Cigarettes."

57. B. Beland, "On the Measurement of Temperature," *Fire and Arson Investigator* (September 1994); *NFPA 921,* Table 5.3.1.1.

58. G. Damant, lecture, California Conference of Arson Investigators (Sacramento, CA: California Department of Consumer Affairs, Bureau of Home Furnishings, June 1979).

59. D. A. Eichman, "Cigarette Burning Rates," *Fire and Arson Investigator* (June 1980).

60. J. F. Krasny, W. J. Parker, and V. Babrauskas, *Fire Behavior of Upholstered Furniture and Mattresses* (Norwich, NY: William Andrew, 2001); G. Damant, "Cigarette-Induced Smoldering of Uncovered Flexible Polyurethane Foams," *Journal of Consumer Product Flammability* 2 (June 1975).

61. J. L. Sanderson, "Cigarette Fires in Paper Trash," *Fire Findings* 6, no. 1 (Winter 1998).

62. R. Berrett, personal communication, November 1998.

63. R. Holleyhead, "Ignition of Solid Materials and Furniture by Lighted Cigarettes: A Review," *Science and Justice* 39, no. 2 (1999): 75–102.

64. J. M. Shannon, "Fire-Safe Cigarettes; Keep Fighting," *NFPA Journal* (March–April 2009): 6; R. G. Gann et al., "Relative Ignition Propensity of Test Market Cigarettes," NIST Technical Note 1436 (Gaithersburg, MD: NIST, January 2001).

65. A. L. Miller, *The U.S. Home Product Report, 1987–91: Forms and Types of Materials First Ignited in Fire* (Quincy, MA: NFPA, February 1994).

66. E. Braun et al., "Cigarette Ignition of Upholstered Chairs," *Journal of Consumer Product Flammability* 9 (1982).

67. J. A. McCormack et al., "Flaming Combustion of Upholstered Furniture Ignited by Smoldering Cigarettes" (North Highlands, CA: California Department of Consumer Affairs, Bureau of Home Furnishings, 1986).

68. Krasny et al., *Fire Behavior of Upholstered Furniture and Mattresses.*

69. G. H. Damant, "Cigarette Ignition of Upholstered Furniture" (Sacramento, CA: Inter-City Testing and Consulting, 1994); R. Holleyhead, "Ignition of Solid Materials and Furniture by Lighted Cigarettes."

70. R. Holleyhead, "Ignition of Flammable Gases and Liquids by Cigarettes: A Review," *Science & Justice* 36, no. 4 (1996): 262–66.

71. Ibid.

72. Babrauskas, V., *Ignition Handbook,* 2003, 717.

73. J. E Malooly and K. Steckler, "Ignition of Gasoline by Cigarette," ATF (May 2005); see also R. E. Tontarski

et al., "Who Knew? Cigarettes and Gasoline Do Mix," presentation B84 at American Academy of Forensic Sciences 59th Annual Meeting, San Antonio, TX, February 19–24, 2007.

74. J. L. Sanderson, "Cigarette Fires in Fabrics," *Fire Findings,* 15, no. 3 (Summer 2007): 1–3.

75. D. A. Schuh and J. L. Sanderson, "Cigarette Testing Reveals Dry Potting Soil, Peat Moss Can Be Viable Fuel Sources," *Fire Findings* 17, no. 2 (Spring 2009): 1–3.

76. J. D. DeHaan, "Spontaneous Combustion: What Really Happens," *Fire and Arson Investigator* (January and April 1996).

77. F. C. Wolters, P. J. Pagni, T. R. Frost, and D. W. Vanderhoot, *Size Constraints on Self-Ignition of Charcoal Briquets* (Pleasanton, CA: Clorox Technical Center, 1987).

78. N. D. Reese, G. J. Kloock, and D. J. Brien, "Clothes Dryer Fires," *Fire and Arson Investigator* (July 1998): 17–19; J. L. Sanderson and D. Schudel, "Clothes Dryer Lint: Spontaneous Heating Doesn't Occur in Any of 16 Tests," *Fire Findings* 6, no. 4 (Fall 1998): 1–3.

79. D. C. Mann and M. Fitz, "Washing Machine Effluent May Provide Clues in Dryer Fire Investigations," *Fire Findings* 7, no. 4 (Fall 1999): 4.

80. W. M. Mackey, *Journal of the Society of Chemical Industries* 14, no. 940 (1895); D. A. Frank-Kamenetski, *Diffusion and Heat Transfer in Chemical Kinetics* (New York: Plenum, 1969).

81. G. Helmiss and W. Schwanebeck, "Aspects of the Investigation of the Chemical Processes of Self-Heating by Means of Quantitative Thermal Analysis," *Journal of Forensic Sciences* 30, no. 535 (April 1985); A. Baxton, E. Stauffer, and O. Delemonte, "Evaluation of the Self-Heating Tendency of Vegetable Oils by Differential Scanning Calorimetry," *Journal of Forensic Science* 53, no. 6 (November 2008): 1334–46.

82. Babrauskas, *Ignition Handbook,* 405–36.

83. B. F. Gray, "Interpretation of Small Scale Test Data for Industrial Spontaneous Ignition Hazards" in *Proceedings Interflam 2001* (London: Interscience Communications), 719–29.

84. D. Hevesi, "Warehouse Caught Fire When Gloves Combusted," *New York Times,* August 10, 1995.

85. D. Kong, "How to Prevent Self-Heating (Self-Ignition) in Drying Operations," *Fire and Arson Investigator* (October 2003): 45–48.

86. D. G. Howitt, E. Zhang, and B. R. Sanders, "The Spontaneous Combustion of Linseed Oil" in *Proceedings International Conference of Fire Safety,* San Francisco, CA, January 1995.

87. J. D. DeHaan, unpublished data, 1993.

88. Howitt, Zhang, and Sanders, "The Spontaneous Combustion of Linseed Oil."

89. Ibid.

90. B. Dixon, "Spontaneous Combustion," *Journal of the Canadian Association of Fire Investigators* (March 1992); Ziegler, D. L. "The One Meridian Plaza Fire:

A Team Response." *Fire and Arson Investigator,* September 1993.

91. B. M. Dixon, "The Potential for Self-Heating of Deep-Fried Food Products" in *Proceedings FBI International Symposium on the Forensic Aspects of Arson Investigators,* Fairfax, VA, 1995.

92. Babrauskas, *Ignition Handbook,* 403–4; see also Bowes, *Self-Heating.*

93. K. Gaw, "Autoignition Behavior of Oiled and Washed Cotton Towels" in *Proceedings Fire and Materials,* San Francisco, CA, 2005.

94. J. D. DeHaan, "Spontaneous Ignition: What Really Happens, Part I," *Fire and Arson Investigator* 46 (March 1996); "Part II," *Fire and Arson Investigator* 46 (June 1996).

95. W. Vlachos and C. A. Vlachos, *The Fire and Explosion Hazards of Commercial Oils* (Philadelphia PA: Vlachos & Co., 1921).

96. NFPA, *Fire Protection Handbook,* 20th ed. (Quincy, MA: NFPA, 2008), table 6.17.11, pp. 6–288 to 6–292.

97. E. Stauffer, "A Review of the Analysis of Vegetable Oil Residues from Fire Debris Samples," *Journal of Forensic Science* 50, no. 5 (September 2005): 1091–1100; E. Stauffer, "A Review of the Analysis of Vegetable Oil Residues from Fire Debris Samples," *Journal of Forensic Science* 51, no. 5 (Septemeber 2006): 1016–32.

98. E. J. Hoffman, *Journal of Agricultural Research* 61, no. 241 (1941); J. B. Firth and R. E. Stuckey, *Society of the Chemical Industry* 64, no. 13 (1945); and 65, no. 275 (1946); *Proceedings International Symposium: Self-Heating of Organic Materials*, Delft, Netherlands, February 1971.

99. H. Ranke, *Liebig's Annals of Chemistry* 167 (1873): 361–68.

100. Firth and Stuckey, *Society of the Chemical Industry.*

101. C. A. Browne, "The Ignition Temperature of Solid Materials," *NFPA Quarterly* 28, no. 2 (1935).

102. Bowes, *Self-Heating.*

103. H. P. Rathbaum, "Spontaneous Combustion of Hay," *Journal of Applied Chemistry* 13 (July 1963): 291–302.

104. B. F. Gray, "Spontaneous Combustion and Its Relevance to Arson Investigation," paper presented at IAAI-NSW Annual Conference, Sydney, Australia, September 1990.

105. L. Lowenweent, "Fire Investigation," *International Criminal Police Review,* no. 344 (January 1981): 2–3; see also *Proceedings International Symposium: Self-Heating of Organic Materials.*

106. Firth and Stuckey, *Society of the Chemical Industry.*

107. A. J. Hicks, "Hay Clinkers as Evidence of Spontaneous Combustion," *Fire and Arson Investigator,* July 1998, 10–13.

108. A. T. Tinsley, M. Whaley, and D. J. Icove, "Analysis of Hay Clinker as an Indicator of Fire Cause," paper presented at *ISFI 2010,* Adelphi, MD, September 27–29, 2010.

109. D. J. Hodges, "Spontaneous Combustion: The Influence of Moisture in the Spontaneous Ignition of Coal," *Colliery Guardian* 207, no. 678 (1973); N. Berkowitz, "Heats of Wetting and Spontaneous Ignition of Coal," *Fuel* 30, no. 94 (1951).

110. A. V. Smirnova and A. K. Shubnikov, "Effect of Moisture on the Oxidative Processes of Coals," *Chemical Abstracts* 51, no. 18548 (November 25, 1951).

111. V. Fuchs, *Forces of Nature* (London: Thames and Hudson, 1977), 62–63.

112. D. J. Latham and J. A. Schlieter, *Ignition Probabilities of Wildland Fuels Based on Simulated Lightning Discharges,* USD Research Paper INT-411 (Ogden, UT: USDA Forest Service, September 1989).

113. General Electric Co., Incandescent Lamps, General Electric Publication TP-110.

114. R. L. Gruehn, "To Be Prepared Is Everything," *Fire and Arson Investigator* 39, no. 1 (September 1988): 54.

115. J. D. DeHaan, and J. C. Albers, Low-Energy Ignition Tests, CCAI, November 7, 2007 (unpublished data).

116. R. A. Cooke and R. H. Ide, *Principles of Fire Investigation* (Leicester, UK: Institution of Fire Engineers, 1985).

117. Lowe and Lowe, "Halogen Lamps II."

118. Ibid.

119. Cooke and Ide, *Principles of Fire Investigation.*

120. L. Bilancia, "The Ignition Matrix," CCAI, November 2007.

7

Structure Fires and Their Investigation

Courtesy of David J. Icove, Ph.D.,
University of Tennessee.

KEY TERMS

alligatoring, *p. 281*

annealing, *p. 289*

area of origin, *p. 250*

calcination, *p. 288*

chain of evidence (chain of custody), *p. 320*

char depth, *p. 279*

clean burn, *p. 281*

crazing, *p. 293*

dropdown, *p. 264*

fire patterns, *p. 257*

ghost marks, *p. 286*

heat horizon, *p. 268*

ignitable liquid, *p. 272*

indicators, *p. 257*

overhaul, *p. 234*

salvage, *p. 236*

smoke horizon, *p. 268*

spalling, *p. 283*

trailers, *p. 271*

OBJECTIVES

After reading this chapter, you should be able to:

- Explain the comprehensive steps necessary in conducting structure fire investigations.
- Recognize the necessary investigative information to be identified, collected, and preserved during fire suppression.
- Understand the general principles of fire behavior that combine combustion of common fuels with the dynamics of heat transfer.
- Explain the methodology for examining a structure fire scene.
- Recognize the myths and misconceptions traditionally linked to incendiary fires that have been shown to be unreliable.
- Explain the systematic documentation of the fire scene through photography, sketching, evidence collection and preservation, and use of timelines.
- Understand how to state correctly the conclusions about a fire's origin and cause.
- Recognize and interpret damage as indicators of fire travel (i.e., fire patterns).

Before attempting to investigate a fire, the investigator should be reasonably familiar with the principles of fuels, ignition, and fire behavior outlined earlier. He or she must have a clear understanding of the purposes and goals of the investigation and a rational, orderly plan for carrying it out to meet those purposes. Having an investigative protocol and following it are now major expectations of courts following on *Daubert, Kumho Tire, Benfield,* and other decisions regarding admissibility of expert witnesses and their opinions. Critical to evaluating fire patterns accurately is reconstructing the pre-fire scene as completely as possible. This includes not only physically re-locating fuel packages in their original locations but also identifying wall, floor and ceiling coverings, determining window and door openings, and conducting a fire engineering analysis for a reconstruction of fire spread. No two fire scenes are ever alike, and no routine can be applied in the same way to each fire. A carefully reasoned plan of attack can be applied as each case requires.

The purposes of the investigation can include the following:

■ To determine the origin of the fire. Where did it start?
■ To determine the cause—the nature of the initial fuel and means of ignition. What was ignited? What ignited it? How did first fuel and ignition source come together?
■ Was it an accidental fire or an intentionally set fire?
■ If the fire was intentionally set, was it set with willful malice to destroy property, or was it set to destroy unwanted personal property without intent to harm or defraud another?
■ If the fire was accidental, what factors contributed to its ignition or spread?
■ If life was lost, what contributed to the death? Was it the rapidity of the fire spread or the toxic gases and smoke? Why or how did others survive?
■ How did fire protection equipment or systems function to minimize losses or fail to protect the property?

As recommended in Icove and DeHaan, the full forensic fire scene reconstruction involves the following steps (not necessarily in a fixed order).[1]

1. Document the fire patterns, scene processing, and fire growth/behavior if possible.
2. Evaluate heat transfer damage.
3. Establish the starting conditions—structure, fuels, heat sources, ventilation, and environment.
4. Collect and correlate human observations and factors.
5. Conduct a fire engineering analysis.
6. Formulate, evaluate, and test hypotheses and form conclusion.

The preceding steps follow the precepts of the scientific method as described in Chapter 1, and they also outline the best approach to any fire investigation, as follows:

1. Collect data—observe and document the scene as it exists, then document the excavation and examination of the scene. Interview firefighters, occupants, reporting party(ies), and other witnesses.
2. Establish starting conditions—based on the data collected, establish what the structure and its contents looked like at the start of the fire.
3. Assess likely direction(s) of spread by observing patterns of heat and fire damage, smoke deposits, structural collapse, and detector activation.
4. Ascertain whether there were any human (witness) observations or video or photographic recordings. These help establish likely areas of origin (or eliminate possible ones), starting conditions, and time frames.
5. Analyze the data by applying fire engineering principles to explain and interpret the information collected to lead to possible areas of origin and ignition sources. Humans affected by the fire provide information about the fire through tenability analysis.
6. Formulate and test scenarios about the fire's origin and causation against available data (including case histories, laboratory tests, and peer-reviewed scientific papers). Some hypotheses may require additional data that have not been collected and published to permit their evaluation. In that case, tests must be conducted to address particular issues. For example, when fire death cases involved the combustion of human bodies, there were no reliable data about the heat of combustion, mass loss rates, or combustion mechanisms of human tissue, so experiments were carried out and published in peer-reviewed scientific journals to provide reliable data.

These steps are summarized in Table 7-1 Similar protocols are published in *NFPA 921* and the *NIJ Guide* previously cited (as well as IFSTA's *Fire Investigator*).[2]

Beginning the Investigation

With respect to the origin of the fire, the investigation may begin at any time. Because every fire should be investigated to detect the crime of arson, or to seek out dangerous products or practices to preserve the safety of the community, the investigation may well begin while the fire is still burning. In other cases, because of the circumstances of the fire or the criminal or civil litigation surrounding it, the investigation may not begin until many weeks after the fact. Let's examine the sequence of events. If feasible, every good fire scene investigation begins with an initial "noninvasive" survey of the scene and its immediate surroundings.

DURING THE FIRE

When possible, it is often helpful for the investigator to be present during the fire and its suppression. The region of the structure in which the fire started may readily be apparent or may be obscured by flames and smoke that prevent close inspection. The investigator can profit from observing the suppression activities of the firefighters, since these can alter the course and intensity of a fire as well as cause unusual burn patterns. The investigator can also photograph or videotape the sequence of the fire and begin the search for witnesses. Preliminary interviews with witnesses to establish what they saw or heard, and where and when, is an important part of nearly every fire investigation.

The responsibilities of the firefighters with respect to investigations go far beyond merely extinguishing the fire. They are often the first witnesses on the scene and have

TABLE 7-1	General Protocol for Fire Scene Examination

Survey and Assessment
    Safety
    Exterior

Find and Interview Witnesses
    Fire personnel
    Police
    Neighbors, occupants, bystanders, etc.

Documentation
    Notes, sketches, diagrams, photos
    Exterior
    Interior
    Evidence
    Specialized (X-rays, panoramic photos)

Establish Starting Conditions

Examine Fire Patterns

Conduct Fire Engineering Analysis (formulating and testing hypotheses) Using Ignition Matrix

Collect and Evaluate New Data as Needed

Establish Likely Area(s) of Origin

Evaluate Possible First Fuels and Ignition Sources

Test Hypotheses about Ignition Mechanisms

Formulate Conclusions (final hypothesis) about Origin, Cause, and Responsibility (as appropriate)

Prepare Report

enormous control over the fate of all types of evidence at the scene—from burn patterns, to human remains, to residues of the ignition source. Their involvement in the investigation is detailed later in this chapter.

IMMEDIATELY AFTER THE FIRE IS EXTINGUISHED

So that access to the scene remains secure, it is important for the investigator to be on the scene of the fire as soon as possible after extinguishment. Although the condition of the hot remains is often not suitable for anything approaching a complete investigation, some exterior surveys may be carried out. The general consequences of the fire may be more apparent at this time than will be true later when the rubble is cleared out. In some fires, it is virtually impossible to make an examination until some clearing has occurred because of the large amount of collapsed rubble usually present. In small fires, however, much can be accomplished at this stage, because when the fire is localized, so is the damage, and reasonable access is available. In the largest fires, it may be days before a complete investigation may even be initiated, much less finished. Potential hazard areas (e.g., structural collapse, chemical or electrical hazards) can be identified.

DURING THE CLEARING OF THE SCENE

overhaul ■ The fire-fighting operation of eliminating hidden flames, glowing embers, or sparks that may rekindle the fire, usually accompanied by the removal of structural contents.

It is critical that **overhaul** (the removal of potential seats of smoldering fire for the prevention of rekindles) is kept to an absolute minimum until the investigator arrives and examines the scene. In scenes where there has been a death, overhaul in the vicinity of the body should be strictly limited to dousing of smoldering furnishings with a hand line to prevent reignition.

It is most important for the investigator to be on hand during cleanup and to assume some supervision of the overhaul operation. It is during this phase that most of the remaining evidence is destroyed or displaced, thereby reducing its interpretive value. Significant items of physical evidence are frequently disturbed or lost during cleanup. Such problems can be avoided if the investigator is present and able to take the necessary steps to preserve the evidence. Many fire service personnel are instructed in recognizing evidence so that they can call the investigator's attention to relevant items as they are found. This interdependence is most valuable when the investigator cannot supervise all areas of a large fire scene and must rely on capable and trained assistants. When it is necessary to remove furnishings during cleanup, it is critical that the location of major fuel packages be documented. It was once thought that the only fire patterns of value were on the floors, so all "debris could be shoveled or flushed out leaving a clean scene for the investigator." Nothing could be further from the truth. It is essential that *no* debris be removed by fire crews prior to the arrival of fire investigators (see Wallace and DeHaan).[3]

AFTER CLEANUP

The investigator now has a chance to examine burn patterns in full, provided they have not been destroyed by the demolition of unsafe portions of the building during cleanup. Dismantling and removal of the building (and its contents) should be prohibited or, at worst, kept to the absolute minimum required for public safety, until the investigation is completed. The investigator must not be less thorough because much of the building has been removed. Nor must any remaining rubble or specific items removed from the structure be overlooked. It is sometimes possible to recover highly significant evidence from such debris. The documentation of their original location will be more difficult, however.

Investigation after extensive demolition or repair is necessarily handicapped to a considerable degree. Often, photographs or videos are available that will assist in evaluating the limited remains of the fire scene. These can be very helpful, but they are no substitute for original investigation at the scene. If photographic, diagrammatic, and interview documentation is sufficiently comprehensive, most fires can be reconstructed and accurately investigated long after the fact, but in some cases it is simply not possible to arrive at firm conclusions as to the cause of the fire if the scene has been extensively cleared. If fire destruction is nearly complete, very little can be done in the way of fire pattern analysis, as depicted in Figure 7-1. Only rarely can the investigator hope to prove the use of liquid flammables from the pattern of the fire, and then only when extensive post-flashover

FIGURE 7-1 This office complex was destroyed when an arson fire was set in a vacant office, and the fire penetrated into an unprotected common attic space that included 10 separate buildings. This particular office was the Arson and Bomb Investigation Unit of the State Fire Marshal. *Courtesy of John D. DeHaan.*

burning has *not* occurred and when the pattern is so distinctive as to leave no doubt as to whether they were used. Many investigators working under these circumstances are tempted to limit their activities to the questioning of witnesses, almost to the exclusion of examining any physical evidence still remaining. It is vital to the success of the inquiry to remember that the most productive investigation will be done right at the fire scene, regardless of the time sequence of the investigation as related to the fire itself. However useful interviews may be for suggesting lines of inquiry or corroborating conclusions, the irrefutable physical evidence of the fire's origin and cause will be found only by careful examination of the fire scene. The systematic examination of a typical structure fire scene can follow a summary guideline such as that in Table 7-1.

Investigative Information during Suppression

One of the major difficulties of fire investigation is that the fire scene is everything that the good crime scene investigator would like a scene *not* to be. The ideal crime scene has no personnel present besides the investigator and any coworkers. It is invisible to the outside world to minimize sightseers and their complications, and it is relatively clean and orderly with items of evidence readily visible. Fire scenes are, of course, just the opposite. The fire is usually detected by an outsider who may compromise what one would consider to be ideal "crime scene" security right at the outset. Fire scenes are almost always peopled by large numbers of firefighters and their supervisors, onlookers, and the press. A great deal of damage and disorder is caused by the fire and its suppression. Most significant of all is that, unlike at other crime scenes where a body, broken window, or a gunshot indicates that a crime has been committed, suspicions of criminal fire activity are often not raised until long after the efforts of suppression. Arson is unique in that, because of the nature of fire, the crime destroys rather than creates evidence as it progresses. The firefighter's prime responsibility, of course, is to protect life and property. The responsibility to protect public safety by the detection of criminal activity is secondary to other functions and, in fact, does not fall within the official purview of the fire service at all in some countries. Although it is recognized that virtually all fire scenes start with the complications of a foreign influence or presence, the fire investigator with proper forethought can make the most of the shortcomings.

RESPONSIBILITY OF THE FIREFIGHTERS

In police work, the "first person on the scene" is the most valuable witness as to the nature of the undisturbed scene and the circumstances surrounding it. If investigators can familiarize line firefighters with some aspects of fire investigations and physical evidence, these first people on the scene can be a valuable asset to the fire investigator from the outset. In some progressive jurisdictions, the firefighters are instructed that in the event of any suspicious nature of a fire, they are to use extra caution in the suppression, overhaul, and **salvage** of that fire scene. All three of these activities are, of course, part of the duties of the firefighter, but each of them can vastly complicate the scene, if not make it completely worthless. With the proper precautions, unnecessary damage or disruption may be prevented. For instance, the use of excessive quantities of water at high pressures should be avoided. Large quantities of water can literally flush important evidence away from the scene or can, at best, complicate the scene investigation by causing collapse and making extensive cleanup efforts necessary. Using excess water in suppression can cause collapse of plaster and sheetrock forcing investigators to spend many hours removing hundreds of pounds of sodden plasterboard and waterlogged fiberglass insulation from the scene. In smaller structures, the use of straight streams of water should be avoided. "Fog" streams are not only less destructive but usually more effective as a firefighting technique.

salvage ■ Procedures to reduce incidental losses from smoke, water, and weather following fires, generally by removal or covering of contents.

Unnecessary structural damage, such as removal or destruction of windows, walls, and doors, should be avoided. The use of straight hose streams to break windows or strip siding off frame buildings may be necessary for suppression, but excessive use (sometimes out of juvenile destructiveness) is to be strongly discouraged. The use of such high-pressure streams inside small buildings can strip interior walls and ceilings of their gypsum board or plaster coverings and destroy invaluable burn patterns on those surfaces. High-pressure streams have also been seen to disrupt badly burned bodies and destroy the fragile remains of incendiary devices. Fog or modified fog streams are more effective in most interior attacks and are much less destructive. Fire crews have been known to push the walls of buildings into a burning pile to minimize postfire cleanup. This practice is a complete antithesis of the preservation of a fire scene. It is understood that rapid entry to a fire scene may, on many occasions, be gained only by forcible entry and that subsequent requirements to vent the scene may result in the breakage of many of the windows. Firefighters should try to note mentally which doors or windows were broken or open as well as those that were locked on their arrival. The recognition of signs of forced entry, such as broken door frames or torn window screens, is also very valuable to the investigation, and such observations should be related to the investigator. During the early stages of suppression, firefighters are in the best position to detect possible multiple points of fire origin. The realization that several fires simultaneously established themselves in *separate* areas of a building is an important step in the detection of multiple arson sets, a common technique used by fire setters. The fire investigator will be interested in not only the location and extent of the fire but also the direction of suppression efforts, the points at which the fire was attacked, and the manner in which the structure was vented, since all these can affect the fire patterns within. The firefighters should note any unusual flames or smoke colors, odors, and weather conditions (wind direction and intensity, temperature, etc.), and condition (operability) of fire detection and suppression systems in the building. These guidelines are emphasized in greater detail in the *NIJ Guide* published in 2000.

MINIMIZING POST-FIRE DAMAGE

The primary sources of post-fire damage to the scene are firefighting, overhaul, and salvage operations.

Overhaul

Overhaul of the fire scene generally entails the removal and destruction of possible pockets of smoldering fire in walls, floors, and upholstered furniture. These overhaul activities should be limited to that necessary to prevent a recurrence of the fire.[4] For example:

- When walls or floors are torn up, mental note should be made of their condition.
- Were the rugs in place?
- Or were they folded up and thrown in a corner?
- Was the furniture in place?
- Were the walls and floors intact or had someone removed their covering materials previously?

Sometimes, an arsonist will intentionally open ceilings, remove attic access panels, or open holes in fire-resistant plaster or gypsum board walls to expose the internal studding and improve the rate of flame spread from room to room. Large items of fire-damaged upholstered furniture, such as sofas and beds, are normally removed to the outside of the building during overhaul. The dumping of all such materials from an entire house into a single pile can make the subsequent investigation impossible. When furniture such as this is removed, someone should remember or note the locations from which various pieces were removed. When conditions permit, all contents from all the rooms should not be dumped into a single pile but in separate piles with a location for the contents of each room established. Only the minimum amount of overhaul necessary for safety should be allowed until the origin of the fire has been determined.

Gasoline-powered fans, pumps, and saws are often used during overhaul. Use of such possible sources of contamination should be restricted to exterior placement and then carefully documented. All fueling of such tools should be carried out in a designated area remote from the fire. Electric or hydraulic tools are preferable if they are available.

Salvage and Security Concerns

During the salvage portion of the fire scene cleanup, owners, tenants, and employees are sometimes allowed to enter the fire scene to remove items of value. The fire official responsible for the scene should not allow any unauthorized persons to enter a fire scene and especially should not allow materials to be removed from the scene until the investigator has had a chance to evaluate those items. Cases have been reported in which the "victim" returned to the fire scene to pick up some valuables and then proceeded to destroy the evidence of one or more arson sets for which he or she was responsible. The fire crews therefore have the best opportunity to detect suspicious circumstances because, on entry, they can be aware of unusual conditions, contents, or fire behavior. Smith has published an excellent in-depth review of firefighters' responsibilities and their important role in successful fire investigation.[5] It is the responsibility of the fire department to maintain custody and control of the scene until the investigation is complete, even if that entails hiring a private guard. The legal requirements for maintaining control of a scene will be discussed in Chapter 17.

Documenting the Fire Scene

The general approach to scene documentation is outlined in Table 7-2. Thorough documentation using photographs, sketches, diagrams, and notes is the foundation of good investigative process. Such information can be critical in demonstrating to the court what was relied upon by the expert. It will also be invaluable to the original investigator when called on years later to recall details of the investigation.

Throughout this chapter it has been repeatedly indicated that the appearance of the fire scene should be documented with both photographs and sketches. The importance of these steps cannot be overemphasized. Demonstrating to a court and jury the condition and exact location of evidence is as important as the evidence itself. This documentation is the beginning of the chain of custody and can be accomplished only with the use of a camera and sketch pad.

PHOTOGRAPHY AND PHOTOGRAPHIC EQUIPMENT

Camera equipment need not be terribly complex or expensive, although investing in better equipment will usually result in better pictures as a product. A good-quality 35-mm camera or digital equivalent, preferably one that will accept a variety of special-purpose (close-up and wide-angle) lenses, is the best all-around choice for fire photography. A variety of film is available in 35 mm, and processing is less costly per frame than with other film sizes. Smaller pocket-type cameras with plastic lens elements are to be avoided except as an absolute last resort. Plastic lenses can distort or diffuse the image captured by the film or digital array. The problems may not be apparent with snapshots but may make an enlargement so fuzzy that it is an embarrassment to show it in court. Instant-print cameras, although handy for some investigative purposes (such as witness interviews), are generally more costly to purchase. The cost per photograph is many times that of conventional photographs, and most investigators discover that their convenience does not compensate for the high costs and sometimes less-than-ideal picture quality.

Because regular film is so low in cost, and permanent in format once processed, some investigators prefer that all critical scene photos be taken with traditional film, then scanned

TABLE 7-2 | Systematic Documentation

STEP	TYPE	GUIDANCE	PURPOSE
1	Exterior	• Photograph exterior and perimeter of property. Sketch exterior. Use GPS to obtain location.	• Establishes venue and location of fire scene in relation to surrounding visual landmarks • Documents exposure damage to adjacent properties • Reveals structural conditions, failures, violations, or deficiencies • Establishes extent of fire damage to exterior of scene • Establishes egress and condition of doors and windows
2	Interior	• Record and sketch extent of fire damages, potential ignition sources, and data needed for fire reconstruction and modeling.	• Documents heat transfer damage, heat and smoke stratification levels, and breaches of structural element • Traces fire travel and development from exterior to suspected point(s) of fire origin • Documents heat transfer damage, heat and smoke stratification levels, and breaches of structural element • Documents condition of power utility and distribution, furnace, water heater, and heat-producing appliances • Documents positions and damage to windows, doors, stairwells, and crawl space access points • Documents fire protection equipment locations and operation (sprinklers, heat and smoke detectors, extinguishers) • Documents readings on clocks and utility equipment • Documents alarm and arc-fault information
3	Investigative	• Document clearing of debris and evidence prior to removal; smoke, burn, and charring patterns; packaged evidence.	• Assists in recording fire pattern and plume damage, isochar lines • Establishes condition of physical evidence, utilities, distribution, and protection equipment (breakers, relief valves) • Documents integrity of the chain of custody of evidence
4	Panoramic	• Produce multidimensional photos and sketches.	• Clearer peripheral views of exterior and interiors • Establishes photograph viewpoints of witnesses

Source: D. J. Icove and J. D. DeHaan, *Forensic Fire Scene Reconstruction,* 2nd ed. Upper Saddle River, NJ: Pearson Education/Brady Publishing, 2009.

for electronic manipulation. This option also precludes legal arguments as to the originality and content of scene photos. Some agencies use two cameras, one film and one digital, for critical scenes. In addition to a camera and film, a flash unit is essential because most photos will be taken inside burned buildings with no lights of their own. A high-intensity electronic flash unit is often needed for photographing charred surfaces in dark interior scenes. (This is a limitation of built-in flash units on most digital cameras.) The authors have discovered that the heat flux from high-power electronic flash guns can ignite freshly deposited, susceptible soot and pyrolysis accumulations on walls if discharged close by.

Digital Images

Digital cameras are now in very common use. The quality of the image produced should be the primary criterion for selection. Some low-cost digital cameras offer a pixel count that produces adequate small prints but do not have the resolution for any sort of enlargements. A camera that provides $1,200 \times 2,500$ pixels (3 megapixels/image) should be considered a *minimum* requirement for ensuring courtroom-quality prints. Cameras that produce 8-megapixel photos are very common at a moderate price. Storage of images containing 3 megapixels or more each requires considerable computer capacity.

Digital images provide flexibility in final format and simplify preparation of illustrated reports. Some digital cameras offer security devices that identify "original" images. Files copied directly onto a computer or disk must retain their original format without compression. Many fire departments have adopted formal protocols for preserving digital images.

Agencies or individuals using digital photography, both simple and complex, should develop a standard operating procedure and departmental policy for preservation of digital images. Many departments have adopted formal protocols for the preservation of digital images consistent with the guidelines recommended by the Scientific Working Group on Imaging Technology (SWGIT).[6] These guidelines cover imaging technologies, recommended policies, best practices, and techniques for photographing specific impression evidence. These resource documents are available on the FBI (www.fbi.gov) and International Association for Identification (www.theiai.org) websites. The SWGIT procedures recommend preserving original images by storing and maintaining them in an unaltered state and in their native file formats. Duplicates or copies should be used for working images. Original images should be preserved in one of the following durable formats: silver based (non-instant) file, write-once compact recordable disks (CDR), or digital versatile recordable disks (DVD-R).

Enlargements and Film

Enlargements are often necessary for court display. Color prints of sufficient size are very costly and may be prohibitive in cost if full documentation is to be offered. Many investigators prefer to take all their scene photos on color transparency (color slide) film such as Ektachrome and then project these in the court during testimony. This option offers the lowest cost per photograph and the means of re-creating the scene at life size for the jury. Its drawbacks are that some courtrooms are not easily darkened for proper projection and that it is awkward for the jury to review them in the absence of printed labels and projection equipment. The ideal presentation would probably include transparencies for courtroom testimony with a duplicate set of small color prints to which the jury could refer. Many courtrooms today have digital image projection capability. If the format of the storage medium can be guaranteed to be compatible with the court's equipment, high-quality digital images can be shown in open court and later used in print format for record-keeping or jury review. Color photographs or images are the standard medium. For court displays involving fatalities, however, the court may insist on black-and-white images only. For the most part, however, the subtle coloring of burn patterns is lost on black-and-white images, and the photos contain less information. The investigator must be careful that the color rendition of the final images used correctly and accurately duplicates the original scene.

Photography for the Fire Investigator

If the fire is still in progress when the investigator arrives, he or she should photograph the building from various angles and at various times. These photos will help document the extent of damage to the building on arrival. If the building is then totally destroyed by fire or collapse, information is available to indicate in which direction the fire was progressing. Other potentially valuable information in such early photos will include conditions of doors or windows, presence of suspicious vehicles, and the appearance of onlookers. (The photos of spectators from a series of similar fires should be compared to detect faces that frequently appear at fires.)

When the fire has been extinguished, photographic documentation should follow the same general pattern as the scene search (exterior, interior, investigative, specialized). The exteriors of all remaining walls of the structure should be photographed to document the extent of the damage. The interior of the structure, including ceilings and floors, should then be photographed to preserve their appearance before the investigation or overhaul. Prior to exterior photography, the windows and doors on the ground floor (or other openings that could have provided access) should be numbered sequentially (e.g., clockwise starting at front door) and numbered markers positioned at each one so that they are visible as each door and window is documented as to condition and security. The markers should then be moved inside to help ensure that all windows, doors, and possible entry points are documented.

During the investigation, many points in the burn pattern; partially destroyed appliances; utility materials; the remains of incendiary devices; possible sources of accidental ignition; signs of forced entry, burglary, or other crime; alarm or sprinkler systems; and other items of possible importance will have been noted. These should be photographed carefully so as to record their general features, condition as found, and their relationships to other objects nearby. These photographs can be most important in determining whether there is evidence that the fire spread from possible ignition sources to other combustible objects or to adjacent structures. Such a fire spread must cast suspicion on the object that initiated it, even though it may ultimately be identified as simply an agent of normal fire spread.

The largest single activity related to most investigations after a fire is moving debris so that critical indicators and items can be uncovered. In many instances this removal must be done by a work crew, or even machinery, and may require days to accomplish. If possible, the investigator should be present, with camera, because an immediate record can be obtained of anything suspicious or informative that may be uncovered before further clearing activities may destroy or damage the item in question. Removal of the room contents and their reassembly outside the structure on a proportioned full-scale map on a plastic sheet may aid in visualizing the pattern, as illustrated in Figures 7-2 and 7-3.

(a)

FIGURE 7-2A The investigators of this trailer fire chalked an outline on the soil and transferred items to the outline as they were removed. *Courtesy of Jamie Novak, Novak Investigations and St. Paul Fire Dept.*

(b)

(c)

FIGURES 7-2B AND C (b) Investigators dismantled the garden shed in this fatal fire, leaving the body and debris undisturbed. (c) They were removed to a plastic sheet marked with the same grid as the structure. *Courtesy of Jamie Novak, Novak Investigations and St. Paul Fire Dept.*

FIGURE 7-3A This mobile home fire required a large plastic tarp with equal-scale grids overlaid on a floor plan. *Courtesy of Don Perkins, Fire Cause Analysis.*

FIGURE 7-3B As debris was removed from each grid within the scene it was placed in the corresponding grid on the tarp. *Courtesy of Don Perkins, Fire Cause Analysis.*

After a room or area has been cleared of debris, it is often of value to reassemble the furnishings in their original locations (using protected areas to reposition them) and rephotograph them in place.

In large rooms, single photos cannot capture all relevant information. With computer scanning of photos readily accessible, it is possible to scan a series of overlapping photos and then, with specialized programs, stitch the digital images together to form a panorama of the room (such as with the Photo Stitch link in Roxio's PhotoSuite). In lieu of such programs, photos can be physically overlapped in a mosaic of images to re-create a view (because the human eye has a wider field of view than the typical 50-mm focal length of a camera lens).[7]

It is expected that ultimately the point of origin will be revealed and, it is hoped, recognized. At this point the investigator should carefully supervise the clearing and record photographically all the evidence that gave rise to the choice of this as the point of origin. Close-up photos should be taken of any *possible* source of ignition or similarly critical evidence in the vicinity of the suspected origin. This includes possible ignition sources that were considered and eliminated, because that is part of the investigator's analysis and opinion.

Whether or not the investigator is later called as a witness in an arson trial or in a civil action, the photographs will be the best evidence of what was done, what was seen, and why he or she reached the stated conclusion. The evidentiary value of photos can be compromised if they are not documented properly with a photographic log. A typical log is included in Appendix I. Note that the log provides for a description of each photo as it is taken, including the subject matter and direction in which it was taken. This data can be recorded on paper as the photos are taken or dictated into a cassette or digital recorder for later transcription. This log can be in the list format of Appendix I or as a floor plan with numbered circles and arrows, as depicted in Figure 7-4.

Neither the human memory nor the notebook, important as both are, will ever match the camera as a means of recording observations. Fire scenes are some of the more difficult photographic subjects because of the properties of burned wood in relation to light. In addition, large structures offer special problems because many such scenes have insufficient light to focus a camera accurately, and flash lamps are ineffective at the distances that must be used in many instances. Another source of difficulty is that the general reflectivity of a burned scene is so low that it is recommended that the lens diaphragm or f-stop be opened one or two stops, even compared with a meter reading. Otherwise, photographs tend to be underexposed to the point of being useless. The "automatic" settings of digital cameras can be overridden with programmed settings to accommodate such scenes. The investigator should become familiar with the manual setting of any camera used to maximize its usefulness.

Burned wood, although black, is highly reflective at certain angles of light incidence. These glare spots photograph white, appearing lighter in many instances than unburned wood nearby. The use of a polarizing filter will assist in reducing these glare spots and giving a truer representation of color intensity. Also, critical areas may be reasonably reproduced by illuminating them head-on so as to avoid glare angles.

Some investigators use video recorders to preserve an image of a fire scene. Although such media are convenient to use, they are very limited in their capability for preserving fine detail, and review of specific images is difficult. Video should not be substituted for good sketches and comprehensive photography at a scene, but it may be useful for supplementing documentation of a fire scene. Video of the fire in progress can be very useful in retracing its path. Some digital video recorders have the capability to record still photos (or video with "still-photo" resolution) and audio "notes," allowing several means of information to be captured in one tool.

There are many tricks that photographers learn to allow them to take pictures under even the most difficult conditions. Space does not permit us to examine them here. The reader is referred to specialty texts in this area (e.g., Lyons or Miller and McEvoy or cfitrainer.net) for further guidance.

SKETCHING

Fire scene sketches and diagrams should be considered important supplements to photographs and written notes, not replacements for them. A sketch is a drawing or layout that graphically portrays the scene and items of evidence within it. Only sketches record and reveal spatial relationships between items of evidence found at the scene. They offer the visual advantage of illustrative evidence while allowing the elimination of distracting or obscuring detail found in some photographs. A simple floor plan of a structure can capture more information more clearly than many pages of narrative description. Accurate and detailed scene sketches allow reconstruction of a scene weeks, months, or years after the incident, by indicating the placement of various items of evidence. They help refresh the memory of investigators or witnesses, and they can indicate the perimeter of a crime scene search with respect to a larger geographical area. A simple plan (overhead) view of a room or building (as shown in Appendix F) can capture as much information as pages of narrative and convey it quickly and accurately to a reader. Although careful execution and neatness certainly improve the visual quality of sketches, the most important requirement is that they portray evidence accurately. Because their artistic merit is of secondary consequence, good scene sketches do not require a high degree of artistic ability. Composing a detailed room diagram can focus an investigator's attention on identifying and locating fuel packages and all potential ignition sources in a room. Such a diagram will aid in the preparation of an ignition matrix, as described in Chapter 6.

To be useful as overall evidence, fire scene sketches must depict the following:

- The outline and dimensions of the room, building, or area of interest
- The locations and relative positions of pertinent evidence found at the scene or critical features of the scene itself, including significant burn patterns
- The locations of approaches or possible points of entry such as doors, windows, stairways, skylights, paths, and roadways (including dimensions of windows and doors)

Fire scene sketching requires only paper, pencils, pens, tape measures, a ruler, and a clipboard for the rough original "on-the-scene" sketch. Paper ruled into a grid of 8 to 12 squares per inch (graph paper) makes scene sketching easier, since it is easier to draw outlines of rooms or structures by tracing along the grid, counting off the divisions corresponding to the measurements. Drawing aids such as the Acculine device (see Appendix F), which helps produce straight lines on a hand-drawn diagram, may be useful. The finished drawing, to be filed with the report or used in court, requires unlined paper, pens, rulers, drawing board and, preferably, T-square and triangle or, more commonly today, the use of a drafting system such as Chief Architect or The Crime Zone, discussed later in this chapter. True "scale" drawings are done to architectural standards and are done only in special cases.

The best rule to keep in mind is that the final sketch must be completely intelligible to the viewer without detailed study. If many types of evidence are to be denoted on a sketch, a single sketch may not be adequate, and sketches for each individual purpose would be a good idea. If too much detail is included in a single sketch, the sketch's major advantage over a photograph—its simplicity—is lost.

To ensure intelligibility and acceptability as evidence, every sketch or diagram should always include the following:

- The investigator's full name and agency, and the date and time at which the sketch was prepared
- The case number of the incident, if available
- The full names of all persons involved in making the sketch or assisting in measurements
- The address or location of the fire scene
- The scene's geographic orientation (compass marking) and GPS location, if available
- A legend that includes the meaning of all symbols used, a numbered list of numbers or letters used to denote items of interest and their meanings, and a scale of the drawing (for example, 1 inch equals 1 foot; $1'' = 1'$)

Because the sketch may be used at a later time to help place various pieces of evidence at a reconstructed scene, the accuracy of all locations and dimension measurements is critical to its acceptability as evidence. Use a conventional system of measurement (feet and inches, meters and centimeters). Use a second person to verify all measurements, if possible. Do not use paces or "steps" on the final drawings, since they are not standard measurement units. Do not measure a distance in paces and then convert to feet and inches because this implies accuracy that is not obtainable by a pace measurement. Remember, the method of sketching and measurement chosen should always give someone viewing the result the best information possible. If a body is involved, be sure the sketch accurately depicts the orientation and position of the torso as well as the orientation of all extremities. For further details on proper scene sketching techniques, see Appendix F. Sketching can be combined with a photograph to enhance the interpretive value of an exhibit.

Some investigators may find it useful to use different colored pens or markers when fire patterns to coincide with the "depth" designations described earlier. For instance, blue could be used to outline (or crosshatch) an area that has only surface deposits, green to denote areas where there are some thermal surface effects, yellow to denote areas where there have been some deeper effects, and red to outline areas where the material is penetrated completely or consumed. This technique can be used with both manual and computer-aided sketching and offers a supplement to scene photographs or notes that may not conveniently record depth factors.

Computer-aided drafting and drawing programs specially designed for crime scene documentation are available now, although one found to be most useful is "Home Designer Deluxe" offered by Better Homes & Gardens. BH&G is planning to offer a special "fire investigation" package that will include a set of graphic symbols and features for that application.) Microsoft Visio 2000 products for crime scene applications are available from Microsoft's website. There also are specialized consumer diagram programs such as The CrimeZone.[9] See Appendix F for an example of Home Architect capability.

Measurement and Scanning Systems

Measurements of rooms, doors, and windows are essential data for the accurate analysis and reconstruction of most fire scenes. Although they do not have to be precise, they should be within ±5 percent of the actual dimension. A variety of manual and optical tape measurements (available in Appendix F) are essential.

Investigators who frequently have to diagram large, complex scenes (particularly outdoors) have used a laser-based surveying method (called Total Station) that measures and records the distance as well as the angles of position and inclination or declination from a reference point to each item or point of interest onto a computer program. The program then plots out all recorded points, designating walls, fences, curbs, locations of evidence, and the like. Computer-driven laser scanners are now finding more use in rapidly capturing fire scenes with tremendous accuracy. The newest imaging techniques combine wide-field image capture with measurement data. Laser scanning cameras use an automatic rotating laser measuring (survey) system and a synchronized digital camera to produce a fully explorable, 3D virtual reality image that captures dimensional detail with great accuracy.[10] They can be used to quickly document even large fire and explosion scenes. Systems by Leica Geosystems and Delta Sphere have been shown to be usable in a variety of scenes, even in small rooms.[11] An example of a laser scan of a burned room is shown in Figures 7-5a and b. An alternative is a digital scanning camera such as the Panoscan that captures a 360° image with measurement capability on the virtual reality image, as shown in Figures 7-6a and b.

NOTES

Notes are the final element in comprehensive fire scene documentation. Information that is not visual cannot be captured in a photo or diagram. Weather data, witness

FIGURE 7-5A Laser scan of fire scene. *Courtesy of Precision Simulations.*

FIGURE 7-5B Standard photograph of same scene. *Courtesy of Don Perkins, Fire Scene Investigation.*

Object

FIGURE 7-6A The Panoscan digital imaging system can take two 360° scans from top and bottom positions on the tripod, allowing a stereo-optical re-creation of distances. *Courtesy of Panoscan.*

FIGURE 7-6B Panoscan image of large fire scene in a commercial kitchen. (When viewed as a computer image, the spherical distortions are compensated.) *Courtesy of Robert Toth, IRIS Fire Investigations.*

FIGURE 7-7 Data form for complete documentation of room fire.

ROOM FIRE DATA

Room # _____

Room _____

Length _____

Width _____

Height _____

Note ceiling height changes _____

Floor Plan

Walls: Structure/material _____ Thickness _____ Covering _____ Sample? Y/N

Structure/material _____ Thickness _____ Covering _____ Sample? Y/N

Ceiling: Structure/material _____ Thickness _____ Covering _____ Sample? Y/N

Floor: Structure/material _____ Thickness _____ Covering _____ Sample? Y/N

Openings (door, window, other vents) into room (number on plan above):

Height (bottom to top of opening)	Sill Height	Soffit Depth (above opening)	Width	Open or Closed? Changes During Fire?
1.				
2.				
3.				
4.				
5.				
6.				
7.				

HVAC System:

Description _____

On/off prior to/during fire? _____

Furnishings (descriptions of major fuel items, including floor and wall coverings, draperies):

_____

_____

_____

_____

Time Line: Alarm time _____ FD arr. time _____ Control time _____

Detection: _____

Pre-Fire Events: _____

interviews, odors, time events, contents, and materials are captured as notes. These may be written on plain paper or recorded on a protocol form such as the one shown in Figure 7-7. Note that this form reminds the investigator not only to measure dimensions of rooms, doors, and windows but also to identify and collect samples of wall, ceiling, and floor coverings.

No matter when the scene investigation is begun, documentation is essential, for it forms the basis for all that follows—fire pattern analysis, engineering analysis, hypothesis formulation and testing, reporting, and testimony. The days when an investigator could walk through a scene, visually evaluate fire patterns on the fly, reach a conclusion, and then testify to that opinion without documentation are gone. The expectations of both the courts and the profession are much higher today. The various investigation

FIGURE 7-8A Coffeemaker fire involving other fuels (cereal boxes, towels). *Courtesy of Jamie Novak, Novak Investigations and St. Paul Fire Dept.*

FIGURE 7-8B Fire spreads into cabinets. Plume entering air return grille in suspended ceiling. *Courtesy of Jamie Novak, Novak Investigations and St. Paul Fire Dept.*

FIGURE 7-8C Extension above suspended ceiling very rapid. Light fixture diffuser softens and falls out. *Courtesy of Jamie Novak, Novak Investigations and St. Paul Fire Dept.*

protocols describe the preferred professional standards and practices, such as *ASTM E620*,[12] *ASTM E678*,[13] *ASTM E860*,[14] *ASTM E1020*,[15] and *ASTM E1188*.[16] *NFPA 921* contains forms for field data collection. Equivalent forms are included in Appendix I.

The photographic and diagrammatic documentation of a typical fire scene employ the techniques described in Table 7-2. They follow a logical flow that maximizes evidence detection and preservation while minimizing the danger of loss or contamination. This topic will be examined in greater detail in a later section of this chapter.

The purpose of the exterior examination and documentation is to establish the location of the fire in relation to other landmarks and record the condition of the structure prior to any scene disturbance. It also encourages a scene search for physical evidence and paths of possible entry and exit. Hazardous conditions can be assessed and needed resources can be requested (e.g., scaffolding for an unstable wall.)

The interior documentation includes recording all rooms of a structure (not just the fire-damaged areas). This ensures that some information will be available later about pre-fire conditions such as furnishings; wall, floor, and ceiling finishes; and maintenance conditions. Heat, smoke, and explosion effects are, of course, documented in detail. Interior documentation also includes the existence and condition of fire protection systems and possible ignition sources.

Investigative documentation includes anything that will be of assistance in conducting the analysis and hypothesis testing. This includes ignition sources, fuel loads, witness observation, measurements, specific fire patterns, and electrical systems (arc fault mapping). This documentation will be in the form of physical evidence, photos, diagrams, sketches, notes, and even video or audio recordings.

Specialized documentation may include X-rays; thermal, infrared (IR) or ultraviolet (UV) images; or panoramic photos. Some of these specialized techniques will be described later in the text and in the appendices.

RECONSTRUCTING THE PRE-FIRE CONDITIONS

Before meaningful analysis of fire behavior and fire patterns can be carried out, the investigator must be familiar with the conditions that existed at the start (or immediately before) the fire. This means knowing the room dimensions (including ceiling height), interior finish, fuel packages (size, type, and location), and conditions of windows and doors (open or closed, sizes, and sill and soffit heights). This is why documentation of unburned areas of a building is so important—to give some idea of ceiling and wall finishes, the nature of furnishings, housekeeping standards, and the like. Previous chapters have discussed the roles that fuel type, physical state, orientation, and ventilation play in a fire. Even the orientation of room contents and lines of sight can affect what witnesses can see and when during a fire's development. If one is going to attempt any meaningful, accurate, defensible fire engineering analysis or computer modeling, reliable data on the starting conditions are essential. Figures 7-8a–c show a test fire in which rapid extension into the ceiling void occurred because of a ceiling vent above the origin.

General Principles of Fire Behavior

You should now have some appreciation of the basic principles of combustion of common fuels and the dynamics of heat transfer, fire, and flame. We now apply those principles to more complex situations involving combinations of fuels and their structural arrangement.

FIRE PATTERNS

The effects of smoke, hot gases, heat, and flames on materials are what form the fire patterns that investigators use to trace the course of the fire and, by reversing that process, establish the **area of origin** of the fire. These effects can be generally categorized as follows:

- Surface deposits (with no irreversible effects on substrate)
- Surface thermal effects—scorching or melting, discoloration
- Charring—surface
- Penetration—charring below surface
- Consumption—charring completely through a material or actual destruction

These effects are the foundation of the fire patterns investigators use. They reflect the intensity and duration of thermal exposure.

If no thermal effect is observed on a surface, the investigator can conclude it was not exposed to enough heat to induce any chemical change. In general, the observed changes reflect the intensity of the applied heat flux—the higher the heat flux, the higher the surface temperature (dependent, of course, on the thermal inertia of the surface). In general, the deeper the penetration of the changes, the longer the *duration* of the exposure. However, heat penetration is not linear with time (although it becomes nearly so for long exposures). The higher the heat flux (or the higher the temperature of the exposed surface), the faster heat is conducted into the material. The total observed damage is therefore the result of both intensity and duration, so both must be taken into account.

TRACING THE COURSE OF THE FIRE

The fire investigator's task would be easy indeed if one were present for the duration of the fire, from ignition to extinguishment. Unfortunately, fire scenes are nearly always cold and badly burned, and the investigator must reconstruct the sequence of the fire backward from what is visible afterward to its point of origin. Fortunately, there are some general guidelines that can be applied to all structural fires to assist in this reconstruction. With few exceptions, fires, however large, start with a small flame, such as a match, candle, or lighter, or a spark of some type. The point and source of ignition are two of the key elements to be determined by the investigator, regardless of the size of the conflagration that followed this tiny flame (the others being first fuel ignited and circumstances of ignition). These elements determine the origin of a fire and must be of primary importance in any fire investigation. With very few exceptions, no defensible conclusions can be reached about the cause of a fire if the area of origin cannot be established. The primary exception are fires where the first fuel ignited is a flammable gas or vapor.

An investigator who spends a large proportion of time surveying what are obviously the later stages of the fire is wasting much of that time, because such a survey cannot answer the fundamental question of nearly every investigation—what started the fire? The insurance adjustor may be interested in the final stages of the fire in which most of the property damage occurred. The arson investigator may be interested in what contents were destroyed from the standpoint of helping uncover a motive for a malicious fire. Otherwise, the predominant focus is on why and how the fire occurred, and the investigation must similarly focus on the tiny initial flame and the first fuel ignited.

Because it is of the greatest importance to trace the behavior of this small flame as it grows into the large fire, the following rules of fire behavior are offered:

- Hot gases (including flames) are much lighter than the surrounding air and therefore rise. In the absence of strong winds or physical barriers (such as ceilings), which force hot gases to travel elsewhere, smoke and hot gases will always preferentially move upward.
- Due to heat transfer processes in the buoyant flame plume, fire will always preferentially move upward much more rapidly than horizontally or downward (with some downward progression as a result of radiant heat and falldown).

- Combustible materials in the path of the flames will be ignited, thereby increasing the extent and intensity of the fire. The more intense (higher heat release rate) the fire, the faster it will rise and spread.

- If there is not more fuel above or beside the initial flame to be ignited by convected or radiated heat, or if the initial fire is too small to create the necessary heat flux on those fuels, the fire will be self-limiting and often will burn itself out. A flame plume that is large enough to reach the ceiling of a compartment is likely to trigger full involvement of a room, for it is charging the upper gas layer in the room with gases that have not cooled by mixing or radiative losses as much as gases in a shorter plume. As a result, the hot gas layer maintains a higher temperature and is more likely to reach its critical temperature (~600°C or 1,150°F in a typical room), which then induces flashover or full room involvement.

- The fuel load of the room or structure is of paramount importance in the development of a fire. This fuel load is not only the structure itself but also the furnishings and contents, and the wall, floor, and ceiling coverings that feed a fire and offer it paths and directions with optimum fuel conditions. In evaluating a fire's progress through a room or structure, the investigator must establish what fuels were present and where they were located. The chemical nature of the fuels and their physical forms will affect their ignitability and their expected heat release rate. In the reconstruction of a fire, the fuel load is not just the total number of joules or Btu's of heat that can be generated but the rate at which that heat is released.

- Variations on the upward spread of fire or smoke will occur when air currents deflect the flame, when horizontal surfaces block the vertical travel, or when radiation from established flames ignites nearby surfaces. If fuel is present in these "new" areas, it will ignite and spread the flames laterally.

- Upward, vertical spread is enhanced when the fire finds chimneylike configurations. Stairways, elevators, utility shafts, air ducts, and interiors of walls and hollow support columns all offer openings for carrying flames generated elsewhere, and fires may burn more intensely because of the enhanced draft.

- Downward spread will occur whenever there is suitable fuel in the area. Fire will burn downward across a solid fuel but at a rate that is a tiny fraction of its upward spread rate. Combustible wall coverings, particularly paneling, encourage the travel of fire downward as well as outward. Burning portions of ceiling and roof coverings, draperies, and lighting fixtures can fall onto ignitable fuels below and start new fires that quickly join the main fire overhead (called "falldown" or "drop down" fires). Radiation from rollover fires or very hot gas layers can ignite floor coverings, furniture, and walls even at some distance, creating new areas of fire. The dynamics of such fires were treated in Chapter 5. The resulting fire patterns can be complex to interpret, and, once again, the investigator must remember to take into account what fuel packages were present from the standpoint of their potential heat release rate contributions.

- Suppression efforts can also greatly influence fire spread. Positive-pressure ventilation (PPV) or an active hose stream attack on one face of a fire may force it back into other areas that may or may not have already been involved, and push fire down, even under obstructions such as doors or cabinets. The investigator must remember those "unusual" conditions and check with the firefighters present.

- Fire tends to flow through a room or structure much like a liquid—upward in relatively straight paths and outward and around barriers.

- Ventilation from open doors, windows, or vents, or forced by heating, ventilating, and air conditioning (HVAC) systems or PPV fans can affect fire growth and movement as much as fuel arrangement. All ventilation must be documented.

- The development of a fire in a compartment (which may be a room or a portion of a room) can be likened to filling a leaky bucket from a water hose. One can fill even a leaky bucket if the flow of water is great enough. Similarly, the hot gas layer in a room can reach its critical intensity (producing a heat flux of >20 kW/m^2) if the fire is large enough to overcome the "leaks" of hot gases due to openings or due to radiant and conductive losses. As seen in the 1985 Bradford (UK) football stadium fire, even a compartment that is open on three sides can develop flashover conditions when the fire gets big enough.

- Fire intensity in a post-flashover room is often greatest around ventilation sources, so damage accumulates faster around door sills, windows, or other openings where fresh air could be drawn into the room. These ventilation effects can even reach walls opposite the opening.

- The total fire damage to an object observed after a fire is the result of both the intensity of heat applied to that object and the duration of that exposure (with the realization that intensity of heat varies considerably during a fire and that all the exposure may not have occurred at the same time).

IMPLICATIONS FOR THE FIRE INVESTIGATOR

Fire investigation and the reconstruction of a fire's growth pattern back to its origin are based on the fact that fires form patterns of damage that are, to a large extent, predictable. With an understanding of combustion properties, heat transfer, and general fire behavior, the investigator can reliably examine a fire scene for particular indicators. Using each indicator as an independent test for direction of travel, intensity or duration of heat application, or point of origin, the investigator can approximate the fire's development. It must be remembered that every fire forms a pattern that is chiefly determined by the configuration of the room or structure, the availability of fuel, and ventilation effects. Because of the upward-reaching tendency of flames, some conical form of fire damage is most common, with the apex at the bottom being a source of fire (possibly a point of origin) and the fire rising and spreading out from there. Naturally, this pattern will be altered by the presence of obstructions or the availability (or lack) of fuel in an area. Thus, interior fires often appear to have very complex patterns as the result of complex structural arrangement rather than any unusual property of the fire. These complications have already been noted in part and will be further discussed as needed in the relevant section on diagnostic signs.

In its simplest terms, the requirement is to identify the origin (the point or space where the fire began) and then to examine that space for possible ignition sources. The fuel first ignited at the origin must be identified (even by elimination) for a causal link to be established between the ignition source and the final destructive fire. If the investigator cannot identify the fuel first ignited, the ignition source competent to ignite that fuel, and the circumstances that brought that fuel and that ignition source together, the cause cannot be conclusively established. This determination is not limited to the identification of physical residues of both fuel and ignition source but may depend on the hypothesis testing of observed fire behavior and damage.

Examination of a Structure Fire Scene

GENERAL CONSIDERATIONS

The successful fire investigation begins with the proper attitude on the part of the investigator. In the authors' experience, a significant percentage of all structure fires today are maliciously ignited. Therefore, every fire scene must be approached as a potential arson fire from the standpoint of preparation, control, search, documentation, and collection of evidence. The investigator must be careful, however, not to prejudge the cause of any fire and should consider every one to be undetermined in causation until sufficient proof is found. Failure to consider a presumption of innocence may lead to a tragic wrongful prosecution. Arson is a most difficult crime to prove because it is often not apparent until the investigation is nearly complete. As a result, every fire investigation must be treated with the caution and thoroughness of a crime scene until it can be clearly demonstrated that the fire was accidental or natural in origin. Even accidental fires may have far-reaching consequences from a liability standpoint, and the physical evidence may be just as crucial to investigators from regulatory agencies or the insurance industry. (See *ASTM E860, E1188*, and *E1459*[17] for guidance on preserving evidence.) There are also concerns of spoliation of evidence in civil cases where evidence critical to establishing what caused a fire and what did not has been destroyed or removed by the public-sector investigator. See Chapter 17 for a discussion of spoliation issues.

This caution extends *both* to the security of the scene and to the initial dealings with occupants and all witnesses in that individual accounts of what happened should be carefully cross-checked and verified. The scene *must* remain in the custody of the fire or police department from the time of the fire to ensure the integrity of the scene and its evidence. It is the responsibility of the fire department to ensure continuity of possession. If that custody has not been maintained, a search warrant must be obtained or the owner's permission secured before any evidence can be rightfully recovered. Some of the legal considerations of fire scenes are discussed in Chapter 17.

It is vital that the investigator be willing to do a complete investigation before he or she begins. It must be appreciated that this may entail many hours of difficult and unpleasant effort, sometimes under less-than-ideal conditions. Custody and control of the scene must be retained until the investigation is complete. Supreme Court decisions have made it clear that security of the scene must be strictly maintained both before and during the investigation. The investigator must be willing to examine *every* fire systematically and thoroughly, paying close attention to small details. The scene must be sketched and photographed to preserve its appearance for later court inspection. These sketches and photos must be accompanied by careful notes and an adequate report to permit an accurate reconstruction at any later time. Because so many fires are arson caused, the fire investigator must be careful to examine every fire scene for things that look out of place—objects that do not belong in the area in which they are found or are in positions that do not appear to be normal or logical. Because of the tremendous damage that many structure fires do, it is attention to small details and such vague properties as "out of place" that can make the difference between a successfully identified fire origin and the frustrating opinion "undetermined cause."

INTERVIEWS WITH FIREFIGHTERS

Every fire scene investigation follows a certain protocol or logical sequence (as outlined in Table 7-1) that will vary slightly depending on the circumstances of the individual fire and its investigation. If a fire investigator is fortunate enough to arrive at the scene while the fire is still underway, first contact should be made with the fire officials present. The ranking fire official can provide information as to the time of alarm, time of arrival, and time required for control, as well as the names of owners, occupants, and witnesses. Copies of all relevant departmental reports should be requested. The firefighters will be able to provide valuable information as to the extent of involvement of the structure, behavior of the fire, unusual odors, and whether any explosions took place. Because the suppression activities can affect fire patterns inside, the investigator should determine where the structure was entered and what attacks were made on the fire. Were windows and doors open, closed, or locked upon arrival? Were entry paths blocked? Were doors or windows blocked open (to improve draft)? Were doors locked? If so, how was entry made? Did anyone try the lock before forcing the door? It must be remembered that an arsonist usually has to gain entry to a building in the same way as a burglar or other criminal. Signs of forced entry such as damaged locks or doorjambs, or broken windows are sometimes overlooked by fire investigators who are focused on the fire and its aftermath. Where were attempts made to ventilate the structure to aid in suppression? Were tactics like positive-pressure ventilation used? Because such an operation can dramatically affect the course, intensity, and time factors of a fire, the investigator must be made aware of their use and of the response of the fire.

The observations of the first-in firefighters are often crucial to the reconstruction of a fire. What was on fire and what was not? Was the fire localized on one side of the room, just at floor level, just at ceiling level, or was everything on fire (i.e., post-flashover)? Was the smoke so thick one could not tell? More departments are using thermal imaging cameras, and that information may be useful in establishing where the hottest part of the fire was. With the use of protective clothing and self-contained breathing apparatus (SCBA), observations of temperature and odor will usually be very limited, so visual observations will be most important. If the time of entry can be established from the radio log or arrival times of other units, a timeline may be reliably established to at least trace the development of the fire from that point. Fire crews should also be asked about their use or refueling of any gasoline-powered tools or equipment, precautions taken, and locations of use or refueling.

The fire crews can also offer important information as to the weather conditions (wind, temperature, humidity) and the nature of the utilities. Were the gas, electricity, phone, and cable TV in service? Who disconnected them? Did any of the occupants or witnesses speak to the firefighters? What did they say regarding the fire, the building, its contents, or the activities of people around the building recently? The sum total of information gathered by the firefighters combined with their own observations often offers the key to the entire investigation. The skill is to gather the details, compare them with other witnesses' accounts, and make accurate assessments and records in order that conclusions can be properly presented. Many fire departments today request a "debriefing" report from every firefighter who attended, describing what was seen and done.

INTERVIEWS WITH WITNESSES

While the investigation relies on application of the scientific method, the gathering of information from witnesses requires the art of communication. Just like fires, no two witnesses are exactly the same. They may be traumatized or excited, helpful or obstructive. The investigator must be skillful in adopting a demeanor and approach that will encourage the free flow of information and enable an accurate picture to be drawn. It is as important to identify and properly react to the genuine concerns of fire victims as it is to recognize suspicious behavior.

The investigator should attempt to discreetly canvass the neighborhood for witnesses to the fire or people with knowledge as to the structure, its owners, occupants, or its past history. He or she should determine who was in the building last and when and, most important, what the furnishings and contents of the building were and how they were distributed. The combustible contents of a structure have the largest single effect on burn patterns, and the pre-fire fuel load must be identified. Appendix H includes a field interview form developed by the Chicago Fire Department to ensure that critical information is not overlooked in any interview with firefighters or police officers. Also, an incident report form used by the same department illustrates comprehensive documentation. *NFPA 921* and *Forensic Fire Scene Reconstruction* also include witness interview forms and many other useful forms.

Civilian witnesses to the earliest stages of a fire or explosion can offer information to the investigator that may be impossible to establish in any other way, especially in cases where there is extensive destruction before the fire can be reported or where fire service response is delayed. Their observations could include the following: what was on fire when the fire was noticed, where in the building they first saw fire and/or smoke, how they first become aware of the fire (explosion, smell, sound of breaking glass, smoke or fire alarm, etc.), what unusual smells, sounds, vehicles, or people they noticed just before the fire. When these interviews are conducted, it may help to walk with the witnesses to where they said they were to see if lines of vision are unobstructed. Sometimes the process of re-creating witness position and path triggers additional memories. These interviews will go a long way to helping the investigator establish the starting conditions, that is, what the scene looked like at the start of the fire, a step essential to accurately interpreting the indicators.

SEARCH PATTERNS AND PRACTICES

As described earlier in this chapter, before beginning the actual search for evidence, the investigator should examine the exterior of the structure from all sides. This walk-around survey will give a fair idea of the limits of the fire damage, the areas of heaviest damage, and the possibilities of an external source of ignition (such as a trash fire) or reveal obvious forcible entry. View lines can be established at this stage. What was visible to passersby or neighbors? Which neighbors have a view of the scene? Once the survey is completed, it is useful for the investigator to step back from the immediate scene for a moment and determine what areas *surrounding* the structure must be searched as well—for example, along likely paths of entry or exit and the areas under windows or around doors. If a tall adjoining building is not available, it is often helpful to use an aerial snorkel or ladder unit to get immediately above the fire scene to facilitate observing, charting, and photographing the damage. Figure 7-9 illustrates the value of an overhead photo from an aerial ladder—showing the location of the most damage.

Once an accurate idea of the scene limits is in mind, a perimeter is established, preferably by means of a physical barrier such as rope, banner, or barricades. All personnel not essential to the search are then kept outside that perimeter for the duration of the scene examination. This perimeter method provides both scene security and a good starting place for the search that follows, since it sets the outer limits of the search pattern. Documentation also begins before the search, normally at the outer perimeter, in the form of photos, notes, and sketches. Some investigators find it useful to prepare a rough floor plan at this stage to help plan the investigation and begin to put the scene features into context.

Most structure fires lend themselves to a spiral search pattern, beginning at the outer perimeter and working toward the center of the fire. The search pattern will, of course, depend on the size and shape of the area to be searched and the time and number of people

FIGURE 7-9 Photo from aerial ladder shows extent of damage. *Courtesy of Greg Lampkin, Knox County, TN, Fire Investigation Unit.*

available to do it. The exterior areas of the scene must be carefully examined for shoe and tire impressions that do not appear to be those of fire or police personnel or their vehicles, discarded tools (e.g., from a forced entry), fuel containers, beer or soft drink containers, matchbooks, or shards of glass from windows. If an explosion preceded or accompanied the fire, the exterior search will focus on any fragments of the structure or explosive device. Because the trajectory these fragments followed must be reconstructed to establish the seat of the explosion, it is vital that the location and distribution of any such fragments be carefully charted on the crime scene sketch (see Chapter 12). (Special attention should be paid to the distribution and distance of dispersal of all window glass.) This process minimizes the chances of destroying evidence that was damaged by the fire or explosion before its relationship to the scene has been assessed. As the investigator works through the scene, the process is one of tracing the flow of the fire back to its origin and cause, which is often (but not always, for reasons to be discussed) in the area most damaged. Indicators of direction of movement will be discussed in an upcoming section. Large scenes may require division into sectors that are then searched in sequence, as dictated by access or exigencies (limitations). An example is shown in Figure 7-10.

As a general rule, anything that looks out of place should be checked. Does the item belong to the scene? Is it clean and fresh, as if recently dropped there, or stained and weathered, as if exposed for many weeks? Is the grass underneath it still green? Anything that looks odd or out of place should be charted on the scene sketch, photographed in place, and then collected. It is always easier to disregard collected items if proven later to be unrelated to the fire than to try to recover important items once they have been left behind. *All* collected objects should be considered potential sources of latent fingerprints and therefore must not be handled directly by more than one person, and packaged to minimize damage and contamination by others' handling them. Vinyl or latex gloves prevent contact with chemical or biological hazards and minimize (but not always prevent) deposition of fingerprints or DNA.

Proceeding toward the building itself, the investigator looks for signs of attempted forced entry, such as tool marks, fresh damage to door or window frames, broken locks, or broken windows. Are there signs of prior efforts at extinguishing the fire—a garden

FIGURE 7-10 String grid divides this large scene into sectors for searching. *Courtesy of Don Perkins, Fire Cause Analysis.*

hose or an empty fire extinguisher? If so, an attempt should be made to find the people responsible, since they may have valuable eyewitness information to offer about the early stages of the fire. Documentation should include photos supported by notes or diagrams showing the presence or *absence* of signs of forced entry at all possible access areas, as well as signs of activity—tire impressions, shoeprints, extinguishers, and the like. All external areas of the structure must be photographed before disturbing them.

No matter what the nature of the structure or the extent of the fire damage, nearly every thorough fire investigation starts in the least damaged areas and progresses toward the most heavily damaged areas—starts with the areas of *least* confusion and disruption and proceeds to those *most* disrupted by fire or explosion. This process in part helps establish the starting conditions in the fire damaged areas by showing the investigator what wall and ceiling finishes were like, as well as furnishings. As the investigator works through the scene, the process is one of tracing the flow of the fire back to its origin and cause. It can be thought of as a search for what material was first ignited and the source of heat that ignited it. Logic and knowledge of basic fire dynamics are the most important tools to keep in hand. Materials exposed to a violent fire can do some unusual things and exhibit unexpected properties, but the fundamental precepts of matter, energy, and heat transfer cannot be violated. There must be fuel, oxidizer, and energy in appropriate forms and amounts before there will be fire. The combination of fuel and energy that the investigation indicates as the source of the fire must meet the scientific and fire engineering requirements of energy release, fuel, and oxygen supply. As fuel packages and potential ignition sources are discovered they should be plotted on a detailed scene diagram. For the reasons we have discussed, the area most heavily damaged is not always the area of origin, since the intensity and duration factors will be influenced by fuel load, manner of ignition, ventilation, and fire suppression. The investigator must take all these into account when evaluating the damage pattern in the search for the origin.

FIRE BEHAVIOR INDICATORS

Indicators are the visible, and usually measurable, changes to the surfaces and materials of a fire scene produced by smoke, heat, or flames. Because they are produced by predictable physical processes, they can be used to re-create the path of the fire's development, the location of fuel packages, the nature of ventilation, and sometimes the height, intensity, and duration of the flames. There are many indicators that the investigator can use in reconstructing the path of the fire. None of them are precise. None of them are foolproof. None should be relied on to the exclusion of all others. Depending on the nature of the structure and the extent of the fire, it may not be possible to apply some indicators. The most accurate origin determinations are accomplished by applying a number of different indicators independently and then combining the results as a series of "overlays." The investigator examines the structure methodically, checking and recording one fire indicator at a time. Each search will reveal some useful information about the direction of fire travel (sometimes called "arrows") and the relative intensity in each area. After several such surveys, stacking the indicators up against each other will reveal points of similarity and agreement. Further searches can then be concentrated in the general area of suspected origin indicated by the combination of indicators.

Some indicators of fire travel result from the deposit of soot and other combustion products on exposed surfaces without any chemical changes. Most other indicators are produced by heat-induced changes to surfaces. The intensity and duration of the heat applied will dictate the nature and depth of those changes. Because indicators of damage (or relative lack of damage) form patterns, and are created by a combination of smoke, heat, and flames, they are sometimes called **fire patterns**. The diagnostic signs most often used to indicate fire intensity and direction of travel are outlined in the following sections.

indicators ■ Observable (and usually measurable) changes in appearance caused by heat, flame or smoke.

fire pattern ■ Indicators of damage (or relative lack of damage) created by a combination of smoke, flames, and heat effects.

Burn Patterns

Burn patterns are created when applied heat fluxes are above the critical thresholds to scorch, melt, char or ignite a surface. Where the fire's destruction did and did not take place constitutes the burn pattern. Burn patterns are the cornerstone of all fire investigation because of their universal applicability. Data from a fire scene include the evaluation of any and all heat transfer patterns. In seriously damaged structures, the question, Where was the destruction most complete? may accomplish the same end. We know that fire *generally* travels upward and outward, leaving, in the absence of barriers or unusual fuel conditions, a V-shaped or conical pattern in the structure it leaves behind. In the absence of active suppression, fire will usually have time to burn longer at its point of origin than at places subsequently involved by spreading flames. As a result, destruction will be more extensive in that area, all other things being equal. These include fuel loads and ventilation, however, that are rarely equal throughout an entire scene. Figures 7-11a–d illustrate patterns in a bedroom fire originating from a wastebasket fire.

FIGURE 7-11A Small waste-basket fire quickly spreads to drapery and bedding. *Courtesy of Jamie Novak, Novak Investigations and St. Paul Fire Dept.*

FIGURE 7-11B "V" patterns on wall and headboard point back toward wastebasket. Note thermal patterns at margins of pattern (charring to scorching to no damage). *Courtesy of Jamie Novak, Novak Investigations and St. Paul Fire Dept.*

FIGURE 7-11C Damage pattern on bed (including partially collapsed springs) becomes more intense toward origin (lower right corner). *Courtesy of Jamie Novak, Novak Investigations and St. Paul Fire Dept.*

FIGURE 7-11D TV set shows more collapse and glass fracturing on side facing growing fire as fire grows and spreads around corner from bed. *Courtesy of Jamie Novak, Novak Investigations and St. Paul Fire Dept.*

FIGURE 7-12
Temperature distribution of gases in a buoyant plume (a) unrestricted (b) under ceiling (temperatures in degrees Fahrenheit).

(a)

(b)

One of the factors that is not always equal is fire suppression. Fires are attacked at what appear to be their most critical areas—where other buildings are threatened, where victims are trapped, where dangerous contents are to be protected, and so forth. Occasionally, suppression will be focused on the very area of a structure in which the fire was started, and even though the rest of the building is destroyed, the area of origin may be the most intact. That is why interviews of the suppression crews are so important to revealing where the fire was attacked and with what tactic.

Burn patterns that can be used to indicate the direction of fire spread can be observed on structural elements, wall surfaces, or even furnishings remaining in the building. Because these elements are of most value when they remain undisturbed, it is most important that overhaul of the structure be kept to an absolute minimum to prevent their destruction as evidence. Application of heat to a material causes changes first only to its surface and if sustained, changes to its bulk. If we apply the principles of heat transfer and the structure of flame plumes from Chapter 3, we can understand how these patterns are formed. The temperatures in a flame and smoke plume decrease with height in the plume or distance from the centerline, as illustrated in Figure 7-12. Because the radiant heat flux intensity is controlled by the temperature of the source, we can see why the pattern of damage from a moderate-sized fire to an adjacent noncombustible vertical surface is often in the form of an inverted "V" (as depicted in Figure 7-13). As room air is entrained into the rising flame and smoke plume it will be spread out as it rises, but is will also be diluted and cooled. If the adjacent surface is very susceptible to thermal changes or if combustion products are deposited as the gases cool, there may be a pattern resembling an hourglass or a "V" in outline.[18] The overall pattern most often resembles a "V"or "U" with its apex pointing toward the heat source.

Wherever the total heat flux onto the wall is below the critical intensity needed to affect the surface, there will be a *line of demarcation* for that heat pattern. Areas outside that demarcation will not be visibly affected. Note that this process is driven by the buoyant flow of the hot gases, so the "V" (or inverted "V") is produced on vertical surfaces, not on horizontal ones. Because the gases cool as they move away from the centerline of the plume, damage patterns to horizontal surfaces above the fire will tend to be circular, with a "bull's-eye" of the most damage directly above the fire.

A typical "V" pattern on a wall as a result of a localized fire is shown in Figure 7-14. Notice how the pattern spreads out naturally, with the apex of the "V" right at the origin

FIGURE 7-13 Energetic fire on chair in corner produced an hourglass pattern on a noncombustible and on the ceiling as a result of the corner configuration. Note gradations of thermal damage. *Courtesy of Jamie Novak, Novak Investigations and St. Paul Fire Dept.*

of the fire. Although it is sometimes claimed that the more vertical the sides of the "V," the faster the initial fire (and therefore the more suspicious), one can appreciate that the nature of any wall covering or vertical target surface, size of the fuel package, height of the ceiling, and conditions of ventilation determine the shape of the pattern. The angle and width of the "V" are not dependent on the rapidity of ignition of the initial fuel but on the fire properties of the target surfaces (see Figure 3-29). The "V" pattern associated

FIGURE 7-14 Fire that started at end of sofa produced "V" pattern of damage on wall behind sofa and on sofa. Note clean burn on drywall, thermal transfer pattern on side wall, and enhancement of ceiling damage as a result of corner effect. *Courtesy of Greg Lampkin, Knox County, TN, Fire Investigation Unit.*

FIGURE 7-15 The wide, shallow "V" pattern on the wall is due to a very large plume flattened by the low ceiling. The room approached flashover conditions in under 6 minutes. *Courtesy of John D. DeHaan.*

with most fires is a result of several factors, one being that the natural plume shape of a large fire is a cone; the interaction of this plume with a wall or a ceiling is the major factor in producing a pattern. The shape of this cone depends on the size of the fire with respect to a ceiling that tends to flatten and spread it horizontally, as shown in Figure 7-15. A vertical wall intersecting this cone may then display a vertical cross section of the cone. If a fire is located some distance away from the wall, the cone does not intersect the fire cone at its base, so a shallow "U" or "V" pattern may appear, as in Figure 7-16. The ceiling directly above a large fuel package may exhibit a circular bull's-eye pattern of damage (unless bounded by nearby walls), with the center reflecting the effects of the highest temperature gases, and succeeding rings of less damage. Because hot gases moving across a surface lose some of their heat by convective transfer, they cool, producing less damage as they move away. If the fire is located against a wall, the circular pattern is interrupted by the wall to form a semicircle (as in Figure 7-16).

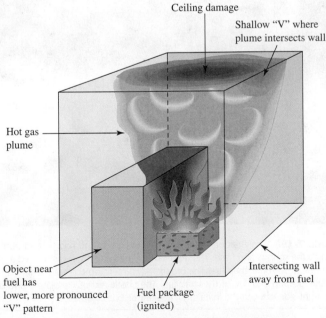

Ceiling damage

Shallow "V" where plume intersects wall

Hot gas plume

Object near fuel has lower, more pronounced "V" pattern

Fuel package (ignited)

Intersecting wall away from fuel

FIGURE 7-16 A fuel package burning in a room may produce a circular burn on the ceiling if the plume is large and energetic enough. Vertical surfaces intersecting the plume may have areas of plume-induced charring.

FIGURE 7-17 Even if the wall covering is destroyed, the "V" pattern is often reflected in the remaining structural members.

FIGURE 7-18 Beveling of exposed combustible surfaces indicates that fire moved from right to left.

A small fire of low energy and limited duration will typically display a triangular, inverted-"V" or "A" pattern of damage to a nearby adjoining wall. This is especially true if the wall or its covering is not readily ignitable. If the wall covering is very combustible, the flame travel can be almost vertical, as shown in Figure 3-29g. If the wall covering has been destroyed, the "V" pattern is often recorded on the wall studs still remaining, as in Figure 7-17. Because the sides of an object exposed to an encroaching flame will be more heavily burned than the protected sides, and the buoyancy of the hot gases causes them to rise as they flow horizontally, the object will often have a beveled appearance that is more pronounced on the side and edges facing the oncoming fire, as illustrated in Figure 7-18. If a fire is traveling very rapidly, the hot gases will flow over recessed areas on the downwind side, leaving relatively undamaged areas on the side away from the fire.

Furniture will also exhibit burn patterns of use in determining point of origin and direction of spread, since an object, whether upholstered or not, will generally be burned more extensively on the surfaces exposed to an oncoming fire. In large combustible objects, significant portions of a "V" pattern may be visible, as in Figures 7-19a and b.

FIGURE 7-19A The "V" pattern can be reflected in large items of furniture.

FIGURE 7-19B The chair on the right was ignited first and had time to burn completely down to the springs and base. The right side of the wine rack was burned away by the radiant heat from the chair fire. The broad "A" pattern on the wall behind is consistent with the location and size of the chair fire. The plastic receptacle plate is also burned more on its right side (toward the initial fire). *Courtesy of John D. DeHaan.*

If the original location of the furniture can be established by observation or by witnesses, the patterns can be incorporated into the overall pattern observed in the room.

The before-and-after photos of a nonaccelerated test fire in Figures 7-20a–d demonstrate a number of these features, including the "V" pattern on the corner walls, as well as on both sofas. The paint and paper have been burned off the walls and the ceiling in the corner around the origin. Radiant heat damage to the left side of the TV stand and the armchair shows more intense heat coming from the left. The fire was approaching flashover when it was extinguished, so the carpet in the corner where the fire began and the wood floor beneath it were very badly burned by radiant heat. The carpet and furnishings closer

(a)

(b)

FIGURES 7-20A AND B
(a) Before and (b) after photos of a test fire ignited by setting fire to right front corner of the left sofa. The "V" pattern extends up and back from the origin, across left sofa, and onto the wall behind. Damage to right-hand sofa is most severe in the front center (directly across from the origin) and decreases to right and left. *Courtesy of John D. DeHaan*

FIGURE 7-20C Overall photo of scene shows floor-to-ceiling plume damage in the corner, floor-level damage between the sofas, and higher damage farther away. *Courtesy of Keith Parker, Marin County Fire Department, Woodacre, CA.*

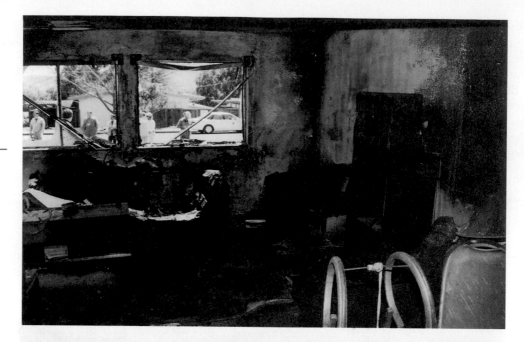

FIGURE 7-20D After removal of the sofas, the protected area on the wall shows the location of the sofa back. Radiant heat damage to the left front corner of the chair was produced by the sofa fire. The "V" pattern from the sofa includes the left side of the TV and stand. Note the damage to the floor caused by radiant heat ignition of all-synthetic carpet and pad. No ignitable liquids were used. *Courtesy of Keith Parker, Marin County Fire Department, Woodacre, CA.*

to the photographer were just being affected by the radiant heat from the hot gas layer. The header formed by a structural beam restrained the hot gas layer from flowing into the dining room. The protected area on the right (far) wall shows where the back of the smaller sofa was in contact with it.

In another test fire, depicted in Figure 7-19b, the fire damage shows the beveling effect on the wine rack resulting from the fire started in the chair in the corner. The chair was reduced to its springs and metal base by the fire, and the floor around it was charred by the radiant heat and from flaming **dropdown** of the synthetic upholstery falling on it during the fire. The thin wooden top of the small table in the foreground is burned through by radiant heat.

Beveling of exposed edges occurs when heat is applied and the temperatures of edges and corners rise faster than those of adjacent flat surfaces (because of their more limited

dropdown ■ The collapse of burning material in a room that induces separate, low-level ignition; falldown.

FIGURE 7-21 Beveled areas (in cross section) reveal the direction of the fire development through the floor. *Courtesy of Lamont "Monty" McGill, Fire/Explosion Investigator.*

Fire direction

thermal capacity, they cannot lose heat as quickly to the interior). That is why such edges are easier to ignite than a flat surface and, once ignited, burn more quickly. This predictable behavior allows the investigator to use such beveling to re-create the direction of travel of the fire gases. Directional travel can produce cross-sectional beveling of exposed wood elements that can indicate the direction of the flow of hot gases around them. Such beveling, then, can be used to plot pointers or "arrows" of fire spread direction on a plan of the building.

This beveling effect can be very useful in establishing the direction of burn-through of floors. Because of the complexity of most structure fires, it is not enough to record the fact that the floor has been burned through in a particular area. It must be determined whether the floor has been burned from underneath or from the top. Figure 7-21 illustrates the typical appearance, in cross section, of the two cases, that is, burned from underneath and from the top. It is obvious that joists will protect the flooring immediately above them when the fire burns from underneath, but all the flooring surrounding the hole will be destroyed first if the fire is burning downward.

Once these indicators of direction of fire spread (sometimes called *vector* or *arrow patterns*) have been observed throughout the structure, they can be combined with observations of relative char depth and the heights of heat and smoke horizons in various rooms to map out the fire's approximate course of travel.[19] The combination of direction and intensity indicators is sometimes referred to as *vector analysis*. These vectors can be mapped out on a floor plan to help establish a likely area of origin. This information, of course, must be considered in light of what is known about the suppression efforts that occurred at the scene. It is well known that venting a burning building or actively attacking it on one side can force fire to reverse its course and attack new areas or reburn previously affected areas. It has been observed that if a fire starts in one room and spreads to another with a larger fuel load, the subsequent extension of the second fire can obliterate the direction indicators on adjoining walls, making it look as if the fire started in the second room. Ventilation through windows, whether wind-driven or from PPV fans during suppression, is a critical factor in evaluating burn patterns and fire behavior.[20]

Heat Level (Heat Horizon)

As hot gases are released into a room their buoyancy causes them to rise to the ceiling, and then their accumulating mass and momentum causes them to start to flow outward. When hot gases flow across a surface, the flame front is always at a marked acute angle to the surface, as depicted in Figure 7-22, because of the physical properties of the density-driven flow. In the same manner as when a dam bursts, and the gravity-driven flow of the "wall" of water causes it to flow with its face at an angle to the ground (as illustrated in Figure 7-23), the same occurs with hot gases. If the flame front comes into contact with vertical surfaces, it produces a characteristic "front" that will tell the investigator the direction of flame spread at that point at some time during the fire. The effect of such progression across a thin-paneled, hollow-core door is shown in Figure 7-24.

FIGURE 7-22 Hot gases entering this room produced a front at an angle of about 45° from the ceiling. A layer of gases moving across a surface will not assume an angle of 90° or more because of the momentum flow pushing it across the ceiling. *Courtesy of Lamont "Monty" McGill, Fire/Explosion Investigator.*

FIGURE 7-23 Advective (gravity-driven) flow of water released from a collapsing dam assumes an acute angle from the ground.

FIGURE 7-24 Hot gases flowing from left to right (out door) produced characteristic "beveling" damage to a hollow-core door. *Courtesy of John D. DeHaan.*

FIGURE 7-25 A uniform hot gas layer ignited paneling near its entry to the room and scorched the paneling across the room. *Courtesy of Detective Mike Dalton, Knox County Sheriff's Dept., Knoxville, TN.*

As the hot gases continue to accumulate they fill the room from the top down, producing a hot gas layer. In large rooms or with very large (high HRR) fires, there may be times when the gas accumulates (in a corner, for instance) in a deeper layer immediately above the fire than it does in the rest of the room; eventually the layer will fill the room to a uniform height throughout (except near ventilation openings, where it rises) (see Figure 7-25). The larger the fire or the smaller the room, the faster this hot gas layer deepens until it can begin spilling out windows or under headers into adjacent rooms. As the hot gas layer spills out into adjacent rooms it begins to fill those rooms from the ceiling downward. This produces "shallower" layers in other rooms, with the "deepest" layer in the room of origin. This process produces a characteristic stair-step pattern of smoke or heat damage that will help trace the flow of the fire, as shown in Figure 7-26.

FIGURE 7-26 Fire moving down this hall from far end produced a stair-step or rising heat horizon pattern. Note damage to paneling in next room on left. *Courtesy of Greg Lampkin, Knox County, TN, Fire Investigation Unit.*

The height of significant levels of heat (i.e., the depth of the hot gas layer described earlier) in a room is revealed by the charring, burning, blistering, or discoloration of paint or wall coverings. The **heat horizon** will generally be level, that is, of uniform height from the floor when the atmosphere in the room is fairly static. Changes in level indicate points of ventilation, such as open windows or doors or breaks in the walls or floor. Typical questions for the fire investigator include the following:

■ Do such points of ventilation agree with the known history of the fire, or do they indicate that an attempt was made to encourage fire spread by opening walls, ceilings, or floors prior to the fire?
■ Are there signs of such intentional venting?
■ If windows or doors were open, were they known to be open before the fire?
■ Were the windows or doors damaged before the fire or during it?

The heat level will often drop markedly in the vicinity of the point of origin or when there is combustible material ignited lower down or along the base of the wall. Low burns can also be produced by combustion of floor coverings during full involvement of the room. As the hot gas layer deepens, the heat level will sink from near the ceiling as the fire grows. At flashover with involvement of floor coverings, the heat level will be at or near floor level (bases of walls). Ignition of a premixed quantity of natural gas, LP gas, or flammable liquid vapors in a room tends to produce a *scorching* of wall surfaces that is uniform from floor to ceiling, while post-flashover burning usually produces deeper burning with its longer duration. In rooms not involved in fire, the heat level may be revealed by the distribution of melted objects—candles, plastic fixtures, even appliances—in the room.

The temperature of the heat source determines the intensity of heat flux it produces. Therefore, the temperature profiles in Figure 7-12 will produce commensurate effects. As hot gases move across a solid surface some of their heat is lost by convective transfer, so they cool as they move away from the primary contact point. This produces a gradation of effects, as shown in Figure 7-27, which allows testing of hypotheses about contact and movement.

Smoke Level (Smoke Horizon)

Even when there is no heat effect on a surface, surface deposits can reveal the height at which smoke has stained the walls and windows of a room (the **smoke horizon**), which reflects the nature and extent of the fire and the amount of ventilation. Smoke staining can occur even in rooms that have not been damaged by the flames themselves. Such staining is produced by the condensation of liquid products of incomplete combustion (pyrolysis products) and the deposition of soot. As with heat levels, smoke levels will change at points of ventilation and therefore may be useful in estimating which doors and windows were open during the fire.

The smoke level may also be useful in assessing the cause of fire deaths. For instance, it may be possible to estimate if a person of a certain height would have been able to see an exit if standing in the room during the fire or whether he or she would have been overcome by the cloud of hot toxic gases accompanying the smoke. Occasionally, victims who did not stand up during a smoky fire escaped, while people who did were quickly overcome. The heat and smoke levels throughout a building are most conveniently charted on a separate scene sketch for later comparison with other data.

Buoyant flow of hot smoke can produce a "backsplash" or "mushrooming" effect if the hot gases have sufficient momentum when they encounter a wall. Just as water waves pile up against a breakwater, hot smoke can "pile up" against a wall producing a dropdown of the smoke horizon (sometimes called *momentum flow*). Flow through a door often leaves a void or clear area on the "downstream" side of the header, as illustrated in Figure 7-28.

Low Burns and Penetration

We examined the dynamics of the fire environment in previous chapters and can now appreciate the reasons why low burns on walls or burned floor materials may be significant.

FIGURE 7-27
Gradations of thermal damage on this wall show the intensity of the plume as it cooled from the centerline out and vertically. Smoke deposits (nonthermal) surround the upper portions of the pattern. *Courtesy of Jamie Novak, Novak Investigations and St. Paul Fire Dept.*

The lowest point of burning in a structure or room must always be examined carefully, for it may well bear evidence of the cause of the fire, as well as its origin. Fires ignited in a room's normal fuel load most rapidly burn vertically upward from their point of ignition, leaving the floors and lower reaches of the walls much less damaged than the ceiling and upper walls. Any localized area with a floor burned or a wall burned right down to the

FIGURE 7-28
Momentum flow and doorway flow, leaving clear area "downstream." Deeper layer exists in room of origin.

floor should be considered deserving of further investigation. Such a burn by itself does not mean that the fire is incendiary in origin. It means that high temperatures or high heat fluxes were encountered at floor level. Something normal to the room may have caused it to burn in this fashion, such as combustible floor coverings, post-flashover burning, falldown, or collapse. Each possible hypothesis has to be tested and eliminated.

During the free-burning phase of a room fire after flashover, the radiant heat is extremely intense (120 to 150 kW/m^2), and the combustion in the room becomes unstable and extremely turbulent. The intense radiant heat causes ignition of floor coverings and the floors themselves, producing a generalized (although not always uniform) burning. The synthetic carpet and pad materials in common use today melt and ignite readily upon exposure to these post-flashover radiant heat fluxes. In a post-flashover fire, when the size of the fire is controlled by ventilation, the most intense burning will be in the vicinity of ventilation openings (see Figures 7-29a and b). The very high flame temperatures

FIGURE 7-29A
Extensive melting of synthetic carpet caused by radiant heat (pre-flashover). Notice the pattern is very irregular. "Clean burn" on right wall is due to ventilation (fresh air drawn into fire at floor level). *Courtesy of SA/CFI Michael A. Marquardt, ATF, Department of the Treasury, Grand Rapids, MI.*

FIGURE 7-29B The same carpet in a room fire ignited with gasoline shows similar irregular melting. Charred and scorched area shows location of gasoline pour. "Clean burn" beside chair is a ventilation effect as the energetic gasoline fire beside chair entrains fresh air from front of room. *Courtesy of SA/CFI Michael A. Marquardt, ATF, Department of the Treasury, Grand Rapids, MI.*

of burning carpet, pad, and floor materials can be sustained as long as there is fuel—much longer than the typical flammable liquid fire—and can induce fire damage down to floor level. Even before flashover occurs, the room is exposed to high heat fluxes across a wide area, producing a generalized (although not always uniform) scorching and melting. Localized low burns surrounded by large areas of undamaged floor are unusual and deserve close attention. The pre-fire conditions of the floor are important because even combustible trash or clothing will protect the floor covering from radiant heat, leaving the exposed areas to be damaged by the full brunt of the post flashover fire.

Figure 7-29 shows two identically furnished rooms, one accelerated with gasoline, the other ignited by open flame, that were extinguished just as flashover occurred (with ceiling and floor temperatures of 815°C (1,500°F). Note that the synthetic carpets melted in both rooms in an irregular fashion, probably due to their nonuniform contact with the floor (from curling). The pour pattern of the gasoline is still visible but would be obscured once the carpet and floor were fully ignited by post-flashover radiant heat. Figures 7-30a–d illustrate the flames and post-fire patterns of a gasoline **trailer** in a carpeted room.

trailers ■ Long trails of fast-burning materials used to spread a fire throughout a structure.

FIGURE 7-30A Ignition of gasoline trailer from door to sofa. Note flash (blue) flame extending through vapor layer from pool. *Courtesy of Jamie Novak, Novak Investigations.*

FIGURE 7-30B Seconds later, flames have traveled along trailer against the base of the wall to recliner. *Courtesy of Jamie Novak, Novak Investigations and St. Paul Fire Dept.*

FIGURE 7-30C Fire extinguished just before flashover. Post-fire (after removal of some of suspended ceiling). Trailer pattern on carpet barely visible on left. Uniform damage across sofa and to floor level to walls to left and right is not consistent with limited damage to other furnishings. *Courtesy of Jamie Novak, Novak Investigations and St. Paul Fire Dept.*

FIGURE 7-30D Damage to left side of recliner down to floor level should prompt further exam. Drapery collapse may have caused this, but residues of the gasoline trailer in the carpet would betray an arson cause. *Courtesy of Jamie Novak, Novak Investigations and St. Paul Fire Dept.*

One possible cause of such low burns is the ignition of a pool or trailer of **ignitable liquid** on the floor. When localized burn damage extends right down a wall to the floor surface, it may indicate that an ignitable liquid was used, since other sources of intense local combustion, such as trash, usually do not burn right to floor level. The flames from a pile of flammable trash burn upward in the typical convective cone from the fuel surface and contact the wall higher than floor level. Sustained combustion of a pile of cellulosic clothing, bedding, or furniture that has collapsed onto a wood floor, however, has been seen to cause penetrations through floors. Ignitable liquids alone will not burn long enough to achieve generalized penetrations of wood floors. However, a liquid can soak

FIGURE 7-31A Post-flashover fire deeply charred all exposed surfaces down to floor level. *Courtesy of Jamie Novak, Novak Investigations and St. Paul Fire Dept.*

FIGURE 7-31B View looking up the stairwell shows ignition of top step. *Courtesy of Jamie Novak, Novak Investigations and St. Paul Fire Dept.*

FIGURE 7-31C View looking up the stairwell shows ignition of lower steps. *Courtesy of Jamie Novak, Novak Investigations and St. Paul Fire Dept.*

into the crack that separates the wall from the floor and take the fire to the very bottom of the wall. A puddle or trail of flammable liquid is the most commonly detected arson device. The use of these liquids often produces effects that are most characteristic if the room has not gone to flashover. If the corners of a room are burned out *before* other exposed areas of the room are damaged, that is unusual fire behavior, because corners are usually dead air spaces that are destroyed last in accidental fires. Post-flashover fires, however, can generate so much radiant heat and be so turbulent that destruction can extend into the corners and even down stairs, as shown in Figures 7-31a–c. Older structures may have uneven floors, and any liquid poured on them in quantity seeks out the lowest surface. Such pools of poured liquid provide a longer-duration fire with potential to create more localized damage.

The behavior of ignitable liquids in a fire follows certain physical rules, which investigators must appreciate if they are to detect such accelerated fires. These physical rules include the following:

■ Liquids (and their vapors) flow downhill, seeking out the lowest available surface. They can flow under doors and even under the toeplate of wood-frame partition walls.

■ Liquids will seep into seams or cracks in a floor and, from there, provide a reservoir of liquid to sustain more prolonged burning and localized charring.

■ Very volatile liquids (camp fuel, acetone, alcohols, lacquer thinners, and even gasoline) will flash quickly off a nonporous surface. [In burn tests of 1- to 4-L (1- to 4-qt) quantities, which produced a freestanding pool less than 0.5 mm deep, a regression rate of 0.5 to 1 mm/min (0.02–0.04 in./min) and a total burn time of about 1 min was expected and observed].[21] Because of the low-boiling-point liquids involved, such fires produce minimal surface scorching (see Figures 7-32a and b) compared with the relatively deep charring induced by burning an overturned cardboard box filled with crumpled newspaper (Figure 7-32c). Only the reservoirs along cracks and seams provide fuel for deep charring (as in Figure 4-13e). Scorching and some sooting of the floor surfaces surrounding the pool are likely, particularly on the downwind side of a pool burning in a draft (as in Figure 7-33). The floor under the pool may be discolored by the solvent action of the liquid or damaged by the brief exposure to flames as the last of the liquid boils off. Flooring with very low thermal damage thresholds such as modern vinyl sheet flooring may be scorched or distorted by a burning gasoline pool.

■ Less-volatile fuels with a higher boiling-point range such as paint thinner or kerosene will exhibit different effects. Recall that it is the vapors of liquid fuels that burn rather than the liquids themselves. The more radiant flames from these fuels produce a more pronounced

FIGURE 7-32A Gasoline burn pattern on plywood. *Courtesy of John D. DeHaan.*

FIGURE 7-32B Paint thinner burn pattern on plywood. *Courtesy of John D. DeHaan.*

FIGURE 7-32C
Cardboard box with
crumpled newspaper
burn pattern on plywood.
Courtesy of John D. DeHaan.

FIGURE 7-33 Sooting
and scorching around the
margins of the discol-
oration pattern on a vinyl
floor reveal the presence
of a flammable liquid.
*Courtesy of Lamont "Monty"
McGill, Fire/Explosion
Investigator.*

charring on the floor surface *surrounding* the liquid pool, often producing a halo or ring of damage. The higher boiling point of these liquids allows the temperature of the liquid pool to rise enough that the floor surface beneath the pool may be scorched (see Figure 7-34). These liquids can form deeper pools on nonporous surfaces and burn more slowly than the volatile fuels and have more time to spread into floors and penetrate crevices and porous edges. The resulting pattern can be a deep ring of char damage around the edges of a lightly charred area that defines, to some extent, the liquid pool (as in Figure 7-35). However, a similar pattern can be created by the burning of thermoplastic materials (including uphol-stery) that melted and burned as a semimolten pool. Residues of unburned polymer should be found adhered to the floor in such cases.

■ When first ignited, any liquid fuel burning on a carpet will produce a pronounced ring or halo of damage on the adjacent carpet pile, leaving the carpet pile in the center of the pour undamaged (the pile acts as a wick) until the liquid is nearly consumed, as depicted in Figure 7-36.[22] When that last liquid is burned, the center of the pool will scorch, char, and ignite. On some carpets, this sequence will leave a charred area sur-rounded by a more deeply charred halo. On carpet, the size of the damaged area is not

FIGURE 7-34 Test burns on wood plank floor. Foreground burn is burn pattern resulting from lacquer thinner: The center pattern is kerosene, and the rear pattern is a gasoline and diesel fuel mixture. *Courtesy of John D. DeHaan.*

FIGURE 7-35 Halo pattern around a pour of gasoline on a carpet in an arson fire. Note the unburned carpet in the center of the pool. *Courtesy of Jamie Novak, Novak Investigations and St. Paul Fire Dept.*

FIGURE 7-36 This polypropylene carpet continued to burn for more than 45 minutes after being ignited by a small, localized pool of gasoline (located near the center of the carpet panel). *Courtesy of John D. DeHaan.*

FIGURE 7-37 Linear burn pattern on carpet from burning oxygen cannula tubing.
Courtesy of Detective Mike Dalton, Knox County Sheriff's Office, Knoxville, TN.

FIGURE 7-38 Floor after a test fire (see Figure 7-20). The carpet was destroyed by radiant heat in the corner between the sofas, and the floor was deeply charred by radiant heat and the falldown of burning upholstery materials (no accelerant).
Courtesy of John D. DeHaan.

always a reliable indicator of how much fuel was involved or even how large the liquid pool was prior to ignition because carpets vary greatly in the amount of penetration (into the pad) and the response of the carpet to radiant heat from the flames.[23] Data and observations from tests of similar carpet and pad can support conclusions about pool size versus quantity and fire behavior. New all-synthetic carpets can self-sustain combustion and produce extensive burn patterns that bear no similarity to the original pool of liquid fuel that ignited them, as in Figure 7-36.[24]

Oxygen masks and cannulas fed by clear plastic tubing are becoming common in residences today. If the oxygen-saturated tubing is ignited, a blowtorch effect rapidly burns the tubing away, leaving behind a narrow trailer of burn damage (see Figure 7-37).

Floor Burns and Penetrations

The investigator must be aware of the effects of radiant heat from burning fuel packages, or from masses of hot gases in a post-flashover room or moving through ventilation openings such as doors and windows.[25] The mixing of fuel-rich hot gases with fresh oxygen-rich air as they leave such openings produces very high temperature flames and enhanced radiant heat output, causing enhanced thermal damage. Floors can be charred and even ignited at heat fluxes of 20 kW/m^2 or greater even if no fuel is on fire in the immediate vicinity. Figure 7-38 shows the charred floor produced by the radiant heat produced by two sofas (Figure 7-20) and the total combustion of the all-synthetic carpet and pad. Areas under the two sofas were burned as the synthetic upholstery materials ignited and collapsed to the floor. Due to the effects of radiant heat from burning surfaces and combustion of falldown fuels, it is not sound practice merely to clear a room of its burned contents and rely on the burn patterns on the floor as indicators of fire location. Post-flashover burning is very intense but nonuniform and unstable and can produce floor burns that bear no relationship to fuel locations, as illustrated in Figure 7-39. Synthetic carpets and pads melt in such exposures, leaving irregular pools of melting and burning residue and equally irregular areas of exposed floor.[26] The ignition of floors and floor coverings can melt aluminum threshold strips that tests have shown will not be melted by ignitable liquid fires alone. The sustained flames of the burning floor coverings add to the radiant heat input from the room fire and can overwhelm the high thermal conductivity of the aluminum. Figure 7-40 shows the difference in effect on an aluminum threshold between exposure to gasoline and diesel fuel fire and a post-flashover fire (<10-min duration) with no ignitable liquids.

Often, the investigator is tempted to identify any burning through a wooden floor as having been caused by the ignition of a flammable liquid. As discussed previously, fire burns downward into a solid fuel very slowly. While penetrations can be caused by a liquid's seeping into the seams and joints of a leaky floor and then being ignited along the seams or underneath, this is usually not the cause. Penetration is more likely to occur due

FIGURE 7-40 Aluminum carpet strips after fire exposure. Top: Exposed to post-flashover fire (with carpet beneath). The strip is beginning to melt. Duration of fire approximately 10 minutes. Bottom: Exposed to gasoline–diesel fuel fire on plywood. Note the strip is sooted and scorched but the metal is still intact. Duration of fire about 2 minutes. *Courtesy of John D. DeHaan.*

to the radiant heat from a prolonged post-flashover room fire; from the sustained burning of a substantial fuel package like a sofa, bed, or overstuffed chair; or from prolonged burning of other fuels that fall onto the floor from above while still burning energetically. Sustained smoldering combustion of such fuels as draperies or upholstered furniture has been observed to cause isolated, localized penetration down through a wood floor.

Other fuels burning from underneath must also be considered. Unusual floor construction, fuel materials stored on the next floor down, or unusual air draft conditions can all contribute to floor penetration. Such penetrations are more common in post-flashover room fires in the vicinity of ventilation openings. The beveling effect of the burn pattern discussed earlier will be of great help in establishing whether the fire came from above or below as it penetrated the floor.

An unusual pattern of fire penetration was observed in a test one of the authors conducted on an all-wood structure. The room had a tight-fitting tongue-and-groove floor over a normal wood subfloor. A substantial fire was ignited without the use of an accelerant, allowed to burn (post-flashover), and then extinguished before the floor was significantly penetrated. The burn pattern on the subfloor (diagonal planks) is shown in Figure 7-41. The diamond-shaped holes correspond to the overlap of gaps between the subfloor planks and the finished flooring.

FIGURE 7-41 The burn pattern on the wood subflooring (on the left) shows unusual penetration in spite of the absence of any flammable liquids. This room sustained extensive post-flashover burning. *Courtesy of Lamont "Monty" McGill, Fire/Explosion Investigator.*

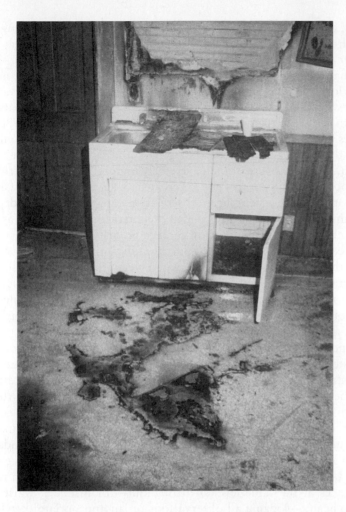

FIGURE 7-42 Isolated burns on kitchen floor resulting from falldown of cabinet (and its contents) above the sink. Note "low burn" on front of cabinet from falldown fire. *Courtesy of Jamie Novak, Novak Investigations and St. Paul Fire Dept.*

Isolated burns can be produced by burning falldown fuels (as in Figure 7-42), so the investigator must be careful about ascribing an ignitable liquid cause. Once again, a sketch of the fire scene is made with the extent and distribution of any low burns or floor penetrations recorded on it. This is retained for later comparison with the other indicators. Observations must also be made of any possible falldown sources—cabinets, draperies, light fixtures, wall paneling, and ceiling materials.

Char Depth

The depth to which the pyrolysis action of fire has converted an organic material (usually wood) to its volatile fractions and charcoal is called the **char depth**. As we saw in Chapter 5, heat applied to wood surfaces causes progressive destruction of the cellulose and other organic constituents of the wood. This charring will progress through a layer of wood at a somewhat predictable rate. Some investigators have been known to place considerable emphasis on the time dependence of char depth, thereby claiming an ability to precisely time the duration of fire exposure throughout a structure. The char rate of structural wood [which is often cited as being on the order of 2.5 cm (1 in.) in 45 minutes] is, first of all, *not* a linear phenomenon. That is, wood chars very quickly for the first few fractions of an inch and then at a much slower rate, presumably due to an insulation effect of the newly formed char and to reduced air/fuel interaction. It is only an average rate that can be projected, so any attempt to interpolate short times of fire exposure precisely based on fractions of an inch of char depth is seriously misleading.

char depth ▪ The measurement of pyrolytic or combustion damage to a wood surface compared with its original surface height.

Most significantly, it is not the time of burning, but the *intensity* of the heating (which varies locally and with time to a great degree) that most affects the char rate, as seen in Figure 5-9. The higher the *intensity* of the applied heat, the faster the charring, and the deeper the resulting char in a given amount of time.[27] Anything that affects the intensity, such as ventilation or localized fuel, will affect the char depth. Different species of wood will also char at somewhat different rates. The density of the wood is a major factor in char depth, as it is with regard to the rate of spread of a fire. The same wood will char at different rates on faces cut across its grain than on faces cut with it (and create different char patterns) due to the differences in heat conductivity and transport of fluids within the wood. The adherency of the char is also an extremely important variable with certain species such as redwood. Variations of exposure created when char flakes off in some areas and adheres in others further complicate the relationship between char depth and fire exposure. Edges and corners cannot lose heat to the interior of a fuel mass as well as flat surfaces and will char more quickly. This difference in heat loss can produce cross-sectional beveling on exposed wood members that can indicate the direction of flow of hot fire gases around them. Even the dimensions may have an influence because of their variable effects on heat capacity and conductivity. To complicate the practical picture further, misinterpretation may follow because water may have been applied in one area in fighting the fire, inhibiting charring there, while water may not have been applied in another area close by, although the general fire intensity was similar.

Although char depth cannot be used to establish precise times of fire exposure, it can be used, if applied systematically, to assess the *relative* fire exposure throughout a structure. Figure 7-43 illustrates a charred wooden beam in cross section. When testing for depth of char, remember to measure depth from the original wood surface, not the existing char layer (which may often be impossible). Use whatever small-diameter, blunt-tipped tool is most convenient to test for the depth at which intact wood remains. (A sharp knife point should be avoided, since it tends to easily penetrate into undamaged wood; some investigators prefer a tire tread-depth gauge or plastic caliper, since it has a blunt wire indicator.) Use the same tool throughout the scene. Take char depth readings on similar wood surfaces at the same level(s) throughout—for instance, waist height, head height, and knee height—logging all observations on an appropriate sketch of the scene. Some investigators prefer to plot an *isochar* diagram of the scene by connecting points of similar char depth. The resulting contours (resembling the isotherms of a weather map) outline the areas of deepest charring. These areas can then be correlated to fuel packages and ventilation sources. Remember, with *all* else being equal, char depth is most likely to be at its maximum at or near the fire's point of origin (where it burned longer) *or* near another localized source of fuel or ventilation (where it could burn more intensely). Distribution of char depth *must* be correlated to the fuel load and ventilation to be of any value.

FIGURE 7-43 Proper use of blunt-tipped measuring tool to estimate char depth. *Courtesy of Lamont "Monty" McGill, Fire/Explosion Investigator.*

Variations in char depth may be of use in positioning fuel packages and establishing ventilation conditions. Due to the rapid combustion of wood in high heat flux exposures and the production of such exposures around ventilation points such as doors or windows in post-flashover fires, floor penetrations in or near door openings in such fires have no value in indicating placement of accelerants. In a post-flashover room fire, the interior of the room may be burning in an oxygen-deficient condition except around ventilation sources. There may be deeper charring or **clean burn** patterns around such vent openings or in the flow path of incoming fresh air. The analysis of post-flashover rooms by Carman has revealed that ventilation effects not only enhance combustion around ventilation points but can produce clean-burn patterns of fire damage across the room from the ventilation source. This may give the appearance of a "V" pattern where there was no localized fuel package.[28]

Appearance of Char Surface

The appearance of the char surface may be indicative of the fire conditions. Many investigators rely on a rule that says that large, shiny blisters of char (**alligatoring**) indicate a fast-moving, rapidly burning fire, while small, dull checkering indicates a slow-burning fire—without realizing that many of the factors mentioned previously affect surface appearance as well as char depth. While a deeply seated, flat, or "baked"-appearing char is most often produced by prolonged heating, as was seen in Figure 6.15c, it can also be produced by a short-duration fire under some fire conditions (very intense fire with limited ventilation).

One observation of the authors (which awaits systematic testing) is that charring of wood surfaces in oxygen-deficient conditions (such as the underside of a wood floor deck) is more likely to produce flattened, dull, sharp-edged char. This char formation is caused by heat-driven shrinkage. Insufficient oxygen reduces the combustion rate of the exposed margins. The undependability of the size of the checkering or alligatoring is illustrated in Figure 7-44, where 1-in. by 6-in. wood planks in the same wall demonstrate remarkable variation in char appearance. One study by Ettling showed a relationship between temperature of exposure and size of char blisters (on one species of wood)—the higher the temperature, the smaller the blisters. However, oven exposure at a constant temperature for various times caused differences.[29] In a real fire where both times and temperatures are unknown, one cannot reliably assess the effects of both. Exposure to low (below 300°C; 600°F) oven temperatures did produce a characteristic sunken appearance of the wood surface with infrequent cracks or fissures that wandered over the surface in Ettling's study.

FIGURE 7-44 Fir boards on the wall illustrate the variability of alligatoring dimensions. *Courtesy of Lamont "Monty" McGill, Fire/Explosion Investigator.*

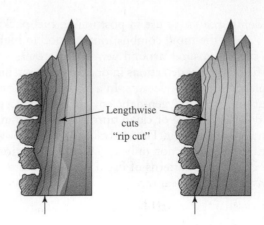

FIGURE 7-45 The difference in appearance in char and pyrolysis zone demarcations between (left) slow- and (right) fast-developing fires. *Courtesy of Lamont "Monty" McGill, Fire/Explosion Investigator.*

Lengthwise cuts "rip cut"

One indicator that may be more useful is the appearance of the charred wood in cross section. As we saw in Chapter 5, the thermal changes in wood occur at characteristic temperatures of short duration. As we saw in Chapter 3, an intense heat exposure to a high-intensity heat produces a sharp, steep gradient between a high-temperature layer and unaffected material below in a poor conductor like wood. Spearpoint and Quintiere confirmed such gradients in wood.[30] The transmittal of heat through the wood has less time to occur in a sudden, fast-developing fire, and the resulting pyrolysis zone is very narrow. As a result, there is often a sharp line of demarcation between the charred and unburned layers of the wood. When a charred beam is cut lengthwise, the gradation between the charred layer and underlying undamaged wood is more gradual with a prolonged, slowly developing fire. Figure 7-45 illustrates the contrast. No matter how reliable an indicator such gradients may be regarding the speed of development of a fire, it must be noted that a fast-developing fire may or may not be accelerated (i.e., arson) depending on what fuels are found in the room and how it was ignited. An accidental fire involving direct-flame ignition of a polyurethane mattress can produce flashover in a room just as quickly as intentionally ignited gasoline.

Surface Effects
Radiant heat fluxes on the order of 10 kW/m^2 (or prolonged contact with hot gases over 150°C; 300°F) will cause surface thermal effects such as scorching of wood or thermosetting polymers or melting of thermoplastic coatings. Higher heat fluxes or high temperatures will induce charring even without ignition. Such indicators need to be documented as to type, location, and depth. Comparison samples of undamaged materials are needed to allow identification and establishment of critical heat fluxes or temperatures.

Displacement of Walls and Floors
Once again, it is vital that extensive overhaul and dismantling of the structure not have been conducted, because such activities may make it impossible to determine whether walls, floors, or ceilings were moved by the effects of an explosion prior to or during the fire. It is a fairly obvious observation to make if the structure is still standing, as in Figure 7-46. The mechanisms and effects of explosions will be treated in full in Chapter 12. The amount of material moved and the degree of fragmentation of the resulting pieces are the critical observations. Walls may be dislodged at their tops, at their bottoms, or both, depending on the directional effects of the developing blast or its reflections, or simply because one edge of the wall was fastened into the rest of the structure more rigidly than another.

The majority of the examination for displacement is easiest to carry out during the examination of the exterior, but damage is sometimes readily detected upon checking the interior. Window glass is the most readily broken structural material and may be thrown large distances by an explosion. Diffuse vapor/air explosions may not break window panes but may knock them free of their frames and toss them intact some distance away. All

FIGURE 7-46 A gasoline vapor explosion in this frame structure produced this displacement of walls and structural members. *Courtesy of John D. DeHaan.*

materials moved by an explosion should be carefully charted and compared against their original positions so that the direction of force can be assessed. If the structure collapsed or was razed, careful examination of the floors or foundations where the sole plates of the walls were attached sometimes may yield indications of sudden disruption. The investigator must measure and document the displacement of all structural elements.

Spalling

The chipping or crumbling of a concrete, plaster, or masonry surface is called **spalling**. It can be caused by the effects of heating, by mechanical pressure, or by a combination of both. Concrete is an aggregate of sand and gravel held together by a crystalline matrix (cement) containing a considerable quantity of water of hydration. When exposed to heat, some of this water is released from both the crystalline matrix and the aggregate (gravel), and the material loses its integrity and crumbles. While concrete is very strong in compression, it is relatively weak in tension. As a result, heating may cause some types of sand or gravel aggregate to expand more than others, which puts the cement matrix into tension, resulting in failure. This material property is a variable that many experiments in inducing spalling have overlooked.[31]

Moisture trapped in the porous structure of concrete or masonry is also evaporated into steam by heating. Forces generated by the expanding steam then can disrupt the cement, causing it to crumble. (The inclusion of polypropylene fibers in concrete can provide pressure relief as the fibers melt during fire exposure, reducing spalling.) This steam expansion is especially true of recently poured (green) concrete and masonry. The authors know of one case where a large fire was built in a just-completed masonry fireplace, and the resulting steam explosion created enough fragmented stone to fatally injure the installer. Heat expansion of steel reinforcing components within a concrete floor, beam, or column can also cause failure of the covering concrete. Steel reinforcing rods have a higher rate of thermal expansion than concrete, so steel components, if not buried deeply

spalling ■ Crumbling or fracturing of a concrete or brick surface as a result of exposure to thermal or mechanical stress.

FIGURE 7-47 Radiant heat from ignitable liquid fire on a concrete surface causes evaporation of liquid fuel and warms the surface beneath. The temperature of the surface outside the pool will rise as radiant heat is absorbed by the concrete. The temperature of the surface beneath the pool is limited by the boiling point of the liquid in contact with it.

Radiant heat

enough, can heat, expand, and distort to displace overlying concrete, compromising the mechanical strength of the beam, column, or floor. Other forms of mechanical stress can cause spalling of concrete or masonry, especially high mechanical pressures or the shock of rapid cooling of a superheated concrete surface by a stream of cold water. Although these other causes can contribute to spalling, by far the most common cause of spalling of structural concrete in structure fires is localized heating as a result of some ordinary fuel load in the immediate area. The formation of the crumbling surface is quite variable and depends on the nature of the aggregate used, the age of the concrete, the amount of mechanical restraint (as in prestressed panels or reinforced slabs), and the rapidity of the rise in temperature, as well as the temperature itself.[32] Because spalling can be triggered by thermal expansion, it is important to consider the physics of heat transfer from a fire on a concrete surface into that concrete. As shown in Figure 7-47, the radiant heat from the flames first raises the temperature of the fuel, causing faster evaporation. Some heat will pass through the liquid and raise the temperature of the surface beneath the pool. Excess heat from the surface is quickly transferred to the liquid, keeping the maximum temperature of the surface at or just above the boiling point of the liquid. As we have seen, the boiling point of gasoline (a mixture) is from 40°C to 150°C (100°F to 300°F); therefore, the temperature of the concrete beneath a burning pool of gasoline can never be much higher than 150°C (300°F), which is not high enough to trigger the thermal expansion *usually* needed for spalling. The duration of a flammable liquid fire on concrete is also very short (typically 1 to 2 minutes). That is not enough time for a significant amount of heat to penetrate concrete with its limited thermal conductivity.

As a fire indicator, spalling can indicate the presence of a significant fuel load of ordinary combustibles, as well as sources of intense localized heating such as a chemical incendiary.[33] Experiments by one of the authors and James Novak have shown that concrete spalling is very unlikely (although it can take place) when a flammable liquid alone is ignited on the surface. It must be remembered that a flammable liquid burns only at its top surface and along its edges while protecting the area immediately beneath the pool from high temperatures.[34] As a result, spalling (if it occurs) is often more pronounced along the outside edges of the burning pool. Freestanding pools rarely burn long enough to induce much radiant heat into the surface, with minimal expansion effects. Once the main pool of liquid is consumed, flames may continue at cracks, holes, seams, and expansion joints where pockets of fuel have collected. As a result, spalling tends to be more pronounced in the vicinity of such irregularities. If a heavier petroleum distillate fuel such as kerosene is burning, its higher boiling-point range will permit higher temperatures beneath the pool, and more persistent, more radiant flames will result. Such flames *may* have more pronounced effects on the concrete, but such fuels are also unlikely to be used as accelerants. Edges of spall patterns may be stained or discolored from the tarry residues

FIGURE 7-48A Spalling test on interior concrete floor. Pool fire of white gas. *Courtesy of Jamie Novak, Novak Investigations and St. Paul Fire Dept.*

FIGURE 7-48B After self-extinguishment, some staining around margins, but no spalling. *Courtesy of Jamie Novak, Novak Investigations and St. Paul Fire Dept.*

of such combustion, but such discoloration is not proof of the involvement of flammable liquids, since such staining has been observed in tests where only normal combustibles were burned. In tests by one of the authors, localized spalling was much more likely if the floor was covered with carpet or tile. In such instances, the localized burning induced by a gasoline spill was more likely to involve prolonged burning of the combustible floor covering. This combustion would involve no protective pool of liquid and was more likely to induce spalling than a pour of liquid alone. Note that the depth of spalling has no relationship to the quantity or type of liquid fuel involved.

Spalling is more likely to be produced over large, unprotected areas of concrete floors by radiant heat developed by normal fire progression, especially where there is extensive collapse or falldown. See Figures 7-48 and 7-49 for the results of an experiment contrasting the effects of a gasoline fire and a wood bonfire on the same type of concrete (note the staining around spalled areas on the wood-fire concrete). The symmetry and location of spalled areas must be carefully considered. Even localized spalling, when accompanied by staining, may not be the result of a flammable liquid. Figure 7-50 shows several pronounced spalled areas that were produced by burning asphalt falling from a collapsing

FIGURE 7-49A Spalling test on interior concrete floor. Large wood-fueled bonfire. No ignitable liquids. *Courtesy of Jamie Novak, Novak Investigations and St. Paul Fire Dept.*

FIGURE 7-49B Prolonged burning promoted explosive spalling during fire. *Courtesy of Jamie Novak, Novak Investigations and St. Paul Fire Dept.*

FIGURE 7-49C Post-fire appearance. Note extensive spalling (especially nearest ventilation source—door to left). Note "pool-shaped" discoloration on concrete. *Courtesy of Jamie Novak, Novak Investigations and St. Paul Fire Dept.*

roof onto a concrete floor. This is exactly the mechanism most likely to induce spalling—sudden contact with a very hot, molten, or burning material.

Some investigators have reported that an ultraviolet light (or various wavelengths of visible light) from a forensic light source played on a concrete surface aids in detecting ignitable liquid residues as a result of their fluorescence. An early reference addressed this technique.[35] While this suggestion has not been investigated in depth by the authors, the presence of fluorescent compounds in asphalts, greases, and other petroleum products indicates that they would be present in the pyrolysis products of many complex synthetics. Such compounds would be expected to fluoresce in the same manner as the petroleum distillates and thereby interfere with their detection by producing false positives. Preliminary field tests observed by the authors have revealed few useful patterns on real-world surfaces that would indicate the presence of ignitable liquids. A new generation of pulsed-laser fluorescence detectors holds promise for distinguishing pyrolysis products from ignitable liquid residues, but they await validation.[36]

Ghost Marks

Localized spalling of concrete surfaces often takes place in conjunction with another diagnostic sign of flammable liquid fires on floors covered with vinyl or asphalt tiles—called **ghost marks**. When asphalt tile squares are affixed to concrete surfaces, they are glued down with a more-or-less continuous layer of tarry adhesive. When a petroleum-based flammable liquid is poured on such a floor and set afire, the heat and solvent action tends to curl the tiles up

ghost marks ■ Stained outlines of floor tiles produced by the dissolution and combustion of tile adhesive.

FIGURE 7-50 Spalling of concrete floor caused by the combustion of extremely hot asphalt from a collapsing roof. Note the piles of tar and gravel near each spalled area and the staining of adjacent concrete surfaces. *Courtesy of John D. DeHaan.*

FIGURE 7-51 Ghost marks on concrete indicate pour pattern through a doorway. The damage of the threshold of the door frame is in direct line with the apparent pour. *Courtesy of Lamont "Monty" McGill, Fire/Explosion Investigator.*

at the edges. Liquid on the surface then seeps between the tiles and, in some cases, partially dissolves the adhesive. As the fire progresses toward full room involvement, the combination of flammable liquid and adhesive provides fuel for the flames, which tend to be more intense at the seams than in surrounding areas. As a result, more of the adhesive is destroyed, and the underlying concrete floor may be locally discolored and is more likely to be spalled. The adhesive in these areas will be burned off the surface leaving *subsurface* staining in the concrete along a grid pattern corresponding to the seams between the tiles, as depicted in Figure 7-51. Depending on the fire conditions and the nature of the floor tile, the authors have observed in experimental room fires that tiles will shrink even in the absence of flammable liquids, exposing the floor at the joints to higher heat fluxes and producing similar *general* effects. In these cases, the residues of the adhesive tend to remain on the surface of the concrete. Very faint subsurface staining has been observed on concrete that had a tile floor over it but no fire damage (as in Figure 7-52). This staining was thought to be a result of the solvent

FIGURE 7-52 Faint stain lines in concrete removed from a concrete floor that was undamaged by fire. The stains align with the tile margins, apparently resulting from wax-stripping solvents used on the floor. *Courtesy of John D. DeHaan*

action of the floor-wax-stripping solvent used for many years in the room as it soaked between the tiles. Once again, the observation of such a feature across a wide, general expanse of floor rather than in small, isolated or discrete areas should indicate by hypothesis testing that it was the result of some general feature of the room and not the result of a release of massive quantities of a flammable liquid. In the authors' experience, large quantities (more than a gallon or two) are used only in a very small percentage of cases, and they usually result in physical destruction of the building by deflagrating explosion.

Ghost marks, then, should not be considered to be proof of the use of a flammable liquid, but taken in proper context may be an indicator of such an accelerant. Samples of flooring and adjacent framing should always be taken for lab analysis.

Calcination of Gypsum Board

calcination ■ Loss of water of crystallization caused by heating.

As we have seen previously, gypsum wallboard consists of a layer of calcium sulfate (plaster) between layers of heavy paper. The calcium sulfate is normally in the dihydrate form (two molecules of water per molecule). When exposed to heat, this water of crystallization is lost through **calcination** in two major steps at predictable temperatures (52°C–80°C; 125°F–175°F and 130°C–190°C; 266°F–375°F). In wallboard each dehydration is detectable as a loss of mechanical strength (penetration resistance) when a probe is pushed into the surface. In addition, as the paper and paint burn off the exposed surface, the combustion products appear to condense in the outermost portion of the gypsum. As heating continues, this layer migrates through the matrix, resulting in a visible color change from gray to white, as seen in Figure 7-53. As the gypsum is heated and dehydrated it shrinks, cracks, and loses its mechanical strength until it is completely dehydrated to a fragile powdery substance. In structure fire experiments conducted by the authors, the detectable changes (mechanical and visible) were proportional to the intensity and duration of the exposure to fire.

Mann and Putaansuu recently reported on a series of tests using a cone calorimeter to reproducibly heat samples of gypsum wallboard, after which the amount of dehydration was measured using FTIR, and penetration depth with a modified tire tread-depth gauge.[37] These tests indicated that probe penetration with 0.87 kg/mm^2 (6 lb) of force followed the depth of the interface between the fully hydrated layer and the hemihydrate

FIGURE 7-53
Calcination of gypsum drywall advancing from fire exposure (fire on right side). *Courtesy of Jamie Novak, Novak Investigations and St. Paul Fire Dept.*

form, as the anhydrous and hemihydrated forms offered less resistance than did the fully hydrated form. Color changes, however, did not. It was suggested that the deposition and migration of colored combustion products have numerous variables. Mann and Pataansuu reported that similar testing of fiberglass-reinforced (X-rated) gypsum board revealed similar linear results between penetration depth, exposure time, and depth of dehydration. The presence of the fibers or vermiculite reduces the shrinkage of the gypsum and the formation of cracks that weaken the material. If applied systematically with some care to ensure reproducible force applications (0.87–1.03 kg/mm², 6–10 lb), this technique does yield reliable data on the duration of typical fire exposures for these products. Mann and Pataansuu reported that the visible "smoke layer" position correlates well to the depth of the anhydrous/hemihydrate interface due to changes in porosity caused by dehydration (but not the actual depth of the hydrated layer).

There may be two or three discrete changes in color (white to gray to white), possibly as a result of the two steps of the dehydration or the chemistry of the condensed pyrolosis products. The colors of interest are transitions between gray and white. Other colors such as pink, blue, or green have no causal link to the fire, since they have been linked to impurities in the gypsum. Data on this color phenomenon are still fairly limited.[38] Because gypsum is usually a uniform manufactured product with a predictable and uniform thermal conductivity, the depth of thermal effects should be directly related to the duration of exposure with the caveat that extreme radiant heat fluxes will cause faster buildup of surface effects. Air pockets or voids in gypsum may also give rise to erroneous estimates. A thorough survey of a room can allow the mapping of areas with the highest "penetrations" corresponding to the locations of the most severe heating (from fuel packages, plumes, or ventilation effects). Kennedy and others have examined the processes of heat effects on drywall and have also demonstrated the reliability of calcination observations.[39]

Annealed Furniture Springs

As a steel wire spring is heated it will reach a temperature, well below its melting point, at which it loses its springiness, or its *temper*. This temperature is called its **annealing** temperature. At temperatures of about 600°C to 650°C (1,100°D to 1,200°F) the spring will collapse if there is any weight on it and may even collapse of its own weight. This temperature-related indicator may be of assistance to the investigator in establishing whether a fire developed within an upholstered sofa or bed or enveloped it from without. A smoldering fire developing slowly within the upholstery of a sofa or bed often produces a localized area of intense heat, as pictured in Figure 7-54. The springs in this localized

annealing ■ Loss of temper in metal caused by heating.

FIGURE 7-54 Intense fire inside this mattress reveals high temperatures (greater than 1,000°C; 1,800°F) with open flames just beginning to show (ignition by cigarette). *Courtesy of Lamont "Monty" McGill, Fire/Explosion Investigator.*

area will often be more collapsed (or extended if they are in suspension) than those springs around the margins. When furniture is enveloped by an external fire, heating may be less intense, and the general degree of spring collapse is often less marked than from a long-burning internal fire. In a prolonged post-flashover fire, however, collapse will often be total no matter what the origin of the fire.

One useful indicator of spring collapse is directional, in that the collapse of springs may be more pronounced in the corner or side facing an oncoming fire, if the resulting destruction is not complete and ventilation is fairly uniform. Temperatures adequate to cause collapse of springs by loss of temper can be produced by both flammable liquids and by ordinary combustibles, so their collapse cannot be reliably used to demonstrate such an incendiary origin.[40] It has also been shown that such springs often taper away, appearing to have been melted by fire exposure. This tapering is due instead to repeated exposure to turbulent flames and not to intense heat from an incendiary mixture. Such appearances are more often noted in structures gutted by fire.[41] Objects lying on the springs or debris falling onto them during the fire will cause more collapse, and any interpretation of indicators from spring collapse must keep such limitations in mind. Like other fire-direction indicators, spring collapse should be noted and photographed as part of the general scene search.

Glass

Glass is a structural material found in virtually all structures. It has some properties that make it a very valuable indicator to the investigator. First, although it looks and generally acts like a solid, it is in fact a supercooled liquid. As a result, when exposed to stress, glass is elastic and will actually bend to a substantial degree before breaking. When a large object strikes a normal (i.e., nonreinforced, nontempered) glass window, the pane bows out until its elastic limit is reached, whereupon it shatters. This reaction results in two sets of fracture lines, radial and concentric, as shown in Figure 7-55. The edges of mechanically broken glass will almost always bear a series of curved, *conchoidal* fracture lines, which are produced by the release of internal stress within the glass when it breaks. These conchoidal fracture lines are roughly perpendicular to one side of the glass and curve around to be roughly parallel to the other, as shown in Figure 7-56. These lines can reveal the side of the pane from which the breaking force came. Note that it is vital to know whether one is dealing with a radial fracture or a concentric fracture. For this reason, it is best that as much of the broken window as is practical be recovered and submitted to the laboratory for reconstruction.

FIGURE 7-55 Typical mechanical impact fracture allowing the development of radial and concentric fracture pattern.

FIGURE 7-56
Conchoidal fracture lines
on the edges of glass
fragments can reveal the
direction of the impact.
*Courtesy of Jamie Novak,
Novak Investigations and
St. Paul Fire Dept.*

Glass can also break as a result of stress produced by nonuniform heating. The resulting fractures tend to curve randomly about the pane (as opposed to mechanically caused fractures, which are almost always straight lines), and the fracture edges bear little or no conchoidal lines, usually being almost mirror smooth, as in Figure 7-57. Glass expands when heated, so convected or radiant heat raises the temperature of the exposed portion of the glass pane, and it expands. The portion of the pane behind (or in) the frame or molding is protected or shadowed and "sees" no heat, and because of its poor thermal conductivity, maintains its starting temperature and size. Tensile stress builds up until the tensile strength of the glass is exceeded, and the pane cracks, usually roughly parallel to the framed edges

FIGURE 7-57 Edges of
thermally fractured glass
reveal virtually no con-
choidal fracture lines.
*Courtesy of Jamie Novak,
Novak Investigations and
St. Paul Fire Dept.*

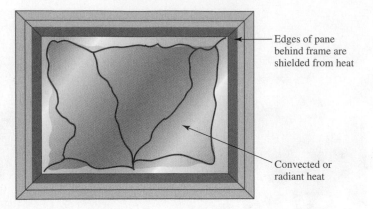

FIGURE 7-58 Radiant or convected heat on the central portion of the glass pane causes thermal expansion of heated portion. The portion behind the molding or frame is in shadow and "sees" no heat and does not expand. When tensile stress becomes sufficient, the glass breaks, typically in fractures roughly parallel to the frame.

Edges of pane behind frame are shielded from heat

Convected or radiant heat

(see Figure 7-58). This process can be predicted by mathematical modeling in terms of temperature differences.[42] Studies of this process have shown that it is dependent on the strength and type of glass; ordinary window glass breaks when there is a temperature difference of about 70°C (126°F).[43] This breakage can occur in any glass object that is subjected to uneven heating and may reflect the direction from which heat approached. Thick glass can shatter due to a sufficient temperature difference between the side facing the fire and the one facing away from it. Double-glazed windows fail in sequence; the pane facing the fire fails first, then shields the other pane until it falls out of the frame. If the glass simply cracks and does not fall away, the outer pane may survive even through flashover.[44]

Not only can one determine whether the glass was broken by mechanical or thermal stress, from the presence or absence of soot, char, or ash deposits on the broken edges, one can tell if the breakage occurred before or after the fire (Figure 7-59). The location of the glass in the fire debris is of utmost importance in interpreting its evidential information. Glass with relatively little soot found on the carpet of a room, for instance, with all fire debris on top of it would be a strong indication that the window was broken before or very early on in the fire. Glass found lying on top of the debris most probably resulted from breakage during suppression or overhaul. The appearance of deposits on the inside surface of the window glass is of less evidential value than was previously thought. Because of the common use of synthetic materials in upholstery and in wall and floor coverings, the presence of thick, oily soot can no longer be used as an indicator of the use of a petroleum distillate as an accelerant. Unfortunately, the color, density, and tenacity of smoke deposits are more closely related to the degree of ventilation of the fire than to the type of fuels involved.

FIGURE 7-59 Broken edges of glass fragments reveal deposits of soot and ash consistent with having been broken before the onset of the fire. *Courtesy of the State of California, Department of Justice, Bureau of Forensic Services.*

FIGURE 7-60A A glass window fractured and softened by heat exposure during an ordinary combustible fire (post-flashover). Crazing was caused by suppression water spray. *Courtesy of Jamie Novak, Novak Investigations and St. Paul Fire Dept.*

FIGURE 7-60B Close-up of crazed glass. *Courtesy of Jamie Novak, Novak Investigations and St. Paul Fire Dept.*

Glass breakage often occurs just as the room approaches flashover. If enough cracks develop that the glass can actually fall out, fresh air can be admitted suddenly, possibly triggering a backdraft coincident with flashover. Sudden heating (as from a flammable liquid accelerant) may induce more cracking than a slow fire but not so much more that it will be detectable by post-fire examination of the fragments. If such heating occurs very quickly, the glass pane usually simply shatters.[45] Crazed glass, in which the pattern of partial-thickness fractures or cracks (Figures 7-60a and b), resembling a complex road map in the glass, was once thought to be indicative of a very rapid buildup of heat sometime during the fire. Laboratory tests have shown that **crazing**, like cratering (pitting of the surface in small circular defects), is caused by a spray of water hitting a window already heated by fire exposure, usually during suppression.[46] Narrow slivers or long shards of glass indicate that an explosion of some type took place, as pictured in Figure 7-61. The presence of heavy soot or varnish on splinters of glass indicates that the explosion took place late in the fire, most probably as a result of a smoke or backdraft explosion. (Glass can also be broken and projected some distance from the window if struck by a straight-stream

crazing ■ Stress cracks in glass as the result of rapid cooling.

FIGURE 7-61 Glass shards caused by deflagration of flammable liquid vapors. Parallel, partial-thickness fractures are typical. *Courtesy of John D. DeHaan.*

hose.) Absence of soot, or a very faint coating, indicates an explosion early in the fire, possibly caused by an explosive device or fuel/vapor deflagration.

Thus, the appearance of the glass can tell the careful investigator a great deal about the nature of the fire or explosion; it can tell even more if its location in the debris is carefully noted and photographed before it is disturbed. Broken glass scattered by an explosion should be charted (with dimensions) and photographed during the exterior examination, and this documentation can be supplemented with observations of the glass remaining in and around the window frame. Tempered glass, found in vehicles and many structure doors, is made by treating the glass with chemicals or heat during manufacture, which results in glass sheet that is much more resistant to mechanical breakage. When it does break, it shatters into thousands of small cubic blocks. In general, it is not possible to reconstruct such a window or to determine whether it was broken by mechanical or thermal shock. One exception is in vehicles or metal-framed glass doors where there is often a rigid frame that holds the peripheral fragments in their original configuration even after the collapse of the majority of the glass. The fracture pattern remaining can sometimes allow discrimination between mechanical and thermal fractures.

Melting glass, whether it is part of a window, furnishings, or electrical lamps, can be useful in indicating the direction of fire travel. Heat causes glass to expand on the surface being heated, which often causes a window to bend or "belly-in" toward the source of heat. When it finally melts at a temperature of approximately 750°C (1,400°F), glass will soften on the side facing the heat first, and so it will sag or flow in that direction as it loses structural strength. Note that windows in aluminum frames may react to heat-induced distortion of the frame rather than the glass. Observed tests show that normal single-glazed windows more often fall inward toward the heat but *can* fall outward as well.

An electric lightbulb is first evacuated and then filled with an inert gas, often nitrogen (except for very low wattage lamps, which are evacuated). When a lightbulb is heated, the expanding gas inside creates overpressure, which balloons out the first area of the bulb to soften in the advancing flame. This "blowout" points back in the direction from which the fire came, since the side facing the oncoming flame will usually be the first to soften. The bulb in Figure 7-62 is a good example of this effect. Other glass objects in the

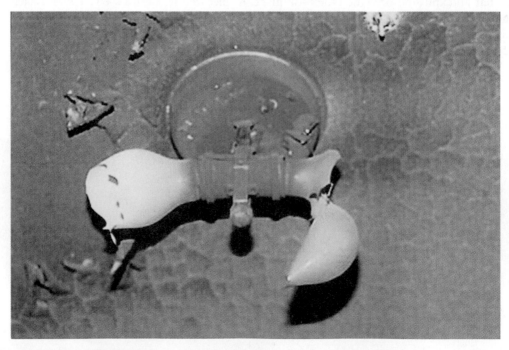

FIGURE 7-62 The "blowout" on this lightbulb faces the oncoming flames and indicates the left to right direction of fire travel. *Courtesy of Detective Mike Dalton, Knox County Sheriff's Office, Knoxville, TN.*

structure—containers, mirrors, and the like—will indicate to the observant investigator not only what temperatures were developed but also from which direction the flames came—but only if they remain undisturbed until they are photographed and examined. Glass bottles, for instance, may be much more intricately fractured on the side facing the heat source even if they appear to be intact. They can shatter, however, if any attempt is made to move them.

Melting Points of Materials

Plastics are often the materials first affected in a fire. As we saw in Chapter 5, many plastics begin to soften and melt at low temperatures of 100°C–200°C (212°F–400°F), and others decompose at similar temperatures. The melting behavior of glass was mentioned in the previous section as being diagnostic of the conditions created within a structure. Glass can vary in its melting point over a wide range, depending on its composition. Ordinary soda-lime glass used in most windows has a melting point of about 700°C to 800°C (1,300°F to 1,400°F), a temperature that is reached in many wood-fueled structure fires.

As additional indicators of the intensity of a fire, metals are often very useful. When a fire scene is examined, previously molten aluminum residues are often found. Because this metal melts at a comparatively low temperature, 660°C (1,220°F), this is not unexpected or indicative of a particularly unusual fire when found. Its high thermal conductivity means that localized flames are unlikely to melt thick metal components. This is especially true of flammable liquid fires, whose typical durations are so short that even low-melting-point metals are likely to be affected (see Figure 7-40). Zinc/aluminum die-cast fittings (sometimes called "pot metal") have a very low melting point (380°C; 718°F) and are very common in fire scenes.[47] A more useful indicator is melting brass or copper. Depending on its composition, brass melts over a range of 875°C to 980°C (1,625°F to 1,780°F). Its presence in a burned structure is not unusual, but it may assist in establishing which parts of a room reached higher temperatures than other parts. Copper, commonly present as water piping and electrical wiring, will occasionally melt in the hottest structure fires, its melting point being 1,080°C (1,981°F). In their extensive study of buildings totally burned during an urban firestorm, Lentini, Smith, and Henderson discovered melted brass and copper fittings in many of the ruins. These included copper water pipe (although water service to the area was interrupted during the fire, water should still have been present in these lines). This demonstrated that such melted materials in individual fire-gutted scenes (so-called black holes) cannot be linked to the use of an accelerant.[48] (See Appendix C for melting points of some common materials).

As we saw in Chapter 3, the flame temperatures produced by flammable liquids burning as a pool in air are not necessarily higher than those produced by ordinary combustible fuels, so the presence of melted metals should not be taken as proof of the presence of liquid accelerants. It is of critical importance to the fire investigator to note that molten aluminum can interact with copper and dissolve in it to form a bimetallic alloy. This alloy can have any percentage of copper and aluminum, but at a ratio of 67 percent aluminum and 33 percent copper, it forms a eutectic metal alloy with a melting point of only 548°C (1,013°F), below that of either original metal. Molten aluminum can therefore attack and dissolve away heated copper conductors in electrical systems and appliances. The reader is referred to the article by Beland, Ray, and Tremblay for a complete discussion of this phenomenon.[49]

Zinc falling onto heated steel can cause penetrations by a similar alloying process. Cast iron, which melts in the range of 1,200°C to 1,450°C (2,200°F to 2,700°F), is sometimes found melted in the ashes and rubble of heavily burned heavy-timber structures. Because well-ventilated wood and charcoal fires can produce temperatures on the order of 1,300°C (2,400°F), such melting may occur by strictly thermal processes.[50] It has also been suggested that carbon monoxide combustion and actual reductive smelting of the cast iron may play a part. It has recently been suggested that black iron can alloy with calcium salts to form a eutectic with a relatively low melting point. The presence of plaster or gypsum residues in contact with melted iron may be significant.[51]

Whenever any residues of a molten metal are found at a fire scene, they can help establish the minimum temperature occurring at that spot in the structure. While such melted metals cannot and should not be used as proof that the fire was incendiary, the fire investigator should note their presence, extent, and distribution. Such information can be of help in establishing differences between temperatures produced by normally fueled and ventilated accidental fires and those produced by enhanced draft conditions or unusual fuel loads from chemical accelerants in incendiary fires.

Copper and, to a lesser extent, copper alloys are very susceptible to oxidation when heated in air. This reaction results in the formation of black copper oxide on the surface of the metal. If the heating and exposure to air are continued long enough, ultimately the entire object, for instance, wire, may be changed to oxide. The degree of oxidation of exposed copper wiring can serve as an indicator of the conditions existing at the point of exposure. Copper wire carried in conduit is not so subject to oxidation because of the limited access to air and the general reducing (rather than oxidizing) atmosphere resulting from the decomposition of insulation materials. See Chapter 14 for a discussion of the analysis of copper wire residues. The surface of copper wire and other conductors may turn green as the hydrogen chloride gas (HCl) released by the pyrolysis of the PVC insulation turns it to copper chloride.

Clean Burn

When a noncombustible surface is exposed to soot and pyrolysis products, they condense on it. If that surface is heated sufficiently, those materials can be burned off, leaving a "clean burn." The surface temperatures needed to accomplish this are flame temperatures (>500°C; 1,000°F). Therefore, a clean burn indicates direct flame contact of some duration. The flames may be from a fuel package playing on the surface or from incursion of fresh air into an underventilated post-flashover room, as shown in Figure 7-63.

MYTHS AND MISCONCEPTIONS ABOUT INDICATORS

A number of indicators have been traditionally linked to incendiary fires that have been shown to be unreliable. Most of them have been dealt with in this chapter or previous ones but are so pervasive that they deserve to be specifically numerated here.

Evidence (Documentary or Witnesses) of an Abnormally Fast Rate of Fire Spread or Collapse

Fact: Fires are rarely seen from the moment of their ignition, and so their actual rate of fire development is rarely known. The furnishings in use today will also support more energetic, much faster growing fires than most people appreciate. The pre-fire fuel loads and their placement are critical to evaluating fire spread. Lightweight construction techniques often used today in commercial and residential structures can lead to much more rapid penetration and collapse of floors, ceilings, and roofs than occur with traditional techniques.[52]

Evidence of Abnormally High Temperatures (Melted Metals, etc.)

Fact: Flame temperatures of flammable liquid fires are about the same as temperatures of wood-fueled fires. Higher temperatures (in excess of 1,000°C; 2,000°F) can be produced by the combustion of many synthetic materials, and post-flashover fires can create temperatures of that magnitude from floor to ceiling in modern furnished rooms in the absence of any flammable liquids.

Spalling of Concrete

Fact: Radiant heat from sustained fires in ordinary fuels and structural collapse are much more likely to transfer enough heat into concrete to cause it to spall than is the heat from short-lived fires of pools of flammable liquids. Temperatures and heat fluxes under pools

FIGURE 7-63 The "clean burn" on the back wall of this post-flashover test cubicle resulted from the entrance of air *after* the glass viewport shattered. *Courtesy of John D. DeHaan.*

of flammable liquids are too low to induce sufficient heating to cause most concrete to spall. Flammable liquids on floor coverings can induce sufficiently energetic burning of those coverings to induce spalling.

Crazing of Glass

Fact: Thermal stress can cause glass to fracture, but rapid buildup of such thermal stress (thermal shock) causes rapid failure, often producing shattering of the glass, rather than the intricate partial-thickness patterns associated with crazing. Such patterns are produced when glass heated by fire exposure is suddenly cooled (usually by hose stream during suppression).

Irregular Damage to Floors and Floor Coverings

Fact: Heat fluxes on floors in post-flashover room fires are not only very high, they are not constant, driven by extremely turbulent mixing between the combustion gases and air coming into the room through doors and windows. As a result, the post-fire damage is often more intense around these ventilation openings. The reaction of synthetic carpets and pads to such radiant heat causes irregular melting and combustion. Falldown or collapse of overhead structural materials can cause irregular, intense floor-level damage even in the absence of flashover conditions.

Black, Heavy, Oily Soot on Windows/Black Dense Smoke

Fact: Black dense smoke is a result of the type of fuel burning in a room and the efficiency of the combustion. Modern synthetic materials are almost always petroleum-based and when burned in a typical fire, release large quantities of black soot and dark pyrolysis products.

Annealing of Steel Springs and Steel Structural Materials

Fact: Steel loses its temper and tensile strength at temperatures above 500°C (1,000°F). Such temperatures can be produced by smoldering or flaming combustion of ordinary upholstery materials. Ignitable liquid accelerants are not required.

Floor-to-Ceiling Heat Damage

Fact: As the temperature of the hot gas layer in a room exceeds 600°C (1,150°F), the radiant heat flux exceeds that needed to char and ignite all exposed fuel surfaces. Due to the large surface area of the hot gas layer, shadowed or protected areas beneath furniture may not be adequate to prevent ignition of carpets under furniture.

Deep Char

Fact: The rate at which wood chars is dependent on many factors (type of wood, nature of finish, etc.) but especially on the intensity of the radiant heat (heat flux) falling on it. Direct flame exposure can cause wood to char at twice the rate at which it chars at ignition, and wood in a post-flashover fire can char at 10 times the rate at which it chars at ignition.

Progression to Flashover

Fact: Today's furniture with its synthetic fabrics and padding is capable of producing very high heat release rates. (A single modern recliner fully involved produces as much heat per second as a half gallon of gasoline poured on the floor and ignited.) Although modern furnishings are more resistant to smoldering heat sources, they are more susceptible to open-flame ignition and, once alight, can cause the spread of a fire in a 3-m × 4-m (10-ft × 14-ft) room to flashover in 5 to 8 minutes in the absence of any ignitable liquids.

Alligatoring (Shiny or White)

Fact: The surface appearance of charred wood is dependent on the nature of the wood and ventilation conditions during its combustion. The presence of white ash on charred surfaces is the result of prolonged burning in a well-ventilated fire, resulting in complete combustion of carbonaceous residues, leaving behind the light-colored oxides of inorganic compounds.

ARSON EVIDENCE

There are many other observations that will be made in the course of the scene examination that may reveal the fire's origin and cause. Incendiary fire cause may be revealed by obvious indicators such as separate multiple points of origin, the presence of trailers of flammable liquid or paper or rags connecting areas of intense burning, or the remains of timing or ignition devices.

Trailers

What appear to be trailers may be artifacts of normal fuel combustion, as illustrated in Figures 7-64 and 7-65. Arsonists often use multiple "sets" or ignition sources to ensure maximum destruction. Very often, unless the devices are carefully synchronized, the first devices ignite a fire that interferes with the operation of later ones. They may be found incompletely burned and readily identified. What appear to be multiple points of origin must be carefully evaluated so as to eliminate their cause by falldown of draperies or lighting fixtures, by lightning or power surges (causing multiple failures of appliances), or even by radiant heat or hot fire gases traveling across a room or between rooms.

Containers

Containers are commonly found at fire scenes. While most of them will be normal contents of the house, such as paint cans and household supplies, they may have contained flammables that could be associated with arson and, if not, may still provide useful information. If the can contained a liquid, from a small amount up to a full can, and it was tightly sealed and heated, exposure to heat would cause it to bulge and split around one or more seams. If the liquid was flammable, it would have boiled out of the can and contributed substantially to the total fuel. If such a can is not bulged or burst, it is an indication that it was either totally empty or that the cap was loose. Under some circumstances, lightly soldered cans may selectively heat from radiant fire above them, so that the solder softens and allows escape of the contents without showing a bulge. When arson is suspected, it is essential that all such containers be collected and submitted for laboratory examination, with care being taken to preserve possible latent fingerprints.

Contents Inventory

All portions of the structure should be carefully inventoried and documented by the investigator, whether or not the contents were damaged by the fire. This process serves two

FIGURE 7-65 Intense post-flashover fire produced irregular deep char in this test fire with no ignitable liquids. *Courtesy of John D. DeHaan.*

purposes. First, it is for the protection of the investigators, who may be called on to prove the appearance and contents of the scene at the time of their possession. This is especially important when the contents are of high monetary value. Second, it ascertains the contents of the structure at the time of the fire. A post-fire inventory should be carefully checked against an inventory prepared by the owner or tenant. Removal of a building's contents before a fire is a strong indication that the fire was planned. The investigator must be sensitive to the absence of major appliances, tools, equipment, stock, guns, clothing, jewelry, memorabilia, antiques, sports equipment, and pets. In cases of insurance fraud, junk or cheap goods may be substituted for valuable goods or stock (which has been sold or stored elsewhere). Many of the other signs of such manipulation are discussed in Chapter 16, on arson fires.

We have illustrated in a number of instances how the fuel load in a structure affects the fire pattern. A thorough inventory of the burned structure provides the investigator with a clear and complete picture of the fuel load in the building at the time of the fire. Objects that could contribute significantly to the fuel load of each room—sofas, upholstered chairs, drapes, carpets, and the like—should be marked on a sketch of the scene. If the contents are removed during overhaul, the investigator may seek the assistance of the owner or tenant in identifying what furniture was present and where it was located before the fire.

Ignitable Liquids

Although it is possible to set large, very destructive structure fires without the use of ignitable liquids, by far the most commonly *detected* arson means is the pouring or spilling of an ignitable liquid (usually a petroleum product) onto the floor and contents of a room followed by direct ignition with a match. Because of the frequent use of such materials, the investigator who is able to locate identifiable residues of such liquids will be more effective in detecting incendiary fires. Ignitable liquids are used as accelerants because they are cheap, easily available, and are volatile and therefore are easily ignited. Whenever such a liquid is exposed to the high temperatures of a fire, it is going to vaporize quickly even if it does not burn. When the post-fire debris is being examined, only that liquid that has been protected from the flames and high temperatures will remain. This protection can be afforded by absorbent materials such as carpet, upholstery, untreated wood, plaster, or soil or by cracks or crevices in floors or between floors and baseboards, as shown in Figure 7-66. The water used to extinguish most fires will then tend to protect these residues from evaporation for some time. Whenever localized burn patterns are discovered on walls, floors, or floor coverings, nearby debris should be checked for the

FIGURE 7-66 Volatile liquids are protected from evaporation by joints in floors, baseboards, and the like. A close-up of burn pattern from gasoline–diesel mixture shows the percolation of unburned fuel oil from pockets along the edges and end grain of planks. This effect was most pronounced the day after the fire. *Courtesy of John D. DeHaan.*

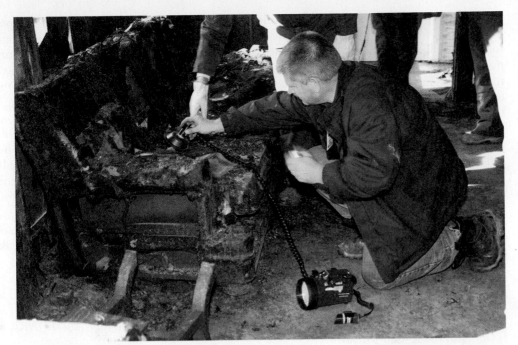

FIGURE 7-67 Electric hydrocarbon detector uses a solid-state sensor that vapors enter by passive diffusion. Operator must move the sensor slowly. *Courtesy of Cameron Novak, Novak Investigations.*

odors of common flammable liquids. The appearance of such burn patterns is often enhanced by cleaning the floor with a broom and then flushing with water. The pattern can then be photographed and sketched as needed. Seams, crevices, and other protected areas can then be checked for distinctive smells by nose or portable hydrocarbon detector (Figure 7-67). Many investigators rely on olfactory detection of volatile fuels, and some have very sensitive and accurate senses of smell. There are several drawbacks, however. First, the human nose is very subjective in its perception of odors and can be "fooled" into false reactions. The human nose fatigues quickly, and inhalation of some vapors can render the nose incapable of detecting other odors. The fire scene also contains toxic gases and vapors that should not be inhaled in high concentrations.

If a floor is porous or leaky, some of the accelerant may have seeped through and penetrated the soil underneath. Water and debris falling onto the soil then trap the volatiles for later detection. In recent years, specially trained canines have proved their value in locating areas of debris where ignitable liquid residues can be identified by laboratory analysis. While not substitutes for a complete, professional investigation, trained canines can help expedite the scene examination for likely areas to sample. Their role will be examined in greater detail later in this chapter.

As we saw in Chapter 4, the flame temperatures produced by accelerants such as gasoline burning in air are not significantly higher than those produced by ordinary combustibles (and in some cases are lower). Ignitable liquid accelerants are noteworthy because of their potential for rapid combustion, localized damage, and unusual distribution. Liquid accelerants can burn at low levels in a room and often in protected areas, leaving burn patterns that are often distinguishable from those produced by ordinary fuels (barring overwhelming subsequent fire damage such as post-flashover burning or collapse).

Remember that some ignitable liquids do not have a characteristic odor, and some do not cause an electronic detector (or canine) to react. It is also possible for a sustained fire to consume all liquid fuels used as accelerants. Other materials, such as crumpled newspaper, waxed paper, plastic packing materials, and the like are used very easily to accelerate a fire and leave very little residue detectable by chemical or even visual examination. In the final analysis it is *always* the experience of the investigator (backed by appropriate

Chapter 7 Structure Fires and Their Investigation   **301**

published experimental data) that determines what importance is to be made of such patterns. When indications are that an ignitable liquid has been used, the prudent investigator will recover samples of the flooring and nearby associated debris for laboratory testing no matter what odors are present. Water-soluble flammable liquids such as methanol, acetone, or ethanol are diluted and may be washed away by the application of hose streams during suppression or debris removal. Empty containers of such solvents should warn of their possible use as accelerants.

For all the reasons described in this text, the investigator should not rely on the presence of "unusual," "pour," or "pool" patterns of floor damage as proof that ignitable liquid accelerants were used in the absence of positive laboratory findings. This is especially true if there has been structural collapse or post-flashover burning. Time and exposure are the enemies with which the investigator must always contend when seeking out volatile residues. When the debris dries out, sun and high temperatures will evaporate even protected residues in a matter of days. (Some residues have been detected in soil under houses a few weeks after a fire, but the finding of liquids under the house bespeaks of very large quantities used.) While cold, wet weather may aid in the retention of ignitable liquid residues, it may adversely affect burn patterns and other evidence exposed to the elements. The fire scene investigation must be carried out as soon as is practical to provide the best chance of detection. It also minimizes the chances of contamination of the scene (accidental or intentional) from spurious sources of ignitable liquids (such as from equipment used for salvage).

If the burn pattern is on a carpet or rug, it is often useful to brush it vigorously with a clean, stiff broom—after carefully visually examining it for the presence of trace evidence (adhering glass fragments or plastic, paper, or fabric from an incendiary device)—to remove loose debris and enhance the burn pattern for examination. The carpet and any pad can then be sampled for lab testing. The resulting burn pattern may be enhanced by sweeping away loose debris and then flushing with clean water. Residues may be found between the tiles remaining, under baseboards, or soaked into the wood subfloor. In a multiple-layer floor covering, residues of ignitable liquids (with evidentiary significance) are not going to be found in the lowest layers if they are not found in the ones above.

No matter from where laboratory samples are to be taken, two things must be kept in mind: First, volatile liquids are driven off by high temperatures; therefore, they are *not* likely to be found in the areas of deepest char. Rather, they are more likely to be found in places where the liquids would be protected from the heat. Second, many building and upholstery products involved in fires today are synthetic materials that share the same petroleum-based origin as most accelerants. To ensure that the lab will be able to distinguish minute traces of accelerants from the semivolatile decomposition products of synthetics, comparison samples are very important. A quantity of undamaged material should be collected from a nearby area that does not appear to have been contaminated with accelerant. Even in badly damaged structures, comparison samples of floor coverings may be found under appliances, bookcases, or large furniture. Because these scene samples cannot be guaranteed to be free from accelerants, they cannot be considered to be *control blanks* and should be considered and labeled as *comparison samples*. True control blanks would have to come from a noncontaminated source such as a retail store, manufacturer, or identical unburned vehicle.

Detectors—Electronic and Canine

"Sniffers" or, more properly, portable hydrocarbon detectors are used by many successful investigators. Many varieties are available, with any degree of sophistication desired (the advantages and disadvantages of all of them will not be treated here). Suffice it to say that, no matter which hydrocarbon detector is used, it should not be used to the exclusion of other tests examining appearance or odor. Even the best electronic detectors have a sensitivity on the same order as that of the human nose and therefore offer no complete advantage in that respect. Like the human olfactory system, detectors can be overwhelmed or fatigued by continued exposure to high levels of volatile fuels. They may not be sensitive at all to some common accelerants (odorless charcoal lighter seems to be the most elusive) and thus yield false negatives when there really is an ignitable liquid present. Most of them are sensitive to an oxidizable gas or vapor, including the carbon monoxide (CO), ammonia (NH_3), methane (CH_4), and methyl alcohol (CH_3OH) vapors produced by the destructive distillation of wood and give false positives. Many modern sniffers rely on an electronic detection element that is not as sensitive to pyrolysis products or water vapor as were older designs. One successful approach is to use two matched detectors, one of which is located in a burned "reference" area while the other is used to examine a burned "target" area. The difference between the two signals is then more likely to be significant and less likely to indicate simple volatile pyrolysis products.[53]

Accelerant-detection canines, typically trained to detect one drop of partially evaporated gasoline or paint thinner, have a detection limit on the order of 0.1 part per million (ppm) [0.1 microliter (μL) in a 1-L can] in samples with a clean background. The presence of burned debris (particularly carpet and pad) can reduce this sensitivity significantly, but at scenes dogs have been able to discriminate traces of ignitable liquid from fire debris more successfully than the human nose, and they carry out a search of a scene much more quickly than a person.[54] As with drug-detection dogs, their positive alerts can be used as probable cause to search a person or take a sample from a fire scene, but their positive indications should *not* be accepted as proof of the presence of accelerants. Dogs cannot discriminate between gasoline used as an accelerant and volatile traces of carpet adhesives or insecticide sprays. Only a proper laboratory analysis can specifically identify the material, and only then can it be put into context with the scene findings, for an accelerant must be identified by its intended use, not by its mere detection.[55] Because of the variety of petroleum products that can be detected at trace levels in normal household items, the investigator must be careful not to put too much value on detection at very low concentrations unless proper comparison samples are collected.[56]

Figure 7-68 illustrates the areas of wood surrounding a char pattern that should be chiseled or pried up and tested with a sniffer or the nose. Areas yielding any sort of response can then be cut out and sealed in appropriate containers for laboratory testing. When all the areas have been checked and sampled, the containers can be left in place, labeled with numbered cards, and the entire area then photographed to record graphically the disposition of the samples recovered. Such a process is shown in Figure 7-69.

Liquid pattern

Tongue-and-groove flooring seam

FIGURE 7-68 Drawing showing the areas of wood surrounding a char pattern that should be sampled for laboratory testing (including the central area). *Courtesy of Lamont "Monty" McGill Fire/Explosion Investigator.*

FIGURE 7-69 After
collecting samples, it is a
good idea to photograph
the area sampled with
cans in place. *Courtesy of
Lamont "Monty" McGill,
Fire/Explosion Investigator.*

General Considerations

PROTECTED AREAS

Areas that have not been damaged by the flames are sometimes as significant as areas that
are badly charred. In any fire, the greatest damage possible will be at points where active
flame or radiant heat existed for an appreciable time. Because flames burn upward, it follows that floor areas do not receive as much active flame as other fuels. They will char by
radiant heat in most instances, but it is often noted that areas of the floor are totally
undamaged though surrounded by damaged areas.[57] Such areas tend to have a definite
form—circular, rectangular, and so on—and invariably show clearly the location of an
object sitting on the floor at the time of the fire (see Figure 7-70). The object may be combustible and show the effect as well as though it were noncombustible. Cardboard cartons often serve to protect the floor under them, because the bottom of the carton does
not burn well due to poor ventilation, even when the remainder of the carton is totally
destroyed.

The importance of such protected areas lies in two factors: (1) locations of objects are
definitely determined, and (2) the general type of the object is defined. For example, a
bucket that originally contained liquid accelerant (or a metal wastebasket where a discarded cigarette ignited the contents) might be placed on the floor in the fire area, and its
location would be proved. Delineation is generally so accurate that the size and shape of
the bottom of the object is readily determined and compared with any suspect item that
is located.

In fires that approached flashover, protected areas on the floor can reveal the locations of furniture or victims that were moved, or even combustibles like paper or cardboard that were consumed during the later stages of the fire, as in Figures 7-71 and 7-72.
These objects are frequently removed during the firefighting or in clearing the area afterward and thus will only occasionally be found still sitting in the original position. When

FIGURE 7-70 Protected areas of the carpet show the locations of furniture during the fire. *Courtesy of Jamie Novak, Novak Investigations and St. Paul Fire Dept.*

such undamaged areas are found, it is up to the investigator to determine what object protected the floor and where it is now. Such objects are most often removed during suppression or overhaul, but more than once have the remains of a container or delay device been removed after the fire by the perpetrator. This is a primary reason why the scene must be secured from all persons until the investigation is complete. Such protected areas may help

FIGURE 7-71A Protected areas of carpet show the locations of furniture during the fire. *Courtesy of John D. DeHaan.*

FIGURE 7-71B

establish the position of doors, a feature that is critical to estimating ventilation conditions and access possibilities at the time of the fire (see Figure 7-73).

Another situation in which a protected area is important is in the claimed fire in a metal wastebasket or trash can. If there is not much fuel in the can or the fire is prolonged, fire may heat the bottom of the can and scorch the floor beneath. In most cases, the bottom of the container will protect the area under it as well as retain manufacturer

FIGURE 7-72A Charred "path" on carpet from door (to left) to broken windows (to right) after preliminary debris removal from a post-flashover fire. *Courtesy of Det. Rich Edwards, Los Angeles Sheriff's Office (retired).*

FIGURE 7-72B Piles of clothing and bedding protected irregular areas of carpet. (Photo taken just after extinguishment shows piles of clothing and gray firefighting foam. *Courtesy of Det. Rich Edwards, Los Angeles Sheriff's Office (retired).*

data. Its position may then be accurately determined if the floor is discolored or charred around it. A fire starting inside a plastic wastebasket will tend to cause collapse uniformly around the perimeter, as pictured in Figures 7-74a and b. A fire involving such a container from outside will usually cause asymmetric melting (greater on the side facing the oncoming fire). By examining the arrangement adjacent to the trash container and the extent of fire in this region, the investigator may be able to assign the source of the fire

FIGURE 7-73 Protected area of carpet shows the position of the door during the radiant heat stage of fire. *Courtesy of Lamont "Monty" McGill, Fire/Explosion Investigator.*

FIGURE 7-74A Plastic trash can was melted uniformly around its perimeter when the fire was ignited inside. *Courtesy of John D. DeHaan.*

FIGURE 7-74B Bottom of can still bears identification information. *Courtesy of John D. DeHaan.*

to this origin or to rule it out. Without the knowledge of the exact location of the container, this is likely to be impossible or very difficult. Often, a fan-shaped char pattern will occur on a wall just over such a point, with its lower boundary at the height of such a trash container. This circumstance is frequently the very best evidence of the trash container as the origin of the blaze. When a plastic gasoline container is ignited in a fire, it burns or melts from the top down. If there is liquid fuel left in it, the fuel will be trapped between the 'folds' of the collapsing container as in Figure 4-5g. Some liquid may leak out when the mass is overturned, or it may be detected by a visible or audible "sloshing."

Protected areas on walls are also frequently observed for similar reasons. The location of an item of furniture that is no longer present can often be determined accurately from such an area, as shown in Figure 7-75. The opposite effect is also observed occasionally in

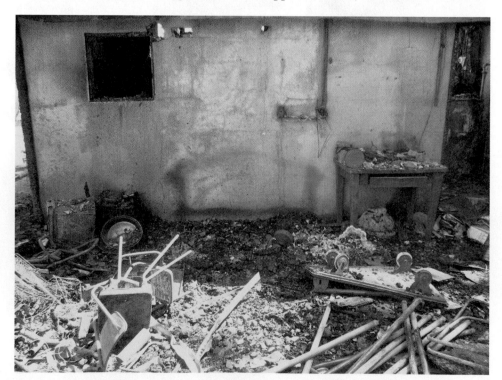

FIGURE 7-75 This fire occurred on a covered patio (assuring plenty of ventilation) and burned post-flashover. The outline of the completely destroyed sofa (and likely origin) is still visible on the stucco wall. *Courtesy of Derrick Fire Investigations.*

FIGURE 7-76A Protected area on left wall shows there was a mattress in place at the time of the fire. Damaged area on right wall and damaged area of mattress in foreground shows the mattress was overturned during firefighting. *Courtesy of Jamie Novak, Novak Investigations and St. Paul Fire Dept.*

instances where the furniture burned and collapsed; instead of protecting the wall behind it, it enhanced the burn (as in Figures 7-76 a and b). Such protection effects will not necessarily serve to locate the position of a specific piece of furniture, because the marginal dimensions are lost, but they can, at times, contribute significant knowledge of the fire origin, when considered along with the general fire pattern. Localized floor penetrations are often

FIGURE 7-76B Broken window frame caused localized damage to the wall behind it and allowed smoke to vent during the fire. *Courtesy of John D. DeHaan.*

FIGURE 7-77 Deep localized charring and penetration of the wood floor was the result of a wood-and-cardboard "bonfire" in the center of this room with no ignitable liquids. *Courtesy of John D. DeHaan.*

FIGURE 7-78 Flashover fire briefly forced flames out under door, producing charred door and floor (repeated several times). *Courtesy of John D. DeHaan.*

caused by combustible furniture burning for prolonged periods, as seen in Figure 7-77. A combination of the two effects is also sometimes seen. When a piece of furniture is afire, it may both protect the wall behind it and, by intensifying the local fire, increase the amount of wall damage on the sides and above the item. This is especially important when the fire started in the furniture in question. Ventilation effects can also produce low burns, even in the absence of localized fuel packages. Figure 7-78 shows the flames of a fully involved room fire being pushed out under the door as the room briefly reached an overpressure condition. This condition could be maintained for only a few seconds, but the flames "pulsed" from beneath the door several times before the fire found its ventilation limit.

UTILITIES

The condition of the utilities servicing the building should always be inspected and documented. When utility companies respond to a scene to remove gas or electric meters (to ensure the safety of scene investigators), they should be asked to evaluate the service for signs of tampering, bypassing, damage, or malfunction. The investigator should determine when power, gas, telephone, cable TV, and the like were cut off and by whom. Sometimes a planned fire will be betrayed by the owner's cutting off costly services several days before the structure was destroyed.

ELIMINATION OF ELECTRICAL IGNITION SOURCES

Remember that in accordance with the scientific method, all logically possible sources of ignition from natural and accidental causes must be eliminated if there is to be proof of the crime of arson. The use of an ignition matrix as a means of structuring and documenting the elimination process was illustrated in Chapter 6. This will, of course, include the utilities, appliances, and electrical wiring (both fixed and power cords). To check out an electrical system in a burned structure is often the most frustrating portion of the entire examination. Wires, conduits, and remains of fixtures are often scattered throughout much of the bulk of the rubble. Wires will generally lack insulation, this having been burned off in the fire. Because wiring is most commonly of copper, and copper when heated is readily oxidized to copper oxide, the blackened and brittle condition of the wires will make them difficult to handle when attempting to trace them. The most significant item to look for in such an instance is not just a mass of damaged wire but rather fused copper in localized areas. Such fused masses are most likely to occur in the neighborhood of switches, outlets, and similar connections, and at the fuse box. At any of these places, it will be common to find wires with rounded, fused ends, demonstrating that there was arcing, almost certainly by short-circuiting. When there are numerous such fused ends, it is unlikely that a fire origin has been found but rather that there was failure of the breakers to shut off current as wires farther away from the breaker or fuse box shorted together as the insulation failed in the fire. When there is a single heavy fusion, but other portions of the wiring are generally intact, this single point becomes highly suspect as a *possible* source of ignition. Normally, in such an instance, the fuse is blown or the breaker is open, indicating that this short circuit occurred and triggered the safety mechanism. Such protective devices respond to a direct short and usually interrupt power before a fire can be started. High-resistance connections, on the other hand, can create enough heat for long periods of time without drawing enough current to trip protective devices. These long-term heat sources are much more likely to ignite ordinary fuels than momentary short-circuit arcs. Examination of the fuse box or circuit breaker panel to determine which circuits have been tripped may be helpful in eliminating areas and the circuits that serve them.

It should be noted also that wire within a metal conduit will rarely short except when the conduit is heated sufficiently by an external fire to char the wire's insulation. If a short circuit is formed within a conduit from any cause, it must be a high-energy one to penetrate to the exterior and start a fire, although such events are not unknown (involving unprotected circuits, high-current applications [>40 A], failed overcurrent protection devices, or three-phase wiring). To start a fire requires that the conduit itself be heated to the point of fusion. In a very rare instance, it might be heated above the ignition temperature of some material in contact with it, which could become ignited. The high heat conductivity of the metal conduit is actually a very good protection against starting of fires from wires enclosed in the conduit. See Chapter 10 for more information.

One additional effect that has been noted in more than one instance involves the zinc die cast used in conduit connectors. Zinc die cast has a relatively low melting point and will melt in a fire surrounding the conduit. The heat of the fire may also remove the insulation, so that molten zinc will contact copper. When this occurs, the zinc and copper will form brass. Either the molten zinc or the brass formed may complete a short circuit between wires that were not previously in contact. Such an effect is not a cause of the fire but the result of an external fire, and it should not be interpreted otherwise. The presence of brass within a conduit or spilled from a broken conduit is actually very good evidence of an external fire that led to its formation. The diagnostic signs of electrically related fires are discussed in great detail in Chapter 10 of this volume.

ARC MAPPING

The plotting of electric arcing in electrical wiring found at a fire scene has been found to be of considerable value in testing hypotheses about possible areas of origin. A map of arc faults can be overlaid on diagrams of other indicators. If the electric circuit involved can be traced back to its power source, the location of the arc fault (which caused the overcurrent device to trip or the conductors to part) farthest from the power source is very likely to be the area closest to the origin. This assumes, of course, that the wiring was equally exposed to the fire. Wiring concealed in walls will be affected much later than will exposed power cords. Arc-fault mapping is not used to establish whether an electrical fault caused the fire but as an indicator of area of origin. The technique is described in more detail in Chapter 10 of this text.

APPLIANCE CONDITION

The condition of an appliance that started a fire is not expected to be identical with one that was in a fire external to it that it did not start, but the differences are sometimes difficult to spot. The external fire will damage any appliance by burning the enamel, discoloring and corroding the steel, and otherwise distorting or warping it. If the fire is external to the appliance, whatever damage it suffers will show uniformity, or at least agreement, with the fire patterns on walls, furniture, or other appliances in the vicinity. The appliance is much less likely to be damaged internally if the cabinet or housing is noncombustible. If the appliance started the fire, it will be damaged more internally than externally and usually have some local damage or effect that is quite distinct from the general damage pattern. This will be a point of consequence to the investigator. Because fires are easily blamed on a faulty oven or a maladjusted furnace, incendiary fires are sometimes ignited in the vicinity of such appliances to throw the investigator off the track.

The key to uncovering this situation is again a study of the fire pattern. In most such instances, the arsonist will use a flammable liquid that is poured around and under the appliance and then ignited. In such instances, it is normal for the liquid to penetrate well below the appliance in question and to give a burn far below that which is possible for the appliance itself. Gas appliances and fireplaces will usually lead to fire above the base level of the normal combustion, not to a region under the floor that supports it. Burns below this level of support may be traceable to a failure or defect of the gas supply or firebox or to improper installation (as seen in Figure 6-4c). In some instances, there may be penetration of the floor under a fireplace or stove/heater due to the dropping of hot materials onto the combustible floor through gaps in the firebox. Burning clear to the floor level is not normal for a fire escaping from an appliance unless there is very obvious and severe damage to the unit itself (see Figure 6-4a for an example). Naturally, the investigator will need to check carefully that no such abnormality exists in the appliance. Misuse of any heat-producing appliance can cause a fire, and any fire in the vicinity should be considered suspect and investigated with extraordinary care. A portable X-ray system, such as those used by bomb squads, can be used to examine appliances and connectors at the scene (see Chapter 15). Such examinations may reveal useful information without destructive dismantling.

One cause of accidental fires that *is* often related to such appliances is the formation of "self-heating charcoal" by prolonged exposure of a wood surface to heat, as discussed in Chapter 6. The investigator should be aware of such processes wherever wood is exposed to even moderate temperatures [over 105°C (220°F), or so)] in conditions where ventilation is excluded or very limited. It must be remembered that the lower the temperature, the longer it takes for a sufficient mass of such charring to form, and the less likely its formation will occur at all. Wood exposed to temperatures below 300°C (600°F) for such long terms is characteristically dull- or flat-finished, with a sunken appearance. Char that has penetrated completely through the wood structures involved and is much deeper than the known circumstances of the fire itself would warrant should prompt further examination. This phenomenon has been thought to turn even large wooden structural members completely to charcoal over a period of months or years before finally igniting them (see Chapter 6). Post-suppression smoldering can, of course, produce deep destruction if prolonged.

TRASH

Trash accumulations that are apparent as a portion of the fire scene will ordinarily be scrutinized very carefully, because they provide a suitable place for igniting a fire, especially when doused with a liquid accelerant. However, they also provide an inherent fire risk from accidental causes. Both possibilities must be carefully weighed before considering that a trash pile was used by an arsonist. If liquid flammables were poured on the trash, some of the liquid may be recovered and identified, in which case, strong evidence of arson is available. When a trash pile appears to be an origin or possible origin of the fire, it is highly important to search for ignitable liquids, as well as ignition sources. It is important to search for means of ignition that could be accidental in nature. Common sense sometimes does not dictate what accumulations of trash some people keep, even in their kitchens and bathrooms. Even the most elementary safety measures of protecting such trash from open flames such as candles, fireplaces, or room heaters are not taken by some people. It must not be taken for granted that "no one would have that much trash around" or "no one would light a candle near all that junk," because time and again, people act unsafely.

DETECTION SYSTEMS MAPPING

In a large compartment or a building protected by fire alarm or sprinkler systems, the detectors or sprinkler heads that were triggered during the fire should be mapped out on a plot of the building. This map can reveal the most energetic (if not the first) stages of the fire. When the times of their responses are included, a sequence of the fire's spread may be reconstructed. Many alarm systems have a data chip that can be read after the fire that reveals the time of functioning of connected systems and the sequence of various alarm functions. Alarm companies should always be contacted for information about alarm settings, functions, and history.

The sprinkler system and its connections should be examined and documented. Fire investigators should be familiar with system components, such as the one in Figure 7-79, and signs of tampering or inoperability.

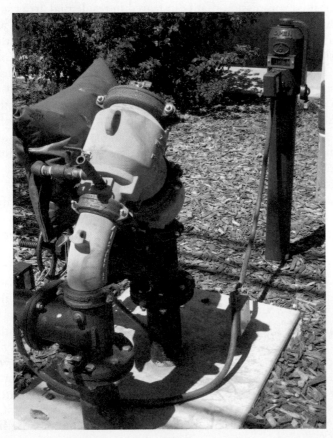

FIGURE 7-79 Typical sprinkler system components—OS&Y valves (with chains and locks) backflow preventer, PIV (post-indicator valve), and water-flow alarm connection. *Courtesy of John D. DeHaan.*

INTERIOR FIRES FROM EXTERIOR SOURCES

Fires within structures need not invariably have their origin in the interior. Buildings are frequently consumed as a result of a fire that is initially outside the structure. Such sources of fire may be roughly classified as follows:

- fire from an adjacent building spread by radiant heat or burning debris
- grass or brush fires exterior to the building
- trash fires that are out of control
- arson by means of trash or accelerant ignition against the side of the structure, under porches, around doors, and the like

When the fire is initially outside the building, it may involve exterior surfaces and never penetrate to the interior. Such a fire pattern is readily evident and the cause simple to determine. However, special circumstances may exist. For example, an open door or window, combined with wind and burning fragments from a nearby fire, may start a fire on the interior with no burning of the exterior structure. This fire would have an interior origin that would be traced in the conventional manner. Radiation from a large nearby fire can ignite the interior of a building through windows and doors before the exterior is ignited. The determination of cause in such an instance will have to rest on demonstrating the existence of a nearby fire, the wind, the opening, and the absence of any other reasonable interior cause. While such instances are not common, they can be quite troublesome to the investigator, since they prompt reports of multiple fire origins from witnesses and firefighters.

ROOF AND ATTIC FIRES

The most common type of transmission of fire from an exterior source is by burning material that settles on a combustible roof. The resulting fire is not initially an interior fire. This is the mode of fire transmission that is so feared in residential areas surrounded by brush or timber. Burning embers as large as whole branches are carried aloft from major wildland fires to land on cedar or redwood shake roofs, quickly igniting them. Such fires caused heavy destruction in San Diego, San Bernardino, Orange, and Los Angeles counties in California in recent years, as well as being one vector of fire spread for the destructive Oakland Hills fire that destroyed 2,500 homes and 500 apartment units in a single day in 1991. So many variations in the mode of transmission of fire from point to point (depending on the ambient conditions) are encountered that care and thoroughness on the part of the investigator are essential. All possible factors must be considered and each weighed comparatively and in light of the established environmental conditions. (See Chapter 6 for an example of wind-drawn urban fire.)

In residences, especially, it is not uncommon for a roof to burn away or for a combined attic-roof fire to occur. For this reason, attic and roof fires are often confused, and this can lead to misinterpretation. Actually, a roof-started fire is totally different in behavior from an attic-started fire. In both instances, the absence of fire on a lower level will define the fire as being one or the other, but often it is difficult to determine which.

An attic fire is started under the roof and will usually involve all the roof that is accessible to the spread of flames beneath it. Thus, it is likely that the entire roof is either burned away or that destruction is very widespread. Widespread burning in the attic is always indicative of this type of fire, as in Figure 6-13.

A true roof fire, in contrast, must have been ignited from the exterior, otherwise it would have developed as an attic fire. Sparks from chimneys, or similar causes, may ignite leaf debris, shingles, or shakes on a roof, resulting in an exterior fire. This fire will follow the usual inverted-cone pattern on an inclined roof and will ultimately penetrate the roof into the attic below. It is at this point that the main distinction between an attic-started and a roof-started fire becomes apparent. The first hole that burns through the roof ventilates the fire, drawing air upward through the hole, which generally confines the fire to the exterior and to parts of the roof immediately adjacent to the hole. Here, air is being drawn from the attic to the fire, which in some instances suffices to prevent a secondary spread as an attic fire. The major burn is usually on the outside of the roof, not under it, although in some steep roofs fire can spread internally up along the inside pitch of the roof. If enough of the roof is involved, of course, structural collapse can spread the fire down into the living space, eventually causing a total burnout.

In an attic fire, there is almost always a great deal of burning under all parts of the roof before the latter is finally breached. Because upward ventilation is limited, spread of the fire is inevitable. Fire spread on the underside of a combustible roof assembly (plywood or wood lattice) can be very rapid and difficult to extinguish. After a hole is chopped in the roof to ventilate the fire under it, that fire can then be controlled much more easily because the flames are funneled toward the hole that ventilates them. Figure 7-80 illustrates the "burned-from-the-top-down" appearance of ceiling joists after a typical attic fire.

Although the destruction may not be significantly different in the two cases, the investigation is greatly affected, because the roof fire is caused by external ignition, while the attic fire is ignited within the structure.

TIMELINES

Every fire has a life cycle from the time a heat source first comes into contact with the first fuel to its extinguishment. The timeline for this cycle is defined by several time factors: ignition time, incipient fire duration, response time, extinguishment time. As we saw in Chapter 3, the nature of the ignition source and the nature of the first fuel control the ignition time and incipient fire. The fire growth is controlled by the fuel, its arrangement, and the ventilation conditions. These factors can be estimated from test data (from existing research or specialized tests) or case history as part of the hypothesis testing. As seen on the modeling data form (Figure 7-7), establishing a timeline includes both "hard" times and "soft" times. "Hard" times are linked directly or indirectly to an accurate and reliable clock or timing device. They include the time someone was last in the vicinity, alarm set or activation times, time of detection by witnesses, time of 9-1-1 call, arrival time, and sometimes time of control or extinguishment. "Soft" times are *estimates* of clock time or duration by witnesses or engineering analysis of known fire behaviors. Such timelines can be used to test and exclude some scenarios (such as smoky self-heating processes or "instantaneous" ignition by discarded cigarette) and suggest others. The successful use of a timeline depends on the investigator's knowledge of time factors in all manner of fires. Timelines can be reconstructed from fire or police dispatch recordings, 9-1-1 calls, or videos of events. Security and surveillance cameras are everywhere today. Videos from cell phones or police, fire, or public transport "windshield" cameras should not be overlooked.

Collection and Preservation of Evidence

The contents of a complete fire scene evidence kit are detailed in Appendix D. Briefly, an evidence kit has a number of containers for various kinds of physical evidence and the tools that may be required to recover them. Evidence recognition, recovery, and preservation all follow rules of plain common sense. Preservation of evidence means just that—*preserve* it until it reaches the laboratory or the court. Preservation, via proper collection and packaging, means the evidence is not allowed to change by evaporating, breaking, spoiling, or being contaminated. Think about the limitations of the evidence being picked up and package it accordingly. If it is a volatile liquid you suspect, you do not want it to evaporate, so package it in an airtight can or jar and keep it cool and sealed. If it is an organic—blood, tissue, soil, or hair—you do not want to let it decompose by mildewing or developing mold. If you allow such material to air dry at room temperature, package it in clean paper containers and keep it dry and cool, it will stay intact. With fragile objects, avoid further breakage by packing in suitable containers and hand-carry to the lab. Minute trace evidence, which is easily lost or contaminated, is best preserved by placing in a clean container of suitable size and sealing it. Comparison samples of any floor coverings involved in the fire's origin should be taken (preferably from areas not damaged by fire) not only to aid laboratory analysis for volatiles but also to help reconstruct the potential contributions the carpet and pad could make to the fuel load in the room and the spread of fire.

The procedures for other kinds of evidence are just as logical and straightforward. Table 7-3 lists some common evidence types and collection and preservation methods. Guidelines for evidence collection are the subject of *ASTM E1188* and *E1459.*[58]

DEBRIS SUSPECTED OF CONTAINING VOLATILES

The most commonly submitted form of arson evidence is solid debris suspected of containing the residue of ignitable liquids. The major concern here is to prevent the evaporation of volatile liquids. Collection of all physical evidence must be carried out with full realization of the chances of contamination. Due to the sensitivity of laboratory methods, invisible traces of volatiles transferred by the tools, gloves, or even the boots of the investigator can compromise the value of such trace evidence. Carefully clean tools using detergent and water between scenes and between collection sites at the same scene. When collecting debris, you must wear clean disposable plastic or latex gloves and then discard them before collecting other samples. Clean trowels, scoops, brooms, and brushes in clean running water and wipe them with a clean paper towel between samples. Always use a container similar in size to the amount of debris and collect *more* than you think the laboratory will need. No matter how strongly it smells of accelerant, more debris is always preferred.

Such debris can best be preserved in clean metal friction-lid paint cans, which are reasonably airtight and unbreakable. The cans come in a variety of sizes, they are easily opened

EVIDENCE TYPE	SAMPLE AMOUNT DESIRED	PACKAGING	SPECIAL HANDLING
Checks, letters, and other documents	All	Place in manila envelope; do not fold; wrap securely.	Do not handle with bare hands.
Fire debris with suspected accelerants	All	Seal in clean empty metal paint can, Kapak (by Ampac) or nylon bag, or glass jar; fill container no more than 3/4 full.	Label as to origin and collecting investigator; keep cool see text.
Flammable liquids	All (up to 500 ml; 1 pt)	Seal in glass bottle, jar with Bakelite or metal top with Teflon liner, or metal can.	Do not use rubber stoppers or jars with rubber seals; do not use plastic bottles.
Charred or burned paper or cardboard	All	Pack loosely on soft cotton; hand-carry to laboratory.	Pack in rigid container; mark FRAGILE; do not treat with lacquer or other coating.
Clothing	All	Mark directly on clothing in waistband, pocket, or coat collar with initials and date; pack in clean paper bags; wrap each item separately in paper; do not use plastic bags; on outside of container indicate nature of item, date obtained, and investigator's name.	If clothing is wet with water or blood, hang to air dry before packaging to prevent putrefaction; fold neatly with clean paper between folds, and refrigerate if possible. If there are residues of ignitable liquids, seal in metal can or Kapak bag. Freeze if possible.
Hairs and fibers	Any loose hairs or fibers. Comparisons: clumps of 10 to 20 closely cut pubic or head hairs from various areas	Pill box, paper envelope or bindle, cellophane or plastic bag; seal carefully.	On outside of container: label as to type of material, where found, date, investigator's name.
Paint chips	At least 100 mm^2 (1/2 in.^{2}) of flake; entire object if small	Pill box, paper bindle, or envelope; seal carefully.	Obtain control sample from outside transfer area.
Soil	At least 25 g (1 oz)	Plastic bag or jar. If suspected of containing volatiles, seal in glass jar or metal can and freeze.	Collect several samples from area.
Tools	All	Wrap each tool separately to prevent shifting and loss of trace evidence	Secure envelope or folded sheet of paper carefully over end of tool to prevent damage and loss of adhering paints, etc.
Tool marks	Send in tool if found; submit all items having tool mark scratches	After marks have been protected, wrap with wrapping paper and place in envelope or box.	Cover marks with soft paper, tape in place; keep from rusting.
Dry blood stains	As much as possible; submit object if possible	Use paper bags, boxes, bindles, or envelopes; sealed to prevent loss of scrapings; keep dry; do not use plastic bags; refrigerate if possible.	On outside of container: type of specimen, date secured, investigator's name and where found.

(continued)

| TABLE 7-3 | | Physical Evidence Collection and Preservation (*continued*) | |

EVIDENCE TYPE	SAMPLE AMOUNT DESIRED	PACKAGING	SPECIAL HANDLING
Glass fragments	As much as possible	Box, paper bags, bindle, or envelope; pack and seal to avoid movement in container.	On outside of container: type of specimen, date, investigator's name and number. Keep evidence separated from control sample.
Firearms	All	Place revolvers and automatics in cardboard box or manila envelopes; tag rifles; hand-carry to laboratory.	Unload: note cylinder position; if automatic, do not handle side of magazine—possible fingerprints in this location; note serial number and make of weapon.
Latent fingerprints	All	Secure in rigid container so surfaces do not rub on packaging. Clean paper envelope OK for porous objects. Keep dry and cool.	Allow to air-dry if damp. Keep immersed in water if recovered from water.

and resealed, and the top can easily be punched for insertion of a sampling syringe. They can be lined or unlined, but unlined cans tend to rust very quickly, so lined cans (used for water-base paints) are preferable. Today's linings contain no significant volatile residues. The second choice for containing debris suspected of containing volatiles is the familiar Mason jar, which is airtight and transparent, permitting ready visual examination of the contents, and has a metal top for easy sampling. The third choice is a glass jar or bottle with a metal or Bakelite cap. Plastic caps, rubber seals, or stoppers should never be used when a liquid sample of suspected material is submitted. Gasolines, paint thinners, and other common volatile hydrocarbons will attack and dissolve such closures. Do not fill any container more than three-quarters full of debris to leave a headspace for sampling.

Two types of polymeric bags have been found suitable for packaging most volatile evidence. Bags made of nylon film (sold by Grand River) or special polyester film (Kapak sold by Ampac) have been found to retain even the lightest hydrocarbons for moderate lengths of time. They will allow the loss of methyl and ethyl alcohols, however, and debris suspected of containing these unusual accelerants should be sealed in metal cans or glass jars. For most debris, however, such bags provide a convenient, low-cost, and unbreakable alternative container.[59] The polyester bags are heat-sealable and provide even greater integrity to the evidence container than a tape seal.[60] Any polymeric bag can be penetrated by sharp debris, so care must be used.[61] Because of a minute chance of contamination during manufacturing of some cans and bags, a sample container from each batch should be tested for the presence of trace hydrocarbons before they are issued for use.[62]

A common household plastic container should be used as a last resort only. Polypropylene jars are permissible, but never polystyrene, which is readily soluble in gasoline. Polyethylene plastic bags, caps for coffee cans, and wrapping films are permeable to some hydrocarbons and allow them to escape (or cross-contamination to occur) before analysis can be attempted and should not be used for arson debris.[63] Paper and polyethylene bags also allow cross-contamination, as external volatiles can permeate them and be absorbed by evidence within. For large, awkward items, large plastic bags may be the only packaging alternative. In such cases, the evidence should be double-bagged and taken to the laboratory as soon as possible, and be kept as cool as possible in the interim.

Refrigeration and even freezing is recommended for all arson fire debris except for glass jars containing wet debris that can expand and shatter the container if frozen. Freezing is especially important when soil is collected for ignitable liquid residues, as the bacteria in garden soil can decompose petroleum products if stored at room temperatures.[64]

OTHER SOLID EVIDENCE

Chemicals from clandestine drug laboratories and most chemical incendiary mixtures are corrosive and care must be taken to ensure they are *not* packaged in metal cans, which can be destroyed by the chemical action. When such corrosives are suspected, the debris may be packaged in glass jars (then kept upright to avoid contact between the lid and debris) or in nylon or polyester bags, which are then sealed in cans.

Nonporous materials such as glass do not retain volatile petroleum distillates, but wrappers, labels, and associated debris may offer absorbent surfaces to retain enough for characterization in the lab. If bottle fragments are found from a suspected Molotov cocktail, locate the neck and base of the bottle and package all of it to prevent further breakage. The neck often retains portions of the cloth wick, the base bears manufacturer's data to identify the bottle, and the sides and label may bear usable latent fingerprints. The neck may also bear saliva with identifiable DNA.

Newly developed techniques using lasers, chemicals, and physical developers are now making it possible to recover identifiable latent prints from objects that have been exposed to gasoline, water, or even fire. Unless an object has been heavily damaged by direct exposure to fire, there is a chance that latents can be recovered. The crime laboratory should be contacted for advice before dismissing the possibility of prints, because prints have been identified on glass bottles, plastic jugs, and metal cans even after fire exposure.[65] It has recently been reported that DNA-typable materials were recovered from unburned towel fragments from the neck of a broken and burned Molotov cocktail.[66] More discussion is presented in Chapter 14.

LIQUIDS

Uncontaminated flammable liquids from the scene are the best kind of evidence for laboratory analysis. If the liquid is still in its original container, as long as that container can be properly sealed, it may be left there. Otherwise, transfer liquids by eyedropper (pipette) or syringe to a clean sealable container. Do *not* use a bottle with a rubber stopper or a jar with a rubber sealing ring for any liquid hydrocarbon. It will dissolve the rubber and leak out. A small quantity of liquid in debris may be recovered directly with an eyedropper or by absorbing it into a pad of clean absorbent cloth, gauze, or cotton wool, and then sealing it in a suitable airtight container.

When flammable liquids are suspected in concrete, samples can be collected from cracks or expansion joints, but often, a better method is to wet-broom the suspected area and then spread a quantity of absorbent material. The best absorbent is clay-type kitty litter, although diatomaceous earth or even flour can be used. This layer is allowed to stand for 20 to 30 minutes and then recovered into a clean metal can using a clean putty knife and brush for laboratory analysis.[67] A control sample of the unused absorbent must be sealed in a separate can. The cleanliness of the absorbent (and any collection tools used) is critical to the value of this evidence. The use of hazardous material recovery pads is not recommended due to the possible contamination of such pads by exposure to volatile materials in the environment. Chemically aggressive absorbents such as activated charcoal are not recommended due to their susceptibility for external contamination.

TESTING OF HANDS

Liquid hydrocarbon fuels can be absorbed into the skin after even brief exposure, but the warmth of living tissue causes rapid desorption of such volatiles. Over the years, a number of methods have been suggested for testing of hands of suspects in arson investigations. Canines have often alerted to hands contaminated with gasoline even after some time has elapsed, but the trace amounts present make recovery for laboratory analysis very difficult. Methods such as swabbing with a dry or solvent-wetted cotton gauze are useless for recovering such traces. The best system is to place a clean disposable PVC or latex glove on each of the suspect's hands or to secure a polyester or nylon fire debris bag around the hand with a rubber band at the wrist, enclosing an activated carbon recovery device (such as a carbon strip or solid-phase microextraction (SPME) device) in the bag for 15–20 minutes. The warmth of the skin will cause the absorbed volatiles to evaporate into the glove or bag (assisted by exposure to a heating lamp), where they are sampled by the activated carbon. After the subject removes his or her hand, the glove or bag is sealed with the sampling device inside and forwarded to the lab for analysis. Tests indicate that for this testing to yield positive results, it must be done within an hour or two of contact with gasoline.[68]

Swabbing of the hands of suspects in explosives cases using dry cotton-tipped swabs [such as those in gunshot residue (GSR) kits] or ones wetted with distilled water or methyl or isopropyl alcohol (but *not* with the acid supplied with the GSR kit) can capture traces of either organic or inorganic explosives. Fingernail scrapings using clean flat wooden toothpicks are collected and sealed in clean glass or plastic vials. This technique is particularly valuable for soft organic explosives such as dynamite or C-4.

TESTING OF CLOTHING

Rapid pouring of an ignitable liquid onto a hard surface often causes spattering onto nearby shoes and pants cuffs. Arsonists may also step into the liquid after it is poured. The gas chromatographic profile of such transfers will be different from that from vapor contact alone (see Chapter 14). Because such volatiles are lost within a few hours (1–6) if the garment is worn, residues on clothing can help identify the perpetrator. Clothing of a suspect should be seized as soon as possible and packaged *separately* in appropriate sealed containers until tested by the lab.

A New Zealand study illustrated the desirability of analyzing clothing and shoes of suspected arsonists for the presence of ignitable liquids. The study found that detectable amounts of petrol were transferred to the clothing and shoes of a person during the action of pouring it around the room, with various pouring heights and floor surfaces. Results of the study showed that petrol was always transferred to the shoes and often transferred to both the upper and lower clothing but that the seizure and proper packaging must take place quickly.[69] In tests using a carbon strip extraction method, it was found that 10 ml (0.34 oz or 2 tsp) of gasoline on clothing worn continuously evaporated much more quickly than on the same clothing left on the lab bench at 20°C–25°C (68°F–77°F) and was barely detectable after only 4 hours of wear.[70] Clothing must be recovered and packaged soon after exposure.

A second New Zealand study further examined a potential defense argument that the gasoline residues located on a suspect's clothing came from a legitimate source. The study showed that it is extremely unlikely to have detectable levels of gasoline on clothing and shoes following the activity of filling a vehicle with fuel, without the spillage of any noticeable amount of fuel onto the clothing and shoes. The study found it possible for gasoline to be detected on the shoes of people after using a gasoline-powered mower to cut their lawns. The study further underscored that the time delays between commission of the crime and seizure of the suspect's clothing need to be taken into account when considering the possibility of evaporation of any volatile residues from the suspect's clothing.[71]

CHAIN OF EVIDENCE

chain of evidence (chain of custody) ■ Written documentation of possession of items of physical evidence from their recovery to their submission in court.

No matter what kind of material is recovered or how carefully it is preserved, it is valueless as evidence if it has not been maintained and documented completely. The **chain of evidence (chain of custody)** is a record of the existence and the whereabouts of the item from the time it is discovered at the scene to the moment it is yielded to the court. This documentation of an item is in the form of a written record, usually on the sealed evidence package, that lists every person who has had custody or possession of the item. The list must include the dates and times of transfer and the signature of anyone involved in that transfer (see *ASTM E1459*). The documentation starts with the photography of the item in place at the scene and records *all* people who handle the evidence or its sealed container. The number of people handling evidence should be kept to a minimum to minimize chances for damage and to reduce the number of people who must be ready to testify in court as to the existence of the evidence, for each person is then a link in this crucial chain.

An evidence log is an accurate, complete list of all items of evidence recovered from a scene or location. Many agencies prefer to use a preprinted form such as that in Appendix I (or see *NFPA 921*). Note that the form provides a checklist for recording what was recovered, from where, and by whom, and also that the item was documented by photograph and diagram before recovery. If more than one person is recovering evidence at a scene, it is preferable that the evidence log for a scene be maintained by one person to avoid duplicate or conflicting numbering. It is the responsibility of the person recovering the item to ensure that each item is properly packaged, labeled, or tagged, and that the photos and diagram entries have been made. Such a log is a useful supplement to the custody list on the item itself.

The chain of custody is much like a ship's anchor chain. If the anchor chain is broken, it cannot be recovered, and the ship will probably be lost. If the chain of custody is broken, it cannot be restored, the evidence is valueless to the court, and the case may well be lost. A complete, well-documented chain allows the court to admit evidence that can be verified, allowing the investigator to say, "Yes, this is the device I recovered from the scene. I know that because here is where I marked it, here is the container in which I sealed it." The importance of the chain of evidence to a successful fire investigation cannot be overemphasized.

Analysis and Hypothesis Testing

The investigator must be aware of any tendency to prejudge the origin and cause of any fire. The fire scene investigation is carried out to identify the origin of the fire—where it started—and its cause—what burned at the area of origin, and what caused the fuel to ignite. Careful evaluation of the indicators should result in an accurate determination of at least an area, if not a point, of origin and a hypothesis of the direction, intensity, and duration of the fire. Without at least a definable general area or origin, no reliable evaluation of ignition sources can be carried out. In the vicinity of the origin, the presence of the first fuel ignited and a possible ignition source should have been detected. The investigator must be able to define and defend the chain of events from the event that brought the first fuel and ignition source together to the spread of the fire, based on knowledge of the location and size of major fuel packages and relevant fire dynamics. The amount of damage visible must correspond to the heat release rate that would be expected from those packages.

The scientific method demands that all reasonably possible alternative hypotheses be evaluated, tested (against test data, scene evidence, or witness information), and eliminated to demonstrate the reliability of the final conclusion. To be sure all data critical to such testing are collected, a form such as the one shown in Figure 7-7 is strongly suggested.

If there are indications of a natural or accidental cause, do all the other indicators agree? Accidental sources include lightning, trash fires, shorted or overheated electrical connections, smoking materials, cooking accidents, chemical reactions, faulty appliances, or even children playing with matches. Is there identifiable contact between a possible heat source and a susceptible fuel, and is that contact enough to promote ignition? During the course of the investigation, any indication of incendiary origin found will prompt other inquiries—Who set it and why? What other signs of criminal activity such as burglary, robbery, fraud, or murder are present? Even in the absence of an incendiary device, the crime of arson can be proven in the absence of *all* logically possible accidental and natural causes at the point of origin. This methodology is sometimes called a *negative corpus,* since a complete positive corpus for the crime of arson includes proof of burning, proof of incendiary cause, and proof of absence of natural and accidental causes. The term *negative corpus* has no recognized definition but *implies* that the corpus of the crime (or event) has not been proven. This is *not* to say it is permissible to conclude a fire is an arson simply because "too much damage was done" or because "no accidental source could be identified." While some investigators treat negative corpus as a "catchall," it is a very difficult case to *prove* and should be relied upon only in the most special circumstances. Negative corpus cases are accepted in some jurisdictions and not in others, so the investigator should confer with local prosecutors before pursuing such cases.

The thorough investigator also considers all possible alternative explanations for the fire and evaluates the evidence (whether physical or testimonial) in deciding whether those alternatives are excluded. Diligent application of the ignition matrix concept described in Chapter 6 will result in consideration of likely "first fires" and ignition sources. Such consideration and elimination of alternative scenarios is a primary element of the scientific method of fire investigation. Careful engineering analysis (including fire modeling) may be useful in testing and eliminating hypotheses. If alternative explanations cannot be excluded, additional information either from the scene or from research or testing can throw light on the energies, times, or conditions needed for each. Based on these new data, it may be possible to exclude all the other possible hypotheses and make it easier to defend the final conclusion. (This process gives rise to the term *affirmative exclusion* to be used instead of *negative corpus.*) Such reasoning can be applied in cases where the first fuel ignited and source of ignition cannot be positively demonstrated if the scientific reasoning and assessment can be shown to be sound. If all possible explanations cannot be excluded, then the investigator must admit so. It is not possible to identify the cause of every fire. Depending on the extent of damage to the structure, there may not be enough information to reliably conclude the actual cause. The professional investigator must realize his or her limitations, as well as the limitations of the evidence, and know when to admit that the cause is "undetermined."

Summary

Before reaching a final decision, it is always a good idea to step back, mentally if not physically, and reassess what has been revealed. This process involves a resurvey of the entire site in view of the presumed finding of the point of origin, source of fuel, ventilation, and ignition mechanism, including heat release rates of the fuel packages involved. This is an important portion of the investigation, because if there are discrepancies in the theory that is presently offered, they can be found and reconciled, new data collected, or the theory modified. Such disparities can be used by the opposing attorney in any subsequent trial that may occur; failure to resolve all the contradictions can be highly embarrassing. This is the process of testing the final hypothesis.

It is wise to note specifically any types of building material that may have enhanced the development of the fire and to make certain that one clearly understands the plan of the structure. The type and location of ventilation sources, as well as those of fuel packages, must also be fully considered. Although these details may have all been noted in the earlier investigation, failure to fix them in the memory and the documentation can later raise problems that are better avoided. It is also possible that this review of the inspection will turn up unclear details that should be known, although they may have little bearing on the present theory or interpretation of the total fire.

As has been previously discussed, the accuracy of the determination of origin and cause depend upon the observational and analytical skills of the investigator. With so much depending on this determination, it should be obvious that the investigator must carefully assess the indications and "proof" available before reaching a final conclusion about a fire. It is not a dishonor to conclude that no origin or cause can be determined if there is any doubt in the mind of the expert investigator. If that trained expert investigator cannot identify an origin or cause, a less-qualified person should not. In accordance with the scientific method, if new data comes to light or the unreliability of previous information is shown, the professional investigator must reassess the conclusions previously reached. This reevaluation may result in a modification of the earlier conclusion. Certainly, in light of the destructive force of the event itself, it should not be surprising that the destruction of some structures is so complete that no determination can be made in good conscience. No matter how skilled an investigator may be, in the absence of adequate indicators of fire behavior or divine intervention, some fire causes simply cannot be reliably identified.

Review Questions

1. What is overhaul and why is it critical that it be kept to a minimum?
2. What basic fire behaviors are revealed by post-fire indicators?
3. Name three ways suppression efforts can affect fire spread and post-fire indicators.
4. Name five types of information that can be gathered from firefighters.
5. What is the most common search pattern for structure fires and what are four of its advantages?
6. Why is the documentation and reconstruction of fuel load so important?
7. What does the angle of a burn pattern on a wall or furniture reveal about the fire's development?
8. What is a heat horizon and how does it help establish direction of development of a structure fire?
9. Name three ways ventilation openings can affect fire patterns in a room fire.
10. What types of fuels produce the deepest burn pattrns on floors?
11. How does the radiant heat flux on a wood surface affect the char rate of the wood?
12. How does window glass provide useful information to a fire investigator?
13. Why is a contents inventory important in a fire investigation?
14. Name four ways a fire scene should be documented.
15. What is the chain of custody and why is it important?

References

1. D. J. Icove and J. D. DeHaan, *Forensic Fire Scene Reconstruction*, 2nd ed. (Upper Saddle River, NJ: Pearson Education/Brady, 2009), 22–27.
2. *NFPA 921: Guide for Fire and Explosion Investigation* (Quincy, MA: NFPA, 2011); *NIJ Guide for Fire and Arson Scene Evidence* (Washington, DC: Department of Justice, 2000); IFSTA, *Fire Investigator*, 2nd ed. (Oklahoma State University, Stillwater, OK: Fire Protection Publications, 2010).
3. M. Wallace and J. D. DeHaan "Overhauling for Successful Fire Investigation," *Fire Engineering* (December 2000), 73–75.
4. Ibid.
5. D. Smith, "Firefighter's Role and Responsibility in Preserving the Fire Scene and Physical Evidence," *The Times: Risk Management for Emergency Medical Services* 3 (September 1995).
6. Federal Bureau of Investigation, "Definitions and Guidelines for the Use of Imaging Technologies in the Criminal Justice System," *Forensic Science Communications* 1, no. 3 (October 1999).
7. J. D. DeHaan, "Advanced Tools for Use in Forensic Fire Scene Investigation, Reconstruction and Documentation" in *Proceedings Interflam 2004* (London: Interscience Communications, 2004), 1079–1100; *NFPA 921: Guide for Fire and Explosion Investigations* (Quincy, MA: NFPA, 2008 and 2011).
8. Better Homes and Gardens, Chief Architect, Chief Architect, Inc., Coeur d'Alene, ID; retrieved November 21, 2010, from http://www.chiefarchitect.com/company/location.html.
9. The Crime Zone (Beaverton, OR: The CAD Zone Inc.).
10. T. G. Gersbeck, "Advancing the Process of Post-Investigation," paper presented at AAFS, Washington DC, February 2008. See also: M. G. Haag M. G. and T. Grissim, "Technical Overview and Application of 3-D Laser Scanning for Shooting Reconstruction and Fire Scene Reconstruction," paper presented at AAFS, Washington DC, February 2008.
11. Icove and DeHaan, *Forensic Fire Scene Reconstruction*.
12. *ASTM E620: Reporting Opinions of Scientific or Technical Experts* (West Conshohocken, PA: ASTM, 2004).
13. *ASTM E678: Evaluation of Technical Data* (West Conshohocken, PA: ASTM, 2007).
14. *ASTM E860: Standard Practice for Examining and Testing Items That Are or May Become Involved in Litigation* (West Conshohocken PA: ASTM, 2006).
15. *ASTM E1020: Reporting Incidents* (West Conshohocken, PA: ASTM, 2006).
16. *ASTM E1188: Collection and Presentation of Information and Physical Items by a Technical Investigator* (West Conshohocken, PA: ASTM, 2005).
17. *ASTM E1459: Standard Guide for Physical Evidence Labeling and Related Documentation* (West Conshohocken, PA: ASTM, 2005).
18. D. I. Icove and J. D. DeHaan, "Hourglass Patterns" in *Proceedings of IFSI Conference*, Cincinnati, OH, May 2006.
19. *NFPA 921: Guide for Fire and Explosion Investigations* (Quincy, MA: NFPA, 2008 and 2011).
20. D. Madrzykowski and S. I. Kerber, "Fire Fighting Tactics under Wind Driven Conditions: Laboratory Experiments," NIST Technical Note 1618, 413 pp. (January 2009); S. I. Kerber and D. Madrzykowski, "Fire Fighting Tactics under Wind Driven Fire Conditions: 7-Story Building Experiments," NIST, TN 1629: NIST Technical Note 1629, 593 pp. (April, 2009).
21. J. D. DeHaan, "The Reconstruction of Fires Involving Highly Flammable Hydrocarbon Liquids," Ph.D. dissertation, Strathclyde University, Glasgow, Scotland, July 1995.
22. Ibid.
23. A. Putorti, J. A. McElroy, and D. Madrzykowski, *Flammable and Combustible Liquid Spill/Burn Patterns*, NIJ Report 604-00 (Washington, DC: NIJ, 2001).
24. J. D. DeHaan, "The Consumption of Animal Carcasses and Its Implication for the Combustion of Human Bodies in Fire," *Science and Justice* (January 1999).
25. J. D. DeHaan, "Are Localized Burns Proof of Flammable Liquid Accelerants?" *Fire and Arson Investigator* 38 (September 1987); J. C. Gudmann J. C., and D. Dillon, "Multiple Seats of Fire: The Hot Gas Layer," *Fire and Arson Investigator* 38, no. 3 (1988): 61–62; J. Albers, "Pour Pattern or Product of Combustion?" *California Fire/Arson Investigator* (July 2002): 5–10.
26. R. J. Powell, "Testimony Tested by Fire, "*Fire and Arson Investigator* (September 1992): 42–51; J. J. Lentini, "The Lime Street Fire: Another Perspective," *Fire and Arson Investigator* (September 1992): 52–54; J. D. DeHaan, "Fire: Fatal Intensity," *Fire and Arson Investigator* (September 1992): 55–59.
27. D. D. Drysdale, *An Introduction to Fire Dynamics*, 2nd ed. (Chichester, UK: Wiley, 1999), 189; V. Babrauskas, "Charring Rate of Wood as a Tool for Fire Investigators" in *Proceedings Interflam 2004* (London: Interscience Communications), 1155–70.
28. S. Carman, "Progressive Burn Pattern Development in Post-Flashover Fires" in *Proceedings Fire and Materials*, 789–800, San Francisco, CA, 2009.
29. B. V. Ettling, "The Significance of Alligatoring of Wood Char," *Fire and Arson Investigator* 41 (December 1990).
30. M. J. Spearpoint and J. G. Quintiere, "Predicting the Ignition of Wood in the Cone Calorimeter," *Fire Safety Journal* 36, no. 4 (2001): 391–415.

31. J. L. Sanderson, "Tests Add Further Doubt to Concrete Spalling Theories," *Fire Findings* 3 (Fall 1995); F. P. Smith, "Concrete Spalling: Controlled Fire Tests and Review," *Fire and Arson Investigator* 44 (September 1993): 43; C. R. Midkiff, "Spalling of Concrete as an Indicator of Arson," *Fire and Arson Investigator* 41 (December 1990): 42.

32. D. V. Canfield, "Causes of Spalling of Concrete at Elevated Temperatures," *Fire and Arson Investigator* 34 (June, 1984).

33. B. Béland, "Spalling of Concrete," *Fire and Arson Investigator* 44 (September 1993): 26.

34. DeHaan, "The Reconstruction of Fires"; J. Novak, personal communication, 2008.

35. R. N. Thaman, "Laboratory Can Help Investigator Pluck Arson Evidence Out of Debris," *Fire Engineering*, 131, no. 8 (August 197): 48–50, 52.

36. N. Saitoh and S. Takeuchi, "Floor Scene Imaging of Petroleum Accelerants by Time-Resolved Spectroscopy with a Pulsed Nd-YAG Laser," *Forensic Science International* 163 (2006): 38–50.

37. D. C. Mann and N. D. Pataansuu, "Studies of the Dehydration/Calcination of Gypsum Wallboard" in *Proceedings Fire & Materials 2009* (London: Interscience Communications, January 2009), 827–38.

38. J. E. Posey and E. P. Posey "Using Calcination of Gypsum Wallboard to Reveal Burn Patterns," *Fire and Arson Investigator* 33 (March 1983).

39. P. M. Kennedy, K. C. Kennedy, and R. L. Hopkins, "Depth of Calcination Measurement in Fire Origin Analysis" in *Proceedings Fire and Materials 2003* (London: Interscience Communications, 2003).

40. W. A. Tobin and K. L. Monson, "Collapsed Springs in Arson Investigations: A Critical Metallurgical Evaluation," *Fire Technology* 25 (November 1989).

41. J. J. Lentini, D. M. Smith, and R. W. Henderson, "Baseline Characteristics of Residential Structures Which Have Burned to Completion: The Oakland Experience," *Fire Technology* 28 (August, 1992).

42. M. J. Skelly, "An Experimental Investigation of Glass Breakage in Compartment Fires," NIST-GCR-90-578, U.S. Department of Commerce, 1990.

43. P. J. Pagni and A. A. Joshi, "Glass Breaking in Fires" in *Proceedings Third International Symposium on Fire Safety Science* (London: Elsevier Applied Science, 1991), 791–802.

44. N. Nic Daéid, "The ENFSI Fire and Explosion Investigation Working Group and the European Live Burn Tests at Cardington," *Science and Justice* 43, no. 1 (January–March 2003).

45. J. J. Lentini, "Behavior of Glass at Elevated Temperatures," *Journal of Forensic Sciences* 37 (September 1992).

46. Ibid.

47. R. J. Gilchrist, "Die Casting," *The Ohio Engineer*, (April 1937).

48. Lentini, Smith, and Henderson, "Baseline Characteristics."

49. B. Béland, C. Ray, and M. Tremblay, "Copper-Aluminum Interaction in Fire Environments," *Fire and Arson Investigator* 37 (March 1987). Reprinted from *Fire Technology* 19 (February 1983): 22–30.

50. R. W. Henderson, "Thermodynamics of Ferrous Metals," *National Fire and Arson Report* 4 (1986).

51. Dennis Farrell, Minneapolis, MN, personal communication, 1996.

52. A. R. Earls, "It's Not Lightweight Construction. It's What Happens When Lightweight Construction Meets Fire," *NFPA Journal* (July–August 2009): 38–45.

53. KTW Corp., Newark, CA, personal correspondence, 2008.

54. J. D. DeHaan, "Canine Accelerant Detection Teams: Validation and Certification," *CAC News* (Summer 1994); M. E. Kurz et al., "Evaluation of Canines for Accelerant Detection at Fire Scenes," *Journal of Forensic Sciences* 39 (November 1994); R. Tindall and K. Lothridge, "An Evaluation of 42 Accelerant Detection Canine Teams," *Journal of Forensic Sciences* 40 (July 1995); D. M. Gialamas, "Enhancement of Fire Scene Investigations Using Accelerant Detection Canines," *Science and Justice* 36, no.1 (1996): 51–54.

55. D. M. Gialamas, "Is It Gasoline or Insecticide?" *CAC News* (Spring 1994): 16–20.

56. J. Lentini, C. Cherry, and J. Dolan, "Petroleum Laced Background," *Journal of Forensic Sciences* 45, no. 5 (2000): 965–89; J. D. DeHaan, "Our Changing World: Part 3 Is More Sensitive Necessarily More Better?" *Fire and Arson Investigator* 53, no. 2 (January 2003).

57. J. L. Sanderson, "Science Reconstruction," *Fire Findings* 9, no. 1 (Winter 2001): 12; DeHaan, "Are Localized Burns Proof of Flammable Liquid Accelerants?" 90.

58. *ASTM E1188; ASTM E1459.*

59. J. D. DeHaan and F. A. Skalsky, "Evaluation of Kapak Plastic Pouches," *Arson Analysis Newsletter* 5, no. 1 (1981).

60. D. C. Mann, "In Search of the Perfect Container for Fire Debris Evidence," *Fire and Arson Investigator* (April 2000): 21–25.

61. J. Demers-Kohl et al., "An Evaluation of Different Evidence Bags Used for Sampling and Storage of Fire Debris," *Canadian Society of Forensic Science Journal* 27, no. 3 (1994).

62. W. D. Kinard and D. R. Midkiff, "Arson Evidence Container Evaluation II: Kapak Bags, A New Generation," *Journal of Forensic Sciences* (November 1991).

63. R. E. Tontarski, "Evaluation of Polyethylene Containers Used to Collect Evidence for Accelerant Detection," *Journal of Forensic Sciences* 28, no. 2 (1983).

64. D. C. Mann and W. R. Gresham, "Microbial Degeneration of Gasoline in Soil," *Journal of Forensic Sciences* 35, no. 4 (1990).

65. J. Deans, "Recovery of Fingerprints from Fire Scenes and Associated Evidence," *Science and Justice* 40, no. 3 (2006): 153–68; N. Nic Daéid,

Fire Investigation (London: Taylor and Francis and CRC Press, 2004).

66. "Testing Molotov Cocktails for DNA," *Fire Findings* 12, no. 4 (Fall 2004): 14.

67. R. E. Tontarski, "Using Absorbents to Collect Hydrocarbon Accelerants from Concrete," *Journal of Forensic Sciences* 30, no. 4 (1985); D. Mann and N. Putaansuu, "Alternative Sampling Methods to Collect Ignitable Liquid Residues from Non-porous Areas Such as Concrete," *Fire and Arson Investigator* 57, no. 1 (July 2006): 43–46.

68. J. R. Almirall et al., "The Detection and Analysis of Ignitable Liquid Residues Extracted from Human Skin Using SPME/GC," *Journal of Forensic Sciences* 45, no. 2 (March 2000): 453–61; M. Darrer et al., "Collection and Persistence of Gasoline on Hands," paper presented at EAFS, Cracow, Poland, September 2000.

69. S. A. Coulson and R. K. Morgan-Smith, "The Transfer of Petrol onto Clothing and Shoes While Pouring Petrol around a Room," *Forensic Science International* 112, no. 2 (August 14, 2000): 135–41.

70. R. Morgan-Smith, "Persistence of Petrol on Clothing," paper presented at ANZFSS Symposium of Forensic Sciences, Gold Coast, Queensland, Australia, March 2000.

71. S. Coulson et al., "An Investigation into the Presence of Petrol on the Clothing and Shoes of Members of the Public," *Forensic Science International* 175 (2008): 44–54.

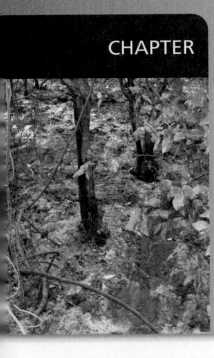

8

Wildland Fires and Their Investigation

OBJECTIVES

After reading this chapter, you should be able to:

- Identify the common causes of wildland fires.
- Explain the basic concepts of fire behavior and spread of wildland and grass fires over uniform ground cover with varying degrees of slope and wind conditions.
- Describe the investigative methodology for determining the origin of wildland fires.
- Explain how to conduct a fire scene search of a typical "V" pattern on a hillside.
- List the typical burn patterns seen at wildland fire scenes.
- Explain several techniques used to document wildland fire investigations.
- List several examples of common sources of ignition for wildland fires.
- Understand the concepts of the collection and preservation of physical evidence, including documentation.

Fires in open land covered with grass, brush, or timber are often termed *wildland fires*. Although they are often terrifying in their destructive power and intimidating in their coverage, they begin, like almost every other fire, with suitable fuel and a small, localized source of ignition. In some respects, the investigation of a wildland fire is simpler than that of a structure fire, because the fuels involved are generally limited to naturally occurring vegetation. They are ignited by some finite source of heat—natural, such as lightning; accidental, such as a discarded match; or incendiary, with an intentional ignition device. Some evidence of the source of ignition will remain unless, as in a fire started with a lighter, the source has been removed from the scene.

The causes of wildland fires are more varied than for structure fires. The California Department of Forestry and Fire Protection (CDF), which annually responds to an average of over 5,500 wildland fires in its direct protection area, reported 11 major categories for identified causes, as shown in Table 8-1.[1]

TABLE 8-1	Wildland Fire Causes in California, 2006–2008			
CAUSE	2006	2007	2008	PERCENT (3-yr)
Arson	319	227	220	6.4
Campfire	113	41	23	1.5
Debris burning	455	490	431	11.5
Equipment use	1,237	489	401	17.7
Lightning	237	126	332	5.8
Miscellaneous	627	1,061	1,115	23.3
Playing with fire	40	12	8	0.5
Power line	130	27	20	1.5
Railroad	19	0	4	0.2
Smoking	87	84	62	1.9
Undetermined	986	969	908	23.8
Vehicle	555	84	69	5.9
Total	4,805	3,610	3,593	100.0

Source: 2006–2008 Historical Wildfire Activity Statistics (Redbooks) (Sacramento, CA: California Department of Forestry and Fire Protection, 2008).

Unlike with structure fires, however, there are complications in retracing the progression of the fire by variable environmental conditions—fuels, wind, weather, and terrain. The large scale of many wildland fires results in major firefighting efforts that can disturb or destroy delicate evidence. Because of the fire, an area of interest may remain inaccessible to the investigator for days, during which time the evidence is exposed to the elements, further diminishing its value. All fire investigations require thorough and systematic examination of the suspected area of origin and logical and analytical assessment of the evidence found, and wildland fires are no exception. The investigator who understands fuels, fire behavior, and the effects of environmental conditions is in a better position to interpret the subtle and sometimes delicate signs of fire patterns in wildland fires, and therefore is better able to identify the origin and cause, no matter what type of fire is involved.

Fire Spread

The spread of a wildland fire (and thereby the means of retracing its development back to a point of origin) involves consideration of many more factors than the typical structure fire. These factors include the following:

- Topography
  - Slope (and the local winds slopes can engender) and
  - Aspect (fuels on south-facing slopes in the Northern Hemisphere are exposed to more sun than those on the north-facing slopes and are therefore drier and able to support larger, faster-moving fires)
- Terrain (natural noncombustible barriers of rock or bare soil versus chimneys (defiles) that can draw fires faster upward
- Weather
  The weather obviously plays a much more important role than in most structure fires, both before a fire in drying out fuels and during the fire. Weather conditions include wind direction, speed, and variability, as well as temperatures and relative humidity (as discussed in Chapter 3)[2]. Since these conditions affect both ignition and spread, they must be documented for times before and during the fires as thoroughly as possible
- Fuels
  Fuels can be extremely varied and must be documented (both in the fire area and in adjoining areas). While the names can vary, the general categories are displayed in Table 8-2.

Each category supports a different type of fire. The duff and surface litter most often burns as a slow smoldering fire, only contributing when strong winds can kick up burning embers to produce airborne fire brands. Grasses, shrubs, and slash can all support fast-moving flaming fires that can ignite the **ladder fuels**.[3]

The crown, being a porous array of fine fuel elements surrounded by lots of air, can support extremely fast burning, intense, flaming fires. The canopy can be nonhomogenous, as well. Ladder fuels constitute the bridge that can connect fires started in ground litter to the canopy or connect a crown fire to the next susceptible fuel array. The oncoming fire first evaporates (by radiant or convective transfer) the water in the exposed foliage, then evaporates the volatiles present, and then pyrolyzes the hemicellulose, cellulose, and lignin components into combustible gases.[4] The seasonal stage of the foliage (green or live, mature, cured or dead) has a significant effect on the fuel moisture and thereby on the fuel's ignitability and rate of spread of the fire. The volumes and arrangements of fuels (continuous, intermittent, or isolated) also are important variables that the investigator must consider. Rothermel has offered guidance regarding the interpretive

ladder fuels ■ Intermediate height fuels between the ground litter and the crowns of trees overhead.

TABLE 8-2	Fuel Categories in Wildland Fires
CATEGORY	**EXAMPLE**
Duff	Decomposed organic layer (humus) on the soil
Ground or surface litter	Dropped leaves Dropped branches Needles and twigs
Ladder fuels	Intermediate-height grasses Shrubs [less than 2 m (6 ft) in height] Seedlings Small trees Lower branches Loose bark
Crown	Canopy of foliage leaves, needles, and fine twigs (more than 2 m above ground)
Slash	Branches and foliage left as heavy ground cover after logging operations[a]

[a]From R. C. Rothermel, *How to Predict the Spread and Intensity of Forest and Range Fires*, Gen. Tech. Rep. INT-143 (Ogden, UT: U.S. Dept. of Agriculture, National Wildfire Coordinating Group, June 1983), 10–12.

complexities of these factors and has developed nomograms for predicting the effects of numerous such variables on fire behavior.[5] Anderson has published a photographic guide to identify the fuel model of various fuel arrays.[6]

The wildland fire investigator must be prepared to take all these factors into account when evaluating the spread of a large fire with the intent of backtracking it to its origin and then evaluating various ignition mechanisms there. The proper investigation therefore entails the determination and documentation of as many of these factors as possible.

Fuels

By definition, wildland fires consume grass, brush, and timber but will also involve structures and even vehicles and equipment as subsidiary fuels. Although the basic fuels are all cellulosic in nature, they can vary considerably in their fire behavior because of their volume, density, arrangement, and moisture content. Short dry grasses, for instance, tend to flash over once, contributing little to the fire (except to carry it into denser fuels), while short green grasses may not burn at all during their first exposure to the fire but will dry out, ready to provide fuel for a reburn of the same area. Long dry grasses may be burned off at their tops if the fire is passing quickly across dense growth. The remaining grass stems then provide fuel for a second or "following" burn. If a slow-moving fire moves into tall grass, it is more likely to burn the stems away near the ground, allowing the tops to fall. This behavior is most often seen in areas where the fire is **backing** into the wind or down a hill or **flanking** or moving laterally at roughly right angles to its major direction of progression.

Brush and heavy timber may also burn in the same ways. Fire may flash across and upward through the leaves and finer twigs (called **crowning**), leaving heavier portions to ignite later as fire conditions change. This condition is especially notable downwind and uphill from a fast-moving fire, in the area sometimes called the *head* for obvious reasons. Radiant and convective heat from an intense nearby fire can raise the temperature of an entire bush or tree or its volatile resins to its ignition temperature, whereupon it "explodes" in flames.

backing ▪ Slow extension of a fire downslope or into wind in the opposite direction of its main spread.

flanking ▪ Lateral spread of fire in a direction at right angles to the main direction of fire growth.

crowning ▪ Rapid extension of fire through the porous array of leaves, needles, and fine fuels above 2 m.

FIRE SPREAD

The nature of the fuels will have considerable effect on the spread of the fire and its intensity. The higher resin content of firs and pines causes them to burn faster and more fiercely than many hardwoods. Few trees are resistant to the intense flames of a fully developed forest fire. Some, like the firs, will die when the *cambium*, the living layer of cells just beneath the bark, is breached even though damage to the rest of the tree may be slight. The critical temperature for cambium death is about 60°C (140°F), so as a general rule, the thicker the bark, the greater the tree's resistance to fire death.[7] A few trees, like the giant redwoods of the Pacific Coast, resist even serious fires with a thick, almost noncombustible bark. (Some species actually *require* a fire to open their cones to allow the sowing of new seedlings.)

Finely divided cellulosic fuels, such as grasses with a high surface-to-mass ratio, will, of course, ignite easily when dry and allow the spread of fire by fast flashing across the tops or by slow, creeping, smoldering spread along the ground. In semiarid regions, the brush (sometimes generically termed *chaparral*) protects its living cells from death via evaporation by developing a high oil or grease content in the wood. The high oil or grease content shields the cells from the hot, dry environment allowing the cells to maintain a normal moisture level. The oils or greases in this brush, however, provide an exceptionally high-quality fuel for a fire. A fire in this type of fuel develops extremely high temperatures that cause the oils of nearby plants to volatilize at a rapid rate, generating clouds of flammable vapors. The passage of a fire through such brush is awesome in its ferocity.

MOISTURE CONTENT

The general response of wildland fuels to changes in relative humidity is categorized in terms of the time required to equilibrate. Small-diameter fuels (grasses, foliage, twigs) are considered to be "<1-hour" fuels. Larger diameter branches are "1–10 hour" fuels. Timber is rated 10–100 hours.

The moisture content of fuels is of more interest to the wildland fire investigator than to the structure fire specialist, because it can vary considerably with the environment. The effects of surface moisture and internal moisture content were treated in a previous chapter on solid fuels. Fuels having excessive surface moisture will, of course, be more resistant to ignition on the first pass of a fire. The heat generated can evaporate the moisture, however, preparing the fuel for more ready ignition should the fire alter its course and reburn the same area. The same holds true to a lesser extent for internal moisture content. This is water trapped within the plant (fuel moisture) that reaches equilibrium with the humidity of the surrounding air through the life processes of the plant's living cells. Dramatic changes in the temperature and humidity in the surrounding air can affect this cellular water, only at a much slower rate than the surface moisture.

The passage of fire or even hot gases can, however, reduce the moisture content of leaves, brush, and needles to materially increase their combustibility in a short time. The relationship of atmospheric humidity to fuel moisture content was examined with regard to nonliving fuels in Chapter 3. The relationship is more complex with living cellulosic materials, since wind and sun exposure can accelerate moisture losses. Fuels that are difficult to ignite with "weak" sources such as cigarettes in still air at moderate temperatures and humidity become much more susceptible with increases in atmospheric temperature and wind speed or decreases in humidity.[8] Weather and exposure conditions in the area of origin just prior to the start of the fire are therefore very important data to have.

INTENSITY OF WILDLAND FIRE

The intensity of a wildland fire is often measured in kilowatts per meter of a fire line. This intensity can range from 40 kW/m for a smoldering fire in duff to 100–800 kW/m for backing fires in surface or ladder fuels, 200–15,000 kW/m for advancing "head" fires, and up to 30,000 kW/m for a single-front crown fire in timber.[9] Crown fires have been measured to produce peak incident heat fluxes of 40–150 kW/m^2 at distances of 10 m (33 ft) from experimental fires in stands of 10-m tall jack pine–black spruce trees.[10] The durations of such peak exposures were typically 30 to 90 seconds.

Fire Behavior

Virtually all wildland and grass fires begin as a small source of flame or smoldering fire—match, cigarette, electrical arcs, hot fragments or embers, lightning, and the like—that kindles easily ignitable fuels in the area. As the small fire aimlessly explores sources of fuel its growth and direction of travel will depend on availability of fuel, direction and strength of wind, and the slope of the surrounding terrain. A small fire burning through uniformly distributed fuel on level ground will grow slowly in every direction in which fuel is available if there is no external wind, as illustrated in Figure 8-1a. Hot gases generated by the fire rise at the center of the fire and draw in air from all directions along the ground (as shown in Chapter 3). The fire must then "back" into surrounding fuel against its own induced draft, and its growth rate is limited as a result.

If this same small fire and fuel are arranged on a slope, convective entrainment draws more air from the bottom of the incline, and the spread of the fire is then enhanced by both this draft and the exposure of more fuel to hot gases and flames on the uphill side of the fire. The fire progresses more rapidly up the slope and travels predominantly in this direction, following the available fuel. This gives the fire direction, and the head of this advancing flame front grows in energy as it moves into more and more fuel. The draft caused by the fire grows in strength as the fire grows, thus perpetuating the development in a chain reaction. Fire winds are those induced by the buoyancy or convective entrainment of the fire plume. In a large wildland fire, this draft or entrainment effect is great, and a very strong wind is created by the fire itself. In a fully developed fire, this can become a firestorm of superheated gases and flames igniting all fuels in its path. The fire not only spreads uphill in a fan-shaped pattern as a result of these dynamics, it can progress downhill as well (by radiant heating or direct flame contact). If sufficient suitable fuel is available, the fire will back slowly down the slope, even against a fire-caused draft. Burning

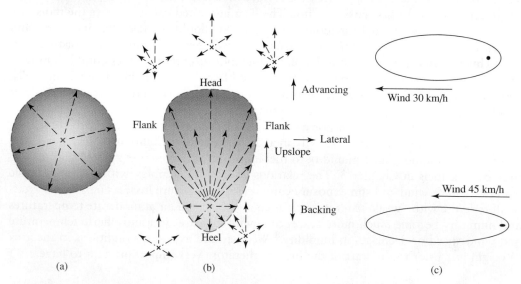

FIGURE 8-1 (a) Typical fire spread in uniform ground cover with no wind on level ground. (b) Fire spread in uniform fuel on moderate slope with no prevailing wind. Although some spot fires may occur downslope from the origin, they will predominantly be upslope. (c) The higher the wind velocity, the more elongated the fire area becomes and the closer the origin to the upwind heel.

embers, falling timber, and loose, burning debris can also fall downhill, starting secondary fires, sometimes at a significant distance from the actual fire (see Figure 8-1b).

EFFECT OF WIND

Natural winds driven by high- and low-pressure atmospheric areas are called *meteorological winds*. Any prevailing atmospheric wind will affect the progress of the fire, particularly during its erratic, early development stages. It will force the flames and hot gases in one direction or spread an uphill fire sideways across the slope, modifying its fan-shaped pattern of general development. The higher the wind speed, the more narrow the elliptical burn pattern becomes, with the origin becoming closer to the downwind end (as shown in Figure 8-1c).[11] Very strong winds can force a fire downhill or cause a fire to reburn previously flashed-over areas where unburned fuel remains. An external wind will have the greatest effect on the downwind edges and tops of the forest, as it is quickly dissipated by the porous canopy.[12]

Warm, dry winds flowing across a mountain ridge can produce very strong downslope winds on the leeward side (especially in valleys and passes). These are sometimes called *foehn winds* (referred to locally as Chinook or Santa Ana winds) and have been responsible for many major firestorms.[13] Diurnal winds are caused by daytime solar heating and night-time cooling. They flow upslope during sunny days and downslope after sunset.

EFFECT OF TALL FUELS

The growth of fires in tall fuels (>2 m; 7 ft) becomes more complex than growth in structures fires because the fire grows through a porous array of small fuel elements—leaves, needles, twigs—rather than across a simple surface. As the buoyant (upward) flow and wind-driven flow carries flames through the foliage the fire can quickly grow to enormous intensity, the process called *crowning*. Fully developed crown fires in timber can reach heat release rates of 30 MW/m of fire line.[14] This intensity causes remote ignition by radiant heating and generation of wind-driven flaming brands.

- These slow-moving (essentially backing) fires are low in energy and often leave a lot of unburned or slightly burned material on the ground and on vertically arranged fuels above the average flame heights. These lightly burned materials are sometimes referred to as *color*. This phenomenon is commonly seen in an origin area, with the fire starting with low intensity at ground level and then transitioning into higher fuels as it moves away from its origin.

OTHER EFFECTS

- Ambient wind will modify the pattern by adding an additional spreading component, so that the fan-shaped pattern on the hillside will deflect to one direction or the other, and a predominant direction of travel will be created on level ground.
- In the circumstance of a strong downhill wind, the fire will burn downhill only to the degree that the ambient wind can overcome the fire's own tendency to burn uphill.
- In a fire having an extended perimeter, the direction of burning may vary locally in almost any direction, depending on the interdependence of the fuel, topography of the terrain, the air currents created by the fire itself, and the ambient wind.
- High winds tend to produce long, narrow burn patterns.

Determination of Origin

The basic premise of wildland fire investigation is the same as that for structure fires—determine the area of origin first, then establish what fuel was present and what source of heat ignited it. Once these factors are identified, the investigation can then focus on what actions and series of events brought the ignition source and first fuel together to cause the fire and who the responsible party was (i.e., was the fire human caused?). Because terrain, weather conditions, and fuel distribution all play important roles in wildland fire behavior, those must be documented as part of the investigation. The determination of area of origin is generally carried out by examining the fire scene for indicators of direction of fire travel. Once a pattern of these directional indicators can be established, it can be used to point back toward the area of origin. Such interpretive signs should be used strictly as indicators—none are foolproof, none are fail-safe. With the complexity of exterior fire behavior, modified as it is by weather, fuel distribution, terrain, and suppression activities, it is the *predominant pattern* of indicators that must be sought out, not a single isolated one here or there.

INVESTIGATION METHODOLOGY

Just as with structure fire examination, wildland fire scene examination proceeds in accordance with a methodology that follows the principles of the scientific method.

- Receive the assignment: know role and responsibility (usually origin and cause with report); be aware of civil and criminal outcomes and type of report needed.
- Prepare for the investigation.
  - After initial survey of scene and interviews, establish most suitable investigation plan.
  - Obtain necessary personnel and equipment.
- Conduct the examination.
  - Interview witnesses.
  - Examine scene to collect data.
  - Document investigation via photographs, diagrams, maps, evidence collection.
- Collect and preserve necessary physical evidence and get analysis as needed.
- Evaluate and analyze available data, test possible hypotheses, and establish final conclusions. These conclusions will usually entail establishing origin, cause, mechanism(s) of fire spread, and responsibility for the fire's ignition.

The most widely accepted protocol for wildland fire investigation today is described in the NWCG manual *Wildfire Origin and Cause Determination Handbook*.[15] It is the foundation for the FI 210 course, "Wildfire Origin and Cause Determination," offered nationally by a wide variety of state and federal agencies in the United States, Canada, Australia, and other countries.

Prior to the start of the scene examination, the investigator must be prepared with knowledge of the local area including topographical maps, people with reliable local knowledge, and even fire histories of previous events in the area. Prior to scene arrival, the investigator can record observations (weather conditions, fire and smoke conditions, equipment, vehicles, and people in the area) and assess available manpower and expertise. On arrival, the investigator must assure that the suspected area of origin is protected (from vehicles, foot traffic, retardant or water drops, water runoff). Interviews with witnesses are a critical first step to minimize wasted search time.

FIRST EVALUATION

It may seem to be a daunting task to seek out a small area of origin, say of a few square meters in a whole forest of burned trees and brush, which may be hundreds or thousands of hectares (acres) in size. The wildlands fire investigator is no more interested in the damage wrought by the later stages of a major fire than is the structure fire specialist concerned about buildings or vehicles burned by later stages of a large structure fire. Almost every wildland fire will have been observed in the early stages of its development by someone—camper, sightseer, hiker, ranger, pilot, or firefighter. Although these people may not be easy to find, it will be much easier than spending days trekking over the later-burned areas spotting indicators of fire direction. All subsequent investigation is then limited to the general locality that witnesses indicate was first burned regardless of how far it spread later. The investigation should follow a logical scheme that allows thorough and accurate determination while preserving the evidence remaining (not to mention the investigator's time). Several of the first firefighters present should be questioned as to wind direction and weather conditions, as well as to the extent, location, and behavior of the fire. First-in fire companies are often aware of where the fire appears to have begun. They are often trained to protect such areas (which may have already burned out) with flags or barrier tape.[16] Interviews with local residents or law enforcement witnesses may help establish an area of origin or identify unusual vehicular or pedestrian traffic or equipment use in the area.

OTHER SOURCES OF INFORMATION

Infrared photography or video imaging from aircraft has been used to map wildland fires and detect hot spots within a fire area. If these are taken over a time interval, the progression of the fire can be retraced via these images. Images from earth-resources satellites can show smoke movement as well as thermal images. These images, when combined with topographic and weather maps of the fire area, may permit retracing the fire's development back toward its origin and eliminating certain mechanisms of fire spread. Establishing multiple origins that are separate in place but coincidental in time and that could not be the result of natural fire progression by airborne embers would go a long way to demonstrating an incendiary origin for those fires. In the Oakland Hills fire of 1991, infrared imaging from a C130 aircraft and mapping via global positioning satellite (GPS) and transponders allowed the rapid assessment of the fire's extension (through fog, smoke, and darkness) and its damage.[17] American Defense Support Program (DSP) satellites today offer a 10-second "revisit" rate of a target area, allowing for minute-by-minute detection and monitoring of wildland fire events.[18] The use of remote-controlled surveillance drones is also reportedly being evaluated by DSP researchers.

THE SCENE SEARCH

The scene search involves a perimeter search and then a more focused search of areas closer to the apparent area of origin.

Perimeter Search

First, the general area of the first-observed fire is surveyed to determine the outer perimeter of the search area. The general shape of the burn pattern often helps investigators focus on a likely area of origin. Depending on the nature of the terrain and the extent of the fire, a fan-shaped or V-shaped pattern may be visible from a short distance away. The apex of this pattern can then be selected as the focal point for the search pattern (see Figure 8-2). A barricade of rope, twine, or plastic banner is arranged to control access and limit further damage to the physical evidence from personnel, vehicles, or the heavy equipment so common to wildland fires. The general area is mapped out on a rough sketch; a topographic map showing hills, valleys, and similar features of the area may be very helpful at this time. This general origin area can be identified through witness statements, fire behavior observations, or gross indicators of direction. The outer portions of the secured area (away from the apparent origin) are searched by whatever systematic pattern best suits the terrain. These early sweeps are for general indicators, observing the larger, more obvious signs (sometimes called *macro-scale* indicators, to be described later), if possible, working from one side or flank of the burn pattern to the other. Indicators can then be mapped onto a drawing of the scene as they are found. It is recommended that the perimeter of this general area of origin be searched twice from opposite directions.

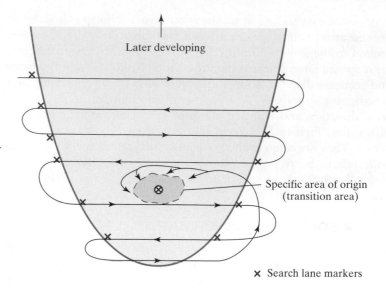

FIGURE 8-2 Typical "V" pattern of burn on hillside with suggested search pattern. This pattern starts from outside the margins of the burn pattern and works progressively toward the earliest portions of the fire.

Later developing

Specific area of origin (transition area)

× Search lane markers

This perimeter search should focus on the macro indicators and examination of adjacent *unburned* areas to establish pre-fire fuel type and distribution. Unless the scene is very large, this search should be done on foot, but always slowly, methodically, observing and recording. Photographs should be taken of all significant indicators (macro and micro). It may be helpful to keep track of this searching process by using visible markers or flags (color-coded to reflect advancing fire progression, lateral progression, flanking, backing, and items of evidence). A standardized color flag system has been developed to help denote fire movement and evidence location, as seen in Table 8-3.

These markers can help keep track of the fire spread and provide a visual reference of the overall fire scene.[19] Keeping a written fire-spread map that corresponds to this flagging system is also a good idea. Remember that the direction determinations you make are based on the indicators you have found, not on the flags added.

Focused Search

The search can then progress to areas of interest closer to the apparent area of origin. As the investigator's search approaches the specific area of origin, there is often a change in the burn pattern indicators being examined, from the larger-scale indicators to the indicators (called *micro-scale* indicators) characteristic of the origin area. Note that the focus is not on a single spot but on a small area of several square feet or yards. Wherever fires of multiple origin merged, the areas of erratic fire travel around each will remain. These areas contain the smallest and most delicate indicators and, of course, any remaining physical evidence. It is always best to minimize all traffic in these areas until an inch-by-inch examination can be carried out. Burn indicators on larger brush and trees will usually not be usable unless the fire took on a direction and some intensity. In the area surrounding

TABLE 8-3	**Standardized Color Flag System**
FLAG COLOR	**INDICATOR**
Red	Advancing fire spread
Yellow	Lateral (flanking) fire spread
Blue	Backing fire spread
Green, white, other	Physical evidence

the point of origin, the smaller indicators in this specific origin area may show considerable variation as the fire sought fuel and responded to the conditions of wind and topography. For this reason, once the area of origin has been located, it is usually necessary for the examination to be carried out on hands and knees looking at individual pebbles, leaves, and twigs for the resulting signs, and, ultimately, for the cause of the original ignition.

As the **area of transition** (to directional spread) is detected, the investigator should proceed slowly *around* this area, but doing so from outside the burn itself or the suspected area of origin. It is best to do the hands-and-knees search working in from the advancing fire area, continuing to read the micro-scale indicators. The most common method is to grid or lane this area off, using twine on stakes. Because the transition from random to directional (driven by wind, fuel or terrain) may be very subtle, it is easier to overrun the origin if it is approached from the backing or heel area; approaching from the advancing area is recommended. Each lane or grid can then be searched visually, with magnification and a magnet. The detail search is often carried out using a straightedge (ruler) as in Figure 8-3 to help focus the visual examination. A handheld or headband magnifier or 2X–3X magnifying glasses are often useful at this point. The debris is then swept with a handheld magnet in a plastic bag to collect iron-containing debris. A sensitive metal detector may be swept over the suspected area to locate metal vehicle or tool components, or ignition devices. A small kitchen baster, photographer's bulb, or blowing by mouth can be used to direct small puffs of air to move surface debris and light ash. Heavier covering debris can be moved using a soft paint brush, tweezers, or long forceps. If the ash layer is deep, it will need to be removed by gently blowing it away after the first search, possibly revealing additional indicators beneath.

area of transition ■ Mixture of fire directional indicators in a wildlands fire.

FIGURE 8-3 Detail search for micro-scale indicators using a straightedge. *Courtesy of Lamont "Monty" McGill, Fire/Explosion Investigator.*

FIGURE 8-4 Char depth will be deeper on the side (with uniform fuel) facing the fire.

Fire travel →

Char depth deeper on side facing fire

Remember that suppression and control efforts can vastly complicate the reconstruction of a wildland fire. Bulldozers can build firebreaks to divert "normal" progress. Tanker air drops can cool off areas, allowing nearby areas to burn. Most important, secondary "backfires" or "burnouts" may be set deliberately to deprive the main fire of fuel. All these can dramatically affect the appearance of burn indicators. Wind and rain may quickly compromise many useful indicators, so the investigation should be carried as soon as it can safely begin.

BURN INDICATORS

Because the effects of the nature of fuels and fire conditions in a wildlands fire are fairly predictable, there are a number of fairly reliable indicators that can be used to determine the direction of fire spread past a given point.

■ Charring of a tree trunk, plant stem, or fence post is deeper on the side facing the oncoming fire. Char depth can be checked with a knife blade, pencil point, ice pick, or screwdriver. It is strictly a comparative indicator, so absolute depth is of little consequence (Figure 8-4).

■ Destruction of a bush or tree may be more extensive on the side facing the fire (Figure 8-5). Fire will progress upward as it moves forward through the foliage, leaving an interface between burned and unburned portions as the buoyancy of hot gases causes the flame front to rise as it moves forward.

■ The beveling effect of a fast-moving fire may influence the appearance of branches or twigs remaining upright. Twigs and branches facing the fire may have flat or rounded ends, while those facing away from it (downwind) will be tapered or pointed, as shown in Figure 8-6.

FIGURE 8-5 Partial combustion of trees and bushes can reveal the fire spread. Fire will progress upward as it moves forward through the foliage, leaving an angled interface between burned and unburned portions. Remaining leaves or needles may be frozen in position in the direction of the prevailing wind.

Remaining leaves or needles may be "frozen" in position in the direction of the prevailing wind

Fire spread →

Bare branches

Fire spread →

or less commonly →

FIGURE 8-6 Beveling effects of oncoming fire blunt the ends of branches facing the fire and taper those away from it.

FIGURE 8-7 In a fast-moving fire, the fire pattern on a tree will start lower on the trunk side facing the oncoming fire, even on a slope. The charred area *may* wrap completely around and extend up the "back" side. (See Figure 8-8).

■ A fast-moving fire creates a draft around large objects, which creates an angled pattern around tree trunks, posts, plant stems, and the like (Figure 8-7). On foliage crowns, this angle of fire pattern may also be evident when the tree or burn is viewed from the side. It may reflect the fire behavior of the flames "coming in low and exiting high."

■ Fast-moving (either fire- or wind-driven) fires moving past large-diameter vertical tree trunks or power poles create a low-pressure area on the downwind side. This low-pressure area draws the passing flames from local fuels into a vertical vortex, creating a laminar flame. The laminar flame lasts longer than the passing flame front and can induce significant fire damage to the downwind side, extending several times the height of the flame front.[20] (See Figure 8-8.) A very slow moving fire, especially one that is "backing"

FIGURE 8-8 This fire (moving from right to left) induced a low-pressure zone on the leeward side of the center tree trunk which caused the formation of a laminar flame entraining the flames in the ground cover at its base. This flame lasted longer than the passing flame front and produced much higher burn damage on the downwind side. *Used with permission of Donna Deaton, Deaton Investigation, LLC.*

FIGURE 8-9A In a slow-moving or "backing" fire, the burn pattern will be *roughly* parallel to the ground. Wind-driven patterns often exhibit a rising char pattern on the leeward side. Accumulations of ground litter may produce char patterns on the upslope side of tree trunks and utility poles.

FIGURE 8-9B Observed low-intensity test fire moving from right to left in light ground fuel with a light (5 mph) wind. Char pattern is higher on the leeward (downwind) side of the trunks. *Courtesy of Marion Matthews U.S. Forest Service.*

against the wind or down a slope, will create a burn pattern *approximately* parallel to the ground (as depicted in Figure 8-9a). Note that such patterns can be influenced by local fuel load such as needles, leaves, and debris around the base of the tree, leaving a pattern higher on the side facing the oncoming fire or the upslope side (which catches such debris). This may also be the result of momentum flow of the fire against the vertical surface. Wind-driven (either from fire plume convection or atmospheric wind) fires often produce a rising char pattern on the leeward side of the trunk or pole, as in Figure 8-9b.

■ If the vertical stem is burned away, the remaining stump will be beveled or cupped on the side facing the fire, as shown in Figure 8-10. This will be true even for stems of weeds. When the back of the hand is brushed lightly against such stubble, the beveled tops allow the skin to pass smoothly across in the same direction as the fire but resist with sharp points a hand passing in the opposite direction.

FIGURE 8-10 Beveling or cupping on vertical stems or stumps will point back in the direction of the fire's origin.

FIGURE 8-11A Deposits left on upper left sides of large rocks as a result of observed fire spread (through coarse grass) from top to bottom. *Courtesy of Marion Matthews U.S. Forest Service.*

FIGURE 8-11B Micro-scale indicator deposits on small stone amid burned grass confirm fire was moving from right to left. *Used with permission of Donna Deaton, Deaton Investigations, LLC.*

■ Rocks, signs, and other noncombustibles will show greater heat discoloration (staining) on the side facing the fire, as pictured in Figures 8-11a and b. Lichen, moss, and close-growing grass may survive on the side away from the fire. Extensive exposure may burn all soot off (leaving a light-colored or white surface) on the side facing the oncoming fire.

■ Large rocks, walls, and large objects provide a barrier to the flow of the flames. The area downwind of the obstacle may escape damage completely or be burned later as the fire changes direction (Figure 8-12). Intense flame contact may cause spalling of stone or concrete on the upwind or facing side.

■ Tall weeds and grasses, when burned by a slow-moving fire (particularly as it spreads laterally along the flanks or backs downslope or against the wind), will be undercut by the fire moving along the ground. The stems, if vertical, will then fall toward the fire with the heads of grass stems pointing back toward the fire origin, as in Figures 8-13a and b. This effect is highly dependent on the wind conditions prior to and during the fire. Tall grass that is already matted down pointing away from the fire will fall in that direction, so there may be conflicting directions from such indicators.

■ The height of remaining stems and grasses is roughly proportional to the speed of the fire. The effect is dependent on the moisture content of the plants and will not be visible in areas that have reburned.

■ In the immediate vicinity of the origin, the fire will not have developed any particular direction, and so the indicators will be variable or contradictory. Fuels in the

Fire spread

Rock wall

Unburned "protected" grasses

FIGURE 8-12 The flow of a fast-moving fire around an obstacle often leaves protected vegetation on the downwind or "lee" side of the obstacle.

FIGURE 8-13A Tall stems of grass or weeds point predominantly back toward the source of the fire in the absence of a strong prevailing wind.

FIGURE 8-13B Fire is backing to left (into wind). Note the long stems have burned off at the base and stems have fallen to the right. *Courtesy of John D. DeHaan.*

spot fires ■ Fires started by airborne embers some distance away from the main body of the fire.

form of grasses or weeds may still be upright or only partially burned. Large fuel may be burned only at its base with fire extending into upper branches only as the fire moves away from its origin.

■ Extension of a fire beyond a barrier, for example, a road or river, will cause the appearance of a brand new origin. This is particularly true when the "new" fire is started by airborne embers or burning debris (brands) from an established fire. (Such new fires are sometimes called **spot fires**; see Figure 8-14.)

■ By their quantity and distribution, ashes can indicate the direction of wind and the amount and nature of fuels present.

■ If the fire is very recent, the white ash formed by the fire's impact on the facing side (with more complete combustion) can be compared with the darker area of less intense fire on the leeward side. Viewing the overall fire scene from the origin may reveal the white ash deposits on exposed surfaces. Wind or rain will remove such deposits.

It should be noted that most of these indicators (particularly the micro-scale ones) require careful systematic *comparison* of one side of an object with the other, not an assessment of absolute condition. The investigator should also be aware of ventilation

FIGURE 8-14 Remote spot fires caused by airborne embers across natural barriers.

Remote spot fires caused by airborne embers across natural barriers

FIGURE 8-15 Diagram of fire scene search showing indicators for fire spread and evidence. *Courtesy of Lamont "Monty" McGill, Fire/Explosion Investigator.*

and venturi effects in high-wind conditions or around/under objects not in full contact with the ground, as they may produce contradictory indicators. Because "wind" conditions change as the fire grows or responds to slope and terrain, the investigator must consider the totality of indicators observed before concluding direction of fire spread. Reliance on a single indicator is not good practice, as it may be misleading.

Once the area has been thoroughly examined, with direction indicators carefully flagged and noted on the sketches, as shown in Figure 8-15, it is often beneficial for the investigator to stop for a few moments and reevaluate what has been found. Questions such as the following can be helpful:

- What patterns do the indicators reflect?
- What features of the terrain, weather, or the fuel load could affect the development of those patterns?
- Do there appear to be multiple origins that grew into one large fire?
- What area of origin appears to be consistent with those patterns?
- What firefighting tactics were used in the area that could have produced patterns of their own—backfires, or retardant or water drops.

A few minutes' reflection may reveal disparities in the indicators that were not apparent during the actual search. Resolving those disparities at this time will avoid having to repeat tiresome, painstaking, inch-by-inch examination on more than one indicated area of origin. Remember, until the fire develops direction as a result of wind, fuel, or terrain, it will often extend erratically in several directions, producing a pattern of contradictory direction indicators. This specific origin area is worthy of close examination as a possible point of origin. Once the questions have been resolved, one can then proceed to careful examination of the fire indicators on a micro scale in the vicinity of the actual ignition source. This allows documentation (by sketching and photography) of even the smallest details while minimizing the risk of destroying such details by kneeling or stepping on them during the search. Ultimately, the search of the specific area of origin is carried out for one major purpose—to discover the means of ignition of the first small fire.

Once the point of origin has been located and the first fuel identified (and possibly an ignition source detected), the origin needs to be thoroughly documented with photographs. It is recommended that a sampling of the specific indicators by class and category be selected at various intervals throughout the general area of origin. These indicators are photographed, documented in the photo log, and their location measured from a previously selected main reference point. This information can then be used to draw a to-scale scene diagram that corresponds with the photographic record and re-creates the origin determination process. (Refer to Chapter 7 and Appendix F for information on surveying and documenting large scenes.)

DOCUMENTATION

The recommended minimum documentation for wildland fire investigation should include a comprehensive narrative report, photographs of burn pattern indicators used, maps, diagrams, witness statements, weather information, and forensic lab analysis reports (if applicable). Photographs, photographic logs, and diagrams are discussed next.

Photographs

Photographs provide the reader of the report a clear understanding of what was seen, done, and used, and preserve critical data for review later by the investigator.

Photograph log

A photo log is essential for describing the subject of each photo and direction in which the photo was taken. This may be in the form of a scene diagram with symbols (numbered to show frame [image] number, or negative/roll number) with an arrow to show direction, or a numbered list.

Diagrams

Diagrams show the location of indicators used, witnesses, and evidence collected, as illustrated in Figure 8-15. They must include date, location, compass, initials of preparer, and documentation of dimensions and topographic features.[21] Measurements can be in one of three formats, as shown in Figure 8-16. GPS coordinates can be used to establish locations

FIGURE 8-16A Right-angle transect: baselines at right angles with end-points set by GPS and compass. Baselines are marked by string, paint, or metal tape.

FIGURE 8-16B Azimuth/baseline: single baseline (in N–S or E–W orientation) (set via compass and/or GPS). Locations noted by angle and distance from baseline or reference point.

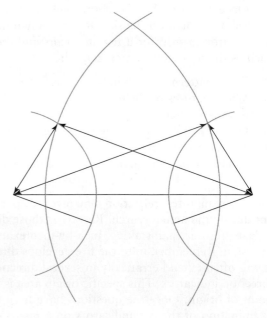

FIGURE 8-16C Intersecting arcs from two fixed reference points.

FIGURE 8-16 Methods for baseline reference at wildland fire scenes.

of evidence and possible areas of origin. These coordinates can be used to create reports and diagrams and to confirm jurisdiction.

There are several computer-aided drawing systems designed for use in wildland fire investigations. These include standardized symbols and plotting methods.

Sources of Ignition

The next step in the investigation of the wildland fire is to determine the cause when possible. In some instances this may be impossible because the evidence has been destroyed. It is always necessary to make a reasonable search for the physical remains of the ignition source and to establish a defensible cause. Ascribing such cause to "unknown factors" should be only a last resort. The causes of outdoor fires are well known in general. The National Wildfire Coordinating Group (NWCG) maintains records regarding fire cause for all wildland fires in the United States. Statistically, human-caused fires account for approximately 80 percent of all wildfires annually.[22] The California Department of Forestry reports that in over 12,000 wildland fires over a three-year period, in the average year, the leading identified causes (in descending order of occurrence) were equipment use (tractors, chain saws, weed cutters, etc.), debris burning, arson, vehicles (trucks and cars), lightning, smoking, campfire, power line, playing with fire, and railroad.[23] (See Table 8-1.) The following are examples of some of the possible causes:

- Unextinguished fires (including discarded unextinguished coals) left by campers, hunters, and others.
- Carelessness with smoking materials, including burning tobacco and matches.
- Trash burning and, in some areas, controlled grass and brush burning (including airborne firebrands).
- Sparks (hot exhaust particles) from vehicles, especially locomotives or other motor-driven equipment.
- Lightning, a major cause of timber fires.
- Power transmission lines and equipment, including birds and mammals that may come into contact with the power, short-circuit it, and catch fire, falling into nearby fuel.
- Arsonists, who may use a variety of devices but generally use a lighter or match.
- Firearms, which under certain circumstances may blow sparks of burning powder into dry vegetation, and steel-core and tracer bullets.
- Spontaneous combustion, limited to very specific types of circumstances. (See Chapter 6.)
- Overheated machinery, which may be in contact with combustibles.
- Sparks from any source, such as impacts of metal with rock.
- Discharges of static electricity.
- **Rekindling** of smoldering embers from previous controlled fires, whether they were barbecues, campfires, or previously burned timber slash or debris piles, or small (possibly unreported) brush fires. These are sometimes referred to as *holdovers*. These have been reported to have time frames of up to and exceeding 12 months.[24]

  rekindle ▪ Reignition of a fire due to latent heat, sparks, or embers.

- *Sleepers,* which are old trees or stumps containing much rotted wood internally in which a smoldering fire has been previously induced. Because of their unique burning characteristics, fire may persist in them for long periods, documented in some instances to be as much as nine months.
- Miscellaneous objects sometimes found in trash, including clear glass (cylindrical or lens-shaped in cross section) or concave reflective surfaces that can focus the sun's rays. (See Chapter 6.) These are rarely responsible for wildland fires but should be considered.

The examination of ground cover for ignition sources can be very tedious, but it is essential. The searcher is encouraged to keep an open mind, look for things that do not belong, and take frequent breaks. Circumstances and proximity of roads, trails, railroads, or signs of recent equipment use (freshly cut weeds, recent gouges in road or stone surfaces) are useful in some cases. A second search from a different direction or a second examination is recommended.

The mechanisms of most of these sources were discussed in Chapters 5 and 6, and the reader is directed there for further information.

The nature of the fuel at the origin and the relevant characteristics of the proposed ignition source must be evaluated together. Because most wildland fires involve cellulosic fuels, either open flames or hot (glowing) sources must be considered as competent.

POWER LINES

Power lines generate fires often enough so that the mere presence of such a line within or close to a fire area renders it suspect as the cause of the fire. There are several methods by which power lines may initiate a fire, including the following:

■ Transformer short-circuits and malfunctions—perhaps one of the more common and dangerous situations, causing overheated oil or molten metal droplets to fall into ignitable fuel
■ Leakage over dirty insulators and supports when moist, giving rise to "pole top" fires
■ Fallen wires that arc with the ground or objects on it
■ Arcing between conductors that accidentally come into contact (typically in high winds)
■ Grounding of in-place conductors by external objects such as fallen trees, branches, or birds
■ Unprotected fuses dropping hot metal
■ Tree limbs' coming into contact with conductors
■ Animals' coming into contact with equipment and shorting it out (see Figure 6-34)

The mere presence of a power line in the neighborhood of a fire should never be taken as the cause without a thorough investigation. It is often easier to blame the obvious than to prove the matter. Despite the inherent hazard of power lines as a source of ignition, some fires may have been blamed on them when careful analytical investigation would have revealed the fire to have been started by other means.

The wildland fire investigator should be familiar with basic power pole equipment such as transformers, fuses, breakers, and switches, as shown in Figure 8-17, and how they fail. High-voltage conductors that come into contact with soil can produce glassy fulgurites, which are produced by the high-temperature fusion of minerals in the soil (see Figure 8-18). They will be found just below the surface of the soil by detail search and sifting.

LIGHTNING

Lightning is another common cause of timber and grass fires. Lightning strikes can shatter or explode trees or limbs, peel bark from trees, split rock, or fuse soil or sand into glassy fulgurites at the point of contact (similar to those in Figure 8-18 but much larger in size.) Such signs should be sought out during the search for the fire origin. In addition, evidence of lightning activity may be available from witnesses. The mechanism of lightning was discussed in full in Chapter 6. The tremendous heat generated by the passage of lightning through wood can ignite even massive timbers under some circumstances. Reports of lightning in an area can be compared against lightning data accumulated by the National Oceanographic and Atmospheric Administration (NOAA), which operates a National Lightning Detection Network reporting system. There is also a private firm, Vaisala Inc., that employs more than 100 sensing stations in the lower 48 United States and a global positioning system. This network can locate lightning strikes (to the earth) to an accuracy of 500 m (1,500 ft). A report recording the time, intensity, polarity, and

FIGURE 8-17 Investigators must be familiar with pole-mounted electrical equipment that can start fires or yield clues as to the start, such as this transformer with lightning arresters (arrows). *Courtesy Pacific Gas & Electric.*

FIGURE 8-18 Small fulgurites of soil fused into a glassy mass by a ground strike of a high-voltage power line. Larger ones can be produced by lightning strikes. *Courtesy of Marion Matthews, U.S. Forest Service.*

FIGURE 8-19 StrikeNet map of lightning strikes in area, color-coded to differentiate those between 0200-0300 and 0300-0400. *Courtesy of Vaisala, Inc., Tucson, AZ.*

location of all strikes for a particular location for a requested time period is available online for a fee.[25] See Figure 8-19 for an example. This service is reportedly 90 percent accurate (multiple reports can sometimes reveal a strike not reported in a single report). Lightning can also strike wire or metal fences, buried pipes or cables, and underground storage tanks, causing direct ignition or ignition remote from the actual strike. See Figure 8-20 for an example of a lightning strike on a tree.

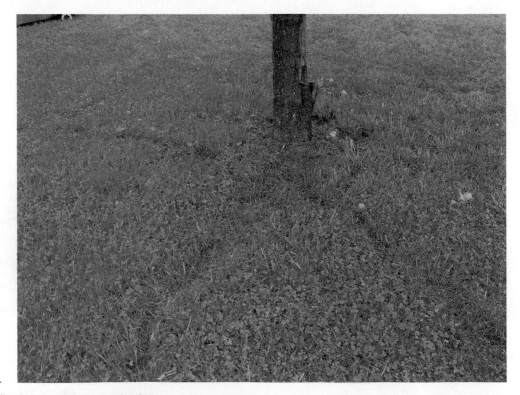

FIGURE 8-20 Lightning striking this tree stripped its bark and charred surface grass along its roots. *Courtesy of Greg Lampkin, Knox County, TN, Fire Investigation Unit.*

BURNING OR HOT FRAGMENTS

Fragments from a disintegrating catalytic converter are small and may escape detection, but they may have a characteristic ceramic "honeycomb" appearance in which the cells may be square, rectangular, or hexagonal in shape. Small spherical pellets are also often used. The rare metals used in them, such as platinum or iridium, can be readily identified in the laboratory. Fragments or "sparks" of burning carbon from the eductor tubes of diesel locomotive engines are nondescript in appearance but tend to be dense and may "burrow" through layers of dry leaves or grass before igniting them. Some exhaust fragments (especially those from gasoline engines) contain iron residues that can permit their recovery using a strong permanent magnet.[26] Fragments of automobile or railroad equipment and welding slag will also be recovered with a magnet. See Chapter 6 for a full discussion of such sources.

CAMPFIRES

Campfires can ignite wildland fires through direct flame contact, aerial sparks (embers), improper disposal or covering of ashes and coals, or even direct ignition of hidden pockets of combustible duff or buried roots. Ignition is usually close to the campfire site, and proof of mechanism and improper care or extinguishment will be an important element of the case. Shoe prints or food containers may reveal recent use. Hot wood or charcoal coals can retain their heat for extended periods of time (1–2 days) and may be buried or discarded away from the main campsite.

CIGARETTES

Cigarettes discarded while still smoldering represent a competent ignition source for grasses under some conditions. The relative humidity has to be low (<22 percent) and ambient temperature over 26°C (80°F) to produce the low fuel moisture necessary.[27] The lower the humidity, the lower the fuel moisture and the more likely that ignition will occur. The fuel has to be loose enough to permit rapid burrowing of the burning cigarette and good contact between the exposed coal and suitable fuel. Surrounding fuel has to be sufficiently dense to sustain the fire spread. Ignition is more likely if the cigarette lands coal-end down or facing into the wind.

INCENDIARY FIRES

Incendiarism cannot be overlooked as a significant cause of wildland fires, especially in rural and urban-wildland interface areas. As in the case of structure fires, all possible causes must be considered. During a thorough search through a fire scene, obvious signs of recent activity should be noted. Cans, bottles, tire and shoe impressions, weapons, and mechanical delay devices may all survive even the most awesome of conflagrations. An eye should be kept open for these during all phases of a scene search. Ignition devices are probably more common in wildland arsons, where one may need many more minutes to flee the area than in urban structure arsons. Unless the fire has been set with a pocket lighter, the remains of the ignition device often are recoverable from the point of origin. Even single paper matches have been found by diligent, trained searchers. Cigarettes have rarely been seen to ignite grasses or brush when the relative humidity of the surrounding air was above 22 percent. Relative humidity lower than 22 percent can produce fuel moistures below 14 to 15 percent, at which the low heat release rate of a smoldering cigarette can ignite fine fuels such as grasses and needles (duff).[28] (See Chapter 6 for the relationship of humidity to fuel moisture.) Weather data for several days prior to the fire are critical for proper testing of hypothesis involving any wildland ignition source.

Multiple points of origin are certainly strong indicators of incendiarism, although there are accidental circumstances that could account for such fires. An overheated catalytic converter spitting glowing remnants of its matrix or a stuck brake shoe or failed bearing on a railroad car spewing bits of hot metal over some distances are examples. Glowing particles of carbon from diesel locomotive engines can be thrown some distance from their moving source to ignite multiple fires along the right-of-way. Witnesses may see such small fires along the right-of-way following the passage of railroad equipment.

Because readily ignitable fuels are present in such quantities in wildland, the use of a flammable liquid such as gasoline or kerosene as an accelerant is uncommon. Unusual odors or the appearance of trailers between points of origin should always be carefully examined.

Devices may be used to ignite remote spot fires, to delay detection and suppression, or to delay ignition of the set itself. The most effective remote ignition sets consist of a bundle of matches wired or taped to a cigarette. A rock or steel washer may be added for additional weight if the device is to be thrown any distance. Once ignited, the device can be thrown from a passing car deep into a brush-covered ravine or similar inaccessible location. Many such devices can be set in a short period of time. Some will not land on suitable fuel, but those that do can ignite serious fires with little evidence with which to link a suspect to the crime. Searches of nearby roadsides in the vicinity of a fire have revealed similar "serial" devices that were unsuccessful and therefore undamaged by ensuing fire. Such devices may bear fingerprints and trace evidence that can be associated with a suspect. Shoe prints and tire prints from such locations have successfully linked a suspect to a general fire area. Wrapping a rock with paper, tissues, or cloth and throwing the ignited bundle into brush can also kindle fires with little traceable evidence remaining.

Incendiary devices are used by fire services for starting backfires. These can range from commercially made pyrotechnic devices with a cardboard cylinder and a fuse to table-tennis balls filled with gasoline/diesel mixture. They have occasionally been stolen or misappropriated from fire agencies. Kerosene-fueled drip torches (as shown in Figure 8-21) are often used to start intentional backfires but have also been misused.

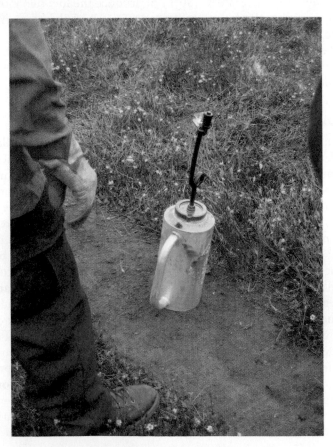

FIGURE 8-21 Hand-carried kerosene drip torch as used by wildland fire fighters to start back-fires as a firefighting tactic. *Courtesy of John D. DeHaan.*

FIGURE 8-22A Cigarette and the match delay of a type often found in wildland fires. *Courtesy of Lamont "Monty" McGill, Fire/Explosion Investigator.*

FIGURE 8-22B Fragile remains of such a set after the passage of the fire should be carefully photographed before any attempt is made to move them. *Courtesy of Lamont "Monty" McGill, Fire/Explosion Investigator.*

Elaborate time-delay devices are very rare in wildland fire arsons. Concave shaving mirrors have been used to start remote fires, sometimes with more than a year's passing before the focused rays by chance fell onto suitable fuel. Other time delays may have been used and have escaped detection and analysis, just as in the case of structural arson sets. Arsonists will generally stay with a device for an entire season if it proves successful, thus creating a signature for linking incidents together and eliminating copycat scenes.

As mentioned previously, the most common time-delay device is a bundle of matches or a matchbook affixed to a burning cigarette, as shown in Figures 8-22a and b. When such a device is found, it is vital that it not be disturbed until it has been fully documented via appropriate sketches and photographs. The appearance and length of ash can indicate if a cigarette was being smoked and discarded, whether it was burned in place by an encroaching fire, or if it was ignited and left deliberately (with malice). The brand of cigarette and the conditions of burning have a great effect on burning time. Tests by the California Department of Forestry demonstrated burn rates of 4–8 mm/min (0.16–0.33 in./min) for various cigarettes burning horizontally in still air. These rates were faster when the cigarette was burning vertically upward and when burning in the direction of an external wind. Total burn times were on the order of 10 to 15 minutes.[29] The brand of cigarette can often be determined if the butt end (with or without a filter) remains unburned. Bourhill's manual for such identifications remains a unique resource.[30] The introduction of "reduced ignition propensity" cigarettes is likely to reduce the number of fires ignited accidentally or deliberately with cigarettes but not to eliminate them from consideration.

Devices that do not provide time-delay ignition are sometimes called **hot sets**. The direct application of flame from a match or lighter to available fuels is almost certainly responsible for a large percentage of wildland arsons. Needless to say, when the source of flame is removed by the arsonist, establishing the cause of the fire will depend on thorough elimination of all other possible causes and extremely careful searching and documentation (including documentation of sources that were considered and eliminated).[31] Hot sets that may leave identifiable remains include dropped matches and cigarettes, "rescue" flares, flaming arrows, or lighted road flares or fusees. The remains of matches and

hot set ■ Direct ignition of available fuels with an open flame (match or lighter).

FIGURE 8-23A Top: Intact debris from a road flare (fusee). The hollow cardboard end cap is at lower right; the plastic cap is melted and burned. *Courtesy of Greg Lampkin, Knox County, TN, Fire Investigation Unit.*

FIGURE 8-23B Lower: After being stepped on, residues are fragmented but still distinctive. *Courtesy of John D. DeHaan.*

cigarettes tend to stay on top of other ash and debris but are easily destroyed by weather or persons searching the scene. Road flares leave a cakey, white, or pastel-colored residue of characteristic appearance (and chemical composition) and usually a wooden end plug and plastic striker cap. The plug and cap often survive even a major fire if they are protected, and the length of the original flare can be estimated from the size of the plug and the amount of residue left behind. Figures 8-23a and b show the characteristically bright-white and pastel residue of a common emergency road flare used to start a grass fire. The components of the flare can vary with manufacturer and age, as pictured in Figures 8-24a and b.

FIGURE 8-24A Components of two varieties of flares (fusees). The top one has paper components with a bottle cap for striker support, cardboard end plugs, and a wooden plug (inside). *Courtesy of John D. DeHaan.*

FIGURE 8-24B The lower one has a plastic cap and striker (low-melting-point plastic is easily destroyed by fire). *Courtesy of John D. DeHaan.*

Figure 8-25 shows a suspected ignition device surrounded by barrier tape until it can be properly documented and recovered.

MODELING

Computer models have been developed for predicting growth and intensity of wildland fires given known fuel, weather, and topography information. The most commonly used programs—Behave, Behave Plus, and FARSITE—can generate predictions of large-scale time lines and behavior patterns.[32] Like computer fire models for room fires, they have been used for hypothesis testing, but they should not be counted on to provide "proof"

FIGURE 8-25 Suspected ignition device and suspected origin protected by barrier tape until they can be photographed. *Courtesy of Lamont "Monty" McGill, Fire/Explosion Investigator.*

of how and where a particular fire was ignited. They await validation for forensic application. Given the complexity of wildland fire dynamics, as described earlier, it is not surprising that accurate modeling has been a difficult challenge. Some programs for two-dimensional, wind-driven spread on grasslands have been validated by fire tests.[33] The advent of ever-more powerful computers has aided the development of limited models of crown fires.[34] NIST is also evaluating the Fire Dynamics Simulator (FDS) for use in predicting fire spread in wildland-urban interface fires where trees and shrubs present the initial fire-spread mechanism, and structures are ignited by radiant heat from those sources.[35] Sullivan has compiled a comprehensive account of wildland fire modeling in the years 1990–2007, and the reader is referred there for more detailed information.[36]

Collection and Preservation of Physical Evidence

The documentation of the appearance of the wildland fire scene and its related evidence by sketching and photography is every bit as important as it is in structure fires. Considering the fragile nature of some important evidence and the vagaries of the environment, it is all the more surprising that some agencies do not routinely document their fire evidence in this way. The same general principles outlined in Chapter 7 apply to sketching and photography of wildland fires, as do the guidelines for collecting and preserving the physical evidence found. Some situations peculiar to wildland fire evidence should be reemphasized, however.

CIGARETTES, MATCHBOOKS, AND OTHER FRAGILE EVIDENCE

When recovering cigarettes or matchbooks, remember that spraying them with lacquer or other preservative will generally make it impossible for the laboratory to test for latent fingerprints or saliva-borne DNA typing results, or to conduct brand identification or physical comparison tests. It is best to photograph the object in place and then package it between layers of loosely fluffed cotton wool in a rigid box for hand transmittal to the laboratory.

One recommended technique for collecting fragile evidence, such as cigarettes or matchbooks, that permits recovery of parts of evidence that are buried in the ash beneath is to dig into the ash and underlying soil using a sharp-edged, flat-bladed garden trowel and loosen the soil under the object in a rectangular area. Once this is done, a clean rectangle of sheet metal (such as sheet metal flashing from the hardware store) is worked carefully through the soil without disturbing the surface debris. The sheet metal is worked all the way under the soil platform and then used to lift the soil, ash, and object together. The entire mass is then transferred to a properly sized and padded evidence box. If the soil is very hard, water can be introduced to the trenches around the target and allowed to stand to soften the soil. On rocky surfaces, it may be possible to lift only the ash layer with the sheet metal.

SHOE AND TIRE IMPRESSIONS

Shoe and tire impressions should be photographed using low-angle oblique lighting and a ruler or scale near (not *in*) the impression. (Placing a scale in or across an impression can obscure critical details.) If at all possible, the impression should then be cast using plaster, dental stone, paraffin, or other suitable casting agent.

CHARRED MATCHES

Matches that are completely charred are probably not useful as evidence (except to demonstrate their presence), but if any unburned cardboard remains, an attempt can be made to match it to a matchbook.

DEBRIS SUSPECTED OF CONTAINING VOLATILES

Debris suspected of containing volatiles should be sealed in appropriately sized cans, jars, or polyester or nylon bags. Polyethylene plastic or paper bags must not be used.

CONTAINERS

All containers much be securely sealed to prevent loss or contamination, and the chain of evidence must be recorded on it in ink. The date, time, and location of recovery and the name of the person recovering the evidence must be on the package as well.

WEATHER DATA

A critical item of evidence that is not found at the scene itself and sometimes overlooked is weather-related data. This should be obtained as soon as possible, either from first-on-scene fire suppression personnel, websites such as Weather Underground (www.wunderground.com) or NOAA, nearby Remote Automated Weather Stations (RAWS), or collected firsthand using either a handheld digital weather recorder or a belt weather kit (shown in Figure 8-26). The RAWS system uses remote sensor stations that automatically send weather data via satellite every hour to the National Weather Service's Weather

FIGURE 8-26 "Belt" weather kit: Pitot-type wind gauge (anemometer), compass, water bottle, psychrometer (wet/dry bulb humidity gauge), and case. *Courtesy of U.S. Forest Service* (photo by author).

Information Management System. From there it is available for spot analysis by meteorologists. Currently, the website www.met.utah.edu/mesowest, operated by the University of Utah, can give investigators acess to RAWS data.

Weather data should be taken from outside the burn but as close to the suspected origin as possible and as close to the estimated/known ignition time as possible. Two hand-held devices that can be used to record temperature, relative (air) humidity, and wind speed are shown in Figure 8-27. Weather data can provide a critical source of information with which to assess fire behavior, rate of spread, and fire intensity through the use of nomogram fire calculations or computer models. Such analysis can support burn pattern interpretation and scientifically corroborate or exclude possible ignition sources.

(a)

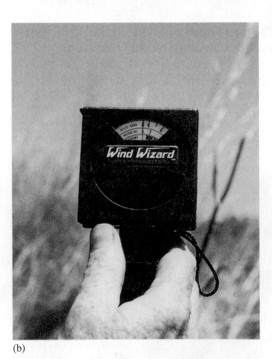

(b)

FIGURE 8-27 Pocket weather tools. (a) Digital thermometer and humidity gauge. (b) Wind-speed indicator (anemometer). *Courtesy of John D. DeHaan.*

Summary

The careful investigator who attempts to locate the exact cause of a fire in wildland areas must, above all, avoid assuming that because some hazard existed in the general area of origin, it was necessarily the cause. Just as with structure fires, the investigator must be sure that the fuel involved was capable of being ignited by the proposed or suspected ignition source. Although a fire along a roadside suggests that a careless smoker may have caused it, without finding the residue of the match or cigarette, it would be unwise to do more than consider this as a possible cause, along with the possibility that sparks from a vehicle exhaust, deliberate hot set, or other source could also have played a part. Conversely, careful search at the origin has sometimes located a partially burned match or the remains of a cigarette that turns out to be the device used by an arsonist.

It is important that the investigator be satisfied that the cause of the fire has been identified and all other reasonably potential causes excluded or satisfactorily addressed. It is the expert opinion of the investigator on which all further proceedings regarding this fire will depend. If there is any doubt as to the cause of the fire, it should not be acted on. In civil cases, of course, the standard of proof is a preponderance of the evidence, as opposed to the "beyond a reasonable doubt" standard in criminal cases.

Every wildland fire investigator must remember that identifying the origin of the fire is only the first step, especially if a human-made cause is found. The person who, through neglect or by an intentional act, was responsible for a destructive fire must be brought to justice. Grass and wildlands are a valuable and irreplaceable resource, just as important as any structure, and their loss to fire must be reduced. This can happen only if the responsible person or agency is identified and civil penalties are assessed for negligence, or criminal charges are pressed for arson.

Review Questions

1. What are four of the major factors that affect the spread of a wildland fire?
2. Why are weather conditions both before and during a wildland fire very important to the investigator?
3. Name three macro-scale indicators and three micro-scale indicators.
4. What is the most common accidental cause and most common natural cause for wildland fires?
5. What is the area of transition and how is it useful to an investigator?
6. Why is a magnet useful in examining suspected areas of origin?
7. What kinds of physical evidence can be found in wildland fires?
8. What is the critical fuel moisture content for cigarette ignition of grasses and other fine fuels?
9. What is a *hot set* in a wildland fire?
10. Name four ways power lines can ignite wildland fires.

References

1. California Dept. of Forestry, "2006–2008 Historical Wildfire Activity Statistics" (Redbooks) (Sacramento, CA: California Dept. of Forestry and Protection, 2008).
2. National Wildfire Coordinating Group, *Wildfire Origin and Cause Determination Handbook* (Boise ID: NWCG, May 2005), 8–10.
3. R. C. Rothermel, *How to Predict the Spread and Intensity of Forest and Range Fires*, Gen. Tech. Rep. INT-143 (Ogden, UT: U.S. Dept. of Agriculture, National Wildfire Coordinating Group, June 1983), 10–12.
4. E. A. Johnson, *Fire and Vegetation Dynamics* (Cambridge: Cambridge University Press, 1992), 40.

5. Rothermel, *How to Predict the Spread and Intensity of Forest and Range Fires*; National Wildfire Coordinating Group, *Fire Behavior Nomograms* (Boise, ID: U.S. Dept. of Agriculture, March 1992).

6. H. E. Anderson, "Aids to Determining Fuel Models for Estimating Fire Behavior" (Boise, ID: National Wildfire Coordinating Group, April 1982).

7. Johnson, *Fire and Vegetation Dynamics*, 56–57.

8. National Wildfire Coordinating Group, *Wildfire Origin and Cause*, 70–71.

9. Johnson, *Fire and Vegetation Dynamics*, 45.

10. B. J. Stocks et al., "Crown Fire Behaviour in a Northern Jack Pine–Black Spruce Forest," *Canadian J. Forestry Research* 34 (2004): 1548–60; J. D. Cohen, "Relating Flame Radiation to Home Ignition Using Modeling and Experimental Crown Fires," *Canadian J. Forestry Research* 34 (2004): 1616–26.

11. Johnson, *Fire and Vegetation Dynamics*, 36.

12. Rothermel, *How to Predict the Spread and Intensity*, 25–34.

13. Ibid., 26.

14. Johnson, *Fire and Vegetation Dynamics*, 45.

15. National Wildfire Coordinating Group, *Wildfire Origin and Cause*, 6.

16. National Wildfire Coordinating Group, NWCG Course FI-110: *Wildfire Observations and Origin Scene Protection for First Responders*, Boise, ID, March 2005.

17. "GPS Technology Aids Oakland Hills Fire Efforts," *Public Works Journal* (May 1992). Reprinted in *Arizona Arson Investigator* (August 1993).

18. D. W. Pack et al., "Civilian Class of Surveillance Satellites," *Crosslink* (El Segundo, CA: Aerospace Corp., January 2000): 2–8.

19. National Wildfire Coordinating Group, *Wildfire Origin and Cause*.

20. S. L. Gutsel and E. A. Johnson, "How Fire Scars Are Formed: Coupling a Disturbance Process to Its Ecological Effect," *Canadian J. Forestry Research* 26, no. 2 (1996): 166–74.

21. G. L. White, "Fire Scene Diagramming for Wildland Fire Investigations," *Fire and Arson Investigator* (October 2004): 31–33.

22. National Wildfire Coordinating Group, Boise, ID, 2005.

23. California Department of Forestry and Fire Protection. "2006–2008 Fires by Cause Statewide," Sacramento, CA, 2006–2008.

24. P. Steensland, "Long-Term Thermal Residency in Woody Debris Piles," paper presented at International Association of Arson Investigators Annual Training Conference, Denver, CO, May 2008.

25. "Lightning Detection Service," *Fire Findings* 5, no. 3 (Summer 1997): 13–14.

26. L. U. Bernardo, *Fire Start Potential of Railroad Equipment* (San Dimas, CA: USDA Forest Services, 1979).

27. C. Countryman, *Ignition of Grass Fuels by Cigarettes* (San Dimas, CA: USDA Forest Services, 1983); P. Steensland, *Cigarettes as a Wildland Fire Cause*, Wildland Fire Origin and Cause Determination FI-210, NFES 2816, instructor guide/student handout material (Boise, ID: National Wildfire Coordinating Group).

28. V. Babrauskas, *Ignition Handbook* (Issaquah, WA: Fire Science Publishers), 2003, 302, 839. See also C. Countryman, "Ignition of Grass Fuels by Cigarette," *Fire Management Notes* 44, 3, 1983, 3–7.

29. D. A. Eichmann, "Cigarette Burning Rates," *Fire and Arson Investigator*, January–March 1980.

30. R. Bourhill, *Cigarette Butt Identification Aid*, 16th ed. (Colorado Springs, CO: Forensic Science Foundation, 1995).

31. R. T. Ford, "Proving Hot Start Arson in Wildfires," *Fire and Arson Investigator* (April 2002): 42–43.

32. P. L. Andrews and C. D. Bevins, "BEHAVE Fire Modeling System: Redesign and Expansion," *Fire Management Notes* 59 (1999): 16–19; M. A. Finney and P. L. Andrews, "FARSITE: A Program for Fire Growth Simulation," *Fire Management Notes* 59 (1999): 13–15; http://fire.org.

33. W. Mell et al., "Numerical Simulations of Grassland Fire Behavior from the LANL-FIRETEC and NIST-WFDS Models" in *Proceeding East Fire Conference*, Fairfax, VA, May 2005; W. Mell, et al., "A Physics-Based Approach to Modeling Grassland Fires" (Gaithersburg, MD: BFRL, NIST, May 2006).

34. R. N. Meroney, "Fires in Porous Media: Natural and Urban Canopies," prepared for NATO Advanced Study Institute: "Flow and Transport Processes in Complex Obstructed Geometries," Kiev, Ukraine, May 5–15, 2004.

35. D. D. Evans et al., "Physics-Based Modeling of Community Fires" in *Proceedings Interflam 2004* (London: Interscience Communications), 1065–76; W. Mell et al., "The Wildland-Urban Interface Fire Problem: Current Approaches and Research Needs," *International J. Wildland Fire* (June 2009).

36. A. L. Sullivan, "Wildland Surface Fire Spread Modelling, 1990–2007: Pt. 1, "Physical and Quasi-physical Models," 349–68; Pt. 2, "Empirical and Quasi-empirical Models," 369–86; Pt. 3, "Simulation and Mathematical Analogue Models," 387–403. *International J. Wildland Fire* 18 (2009).

9

Automobile, Motor Vehicle, and Marine Fires

Courtesy of Jamie Novak, Novak Investigations and St. Paul Fire Dept.

KEY TERMS

Hot-plate ignition (hot surface ignition),
   p. 362

OBJECTIVES

After reading this chapter, you should be able to:

- Identify the major sources of fires in automobiles and motor vehicles.
- List the flash points and ignition temperatures of fluids found in vehicles.
- Explain the systematic process of examining a vehicle for fire pattern damage and other fire-related factors.
- Identify the major problems associated with investigating fires in motor homes, mobile homes, and recreational vehicles.
- Describe the differences between fire spread in ships and in buildings.
- List the motives for vehicle arson.

PEARSON

myfirekit™

For additional review materials, appendices, and suggested readings, visit **www.bradybooks.com** and follow the **MyBradyKit** link to register for book-specific resources.

Register for **MyFireKit** by following directions on the **MyFireKit** student access card provided with this text. If there is no card, go to **www.bradybooks.com** and follow the **MyFireKit** link to Buy Access from there.

All types of motor vehicles—cars, trucks, buses, motorhomes, and the like—are subject to fires of either accidental or incendiary origin. They all contain substantial quantities of ignitable liquids for fuels, electrical and mechanical systems to provide ignition sources, and combustible plastic and metal components and cargo to provide fuel loads. Vehicle fires can start from a variety of accidental causes, and arsonists will often try to simulate such accidental fires to escape detection. It is important for the fire investigator to be familiar with the fuels and sources of ignition present in vehicles so that all such fires can be properly assessed. The NFPA estimated that U.S. fire departments respond to an average of over 270,000 fires in highway vehicles per year (2003–2008), which cause some $1.1 billion in direct property damage each year[1] (representing a slow decline in the number of vehicle fires since 1997) and an estimated 365 deaths. Vehicle fire investigators must be familiar with fuels, ignition sources, and dynamics as encountered in vehicles if they are to carry out correct investigations. We shall describe the fuels, fuel systems, and ignition sources and then discuss the investigation.

Automobiles and Motor Vehicles

FUEL TANKS

The fuel tanks of most autos or light trucks contain a large reservoir of flammable and potentially explosive gasoline and gasoline vapors (diesel, biodiesel, LPG, or CNG are, of course, also encountered). Extensive firefighting experience and laboratory tests reveal that contrary to expectations, fuel tanks very rarely explode during a vehicle fire.[2] Despite movie and television scenes, it is nearly impossible to ignite the gasoline inside. In part, this is because its vapors are much heavier than air and tend to fill the tank completely to the top. Except at extremely low ambient temperatures, the vapor pressure of gasoline is such that it tends to form a mixture with residual air in the tank that is well above the explosive limit and cannot be ignited. It follows that fire will not ordinarily invade an open gasoline tank but will be limited to the outside of any openings where there is sufficient air for combustion. Gasoline vapors outside an open fuel filler neck can usually be ignited with an open flame, to create a steady, torch-like flame longer than a foot (30 cm) or so in length. Propagation into the tank to cause an explosion is almost impossible unless the tank has so little fuel in it that an equilibrium vapor pressure is not reached. (The dynamics of such fires will be explored in more detail in Chapter 12.) A closed vapor recovery system, when exposed to a fire, can release vapor under pressure, creating a torch-like effect, sometimes in locations remote from the tank itself.

When exposed to a general fire, the soldered seams and fittings of a metal tank may melt and separate, releasing fuel (sometimes under considerable pressure) to fuel the fire. Pressure built up by vaporizing fuel can split the tank or blow the filler cap off (of either metal or polymeric tanks). Accidents can, and frequently do, lead to rupture of the fuel tank by mechanical force, producing a very hazardous spill. Since most large trucks use diesel engines, their fuel presents less of a hazard when spilled at ordinary temperatures, but in a general fire, diesel fuel can vaporize to form explosive vapor mixtures. In addition, some vehicles are being fitted with LPG (liquefied petroleum gas) or CNG (compressed natural gas) fuel systems, and a number of fires have been caused by even small leaks or poor filling practices that resulted in explosive vapor mixtures in or around the vehicle.

To minimize fabrication costs and vehicle weight, manufacturers are using polymeric tanks in more and more vehicles today. They are strong and puncture-resistant, but unlike metal tanks, they are subject to cracking and can melt or burn through in an enveloping fire, releasing their contents quickly, giving the appearance of an accelerated fire. Under rare conditions, polymeric tanks can fail explosively as a result of a boiling-liquid, expanding-vapor explosion (BLEVE).

FUEL TANK CONNECTIONS

Because regular fuel tanks are mechanically strong and usually well protected, the tanks themselves are rarely involved in accidental fires (except for severe crashes). Their filling and fuel line connections are a different matter. The filler spouts of many tanks have flexible rubber connections that can crack or loosen and allow fuel to spill into trunks and passenger compartments.

CNG vehicles may have a glass-fiber-wrapped polymeric tank that can fail explosively in a general vehicle fire. Topping off the tank often leads to overfilling, creating hazardous spills beneath the vehicle. (In older vehicles, fuel filler leaks could introduce fuel into the trunk or even the passenger or engine compartment.) Overfilling may cause excess fuel to be siphoned into the vapor recovery canister in the engine compartment. This canister, often filled with charcoal, is designed to trap fuel vapors, which are then drawn into the engine to be burned. Liquid fuel in this canister can pose a fire hazard. The flexible connectors (hoses) on metal fuel lines are subject to splitting or cracking with age or loosening due to abuse or improper service repairs, causing a fuel spill hazard. Many manufacturers have adopted the use of nylon fuel lines and polymeric connectors. Physical damage, exposure to severe heat conditions, or improper service procedures can result in these cracking, splitting, or leaking, which can result in a fire hazard. It was observed that a polymeric fuel line, ignited by accidental contact with an exhaust header, burned the length of the car to the fuel tank, with the fuel in the line supporting the fire.

FUEL PUMPS, FUEL LINES, AND CARBURETORS

The fuel pumps on many older cars are mechanical ones, mounted on the side of the engine, that function only when the engine is turning over. Most imported cars and newer domestic cars are fitted with high-capacity electric pumps that are pumping whenever the ignition is on. Because most electric pumps "push" better than they "pull," they are most often mounted at or near the tank (and many are enclosed in the tank itself). Without a proper interlock (to oil pressure or engine-speed sensors) (or with failure of the interlock safety), such pumps can pump a great deal of fuel out through a loose or ruptured fuel line even though the engine may not be running. Some new vehicles have two pumps—one in the tank and a second one closer to the engine to produce high pressure. If a fuel line is compromised by external fire or mechanical impact (collision), gasoline may siphon out of the tank by gravity even if the fuel pump is inoperative. Although the amount of gasoline in the carburetor itself or in the fuel injection lines (rails) may be small, the fuel lines of some cars are positioned over electrical components or the exhaust manifold, and loose connections pose a hazard for a large fire. If a fire is limited to the fuel in the carburetor or fuel rail, the fire will generally be limited to moderate damage of the air cleaner, hood, and engine wiring and rarely be transmitted to other parts of the vehicle.

FUEL INJECTION SYSTEMS

Most fuel injection systems are electronically controlled and are found on nearly all newer gasoline-fueled vehicles. These systems provide additional fire risks because their fuel delivery lines carry gasoline at pressures from 25 to 100 psi (150 to 700 kPa). Even a pinhole leak in a line or minor crack in a fitting allows gasoline to be atomized and sprayed around the engine compartment at high pressures, quickly producing an explosive vapor mixture requiring only a suitable electrical arc to ignite. The fuel rail is the metal tubing that distributes the fuel to individual injectors on each cylinder (or, in older vehicles, directly to the

TABLE 9-1	Flash Points and Ignition Temperatures of Vehicle Fluids	

FLUID	FLASH POINT	IGNITION TEMPERATURE
Gasoline	−43°C–38°C (−45°F–36°F)	280°C–456°C (536°F–853°F) (varies with grade)
Ethanol	13°C (55°F)	363°C (685°F)
Diesel fuel	52°C (125°F)	260°C (500°F)
Ethylene glycol (100%)	111°C (232°F)	398°C (748°F)
Propylene glycol (100%)	99°C (210°F)	371°C (700°F)
Windshield washer fluid (methanol)	11°C–15°C (52°F–55°F)	464°C–484°C (867°F–903°F)

Hydraulic brake fluid
    DOT 3 (Normal auto service)
    DOT 5 (Synthetic or silicone)
		190°C (375°F)[a]
		260°C (500°F)[b]

Lubricating oil (including power steering and automatic transmission fluid)	121°C–205°C (249°F–402°F)	310°C (590°F)
Compressed natural gas (CNG)	N/A	632°C–650°C (1169°F–1202°F)
Propane (LPG)	N/A	450°C–493°C (842°F–919°F)

Sources: NFPA, *Fire Protection Guide to Hazardous Materials,* 12th ed. (Quincy, MA: NFPA, 2001).
[a]MSDS: Champion 1400 Series, DOT3, rev. 1/13/09.
[b]MSDS Maxima Racing Oils, rev. Sept. 2009

throttle body). It usually has a high-pressure (delivery) side and a low-pressure (return) line. Such systems can leak at the gaskets or seals at the injectors or the connectors.

VEHICLE FUELS

Today, more vehicles are being seen with gaseous fuels rather than liquid ones. Propane (sometimes called LPG) is stored in a vehicle as a liquid in a pressurized metal tank at pressures of approximately 75 psi. The fuel is delivered to the motor as a liquid, which evaporates quickly to a clean-burning gaseous fuel. Natural gas may be encountered (especially in large commercial or passenger vehicles) as compressed gas (CNG) with tank pressures of 2,400–3,600 psi (16,500–24,800 kPa). These tanks may be of steel or glass-fiber composites.

Gasoline is the most abundant fuel in most automobiles and the one with the lowest flash point (see Table 9-1) and is therefore often thought to be the first fuel ignited. However, its properties render it somewhat less likely to ignite upon contact with a hot surface (referred to as **hot-plate ignition**) than are other heavier combustible liquids such as hydraulic fluids and lubricating oils. Tests performed by auto manufacturers and fire investigators have shown that gasoline seldom ignites at hot-plate temperatures near the listed autoignition temperature (AIT) and, in the under-hood environment, requires surface temperatures between 590°C and 730°C (1,100°F to 1,350°F) to reliably ignite.[3] These temperatures may be attained at the exhaust manifolds but only after extreme and sustained engine load or severe mechanical malfunction (timing and mixture) or on the surfaces of turbochargers or catalytic converters. Textures of hot surfaces (including crevices and cracks) may reduce the required temperature (see Chapter 6). The rough undersides of hot cast exhaust manifolds may offer better chances of ignition. Leaks of burning combustion gases around loose spark plugs or faulty manifold gaskets offer a chance of flame ignition.

hot-plate ignition (hot surface ignition) ■ Temperature of a metal test plate at which liquid fuel ignites on contact.

OTHER COMBUSTIBLE LIQUIDS

The flammability properties of some of the other combustible liquid fuels in the engine compartment are shown in Table 9-1. As we saw in Chapter 4, the residence or contact time between a fuel and a hot surface is critical, as are the nature of the surface and the quantity of fuel. Hot-surface ignition of transmission oil (460°C–600°C) (860°F–1112°F) or steering or brake fluid (540°C–570°C) (1000°F–1060°F) is much more likely than "open" hot-surface ignition of gasoline. Of course, in a severe collision, all these fluids may be released. Antifreeze is not often thought to be combustible, but pure ethylene glycol has a flash point of 111°C (232°F) and an AIT of 398°C (748°F) (and the values for propylene glycol are even lower).[4] If antifreeze (often, a 50/50 mix of ethylene glycol and water) is allowed to pool on top of an engine, the water can boil away until the pure glycol can form a vapor ignitable by heated surfaces (530°C–570°C) (986°F–1058°F) or by arcing within the distributor, alternator, faulty spark plug leads, fan motor, or other electrical equipment.[5] The surface temperatures of exhaust manifolds, turbochargers, catalytic converters, and the like may be far in excess of the 398°C (748°F) required to ignite ethylene glycol, and the high boiling point of the glycols slows their evaporation to give more time for energy transfer and mixing). Ignition of a 50/50 ethylene glycol–water mix sprayed directly onto a hot exhaust manifold has been reported, although documented cases of such ignition are very rare.[6]

Brake fluid (which is usually a heavy glycol mix) and automatic transmission fluid (ATF) and power steering fluid (which are light lubricating oil–grade petroleum distillates) also represent potential fuels ignitable by hot surfaces.[7] Such fluids may be present in reservoirs (including polymeric ones that can leak, fracture, or melt) or in pressurized lines. If these lines suffer even a small puncture or leak, the fluid is sprayed out in an easily ignitable aerosol. These fluids have low ignition temperatures and can be ignited if sprayed onto hot exhaust components or disc brake rotors, or if they come into contact with electrical arc sources. Gasoline, by contrast, is not readily ignited by contact with the exterior of a hot exhaust manifold, due in part to its volatility (but other more viscous, less volatile fuels with similar AITs, such as diesel fuel or motor oil, are ignited). As seen in Figure 9-1, the normal maximum temperature for an exhaust manifold of 550°C

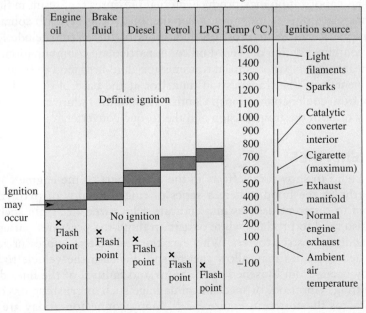

FIGURE 9-1 Representative ignition temperatures of vehicle fluids, temperatures of vehicle fluids, and temperatures of ignition sources. *Courtesy of Fire, October 1991. Reprinted in March, G., "An Investigation and Evaluation of Fire and Explosion Hazards Resulting from Modern Developments in Vehicle Manufacture." Newcastle-upon-Tyne, United Kingdom. Tyne and Wear Fire Brigade, Newcastle, United Kingdom, 1993.*

TABLE 9-2	Typical Operating Pressures for Vehicle Fluids
Power steering with rack and pinion	50–300 psi (345–2,070 kPa) (maximum relief pressures can be over 1,000 psi)
Power steering with conventional steering gear	600–1,200 psi (4,140–8,200 kPa) (maximum relief pressures can be over 1,500 psi)
Automatic transmissions	75–400 psi (518–2,760 kPa)
Fuel pump for carburetors	3–9 psi (21–62 kPa)
Fuel pump for fuel injection	40–120 psi (276–828 kPa)
Engine oil pressure	15–75 psi (103–518 kPa)
Cooling systems	12–17 psi (83–117 kPa)
Hydraulic ride or active ride control systems	125–1,800 psi (862–12,420 kPa)
Operating pressures of brake system non-ABS	0–1,200 psi (0–8,280 kPa)
Operating pressure of ABS brake system	0–2,000 psi (0–13,800 kPa)
Diesel fuel in-line injection or rotary pump (into combustion chamber) (low-pressure side)	3,000–20,000 psi (20,700–138,000 kPa) 40–80 psi (276–552 kPa)

Sources: R. Parsons, Wright Group, Uxbridge, MA; G. Barnett, *Automotive Fire Analysis*, 2nd ed. (Tucson, AZ: Lawyers & Judges Publishing, 2007).

(1,020°F) is not adequate for hot-surface ignition of gasoline (petrol), while temperatures above 395°C (743°F) can ignite DOT 3 brake fluid or automatic transmission fluid dripped onto its exterior.[8]

Many fluids in vehicles are at elevated pressures (see Table 9-2). The release of any ignitable fluid under pressure produces an aerosol mist that is much more readily ignitable than the bulk liquid. Operating fluids are usually at elevated temperatures as well, making them easier to ignite upon release.

Transmission or torque converter failure can produce internal heat sufficient to vaporize transmission fluid. This vaporization may produce internal pressures that result in fluid being expelled from vents, seals, or dipstick tubes. When the fluid (or its aerosol) contacts heated exhaust components, it can ignite. Post-fire indicators of this event include low transmission fluid level combined with a darkened oil color, particulate contamination, or a strong burned odor. Heat transfer patterns from transmission fluid fires tend to be at the bottom of the compartment, near the transmission tunnel or at the base of the radiator (where the fluid can leak from cooler connections). Confirmation of the failure may require removal and disassembly of both the transmission and the torque converter.[9]

ENGINE FUEL SYSTEM FIRES

Engine fuel-system fires tend to show heat effects in the upper part of the engine compartment, and particularly on the hood, where a series of rings or halos of pyrolyzed paint and metal oxidized to different degrees may indicate a localized fire from such a fuel source. Such fires also tend to produce white oxidation (high-temperature) patterns on the firewall, inner fenders, or strut towers. Whenever the fuel system is pressurized, leaks elsewhere in the delivery lines may allow fuel to spill beneath the vehicle to be ignited there. Leaks may occur from loosened connections and failures of the lines due to corrosion, aging, splitting, vibration, or mechanical damage such as crushing or contact against moving parts, as illustrated in Figure 9-2.[10] Many connectors today are of

FIGURE 9-2 A rubber fuel line, connecting a fuel injector to the fuel rail, cracked from age and sprayed gasoline on the top of an engine. *Courtesy of Bloom Fire Investigation, Grants Pass, OR.*

plastic, and some modern vehicles have fuel lines that are entirely of plastic. These can burn away completely, so they may not be detectable post-fire. In-line fuel filters (metal, plastic, or glass and metal) can fracture or separate to produce fuel leaks. O-ring seals in fuel injection systems degrade from heat, pressure, or fuel additives (or improper installation) and release fuel under pressure. Fires on the top of the engine are most likely to be fuel fires, while fires near the firewall (bulkhead) or wheel wells are more likely to be brake fluid or automatic transmission fluid fires. Lines can be cut and even ignited by contact with moving parts such as pulleys, belts, or drive shafts. Because of the complex routing of their hoses, power steering and coolant systems cannot be generalized as to origin location. The trained investigator who is very familiar with the various systems, the fluids they contain, and the variations in construction and routing (model, year, and manufacturer) can sort out the relevant damage patterns.

Once ignited, the myriad hoses, vacuum lines, and plastic-insulated wires of modern vehicles provide a great deal of fuel to support a fire. (Very few of the polymeric materials in engine compartments are fire-resistant.) Many vacuum system and vapor recovery hoses use plastic connectors that can fracture, melt, or simply add to the fuel load. The modern vehicle has many fresh-air, heating, and air-conditioning ducts of synthetic materials that originate in the engine compartment and penetrate the firewall, as shown in Figure 9-3. These are, in turn, connected to polymeric fan enclosures. While they are not ignited by glowing ignition sources, all these can support flaming ignition once ignited and allow flames to readily penetrate into the passenger compartment. Fires in these enclosures and ducts may be ignited by failure of fan motors or by fires in the engine compartment.[11] Fire entering the interior through the large openings created by failure of polymeric enclosures will impinge on plastic dashboard components and spread readily into the interior. If the fire progresses to the point where window glass is exposed to fire, side window glass, being tempered, is

FIGURE 9-3 Pene-
trations of the "firewall"
in modern vehicle.
*Courtesy of Jamie Novak,
Novak Investigations and
St. Paul Fire Dept.*

especially vulnerable to thermal-shock failure, as depicted in Figure 9-4. If the window
glass breaks, there can be sufficient ventilation to allow the fire to *eventually* involve
all combustible interior components. Fire thus transmitted into the passenger compartment may be revealed by fire damage coming from the defroster vents between the
windshield and top of the dashboard. Engine compartment, tire, or brake fluid fires
can ignite the polymeric fender liners used on many cars, and these fires, in turn, can
extend under front fenders and attack side windows, door seals, and the lower front
corners of windshields.[12]

FIGURE 9-4 Tempered
side window glass shat-
ters to cubes. Laminated
safety glass breaks in
curves. *Courtesy John D.
DeHaan.*

ELECTRICAL SYSTEMS

The general principles of electrical fire causation will be discussed in Chapter 10, and they are just as applicable in vehicle circuits as in structural ones. Although the operating voltage of most autos and light trucks today is only 12 V, and the currents (in most fused circuits) are limited to a few amps, electrical causes of vehicle fires cannot be dismissed out of hand. Most power on vehicles is direct current, so arcing, once established will not self-extinguish, as often seen in alternating current events. Trucks and commercial vehicles commonly have 24-V systems for starting and charging systems. Most vehicles today have a single energized conductor connected to the positive battery terminal, with the return path provided by the metal vehicle frame and components. This is called a negative-ground system, since the negative post of the battery is connected to the vehicle's frame (American cars before 1955 often used a 6-V positive-ground system.) This frame-ground system means that an energized conductor coming into contact with any metal component can cause current flow. The high-capacity cables fitted to the battery connections can carry sufficient amperage to start fires of the cable insulation, starter, and related electrical fittings. In short circuits of these or high-current circuits of the horn, cooling fans, cigarette lighter, or lighting equipment, the temperatures developed will cause smoking or even flaming combustion of the insulation and any connectors or terminals. Such circuits are usually energized even when the key is in its off position, but all but the starter are often protected by a fusible link—a special wire connector. There is no overcurrent protection on the starter circuit. If this circuit shorts to start a fire near combustible components (fender underliners, splash shields, ducts, hoses, etc.), a large fire can develop quickly. Attempts to deliberately cause such electrical fires in the protected wiring in a series of modern vehicles, however, invariably caused the overcurrent devices in the vehicles to interrupt the power before ignition could occur.[13] Relays control many of the higher-current accessories (like blower motors) in newer cars, so there is less chance for a high-current fault to occur to initiate a fire.

Ignition switches (which have unfused "hot" wires and ground connections in close proximity) have been linked to fires started when carbon tracking allowed current to flow continuously between contacts in the switch and eventually to degrade and ignite the plastic housing.[14] Some recent vehicle fires have been linked to corrosion of fittings associated with speed control of activation switches on master brake cylinders.[15] When large, energized, unfused conductors can come into contact with grounded metal components, a massive current flow can result, with the accompanying increased likelihood of ignition, as illustrated in Figures 9-5a and b. In one author's experience, a battery cable that came into contact with an exhaust pipe caused a direct short when the insulation melted. The resulting massive current melted and charred the insulation on the cable and interconnected wires within seconds. One form of grounded metal component that is easily overlooked is the steel braid wrapping in high-pressure hydraulic and brake lines. Chafing of these lines against metal edges (such as starter cable terminals) can abrade the protective rubber coating and allow contact. The hot electric arc produced quickly chars the rubber liner, allowing a leak of the ignitable fluid and the potential for a very destructive fire fueled first by the fluid, then by the rubber lines. As with any other kind of fire investigation, the first fuel ignited has to be identified, and the chain of reactions has to be demonstrated. Just having a source of heat, no matter how intense, will not result in a fire unless an adequate supply of a suitable fuel is present and the mechanism for enough heat transfer from that source to that fuel can be shown.

The newest generation of theft protection devices includes keyless entry and starting (with an encoded transponder), remote locking and unlocking, and specially coded ignition keys. Their contribution to fires and investigations is unknown at this time.

FIGURE 9-5A The vibrations of a farm tractor caused chafing to the battery cables, resulting in a dead short. *Courtesy of Tim Yandell, Public Agency Training Council.*

FIGURE 9-5B The main alternator connector on this vehicle failed, causing a significant engine fire. *Courtesy of Tim Yandell, Public Agency Training Council.*

Although the current standard for automotive electrical systems is 12 VDC, negative ground, recently developed technology suggests more efficient higher-voltage systems will be implemented on most automobiles in the near future. These systems will likely employ a 42 V alternator/starter motor combination that will be installed in the flywheel area. Currently, many trucks and other diesel-powered vehicles use 24 V systems. Electric- and hybrid electric-gas-powered vehicles are already in service. The high wattage capabilities of such electrical systems (including batteries that can keep mechanical systems operational) can create fire hazards after crashes. The power for electric propulsion motors is 600 VDC with high-current batteries. This power is supplied in two conductors (with brightly colored insulation—blue, orange, or green) separate from the vehicle's lighting and control systems.[16]

Many modified vehicles use high-current accessories (such as audio systems or hydraulic pumps) in an unsafe fashion, and wiring may be exposed to all manner of crushing and chafing hazards that can compromise the insulation. The same care should be exercised in examining such vehicles as one would employ in examining an older house for dangerous modifications. It is obvious that modern vehicles are often fitted with power accessories such as power seats, power windows, seat heaters, telephones, and even portable computers and video systems. All these add to the complexity of the fire scene even if they do not usually add to the fire hazards when properly installed.

One additional source of electrical fire hazard is the battery. In normal use, it presents no fire hazard, but when it is receiving a charging current, considerable volumes of explosive hydrogen gas are generated. An explosion will result from an arc (even between the cells of the battery) or other source of ignition, and some fire may follow if the charging continues. Explosions unaccompanied by fire are the rule, and even these are rare. Symptoms of malfunction will usually be apparent to the operator. Extra batteries may be concealed in the vehicle to power hydraulic lift systems or high-power sound systems in modified vehicles. Some vehicles have their operational batteries in trunks or beneath cargo floors. Some "accessories" require the use of inverters to convert the 12 VDC of the vehicle to 120 VAC. Such inverters can fail, to cause fires.

MISCELLANEOUS CAUSES

The NFPA estimates that some 50 percent of highway vehicle fires are caused by mechanical failures or malfunctions and 24 percent are caused by electrical failures or malfunctions. Other identified causes include misuse (11 percent), intentional ignition (8 percent), and collision or overturn (3 percent).[17]

Turbochargers

Many vehicles (both gasoline- and diesel-fueled cars and trucks) are fitted with turbochargers that redirect the exhaust gases to a housing to spin an impeller. This impeller then helps push air into the intake system. Such turbocharger housings operate at very high external temperatures and also, because of their very high rotational speeds, are subject to bearing failures that can produce hot-fragment ignition as the bearing, housing, and rotor fail. Bearing failure can also result in the ignition of the lubricating oil supplied to the center bearing under pressure.

Catalytic Converters

Catalytic converters have been fitted to domestic (U.S.) cars since 1975 and to light trucks since 1984. These units convert hydrocarbons, carbon monoxide (CO), and nitrogen oxides (NO_x) to less-toxic compounds by chemical conversion using a metallic catalyst like platinum or palladium in a porous ceramic foam or in loose beads. Under normal operating conditions, these converters operate at internal temperatures on the order of 600°C to 650°C (1,100°F to 1,200°F) with external surface temperatures of approximately 300°C (575°F). If the converter is misused or the vehicle is not maintained properly, the converter can develop internal temperatures that can exceed the melting point of the ceramic matrix, 1,250°C (2,300°F), and surface temperatures on the order of 500°C (900°F).[18] Such radiative sources can ignite combustible materials under the vehicle or in contact with the floorpan above them. Some vehicles have been completely burned when carelessly parked in dry brush or leaves with the hot converter in contact with the combustible litter. Prolonged idling of some converter-equipped vehicles, especially where air circulation under the car is limited, can produce sufficiently high temperatures that floor coverings and upholstery within the vehicle can be ignited by radiant or conducted heat. The newer generation of catalytic converters use smaller, often multiple, reactors whose surface temperatures are much lower than those of the 1980s. They also tend to be shielded better and much less likely to ignite fires inside or below the vehicle by contact with ignitable fuels.

Block and Other Heaters

In cold climates built-in block heaters or accessory block heaters (OEM and aftermarket) have been known to fail and ignite combustibles. Improvised heaters have included hair dryers, electric blankets, and trouble (service) lamps. Considering the damp conditions that may be present, each of these presents new ignition risks that should be considered (see Figures 9-6a and b).

FIGURE 9-6A Fire in engine compartment could have had several possible causes. This one was a challenge because the vehicle was parked and off at the time of the fire. *Courtesy of Peter Brosz, Brosz & Associates Forensic Services, Inc.; Professor Helmut Brosz, Institute of Forensic Electro-Pathology.*

FIGURE 9-6B Engine block heater (above oil filter) failed and caused the fire. *Courtesy of Peter Brosz, Brosz & Associates Forensic Services, Inc.; Professor Helmut Brosz, Institute of Forensic Electro-Pathology.*

Rags

Other types of common carelessness cannot be ignored as fire causes. Rags have been left in the engine compartment or on top of exhaust pipes and mufflers during servicing, only to ignite when the exhaust system reached its operating temperature (as much as 550°C; 1,000°F). The burning rag then spreads the fire to other combustible materials. It matters little whether the rag is oily or not because it is the rag that will be ignited by contact with the hot surface. Improperly installed hydraulic lines can come into contact with moving engine parts to cause penetration and loss of ignitable fluid, as shown in Figure 9-7. Overflows from leaking overfilled or overheated transmissions, power steering, or brake systems can flow or spray onto hot surfaces, causing ignitions.

FIGURE 9-7 An improperly installed power steering line rubbed against an engine fan belt and sprayed its contents, causing an engine compartment fire. *Courtesy of Greg Lampkin, Knox County, TN, Fire Investigation Unit.*

Tires

Drivers can be almost unbelievably insensitive to the condition of their tires, especially those on trailers. A deflated tire, when run against the road surface mile after mile, develops great frictional heat. This heat can build up sufficiently to ignite the tire, the hydraulic system of the brakes, and even the vehicle or its cargo, sometimes with disastrous results. Such a situation is most serious. The movement of the vehicle usually aids in dissipating the heat being developed, but when the driver finally stops to inspect the problem, the heat can ignite the tires to flaming combustion. Although tires under these conditions require a great deal of heat to ignite, once ignited they are extremely difficult to extinguish. Once burning, the oily char of rubber repels water, and the insulative properties of the rubber prevent cooling by conduction. Even a driver prepared with a portable fire extinguisher may be unable to prevent a disastrous fire. Single large tires on industrial (earth-moving) equipment have been measured to produce maximum fires of 3 MW (at 90 minutes after ignition by a diesel fuel fire).[19]

CONSIDERATIONS FOR FIRE INVESTIGATION

One interesting observation is that many accidental vehicle fires occur in the few minutes after being shut off. When the circulation of air, oil, and coolant is stopped, the external temperatures of many engine components (exhaust manifolds and systems, turbocharger housings, and the like) rise. Furthermore, with the cessation of movement and fan circulation, concentrations of ignitable vapors, which had been dispersed or diluted to safe concentrations, can accumulate to above their lower explosive limit. The combination of higher temperatures and higher concentrations of vapors, including pyrolysis products from smoldering plastics, can result in ignition. Such situations may be preceded by unusual odors, noises, mists, or electrical fluctuations, which can paralyze an engine management computer and cause stalling, so the driver should be asked specifically about any peculiarities that may have indicated such problems. These fires normally begin with a smoldering ignition that then goes on to flaming combustion to engulf the car, so the time frame for development is some minutes.

Electrical shorts developing in the maze of wiring and myriad components have a rich source of fuel available once temperatures are reached at which the plastics and resins begin to pyrolyze and ignite. These initial stages take many seconds or minutes to develop. Depending on the initial fuel and ventilation conditions, an accidental fire may not be visible for many minutes to someone outside the vehicle. A fire accelerated with a flammable liquid would be expected to be visible as open flames almost immediately after ignition, especially if a door, window, hatchback, or sunroof was open. However, a fire ignited only by a direct flame in a closed car may not be visible from a distance for some minutes.

The investigator should exercise great care in deciding what amount of damage is consistent with an accelerated fire versus that from an accidental fire. Normally, the time factor of fire development is usually a significant diagnostic sign of accidental origin. If a reliable witness observed the progress of the fire and can establish that the vehicle was totally involved in flames within 5 minutes or so of its arrival or last occupancy, it is a very strong indication that the fire was set with the use of an accelerant. Accidental fires usually require considerably longer to start, develop into flaming combustion, and spread than do accelerated fires. In the absence of a reliable eyewitness or videotape, it is very dangerous to estimate speed of development.

If the vehicle was being driven (or had just been parked) when it caught fire, the driver and any passengers should be interviewed to gain information about possible mechanical problems. Questions should include the following: How long was the car driven just prior to the fire? When was it last fueled? When and where was it last serviced? Were there any problems with starting, steering, or shifting? Were there any unusual smells? Did the car overheat recently? Were there any noises just prior to the fire? Were there any electrical events—lights coming on, instrument readings surging, accessories or controls not working properly, etc.? The owner should also be interviewed about modifications, special equipment, or recent repairs.

COMBUSTIBLE MATERIALS

The upholstery of seats, carpeting, headliners, and side panels represent a significant fuel load in every car or light truck. Modern cars make extensive use of vinyl plastics, polyurethane foam rubber, nylon carpeting, and polyester and cotton seat fabrics. All these products are combustible to a greater or lesser extent and thus provide fuel for fires started by direct accidental ignition with any open flame, as well as for fires started in other portions of the vehicle. It is estimated that some 20 percent (by weight) of a modern auto consists of combustible fabrics, plastics, polymers, and rubber. This constitutes on the order of 228 to 273 kg (500–600 lb) of fuel load, which is responsible for the fire damage to the vehicle, no matter how it is first ignited.[20] Although no recent published data are available, it is clear that today's vehicles use considerably more combustible plastic and synthetic components than these data reflect. Most of the interior (including seat backs, instrument panels, dashboards, and side panels) are readily combustible. Unlike in earlier cars, polymeric materials are used throughout—including exterior body panels, structural components, and even suspension and drivetrain—not just in the upholstery. Before an investigator concludes that a component has been stripped off a car before it was burned, he or she must make sure that it has not simply burned up with the rest of the fuel.

Motor Vehicle Safety Standard 302

It is a common misconception that the plastic upholstery materials in automobiles have to be flame retardant or pass ignition tests. In fact, the only applicable fire performance test is U.S. Motor Vehicle Safety Standard No. 302 (MVSS-302), which applies to all nonmetal parts of vehicle interiors.[21] It is an open-flame, horizontal-spread test that is very easy to pass, as it was designed in response to vehicle fires in which the headliners ignited and dripped flaming residues onto occupants. It has been in effect since 1972 without change and has been found to have had no measurable effect on the incidence or severity of fire injuries to passengers. The MVSS-302 test does not accurately reflect the fire performance of fabrics in real fires in vehicles, and replacement tests are being evaluated.[22] There are no tests for flammability of seat padding, which is almost universally polyurethane foam. Tests by the authors have demonstrated that even a modest open flame applied to the underside of a vehicle seat cushion will result in rapid involvement of the seat and potentially full involvement of the interior of the vehicle.[23] It has been suggested that materials used in higher-value cars today include more fire-resistant compounds, but this has not been confirmed by testing.[24]

MISCELLANEOUS IGNITION MECHANISMS

Accidental fires today rarely start from cigarettes lying in the crevices of seats due to the almost universal use of thermoplastic fabrics and urethane foam padding. (Of course, accumulations of cellulosic trash—facial tissues, newspapers, food wrappers—pose the same risks of smoldering ignition as they do in homes.) Although such trash fires are usually limited to smoldering, they can occasionally produce (if cellulosic fuels are present and if properly ventilated) flaming combustion that spreads quickly through the rest of the car. Pillows, blankets, and other nonstandard accessories provide the same risks to vehicles as they do in dwellings, because they may be subject to both flaming and smoldering ignition sources.[25] Foam rubber in later-model cars is more fire resistant to smoldering (versus open-flame) ignition than the latex foam and cotton padding used in earlier cars. Because of the variation in fire susceptibility, if there is reason to believe that a fire was started accidentally, it may be necessary to obtain a duplicate car from a wrecking yard and test various means of ignition directly. Virtually all synthetic materials used in modern car interiors are susceptible to flame ignition (e.g., a match flame). Most are thermoplastics, which means they melt as they burn, and the burning droplets of material then fall onto other ignitable surfaces and spread the fire very quickly. The flame temperatures and heat fluxes of such plastic fires exceed those of a gasoline fire and can mimic many of the same effects. Studies by Hirschler and Hoffman showed that nearly all plastic vehicle components in passenger compartments had short ignition times and high heat release rates when tested using cone calorimetry.[26]

Vehicle Arson

The NFPA estimated that incendiary fires accounted for 9 percent of all reported highway vehicle fires.[27] (It should be noted that many U.S. jurisdictions do not investigate vehicle fires, so this percentage should be considered very conservative.) Until recently, vehicular arson was quite rare, because a predominant motive, insurance fraud, was of minimal economic benefit to the arsonist. Unlike real property that can be traded back and forth to inflate its insured value prior to the fire, most motor vehicles are insured on the basis of their actual cash value—that is, their replacement cost less a substantial amount for depreciation. If a vehicle is burned, the payoff on the policy is usually based on Kelley Blue Book figures for that make and model, with credits added for accessories and reductions made for high mileage. Because these values are often quite low, compared with the actual resale value of the car or its components, it does not often pay the arsonist to burn his or her own vehicle. Motives are discussed in more detail in a later section of this chapter.

CONSIDERATIONS FOR FIRE INVESTIGATION

Setting fire to a modern vehicle does not require the use of an ignitable liquid accelerant. Any direct flame source of moderate size (a burning crumpled sheet of newspaper) under the seat or under the instrument panel can result in a fast-spreading, destructive fire inside a modern vehicle if there is ventilation. Whether an accelerant is used or not, ventilation is critical. A modern vehicle is sufficiently airtight that if the windows and doors are all closed, a fire in the passenger compartment will usually self-extinguish unless a window (or sunroof) fails from thermal shock early in the fire. As in structure fires, the investigator must be willing to treat every vehicle fire as a possible arson from the standpoint of evidence preservation and documentation, not from investigative bias or preconception.

Protocol for Vehicle Examination

The general protocol for vehicle examination follows the same logical pattern as for structure fires: documentation and examination of exterior, interior, and specialized evidence features. The procedure is outlined in Figure 9-8 as a general checklist protocol.

SAFETY

Safety is an important consideration in vehicle examinations, as it is in structure fires! Before beginning the examination, the investigator must be sure that all batteries are disconnected, air bags (steering wheel, passenger dash, side curtain, side impact, and seat backs) are disabled or discharged, pressurized fuel containers (LP or CNG) are sealed, and hydraulic cylinders (shock absorbers, bumpers, tailgate assists) are cooled. There are also risks from leaking fuel, broken glass, torn metal, and toxic gases from continuing smoldering. More components are made of carbon fiber, and burned carbon fiber has been linked to inhalation injuries.[28] If the vehicle is lifted to allow examination of its underside, it must be securely based on proper jackstands (or service rack) before anyone gets underneath. Although it may be necessary to use a forklift to position a vehicle for examination, it is violation of OSHA standards to have any person under a vehicle being supported by a forklift.

Safety is as important a consideration in vehicle examinations as it is in structure fires!

Motor Vehicle Fires

AGENCY	FILE NUMBER

VEHICLE DESCRIPTION

Year	Make	Model	License # ST Expiration
Color(s)		VIN#	

OWNER/OCCUPANT

Owner's Name	Owner Address	Owner Phone#
Operator Name/License#	Operator Address	Operator Phone#

EXTERIOR

Prior Damage	Fire Damage
Tread Depth	Wheel Type	Lug Nuts Present and Number

Condition of Glass

Tires/Wheels (Missing, Match, Condition)

Parts Missing

FUEL SYSTEM

Prior Damage	Fire Damage

Type Fuel	Condition of Tank	Filler Cap Condition	Fuel Line Condition

ENGINE COMPARTMENT

Prior Damage	Fire Damage

Fluid Levels	OIL	TRANSMISSION	RADIATOR	OTHER

Parts Missing

INTERIOR

Prior Damage	Fire Damage

AirBags Present	Sound systems and accessory

Ignition system	Keys in Ignition Yes or No

Personal Content Missing

Accessories Missing

Odometer Reading	Service Sticker Information

VEHICLE SECURITY

Alarm	Doors and Trunk Locked	Window Position

ORIGIN/IGNITION SEQUENCE

Area

Heat Source

Material Ignitied

Ignition Factor

Odometer Reading	Service Sticker Information

Field Fire Investigation Report Forms v6-10

FIGURE 9-8 Vehicle fire inspection data sheet. *Courtesy of Det. Mike Dalton, Know County, TN, Sheriff's Office.*

Motor Vehicle Fires

AGENCY	FILE NUMBER

Items found in Vehicle

Personal Effects	
Trunk/Cargo Area Contents	
Aftermarket Items	
Airbags In Place/Missing	
Personal Effects	

NOTES

DIAGRAM

Passenger Vehicles Compact–Full size   Pickup Truck   SUV   Van Mini - Fullsize

Field Fire Investigation Report Forms v6-10

FIGURE 9-8 (Continued)

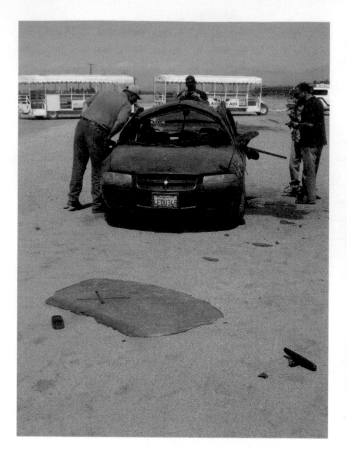

FIGURE 9-9A Vehicle with windshield and mirror blown out by explosion. *Courtesy John D. DeHaan.*

FIGURE 9-9B Partially burned air bag indicates it functioned during the fire. *Courtesy of Jamie Novak, Novak Investigations, and St. Paul Fire Dept.*

PHOTOGRAPHY AND SKETCHES

The vehicle should be thoroughly photographed inside and out before any disturbance. Photos should capture front, rear and both sides, plus quarter views. It is most preferable to initially examine the vehicle at the fire scene if at all possible. There, the distribution of glass can be measured and documented. Air bag explosions are very unlikely to throw glass fragments more than a few feet even to the rear of the car, but the directionality of fragment distribution from such devices should be considered (see Figure 9-9). Glass fragments found more than 2 m (6 ft) away are most likely to have been thrown there by a gas or vapor explosion within the vehicle. Smoke explosions are very rare in vehicle fires, so such glass will be soot-free, indicating an explosion prior to onset of fire, so the condition of the glass should be documented. Components that have fallen off from fire damage or fire suppression (or rescue) can be documented in situ.

The rectangular outline of most vehicles, with two or three "separate" areas of concern, lends itself to dividing the vehicle into quadrants, as shown in Figure 9-10: four in the engine compartment, four to six in the passenger compartment, and one in the trunk (and one for the underside!). Depending on the vehicle, separate quadrants can be considered for the console and dash/instrument panel. A photo showing number tags or a plan-view diagram showing the quadrant designations will make examination and note-taking easier.

Be sure to provide full documentation of the appearance of the vehicle by means of both photographs and sketches at all stages of all the scene examination. If the vehicle has not been gutted, document which portions(s) is more heavily damaged.

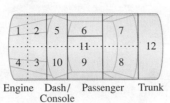

Engine Passenger Trunk

Engine Dash/ Passenger Trunk
        Console

FIGURE 9-10 "Quadrant" system for examining and documenting vehicles: four in engine compartment, four to six in passenger compartment, one or two in trunk, one for the underside. Quadrants can be added for console and instrument panel.

IMPORTANCE OF SCENE PRESERVATION

It is always much more preferable to examine a burned car or truck before it is moved or disturbed in any way. Valuable evidence is often destroyed or relationships lost when the vehicle is towed to another location for examination (see Figure 9-11b). Because most vehicle arsons take place, conveniently enough, in lonely, deserted areas protected from witnesses by trees or heavy brush, leaving the vehicle in situ is rarely a problem of public safety or even inconvenience. Figure 9-11c is a good example of the value of scene preservation. Had the utility trailer (foreground) been moved, the origin and cause of the much

FIGURES 9-11A AND B Even if vehicle has been moved, the scene of the fire should be examined and documented. Evidence of arson may also be found there. *Courtesy of Greg Lampkin, Knox County, TN, Fire Investigation Unit.*

FIGURE 9-11C
Destructive fire in
motorhome caused by
radiant heating from fire
in utility trailer parked
alongside. Notice contigu-
ous "V" pattern on both
vehicles. *Courtesy Chris J.
Bloom, CJB Fire Consultant,
White City, OR.*

more serious and costly fire in the motorhome would have been very difficult to establish. Access to the vehicle should be tightly restricted until the exterior examination is completed. Shoe prints of the perpetrator(s) (and sometimes those of the owner!) may be found around the car if there is soft soil. Another vehicle may have been present to retrieve the arsonists or to tow or push the vehicle to be burned to its final resting place. Tire impressions may be invaluable in reconstructing the crime or identifying the second vehicle and should be preserved by photography and casting, if possible.

EXTERIOR EXAMINATION

Portions of the vehicle that may fall off during any fire—gas cap, door handles, side mirrors, grillwork, headlamps, and the like—should be present in the debris near or under the car if the vehicle was not stripped prior to the fire. It is important to remember that in the interest of saving weight, substantial portions of the exterior trim and major body panels of current vehicles are being made from synthetic materials, which can burn almost completely away. Cans or jars used to carry an accelerant or the remains of initiating devices such as safety flares may be found under the car or nearby. All such evidence must be photographed in place before it is recovered. Latent fingerprints can also survive on the smoother surfaces of jars and cans, and this evidence should not be handled excessively before such an examination is conducted. Plastic side mirrors and door handles may melt off early and should be considered for latent print examination.

EVIDENCE OF STRIPPING

From the outside of the vehicle it should be possible to determine whether it was stripped prior to being burned. Most vehicle arsonists cannot resist removing the more salable items from a car or truck before burning it, thinking all signs of their removal will be destroyed anyway. Of course, many vehicles are burned to destroy the physical evidence relating to the original crime of theft and stripping. Are all mechanical components present and properly installed? Are the wheels and tires correct for the car? Are all the lug nuts present and tight? (Most people will not bother to replace all the lug nuts when putting "junk" tires and wheels on a car.) Even when the tires are badly burned, the portion

FIGURE 9-12 In spite of considerable fire damage to a vehicle, the tread pattern and amount of wear is still determinable from the pads in contact with the ground. *Courtesy of Greg Lampkin, Knox County, TN, Fire Investigation Unit.*

of the tread in contact with the road will not be burned, as shown in Figure 9-12. The tread pattern and depth can be established from these unburned "pads." Do they match the tires that the owner says were on the car?

CONSIDERATIONS FOR FIRE INVESTIGATION

An accidental fire in the engine compartment of a front-engine car is less likely to travel into the passenger compartment than an intentional fire (although it can happen if there are multiple duct penetrations of the firewall or sufficient combustible materials in contact with the firewall).[29] Recent changes in the design of automobiles have resulted in less separation between the luggage and passenger compartments. As a result, fire established in one area is more likely to spread to the other than in earlier cars, no matter what the cause of the fire.

Determine whether the windows were rolled up or down (being aware that position of electric windows may be impossible to establish). (Is the greatest amount of broken window glass down inside the door or inside or outside the car?) Examine the nature of char patterns on the outside of the body. If an accelerant was poured over a vehicle, splash and drip patterns like those in structures may be detected on exterior painted surfaces as discoloration, scorching, checkering, or blistering of the paint, as shown in Figure 9-13 (and Figure 9-4). Weather stripping and rubber seals are usually ignited.

Examination of the vehicle's underside is important. The paint or undercoating on the underside of a footwell floor pan is likely to be blistered or charred if a fire was ignited on the floor with a flammable liquid in that area, as shown in Figure 9-14.[30]

Fire tests have revealed that plastic and pot-metal fuel door latches can fail at relatively low temperatures, allowing spring-loaded fuel doors to pop open. Filler caps are often made of plastic today; these can melt and fall into the filler neck or burn off completely. Internal pressure can push a heat-softened cap off the filler tube. The investigator must be careful to examine the filler tube and tank before concluding the cap was removed deliberately.

FIGURE 9-13 Gasoline poured on the back of the vehicle and then ignited scorched and discolored paint and trim. *Courtesy of Jamie Novak, Novak Investigations and St. Paul Fire Dept.*

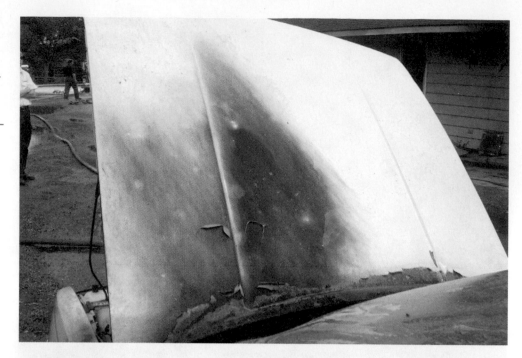

FIGURE 9-14 Gasoline burning in a footwell often produces localized charring or blistering on the underside of the body pan. *Courtesy of Tim Yandell, Public Agency Training Council.*

Char discoloration and oxidization patterns on the painted metal surfaces of the body and fracture patterns to the windshield glass may help reveal the direction of fire spread. If a fire starts in the engine compartment, heat conducted through body panels causes damage to paint layers that decreases in a radial pattern as the heat moves away from its original source. Fire progressing from an engine compartment into the passenger compartment will usually cause the windshield to fail at its lower edge first, with the most intense fracturing and delamination occurring there, as illustrated in Figure 9-15. Fire starting in the passenger compartment will tend to form a hot gas layer near the roof that

FIGURE 9-15 Engine compartment fire pattern. *Courtesy of Jamie Novak, Novak Investigations and St. Paul Fire Dept.*

attacks the windshield from the top and causes patterns of damage that radiate out into adjoining body panels, as shown in Figures 9-16a and b.

During a car fire, electrical connections can be made by the charring of switches, relays, and insulation. With a battery in place, headlights, wipers, horn, gas door releases, and even the starter have been seen to operate. If the vehicle is left in gear, the starter may move the vehicle some distance while burning, unless the handbrake is engaged.[31] Fires ignited in pressurized fluids can produce localized areas of damage from fire plume impact. Fuel leaks from hoses or fuel rails may cause "'hot spots" on engines, firewalls, or frame components. Transmission fluid fires will often be concentrated around the dipstick/filler tube or the housing below. (There may be oily residues adhering to nearby metal surfaces from such fires.) Many petroleum products fluoresce under ultraviolet (UV) or short-wavelength visible light, so examination with a forensic light source *may* be useful in spotting localized deposits.

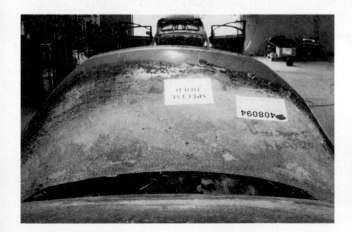

FIGURE 9-16A Concentric passenger compartment fire pattern on hood. *Courtesy of Jamie Novak, Novak Investigations and St. Paul Fire Dept.*

FIGURE 9-16B Concentric passenger compartment fire pattern on hood. *Courtesy of Tim Yandell, Public Agency Training Council.*

FIGURE 9-17 Extensive destruction of light-alloy and plastic components in a nonaccelerated vehicle fire.
Courtesy of Greg Lampkin, Knox County, TN, Fire Investigation Unit.

Because of the large quantity of combustible fuels present in the modern car, the duration and intensity of an extensive fire cannot be related back to the presence or absence of a flammable liquid accelerant. The accelerant serves mainly to ensure ignition of the substantial "natural" fuel load.[32] Figure 9-17 illustrates the extensive combustion of combustible automotive components—aluminum deck lids, plastic trim, composite panels, and light-steel-alloy hoods—in nonaccelerated fires in modern vehicles. Pouring gasoline on the outside of a closed modern car produces surprisingly little damage. Paint will scorch or blister, plastic trim will melt or char, and weather stripping may be ignited. Tests of modern vehicles conducted in 1992, however, showed that even a modest flaming ignition source beneath a seat would result in full involvement of the interior within 2 minutes of ignition, whether the windows were open or not.[33] The interior of a vehicle can be thought of as a compartment that will undergo the same development as the room fire described in Chapter 3. If provided adequate ventilation, fires inside vehicles with their rich fuel load can progress to intense, full post-flashover fires with the same high temperatures (over 1,000°C) as a well-furnished room.[34] Because a vehicle is dimensionally smaller, it will undergo the process even more quickly than a typical room. The maximum heat release rate for a fully involved modern automobile ranges from 1.5 MW to 8 MW, with newer cars exhibiting larger heat release rates.[35] Partial enclosures such as canopies, roofs, or parking garages produce larger fires that can readily spread to adjacent vehicles in as little as 8 minutes.

Direct exposure to fire can cause not only tires to burst but also shock absorbers, gas-filled suspension and hatchback struts, driveshafts, and bumper (collision) impact absorbers to explode, sometimes with dangerous force. Fire-weakened assemblies have also been known to explode as they cool after a fire. Such explosions can inflict damage that could be misinterpreted by an investigator, as well as pose a safety risk to fire suppression personnel and investigators.

Vehicle Identification Number

Before proceeding to the interior examination, the vehicle identification number (VIN) plate should be located and the identity and ownership of the vehicle determined. It is not beyond the realm of imagination that one vehicle was substituted for another to defraud an insurance carrier. The National Insurance Crime Bureau (NICB) and the Motor Vehicle Manufacturers Association (MVMA) produce guides to the location and interpretation of vehicle identification data plates. The NICB or the Canadian Automobile Theft Bureau (CATB) can also assist qualified members with locating "confidential" identification numbers hidden on every modern vehicle if the VIN plate has been destroyed.

FIGURE 9-18A Despite extensive fire damage, the base of the passenger air bag (dashboard) is readily visible. *Courtesy of Jamie Novak, Novak Investigations and St. Paul Fire Dept.*

FIGURE 9-18B The base of the steering column air bag may be on the wheel or on the seat or floor below. *Courtesy of John D. DeHaan.*

Interior Examination of the Vehicle

As previously noted, the vehicle can best be examined in quadrants—four in the engine compartment and at least four in the passenger compartment. Each quadrant can then be evaluated and documented for fire damage, fuel sources, and ignition sources using the ignition matrix concept described in Chapter 6.[36] Fire damage can include total or partial destruction as well as direction and intensity indicators. Because stolen vehicles are often at least partially stripped before they are burned, the most marketable components and accessories should be accounted for in the passenger compartment, including the stereo, CB, CD/DVD player, seats, navigation system, radar detector, air bags, and the like. They may be melted and badly damaged, but some remains should be on the floor beneath the mounting brackets. The seats are very valuable and a frequent target of strip-and-burn auto thieves. The air bag(s) will be triggered by the heat of the fire, but some remains of the deployed bag or its base plate should remain, as shown in Figures 9-18a and b. Because air bags are very expensive, they are often stolen for resale. Air bags functioning during a fire may give bystanders the impression that an explosive device was present. The gas generated in an air bag is nonflammable nitrogen. The remains of an air bag may be found outside the car (see Figure 9-9b).

If the owner is claiming a loss of valuable personal property with the vehicle—tools, clothing, fishing tackle, and so on—the metal pieces of such items should survive even the most complete conflagration. The jack and the spare tire (or at least its metal bead wires, belting, and valve) should still be identifiable. The ignition lock should be located even if it has melted and fallen from its mounting hole. Sometimes an "arranged" theft and fire will have been provided a convenient key, and this will often be found still in the ignition switch (as in Figure 9-19). Signs of tampering with, forcible damage to, or removal of the ignition switch and steering lock should be eliminated.

Upholstery

If the seat upholstery has not burned completely, it will behave in much the same way as household furniture, and its examination for relative consumption and direction of burning may be carried out in the same way. Most vehicles use a pad of foam rubber over a network of metal springs. This pad is covered in turn by a vinyl, leather, or polyester fabric covering. The foam rubber can vary considerably in its flammability characteristics, and if none remains unburned, it may be necessary to secure a control sample from a duplicate vehicle for laboratory testing. It tends to burn completely away in a well-ventilated fire in 10 to 30 minutes, no matter how the fire was started.

FIGURE 9-19 The remains of the ignition key are visible in the lock. *Courtesy of Greg Lampkin, Knox County, TN, Fire Investigation Unit.*

Accelerants and Other Liquids

If a liquid accelerant was used in excess, unburned liquid may be found soaked into the upholstery, carpets, and carpet padding or even pooled in low spots of the body pan.[37] (If the doors and windows are fully closed, the fire may self-extinguish, leaving raw unburned accelerant, as pictured in Figure 9-20.) One study reported that the thick soot on the interior of window glass contained identifiable residues of the ignitable liquids used to burn two vehicles in test burns (but the duration and maximum temperatures reached were not cited).[38] It is often the case that fire went undetected and the vehicle burned to completion, with *all* combustibles within the vehicle completely consumed, including any accelerants. In such cases, flammable liquid that leaked through and soaked into the soil under the vehicle may yet be detectable. When the vehicle is moved, the debris and soil under it should be examined carefully, using a hydrocarbon detector if possible. Samples should be collected and sealed in airtight metal cans or glass jars if there are any indications of flammable liquids. Remember that a leaking fuel tank can simulate such conditions, and a control sample of the fuel should be taken in most instances. Electric fuel pumps *may* continue to pump as long as the ignition is on and will empty the entire tank under the car if the electrical system remains intact and there has been no severe mechanical impact that would have triggered the interrupter fitted to some cars. (In many vehicles there is an interlock, so the loss of oil pressure or electrical signal from a stopped engine will shut the fuel pump off as well.)

Interior Debris

Portions of timing and ignition devices may be discovered in the interior debris. Safety flares make excellent ignition devices (and they are innocuously present in most vehicles) but a careful search can reveal their wooden end plugs and white or gray lumpy residues (see Figures 8-19 and 8-20). Paper matches survive even substantial fires if they are resting on a noncombustible surface and are worth looking for. Crumpled paper makes a good, fast initial fire. Setting fire to a small plastic bag of potato chips provides more than

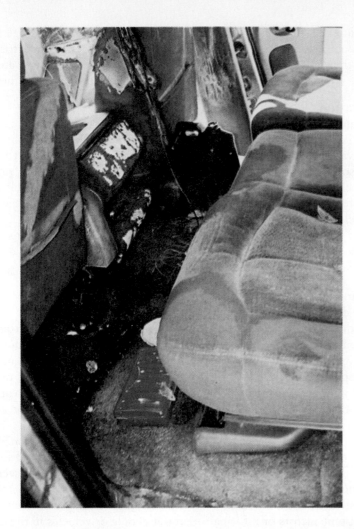

enough heat to ignite the upholstery in a modern car and yet burns away to an amorphous, unrecognizable carbonaceous ash. The oils in the chips produce a prolonged, intense fire because they are distributed on a thin, porous, combustible wick. The residue is easily overlooked and, even if discovered, may be dismissed as accidental. The vegetable oils present, however, may be identifiable in the carpet if it is analyzed by gas chromatography/mass spectrometry (GC/MS) by the same method as used for self-heating oils (see description in Chapter 14.) Fires under seats can also be accidental—from hot work done recently, rags left on exhaust pipes, overheated converters, or service lights left on inside the vehicle. Sometimes, an attempt is made to simulate an accidental fire by loosening fuel lines or drain plugs or disconnecting them completely. All such plugs and connectors should be accounted for and examined for the presence of loosening or fresh toolmarks indicating recent tampering. One multiple-vehicle arson involved the removal of fuel tank caps from a line of diesel tractor-trailer trucks. The fuel tanks were connected by means of fuel-soaked rag trailers, and a series of the tanks were ignited by floating a plastic cup with a small firecracker and blackpowder in each. The resulting fires destroyed a number of the vehicles, despite the failure of most of the rag trailers.

Mechanical Condition

If it is possible, the mechanical condition of the vehicle prior to the fire should be established. Mechanical problems may provide a motive for "selling" the vehicle to the insurance company, as well as a possible cause of an accidental fire. Observations should include whether the battery is in place and connected and whether the engine's electrical connections are still intact. Even if the wires melt or burn away, the metal connectors generally remain. Are all the engine components and accessories present? Do the water, oil, and transmission fluid levels indicate the vehicle was operational at the time of the fire?

Mechanical Examination

Disassembly of the engine or gearbox may reveal the operational status, which can then be compared against the owner's statements. Engine oil and transmission fluid should be drained. Simple observations for condition and contaminants can be carried out by eye, but such fluids can also be laboratory tested to reveal mechanical malfunctions prior to the fire. If the owner claims the vehicle to be complete and operational, a check of the engine compartment may reveal missing, disconnected, or out-of-place components that would render the vehicle inoperative.[39] Accidental fires involving the automatic transmission or power steering unit during operation usually produce severe damage inside the unit as fluid is expelled. External fires do not do the same kind of internal damage. Analysis of the oil or fluid may reveal the mechanical state of the unit.

Theft and Other Non-fire Damage

Non-fire-related body damage should be documented and compared against the owner's description of the condition of the vehicle. Previous damage may be detectable from part numbers or inventory. Body damage that is costly to repair but makes the car unsalable may be reason for destroying the car by an "accidental" fire. As in the investigation of structural fires, it is important that the investigator look for anything that seems unusual and out of place. No matter how clever the vehicle arsonist is, it is very difficult to arrange a destructive fire that can pass for an accident without modifying the vehicle or adding something to it to help it burn.

If the vehicle was found burned and the owner claims the vehicle had been stolen, there are additional avenues of inquiry. The scene where the vehicle was found (even if the vehicle has already been removed) should be examined as a potential crime scene for shoe and tire impressions, tools, accelerant containers, and means of ignition (matchbooks, lighters, flare striker caps). The area should be canvassed for possible witnesses who can help establish time factors or information about people or vehicles in the area at the time. The person who found the vehicle (or reported the fire) should be interviewed as to how they came to be in that location. The owner should be interviewed about the circumstances of the loss (Who last drove the vehicle and when? Was the vehicle locked? Was the alarm set? What were the vehicle contents? At what time and location was the vehicle last seen? At what time was the theft discovered? At what time was it reported?). The status of keys for the vehicle and access by others should be established. Modern theft or GPS/navigational systems may provide data about the movements of the vehicle prior to the fire. Some vehicles are also equipped with event data (black box) recorders that capture vehicle operational data before a crash! These recorders require specialized equipment and expertise for access. A new ASTM monograph describes the examination of vehicle data (black box) recorders from accident vehicles.[40]

Interviews

Wendt has published a comprehensive outline for interviewing the victim of a vehicle theft and fire.[41] Interviews may include the vehicle's owner, driver, occupants, and mechanic. Responding police and fire personnel are also important sources of information.

NHTSA Information

The National Highway Traffic Safety Administration (NHTSA) should be consulted for recalls or problems associated with the type of vehicle involved. One must be very careful not to assume that a recall *proves* the fire in question was caused by the failure reported! Such recalls must be included as a testable hypothesis in complete investigation.[42]

Vehicle Fire Tests

Considerable experience has been gained from laboratory tests involving the burning of complete automobiles under various conditions. These tests reveal that, for the most part, the amount of structural damage done to the exterior of the vehicle does not bear a direct relationship to whether the fire is of incendiary or accidental origin. Some tests have revealed that temperatures near the roof of a burning car reach as high as 1,000°C (1,860°F) at some stage of a fire no matter what the means of ignition.[43] This is more than sufficient to warp sheet metal and melt glass and aluminum trim. As a result, body and roof panels buckle and warp, window glass melts and icicles, and seat springs lose their temper at temperatures developed in slow-burning accidental fires as well as in fast, accelerant-fueled fires. Similarly, the color, density, and appearance of soot and pyrolysis products on remaining windows depend largely on the upholstery materials present and the degree of ventilation present during the fire. Although the amount of structural damage will, of course, be generally related to the intensity and duration of the fire, those features are not necessarily determined by whether or not the fire was incendiary. One series of tests indicated that, in fact, more complete damage can result from a slow-developing accidental fire that has more time to bring more of the entire mass of the vehicle up to high temperatures than from a fast-moving gasoline fire.[44]

Motorhomes and Other Recreational Vehicles

A recreational vehicle is defined as a motor vehicle containing living quarters for recreation, camping, travel, or seasonal use. Recreational vehicles come in different styles, shapes, sizes, and with different accessories. They range from a simple tent trailer to a bus-chassis motorhome. The basic designs are the camping trailer, truck camper, travel trailer, fifth wheel, park trailer, and the motorhome.

CHARACTERISTICS OF MOTORHOMES

Most motorhomes are constructed and assembled from four distinct and important components, often made by different manufacturers: the unit's engine; the transmission; the chassis; and the remaining portions of the unit, including the assembly and installation of appliances and furnishings. It is not uncommon for there to be several fire investigators examining the remaining debris, each one representing the individual interests of each manufacturer involved.

Motorhomes are categorized into three distinct classes: Class A, Class B, and Class C. Class A motorhomes are the largest units and are designed and built from the ground up. Class B motorhomes are van conversions that have been altered to accept appliances and furnishings for the living space. Class C motorhomes are units that incorporate a van cab portion, with an attached body, usually with an over-the-cab sleeping area.

FIRE RISK

Motorhomes are subject to a combination of the risks of accidental fires of vehicular origin with those peculiar to the heating, cooking, and living conditions found in dwellings. Because the interiors of these vehicles are used for day-to-day living, the hazards of cooking and smoking materials are as present as they are in a stationary dwelling. Many such vehicles have moderate-capacity electrical systems of 12 V, so electrical fires, while not completely impossible, are quite rare. Some vehicles have large-capacity, complex electrical systems of 24 V, which provide higher risks. In addition, many of them are fitted with a parallel 120 VAC system to operate lights, heaters, radios, and the like when they are connected to an outside source of AC power or to a portable generator mounted within the vehicle. Many of the same appliances found in the home will be found in the motorhome. High-current devices such as televisions, toasters, and microwave ovens are common, as well as their attendant risks of misuse.

PROPANE TANKS

Most such vehicles also include a propane (or LPG) tank to provide fuel for the heater, stove, refrigerator, and even generator with the attendant hazards of explosions from spills and leaks. If factory equipped, the recreational vehicle will be fitted with either DOT or ASME tanks, steel and/or copper piping, rubber hose connections with brass fittings, valves, and the appliances they serve, including a refrigerator, furnace, stove, and even a generator. The size of the propane tanks can vary from 5-gallon tanks on trailers to upward of 80-gallon tanks on the more expensive motorhomes and bus conversions. There is no guarantee that propane systems will be properly installed in amateur conversions. One explosion, costing 12 lives, involved a modified motorhome. The cause was traced to an improperly installed LP tank that leaked its contents into the vehicle until it was ignited by an electric arc or open flame as the vehicle proceeded down the highway.[45]

All propane tanks are equipped with some type of a pressure-relief valve, which is designed to vent the contents of the tank at a certain pressure. When the propane tank heats up due to flame impingement, the pressure inside increases. When the pressure inside becomes greater than the pressure-relief valve setting, the valve opens and rapidly expels the contents away from the motorhome and tank. This expulsion will continue until either (1) the pressure inside the tank is less than the pressure-relief valve setting, at which time the valve will close again, or (2) the valve expels all the contents of the tank during the fire. See Chapter 4 for a full discussion of propane tanks.

Many motorhomes and travel trailers today are fitted with propane detectors. The location of these must be carefully assessed, for thermal currents in the vehicle may move the fuel/air vapor away from the detector. Recall that the density of an ignitable propane/air mixture is very close to that of air, so it will not settle like pure propane.

Manufacturers make great use of plastics, synthetic fabrics, foam rubber, and particleboard to save both cost and weight. The structural elements are most often a skeleton of aluminum tubing to which is riveted an exterior aluminum or fiberglass skin. Plywood or synthetic paneling is fitted to the inside, and thermal and sound insulation is provided by fiberglass or polystyrene or polyurethane foam sandwiched between those layers. Once the main fabric of such a vehicle is ignited, there is little to prevent it from being completely destroyed. While the vehicle is burning, formaldehyde and various cyanogens are being released from the interior finishes, sometimes with fatal results to any occupants. Although most reputable manufacturers strive to make their vehicles as safe as possible, there is no code applicable to custom conversion facilities, and the fire investigator must be alert to the possibility of finding patently hazardous practices and products in such vehicles. The competition in this area is fierce, and the most cost-effective use of materials, even to taking shortcuts, is the key to survival.

CONSIDERATIONS FOR FIRE INVESTIGATION

Almost every fire involving a recreational vehicle will develop very quickly due to the lightweight construction utilizing foams, resins, plastics, and other highly combustible materials. The federal standards for allowable flame spread rate are on a scale from 25 to 200, with 200 being the maximum allowed. Due to the lightweight construction methods and materials employed, a flame spread rating of 200 is allowed for recreational vehicle construction. Because of the materials involved, when a small fire occurs, it develops rapidly and quickly involves the gasoline or diesel fuel tanks and the propane system. It is not uncommon for a small fire to completely consume a recreational vehicle down to the frame within 12 minutes.[46]

When investigating fires in such vehicles, the investigator should expect substantial destruction of their structural elements, with consequent loss of evidence that would normally permit a reconstruction of the fire sequence. Within the limitations of the construction materials involved, diagnostic signs of fires caused by electrical failures, smoking

FIGURE 9-21 Remains of a Class A motorhome ready for examination. *Courtesy of Greg Lampkin, Knox County, TN, Fire Investigation Unit.*

materials, heating systems, and arson remain largely as they are described elsewhere in this book. The investigator must be aware of the extensive fire damage resulting from the construction of such vehicles, as shown in Figures 9-21 and 9-22.

Photographs should be taken at every step prior to, during, and after the investigation (including the general scene). This is to allow documentation and reconstruction of the scene as found. Valuable information that was not readily apparent during the initial examination could easily become visible after the photographs are developed. When examining the remains of a motorhome, it is also important to examine the scene where the fire occurred, because critical evidence may have dropped off and away from the vehicle

FIGURE 9-22 The remains of a large motorhome being properly transported on a flatbed carrier. Total destruction resulted from an accidental origin. *Courtesy of Joe and Chris J. Bloom, Bloom Fire Investigation, Grants Pass, OR.*

FIGURE 9-23 A passerby recorded this accidental fire in a Class C motorhome, which destroyed the vehicle. Such recordings preserve indicators such as direction of fire travel, time involved, suppression attempts, and wind direction. *Courtesy of Joe and Chris J. Bloom, Bloom Fire Investigation, Grants Pass, OR.*

during the fire and its suppression. It is most important when inspecting the unit, unless it is absolutely necessary to move them, to leave the remains undisturbed. If items or debris need to be moved to allow for an examination, only professionally accepted practices for the documentation and removal of the debris should be employed. When the unit needs to be moved from the scene location to a storage facility, the unit should be moved via a flat-bed towing truck and tarp-wrapped, if at all possible, to minimize potential loss or damage to components. The reader is referred to specialist references for guidance.[47] Because of the enormous variety of styles fittings and interior finishes, it is often helpful to locate and inspect an exemplar unit of the same type, floor plan, and age. Photos or videos taken by owners or passers-by may capture vital data, as illustrated in Figure 9-23.

Mobile Homes (Manufactured Housing)

CONSTRUCTION AND MATERIALS

In many respects, the structure and materials of mobile or manufactured homes (sometimes called *modular* homes), the causes of fires within them, and the problems of investigating such fires are similar to those just discussed in regard to recreational vehicle fires. Structurally, manufactured homes are most often built on a heavy steel frame supported on three axles. A wooden floor structure of plywood or particleboard provides a base for the erection of a wooden or metal framework upon which the exterior aluminum or wooden sheathing is installed. The interior walls are often framed (as are exterior walls) with 2-in. × 3-in. wood studs and then covered with plywood, wood paneling, or more recently, gypsum board specially designed to resist the flexing and vibration induced by moving. Insulation is often fiberglass batting in the roof and exterior walls. The interior finish in many such mobile homes is similar to the paneling of motorhomes and thus poses a high risk of rapid fire development. The occurrence of fully involved (post-flashover) fires seems to be much more common in mobile homes, and the time to flashover is much reduced over that seen in most standard buildings. The small-dimension wood members burn quickly and allow more rapid structural collapse than in a structure of standard construction. With the advent of *NFPA 501B: Manufactured*

Housing (imposed by the U.S. Department of Housing and Urban Development on manufactured homes starting in 1976), the speed and severity of development of fires has been notably reduced.[48] The date of manufacture of the structure being investigated is therefore critical to estimating what the materials and construction would be (as well as smoke detectors and the like). The recent development of gypsum wall finishes suitable for manufactured housing has made fire performance of such housing more closely parallel to that of fixed structures.

The tubular nature of the interior structure uninterrupted by firestops provides for rapid fire spread throughout, and combustible wall finishes and low ceilings (typically 7 ft) provide for extremely rapid buildup and flashover conditions. Davie has reported measuring temperatures at floor level of a mobile home equal to those at the ceiling—750°C to 1,000°C (1,400°F to 1,800°F). This is due to the size and shape of the rooms, the combustible nature of wall and flooring materials used, and the reflective properties of the outer walls and roof.[49]

The development of fires in such structures to post-flashover follows the same pattern as in fixed structures but more quickly. The location, thickness, and spacing of the structural members and the installation of electrical and heating equipment are considerably more variable in mobile homes than in conventional structures. (Walls are often framed and erected after carpet is laid across the entire floor area, for instance.) The motion produced when the vehicle is moved or when fans, ventilators, or air-conditioning units are operated can chafe wire insulation or loosen connections far more than would be expected in fixed structures.

CONSIDERATIONS FOR FIRE INVESTIGATION

It is thought that the small rooms and narrow hallways typically found in these structures contribute to the rapid development of flashover and unusual burn patterns, such as those observed in Figure 9-24, where the most intense fire was to the floor and wall to the right of the bed and not at the origin. The minimal insulation and conductive roof covering (typically aluminum or steel sheathing) may allow fire damage to the roof to record the development of fire in the rooms beneath. An investigator should be wary of placing too

FIGURE 9-24 Fire in the bedroom of this mobile home started near the floor to left of bed. The dynamics of air entrained from the entry door during post-flashover burning caused a false "V" of severe burn to the right of the bed. *Courtesy of Jamie Novak, Novak Investigations and St. Paul Fire Dept.*

FIGURE 9-25 Despite a relatively short burn time of 20 minutes, this furnished mobile home was nearly totally destroyed. Burn patterns on the floor are due to radiant heat. Fire ignition was by means of a small wood crib on the kitchen counter behind and to right of the photographer. Low burns, consumption of the sofa against the far wall, and penetrations of the floor are the result of fire spread. *Courtesy of John J. Lentini, Applied Technical Services Inc., Marietta, GA.*

much importance on the presence of complex burn patterns on the floors of a heavily damaged mobile home. The combustible floor materials and wall coverings and high fuel loads combine to produce intense post-flashover fires that in some cases can penetrate the floors under furniture as well as in exposed floor areas. (Pre-fire degradation of floors from water incursion will accelerate floor combustion and failure.) Rapid failure of single-glazed windows adds ventilation sources. Post-flashover burning (especially when synthetic carpet and pad are involved) will produce irregular burn patterns on floors, as shown in Figure 9-25. Although the fire in a mobile home behaves in a manner consistent with basic scientific and engineering principles, the investigator must be careful not to use the same criteria for evaluating indicators of time and estimating fuel effects as used in evaluating fires in buildings of standard construction.

When investigating fires in mobile homes, the investigator should expect substantial destruction of their structural elements, with consequent loss of evidence that would normally permit a reconstruction of the fire sequence. Within the limitations of the construction materials involved, diagnostic signs of fires caused by electrical failures, smoking materials, heating systems, and arson remain largely as they are described elsewhere in this book.

Heavy Equipment

Motorized equipment such as tractors, loaders, earthmovers, forklifts, harvesters, and the like can pose special fire risks. If not cleaned properly, agricultural equipment can accumulate thick layers of cellulosic dusts that can ignite from hot surfaces (as shown in Figure 9-26). Excess dusts and fibers can accumulate in moving parts to cause frictional heating in belts, bearings, and gearsets. Improperly maintained hydraulic and oil systems cause special problems, as they operate at very high pressures when in use. The extremely fine aerosol mists that are created at such leaks are readily ignited by hot surfaces or electric arcs. Vibration causes loosening of fittings and fracturing of rigid pipes—all leading to loss of fluids under pressure. Fires started from hydraulic or other fluid leaks can spread very quickly to engulf the equipment. Fires started in solid dusts grow more slowly but can be ignited for some time after the equipment is shut off and left unattended. Either

FIGURE 9-26 Fire in harvester caused by accumulation of combustible dust. *Courtesy of Greg Lampkin, Knox County, TN, Fire Investigation Unit.*

ignition can lead to a major fire involving expensive equipment. The size of combustible components is a factor in determining the size of the resulting fire. Heat release rates as high as 3 MW have been observed in tests of single earthmover tires (ignited by a pool of diesel fuel beneath the tire.)[50] Large tanks of diesel fuel or hydraulic oil (or propane cylinders to aid in cold-weather starting) can also contribute to the fire's intensity and duration. The investigator must be very familiar with the specialized equipment used in such vehicles before attempting to determine the origin and cause of major fires.

Boats and Ships

Pleasure boats are more like automobiles in their relative fire hazards and susceptibility than dwellings or larger ships. The boat and the automobile most often have a gasoline-powered motor, gasoline tank, and electrically driven accessories. There is one major difference that makes the boat fire more destructive and much more difficult to investigate than the auto fire, namely, all the flammables are enclosed in a more-or-less tub-shaped container—the hull. Leaking gasoline or its vapors will settle into the bottom of the hull, where they remain unless circulated by drafts from the engine fan or bilge pump. Comparable leakage in an automobile is generally to an open or well-ventilated exterior. This is the basic reason for the greater danger and likelihood of explosion of fuel vapors in a boat. In addition, many boats are made of wood or fiberglass, while the automobile does not usually consist of such a combustible "container."

A gasoline fire in a wooden or fiberglass container is a far more formidable occurrence than the same fire in a steel conveyance. Wooden decks and cabins are likely to be burned completely, and the hull is likely to be burned down to the waterline. Fire patterns are therefore obscured or destroyed, and points of origin are often not identifiable. Smaller rooms and the presence of all-wood interiors will produce more intense, faster growing fires than normally found in the larger, gypsum-wallboard finished compartments of buildings. With larger wooden vessels that have considerable secondary interior structure, patterns will be found that are in every way comparable to those in any other similar wood-framed structure. Such fires can be investigated in accordance with the principles that are detailed throughout this book.

In addition to sources of gasoline vapors, boats may have heaters or stoves that are fueled by kerosene, alcohol, or LP gas. Each of these poses risks of accidental fires and must be considered when investigating these fires; LP gas can settle into the hull like gasoline vapors, and kerosene or alcohol stoves can spill or leak, producing pools of ignitable fuel. The electrical systems may include separate systems for 120 VAC and 12 or 28 VDC power. On newer vessels, these electrical systems must meet the requirements of the U.S. Federal Boat Safety Act of 1971, but earlier ones need meet no requirements, and modifications may or may not have been done correctly. This same act requires flame arrestors, appropriate fuel lines, and other safety equipment. Minimizing the risk of fire and loss of life on such crafts is the focus of *NFPA 302: Fire Protection Standard for Pleasure and Commercial Craft.*[51] Like standards for structures, such guides can only describe recommended (or required) practices, not capture the content and fire risks as seen in real-world occupancies. The investigation of boat fires involves the evaluation not only of the presence of such features but their operability and effectiveness (e.g., flame arrestors may be present but are compromised by a hole punched through them by a careless mechanic). One survey of small-vessel fires reported that 55 percent were caused by wiring and appliances (both AC and DC), 24 percent were caused by overheating of engines or transmissions, 8 percent by fuel leaks, and 1 percent by stoves.[52] Some vessels, such as the houseboat shown in Figure 9-27, combine entirely combustible (exposed wood) structure and linings with large windows and numerous household ignition sources and corroded and contaminated electrical fittings, leading to a high risk of very intense, destructive fires.

SHIPS

Although large steel-framed ships would not be expected to burn, it is fortunate that most fire investigators are not called on to examine such fire scenes frequently, because such

FIGURE 9-27 Houseboat gutted by accidental fire. All-wood structure on concrete "bathtub" hull. Large windows assured excellent ventilation as they failed in the fire. *Courtesy of John D. DeHaan.*

ships and their flammable contents can burn with incredible ferocity and massive destruction. This destructiveness is based largely on three factors: fuel load, spread of fire by conduction, and inhibited access to firefighters, which are all peculiar to large steel ships. Although the steel structure itself is not readily ignitable, a considerable fuel load is present in the ship's interior finishes, furnishings, cargo, and fuel. The interior finishes and furnishings represent the same degree of combustibility hazard as comparable furnishings in a dwelling. A large cargo of freight, however, imposes its own flammability characteristics on the ship. In bulk shipping of products like wood chips, grain, coal, and natural fibers, smoldering fires are common and are very difficult to extinguish. Water tends to wet only the top layers and not penetrate, and extinguishing gases are expensive. In most cases, the holds with the smoldering cargo are sealed and left, hoping the fire will self-extinguish. Unfortunately, due to the low oxygen limits for smoldering combustion, considerable combustion can occur, creating heat and combustible gases that are trapped in the hold. When fresh air is admitted, a backdraft or smoke explosion can result with disastrous effect. Self-heating of bulk cargoes is also a problem, particularly in hot climates. Steam, nitrogen, or carbon dioxide gas injection is sometimes used to suppress cargo compartment fires. The cargo itself can contain flammable or explosive vapors, as illustrated in Figure 9-28.

Explosive dust suspensions are encountered during loading and unloading of flour, sulfur, and similar finely powdered products. Such products as ammonium nitrate, used for chemical and fertilizer production, are well recognized for their hazardous properties, and special precautions are required.[53]

TANKERS

Tankers, often loaded with very large quantities of fuel, generally of petroleum origin, might be expected to be subject to enormously increased hazards. The main cargo of a tanker while contained is no more dangerous than is the gasoline tank on an automobile and for the same reason. Bulk cargo may be contained in freestanding containers or in open holds. Many tanks are provided with inert gas (such as nitrogen) systems to exclude

FIGURE 9-28 Explosion in container midvoyage caused by leakage from used gasoline tanks being shipped for recycling. *Courtesy of Dennis Field, Fire Cause Analysis.*

air from the ullage spaces. However, the danger of a fuel spill above board, where the vapors can mix with air, creates a hazard similar to that produced by a fuel spill in the exterior in any other environment. Ignitable liquid cargoes require that special precautions be taken. Far more serious hazards are associated with a rupture in the hull of a tanker that releases great quantities of fuel or with tanks that have been unloaded, where air can mix with the fuel vapors, forming an explosive mixture.

SHIP CONSTRUCTION AND FIREFIGHTING TECHNIQUES

Once a fire has started, by whatever cause, extinguishment is very difficult and hazardous. The modern welded-steel ship can be thought of as a huge box girder with the steel plates of the sides, bottom, and main deck forming the four sides. This box is strengthened by steel girders forming the keel or "backbone" and the frames or "ribs." It is separated into many smaller compartments by the lower decks and transverse bulkheads.[54] These elements are part of the strengthening substructure and are also all steel (or aluminum in some ships). Unlike a fire in a frame or masonry structure, fire in one compartment of a ship is not thermally insulated from neighboring compartments. The steel plates conduct heat quickly and efficiently to other adjacent or overhead compartments. Combustibles in these compartments quickly reach their ignition points and contribute to the fire spread. Damage may be more extensive in compartments above or alongside the point of origin than in the actual compartment where the fire started. Tunnels and vertical chases for electrical and hydraulic systems provide chimneys for the spread of fire by convection or even direct flame impingement.

TERMINOLOGY

Certain terms peculiar to vessels are helpful to the fire investigator in interviews (see Figure 9-29).
Some of these terms include the following:

Hull: Main body of a vessel—may be steel, aluminum, wood fiberglass, concrete
Deck: Horizontal element separating levels
Bulkhead: Vertical partition separating compartments
Bridge: Location from which the vessel is steered and controlled
Cockpit: In small boats, the opening in the deck from which the vessel is steered
Ladder: Vertical access between decks
Passageway: Horizontal access between compartments (hallway)
Porthole: Transparent windows from a compartment through hull
Galley: Kitchen area
Boat: Smaller water vehicle (sometimes defined as capable of being carried by a ship)
Ship: Larger, oceangoing vessel

Source: F. Herrera, "Investigation of Boat Fires," *California Fire/Arson Investigator* (January 2005): 10–11.

FIGURE 9-29 Terminology for ship construction.

Standard firefighting practice in dwellings often calls for getting above the fire to ventilate it and thereby to cool it. In a steel-compartmented ship, access above the fire is often restricted, and entry may not be achieved by anything less than an oxyacetylene torch. In such conditions, the spread of fire upward through a ship may not be preventable, and efforts will be directed at removing cargo and reducing loss of life.

Fires aboard very large modern cruise ships combine firefighting difficulties with huge life safety concerns in which over 5,000 passengers may be at risk.[55] Fires on such ships have been started by discarded smoking materials, welding repairs, laundry fires, and arson. *NFPA 301* and the International Maritime Organization (IMO) *Convention for Safety of Life at Sea* ((SOLAS) 1992) are the predominant codes for such ships.[56]

Idle vessels are often moored side by side or berthed alongside disused piers, where fire can spread from pier to ship or from ship to ship by radiant heat, flaming debris, or direct flame contact. The investigator must be aware of the mobility of such vessels (before and after the fire) and that the fire may not have been an isolated occurrence. The original location of the vessel must be determined as part of the investigation. Because of the unusual nature of the fire conditions aboard such vessels, the investigator is directed to seek the assistance of specialists in this area.[57]

Motives for Vehicle and Marine Arson

As in any other kind of fire investigation, the intimation of a possible motive for arson may provide a significant turning point for any vehicle fire investigation. Although proof of motivation is not necessary for the conviction of an arsonist, it can pull all the pieces of an investigation together. Many of the same motives discussed in Chapter 16 will apply to vehicle arson. Insurance fraud, either involving the owner directly or the financial institution that is carrying the loan, is certainly a consideration. The fuel crisis of the 1970s and 2006–2009 prompted many owners of large, gasoline-thirsty vehicles to dispose of them. Government-imposed requirements for mandatory emission-control modifications may also force owners to abandon vehicles. Depressions in the general economy in the 1980s (and currently) forced people to acknowledge that they had paid too much for a vehicle or that the monthly payments were too high to maintain. With a substantial increase in the number of leased cars; changes in the tax deduction bases for cars, trucks, and other vehicles; and changes in the marketability of expensive or fuel-inefficient cars of recent vintage, vehicular arson is more common today than ever before. Faced with a substantial payoff due on a leased car or truck or when stuck with a relatively new vehicle whose resale value has plummeted, an individual or a company may see "selling" it to the insurance or leasing company as a convenient way out. Large diesel tractors routinely cost over $150,000 today, and a major engine rebuild can cost over $20,000. If a company or private trucker is operating on a thin margin, the loss of the vehicle to a convenient fire may be the solution to financial problems. Vehicles are also subject to arson motivated by a number of other reasons—spite or hatred, intimidation, vandalism, concealment of a crime, or even pyromania. Vehicles stolen for joyriding, for parts, or for the commission of other crimes are often burned to destroy fingerprints or other associative evidence.

Inquiries as to the purchase price of the vehicle, standing loan balance, and monthly payments should be accompanied by a determination of the fiscal well-being of the owner. Are there indications of reduced income, unemployment, business losses, or gambling debts? If a vehicle is mechanically unreliable or inoperative, it is more likely to be a candidate for arson. The same holds true if the vehicle is no longer used or has outlived its usefulness. Domestic problems are sometimes behind the destruction of an automobile, recreational vehicle, or pleasure boat. A vehicle that has sentimental value to one partner may be destroyed by the other as a gesture of spite or anger. Any indications of marital difficulties should prompt a closer examination of the circumstances of the fire. As with other fires, the investigator must always be careful to keep motive and proof of cause separate and distinct determinations.

With fires in large commercial ships, any of the same reasons that prompt the burning of land-based properties can hold true. Business-related motivations can include intimidation, elimination of competition, labor unrest, or the concealment of other crimes, particularly theft, pilferage of cargo, and smuggling of goods or drugs. The financial motivations are even stronger when entire fleets are idled by international trade or monetary reversals. The thorough investigator must consider all possibilities when the values of vehicles and their cargoes reach into the thousands or millions of dollars, thus providing rich rewards to those clever enough to escape detection. Fires have been set aboard military vessels to delay deployments.[58]

At the same time, the investigator must remember that a motive alone does not change a fire of undetermined cause into an arson. The origin and cause of a fire is established from careful data collection and proper analysis of the evidence at the scene. There are many ignition sources in the average vehicle or vessel, rich fuel loads, and many ways for accidental fires to ignite. The investigator has to be just as diligent in establishing the first fuel ignited (suitable in both type and quantity), a competent heat source, and the mechanism for contact in these fires as in a structure fire before concluding the cause of the fire. If not suppressed shortly after ignition, these fires can result in such complete destruction of the vehicle or vessel and evidence of the fire that an origin and cause cannot be reliably established.

Summary

With their combustible interiors, ignitable liquid heat sources, moving parts, and active electrical systems, it is not surprising that motor vehicles catch fire. Intense, rapidly developing fires can result if sufficient ventilation is available, and they can bring about nearly total destruction in a short time. (Peak heat release rates of modern vehicle fires have been measured at between 1.7 and 7.5 MW.)[59] To carry out a scientific fire investigation and properly create and test a full range of hypotheses, the investigator must be aware of the ignition properties of a range of fuels commonly found in vehicles and the dynamics of fire spread and effects on materials. The investigator must also be very familiar with the wide variety of complex electrical and hydraulic systems and accessories in modern vehicles and how they can ignite and burn. As with structure fires, total destruction and rapid fire development are not proof of an incendiary fire.

Review Questions

1. What are the most readily ignitable liquids found in a modern auto?
2. How do the boiling points of liquid fuels affect their hot-surface ignitability?
3. Which produces the higher internal temperatures in a modern vehicle fire—a flammable liquid arson or direct ignition of the upholstery—and why?
4. Why is it important to examine the scene of a vehicle fire before the vehicle is removed?
5. How can transmission fluid become ignited?
6. Does extensive combustion of body panels and structural and suspension components indicate an accelerated fire in a modern vehicle? Why or why not?
7. Name five of the items most commonly removed before a stolen vehicle is set afire.
8. Why is it nearly impossible to cause an explosion of a gas tank?
9. Why do recreational vehicle fires develop quickly?
10. What makes extinguishment of a ship fire so difficult?

References

1. M. J. Karter, "Fire Loss in the U.S, 2007," *NFPA Journal* (September–October 2008): 46–51.
2. R. J. Hrynchuk, "A Study of Automobile Fires," *Canadian Society of Forensic Science Journal* 11 (1978): 15–22; K. Allen et al., "A Study of Vehicle Fires of Known Ignition Source," *Fire and Arson Investigator* 34 (June 1984): 33–43, 35; (September 1984), 32–44.
3. American Petroleum Institute, "Ignition Risk of Hydrocarbon Vapors by Hot Surfaces in the Open Air," API Publication 2216, 2nd ed., (Washington, DC: API, January 1991); N. R. LaPointe, C. T. Adams, and J. Washington, "Autoignition of Gasoline on Hot Surfaces," *Fire and Arson Investigator* (October 2005): 18–21; J. D. Colwell and A. Reza, "Hot Surface Ignition of Automotive and Aviation Fluids," *Fire Technology* 41 (2005): 105–23.
4. NFPA, *Fire Protection Guide to Hazardous Materials,* 2001 ed. (Quincy, MA: National Fire Protection Association, 2001).
5. M. Higgins, "Vehicle Fires: A Practical Approach," *Fire and Arson Investigator* (Spring 1993).
6. J. H. Potter, Ford Motor Company Specialty Products Engineering, personal communication, 1988.
7. C. Cope, "Vehicle Engine Compartment Fires" in *Proceedings Fire & Materials 2009*, (London: Interscience Communications, 2009), 561–79.
8. G. March, *An Investigation and Evaluation of Fire and Explosion Hazards Resulting from Modern Developments in Vehicle Manufacture*, Newcastle, Tyne and Wear Fire Brigade, UK, 1993.
9. F. J. Suefert, "Automobile Engine Fires," *Fire and Arson Investigator* 45, no. 1 (June 1995): 23–26;

"Automobile Engine Fires (Part 2)," 46, no. 1 (September 1995): 23–26.

10. Ibid.

11. Ibid.

12. R. K. Marley and J. D. DeHaan, *CCAI Vehicle Fire Tests: A Photographic Report* (Sacramento, CA: California Criminalistics Institute, 1994).

13. Ibid.

14. J. Bloom and C. Bloom, "Ford Ignition Switch Overheating and Fires," *Fire and Arson Investigator* 47 (March 1997): 10–11.

15. J. Morrill, "Analysis of a Ford Speed Control Deactivation Switch Fire," *Fire and Arson Investigator* (July 2006): 22–27.

16. *NFPA 921: Guide to Fire and Explosion Investigation* (Quincy, MA: National Fire Protection Association, 2008), 223; (2011, 257).

17. NFPA, "U.S. Vehicle Fire Trends and Patterns" (Quincy, MA: National Fire Protection Association, July 2008).

18. Universal Oil Products, personal communication, 1979.

19. H. Ingason and R. Hammarstrom, "3MW from a Single Tyre," Brand Posten, SP, Boras, Sweden #39 2008, p. 6.

20. Allen et al., "A Study of Vehicle Fires."

21. J. F. Krasny, W. J. Parker, and V. Babrauskas, *Fire Behavior of Upholstered Furniture and Mattresses* (Norwich, NY: William Andrew, 2000), 145.

22. M. Spearpoint et al., "Ignition Performance of New and Used Motor Vehicle Upholstery Fabrics," *Fire and Materials* 29, no. 5 (September–October 2005): 265–82.

23. Marley and DeHaan, *CCAI Vehicle Fire Tests.*

24. G. Barnett, *Automotive Fire Analysis,* 2nd ed. (Tucson, AZ: Lawyers & Judges Publishing, 2007), 96–100.

25. J. D. DeHaan, "A Close Call for a Hot Hudson," *Arizona Arson Investigator* (1993).

26. M. M. Hirschler et al., "Rate of Heat Release for Plastic Materials from Car Interiors," *BCC Flame Retardancy Conference Proceedings,* June 2002.

27. NFPA, "U.S. Vehicle Fire Trends and Patterns."

28. V. L. Bell, "The Potential for Damage from the Accidental Release of Conductive Carbon Fibers from Aircraft Composites," National Aeronautics and Space Administration, Langley Research Center, Hampton, VA, April 1980.

29. Marley and DeHaan, *CCAI Vehicle Fire Tests.*

30. S. Riggs, personal communication, July 23, 2009.

31. Marley and DeHaan, *CCAI Vehicle Fire Tests.*

32. Ibid.; J. D. Nicol and L. Overley, "Combustibility of Automobiles: Results of Total Burning, *Journal of Criminal Law, Criminology and Police Science* 54 (1963): 366–68.

33. Marley and DeHaan, *CCAI Vehicle Fire Tests.*

34. J. D. DeHaan and F. L. Fisher, "The Reconstruction of a Fatal Fire in a Parked Motor Vehicle," *Fire and Arson Investigator* (January 2003): 42–46.

35. J. Mangs, "On the Fire Dynamics of Vehicles and Electrical Equipment," VTT Publication, 521 Helsinki, 2004.

36. Morrill, "Analysis of a Ford Speed Control Deactivation Switch Fire."

37. A. F. Service and R. J. Lewis, "The Forensic Examination of a Fire Damaged Vehicle," *Journal of Forensic Sciences* 46 no. 4 (2001): 950–53.

38. D. Sutherland and K. Byers, "Vehicle Test Burns," *Fire and Arson Investigator* 48 (March 1998): 23–25.

39. F. J. Suefert, "Steal and Burn," *Fire and Arson Investigator* (December 1993): 18–23.

40. ASTM, *Black Box Data from Accident Vehicles: Methods of Retrieval, Translation, and Interpretation,* monograph (W. Conshohocken PA: ASTM, 2009).

41. G. A. Wendt, "Investigating the 'Black Hole' Motor Vehicle Fire," *Fire and Arson Investigator* 49 (October 1998): 11–13.

42. www.nhtsa.gov.

43. Hrynchuk, "A Study of Automobile Fires"; California Department of Forestry, *Vehicle Fire Experiments* (December 1979); see also R. Hrynchuk, "CAFI Vehicle Fire Investigation Techniques: A Report on Vehicle Fire Tests," Canadian Association of Fire Investigators Seminar, Toronto, Ontario, September 30, 1998.

44. Nicol and Overley, "Combustibility of Automobiles."

45. L. McGill, personal communication, 1995.

46. J. Bloom and C. Bloom, personal communication, 1995.

47. J. Bloom and C. Bloom, *Recreational Vehicle Fire Investigation,* 2nd ed. (DVD), (Grants Pass, OR: Bloom Fire Investigation, 2007).

48. A. R. Earls, "Manufactured Homes Revisited," *NFPA Journal* (March–April 2001): 49–51.

49. B. Davie, "The Investigation of Mobile Home Fires," *Fire and Arson Investigator* 39 (September 1988): 51.

50. R. Hammarstrom, "Joint Development of Fire Fighting Systems for Construction Machines," Brand Posten, SP, Boras Sweden, #39, 2008, pp. 4–6.

51. *NFPA 302: Fire Protection Standard for Pleasure and Commercial Craft* (Quincy, MA: National Fire Protection Association, 2010).

52. F. Herrera, "Investigation of Boat Fires," *California Fire/Arson Investigator* (January 2005): 10–11.

53. R. Hadden, F. Jervis, and G. Rein, "Investigation of the Fertilizer Fire aboard the Ostedijk," *Fire Safety Science* 9 (2009): 1091–1101.

54. W. E. Swanson, *Modern Shipfitters Handbook* (New York: Cornell Maritime Press, 1941).

55. J. Nicholson, "Cruise Ship Fires," *NFPA Journal* (January–February 2007): 37–42.

56. *NFPA 301: Code for Safety to Life from Fire on Merchant Vessels* (Quincy, MA: National Fire Protection Association), 2008.

57. D. V. Stuart, "Shipboard Fire Investigation," *Fire and Arson Investigator* 29 (January–March 1979): 55–58.

58. A. D. Sapp et al., "Arson Aboard Naval Ships: Characteristics of Offenses and Offenders," NCAVC, 1995.

59. J. Mangs, VTT Building & Transport, April 2004, pp. 33–35.

10
Electrical Causes of Fires*

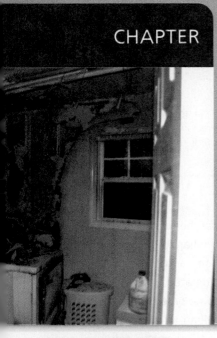

Courtesy of Det. Mike Dalton, Knox County Sheriff's Office, Knoxville, TN.

KEY TERMS

ampacity, *p. 413*	eutectic, *p. 437*	service conductors, *p. 420*
circuit breaker, *p. 411*	ground, *p. 402*	service drop, *p. 437*
conductor, *p. 402*	open neutral, *p. 422*	thermal protector, *p. 419*
duty, *p. 438*	receptacle, *p. 411*	voltage, nominal, *p. 417*

OBJECTIVES

After reading this chapter, you should be able to:

- Identify the most common causes of residential electrical fires.
- Explain the basic concepts of basic electricity: voltage, resistance, current, and Ohm's law.
- Describe the investigative methodology for identifying ignition of fires by electrical means.
- Explain how to identify wiring systems, current-carrying capacity, and overcurrent protection.
- Explain several techniques used to investigate electricity-related fires.
- Understand the concepts of arc-fault mapping.
- List and describe several post-fire indicators used when tracing an electrical system in a burned structure.

*The authors are deeply indebted to Chris W. Korinek, P.E., president and founder of Synergy Technologies LLC, for conducting many demonstrations and contributing much of the material in this chapter. Contributions for the lighting section were made by Richard E. Korinek, P.E., of Synergy Technologies LLC. The material on arc-fault was generously provided by Dr. Robert Svare, Mark Svare, and Mathew Dubbin of Svare Professional Engineering and Chris W. Korinek, P.E. of Synergy Technologies LLC.

conductor ▪ Any material capable of permitting the flow of electrons: (1) bare—a conductor having no covering or electrical insulation whatsoever; (2) covered—a conductor encased within a material whose composition or thickness is not considered insulative; (3) insulated—a conductor encased within a material recognized by the electrical code as insulation.

ground ▪ A conducting connection, whether intentional or accidental, between an electrical circuit or equipment, and the earth, or to some conducting body that serves in place of the earth.

As we saw in Chapter 6, almost any kind of energy can be responsible for igniting a flame. One form of energy—electricity—is both familiar to us because it is in use constantly around us and mysterious because its origins and properties are not well understood, even by many fire investigators. In 2006, an estimated 52,500 home structure fires reported to U.S. fire departments were attributed to some type of electrical failure or malfunction. Such fires resulted in 340 deaths and 1,400 civilian injuries (13 percent of all reported residential fires).[1] Many fires are attributed to electrical causes, fairly or unfairly, and it is therefore important that the investigator have a working knowledge of the basics of electricity, its control, uses, and measurement, so that its effects on fuels in the fire environment can be accurately assessed. That will be the function of this chapter. The post-fire examination of wires, connectors, fuses, and appliances will also be described.

Electricity creates heat as it moves through materials, typically called **conductors**. The term *conductors* refers to anything that conducts electric current and includes wires, terminals, connector lugs, bus bars, metallic enclosures, and plasma (such as in an arc and a flame). A *plasma* is a hot ionized gas that can conduct electric current and will be discussed later in more detail. Even materials normally not thought of as conductors such as charred plastic, electrolytic fluids, and human tissue can act as conductors. Our nerve cells and muscles conduct electricity required for life. Conductors may carry electric current during normal operation (such as the hot and neutral wires and electric device load) and during an electrical fault (such as a short to a metallic enclosure or **ground** wire). Electric current may produce heat as a primary design function, as in the case of heaters, furnaces, range tops, ovens, toasters, and the like. These pose ignition risks when, under normal operating conditions, they are brought into contact with susceptible materials or are misused in such a way that their normal heat production causes ignition. Additionally, in a much larger number of situations heat is an unintended by-product of the passage of electric current where the desired result is the operation of lights, appliances, tools, equipment, or simply the transmission of current from one place to another. In these cases, the current is flowing through its intended or design path, and the heat created is supposed to be dissipated safely so that the device does not degrade or fail. These items pose a risk when they are operated in such a way that the heat created by their normal use cannot be safely dissipated. The user usually is not aware of the heat buildup until it causes failure of the device or ignition of the device or nearby combustibles. There are also situations where the current does not flow through its intended path but instead flows through an unintended path, causing severe localized heating and ignition of the material. When current flows along unintended paths (across "insulating" surfaces or through carbonized materials, for example) the flow is sometimes called *arc tracking*. This flow of current may occur as a result of surface contamination by conductive materials or thermal degradation. We shall examine examples of all three types of electrical fire causation, keeping in mind that it is not simply the creation of heat that brings about ignition but the transfer of sufficient heat to a susceptible first fuel to cause sustained combustion that is the concern of the fire investigator.

Basic Electricity

As we discussed in Chapter 2, all matter is composed of molecules that, in turn, are composed of atoms. The atom is the building block of matter. An atom may be thought of as a positive nucleus surrounded by negative electrons that rotate around the nucleus of positive charges in the same manner that the planets revolve around the sun in our solar system. Electrons and the ease with which they can move about in a material determine the electrical properties of that material.

STATIC ELECTRICITY

Static electricity is an electric charge produced by an actual transfer of electrons from some atoms onto others. If a plastic comb is rubbed with a woolen cloth, electrons will be rubbed from one onto the other, causing extra electrons to be on one piece and fewer electrons on the other. Atoms are so constructed that there are normally the same number of positive charges in the nucleus as there are electrons revolving around them. In other words, atoms are *neutral*—uncharged. If an atom gains electrons, it becomes negatively charged. If it loses electrons, it becomes positively charged. Static electricity is simply an electric charge due to the loss or gain of electrons. When we refer to static electricity, we mean electricity that is not in motion but at rest. Such electricity is not "visible"; since it is not doing any work, it represents only a *potential* source of energy. Static electricity is observable when it is discharged, such as the arc seen and felt when one walks across wool carpeting on a dry day and touches a metal object. In comparison, the effects of electricity in motion are the production of heat and light or magnetic forces, such as those used to drive electric motors. Static electricity has also been known as *frictional electricity*, because it is often generated by the rubbing or merely contact and separation of the surfaces of two bodies of dissimilar materials (rubbing is not necessary).

When static charge accumulates, one body takes a positive charge and the other a negative charge. The charges remain on the outer surface of the bodies unless they come in contact with or near a less-charged, uncharged, or oppositely charged body. At such time, the charge may pass from one body to another and is neutralized. When sufficient charge accumulates to break down the resistance of the insulating atmosphere, an arc of very short duration (an electrical discharge) occurs. (Because this discharge is not sustained, it is often referred to as an electric *spark*. No hot particles are created. It is this resulting arc that is capable of igniting suitable flammable materials such as gases, vapors, or dusts. (An *arc* is defined here as the passage of electric current through a gaseous medium, however brief its duration.) (Under special circumstances passage of current through liquids, plastics, or across a surface can be considered to be an arc.)

The discharge of static electricity contributes to fire and explosion hazards in many industries. Often, it is generated in places where it is not noticed and under circumstances in which it may not ordinarily cause trouble. Charge built up from clothing and discharged to a grounded object is usually harmless, while lightning is often harmless but can be dangerous. It should be recognized, however, that static electricity is an important source of potential ignition, particularly in those processes that involve the handling of flammable liquids, gases, and combustible dusts. Industries in which the hazard of static electricity is relatively common are the petroleum, chemical, textile, plastics, printing, agricultural, powdered metals, mining industries, and service industries such as dry cleaners. Many industries recognize the static electricity hazard and regularly attempt to control it, while others have not yet appreciated its prevalence and danger.

Static electricity is generated in moving parts of equipment such as power transmission belts and by moving columns of dusts, smoke, powders, and even nonconductive fluids in processing and storage equipment. Static electricity arcs can be discharged from such equipment, and proper grounding (as described in the *National Electrical Code*, or *NEC*) is required to prevent any accumulation of charge.

Any process that involves the storage and handling of flammable gases and liquids, combustible fibers and dusts, and similar easily ignitable materials can be subject to the fire hazard of static electricity. Although there is no generation of static electricity in the actual storage of flammable liquids, static electricity charges are produced during their flow or through turbulence in mixing or discharging of liquids into or from tanks or other containers. Movement of clothing (friction or separation) can create static charge. The key to the creation of static charge is that the materials are poor conductors of electricity. Thus, the better the insulative properties, the easier it is for charge to build up to discharge levels. In dry air, the human body can store enough charge to produce a potential of 5–25 kV, which releases about 20 mJ of energy.[2] Discharge of static electricity built up on a vehicle or a person has been identified as the cause of numerous vehicle fires during refueling.[3]

Despite the fact that humid air is actually a better insulator than dry air, arcs from static charges occur more frequently during dry winter weather than in the hot humid months of summer because the moisture accumulated in garments and on surfaces tends to allow the charge to leak away as quickly as it is produced. Lightning, as a special case of static electrical discharge, was described in Chapter 6.

CURRENT ELECTRICITY

Electricity must flow to create heat and do work. Current is a flow of electric charge (electrons or charged particles, called *ions*), through a conductor.

Current electricity may be generated in various ways. One way is through the use of an electric generator. In this device, a coil of wire is brought into a magnetic field. As long as the magnetic field is changing—that is, if either the coil or the magnet is in motion—a *current* will be induced in the coil winding. Another way of setting electricity in motion is by the use of chemical energies in an electric cell (a group of cells is called a *battery*). For the electric energy supplied by an electric cell (or other source) to do useful work, an appropriate electrical path—a circuit—must be provided for the current to flow through. In this respect, an electric circuit is similar to a water circuit, in which water must have a path through which it can pass (a pipe) to perform intended tasks. The pressure applied to the water causes it to flow through the pipe, and the rate at which the flow occurs is measured in gallons (or liters) per minute. *Voltage* can be thought of as the electrical pressure forcing the electrons through the electric circuit, and current as the electron flow rate, as illustrated in Figure 10-1. Current will flow only when the circuit is complete. A circuit can be defined as a complete path that allows current to flow from the source through a load and return to the source. The electrons passing through a conductor constantly collide with other electrons. These collisions causes heat to be produced in the same way a restriction in a pipe causes resistance to the flow of water. The greater the current, the more heat produced. If the conductor offers enough resistance, it can become very hot and thus become a possible ignition source.

Whenever a current of electricity is flowing, as through a wire, a magnetic field is also generated. Winding the wire into a coil concentrates the magnetic field. This

(a)

FIGURE 10-1 Components of an electric circuit (a) and (b) have equivalents in an analogous water circuit (c). A battery provides electromotive force (voltage); a lightbulb offers resistance, and a current flows. Here an applied voltage of 12 V produces a current flow of 4 A through a resistance of 3 Ω. The load (a lightbulb) dissipates 48 W of power. *Courtesy of Chris W. Korinek, P.E., Synergy Technologies LLC, Cedarburg, WI.*

(b)

(c)   Pump

phenomenon is the basis of an electromagnet. As stated earlier, whenever a conductor is moved through a magnetic field, an electric current will be generated in the conductor. An electric generator works on the principle that a coil of wire moving through a magnetic field generates current. An electric motor uses electricity to produce motion. There is a coil of wire inside a magnetic field in the motor. When a current is sent through the coil, the direction of the magnetic field produced by the coil is opposite that of the magnetic field normally present in the motor. This opposition of fields causes the motor to rotate.

A circuit can be designed to make one of these effects greater than the others, but none of them can be completely eliminated (except in the special case of superconductivity). For example, the purpose of electric heaters is to convert electric energy into as much heat as possible. The magnetism that is produced serves no useful purpose, even though it is present around the heating coils and connecting wires. In contrast, in an electric motor, it is the magnetism generated that produces the final desired product—torque; any heat produced is undesirable.

Features of electron current and ionic current are discussed next.

Electron Current

Metallic copper is electrically conductive because its electrons are free to move. If one atom at one end of a copper wire gains electrons, the entire wire becomes negatively charged. As electrons are "pushed" into one end of the wire, electrons migrate from atom to atom, and the charge is found at the other end because electrons pass through the copper wire easily. However, if electrons were pushed into one end of a piece of dry string, they would not be available at the other end; electrons do not migrate easily through string, because it is an electrical insulator.

TABLE 10-1	Materials in Order of Decreasing Electrical Conductivity	
Gold		
Silver		
Copper	Conductors	
Aluminum		
Steel		
Copper oxide		
Nichrome		
Carbon		
Charred wood		
Wet skin		
Dry skin	Semiconductors	
Aluminum oxide		
Salt water		
Paints		
Petroleum products		
Dry wood		
Pure water (deionized)		
Most plastics	Insulators	
Glass		
Porcelain		
Air		
Vacuum		

Some materials are better conductors of electricity than others. In other words, electrons flow with much less resistance in conductors than in insulators. Objects in which static electricity can be induced are usually poor conductors of current electricity. Although these objects can gain or lose electrons and become charged, they remain charged. The current does not readily flow through them because their atoms are arranged in a manner that does not permit the electrons to easily migrate among them.

Metals are generally considered to be good conductors of electricity. At the other extreme, materials such as glass, stone, plastics, and synthetic textiles resist the flow of electricity, being poor conductors (or nonconductors), and are used as insulators. Another variety of materials can be described as *semiconductors;* they are poor conductors at low voltages but will conduct at higher applied voltages, with higher resistances than "normal" conductors. See Table 10-1 for a list of such materials.

Ionic Current

Ionic current is a flow of charged particles, called ions, that occurs only through a conductive liquid or paste or plasma. Ionic current is important in batteries such as the lithium ion type, in current conduction through flames, in wet arc tracking, and in nerve cells and tissue. This phenomenon is more complex than electron current (a flow of electrons) and will not be discussed further in this text, except to describe how it is involved with wet arc tracking.[4]

DIRECT AND ALTERNATING CURRENT (DC AND AC)

A direct current (DC) is a flow of electrons in one direction only. A battery is a good example of a DC power source. An alternating current (AC) is created by an alternating voltage that reverses itself in a periodic manner, rising from zero to maximum strength (voltage), returning to zero, and then going through a similar variation in the opposite direction. One of these complete changes is called a *cycle*. The power supplied to homes in the United States (at 120 or 240 V) varies at 60 cycles per second, often denoted in hertz (Hz) (60 cycles per second = 60 Hz). This means that the cycle—in which the current goes in one direction, back to zero, then in the opposite direction, and then back to zero again— is repeated 60 times every second. In many other countries, the power is supplied at 50 Hz (and at different voltages), which is why appliances from one country require a converter to work properly in another country. Currently, electronic devices such as laptops and cell phones are being sold with "dual-mode" power supplies that will accept 100–250 V at 50–60 Hz. An important reason for using alternating current to supply power to home and industry is the ease with which its voltage can be varied.

A transformer is a device used to convert one voltage to another (higher or lower). However, a transformer operates on alternating current only; it will not work on direct current. AC voltage is often measured as its root-mean-square (rms) value, not its peak voltage. The rms voltage is a numerical "average" of the sinusoidal waveforms and is about 70 percent of the peak voltage.

ELECTRICAL UNITS

The basic units of electrical measurement are the *volt*, which measures differences in potential energy, the *ampere*, which measures the quantity of electrical charge flowing), and the *ohm*, which measures resistance. Actually, the total resistance to current flow in a circuit is called the *impedance*, and it is a sum of the simple (often called real) resistance and more complex components of resistance. Simple resistance will be most commonly used in this text when resistance is discussed. Additional units are *watts* or *joules per second* (which measure power). Each unit will be further defined and explained.

Volt

Voltage (V) is the difference in potential energy between two points of an electrical circuit. This difference provides a pressure or force (sometimes called *electromotive force* and represented by the symbol E), which causes electrons or ions to flow between the two points. This force can be likened to the pressure in the water circuit analogy that forces the water to move through the pipe.

Ampere

The *ampere* (or amp; A) is the measure of the quantity of electricity or charge flowing past a point of a circuit in a given period of time. Current is represented by the symbol I. Because current is a measure of the number of electrons flowing per second, it can be compared to the gallons of water per minute flowing in a pipe circuit. Connecting an ammeter in series into the electrical circuit forces current to pass through the meter, which allows the quantity of electron flow to be measured. An ammeter is analogous to a water flow meter reading in gallons per minute. An ampere is the flow of approximately 6 quintillion (6 billion times a billion, or 6×10^{18}) electrons per second.

Ohm

In the water analogy resistance can be considered as the friction of the water flowing through the pipe. Electrical resistance is measured in ohms (Ω). Mechanical friction produces heat. For instance, if a piece of sandpaper is rubbed on a block of wood, both get warm. When an electric current is sent through a resistance, the resisting element will also get warm. This heat is caused by the "friction" encountered by the electric current as it flows through a conductor and associated connection points. As we saw in Table 10-1, certain materials are better conductors than others. Electrons can "pass" through copper wire quite freely without bumping into too many other electrons. Thus, copper is a good conductor. In contrast, electrons have more difficulty in passing through carbon because the resistance is greater, and therefore more heat is produced. Analogously, water being run through a smooth-walled pipe will encounter less resistance than if it is run through a pipe having rough walls.

Resistance can also be thought of in terms of a pipe with a reduced cross section in the water analogy, because a higher flow (more gallons or liters per minute) can move through a large-diameter pipe than through a small-diameter tube (at the same starting pressure). Thus, a larger conductor offers lower resistance, and therefore more current will flow through it (at the same applied force of voltage) than through a smaller conductor with a higher resistance. An insulator is a material through which electric current will flow very poorly, since it offers very high resistance. It is the opposite of a conductor. Insulators are necessary in electricity and electronics to keep the electric current where it belongs. Most resistances are "positive," meaning that if the voltage is decreased, the current decreases. For certain phases of an arc, the resistance actually is "negative," meaning that when the voltage decreases, the current increases. This helps explain why arcs can be explosive and dangerous.[5]

Impedance

Impedance, or the complex resistance to current flow in an AC circuit, is due to the effects of capacitance or inductance. This complex resistance may be inherent in a component such as a capacitor or inductor or may even be a property of straight wires. Capacitance is a property of a circuit or component where *charge* energy is stored. A capacitor typically consists of two parallel plates separated by an insulator. Inductance is the property of a circuit or component where *magnetic* field energy is stored. An inductor typically consists of a coil of wire.

Watts and Joules

Power is the rate at which electrical energy is dissipated (or work is done). It is measured in watts (W) and is the same unit (J/s) as used in heat release rate measurements. Electric motors are often rated in horsepower (hp), where 1 hp = 746 W.

ELECTRICAL CALCULATIONS

Ohm's law is the basic law of electricity and is very useful in making energy calculations. Ohm's law states that the voltage in any circuit is equivalent to the amperage multiplied by the resistance. These relationships can be expressed as a series of formulas:

	SYMBOL	UNIT	DEFINITION
$V = IR$	V	Volt (V)	Potential (analogous to water pressure)
$I = V/R$	I	Ampere (A)	Current (analogous to water flow rate)
$R = V/I$	R	Ohm (Ω)	Resistance (analogous to water flow restriction)

In the circuit in Figures 10-1a and b, the applied voltage is 12 V, and the lightbulb (load) has a resistance of 3 Ω. Using Ohm's law, $I = V/R$, we can solve for current: $I = 12/3 = 4$ A. In the same way, but using the formula in a different form, we can solve for either the resistance or the voltage, depending on which two of these quantities is known.

	SYMBOL	UNIT	DEFINITION
$W = VI = I^2R$	W	Watt	Power (or, using $W = V \times I = 12 \times 4 = 48$ W) at steady state
$WH = V \times I \times hr$	WH	Watt-hour	Energy used
$kWH = WH/1{,}000$	kWH	Kilowatt-hour	Energy used
$J = W \times s$	J	Joule	Energy used

Power (with simple resistances), is equal to the voltage multiplied by the amperes (or the resistance times the square of the current, expressed as I^2R). For a resistive heater, power can be calculated from I^2R, as the resistances are simple. A motor, however, has complex impedances, so the I^2R calculation cannot be used to calculate power without involving other factors.

The lightbulb shown in Figure 10-1 would then dissipate (or use) $W = I^2 \times R = (4)^2 \times 3 = 48$ W. The surface temperature of such a lightbulb operating in open air will stabilize at approximately 175°C (347°F) (as noted in Figure 6-27).

The watt-hour (WH) is a unit of energy (total work done). It is equal to the watts times the time in hours that an appliance is on. The watt-hour is a relatively small unit, so the kilowatt-hour (kWH; 1,000 WH) is generally used. This is the unit used by utility companies in selling electric energy.

If we substitute a 24-W light in this circuit, it will have a design resistance of 6 Ω, so with a 12-V power supply, the increased resistance causes a decrease in the current to 2 A. The surface temperature of the bulb [about 90°C (194°F) in open air] will then be cooler than that of the 48-W bulb, per Figure 6-27. The temperature of a surface that is radiating heat is a function of the wattage and the area over which the power is generated. This value is most often expressed as "watt density" in units of W/cm². In the preceding lightbulb example, the temperature of a higher–watt density bulb is higher than for a lower–watt density bulb.

For a 120-VAC outlet supplying power to a 48-W lightbulb, the circuit looks like Figure 10-2. In comparison with Figure 10-1, note that although the wattage output has stayed at 48 W, the voltage has increased by a factor of 10 (from 12 V to 120 V), the current has decreased by a factor of 10 (from 4 A to 0.4 A), and the resistance has increased by a factor of 100. Note that the power calculation I^2R still yields 48 W.

(a)

(b)

FIGURES 10-2A–B With 120 V (AC in this case), the 48-W light circuit current drops to 0.4 A, while the lightbulb dissipates the same power. *Courtesy of Chris Korinek, P.E., Synergy Technologies LLC, Cedarburg, WI.*

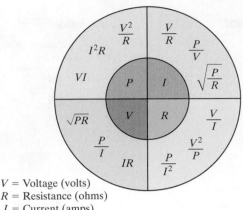

FIGURE 10-3 Ohm's law wheel gives all the combinations of the relationships of voltage, resistance, current, and power for resistive loads such as incandescent lights and heaters. The expressions in the outer ring equal the quantities in the inner circle area. All the values on each segment must be known to calculate the inner quantity.

V = Voltage (volts)
R = Resistance (ohms)
I = Current (amps)
P = Power (watts)

A convenient way to show all the relationships for current, voltage, resistance, and power is the Ohm's law wheel, illustrated in Figure 10-3. All the expressions in each sector of the outer ring are equal to the value in the inner segment. The values in the outer segment must be known to establish the value in the inner sector.

SERIES AND PARALLEL CIRCUITS

Many circuits have more than one load. These loads can be connected in two basic configurations: series and parallel. In a series circuit, the loads are connected so that the same amount of electric charge (current) flows through all of them simultaneously. Figure 10-4 shows a series circuit with two loads (lights). Note that it does not matter whether the switch comes before or after the load in the circuit. No current can flow until the switch contacts close. When the switch closes, the current flows from the battery through each of the loads before returning to the battery. To analyze this circuit, the total load resistance is calculated (3 + 3 = 6 Ω). With a total load resistance of 6 Ω and a 12-V source, the current flow will be 2 A. The water analogy shows that both water restrictions are plumbed into the same line, reducing the total flow rate in the same way two resistances (loads) in series reduce the current flow through both loads. Note that each light uses (dissipates) only 12 W, for a total of 24 W. Series circuits are important when considering modes of electrical overheating such as a poor connection (resistance heating), open/lost neutral, series parting arc, and others. In summary, all resistances in series add up to the total load.

FIGURES 10-4A–B Series electrical resistances are connected in-line, just like the series water resistances shown. Series resistances add to the total circuit resistance. *Courtesy of Chris Korinek, P.E., Synergy Technologies LLC, Cedarburg, WI.*

High pressure   Medium pressure

Water flow

←Parallel→
restrictions

$V = 12\ VDC$

$I = 8\ A$

$I = 4\ A$ $I = 4\ A$

Low pressure

Pump

(a)   (b)

FIGURES 10-5A–B Two parallel electrical resistances are connected such that the load will experience the same voltage, and the two water resistances in parallel will experience the same pressure drop. The two parallel resistors of 3 Ω each reduce the total circuit resistance to 1.5 Ω (resulting in a total current flow of 8A). *Courtesy of Chris Korinek, P.E., Synergy Technologies LLC, Cedarburg, WI.*

In a parallel circuit, seen in Figure 10-5, the loads are connected so that all of them experience the full supply voltage (pressure). The total current will be the sum of all the currents in given portions of the circuit. This is the way in which outlets are wired in normal household systems—one side of each **receptacle** is connected to the "hot" or "live" (pressure) supply and the other to the *neutral* (return side). With parallel circuits it is possible for the current to increase to an amount greater than the rating of an *overcurrent protection device* (OCPD) such as a fuse or **circuit breaker**. The OCPD should interrupt the current flow if the total load current exceeds the rating of the OCPD. In a parallel circuit, each added new resistance (load) adds a new path for current without decreasing the current through other loads. In such a circuit, components are connected so that the total current (flow) can split, allowing a partial flow through more than one conductive path. Note that switches can now control individual loads.

The greatest current will be through the path with the least resistance. This is a very important concept to remember. Given the opportunity, a tremendous current will flow where a path is made available with minimal resistance. Most of the current (but not all of it) will take the shortcut in resistance back to the source instead of going through the higher resistance load (such a short circuit is an unintended current path). This short circuit can produce current flows of hundreds or thousands of amps. In the water analogy, this is like connecting a very large diameter hose between the pressure and return sides. Such a short circuit is shown in Figure 10-6. A wire may have a resistance of only 0.1 Ω, so a parallel circuit like this will create a current flow of 120 A through the short path, dissipating 1,440 W (at a voltage of 12 V). Obviously, this will cause all the wires in the circuit between the battery and the short to overheat tremendously as long as the current flows (without OCPD protection, small-diameter wires may reach fusing or melting temperatures). Parallel circuits are used in most building and vehicle wiring. They help us understand overheating modes such as short circuits, current leakage, excessive current heating, and arc tracking. Note that a current of 2 A continues to flow in the intended branch of the circuit (with resistance of 5 Ω), stabilizing the short-circuit flow in the other branch.

receptacle ▪ A contact device installed as the outlet for the connection of an appliance by means of a plug.

circuit breaker ▪ A device designed to open a circuit automatically at a predetermined overcurrent without injury to itself when properly applied within its ratings.

$V = 12\ VDC$

$I = 122\ A$

$I = 120\ A$

$R = 0.1\ \Omega$
$P = 1440\ W$

$I = 2\ A$

$R = 6\ \Omega$
$P = 24\ W$

FIGURE 10-6 A short-circuiting wire connected across the legs of this circuit (in parallel with the load) has a very low resistance, and as a result, $I = V/R = 120$ A flows through the short-circuit wire, dissipating $I^2R = 1{,}440$ W. The resistance of 0.1 Ω represents the short-circuit wire. *Courtesy of Chris W. Korinek, P.E., Synergy Technologies LLC, Cedarburg, WI.*

Electrical Systems

Vehicle wiring systems (12 to 24 VDC) often use the metal frame of the vehicle as the return path for current. Voltage is supplied to one "hot" or "live" side of lights and accessories, and the other side is connected to a common "ground," (usually denoted by the series of parallel lines in an arrowhead formation on diagrams). In a building, the framework is not (sufficiently) conductive, so wires are provided for a return path, called the *neutral*. The grounding system of building wiring systems extends through the fixed wiring and into the appliance as a non-current-carrying conductor. For a residential 120/240-VAC system, the ground system is intended primarily for occupant safety against shock hazards and does not normally carry current. The grounding conductor of most electrical systems is bonded (firmly connected) between the main service panel and a copper rod or metal water pipe in contact with the earth. The neutral (return) side of the system is usually bonded in turn to the ground at the main service panel, so that there is only one "hot"/ "live" conductor to a 120-V circuit. (At subpanels, the ground and neutral are split.)

Figure 10-7 shows the typical North American wiring system for 120/240-VAC, 60-Hz, single-phase residential and commercial service. Two 120-V loads are shown, each energized by 120 V potential with respect to the neutral conductor. One 240-V load (like a range or dryer) is shown energized by 240 V from one hot line to the other. The neutral is held basically at 0 V by being tied to ground and carries the current only from unbalanced loads on the 120-V legs. For simplicity, no ground wires are shown and only the ground symbol is shown (on the neutral). In Europe and areas of Asia, the typical single-phase system is 230 V (±10 percent), 50 Hz power, as shown in Figure 10-8. Note that 120 V is not available.

CONDUCTORS AND INSULATORS

Three factors determine the resistance of a conductor: the material from which it is made, its length, and its cross-sectional area. The heat produced is dependent on the resistance, current flow, and its operating temperature. The size requirements of a conductor are generally specified by NFPA codes and standards that take into account the following factors:

■ How high a temperature can be tolerated by the insulation or the conductor. This is dependent on the rate of heat generation and the rate at which that heat can be dissipated. The temperature must remain below the point at which the conductor will melt or be otherwise made ineffective. It also must be below the temperature at which the insulation will be damaged. Asbestos insulation was once frequently used on heater and appliance cords. Asbestos has been replaced with higher-temperature-

FIGURE 10-7 Typical normal North American wiring system with a grounded neutral conductor network allows 120-V operation of many appliances as well as 240-V for larger ones. *Courtesy of Chris W. Korinek, P.E., Synergy Technologies LLC, Cedarburg, WI.*

FIGURE 10-8 Typical normal European and Asian wiring system without a grounded neutral conductor network allows only 220- to 240-V operation of appliances. *Courtesy of Chris W. Korinek, P.E., Synergy Technologies LLC, Cedarburg, WI.*

resistant fiberglass or synthetics, which can operate acceptably at higher temperatures and can safely carry greater current than conventionally insulated cords. The resistance of a conductor changes with its temperature, so its normal operating condition must be considered. (For most metals, the higher its temperature, the higher its resistance.)

- The conductor must be of sufficient size to be mechanically sound. A small appliance may require only a very small current, but a very small wire may be too fragile and break.

- The conductor must be of sufficient size that the voltage drop is not excessive. Generally, in North America the size must be adequate to supply the load without exceeding a 2 percent drop in voltage.

Take, for example, a correctly sized wire supplying power to a pump motor located remotely to a dwelling. The wire must be large enough to (1) carry the required current without overheating, (2) withstand mechanical stress such as wind and rain, and (3) be large enough so that the voltage at the pump motor is not lowered significantly.

Metallic conductors are often made of copper or aluminum. Solid wire is used in the smaller sizes and for purposes not requiring flexibility. Larger sizes and wires needing flexibility are constructed of multiple strands of smaller wires. Insulation quality is basically dependent on the type of material and its thickness. In addition, important considerations are the ability of the insulation to withstand elevated temperatures and moisture. Wire insulations usually encountered are rubber, plastic, and asbestos. Other good insulating materials are glass fiber, ceramic, and varnish.

Insulation is rated by voltage that it can withstand or suppress. Conductors with insulation are rated by current and voltage. The current rating is determined by the current that can be carried without a temperature rise above the point at which the insulation could be degraded. The voltage is the maximum voltage that can be applied between the conductor and the outside surface of the insulation without significant current flow.

Wires in U.S. household circuits today usually have thermoplastic insulation that is generally colored according to the following convention:

Black	Line ("hot" or "live")
White or gray	Neutral
Green (or bare)	Grounding (earth) conductor

Other colors are additional line (hot) conductors. They will be found in #10, #12, and #14 gauges in most residences.

Electrical insulation between the hot and ground conductors, as shown in Figure 10-9, is generally extremely high resistance and can be thought of as infinite (at the usual rating voltage of 600 V). In this circuit, the resistance is 10,000,000 Ω (10 MΩ), and the current flow through it is only 0.00001 A. For this reason, electrical insulation resistance is usually not shown on a schematic. If the insulation becomes degraded or contaminated, the level of resistance can become low. As the resistance drops, the insulation can offer a path for higher current flow, and heat production will result.

CURRENT-CARRYING CAPABILITY (AMPACITY)

Ampacity is defined as the amount of current (amperes) that a conductor can continuously carry without overheating (degradation) of its insulation. As with the water circuit, a larger conductor is required to pass a larger current, because forcing a large current through a small conductor will cause it to overheat. The ampacity is dependent on how high a temperature can be tolerated by the insulation and the wire's ability to dissipate heat into

ampacity ■ Current-carrying capacity of electric conductors (expressed in amperes).

FIGURE 10-9 Schematic of good insulation between hot and ground conductors. In this case, insulation is 10 MΩ, and current flow through it is negligible at this voltage. *Courtesy of Chris W. Korinek, P.E., Synergy Technologies LLC, Cedarburg, WI.*

the surroundings. The ratings are usually for a wire at normal ambient temperatures in air. Obviously, packing the wire in building insulation will reduce its ability to dissipate heat and may accelerate the degradation of the wire's insulation. Electrical codes list the maximum allowable current for various sizes of conductors and types of insulation. The generally acceptable current ratings are listed in Table 10-2 for a common grade of insulation in North America but will vary due to special circumstances (high ambient temperatures) or different insulation types.

The ampacity of wire is usually conservatively rated, meaning that the wire generally will not overheat if it carries a small additional current above its rated current capacity, especially for a short amount of time. Over an extended period of time, it is possible for the insulation to degrade if the current is slightly in excess and the wire is thermally well insulated, so the listed rating should be adhered to. The labeled "ratings" of cables and wires are based on what currents can safely be carried under particular conditions, such that the insulation is not compromised by melting or degradation. Flowing twice the rated current, for example, will significantly raise the temperature of the wire and its insulation, and this elevated temperature *may* degrade the insulation over time.

TABLE 10-2	Allowable Ampacities of Common Copper Wire Sizes [for TW Insulation, Moisture-Resistant Thermoplastic at 60°C (140°F)]	
WIRE SIZE (AWG)	**DIAMETER (In./mm)**	**AMPACITY (A)**
18	0.040 (1.02)	6[a]
16	0.051 (1.30)	8[a]
14	0.064 (1.62)	15[b]
12	0.081 (2.05)	20[b]
10	0.102 (2.60)	30[b]
8	0.129 (3.28)	40
6	0.162 (4.11)	70
2	0.258 (6.55)	95
1/0 (0)	0.302 (8.26)	125
2/0 (00)	0.365 (9.27)	145
4/0 (0000)	0.460 (11.684)	195[c]

Source: National Electrical Code (NEC), (Quincy, MA: National Fire Protection Association, 2008), table 310-16, pp. 70–148. Allowable ampacities of insulated conductors rated 0 through 2,000 V, 60°C through 90°C, not more than three current-carrying conductors in raceway, cable, or earth (directly buried) based on ambient temperature of 30°C.

AWG = American Wire Gauge
[a]*NEC 2008,* table 402-5 (fixture wires).
[b]*NEC 2008,* 240.4(D).
[c]At 90°C.

FIGURE 10-10 Schematic of a normally operating circuit with a 1,650-W heater operating at 120 V. In this case, 20 m of NM cable also offers a small resistance of 0.18 Ω, which dissipates 35 W. *Courtesy of Chris W. Korinek, P.E., Synergy Technologies LLC, Cedarburg, WI.*

The fusing current of copper wire is the current at which the wire will melt in a given amount of time and is many times the normal current rating of the wire. For example, if you wanted to fuse (protect) a large-gauge wire by de-energizing it after it carried 3,000 A for 0.1 second, a #12 AWG (American Wire Gauge) solid wire would most likely accomplish this task by melting after 0.1 second at this high current.[6] A current of 3,000 A is 150 times the normal current rating of 20 A for a #12 solid wire.

Also, it should be realized that heat generated in a conductor by current flowing through it is uniform per unit length, provided that the entire conductor is the same size throughout the circuit. Where overcurrent is suspected, examination of the wiring on the same circuit between the suspect area and the service in an unburned location can frequently offer useful information. If the wire insulation in the unburned area of the wire is in good condition, it would demonstrate that the current was not excessive in the questioned area.

The circuit diagram in Figure 10-10 shows the effect of resistance heating in a 20-m (66-ft) length of #14 AWG NM (nonmetallic) cable carrying power to a 1,650-W (1.65- kW) electric heater. Note that the small series resistance of the wire causes a voltage drop, and the resistance heating generates some 35 W. When this cable is out in the open, it allows heat to safely dissipate, but if the cable is buried under thermal insulation or tightly coiled, heat cannot safely dissipate and temperatures can increase to cause the breakdown of the insulation and possibly a fire (as in Figure 10-11).

FIGURE 10-11 Extension cord with 1200-W heater plugged in and buried under a pile of clothing failed and ignited a fatal fire. Note folded loops of cord at center front. *Courtesy of John D. DeHaan.*

Protection—Overcurrent and Short Circuit

Overcurrent protection devices (OCPDs) (fuses and circuit breakers) have two important functions. One is to guard against overheating of conductors from overcurrent. (This protection may be to protect the conductors of the building or to protect appliances and their power cords from overheating.) The other is to provide protection from a *short circuit* (a condition caused by direct contact of the circuit conductors, as shown in Figure 10-6), sometimes called a *dead short*. (A *bolted fault* refers to a short where the conductors are physically restrained so they cannot separate.) A dead short creates the highest current flow and triggers the fastest operation of the typical overcurrent device.

Some equipment, however, requires an initial or starting current that is considerably higher than its operating current. Some motors, for example, draw on the order of 600 percent of rated current during their starting phase. If the overcurrent device triggered immediately any time there was a significant draw, the motor would never start. The protective device on this type of circuit has to be designed to accommodate this brief overload and still provide the needed protection once the motor has started.

FUSES

A fuse basically consists of a short section with a metal of such composition and size that it will melt at a predetermined current flow. In practice, however, a fuse opens when its temperature reaches the melting point of the fuse material. This temperature can be reached by current flow or from the surrounding ambient temperature. A fuse located in the hot sun will therefore function at a lower current, while one located in icy surroundings will pass more current. Two considerations should be noted in regard to fuse functioning:

- Fuses are designed to carry current at their rated current continuously.
- Fuses will carry overcurrents for a time period depending upon the degree of overload (magnitude of the current).

Fuses are made in a variety of forms (as seen in Figures 10-12 and 10-13), varying from a simple piece of fuse wire to the more complicated forms designed to extend the period of time required for the circuit to open. Some fuses have a "slow-blow" feature to allow them to pass starting or inrush currents to motors and other large loads for short periods of time. All fuses, regardless of size and type, have a specified current rating, voltage

(a)

(b)

FIGURE 10-12 Overcurrent protection devices (OCPDs). Left: circuit breakers; Right: fuses. *Courtesy of Chris W. Korinek, P.E., Synergy Technologies LLC, Cedarburg, WI.*

FIGURE 10-13 Appliance protection in the form of a fuse (lower left), automatic-reset thermostat (upper left), manual-reset thermostat (lower right), and TCO or fusible link (upper right). This level of protection is usually more sensitive to appliance overheating conditions than the circuit breaker protecting the building wiring to the outlet feeding the appliance. *Courtesy of Chris W. Korinek, P.E., Synergy Technologies LLC, Cedarburg, WI.*

rating, and thermal characteristics. The correct selection for safe and trouble-free circuit protection relies on thoroughly understanding and considering these factors.

Voltage Rating

The voltage rating of a fuse should be equal to or greater than the voltage of the circuit. This rating is not a measure of its ability to withstand a specified voltage while carrying current. Instead, it is the ability of the fuse to prevent the open-circuit voltage of its system from "restriking" and establishing an arc once the fuse link has parted. North American standard fuse voltage ratings are 600, 300, 250, and 125 V and are usually stamped on the fuse. Fuses are devices that are sensitive to changes in current, not voltage, and remain functional at any voltage from zero to their maximum rating. It is not until the fuse reaches melting temperature and fusing occurs that the circuit voltage and available power influence the fuse performance and determine the safe interruption of the circuit. A fuse may be used at any voltage less than its voltage rating without detriment to its fusing characteristics. A fuse used in a circuit at a voltage far in excess of its rating is potentially subject to a restrike, that is, a renewed flow of current between the fused elements.

Continuous Current Rating

The continuous current rating of a fuse should be equal to or slightly less than the current-carrying capacity of the circuit it protects. Only in special cases in which the connected equipment has unusual characteristics or where the ambient temperature is quite high should the fuse rating be greater than the current-carrying capacity of the circuit.

Examination of Fuses

Examination of functioned fuses can frequently provide some useful information. If the entire link melted, it was very likely caused by external heat. If the fuse link melted only at the center (smallest area), it was probably caused by overcurrent. If the fuse link vaporized and coated the inside of the fuse housing with evaporated material, the cause was very likely an enormous overcurrent (a short circuit). (See Chapter 14 regarding laboratory examination of fuses.)

Nominal voltage is a value assigned to a circuit or system for the purpose of conveniently designating its voltage class (120/240, 480Y/277, 600, etc.).

voltage, nominal ■ A value assigned to a circuit or system for the purpose of conveniently designating its voltage class (120/240, 480Y/277, 600, etc.).

FIGURE 10-14 Schematic of a grounded circuit with a circuit breaker sensing the current through the one energized conductor. If this current exceeds the rating of the OCPD, the OCPD will trip. *Courtesy of Chris W. Korinek, P.E., Synergy Technologies LLC, Cedarburg, WI.*

CIRCUIT BREAKERS

Another common form of overcurrent protection is the circuit breaker (as shown in Figure 10-12). The circuit breaker in its simple form consists of a mechanical means of opening or switching the circuit. Its contacts are "normally open" and are closed by mechanical pressure on an external lever acting against a spring. The contacts are kept closed by a thermally sensitive element. When this element reaches its rated maximum temperature (as a result of overcurrent), it does not melt, as in a fuse, but releases the spring-loaded contacts, opening the circuit, as shown in Figure 10-14. The modern circuit breaker is a dual-function device with either thermal or electromagnetic triggers that allow it to respond to modest overcurrents slowly or to massive overcurrents (short circuits) very quickly. External heating (from fire exposure) can cause circuit breakers to trip.

Circuit breakers are designed to respond in a time-temperature-dependent manner. That is, they can carry 6 to 10 times their rated current for up to four cycles (1/15 second) or twice their rated current for 2 minutes. In contrast, some electronic equipment cannot tolerate any appreciable overcurrent, so the protective device must be a "fast-blow" design to trip immediately. Some varieties of circuit breakers were shown in Figure 10-12. Circuit breakers can be actuated by magnetic or thermal mechanisms and can also be tripped by external heating to temperatures of 135°C to 150°C (275°F to 300°F). A breaker panel exposed to substantial direct flame impingement, internal or external, can be expected to have one or more breakers in the tripped (open) position. European miniature circuit breakers (MCBs) are made to the same European specification. They are classified as types B, C, and D, with C being normally used for discharge lighting and D for motor circuits. These circuit breakers do not have a tripped position; therefore, it is not possible during a scene examination to determine if the breaker was off prior to the fire or tripped during the fire. All European specification circuit breakers incorporate thermal and magnetic devices to protect circuits from overload and short circuit.

Circuit breakers, as opposed to fuses, are mechanical devices and therefore subject to conditions of dirt and corrosion, which can affect their operation. It is possible to perform tests on a suspected breaker, but if laboratory tests are to be made, any attempts to trip the breaker may invalidate any further tests, because a single hand trip could loosen a possible frozen/seized mechanism. X-rays can be taken prior to any testing to verify its condition.

Modern North American and European circuit breakers will function even if the handle is mechanically held, as would be the situation if the handle was secured with a piece of tape. This is not true with some older North American circuit breakers. This also can be determined by laboratory tests on the questioned breaker. Some circuit breakers can also be made to trip by mechanical shock (impact).

Fuses and circuit breakers, although meant for preventing overheating from overloading in an electric circuit, are arc-producing devices and may be hazardous themselves unless designed, installed, and maintained so that the arcs produced cannot ignite flammable atmospheres.

THERMAL PROTECTORS

In modern electronics, other devices called **thermal protectors** provide various thermal, overcurrent, and short-circuit protection. These include manual- and automatic-resetting thermostatic controls (used as high-limit controls) and replaceable thermal cutoffs or cutouts (TCOs) (see Figure 10-13.) A typical TCO consists of a pellet of nonconductor that holds two spring-loaded contacts together. When the designed temperature is reached, the pellet melts and allows the contacts to separate and interrupt the current flow. They are not reusable or resettable and must be replaced if they function. TCOs have been known to fail to operate, and two are often connected in series in heat-producing appliances like coffeemakers. When TCOs are encountered during a fire investigation, they should be carefully preserved for examination by a qualified specialist, such as an electrical engineer.

thermal protector ■
An inherent device against overheating that is responsive to temperature or current and will protect the equipment against overheating due to overload or failure to start.

SURGE PROTECTION DEVICES

With the advent of sophisticated electronic devices such as computers and their peripheral devices came a need to protect these devices from surges or spikes in the power delivered by ordinary wiring systems. "Surge protectors" are available in a number of forms but the most common device for consumer use is based on a "metal oxide varistor" or MOV. This is a small disk of zinc oxide encased in epoxy, with a connector wire affixed to each side. When a normal-service steady (clamping) voltage is applied, the MOV acts as a high-resistance element. When a surge of current or a spike in voltage occurs, the extra energy is dissipated through the zinc oxide. If repeated, such surges can cause the MOV to fail either as a continuous, low-resistance short; an open device (which no longer protects the connected device); or as a high-resistance current path (which heats up as a current flows through it). This heating can melt adjacent plastics, degrade them, and even ignite them, as pictured in Figure 10-15.[7]

Some manufacturers of better-quality power strips bond a TCO to the MOV disk so that when the MOV overheats, the circuit opens. Others use a fine-wire fuse trace or a thermal fuse spring that quickly opens when an overcurrent occurs. Failed MOV disks are rough or melted in appearance, while serviceable disks are smooth-finished and uniform in texture. Failures of MOVs have been linked to fires starting in multiple power tap extension cords (power strips) with plastic cases.[8]

FIGURE 10-15 Failed MOV in this plug strip caused ignition. *Courtesy of Steve Riggs, Public Agency Training Council.*

OVERCURRENT DEVICES AND FIRE INVESTIGATION

The overcurrent device can be of importance from a fire investigation standpoint and should be carefully examined and its condition documented, along with the fixed wiring, appliances, receptacles, and other components. The OCPD is well named the "safety valve" of the system.

The rating of North American fuses and the setting of circuit breakers should not exceed that permitted by the *National Electrical Code* (NEC). Similar guides also apply to other countries. When an overcurrent device operates, it ordinarily indicates an overload or trouble on the circuit. This condition may be due to a momentary overload. However, if the OCPD continues to operate, the cause should be determined and remedied. Repeated tripping of OCPDs may be a clue to the cause of a fire and should be established by interviews of employees or residents.

In North America, as a general rule, the *NEC* requires that every piece of equipment be protected by some form of overcurrent device. Normally, the rating or setting of the overcurrent device should be as low as is practical for the particular equipment concerned. In an electric range, for instance, it should be just large enough to carry the load continuously, but with a large motor, especially one with a variable load, a specialized motor controller is required. Smaller items, such as lamps, and household and office appliances, are grouped under the protection of one overcurrent protective device. On the typical residential circuit used for lighting, overcurrent protection is rated at 15 or 20 A. The mere finding of a slightly larger fuse or breaker, such as a 20-A fuse when a 15-A fuse is proper is most likely not a fire cause, since the extra current flow it will permit will not have a significant effect on the temperatures generated.

The maintenance of good, clean contacts at fuse terminals is important to safety and should be considered by the fire investigator. Overheating at this point from poor or dirty contacts generally results in needless blowing of fuses and, frequently, replacement with larger ones as a stopgap repair.

The following hazards most likely to be found in overcurrent protective devices:

- Fuses of significantly too large a capacity or circuit breakers with too high a rating for the intended load.
- Circuit breakers made inoperative by blocking or taping of the tripping element (old type).
- Plug fuses that have blown and in which pennies have been inserted or where the metal of the fuse holder has been cut back or in which a wire has been inserted.
- Cartridge fuses that have blown and in which nails, wires, or other types of metal have been inserted.
- Refillable or rewirable fuses in which additional strips or wires have been placed.
- Fuses or circuit breakers in poor mechanical condition or that have not been moved and reset in many years.

service conductors ■
The supply conductors that extend from the street main or transformer to the equipment of the premises served.

- The **service conductors** from the utility transformer to the meter and panel have no overcurrent protection, other than a fuse on the primary side of the transformer, so contact with trees, ladders, or tools can draw very high currents.
- Fuses, without enclosures, in the vicinity of combustibles.
- Hazards created by doors or covers of fuse cabinets or breaker panel boxes off or opened.
- Corrosion or contamination of fuses, circuit breakers, or their connectors or sockets.

(Note that all but the last three items on this list can result in very high currents being drawn that can cause fires by a variety of mechanisms, all without tripping the OCPDs.)

GROUND FAULT INTERRUPTERS

Ground fault interrupters (GFIs—also known as ground fault circuit interrupters, GFCIs) are installed to minimize the risk of electric shock rather than to prevent fire-causing situ-

GFCI and AFCI Schematic

GFCI sensing coil

Leakage current
GFCI trips at 0.005 A
AFCI trips at 0.030 A

Leakage current

Ground system connected
to grounded enclosure

FIGURE 10-16A Schematic of a grounded circuit with a GFCI or AFCI sensing the current imbalance through both the hot and neutral conductors. If this current imbalance exceeds the leakage current trip level, the GFCI or AFCI will trip. *Courtesy of Chris W. Korinek, P.E., Synergy Technologies LLC, Cedarburg, WI.*

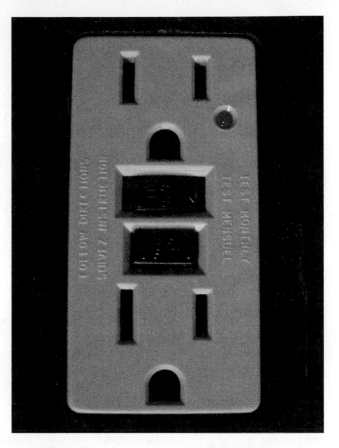

FIGURE 10-16B GFCI Outlet. "Push to test" button actually creates a 0.005-A (5-mA) imbalance and tests the ability of the GFCI to sense the imbalance and trip. *Courtesy of Chris W. Korinek, P.E., Synergy Technologies LLC, Cedarburg, WI.*

ations. They operate on the balancing principle that the current in the line (hot) wire must be equal to and opposite that in the neutral. Any imbalance causes the device to interrupt the flow (see Figure 10-16a.) GFCIs will trip on an imbalance of about 0.005 A (5 mA), which is considered a safe level of current for passage through a human body. For household or light industrial use, GFCIs are commonly in the form of a modified receptacle with manual "test" and "reset" buttons on the faceplate (Figure 10-16b). The permitted currents involved in such devices are generally infinitesimal as compared with that required for ignition of common materials. Some outlets (receptacles) are protected by remote GFCI breakers installed in the distribution panel or by another GFCI outlet "upstream" (toward the power source) from that being examined. Note that GFCIs are designed to prevent electrocution by detecting an imbalance between the hot and neutral, not to prevent short circuits or overcurrent. Fires can be started on connections to a protected circuit. Water contamination in GFCIs has also been linked to fires.[9] In the United Kingdom, GFCIs are called residual current devices (RCDs). The most popular trip rating for general use is 30 mA. A 2009 change in the UK wiring regulations resulted in RCDs' being installed into all new circuits where the wiring is buried (enclosed) in the wall.

ARC-FAULT CIRCUIT INTERRUPTERS

Arc-fault circuit interrupters (AFCIs) are a relatively new type of device that protects conductors from overheating due to arcing faults. As of 2008, AFCIs are required by the *NEC* to be

FIGURE 10-17 Schematic for floating (or open) neutral condition. Note that with the neutral conductor open (not connected to the service ground), the two loads are now in series, instead of being (essentially) in parallel, potentially causing the voltage to be higher than 120 V. Such an overvoltage can overheat some appliances. *Courtesy of Chris W. Korinek, P.E., Synergy Technologies LLC, Cedarburg, WI.*

used in all residential bedroom installations.[10] They will be seen in most living areas of new homes as well (per *NEC 2008*).[11] The operation of the AFCI is similar to that of the GFCI, with the hot and neutral wires connected through a sensing coil (as shown in Figure 10-16a), but the AFCI will detect leakage to the ground (as opposed to imbalances only between the hot and neutral currents). If a current imbalance of more than 30 mA (0.030 A) occurs, the AFCI will trip, de-energizing the circuit. The AFCI is also designed to sense deviations from the sine wave of normal AC current and trip if the voltage curve does not match the "design" curve. Short-duration arcing occurs when normal switches are thrown or a plug is pulled while an appliance is energized, but AFCIs are designed *not* to trip on such events. AFCIs should detect many arc faults, but the mere presence of an AFCI should not be taken as evidence that a fire could not have been started on a circuit protected by one. Fires linked to electrical arcing on AFCI-protected circuits have been reported.

OPEN NEUTRAL

open neutral ■ In an American single-phase, 120/240 V residential system, the neutral return leg is not connected to the service ground.

If an **open neutral** condition exists in a North American electrical household system, the normal return (neutral) leg is missing, and the two legs (each with its normal load) are now in series rather than parallel, as shown in Figure 10-17. The load R_1 expects to "see" 120 V presented to it, but the voltage it sees can be anywhere from nearly zero to nearly 240 V depending on the balance of loads at R_2 and R_3. This higher voltage can cause overheating by excessive current.[12] If the homeowner noticed that some lights were abnormally bright and others dim, or other appliance abnormalities that could indicate high or low voltage, the neutral system should be inspected for continuity from the service through the branch circuits. (Such variations can occur on branch circuits that share a common neutral.) Sometimes, the service neutral is interrupted, which will cause current to find whatever path it can to return to neutral or ground, producing unintended path heating. The presence or absence of a ground (to driven rod or water pipe) at the service panel is critical in determining whether a severed service conductor will cause a voltage imbalance.

In the United Kingdom a lost or open neutral condition will affect only circuits from a 400-V three-phase supply. The majority of UK household electrical supplies are 230 V single-phase, so the lost or open neutral fault would need to be at the substation transformer or a link junction box that supplies a large number of properties. Therefore, lost or open neutral faults on supply substation transformers tend to affect a large number of properties simultaneously.

Electrical Service Distribution

Electricity is delivered to homes and businesses at much higher voltages than are used, for reasons of cost efficiency. In North America, power is typically generated at 12 to 25 kilovolts (kV), transmitted over long distances at 100 to 1,000 kV, and distributed locally at 3 to 20 kV. The voltages are reduced by surface "pad-mounted," underground, or utility pole transformers to 240 V for delivery to residences (as shown in Figures 10-18a–d) and

FIGURE 10-18A Overhead service entrance. *Courtesy of John D. DeHaan.*

FIGURE 10-18B Typical pole-top transformer for 240-V residential service. *Courtesy of John D. DeHaan.*

FIGURE 10-18C Lateral (underground) service entrance. *Courtesy of Greg Lampkin, Knox County TN Fire Investigation Unit.*

FIGURE 10-18D Pad-mounted transformer. *Courtesy of Steve Riggs, Public Agency Training Council.*

240/480 V for businesses. Note that voltages over 350 V are high enough for arcing to initiate through air prior to direct contact with the conductor. (See the work of Pachen, as discussed in Babrauskas's *Ignition Handbook*).

SERVICE ENTRANCE

The service entrance, the point where connection to the utility company is made, usually contains the electric meter, disconnecting means, and overcurrent protection devices. In addition, the system ground is made at this point. It is important for the fire investigator to become familiar with the basic fundamentals of wiring methods and especially the service entrance, because this is the basic electricity distribution center.

An example of a typical North American dwelling service entrance is shown in the block diagram in Figure 10-19. The breakers (or fuses) are identified by size. The required wire sizes are also shown. Each breaker is selected to protect the circuit it supplies from overloads. Therefore, 50-A breakers must be wired to #6 copper conductors (or larger), 20-A to #12, and so on. The usual appliance circuit is wired with #12 wire to a 20-A

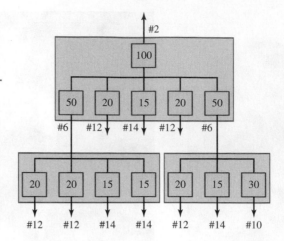

FIGURE 10-19 Typical residential service entrance electrical connections with breaker ratings and wire sizes.

breaker. Branch circuits for bedrooms and similar low-load areas are often wired using #14 wires or cable. Small appliances and lamps are frequently supplied with #18 power cords that are rated at 5 A. It should be noted that when failure occurs in these cords, they are not protected by a 20-A breaker, since four times the cord's rated current must pass (sufficient to cause ignition by overcurrent) before the breaker is even passing its normal rated current.

Frequently, breakers or fuses are found to have operated after a fire. This finding can serve as an indicator of possible electrical fault, but it is just one indicator. Breakers and fuses can be tripped (opened) by fire damage to the electrical system in the absence of any pre-fire fault. Also, fire-causing electrical faults need not draw sufficient current to cause the OCPD to open the circuit.

The service entry may be from an overhead drop from a power pole or via an underground connection (called a *lateral*), as shown in Figure 10-18. The main distribution panel and meter is located at or near the point where power enters the building for maximum safety. This is because any conductor upstream of the main breaker or fuse block (denoted by the 100-A block in Figure 10-19) is capable of delivering the full current of the utility transformer (up to 2,000 A). The fire investigator must identify and document all service entries to a building, and there is often more than one in a commercial or industrial site.

RECEPTACLES

The termination of most fixed wiring in residences familiar to users is the *receptacle* or outlet into which lamp or appliance plugs are inserted. Prior to 1955, electrical plugs were two-pronged for 120-VAC 15-A use (hot and neutral only). Since then they have a third D-shaped opening for a ground. The arrangement of "slots" in the receptacle face is linked to their voltage and current rating (along with specialized outlets for hospital equipment). In typical applications, several outlets will be connected in parallel on a single branch circuit. Receptacles may be "daisy-chained," where the contacts are used to connect "downstream" receptacles. This means that connections in one receptacle will be carrying current even when nothing is plugged into that receptacle. Connections may be made via screw terminals on the sides of each receptacle, or wires may be inserted through holes in the back of each receptacle (push-in connections) (see Figures 10-20a and b).

The security of the connections is a major focus of fire investigators because loose connections cause "high-resistance" connections. The heat dissipated by such high-resistance connections may be on the order of 4–12 W in 120-VAC 15-A service, but that is enough to degrade plastic housings or wire insulation over a prolonged period of use. Such degradation can lead to a flow of leakage current between the hot and neutral sides, even

FIGURE 10-20A Screw-terminal duplex receptacle. Note glowing wire and degradation due to loose connection. *Courtesy of Jamie Novak, Novak Investigations and St. Paul Fire Dept.*

FIGURE 10-20B "Push-in" wiring receptacle. Note stubs of degraded aluminum wiring broken off that caused a fire in the box. *Courtesy of Peter Brosz, Brosz & Associates Forensic Services Inc.: Professor Helmut Brosz, Institute of Forensic Electro-Pathology.*

when nothing is plugged in. This can, in turn lead to fires in receptacle boxes (today often made of plastic instead of steel) and wall cavities.[13]

Ignition by Electrical Means

When evaluating the possible ignition of a fire by electrical means, it is important to consider just how much electrical energy would have been necessary to cause sustained ignition of the first fuel involved in a particular environment. Small areas of overheated wire or small arcs are of little concern in a solid material of high ignition temperature. The smallest discernible arc, however, is a serious hazard in a combustible gas or vapor. As discussed in Chapter 6, ignition occurs only when enough energy is transferred to the fuel to initiate a self-sustaining combustion. It is not the total energy but the concentration of the rate of transfer that determines whether an ongoing reaction will be initiated or will simply cause localized damage such as a char or scorch.

Ignition by electrical means is caused by a number of mechanisms. We have seen how misuse of heat-producing appliances can ignite fires, but here we are going to consider overheating and arc production as a result of failures.

Conduction heating can occur as current travels through a conductor (with the resistance of the conducting medium being the controlling factor) or as it encounters contact resistance (with aluminum wiring causing special problems in that area). Arcs as sources of extremely high temperatures have already been discussed. The degradation of electrical insulation can, of course, contribute to either mechanism. Each of these problems will be examined in turn in the following sections.

CONDUCTION HEATING

The heating of properly installed and used conductors used to carry current is negligible under ordinary circumstances. In North America the *NEC* dictates the maximum current a conductor shall carry and therefore the amount of heat generated. This maximum current rating depends on the size of the conductor, its composition, and the type of insulation covering it. Wherever these specified currents are exceeded (where a conductor is overloaded), the generation of heat becomes a hazard.

The temperature of an electrical device (including cables and cords) is a function of its temperature rise and the ambient temperature, which is the temperature of the surroundings of the device when it is not carrying current. Many conductors and appliances have a temperature rating recommended by the manufacturer to be the maximum the device can safely withstand. Use above this temperature may result in degradation and fire hazard. For example, nonmetallic (NM) cable (such as Romex) insulation has an *NEC* rating of 90°C (194°F). If #14 AWG NM cable inside R38 thermal house insulation carries a current of 14 A, its temperature rise will be on the order of 33°C (60°F). If the cable is in a hot attic with an ambient temperature of 40°C (104°F), the cable temperature will be about 73°C (163°F). If that same 14-gauge NM cable is used to carry 28 A, the power dissipated will be very high, and the cable temperature will exceed its safe operating temperature due to a high watt density. If thermally insulated, the cable can reach temperatures of 172°C (342°F). Note that this is because the current flow has doubled, and since power dissipation goes up as the square of the current ($W = I^2R$), the temperature rise increases by a factor of four [from 33°C to 132°C (60°F to 237°F)].

Overheating can be defined as the permanently destructive heating that has the potential to cause a fire. Overheating degrades the insulation, conductor, surrounding materials, or all three. Overheating causes of fire can be separated into four categories: (1) excessive current, (2) poor connection, (3) insulation breakdown, and (4) induction (rarely seen in nonindustrial applications).

OVERHEATING BY EXCESSIVE CURRENT

On a current-protected circuit, a small overload has only a minor effect on overheating because of the built-in safety factor. For instance, if a 20-A fuse is used instead of a 15-A size, there is a 33 percent overprotection, but even with this amount of overload, some time will be required to cause a significant temperature rise in the conductor, and even then the maximum service temperature may not be exceeded if the conductor is in open air.

Note that the current flow needed to cause a typical conductor to reach its melting (fusing) temperature is very high. For example, a #14 AWG copper conductor in open air will withstand over 100 A before fusing.

Consider the situation in which a 1,200-W appliance is plugged into a 120-V appliance circuit protected at 20 A using a typical #18 AWG extension cord. In case of a short circuit in the extension cord, possibly from frayed or cracked insulation, the excess current will be expected to trip the breaker quickly. In the absence of a short circuit, the normal current draw (10 A) will not cause the breaker to trip (typically at more than 20 A), and the circuit feeding the receptacle will not be overheated because, if properly installed, it will be #12 AWG. The current through the #18 cord, however, is far beyond its rated ampacity (#18 AWG is rated at 5 A), and this will cause overheating of the extension cord even in open air. This is an example of an excessive current overheating mechanism.

When an electric current flows through a single series circuit path, the current (measured in amperes) is the same in all parts of the circuit. The temperature at any point is dependent on the electrical resistance at that point and also on the rate at which heat can be transferred away. Referring to the previous example, the #18 AWG wire has a higher resistance per unit length than the #12 AWG. Therefore, it will operate at a higher temperature

(a)

(b)

(c)

FIGURES 10-21A–C (a) Lab-created short circuit of a battery circuit where a #18 AWG wire carrying current of 120 A degrades as shown by the smoking wire insulation. The light-colored wire is the short-circuit wire. (b and c) Before-and-after photos of cable insulation degraded by overcurrent. *Courtesy of Chris W. Korinek, P.E., Synergy Technologies LLC, Cedarburg, WI.*

even if the same current is flowing through both. Note that the entire length of the extension cord will be heated and so the insulation along its length may soften, melt, and perhaps creep to bring conductors into contact with each other, as shown in Figure 10-21a. At any location along the extension cord where there is thermal insulation (a rug, pillow, blanket), the convective losses will be minimized, and the temperatures will be much higher. These higher temperatures can result in significantly more thermal damage (or faster failure) at those locations. Heating may also produce pyrolysis and charring where the insulation becomes a semiconductor (PVC insulation when heated to 200°C (400°F) begins to conduct current because of its inorganic fillers.)[14] (See Figures 10-21b and c.) Once the insulation breaks down, the current flow through it can heat it to its ignition conditions, with localized arcing able to ignite the pyrolysis vapors. The same process can apply to combustibles in contact with the overheated wire, especially where their contact is impeding heat losses. In a short circuit, where the currents can be very high before the OCPD functions, the conductor may glow along its entire length (if the OCPD is rated at more than 10 times the rating of the wire).

OVERHEATING BY POOR CONNECTION

Arcing can occur as either parallel or series arcs, as illustrated in Figure 10-22. In a parallel arc, the current flows from the hot conductor to ground, causing a short circuit. This short circuit tends to be a brief, high-current event. A series arc takes place between two ends of a broken conductor or through a poor connection. Such arcs can be continuous with a low current flow in the circuit. When current tries to pass through a poor or loose connection (one with high resistance), it can generate heat. In these cases, the heat and its effects on nearby materials will be localized around the poor connection. These failures can occur anywhere two wires connect (splices) or where single wires connect to appliances, power sources, meters, and on the like. A good example of such a failure is where a wire is wrapped around a screw terminal. If the screw is not properly tightened, loosens by expansion and contraction from heating (especially for aluminum conductors), or

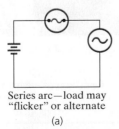
Series arc—load may "flicker" or alternate
(a)

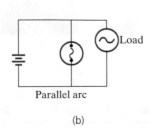
Parallel arc
(b)

FIGURE 10-22 (a) Series arc-load may "flicker" or "operate" not alternate. (b) Parallel arc: load will falter due to insufficient current.

Ohm's law: $V = I \times R$
$2.5\,V = 14\,A \times 0.18\,\Omega$

Power: $P = I^2 \times R$
$35\,W = (14\,A)^2 \times 0.18\,\Omega$

Heat flow: 35 W

Watt density: $35\,W/(0.00006\,m^2) = 550{,}000\,W/m^2$

Temperature: over 400°C (750°F)

(a)   (b)

FIGURE 10-23 (a) Schematic and (b) circuit heat flow analyses of one example of overheating by a poor connection, a glowing screw caused by a loose connection. *Courtesy of Chris W. Korinek, P.E., Synergy Technologies, Cedarburg, WI.*

loosens from vibration, the contact can make and break. The resistance added by the poor connection may be only a fraction of an ohm, but if the circuit is carrying the current to serve a high-wattage appliance, that can mean the creation of considerable heat. In Figure 10-23, a situation is depicted where a poor connection creates a localized resistance of 0.18 Ω. Even at a fraction of an ohm, this resistance is many times the resistance of a good, clean, tight connection, hence the term *high-resistance connection*. In a 120 VAC circuit serving a 1,650-W appliance, this means that nearly 14 A is trying to pass through that connection (with the contact equivalent of 6mm²), creating 35 W of heat. Although this is the same amount of heat created by 20 m (66 ft) of NM cable, the heat is localized at the poor connection. This condition highlights the concept of watt (energy transfer) density. For the same reason that high heat fluxes cause more rapid heating and more likely ignition, high watt densities create high temperatures that degrade nearby insulators and combustibles. Figures 10-24a and b show some of the localized heating effects (sometimes called *glowing connections* for obvious reasons). Ignition occurs when the arc ignites the "cloud" of gaseous pyrolysis products. This ignition may be in the form of a puff of flame, which, if sustained, can ignite the solid fuels to a sustained flame.

(a)   (b)

FIGURES 10-24A AND B A lab-created condition in which a 120-V receptacle with a loose connection carries a current of 14 A. (a) Initially, a series parting arc is seen when the screw and wire are moved. (b) As the heating continues, the screw and wire glow as current travels through the oxides formed and produces a relatively high-resistance connection. *Courtesy of Chris W. Korinek, P.E., Synergy Technologies LLC, Cedarburg, WI.*

These mechanisms often create series parting arcs as the contact between the wire and the connector is made and broken by mechanical movement or vibration. The intermittent arc (current passing through the air gap as the conductors touch and separate) causes very high temperatures at the tip of the wire. This accelerates the localized heating and degradation and ignites the gases and vapors from pyrolysis.[15] The receptacle in Figure 10-25a was ignited by a poor connection fault. The vibration from the washer and dryer caused the screw connector to loosen, and the high current for the dryer caused localized heating of the wire, connectors, and insulating materials. The heated insulation eventually degraded and was ignited by the arcing within the receptacle. The points of high resistance can be where a portion of the conductor size is smaller (as discussed previously), where the conductor material is different (as in an electric heater), and at connection points (ends) where poor electrical contacts or splices are made. See Figure 10-25b for an example. The *NEC* and UK wiring regulations specify that all wiring connections be made in protective enclosures (junction boxes).

To minimize heat buildup at connections, the full cross-sectional area of the conductor must be maintained, and positive contact between the conductors must be ensured. This means that no oxide or other films or foreign material can appear between the conductors. These failures are being seen where volatile sulfide contaminants in Chinese-made wallboard are corroding copper wires and connectors inside walls.[16] The total resistance of a circuit determines the current flow (at a certain voltage) and thus the total rate of heat production (watts). The distribution of the resistance throughout the circuit determines where the heat will be concentrated. An example of the first case would be the use of the inexpensive "zip" cord plugs that are installed by merely pushing the cord into a slot in the opened plug and then pushing the plug back together. The connection to the wire is made by pushing (stabbing) needlelike points through the insulation and into the conductor. The cross-sectional area of contact is diminished at this point, and if such plugs are used for anything other than perhaps a clock or small lamp, excessive heating can occur, as shown in Figure 10-26. Flexing of a zip cord can cause strands to separate and arc, as illustrated in Figures 10-27a and b. This heating can cause ignition of the insulation of the cord or the plug. Frequently, draperies cover wall outlets and can be ignited by the small fire supported by the plug and wire.

INSULATION BREAKDOWN (DEGRADATION)—CARBON TRACKING

With the exception of glass, ceramic, mica, and asbestos insulators, most electrical insulation materials are organic compounds containing carbon. Therefore, degradation of

FIGURE 10-25A Vibration from the clothes dryer plugged into this 240-V receptacle caused a loose connection. The #10 AWG wire involved is sputtered away to a fine point. The phenolic housing pyrolyzed away until it ignited. The resulting flames consumed the exposed wire insulation and the adjacent plug. The fire was confined by a metal receptacle box. *Courtesy of John D. DeHaan.*

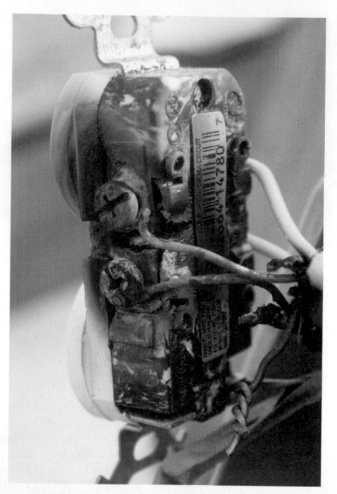

FIGURE 10-25B Ignition of plastic receptacle caused by loose connection shown in Figure 10-20a. *Courtesy of Jamie Novak, Novak Investigations and St. Paul Fire Dept.*

FIGURE 10-26 Zip cord plugs provide minimal contact area. *Courtesy of John D. DeHaan.*

FIGURE 10-27A Cord flexing at strain relief caused shorting.
Courtesy of John D. DeHaan.

FIGURE 10-27B Cord flexing caused a series arc connection failure.
Courtesy of John D. DeHaan.

such insulators by heat produces a carbon char, which is an electrical semiconductor. When this pyrolysis occurs over a large area, it is called *carbonization*. Breakdown can be more subtle, because it applies wherever the insulation has lost its insulating capacity, completely or in part, thereby allowing the current to follow an unintended path. A circuit must be energized for overheating to occur from insulation breakdown; however, the equipment does not need to be operating. The circuit can offer an unintended path for current to flow from hot (energized) to either ground or neutral via conductive contaminants or pyrolysis products by three routes: (1) through the insulating materials, (2) across the surface of the insulating materials, or (3) through air. When current flows in a limited area between conductors, it is often called *arc tracking, arcing through char,* or *carbon tracking*. The result of this carbon tracking is that something that is supposed to be an insulator becomes a semiconductor. This transformation often occurs over an extended time and at such a slow rate that it is not readily detectable until the conductive path it creates can conduct so much current that massive heating can occur. The time frame can be as long as months or years prior to the ignition of a fire.[17] For instance, if an insulator absorbs moisture, a small current can flow, producing some heat. This heat can slowly degrade organic insulation in the vicinity and turn some of it to carbon. Once enough carbon is formed between the conductors, more current can flow along the carbon path, providing localized heating and further degradation. Note that since the current is flowing along a path of limited area, even the passage of 1 A or so can create a very competent ignition source without tripping the circuit breaker protection. As the process continues, the current progressively increases as more and more carbon is formed. Finally, unless a breaker or fuse functions, an arc may be struck, igniting the carbonized insulation or volatile pyrolysis products, possibly resulting in a fire.

Leakage Current
This development of an alternative path is sometimes called *leakage current*. If the surface of an insulator is contaminated, some of the current leaks across the trail of contaminant and begins the process just described. Figure 10-28 shows leakage current heating the surface of a contaminated plastic plug. Note that the plug is energized but not connected to an appliance. The current being drawn by this flow was 0.25 A or less, but at 120 V, this caused some 30 W of heat to be produced in a very small area and created such high temperatures that glowing combustion was visible. This situation is more prevalent in inexpensive electrical appliances in which low-quality insulation is used. This

FIGURE 10-28 Lab-created leakage current across the face of a contaminated plastic plug. *Courtesy of Chris W. Korinek, P.E., Synergy Technologies LLC, Cedarburg, WI.*

occurrence has been reported in appliance switches, from the ignition of the insulation supporting the contacts, and has been seen in many devices—televisions, clothes dryers, coffeemakers, smoke alarms, molded plugs, and the like. Materials used for insulation, even in high-quality appliances, are selected on the basis of the thermal conditions expected under normal use. If an otherwise safe power cord is used in an abnormally hot environment or even buried under carpets or blankets, the heat being generated cannot escape. The heat can then build up, possibly beyond the design limits of the insulation, and the insulation begins to degrade. It should be remembered that overheating conductors first cause degradation of the insulation surrounding them, followed by degradation of plugs, sockets, and other hardware. It is only then that heat from an overheated wire will affect other fuels. Once flame is established, other fuels, whether cellulosic or thermoplastic, can be quickly involved. Figure 10-29 shows the plastic receptacle plate and plug in flames as the result of a simulated insulation degradation failure of the receptacle.

This degradation mechanism has been linked to causation of fires in a variety of assemblies and circumstances: insulators on fluorescent lamp connectors, electrical cords exposed to external heat, plugs, and receptacles. Just as heat exposure causes chemical and physical changes in wood, similar changes can take place when plastic insulators are exposed to heat. This heat can be from high-resistance connections, sustained high-current flows, or external heating. Loose connections in household (15 A) circuits have been found to produce connector temperatures between 150°C and 200°C (300° and 400°F). Such high temperatures (even if cyclic) will cause PVC and urea formaldehyde insulators to degrade and develop tracking capability. Once current begins to flow through the degraded material, localized heating and ignition will follow.[18]

Water and Carbon Tracking

Abuse of an appliance such as soaking it repeatedly with impure water or allowing conductive dusts to accumulate can trigger such carbon tracking. The higher the voltage, the more likely such tracking can take place. It is less likely in low-voltage (<24 V) circuits but becomes more likely at higher voltages.[19] Figure 10-30a shows arc tracking between two terminals energized with a 12-kV neon sign transformer that was triggered by spraying distilled water on the wood between the terminals. The water leached enough materials

FIGURE 10-29 Simulated residential fire after insulation degradation in an outlet produces small flames from receptacle and ignites nearby combustibles such as draperies or upholstery fabrics. *Courtesy of Chris W. Korinek, P.E., Synergy Technologies LLC, Cedarburg, WI.*

out of the wood to support leakage current. The arcing produced as the wood pyrolyzed resulted in the ignition of the wood. Figure 10-30b shows the effect of rainwater that eventually caused a short in a neon sign connector that nearly ignited nearby wood surfaces.

While pure water is an insulator, ordinary tap water contains enough dissolved salts to be somewhat conductive, and the greater the salt content, the better the water conducts. Small amounts of impure water, then, can play a role in triggering arc tracking on insulators, but they can also cause corrosion that can result in high-resistance, heat-producing connections. These connections create localized heat and pyrolysis of their insulative supports, and additional current flows as the char develops. When water enters electrical equipment, it can provide additional paths for the current in parts of the equipment

FIGURE 10-30A Lab-created arc tracking occurring across wood after water exposure between two terminals with a potential difference of 12 kV, a typical neon sign transformer voltage. *Courtesy of Chris W. Korinek, P.E., Synergy Technologies LLC, Cedarburg, WI.*

FIGURE 10-30B Neon sign (PK) connector, which had been exposed to rain, shorted, heating surrounding combustibles, and caused extensive damage to an outdoor sign. *Courtesy of Joe Bloom, Bloom Fire Investigation, Grants Pass, OR, and Chris J. Bloom, CJB Fire Consultant, White City, OR.*

(a)

(b)

(c)

(d)

FIGURES 10-31A–D (a) Lab-created flaming ignition of a switch housing after the switch was exposed to water containing an impurity. The voltage was 120 V, and current did not exceed 1.3 A. (b and c) Plastic between the terminals of this microswitch has been severely degraded, allowing current to flow between "open" switch terminals. (d) New switch. *Courtesy of Chris W. Korinek, P.E., Synergy Technologies LLC, Cedarburg, WI.*

that are not meant to carry such currents. In addition to providing a potential electric shock danger, the presence of impure water has been seen to cause fires by allowing current flows along these unplanned paths. In one case, a storm broke the utility drop connectors and bent the entrance conduit. When repairs were made, the entrance conduit was not repaired, and it had broken loose at the roof, allowing a path for rain to flow directly down into the wall and into the electrical distribution system over time. The water contamination ultimately resulted in ignition and fire destruction.

Fire codes list specific types of electrical equipment for wet or damp locations. When regular equipment is exposed to these conditions, disaster can occur. In the test shown in Figures 10-31a–d, water with an impurity was applied to an electrical switch. Sizzling, smoking, and flaming occurred as the plastic in the switch overheated. Note that the light was in series with the fault and limited the current flow in the circuit. The OCPD was not tripped because the current did not exceed 1.3 A. Wall receptacles under towel dispensers in public restrooms are often affected by water corrosion.

Heating and Carbon Tracking

Heating of an insulated cable (or insulated wires inside a conduit or appliance) will cause the insulation to degrade, sometimes over a substantial area. Energized conductors within can then provide current that flows through the charred insulation (arcing through char) to either the ground or the neutral conductor or the grounded conduit. With conduit, this can "blow" a hole through the metal in the vicinity of the arc (as seen in Figure 10-32). Note that such a hole is almost always a result of a preexisting fire and not an initial cause. The localized heating then can pyrolyze and ignite the insulation. Molten metal particles can also be ejected to fall onto other combustibles below. Very high current

FIGURE 10-32 Lab-created arcing through char (caused by the heat of an external fire) that melted holes in the steel conduit enclosing the wire in a high-current circuit. *Courtesy of Chris W. Korinek, P.E., Synergy Technologies LLC, Cedarburg, WI.*

single-phase or three-phase circuits will allow enough current flow between a damaged insulated cable and a grounded conduit (EMT) or flexible armored cable (BX) to melt a hole through conduit to cause a fire, as shown in Figure 10-33.

A very different mechanism is illustrated in Figure 10-34. There the metal flexible (BX) cable housing became energized due to an open neutral and became a heating element that ignited the adjacent wood joists.

ARCS AND SPARKS

As we saw in Chapter 6, an arc can be defined as the flow of current through a gas (usually air). It can range from a brief discharge (spark) built up by walking on a dry carpet to a bolt of lightning. If an arc is sustained, the atmospheric gases through which the current is passing can be supplemented by ionized (plasma) vapors of the conductors

FIGURE 10-33 Conduit from service mast involved in fire showing arcing and penetration from within. Current would be limited here only by the line transformer (2,000–3,000 A). *Courtesy of E. R. Baker, Fire Investigator, Fresno, CA.*

FIGURE 10-34 Armored cable became energized when open neutral occurred in meter box and acted as a heating element to ignite ceiling joists. *Courtesy of Detective Mike Dalton, Knox County Sheriff's Office, Knoxville, TN.*

involved (or, in the case of welding, by the vapors of a flux or specialized gas). For an arc to establish itself in air between two conductors requires either a high voltage or a very small gap between the conductors, because air is a good insulator. Once an arc is established, however, it can be maintained as the conductors separate via the conductive path offered by the vaporizing metal and the ionized air molecules. This is especially true with DC currents, whose voltage does not cycle through zero periodically as does AC. As described previously, an arc produces very high temperatures, on the order of thousands of degrees, but they are very localized, so a fuel has to be in almost immediate contact with the arc for it to ignite. Gases and vapors, including those from active pyrolysis of insulators, are much more readily ignited to flames than are solid fuels. Unless a solid fuel is very finely divided or thin (like paper), it will resist ignition by all but the most energetic and prolonged arcs. Thin paper, lint, sawdust, and flammable gases and vapors are all ignitable by 120/240-V arcs in AC circuits.

When an excessive current is passed through a metallic conductor, its temperature rises to the melting point of that metal. As the first melting occurs, the wire severs and a *parting arc* occurs at this point (which may be sustained at high enough voltages). This arc creates very high temperatures locally, and melted metal is spattered onto cooler nearby surfaces, where it recondenses as droplets. While these sparks (often incandescent particles) are very hot, they do not have sufficient thermal capacity to ignite massive solid fuels nearby, except under the most unusual conditions. But recalling the information on pyrolysis from Chapters 3 and 5, we realize that sustained heat breaks down most plastic or cellulosic insulators into combustible gases and vapors and conductive char. It is often the gases and vapors that are first ignited by the arc itself.

The breakdown voltage of dry air is about 30,000 V/cm (75 kV/in.), but the minimum voltage at the minimum spacing (~0.01 mm) is still 340 V (DC or RMS AC – 240 VAC peak = 340 V RMS).[20] So, if two bare conductors are kept separate in air by even a very small amount, an arc cannot be started. If the two conductors touch and then are separated, the ionized gases and metallic residues can maintain the arc at some distances until the voltage differential goes to zero. In a 120-V 60-Hz AC circuit, arcing lasts typically 0.5 to 4 cycles (0.008 to 0.05 s).

An AC arc tends to self-extinguish as its cycle goes through zero volts in the AC voltage sine wave. A DC arc can be maintained for as long as there is power. The duration of the arc is critical to its capacity to ignite nearby fuels. While the temperature of gases/plasma in the arc is a function of current and is thousands of degrees Celsius, the duration of the typical arc in an overcurrent-protected circuit will usually be so brief that ignition of solid fuels is nearly impossible, except, for example, when the fuel is finely divided (cotton waste or fine sawdust, for example) and in direct contact with the arc or when the overcurrent protection fails to interrupt the circuit. Flammable gases and vapors, of course, can be ignited by all but the shortest arc due to their low ignition energies and direct contact with the superheated gases of the arc.

ALUMINUM WIRING

Aluminum is sometimes encountered as a substitute for copper because of its lighter weight and lower cost. Aluminum was used for a period of years in homes and mobile homes for all internal wiring; however, today it is mostly used for high-voltage distribution lines and in subpanels. Aluminum cable is routinely used for **service drops** (from pole to weather head). Because aluminum is a poorer conductor than copper, to carry the same current as a copper conductor safely, the aluminum conductor must be larger. However, because aluminum is lighter, less conductor weight is required. Because aluminum costs considerably less per pound than copper, aluminum is cheaper to use overall.

The major problem with aluminum is that of making mechanical connections. Most difficulties with aluminum wiring are found not in the wire span but at the ends where the connections are made. Failures have resulted from directly substituting aluminum wire for copper; for example, using the fittings, receptacles, and switches that were designed for copper wire. The quality of the installer's work is critical to avoiding problems that result from aluminum's peculiar physical and chemical properties. Even the best electrician cannot install small-diameter aluminum wiring to be robust and maintain a good connection indefinitely. The following are some points to keep in mind.

First, the electrical conductivity of aluminum is somewhat poorer than that of copper. As a general principle, aluminum wire should be two sizes larger (two AWG numbers smaller) than its copper equivalent. Wire tables should be consulted for exact size requirements. This requirement may be overlooked by some installers.

Second, aluminum is very chemically active. Exposed surfaces always have an oxide coating. This oxide film acts as an insulator and is a major problem in making good electrical connections. Chemical preservatives (antioxide pastes) are usually applied when making connections to minimize this oxide film formation.

Third, the thermal expansion coefficient of aluminum is different from that of copper, and for any combination of these two metals, allowances must be made for this difference. Connections that are tight at one temperature may be loose at another. Also, the continued tightening and loosening due to changing temperature can result in a permanently loose connection.

Finally, another problem, mainly with the early products, was cold flow. Aluminum creeps like putty when subjected to pressure, and it flows away from the pressure points, resulting in a poor connection.

The basic problems with aluminum wiring have largely been corrected by new aluminum alloy connectors. The connectors, receptacles, and switches appear similar to their old counterparts, but the selection of materials and their design has, for the most part, corrected the problem. The devices acceptable for aluminum are so marked (AL/CU). Aluminum wiring connection failures can still occur even with AL/CU devices.

Aluminum, having a lower melting temperature (660°C; 1,220°F), than copper (1,083°C; 1,981°F), will usually be melted by most fires and therefore may not provide useful information after a substantial fire. As it melts, it can form **eutectic** alloys with

service drop ■ The overhead service conductors from the pole, transformer, or other aerial support to the service entrance equipment on the structure, including any splices.

eutectic ■ An alloy of two materials having special physical or chemical properties, typically having the lowest melting point of any combination of the two.

copper conductors and cause them to melt away, sometimes giving the appearance of a massive failure of the copper conductor. The presence of a gold "brass"-colored "stain" on the remaining copper is a clue that there may have been aluminum alloying.

ELECTRIC TRANSFORMERS AND MOTORS

Electric transformers and motors are similar in that they contain copper wire that is wound on an iron core that creates magnetic fields. Both can overheat from excessive current flow, and this excess heat can cause insulation destruction that, in the absence of proper overload protection, is capable of causing ignition.

Electric Transformers

Transformers contain several windings with some windings that have wires connected to the midpoint of the windings (called *mid-winding taps*). They are used for raising or lowering the voltage (and current) in an AC circuit. Large transformers frequently are oil-filled to insulate them and to provide a cooling bath for the heat normally produced. Large power transformers of this type have a potential fire hazard from the high voltages and the presence of certain types of combustible oils. However, modern transformers are sometimes filled with noncombustible oil or are dry-type devices (air-cooled with no oil).

Fire problems with transformers are similar to those in motors; however, the mechanical and high starting current conditions of motors are not applicable. Some electronic equipment containing transformers has been known to start fires because of the absence of fuses or improper fuse selection by the manufacturer. High-current devices may be able to induce heating by magnetic induction, but this is beyond the scope of the investigations covered by this text.

Electric Motors

In addition to the windings, motors have moving parts. Some have centrifugal switches that operate when the motor attains 75 percent of its normal operating speed. Other motors require an electrical connection to the moving part (armature or rotor), accomplished by brushes riding on a moving contact (commutator or slip ring).

Electrical motors present numerous possibilities for ignition, including the following:

- Overheating because of excessive current
- Defective windings or excessive mechanical load, which prevents attainment of proper operational speed
- Arcs from the centrifugal switch or brushes
- Sparks from frictional contact with moving parts
- Overheating due to excessive use beyond its rated **duty** cycle

duty ■ Conditions of use in electrical service.

These hazards may be overcome by construction of the motor frame so that electrical parts are suitably enclosed, or by keeping combustible material away. The starting equipment of a heavy-duty motor presents hazards in the forms of arcing of current-breaking contacts and the heat produced in some forms of starting equipment. The hazards of overloading a motor may be overcome by proper overcurrent protection.

Fire hazards common to motors include the following:

- Location close to combustible material that may be ignited by arcs or sparks.
- Location in a damp place or where subject to corrosive vapors.
- Linty or dusty condition of environment.
- Burning out (failure of insulation on windings), which may be due to overload, stalling of the rotor, or, in some cases, too low a voltage at the motor terminals.
- Improper overcurrent protection. Except for manually started motors of 1 hp or less, every motor should be protected by a suitable overcurrent device. (Such devices may be internal or external.)

- Starting equipment that produces arcs and that is not located away from combustible material.
- A three-phase motor protected by fuses in which one fuse blows or one leg of the circuit opens while the motor is started or operated.

Post-fire Examination of Motors

Examination of motor remains can frequently provide useful information. When electric motors are found to be in a fire area near a suspected point of origin, a thorough evaluation should be made to determine whether the fire was deep seated in the windings or whether damage resulted from heat conducted to windings from fire outside the motor housing. Figure 10-35b is an example of a failed winding in a motor that ignited the insulating varnish and led to a fire. If the shaft is frozen (will not turn), it may be an indication that the heat was internal. Because small motors have much less mechanical inertia, this symptom is not reliable. If the motor was running during the fire, melted bits of solder thrown about the interior of the motor housing may sometimes be found in motors having a commutator (a rotary electrical connector with soldered leads). Friction-caused overloading is indicated if the motor belt is most heavily damaged where it passes over pulleys. Fire from other sources damages motor belts much more between the pulleys and often does less damage to belts protected by the pulleys. If immediately after a fire the motor housing is too hot to touch, but iron or steel of similar size in the same area is relatively cool, it indicates internal heating of the motor. A stalled motor, or one that is not running up to speed, will overheat from excessive current, which can be caused by a defective motor, failure of the winding's insulation, tight bearings, too small a motor for the purpose, frozen or tight load, and the like.

Frequently, when an electric motor from a fire is completely gutted inside, it does not prove that the motor was faulty. It can also indicate an overload condition or possibly that something jammed the motor and prevented it from turning. If the motor was able to operate with the load partially jammed, a fire could have been started from friction of the belts or pulleys or an external interference (as in Figure 10-36). Fires from electric motors can also be caused by bearings that were not lubricated, faulty starting mechanisms, or excessive dirt in the motor. Excessive dust in a motor is often found where the motor was used to power shop tools. Dusts can be subject to hot-surface ignition if allowed to accumulate on hot motors.

Old motors are generally physically larger, for a given horsepower rating, than modern motors. In addition, modern motors are frequently designed to operate at a higher

FIGURE 10-35A Motor failure when interlock shorted out to armature. *Courtesy of Peter Brosz, Brosz & Associates Forensic Services Inc.: Professor Helmut Brosz, Institute of Forensic Electro-Pathology.*

FIGURE 10-35B Close-up of motor winding failure. *Courtesy of Jean-Claude Martin, Institut de Police Scientifique et Criminologie (IPSC), Lausanne, Switzerland.*

FIGURE 10-36 Chafing of the motor shaft against a piece of clothing wedged in the appliance caused a dryer fire. *Courtesy of Joe Bloom, Bloom Fire Investigation, Grants Pass, OR.*

temperature than old motors because of the use of modern high-temperature insulation. The nameplate or manufacturer's literature should be consulted before a modern motor is suspected of overheating.

FIXED HEATERS

Electric radiant heaters that are fixed installations and hardwired into the electrical system of a building should not be overlooked as potential fire causes. They can be located in walls, in floors, and even in the ceilings of structures. They are designed to operate under certain conditions of extended operation and limited heat dissipation, but because they may be concealed, thought is often not given to the combustibles or the insulation packed too close, beneath, or over them. The Consumer Product Safety Commission (CPSC) estimated that fixed (gas and electrical) heaters and furnaces were responsible for 4,500 fires, 8 percent of all 56,500 accidental residential fires in 2004–2006.[21]

APPLIANCES

It must be remembered that electrical equipment can be divided into two general categories: the fixed conductors, switches, protective devices, and outlets of the distribution system; and all portable, movable lamps, tools, and appliances that can be connected into that system. The fixed distribution system must (in most areas) meet fairly stringent guidelines and is not likely to be changed by the average owner or tenant, so it usually poses less likelihood of unwanted heating. Appliances and all other removable

FIGURE 10-37A Coffeemaker ignited by runaway heating when TCOs were disabled. *Courtesy of Steve Riggs, Public Agency Training Council.*

FIGURE 10-37B Post-fire remains show heating elements, power cord, and TCOs. *Courtesy of Steve Riggs, Public Agency Training Council.*

electric utensils are subject to much more misuse and can be added together to pose all manner of risks. By far the largest percentage of fires are ignited by the appliances, but these fires are also more easily detected and extinguished by the user than are the fires hidden in walls and ceilings. See Figures 10-37 and 10-38 for examples of small-appliance fires.

Electric appliances are found in nearly all structures, and whether they produce quantities of heat by design or by failure or simply represent a convenient source for that all-important arc, their possible contribution to fire causation should not be overlooked. Faults with power cords and plugs may result from poor design or misuse by the consumer. Because power cords and plugs are subject to misuse, faults in them probably represent the majority of fire causes from appliances. The symptoms of poor contact resistance and inadequate conductor sizes were discussed previously in this chapter. A new problem source is Chinese-made appliances, extension cords, or power strips that are made with conductors that are far too small to carry normal currents. These small conductors can overheat quickly and cause ignition of adjacent plastic components.

Some appliances, however, because of their high power load, high voltages, or typically unsafe locations, deserve special mention.

FIGURE 10-38A Runaway clothes iron (TCO disabled). *Courtesy of Steve Riggs, Public Agency Training Council.*

FIGURE 10-38B Plastic components support large flames. *Courtesy of Steve Riggs, Public Agency Training Council.*

FIGURE 10-38C Post-fire remains show melted aluminum base and heating element. *Courtesy of Steve Riggs, Public Agency Training Council.*

FIGURE 10-39A Pattern on walls indicates origin in dryer. *Courtesy of Det. Mike Dalton, Knox County Sheriff's Office, Knoxville, TN.*

FIGURE 10-39B Interior fire burned clothing in drum. *Courtesy of Det. Mike Dalton, Knox County Sheriff's Office, Knoxville, TN.*

Clothes Dryers

High power usage, high temperatures produced even in normal use, poor locations with dust and dampness common, and poor maintenance (such as not cleaning lint filters regularly) are associated with fires in clothes dryers. Lint also can accumulate around motors and heating elements (especially if the dryer is not properly vented) (see Figures 10-39a and b.) Clothes dryers are responsible for an estimated 11.7 percent of all residential fires.[22]

Portable Electric Heaters

High power usage, high temperatures, and careless use (blocked by furniture or in contact with flammable furnishings or clothing, for example) are associated with fires involving portable electric heaters. These devices are responsible for an estimated 2.8 percent of all residential fires.[23]

Electric Ranges, Electric Ovens, and Small Cooking Appliances

High power usage, high temperatures produced, and proximity of fuels (such as foods, utensils, or clothing) are associated with fires involving electric ranges and ovens and small cooking appliances. The CPSC data indicate that these appliances are responsible for an estimated 37.3 percent of all residential fires; however, these are often cooking fires. Small cooking appliances can be dangerous due to the fuel load represented in the food. Figure 10-40 shows a cooking-oil fire in a frying pan. CPSC estimated that cooking equipment was responsible for 7.5 percent of fire deaths in 2004–2006.[24]

Dishwashers

Even though dishwashers use water inside to clean dishes, they have a number of electrical overheating failure modes. These have been known to include overheating of drying elements, lack of prevention of wiring damage during opening and closing, and egress of moisture and other conductive fluids into area where wiring is present, causing arc tracking and ignition of nearby combustibles.[25]

Television Sets

High voltages and poor design (including the frequent use of flammable materials for cabinets) are associated with television fires. According to the CPSC, there was an extremely high incidence of fires in sets made in 1970–1974. In 1979 the CPSC estimated that approximately 2 percent of all residential fires, some 11,000 incidents causing 160 deaths, resulted from television sets. Since 1974 the frequency has been decreasing because of the

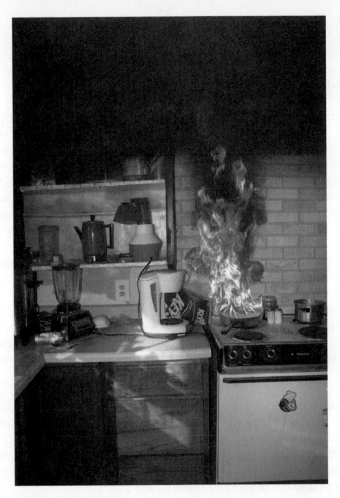

FIGURE 10-40 A cooking-oil fire in this electric skillet involved the cabinets above by convective flame plume. The fire spread from this point was very rapid. *Courtesy of Lamont "Monty" McGill, Fire/Explosion Investigator.*

use of newer sets that have better wiring, lower operating temperatures, and cabinets made from fire-resistant materials.[26] By 2006, that figure had dropped to 700 fires, 0.2 percent of residential fires, with no recorded deaths (including radios and phonographs).[27] Except for televisions sold in North America and Japan, the plastic cabinets are made from non-flame-resistant styrene or similar material. In the UK there have been a large number of TV fires started by unattended candles.

It should be noted that it is possible for fires of both accidental and incendiary origin to destroy all combustible elements within a modern TV. In such cases, even highly experienced laboratory examiners may find it impossible to determine whether the fire started within the set or elsewhere because of the destruction of the diagnostic signs necessary for such a determination.

ELECTRIC LIGHTING

Lighting fixtures and related issues of particular interest to the fire investigator include magnetic and electronic fluorescent light ballasts, overheating in fluorescent lampholders, lamp failures at the end of life, and other lamp considerations, as discussed next. During the period 2003–2006, the NFPA estimated that annually, 2,800 fires in nonresidential structures were attributable to the general category of lighting.[28] The NFPA also estimated that some 5,770 home structure fires were linked to lamps, light fixtures, or lightbulbs, which caused 76 deaths and $181 million in damages in 2006.[29]

Magnetic Ballasts

Ballasts used in fluorescent lighting fixtures create a high voltage to initiate a discharge (current) through the gas inside and to limit the current once the arc is established.[30] While the majority of new ballasts put into service today are electronic, there are still millions of magnetic ballasts in use, especially in fixtures that utilize the older 3.81-cm (1.5-in.) diameter T12 lamps. In Europe, magnetic ballasts are still frequently used in new fluorescent light fittings. Magnetic ballasts generate a considerable amount of heat when in operation. In North America, ballasts manufactured for new fixtures since 1968, and all ballasts manufactured since 1984 have been equipped with autoresetting thermal protectors, which greatly reduce the potential for fires caused by overheating ballasts.[31] However, magnetic ballasts are not failure-proof. If the protector fails to sense an overtemperature condition or its contacts become welded shut, an overheating situation could be allowed to continue. Under ideal conditions, such a ballast may reach temperatures high enough to ignite combustibles above the fixture, such as low-density ceiling tiles or cellulose thermal insulation. Continued overheating can lead to an arcing event that is energetic enough to penetrate both the ballast case and the enclosure of the fixture (as shown in Figure 10-41). Molten metal and burning gasses from the vaporized potting compound ejected out of the fixture as the result of such an event may have sufficient energy to ignite combustible materials nearby.[32] Most fixtures have metal covers to prevent such contact, but covers may be missing due to poor maintenance. The NFPA statistics for 2003–2006 indicate that 26 percent of the fires in nonresidential structures attributable to lighting were caused by fluorescent fixtures or ballasts.[33] (See Figure 10-41.)

Older fluorescent light fixtures with magnetic ballasts can cause fires from a breakdown of the ballast transformers, which contain a flammable potting compound. Some older ballasts have a starting voltage as high as 700 to 900 V and operate at about 400 V; newer units operate at lower voltages. They are designed to operate continuously at over 40°C (100°F), but the temperature can go much higher. This heat sometimes causes melting and vaporizing of the combustible contents, sometimes igniting combustible ceiling material. The vapors can collect and ignite in the housing.

FIGURE 10-41 A field example of an arcing event inside a magnetic ballast that breached the metal fixture body.
Courtesy of David W. Powell, SYTEK Consultants, East Syracuse, NY.

Early ballasts had no thermal protection. Current or newer manufactured units contain a thermal overload switch that is intended to disable the appliance in the event of overheating. These thermal protectors are typically designed to open at 110°C ± 5°C (230°F).

Electronic Instant-Start Ballasts

Electronic ballasts generate much less internal heat and operate at considerably lower temperatures than their magnetic counterparts. As a rule, they are fused internally or are inherently thermally protected. As a result, fire causation due to electronic ballast failure is extremely rare.

Instant-start electronic ballasts use a high voltage (5–600 V) to start the lamps. If the lamp is inserted improperly, arcing in the lampholder, between the lamp pin(s) and the lampholder contacts can occur due to the high voltages present. The extremely high temperatures resulting from a prolonged arc can result in degradation of the lamp holder into combustible materials, melting of the brass contacts, and possible ignition, as shown in Figures 10-42a–c. Because the lamps in the fixture can remain lit while the arcing is occurring in the lampholder, the overheating event can progress, possibly to ignition, before occupants are aware of the problem.[34]

FIGURE 10-42A A lab-created arcing event in progress within a lampholder in a fluorescent light with an instant-start electronic ballast. *Courtesy of Richard E. Korinek, P.E., Synergy Technologies LLC, Arbor Vitae, WI.*

FIGURE 10-42B The result of a brief arcing event at the lampholder contact as a result of improper insertion of the lamp. *Courtesy of Richard E. Korinek, P.E., Synergy Technologies LLC, Arbor Vitae, WI.*

FIGURE 10-42C A field example of arcing at a lampholder contact (second from left) that resulted in a fire. *Courtesy of David W. Powell, SYTEK Consultants, East Syracuse, NY.*

Fixtures for 8-Ft (Commercial) Lamps

Some large lampholders incorporate spring-loaded contacts in which the springs conduct the lamp current. A number of fires have been attributed to thermal relaxation of the springs in the lampholders.[35] Aging and thermal stress can result in poor connection in the lampholder. Arcing can occur, degrading the lampholder while the lamps remain lit. In another variation, improper spacing or mounting of the lampholders on the fixture housing can cause poor contact pressure in the lampholders, resulting in arcing and overheating in the lampholders.

Compact Fluorescent Lamps (CFLs)

Compact fluorescent lamps are becoming very common, and heat-producing failures in the electronics concealed in their plastic bases are being seen, as depicted in Figure 10-43. These lamps are intended for base-down use, but many household lighting fixtures require horizontal or even base-up installation. Some CFLs have already been the subject of CPSC recall, prompting more reported failures.[36] Such lamps can also overheat if used with a dimmer switch.[37]

T5 Fluorescent Lamps

Lighting systems are now available that utilize new thinner fluorescent lamps, such as T5 lamps, with 15-mm (5/8-in.) diameters. At the end of a lamp's life, overheating at the lamp ends can occur, resulting in cracking and melting of the glass envelope, and overheating of the lamp end.[38] Temperatures can approach 700°C (1,335°F), which is high enough to ignite combustibles nearby. Such overheating may also damage the lampholder, resulting in poor connections between the lamp pins and the lampholder, leading to further overheating.[39]

In any situation where overheating occurs at the end of the fluorescent lamps, the possibility exists for such an occurrence to result in flaming ignition if there are combustible materials nearby. For example, many fixture diffusers are made of acrylic plastic, which has a relatively low ignition temperature in the range of 195°C–289°C (383°F–552°F).[40]

FIGURE 10-43 Electronics in the plastic base of a CFL may fail, perforating housing, and cause an external fire. *Courtesy of Steve Riggs, Public Agency Training Council.*

Aquarium Lights

The moist environment near aquariums can result in corrosion of electrical connections in the lampholders. Overheating at a poor connection has been shown to generate heat to ignite the lampholder and then spread to body of the fixture, which is often made of a combustible plastic. Insufficient contact pressure may be a contributing factor in these fires.[41] An example is shown in Figures 10-44a–c.

Metal Halide Lighting

Metal halide (MH) lighting is a type of high-intensity discharge (HID) lighting, along with mercury vapor and high-pressure sodium types. Inside the thin outer glass shell of the MH lamp is an inner arc tube made of fused quartz. The arc tube operates at extremely high temperatures, up to 1,100°C (2,000°F) and pressures ranging from 73 to 441 psi (5–30 atm).[42]

The majority of MH lamps fail passively at their end of life. In some cases, the inner arc tube can break due to the weakening of the quartz arc tube. The breakage can be a violent event, rupturing the outer glass envelope and scattering 1,100°C (2,000°F) particles of quartz glass out of the fixture. Hot quartz particles from such an event have been found to ignite combustible components of the light fixture, materials below, and even wooden gym floors.[43]

Incandescent Lamps

As we saw in Chapter 6, in normal (base-down) use, most incandescent lamps do not produce surface temperatures sufficiently high to ignite most combustible materials in open air. There is no guarantee to the fire investigator, however, that such lamps will have been used properly. Normal use for most lamps entails free circulation of cooling air around the lamp. Any enclosure that restricts the flow of air around a lamp must be of nonflammable materials, and the investigator must determine that no combustible materials came into contact with the surface of the bulb while incandescent. Lamps that are portable by their design—drop cords and inspection lamps—are more likely to be covered by rags, clothes, or wood shavings during their use. If left on in this covered condition, they can readily produce surface temperatures in excess of the ignition temperature of cloth or wood (as discussed in Chapter 6). Lamps and light fixtures were associated with some 4.7 percent of residential fires in 2006 according to the CPSC.[44]

FIGURE 10-44A A field example of an aquarium light in which an overheating lampholder was the cause of a brief residential fire. *Courtesy of Richard E. Korinek, P.E., Synergy Technologies LLC, Arbor Vitae, WI.*

FIGURE 10-44B Corrosion at the end of a lamp that was in an aquarium light that caused a structure fire (field example). The fire occurred at the other end of the lamp. *Courtesy of Richard E. Korinek, P.E., Synergy Technologies LLC, Arbor Vitae, WI.*

FIGURE 10-44C A field example of a lampholder in which an overheating connection at the connector on the right side of the lampholder caused the fire in an aquarium light. *Courtesy of Richard E. Korinek, P.E., Synergy Technologies LLC, Arbor Vitae, WI.*

High-wattage incandescent lamps (over 150 W), due to their higher operating temperatures, can pose an ignition risk if ordinary combustibles come into direct contact with their outer envelopes. As we saw in Chapter 6, halogen lamps have extremely high surface temperatures at their quartz envelope.[45] The lamps can produce high temperatures at their metal shades, and their intensity can melt light diffusers and other plastics set too close in front of them.

ELECTRIC BLANKETS

Electric blankets have been responsible for some fires. The cause of these fires can frequently be attributed to misuse. Some blankets are controlled by a thermostat located at the edge, and when blankets are partially covered with additional blankets or other materials, it is possible for other portions to become overheated. If the main control switch is left on "high" in a cold room, the current can remain on continuously and cause the covered section of the blanket to overheat. Blankets have been known to cause fires when tucked between a mattress and padded box spring, wedged between the bed and a wall, or folded at the foot of the bed. It can be difficult to determine whether a bed fire was caused by a blanket or other cause such as smoking in bed, unless the fire was detected early and the char path is traceable. Sometimes the thermostat and control will remain after a fire, and these can provide information as to whether the blanket was turned on. Some blankets have multiple TCOs embedded in the blanket, so that if even a portion of the blanket reaches a high temperature, the current is interrupted. The location and distribution of the fine heating wires can reveal the use (or abuse) of the blanket at the time of the fire.

Electric blankets have evolved from the continuous-resistance-wire heating elements protected by numerous TCOs to positive temperature coefficient (PTC) cable. This cable, whose resistance varies with its temperature (the hotter the cable, the higher its resistance), has two conductors, each wrapped around a polyester core and then embedded in insulating layers. There is typically an external thermostatic control and a safety circuit board sewn into the blanket. Failures can occur at the connectors or the safety circuit board, or broken PTC wires can cause arcing and arcing through char as the cord degrades.[46]

EXTENSION CORDS

Although they are not part of an appliance, extension cords are a movable element of the electrical system. They are involved in a large number of fires in North America and other countries because of the variety of ways they can be abused. Most often, their multiple outlets offer the chance to plug in so many appliances that their ampacity (typically 5 to 15 A) is readily exceeded. The nature of their use results in their being hidden under furniture or carpets, with the result that their heat dissipation is severely limited. The insulation quickly melts and chars in overcurrent situations, and the char provides a semiconductive path between the conductors that can create even more heat without drawing enough current to trip the protective devices. They can also be used in hidden spaces (behind baseboards, for example) or suffer so much damage during the fire and suppression that they can be overlooked during post-fire examination.

HEAT TAPES AND HEAT CABLE

Buildings in cold climates may have electrical-resistance heaters in the form of tape strips that are designed to be wrapped around water or sewer pipes that might be susceptible to freezing, or laid in gutters to reduce ice/snow loading. The tape may then be wrapped in layers of combustible or noncombustible insulation, concealing it from view. Heat tape typically uses a resistive wire heating element either wrapped helically around a core filament or zigzagging between two bus wires, with an integral thermostat. Heat cable contains two parallel conductors straddling a conductive, carbonaceous matrix whose resistance is dependent on its temperature. At warm temperatures, the heat cable matrix has a high resistance, so no current flows; at lower temperatures, its resistance decreases, allowing current to flow and provide heat.[47]

Maintenance of these installations is often neglected, and the tape insulation can fail (along its length as well as at its connections) from a variety of causes after prolonged use. As the insulation fails, arc tracking can cause prolonged external heating while not exceeding overcurrent protection limits. Such tapes have been linked to fires in many locales.[48]

BATTERIES

Lithium-ion (Li-ion) and lithium ion–polymer (Li-Po) batteries have the potential to overheat and ignite due to their high energy density, flammable internal materials, and susceptibility to generate oxygen.[49] Failure modes include short circuits and overcharging. There have been numerous major recalls of lithium batteries by laptop computer companies in recent years. Lead-acid batteries have been known to explode due to short circuits inside the batteries or when loose external connections ignite hydrogen gas evolved during the charging process.[50] Some battery chargers can develop poor connections, and charging non-rechargeable batteries can cause overheating and ignition.

Investigation of Electricity-Related Fires

When a fire occurs, suspicion of its cause often falls on any electrical equipment present, in part because of the basic fact that electricity can cause heat, and this heat can cause a fire. Because most structures contain electrical wiring or electrical apparatus of some sort, electricity-related causes are often suspected. However, in a notable study in which a number of fires were reexamined by fully qualified scientific investigators, the 20 percent of fire causation attributed to electrical fires was estimated to be on the order of only 9 percent upon further scrutiny.[51] Some investigations focus on electrical causation for no better reason than suspicion based on misunderstanding and mythology. Basically, when considering an electrical cause, the investigation should address the following possibilities:

- Was the installation or appliance designed to produce heat? If so, how did the heat transfer from the unit to ignite the first fuel ignited? (Without knowing what the first fuel was, how can one realistically evaluate the possibilities?) The same analysis applies when heat is a waste product (such as with incandescent lamps).
- If the unit or appliance was not designed to create heat, how could electricity escape from its intended path to produce heating and thereby the ignition?
- Was thermal protection present, and if so, why did it fail to protect the appliance?

When electrical components or wires are found melted after a fire, there is immediate suspicion that the fire and the melting were directly due to an electrical failure of some sort. The following questions should be considered at this point:

- Was the melting caused by the fire and not by an electrical fault?
- Is there any indication that electrical failure occurred? (It is not recommended that an item of evidence be plugged in or energized. This can destroy evidence and should be done only after extensive examination and after the approval of all parties.)
- When an electrical failure is detected, was the failure caused by the fire, or was the fire caused by the failure?
- Is enough energy released by the electrical source to ignite the first fuel in the form in which it was presented to the ignition source?

Fires can be ignited by electrical equipment by one (or more) of the following events:

- *Design errors*—The heat shielding on a resistance heater is not designed correctly and the back of the heater can become dangerously hot, or electrical insulation between two conductors is not designed properly and the unit can flex, bringing the two conductors into contact. Thermal protection devices are absent or poorly located.

■ *Manufacturing defect*—The unit was designed properly, but someone forgot to install the heat shielding; the wrong wire or insulation was used when the unit was built; electrical connections were left loose, and the high resistance caused by the poor connection caused the insulator to pyrolyze and ultimately arc track, with resulting ignition of the plastic insulator housing.

■ *Improper installation or use*—A heater is used inside a closet where its housing comes into contact with combustibles, or it is installed in an insulated enclosure where the convective cooling needed to maintain safe external temperatures is lost.

■ *Use of approved equipment for an unapproved application*—Equipment is used in a high-humidity or high-ambient-temperature environment for which it was not designed; a thermostatically controlled heater is used in an explosive environment.

■ *Accidents*—A heater is properly designed, built, and installed, but rags drop onto its wire shield and ignite; a can of lacquer thinner is used to clean tile spills, and vapors come into contact with thermostat contacts as they open.

If properly designed, installed, and maintained, electrical systems for lighting, power, heating, and other purposes are convenient and safe; but if not installed, used, and maintained in accordance with well-established standards, they may introduce both fire and personal injury hazards. This fact has been recognized since electric power first came into general commercial use. Unfortunately, however, electricity has historically received the blame by fire investigators when they could not determine any other logical method of ignition. There are methods by which useful information can be obtained from electrical debris, and interpretation of this information by a person knowledgeable in the field can very frequently provide valuable testimony.

Over the years, a variety of indicators or guidelines for determining an electrical cause of fire have evolved. Some are useful, but unfortunately, some are not as reliable as generally believed. Little work has been done to substantiate these indicators because they have been based primarily on what appeared to be logical explanations. In fact, some experimental work has shown that some much-used indicators are produced by unexpected mechanisms.[52] When there has been doubt as to the ignition source, reports have named electricity as the sacrificial lamb by noting that "since no other source was determined, the cause probably was electrical." Such speculative comments should never be made, and any opinions should be offered only with sound indications.

Knowledge of the basic principles of fuels and their ignition and a basic knowledge of the principles of electricity are the essential tools of the investigator approaching a fire scene of possible electrical origin. This, of course, includes the majority of instances, because few structures today lack the convenience of electric energy. Electric energy in some form must be included in thorough hypothesis testing.

POST-FIRE INDICATORS

Post-fire indicators to be considered during a fire investigation are numerous. In the case of possible electrical fires, these indicators involve damage to wiring, receptacles and switches, electronic appliances, conductors, heat melting, and electrical failures, as discussed next.

Wiring

Tracing an electrical system in a burned structure can be a tedious and time-consuming task. While any portion of the remaining wiring might yield useful information, the most fruitful places are where connections are made. This includes the service entrance, load center box and other fuse and breaker boxes, switches, receptacles, and fixtures. Particular attention should be given to high-current circuits such as heaters, stoves, and dryers. When part of the structure is undamaged, an examination of this wiring will tend to show the general condition of the wiring before the fire.

Wire inside a metal conduit or box will rarely short-circuit except when the insulation is destroyed by external heat (as in Figure 10-32) or animal damage. If the wire does short, the arcing is generally retained inside, where it cannot cause ignition of thick materials. Thin paper, lint, or sawdust in the vicinity of the arc has been shown to be ignitable by incandescent metal sparks created. Of course, if the arcing in high-current circuits burns through the conduit, it can then cause ignition of nearby combustibles by dropping molten metal sparks, not by the arcing itself.

The following hazards are most likely to be found in the wiring itself:

- Damage to wire insulation caused by chafing against sharp surfaces when pulled through conduit or boxes
- Conductor insulation deteriorated from age or exposure to heat, moisture, or chemical vapors
- Mechanical damage to plastic-insulated cable caused by rodents (with subsequent contamination), sharp edges, or nails or staples driven through the insulator (not by crushing externally)
- Conductors intended for a particular service load being overloaded (via substitution of a significantly larger-capacity overcurrent device)
- Connections between conductors not properly soldered or wire-nutted, allowing a high-resistance connection to create localized heating with subsequent degradation of plastic components
- Wiring (sometimes 18-gauge lamp cord) installed for "temporary" use and never replaced
- Faults in the ground or neutral side of distribution systems that prompt current delivered to an appliance or outlet, instead of returning via the intended conductor path, to return via an alternative path through conduits, flexible metal sheathing, and sometimes even through the plumbing, heating ducts, or wire stucco mesh of the house;[53] covers of outlet or junction boxes removed, allowing dirt or combustibles inside
- Conductors of "open wiring" (knob-and-tube) systems separated from their supports, in contact with each other or with pipes or ductwork
- Cables and power cords carrying substantial but "safe" currents being buried under insulation, carpets, or bedding; coiled in a large mass; or exposed to high ambient temperature conditions, all of which produce "unsafe conditions" with regard to insulation degradation

Receptacles and Switches

In explosive environments, the chief ignition hazard of receptacles and switches is the arcing produced when the switching device is operated or the plug is pulled while current is still flowing (creating a parting arc). Many switches are either so designed or so enclosed as to allow the arc to occur with minimal damage or effects. Ignition of ordinary combustibles by switches is most likely due to one of the following:

- Loose contacts between wires and terminals of switch or receptacle, allowing a high-resistance connection and resulting heat production. These can occur with push-in connectors, in units exposed to mechanical vibration (see Figure 10-25), or with aluminum wires on units not designed for them.
- Overheating due to overload.
- Poor mechanical condition due to extended use or abuse, allowing contact between conductors.
- Enclosures removed or not effective. (This is especially true with the plastic receptacle and switch boxes in common use today.)
- Corrosion or erosion of internal conductors, producing high-resistance connections.

Electronic Appliances

Early electronic equipment contained heat-producing tubes that operated at lightbulb temperatures and could cause ignition if in contact with combustible materials. Old electronic equipment, unless properly maintained, can exhibit frayed or worn-out power cords, improper fuses, and missing cover plates that were intended to retain failures that could cause a fire. Modern electronic equipment uses solid-state devices that generally function at lower temperatures, lower currents, and lower internal voltages.

Some manufacturers have historically been lax in providing proper overcurrent protection. Rectifiers and power transistors have been known to fail and cause power transformers to overheat, thereby causing the insulation to ignite for lack of a proper protective fuse or breaker. Speaker cones have been known to ignite from excess current passing through the voice coil because of a defective power transistor lacking fuse protection, or by being overdriven by a powerful amplifier. Most pieces of household electronic equipment are considered low-power apparatuses, but situations do occur that have not been anticipated by the designer.

Conductors

Over the years, some investigators have erroneously accepted the presence of a conductor with balled or "beaded" ends as a prima facie indication of electrical arcing. Actual burning tests have demonstrated that balled ends can occur from flame exposure without the passage of electric current. Therefore, fire investigators should show extreme caution in reporting the presence of any balled or beaded conductors as symptoms of electrical arcing. The various mechanisms involved in the melting of conductors will be discussed in the following sections.

Heat Melting

Wires with ends showing drop-shaped, balled, and pointed formations are frequently the result of localized heating where some locations along the wire have attained a higher temperature than others (as shown in Figures 10-45a–e). The melting temperature of pure

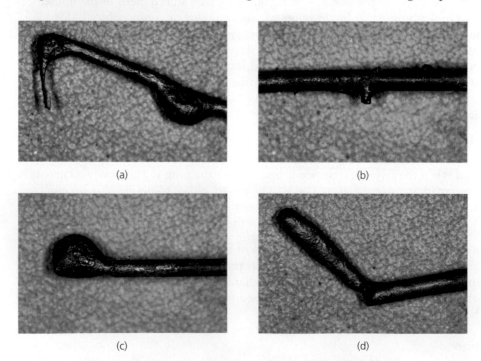

(a)

(b)

(c)

(d)

FIGURES 10-45A–D Melted and offset wire segments resulting from fire exposure. *Courtesy of Chris W. Korinek, P.E., Synergy Technologies LLC, Cedarburg, WI.*

FIGURE 10-45E Microscopic view of the end of a copper wire melted by localized fire heating. This shows the bulging and uniformly oxidized surface of molten and wire globules. *Courtesy of Lamont "Monty" McGill, Fire/Explosion Investigator.*

copper is listed as 1,083°C (1,981°F), which is somewhat above the average temperature of a typical fire. In other words, the copper wire during a typical structure fire is usually below its melting temperature. A small localized area of higher temperature such as that caused by an electrical arc, a concentrated fuel load, or possibly optimal combustion conditions (post-flashover fire with extra ventilation), results in localized melting of the copper wire, as previously characterized.

It has been shown that molten aluminum coming into contact with copper conductors can erode them at surprisingly low temperatures. Aluminum has a melting point of 660°C (1,220°F) and copper 1,083°C (1,981°F), but an alloy of the two metals, called a eutectic, has a melting point as low as 548°C (1,081°F). Aluminum melted from connectors, housings, conductors, or even household goods can literally dissolve copper conductors even without the interaction of electricity.[54]

Electrical Failures

When energized wires make contact and the contact is not a secure connection, such as when two bare wires touch without any mechanical pressure, an electric arc is usually produced. Its magnitude and duration will depend on the overcurrent protection (fuses or breakers) and the physical conditions under which the wires contact each other. The passage of current produces extremely high localized temperatures (as seen in Figure 10-46). If the resulting arc is not confined inside a suitable enclosure, the high temperature of the arc may ignite very thin or finely divided combustible materials in contact. This situation is sometimes referred to as *primary arcing*. Japanese researchers have reported "microdeformation" of stranded conductors in short circuits where the high current flows cause the strands to separate and protrude.[55]

Concurrent with the general origin and cause investigation, the electrical system should be examined to aid in the determination of the area of origin. The electrical system's copper wiring often survives the heat of the fire, leaving points of damage (pitting, beading, or severing) caused by electric arcing as the fire attacks insulation around energized

FIGURE 10-46 Graph of temperature of arcs in ambient-pressure air (experimental data). *Courtesy of Vytenis Babrauskus, Fire Science and Technology Inc., Issaquah, WA., reprinted from the Ignition Handbook.*

FIGURE 10-47 Microscopic close-up of wire showing localized globules of molten copper on surface of conductor caused by electric arc. Striations of wire drawing are still visible just to the left of the globule. *Courtesy of Lamont "Monty" McGill, Fire/Explosion Investigator.*

conductors. The location pattern of the damage (sometimes called *anomalies*) left on the electrical system is often more time-stable and less fragile than the fire patterns left by burning materials. Analysis of this pattern of damage can provide valuable three-dimensional information about the fire's progression and can be carried out on exposed power cords as well as fixed wiring as long as they were energized at the time they were exposed to fire. Arcing can result in either very localized melting on continuous wires, welding the conductors together; severing of the conductors (and spattering droplets of molten copper) (as illustrated in Figures 10-47 and 10-48); and/or blowing of the fuse or circuit breaker protecting the conductors.

MAPPING OF ARC FAULTS

If the wiring in the suspected area of fire origin is traced, and the location of arcing is mapped out on a diagram of the building, there may be an indication of the origin and development of the fire in the building. As a general principle, the arcing found farthest from the electrical source can help identify the area of origin. As a fire grows, the circuit

FIGURE 10-48A Bright copper ball (bead) at tip of this oxidized wire and small adhering globules of molten copper are consistent with this wire's having been exposed to an electric arc. *Courtesy of Lamont "Monty" McGill, Fire/Explosion Investigator.*

may be interrupted when the fire attacks the nearest wiring (either by tripping the OCPD or physically severing the conductors). If the wiring is severed, this means that as the fire progresses, the more distant portions of the circuits will already be de-energized and show no signs of arcing. If the OCPD trips, there will be no additional evidence of arcing.

Arc-fault mapping is initially not a part of *cause* determination but is often useful in determining a possible area of origin of any fire (structural or vehicular). Figure 10-49a shows fire attacking an insulated power cord. The source of electric current in the cord is at the left (plug) end of the wire. The cord is lying on the floor and it feeds a heater at the right side of the room. If a fire starts on the floor in the center of the room by the extension cord (fire #1), it will burn through the cord's insulation, and an arcing fault will occur as the cord's insulation chars and degrades. Depending on the resistance of the degraded insulation, this current flow can range from very low to very high (as high as 150 A has been measured). Because the conductors are not bonded together, they are free to move, so intermittent current flows can delay tripping of an OCPD. The result of this arcing fault is beading left on the conductors (as seen in Figure 10-48b). The beads formed in this case severed both of the conductors but did not blow the circuit breaker.

As fire #1 expands into fire area #2, it attacks the cord "downstream" from the original point of arcing and produces no further beading, since the electrical supply has been cut off to this point (by severing a conductor or tripping the OCPD). Subsequent spread of the fire attack on the cord at point #3, closer to the source, may result in beading of the source, since the cord has not been de-energized by the circuit breaker. The investigator can then determine where the fire first encountered the energized extension cord by looking for the point of beading and severing farthest from its plug, which is the source of its electrical energy.

If the fire started at the heater, there would be a tendency for its line cord to fault and bead in the area of the heater. Conversely, failure in the power cord's plug would leave residue of arcing (beading) in the plug's components but would leave no beading downstream in the line cord or by the heater if the wire was severed. If the fire started in the building wiring, the investigator would not expect to find faulting inside the extension cord's plug or in any of the downstream electrical cords if the fixed wiring was severed.

The process can be described using the water system analogy. This analogy centers on the use of the puddle pattern of a cut and severed garden hose to demonstrate how the point of puddling farthest from the source (faucet) is the most likely point where the hose was first severed. In Figure 10-49b, water will flow from the first cut, leaving a puddle residue. A subsequent second cut downstream will not leave a puddle, since the water source has been interrupted at the first cut. Subsequent cuts upstream from the first cut will leave puddles until the faucet is turned off. If the hose and puddles are examined after the hose has been damaged, the puddle farthest from the faucet is the point where it was first cut and severed. The puddles here are analogous to the arc-fault residue (beading) created in the electrical example.

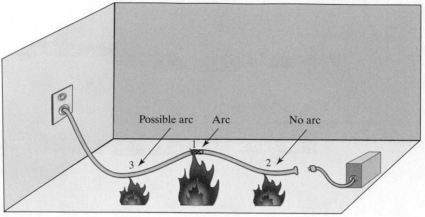

FIGURE 10-49A Fire attacking an energized line cord, resulting in a pattern of arcing.

Possible arc   Arc   No arc

3   1   2

Residue of arcing (beading)

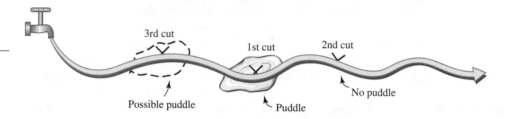

FIGURE 10-49B The water-hose analogy to Figure 10-49a.

3rd cut

1st cut   2nd cut

Possible puddle   Puddle   No puddle

This method of using the arcing damage produced on both single- and three-phase electrical systems has been taught to over 1,000 investigators at the ATF National Academy and is based on hundreds of tests on electrical wiring and components in both laboratory and structural burn situations since the 1980s.[56] It has been accepted by the courts and has also been addressed by *NFPA 921*, Ch. 17. Arc-fault mapping has been experimentally verified as reliable in both North American (120-V branch main distribution) and UK (230-V ring main distribution) systems.[57] It has been extensively tested in exposed surface cable wiring exposed to full-scale compartment fires and found to be quite reliable in indicating a fire's area of origin.[58]

Height of Origin

A "third dimension" of the origin can be determined by examining the vertical and horizontal runs of the electrical wiring. For example (Figure 10-50), the wall receptacle fed from an in-wall building branch circuit feeds a line cord that has signs of beading and severing. This indicates that the early fire was in the room and not in the wall, because an internal wall fire would de-energize the receptacle, so no faulting could occur in the line cord. Conversely, if beading and severing had been found only in the wiring above the ceiling, this would indicate an early fire in the attic. If beading and severing had been found in both the line cord and above the ceiling, then it would indicate that a room fire first attacked the line cord and then progressed into the attic before the branch circuit breaker had time to open. Plotting these points on a three-dimensional diagram that includes the layout of the pertinent sections of the electrical wiring can provide information for the investigator to use in eliminating potential fire causes.

Further Considerations of Arc-Fault Interpretation

The investigator should also be aware that the farthest downstream arcing may not be nearest the origin, because of the effects of wiring placement in relation to other objects in the fire scene and the nonelectrical damage to the wiring caused by the fire and other activities. Some factors to be considered are uneven shielding of the wiring from the fire by

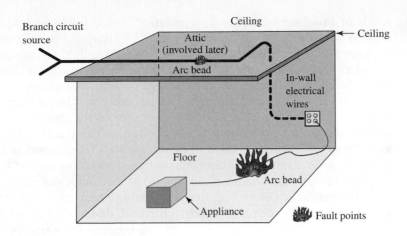

FIGURE 10-50 Plotting the pattern of arc faulting in three dimensions.

Branch circuit source

Ceiling

Attic (involved later)

Arc bead

Ceiling

In-wall electrical wires

Floor

Arc bead

Appliance

Fault points

drywall, types of electrical insulation, available fault current, the size of the OCPDs, and the number and location of sources of electric energy. In larger or multistory structures where there has been significant collapse or remodel, it is often impossible to accurately trace each conductor back to its source, making arc-fault mapping of minimal value. An electrician with experience in wiring can be of great help in considering these factors. This method may also be fuel load–dependent. For example, a very small fire starting in one corner of a room may not have enough energy to melt the insulation on overhead conductors, but a more massive fuel load in the other corner of the room may cause faulting in the conductors in its corner. In the case of a building collapse or explosion, the physical movement or fracturing of structural members, electrical conduit, or other electrical enclosures can expose the conductors inside to arcing due to contact or heating faster than in areas where the conductors are better protected. Such events can then result in arc damage even if they are downstream of more protected areas closer to the area of origin.

Some of the points of electrical faulting may be modified by the subsequent melting and chemical effects of some fires. The investigator should take this into account during the examination. As a general guide, there must be a clear demarcation area between the localized melting (purported arc) and the undamaged area of the conductor. Courtroom battles have been centered on whether an alleged point of beading is really only thermal melting due to the effects of the fire. Experienced metallurgists may be called upon to provide opinions on the source of a bead. As we shall see, the discrimination can be very difficult. The impact of this technique is on location of first major fire, not necessarily on whether the fire was caused by an electrical fault.

The lower melting temperature of aluminum wiring limits its use in this method, since it will undergo extensive gross melting in large fires. However, if gross melting is accounted for, this method of pattern analysis may be useful. Finneran has recently published a thorough description of the approach and its application.[59]

The origin(s) determined by examination of the electrical system should correlate with the origin(s) determined by the other standard investigative methods. It is important that the electrical investigator and the origin and cause investigator reach a consensus as to the point(s) of origin before the cause is conclusively established. The possibility of an electrical cause should then be investigated as part of the continuing examination.

ARCING THROUGH CHAR

Very frequently, signs of electric arcing are not indicative of electrical ignition and are instead a consequence of the fire. Because most wire insulation materials are combustible, arcing is often the result of current transmitted through charred insulation or contact between wires as the insulation was burned away completely. When the condition of the wire indicates arcing has taken place, the arcing could actually have been the result of fire-caused insulation degradation. [Such fire-caused faulting (arcing through char) is sometimes called *secondary arcing*.] This arcing may be between conductors or between an energized conductor and a grounded conduit, box, or appliance housing. The arcing can progress up the wire (toward the power source in single-phase systems) for some distance, producing sustained exterior effects and showers of molten metal, as in Figure 10-32. Such arcing through char can occur and be sustained even in circuits protected by fuses or breakers, since the resistance of the char "connection" may limit the current so that an overcurrent is not produced. The conductors are also free to move away from each other (unlike a "bolted" fault). This means that the arc may be struck and extinguished numerous times in time frames too short for an OCPD to be triggered.[60]

Microscopic examination of questioned wires/conductors may provide useful information. The conditions of melting from external fire heat are different from those of electric arcing. When wires melt from external fire heat, due to high thermal conductivity, there is usually only a small temperature difference between the parts that have melted and the unmelted sections. In this situation, the wire section has been raised to the fire temperature, and only a few additional degrees (from an area of optimized heating) are needed to produce the melt and its balled or pointed symptoms (see Figure 10-45). By contrast, when wires are subjected to arcing (whether that arcing is the cause of the fire or the result of it), the wire itself is at ambient temperature while the arc is at many thousands of degrees. The result is often an abrupt change from undamaged wire to a melted globule, as was shown in Figures 10-47 and 10-48. There may be minute beads of melted copper spattered onto nearby surfaces that will not be present when the melting is caused by thermal damage alone with no current flow. The two different situations are usually discernible by examination. Visual inspection is usually sufficient to detect arcing effects, but running the wire through a wad of cotton wool or even between fingertips may be more effective in detecting very small defects (see next section). Of course, there are gray areas where both situations occurred and also where high temperatures have left the wire in such poor condition that the indications have been minimized or destroyed. The difficult task is to assess what arcing occurred prior to the fire and thereby created an ignition source. Remember that it is not enough just to have an ignition source; there must also be fuel in a suitable physical state in contact with that source for a sufficient length of time to have ignition. The investigator must continuously ask, "Do I have all the elements here, or am I blaming electricity just because there is electrical wiring in the vicinity?"

LABORATORY EXAMINATION

As described more fully in Chapter 14, laboratory examination can assist the investigation in a number of ways. As discussed next, laboratory examination of wiring, insulation, switches, fuses, circuit breakers, and appliances can particularly assist in the investigation of fires of possible electrical origin.

Laboratory Tools and Procedures

In many cases, the diagnostic signs of electrical fires are so subtle that they require laboratory examination. Changes occur as copper wire is fire heated. First, oxidation produces discoloration. As the melting temperature is reached, blistering and bubbling may occur from degassing. During melting, the striations caused by the manufacturing process flow together, and there is a general flow of the molten copper, resulting in balling and areas of erosion with a rough surface. When the wires are oriented in a vertical position, the molten copper can run down to a ball and then drop off, leaving a pointed end. When the ball remains, the portion immediately above it shows a reduced diameter from where the metal flowed. There is no sharp line of demarcation showing points of melting and nonmelting. Stranded wires melt and flow together and form a solid mass. Sometimes the strands are discernible but have been fused together and are not separable. Sometimes several wires inside a common conduit will fuse together. These symptoms will be seen whenever wire has reached its melting temperature. Usually, fusing is localized, because the melting temperature of copper wire is above average fire temperatures, so sufficient generalized heating is uncommon.

Electric arcing, by contrast, does not usually result in random melting, erosion, *generalized* fusion of strands, or other symptoms indicative of flame-exposure heating. The characteristic of electric arcing is the formation of a sharp demarcation line between the arcing area and the unmelted wire, as was shown in Figures 10-47 and 10-48. In arcing situations, stranded wire sometimes exhibits individual strands that have melted together only at the point where the arc occurred, as was shown in Figure 10-48a. Elemental analysis of such spots by scanning electron microscopy/energy dispersive X-ray spectrometry (SEM/EDS) may be of value in establishing contact between conductors of different metals (copper wire against a zinc-plated steel receptacle box, for instance)[61] alloying with aluminum (see Figure 10-51). (See Chapter 14 for descriptions of SEM/EDS technology and its applications to fire investigation. SEM alone can improve the quality of visual examination and allow examination of very fine details, such as the "draw marks" on conductors that disappear when the conductor approaches its melting point.)

SEM/EDS may help confirm a particular accidental ignition mechanism. The presence of metallic contaminants on insulators in switches and relays has been linked to arc tracking and electrical equipment fires. SEM/EDS can help identify the nature and origin of such contamination.

Metallography is the study of a cross section of a metal showing details such as grains, impurities, voids, and other important details which are important in the material's history which may help to show if the material was involved with a fire cause or was a victim of the fire. Some examples of metallographic features are shown in Figure 10-51.

Alloying   vs.   Arcing

FIGURE 10-51
Comparison of a copper wire (heated) alloyed with aluminum (on left) and a copper wire exposed to an electric arc (on right). Top: Color photos. Middle: False-color SEM/EDS elemental maps. On left, green represents copper and red represents aluminum. On right, green represents copper and red is oxygen. Bottom: Metallographic elemental maps. On left, you can see that aluminum (red) has penetrated into the copper as well as on the surface. On the right, the transferred copper is on the surface with a slight amount of copper oxide due to the arc's heat. *Courtesy of Chris W. Korinek, P.E., Synergy Technologies LLC, Cedarburg, WI.*

It has been suggested that the identification of compounds trapped within the copper as it melts and reacts with its environment may permit one to conclusively establish whether the copper failed in an atmosphere of oxygen (open air) or chlorine (from vinyl chloride insulation). In fact, double-blind tests of the Auger electron spectroscopy (AES) method have concluded that the technique does not discriminate between arc beads created during fire exposure and those that initiated a fire.[62] In his very thorough review of the AES analysis issue, Babrauskas concluded that such discrimination is unlikely, since the two processes—electrical and fire—do not produce categorically different residues.[63] Because most post-fire indicators are the summation of the fire's effects, and fire environments are so complex, it would appear that none of the metallurgical or instrumental techniques examined to date can reliably discriminate between copper arc bead residues created as part of the ignition process and those created later in the fire by thermal exposure.

Wiring and Enclosures

Conduction heating (due to excessive current flow for the size of the conductor) heats the full length of the wire all the way back to the junction box or to the supply source. Extreme overheating can cause the wires to reach fusion temperature, and when the first part of the wire melts away, a parting arc can form. If the insulation is destroyed first, the carbonized residue may provide a conduction path, or the bare wires themselves may touch and arc. When these things occur inside metal boxes or conduit, the failures are usually trapped and of little concern. However, if the failure has sufficient energy to burn a hole in the metal enclosure, a fire can result (such penetrations are rare with *protected* 120/240-V circuits). Figure 10-52 shows the failure of a meter service connection that came loose and grounded against the meter base. Frequently, the examination of the wiring in an unburned area can provide useful information, as it can indicate the age and general condition of the wiring. If it has been poorly maintained and the insulation has become brittle and cracked, electrical problems should be considered more likely.

If a section of undamaged wire between the suspected area and the supply can be found, the condition of the conductor can provide some useful information. If the insula-

FIGURE 10-52 Arcing through the meter case caused when power lead came loose and shorted to inside of case. *Courtesy of Greg Lampkin, Knox County, TN, Fire Investigation Unit.*

FIGURE 10-53A The underside of a steel screw head after a glowing connection was lab created on a duplex outlet with a solid #12 AWG wire. Note the melting and discoloration at the bead at the rim of the screw head. *Courtesy of Chris W. Korinek, P.E., Synergy Technologies LLC, Cedarburg, WI.*

FIGURE 10-53B Comparison of wire and screw from normal (tight) connection to oxidized and pitted screw and wire from loose (high resistance) connection. *Courtesy of Chris W. Korinek, P.E., Synergy Technologies LLC, Cedarburg, WI.*

tion and wire are in a "like-new" condition with no tarnishing of the conductor, it is unlikely that overcurrent occurred to the degree that ignition took place. Electric arcing, or arcing through char, however, with its local heating effects, would still have been possible under these conditions. Although failures in mid-run have occurred where the conductor itself was not continuous (causing series arcing), by far the most common failures occur at connections.

Connector Screws

Connector screws can show extensive effects of heat, including arc transfers on the edge of the screw, as shown in Figures 10-53a and b. Prolonged heating in a poor connection condition can cause heating of the entire length of the screw. When a screw is exposed to external heating, the portion in the connector is protected unless fire exposure is very prolonged. Similar localized heating can occur in any connector improperly used—wire nuts, twisted wire ends, "push-in" connectors in receptacles, or stab connectors in plugs. Figure 10-54 shows the heating and erosion of a loose "split-bolt" connector that ignited the adjacent timber.

Insulation

Very old wire was usually insulated with rubber and/or cloth material. Modern insulation for wires is a thermoplastic (polyethylene or PVC being the most common material). Insulation condition can sometimes be used as a guideline to indicate whether the heat was produced internally (overcurrent) or externally (by the fire). Rubber is destroyed by heat. When overcurrent is passed through rubber-covered wire, the internal heat can destroy the rubber nearest the wire, leaving the exterior nearly intact. Afterward, the insulation may remain loose and free to slide back and forth over the wire (sleeving effect). The situation is different with the plastic insulation that is more common today. On an overheated wire, the insulation softens and melts when heated and then rehardens when

FIGURE 10-54 Close-up of split bolt connector showing erosion and arcing to clamp. *Courtesy of Chris J. Bloom, CJB Fire Consultant, White City, OR.*

cooled. Due to its thermal inertia, plastic will discolor or pyrolyze on the surface exposed to the most heat, so this may reveal whether the insulation was heated from the inside (heat from the wire) or from the outside (from an exterior fire). However, any heating is generally accompanied by deformation, sagging, and discoloration of the insulation.[64] Sometimes bubbling also occurs from gas formation as the plastic decomposes from the heat. Bubbling is not as noticeable in rubber insulation because of its porosity, which minimizes the entrapment of gases.

When there is doubt, the laboratory can test a section of undamaged wire by passing a current through it and observing its effects on that particular insulation. Remember, heating from the wire is not necessarily electrical. Heating can result from thermal conduction in a wire exposed to a direct flame some distance away.

Examination of Switches, Fuses, Breakers, and Appliances

The laboratory examination of switches, fuses, breakers, and appliances will be discussed in Chapter 14. X-ray examinations are particularly valuable, since they can reveal internal details without destructive examination.[65] Even modest heating can melt, scorch, or char an electric appliance beyond visual examination (see Figures 10-55a and b). Sometimes the laboratory analysis will address issues other than electrical failure. Interviews with appliance users can sometimes save a lot of investigative effort by suggesting a particular testable hypothesis. The shop vacuum in Figure 10-56 was the cause of the fire, but it was unplugged at the time of the fire. It had been used 5 hours previously to clean up a quantity of dust from a floor-sanding project. Laboratory analysis confirmed that the floor finish on the wood dust was susceptible to self-heating.

FIGURE 10-55A Burned smoke detector, which is difficult to evaluate nondestructively. *Courtesy of John D. DeHaan.*

FIGURE 10-55B X-ray shows battery connections in place. *Courtesy of Dr. David Icove, University of Tennessee.*

FIGURE 10-56 This shop vacuum was the cause of a fire in a utility closet, but close examination showed it was not plugged in at the time of the fire. Lab analyses confirmed that the bag (completely destroyed) contained dust from a floor-sanding operation. The floor finish/sanding dust self-heated to ignition. *Courtesy of Investigator Jeff Weber, San Jose, CA, Fire Dept.*

Summary

If wires, appliances, and other electrical equipment have not been excessively damaged by a fire, a considerable amount of information can be gleaned from them by careful examination in accordance with the principles outlined previously. The evaluation of electrical fire causes includes establishing the means by which electric current could have produced the necessary heat—heat by design, heat by failure or misuse, or heat from unintended current flow. The analysis must also include the amount of heat produced, the rate at which it was produced, and the area over which it was produced. The effect of heat on the first fuel and the mechanism of actual ignition also need to be evaluated. As in other fields, over the years fire investigators have used certain criteria for judging the evidence of electrical fire symptoms. Unfortunately, some are not as reliable or as inviolate as once believed. These include the following:

■ *Ignition due to slightly oversized fuse or breaker.* When a slightly higher rated fuse or breaker is found than the rated capacity of the wire, it probably is not the fire cause. Current-carrying capacities are conservatively rated.

■ *Balled wire ends.* This condition can be the result of fire melting as well as from electrical arcing, neither of which may have anything to do with the fire's ignition.

■ *Loose wire insulation (sleeving).* If the insulation is loose, it probably is from overcurrent. If it is not loose, however, overcurrent should not be ruled out. The particular type of insulation should be tested to determine its thermal characteristics.

■ *Open circuits causing fires.* There must be a flow of current for ignition. Voltage alone on wiring (energized) will not cause ignition—there must be a current flow (amperes). This flow can take many forms—short circuit, overload, leakage current, insulation degradation, and so on. Energized conductors can cause electrical effects during the fire, however.

■ *Arcing—cause or result.* Did the arcing cause the fire, or was the arcing the result of wires that touched because their insulation was burned off by the fire? Could an arc ignite nearby materials? Was there fire exposure that could cause arcing through char later in the fire?

■ *Energy.* Was there sufficient electric energy present to cause ignition of the surrounding materials? The mere presence of some electric energy is not sufficient proof that it could have ignited the fuels in question. Electrical events must be included in a complete scheme of hypothesis testing for nearly all fire investigations because they indicate there was some energy available. The critical test remains: Could this source have caused ignition of that fuel under the conditions present?

■ Remember that, as with other forms of physical evidence, it is possible that there was sufficient fire damage to preclude conclusive analysis. When even a thorough and careful laboratory examination fails to reveal the cause or mechanism of the fire, it should *not* be viewed as motive to invent an electrical origin.

Review Questions

1. Describe Ohm's law and state its basic formula.
2. How can a poor connection cause a fire?
3. How does a fuse function as an overcurrent protection device?
4. How can electric current cause a fire when the circuit is protected by an OCPD?
5. What is arc tracking and how does it contribute to fire ignition?
6. How does arc mapping aid in determination of area of origin?
7. Name three fuel types ignitable by discharge of static electricity.
8. Can arc beads on copper wire that resulted from fire exposure be distinguished from those that were part of the cause of the fire?
9. How does aluminum wiring differ from copper wiring in its conductance, mechanical properties, and melting point?
10. Give three examples of how electric motors cause fires.

References

1. J. R. Hall, "Home Electrical Fires" (Quincy, MA: NFPA Fire Analysis and Research Division, March 2009), 5.
2. V. Babrauskas, *Ignition Handbook* (Issaquah, WA: Fire Science Publishers, 2003), 548.
3. V. Babrauskas, "Ignition Events during Fueling of Motor Vehicles at Filling Stations," *California Fire and Arson Investigator* (April 2005): 16; J. L. Pharr and T. T. Jonas, "Static Electricity Generation and Ignition at Fuel Delivery Systems," *North Carolina IAAI Newsletter*, 2007; (www.nciaai.com/articles).
4. R. C. Lee, E. G. Cravalho, and J. F. Burke, *Electrical Trauma: The Pathophysiology, Manifestations and Clinical Management* (Cambridge: Cambridge University Press, 1992), 33–36; J. Webster, *Medical Instrumentation, Application and Design*, 3rd ed. (New York: Wiley, 1998), 183–86; D. Fink and W. Beaty, *Standard Handbook for Electrical Engineers*, 13th ed. (New York: McGraw-Hill, 1993), chap. 11, p. 59; T. L. Brown and E. H. LeMay Jr., *Chemistry: The Central Science* (Upper Saddle River, NJ: Prentice-Hall, 1977), 356.
5. V. Babrauskas, *Ignition Handbook*, 540–48; International Institute of Welding, *The Physics of Welding*, ed. by J. F. Lancaster (Oxford: Pergamon, 1984), 109–20.
6. Fink and Beaty, *Standard Handbook for Electrical Engineers*, 4-74–4-79.
7. J. Pharr, "Surge Suppressor Fires," *ESD Journal* (Moore, SC: Fowler Associates), www.esdjournal.com.
8. A. Chiste and D. Kladar, "Electronic Systems Protection via Advanced Surge Protection Devices," *Institute of Electrical and Electronic Engineers (IEEE) Industry Applications Conference, IEEE* 13 (2002): 2244–47.
9. B. VanderPas and K. Lewis, "Corroded GFCIs Create Fire Hazard," *Fire and Arson Investigator* (January 1999): 10–13.
10. G. Gregory, "AFCIs Target Residential Electrical Fires," *NFPA Journal* (March–April 2000): 69–71.
11. A. Manche and A. Haun, "Why AFCIs?" *NFPA Journal* (September–October 2008): 74–77.
12. V. Babrauskas, "How Do Electrical Wiring Faults Lead to Structure Ignitions?" in *Proceedings Fire and Materials 2001* (London: Interscience Communications, 2001), 39–51.
13. N. P. Dwyer, "Can a High Resistance Connection Lead to a Fire Even after the Load is Removed?" *Fire Findings* 16, no. 3 (Summer 2008): 12–13.
14. Y. Hagimoto, K. Okamoto, and N. Watanabe, "Short-Circuit Faults in Electrical Cables and Cords Exposed to Radiant Heat" in *Proceedings Fire and Materials 2003* (London: Interscience Communications, 2003), 215–26.
15. K. Okamoto, N. Watanabe, and Y. Hagimoto, "Fire Hazard Caused by Thermal Degradation of Organic Insulating Materials at Plug and Receptacle Connections" in *Proceedings Fire and Materials 2003* (London: Interscience Communications, 2003), 203–14.
16. CPSC, "Corrosion in Homes and Connections to Chinese Drywall," press statement (Washington DC: CPSC, November 23, 2009); see also "CPSC Alert to Fire Safety Professionals," www.cpsc.gov/info/drywall, April 2010.
17. Babrauskas, *Ignition Handbook*, 312–15, 333–34, 734, 774–77, 804–6; Hagimoto et al., "Short-Circuit Faults."
18. Hagimoto et al., "Short-Circuit Faults"; Y. Hagimoto, "Fire Causes in Electrical Plugs" in *Proceedings Interflam 2001* (London: Interscience Communications, 2001); J. Zicherman, "The Fire Performance of Electrical Insulating Materials Used in Fluorescent Light Fixtures," *California Fire and Arson Investigator* (July 2004).
19. Babrauskas, *Ignition Handbook*.
20. V. Babrauskas, "What Voltage Is Required to Cause Arcs and Sparks?" *Fire Findings* 12, no. 1 (Winter 2004): 1–5.
21. D. Miller, R. Chowdury, and M. Greene, "2004–2006 Residential Fire Loss Estimates," (Washington, DC: Consumer Product Safety Commission, October 2009).
22. Ibid.
23. Ibid.
24. Ibid.
25. Babrauskas, *Ignition Handbook*, 732; Consumer Product Safety Commission, "Recall," press release #07-094 (Washington, DC: CPSC, February 1, 2007).
26. B. Harwood, *Fire Incidence in Television Receivers* (Washington, DC: CPSC, 1979).
27. Miller, et al., 5.
28. J. Flynn, "Lamp, Bulb or Lighting Fires in Nonresidential Structures: 2003–2006 Annual Estimates" (Quincy, MA: NFPA Fire Analysis and Research Division, January, 2010).
29. Hall, "Home Electrical Fires," 67.
30. D. Powell, "Fluorescent Lighting," *Fire Findings* 13, no. 3 (Summer 2005), and 13, no. 4 (Fall 2005): 8.
31. *NFPA 70, National Electric Code* (Quincy, MA: National Fire Protection Association, 2008).
32. Powell, "Fluorescent Lighting," 7–8.
33. Flynn, "Lamp, Bulb or Lighting Fires."
34. Zicherman, "The Fire Performance of Electrical Insulating Materials."
35. J. Zicherman, "Technical Note: The Fire Performance of Electrical Insulating Materials Used in Fluorescent Lighting Fixtures," *Fire and Materials* 25 (2001): 209–31.
36. CPSC, "Technical Consumer Products Inc. Recall of Fluorescent Light Bulbs" (Washington, DC: U.S. Consumer Product Safety Commission, December 2004).
37. N. P. Dwyer, "Compact Fluorescent Light Bulbs," *Fire Findings* 16, no. 2 (Spring 2008): 4.
38. Osram Sylvania, *Lamp and Ballast Catalog*, 2008, fn. 74 (Danvers, MA: Osram Sylvania), 163.

39. David Powell, personal communication, November 2009.
40. Babrauskas, *Ignition Handbook*, 1075.
41. P. A. Lynch and C. Bloom, "Slippery Issues on Aquarium Fire Losses," *California Fire/Arson Investigator* (July 2006): 4–10.
42. NEMA, LSD 25-2008, Best Practices for Metal Halide Lighting Systems, Plus "Q&A about Lamp Ruptures in Metal Halide Lighting Systems," (Rosslyn, VA: National Electrical Manufacturers Association, March 10, 2009), 2; NEMA; David Powell, personal communication, November 2009.
43. Babrauskas, *Ignition Handbook*, 750.
44. Miller et al., "2004–2006 Residential Fire Loss Estimates"; see also Hall, "Home Electrical Fires," 14.
45. R. E. Lowe and J. A. Lowe, "Halogen Lamps II," *Fire and Arson Investigator* (April 2001): 39–42.
46. W. Hull and D. L. Pippen, "Electric Blanket Fires," Part I: *Fire Findings* 11, no. 1 (Winter 2003), 7–11; Part II: *Fire Findings* 11, no. 2 (Spring 2003): 1–3.
47. "Heat Tape and Heat Cable Are Not the Same," *Fire Findings* 16, no. 4 (Fall 2008): 5.
48. B. Béland, "Heat Tape Fires," *Fire and Arson Investigator,* September 1994.
49. Brown and LeMay, *Chemistry*, 356.
50. K. Rosenhan, "Acid Tests, *Fire Risk Management* (October 2009): 48–51; (also *Fire & Arson Investigator* (April 2009); V. Babrauskas, *Ignition Handbook*, 742.
51. J. D. Twibell, "Reporting of Electrical Fires Hides True Picture," *Fire Prevention* 282 (September 1995): 11.
52. B. V. Ettling, "The Overdriven Staple as a Fire Cause," *Fire and Arson Investigator* (March 1994).
53. L. Weintraub, "Open Electrical Neutral and Pyrophoric Carbon Phenomenon," *Fire and Arson Investigator* (December 1993).
54. B. Béland, C. Roy, and M. Tremblay, "Copper-Aluminum Interaction in Fire Environments," *Fire Technology* 10 (February 1983).
55. S. Miyoshi, "Micro-Deformation of Stranded Electrical Conductors Caused by Short Circuits" paper presented at IAFS (International Association of Forensic Sciences) 2008, New Orleans, LA, July 2008.
56. L. I. Rothschild, "Some Fundamental Electrical Concepts in Locating the Cause and Origin of a Fire," *Journal of National Academy of Forensic Engineers (NAFE)* 3, no. 2 (December 1986): 37–44; R. H. Kaufman and J. C. Page, "Arcing Fault Protection for the Low-Voltage Power Distribution Systems: Nature of the Problem," *Institute of Electrical and Electronics Engineers (IEEE) Transactions on Power and Apparatus Society* (1960): 160–67; J. R. Dunki-Jacobs, "The Escalating Arcing Ground-Fault Phenomenon," *IEEE Transactions on Industry Applications Society* 1A-22, no. 2 (1986): 1156–61; B. Béland, "Examination of Electrical Conductors Following a Fire," *Fire Technology* 16, no. 4 (1980): 252–56; B. Béland, "Comments on Fire Investigation Procedures," *Journal of Forensic Sciences* 29, no. 1 (1984): 190–97; R. J. Svare, "Determining Fire Point-of-Origin and Progression by Examination of Damage in the Single Phase, Alternating Current Electrical System," invited paper, Ministry of Public Security, Beijing, People's Republic of China. Presented in Shanghai, PRC, April 1988, *Journal of People to People, International Arson Investigation Delegation to the People's Republic of China and Hong Kong, 1988*; M. Deplace and E. Vos, "Electric Short Circuits Help the Investigator Determine Where the Fire Started," *Fire Technology*, 19, no. 3 (1983): 185–91; B. Béland, "Copper Behavior under Fire Conditions," *Fire and Arson Investigator* (June 1994): 40–43; B. V. Ettling, "Arc Marks and Gouges in Wires and Heating at Gouges," *Fire and Arson Investigator* 20, no. 4 (1996): 19–27; J. M. Finneran, "Arc Mapping," *Fire Findings* 13, no. 2 (Spring 2005): 7–11.
57. N. Carey, "Powerful Techniques: Arc Fault Mapping," *Fire Prevention* (March 2002): 47–50.
58. N. J. Carey, *Developing a Reliable Systematic Analysis for Arc Fault Mapping,* PhD dissertation, University of Strathclyde, September 2009; W. Johnson and L. Rich, "Arc Mapping Follow-Up," *Fire Findings* 15, no. 3 (Summer 2007): 7–10. Also see Johnson and Rich, *Proceedings: Fire and Materials* (London: Interscience Communications, January 2007).
59. Finneran, "Arc Mapping."
60. Carey, "*Developing a Reliable Systematic Analysis for Arc Fault Mapping.*"
61. Ibid.
62. D. G. Howitt, "The Chemical Composition of Copper Arc Beads: A Red Herring for the Fire Investigator," *Fire and Arson Investigator* (March 1998): 34–39; R. Henderson, C. Manning, and S. Barnhill, "Questions Concerning the Use of Carbon Content to Identify 'Cause' vs. 'Result' Beads in Fire Investigations," *Fire and Arson Investigator* (March 1998): 26–27; B. Béland, "Examination of Arc Beads by Auger Spectroscopy," *Fire and Arson Investigator* (April 2004): 41–42.
63. V. Babrauskas, "Fires Due to Electrical Arcing: Can 'Cause' Beads Be Distinguished from 'Victim' Beads by Physical or Chemical Testing?" in *Proceedings Fire and Materials 2003*, (London: Interscience Communications, 2003), 189–201.
64. B. Béland and J. Tremblay, "On the Sleeving of Electrical Insulation," *Fire and Arson Investigator* 43, no. 4 (June 1993).
65. J. D. Twibell and S. C. Lomas, "The Examination of Fire-Damaged Electrical Switches," *Science and Justice* 35, no. 2 (April 1995): 113–16.
66. J. D. Twibell and C. C. Christie, "The Forensic Examination of Fuses," *Science and Justice* 35, no. 2 (April 1995): 141–49.

CHAPTER

11
Clothing and Fabric Fires

Courtesy of John Jerome and Jim Albers.

KEY TERMS

elastomers, *p 470*

furniture calorimeter, *p. 478*

natural fibers, *p. 469*

synthetic fibers, *p. 469*

OBJECTIVES

After reading this chapter, you should be able to:

- Identify the common causes for fires in fabrics, upholstered furniture, and mattresses.
- Identify the common types of cloth.
- Explain the basic concepts of the Flammable Fabrics Act.
- Explain how furniture testing is conducted.
- List and explain common tests used to measure cigarette ignition, flame ignition, heat release rate, smoke, and toxicity.
- List and separate describe several flammability tests (examples: clothing, textiles, plastic films, carpets and rugs, mattresses and pads, children's sleepwear, upholstered furniture).
- Search the Internet for current and proposed governmental policies and legislation.

PEARSON
myfirekit™

For additional review materials, appendices, and suggested readings, visit **www.bradybooks.com** and follow the **MyBradyKit** link to register for book-specific resources.

Register for **MyFireKit** by following directions on the **MyFireKit** student access card provided with this text. If there is no card, go to **www.bradybooks.com** and follow the **MyFireKit** link to Buy Access from there.

Although the total number or residential fires in the United States continues to decrease each year, fires in which some fabric or fabric-covered household material was the first fuel to be ignited accounted for 14.3 percent of all residential fires in the United States for the period 2004–2006 but were involved in approximately 40 percent of all residential fire deaths. This means that upholstered furniture, bedding, mattresses, garments, and draperies accounted for an estimated 26,900 residential fires (out of 384,100 total) and 1,030 deaths each year (2004–2006).[1]

The NFPA estimates that upholstered furniture alone was the item first ignited in an average of 7,630 home structure fires per year, causing an annual average of 600 civilian fire deaths, 920 civilian fire injuries, and $309 million in direct property damage (2002–2005).[2] Fires beginning with upholstered furniture accounted for only 2 percent of *reported* home fires but 21 per cent of home fire deaths. During the same time frame, a mattress or bedding was the first fuel ignited in an average of 11,520 home fires. Such fires accounted for 3 percent of the fires but 13 percent of home fire deaths (378 deaths, 1,286 civilian injuries, and $357 million in direct property damage).[3] According to the CPSC, ignition of upholstered furniture was caused by cigarettes or other smoking materials in an estimated 29 percent of home fires and by open flames in about 16 per cent. Ignition of mattresses and bedding was caused in about 22 percent of home fires by smoking materials and in 26 percent by open flames.[4] The NFPA estimated (in 2008) that 14 percent of mattress/bedding home fires and 12 percent of upholstered furniture fires were ignited by smoking materials.[5]

Although garment fires may not frequently go on to cause structural damage, the NFPA estimated that in the years 1999–2000, 120 people died and 150 people were injured (per year) because their clothing caught fire.[6] A more recent study by Hoebel et al. used burn injury data from the National Electronic Injury Surveillance System (NEISS) and death data from the National Center for Health Statistics (NCHS) of the Centers for Disease Control and Prevention (CDC). This study revealed that a great many clothing fires are not reported to the local fire departments (and therefore do not appear in the NFIRS or NFPA data). This analysis revealed that nearly 4,300 injuries and 120 deaths per year were caused (1997–2006) when peoples' clothes caught fire.[7]

The overall number of upholstered furniture, mattress, and bedding fires in homes has decreased dramatically since the 1980s as fire-retardant materials have become more commonly used in some furnishings and as public awareness of the hazards has improved.[8] In spite of this progress, fires in which fabrics are the first materials to be ignited are still a very common occurrence. The fire investigator should be familiar with the nature of common fabrics and upholstery materials and their contributions to both ignition hazard and fuel load in today's studies.

Types of Fabric

Fabrics differ greatly in their properties of combustion, ranging from some that are intrinsically very hazardous to others that cannot be burned at all. A few extremely flammable fabrics are prohibited in the United States by the 1953 Flammable Fabrics Act (16 CFR 1610), however, fabrics that "pass" may ignite and burn easily. Fabrics that are the most fire resistant are, in general, used only for special purposes such as fire-resistant clothing for pilots, firefighters, and drivers of racing cars. The more common types of fabrics made from natural and synthetic fibers and their properties are outlined next.

NATURAL FIBERS

Common **natural fibers** include wool, cotton, linen, kapok, and silk.

natural fibers ■
Fibers of animal or vegetable origin without chemical manipulation.

Wool

Wool is used in many fabrics utilized for clothing as well as for bedding, upholstery, and miscellaneous purposes. Its fire hazard is minimal because of its high nitrogen and protein content. It can be burned but with considerable difficulty, because it has a high ignition temperature and low heat of combustion. It will not normally sustain a flame. Thus, wool blankets are used to wrap a person whose clothing is on fire and is being tested as a barrier layer in furniture. Exposing wool cloth to a flame will char a hole in it, but it will self-extinguish when the outside heat source is removed. When it does burn, wool generates toxic hydrogen cyanide fumes in addition to other combustion products.

Cotton

Cotton is used extensively in cloth for garments, bedding, and upholstery. It is possibly the most combustible of the common natural fibers. It is cellulosic in composition, like wood and paper. Having a very large surface-to-volume ratio, that is, a large surface for combustion, cotton is ignited with relative ease and will sustain a fire or allow it to increase greatly after ignition as long as it receives adequate ventilation. Cotton's ease of burning is conditioned by several factors, which will be discussed later. Cotton batting was once used universally as padding in garments and furniture and also as filler in cushions. It may still be found in some modern furniture, but it has largely been supplanted by polyester fiberfill or other synthetics.

Linen

Linen is being used much less than formerly, especially in clothing, and only occasionally is it seriously involved in major fires. Its cellulosic fibers are derived from stems (bast) rather than seed pods, as is cotton. Because it is a bast fiber, linen is much more variable than cotton, and this difference is expected to have some influence on combustion rates. Cloth from predominantly fine fibers is expected to burn more freely than that from coarser fibers. Like cotton, linen can be ignited by a smoldering cigarette or contact with a hot surface to support a smoldering or flaming fire.

Kapok

Kapok has been used as a filling for pillows, bedding, and insulated clothing but is not normally used as a cloth fiber. Kapok is involved in fires with sufficient frequency to make its fire hazards important. Like cotton, kapok is a cellulosic fiber, but because of its cell structure and its fineness, it exceeds cotton significantly in its flammability. It will ignite with ease, and when ventilation is available, kapok will burn with great intensity. It will also support smoldering fire very well. It is fortunate that kapok's uses are limited, so that it is not often involved in clothing fires. It can, however, pose a significant hazard in industrial areas, such as life preserver and pillow manufacture, where it is utilized. California's strict upholstered furniture flammability laws do not permit the use of kapok.

Silk

Silk is made by unwinding the cocoon of the silkworm and spinning the filament obtained into thread. It is a proteinaceous material like wool and although not easily ignited in bulk, it is often woven into a thin or sheer fabric that can burn quickly once ignited. It generates toxic gases, including hydrogen cyanide, upon combustion.

PETROLEUM-BASED SYNTHETIC FIBERS

Synthetic fibers, because of their versatility, constitute the largest percentage of fabrics in upholstery and clothing in the United States today. These fabrics, which bear names such

synthetic fibers ■
Manmade fibers of chemical origin.

FOAM RUBBER

Butyl, nitrile polybutadiene, and polyurethane rubbers are examples of common synthetic materials used as coatings, padding, and insulation in household goods. They will resist ignition from a glowing source, but without fire-retardant additives they are easily often ignited with a small open flame and will burn with considerable energy.

Polyurethane foam rubber, because of its widespread use in upholstery (and occasionally clothing), is the most commonly encountered synthetic with a substantial fire hazard. (Today, nearly all upholstered furniture sold in the United States contains some polyurethane rubber padding). More and more of the foam being used today is of the flame-resistant type. There are many types of flame-retardant foams; most of them are resistant only to small open flames and will burn fiercely when ignited by a larger flame. Others resist ignition even by much larger flame sources. It has been estimated that the percentage of upholstered furniture using flame-retardant grades of flexible urethane foams rose from 1 percent in 1970 to 50 percent in 1994 as those products were improved in strength, durability, and cost.[9] Flame-retardant grades of foam still remain very underused in residential furniture.[10]

As noted in Chapter 5, not all "foam rubber" is polyurethane. There are other **elastomers** such as isoprene isobutylene or polyester foams. These may have different ignition and flammability properties than polyurethane and usually require lab identification. Most elastomers do not melt but decompose when heated.

elastomers ■
Fibers or materials having elastic properties.

---

as nylon, Dacron, Orlon, Dynel, and Acrilan, can vary considerably in their ignition properties and burning behavior. Some of these variations are the effect of the chemistry of the fiber itself, but they usually are the result of additives and modifiers used to improve particular properties of the fibers. Most synthetic fibers are made from chemicals that originate from petroleum products. As a class, they tend to be much more susceptible to ignition by open flame than by a hot surface or glowing ember. Nylon, polyester, polyethylene, polystyrene, and polypropylene will ignite when in contact with a flame, especially when in the form of thin fabrics or coatings. Fluffy polyester fiberfill (as seen in Figure 5-21) is very commonly found in cushions, pillows, and comforters.

In addition to being flammable, most (but not all) synthetic fabrics have a tendency to melt and flow as they burn. When this occurs, burning droplets may fall into other fuel and start secondary fires. If the fabric is in the form of a garment, the melting material may fall away rapidly with only burning droplets clinging to the skin of the wearer, with their difficult-to-extinguish flames inflicting localized burns.[11] Interestingly, the fabric of choice for the past 35 years for children's sleepwear has been 100% polyester. Its melting and shrinking behavior actually prevents the propagation of small flames. This permits such fabrics to pass the CPSC sleepwear requirements of 16 CFR 1615 and 16 CFR 1616. When exposed to large flames, polyester will ignite and melt, possibly causing serious burns. The ignition and burning properties of synthetics change when they are blended with natural fibers. For instance, cotton-polyester blended fabrics cause more injuries than polyester alone because the cotton keeps the burned fabric together to support a rapidly spreading flame, and the fabric remains in contact with the wearer.[12] Pure thermoplastic synthetic fibers do not smolder, and burn only in flames.

Nomex

One synthetic fiber, Nomex (an aromatic polyamide), resists all types of ignition and is used to weave fabrics for fire protective gear for firefighters, aircraft pilots, and drivers of racing cars. A single layer of Nomex cloth will protect human tissue from burning in contact with a flame of 600°C (1,200°F) for up to 30 seconds. Other fire-retardant fabrics available today include modacrylics, viscose rayon, polybenzimidazole (PBI), Kevlar, Kynol, and some PVC (polyvinyl chloride) fabrics.[13]

NON-PETROLEUM-BASED SYNTHETIC FIBERS

Other synthetic fibers do exist that are not petroleum based—for example, rayon and cellulose acetate. Rayon (regenerated cellulose) is chemically very nearly the same as cotton because it is made by dissolving cotton or other cellulosic material and regenerating the cellulose in the form of an extruded fiber by forcing the solution through a spinneret into the regenerating bath. Because of the similarity to cotton, the main difference in fire hazard is due to the different fabric weave and garment configuration and possibly size of the fibers.

Acetate (acetylated cellulose) is similar to rayon in its manufacture and appearance, but the chemical modification of the cellulose diminishes its flammability. It melts rather easily but burns with some difficulty, except in lightweight fabrics, where it will burn relatively rapidly.

Other synthetic fibers may be encountered in some unusual fabrics. Glass-fiber cloths are manufactured and used mostly for nonapparel purposes, although clothing has been made from them. Drapery is one of the common uses in which their obvious nonflammability and fire resistance would be of importance. Nearly all draperies sold for residential use, however, are readily ignitable and will support an energetic fire. Curtains for public occupancies are fire resistant and are often required to comply with *NFPA 701*.

Asbestos, a naturally fibrous mineral, has been used in special applications such as theater curtains because of its great fire resistance and absolute nonflammability. It is unlikely to be found in clothing or household textiles; it is no longer permitted because of its potential carcinogenicity.

Fire Hazards

It will be noted that few of the fabrics encountered present an extreme fire hazard, especially those found in garments. Most common fabrics are readily ignitable and can support significant fires. Of those that remain, it appears that one of the most common fibers, cotton, is also the most hazardous. The susceptibility of a fabric depends not only on what fibers are used to make it but on how the fabric is woven and the design of the garment in which it is used. Yarn or thread that is tightly twisted will burn with less intensity than loosely spun yarns. In the manufacture of some cloths, the threads comprise tightly twisted fibers; in general, these are the fabrics that are most tightly woven as well. The degree of tightness in spinning is a definite factor in determining the flammability of the fabric when finished. The weight of the fabric (density) is also a key factor with regard to its ease of ignition. Generally, the heavier a fabric is, the more it will resist ignition by flaming sources. However, while heavier fabrics may be harder to ignite, once on fire, the larger fuel load can sustain a fire more likely to cause injuries if the garment cannot be removed or extinguished.

INFLUENCE OF WEAVE AND FIBER

The weave and finish influence the burning behavior of a fabric because they control the amount of its fuel-air interface. A loosely woven cloth, with spacing between the threads, will always burn much more rapidly than a tightly woven fabric of the same thread. For example, a cotton canvas burns relatively poorly, while cotton gauze is rapidly consumed (but canvas is more easily ignited by a cigarette), although both are made from the same fiber. A tightly woven, heavy fabric like denim will ignite and burn much more slowly than a sheer, lightweight cotton fabric such as the broadcloth used in shirts. A fabric with a nap, such as cotton flannel, has air spaces between the loose fibers on the surface and will ignite much more quickly than cotton denim, which is smooth surfaced. Cotton flannel and terry-cloth fabrics have been shown to be particularly susceptible to ignition. A fluffy high-pile fabric such as that used in some sweaters will ignite and burn faster than a close-knit low-pile fabric of the same fiber. Recent tests conducted to replicate fatal fire cases have revealed the rapidity with which cotton shirt fabrics and synthetic windbreakers (anoraks) can become fully involved from a single point of ignition with a flame.[14] The design of the finished garment has a bearing on its fire hazard because of the increased likelihood of accidental ignition of loose-fitting garments such as robes, blouses, and nightgowns. Garments with ruffles, pleats, loose and flowing sleeves, and similar features are more prone to come into contact with a heat source during the course of the wearer's everyday activities. Excesses of material in such loose garments enhance the fuel load as well as the fuel-air ratio and promote rapid burning. From a fire-safety standpoint, the ease with which a burning garment can be removed is also a key design feature that influences the severity of possible injuries.

CLOTHING IGNITION

Ignition of clothing may originate in a variety of ways, most of which are identical with the manner of kindling any other type of fire. Cigarette lighters in homes, as well as utility or fireplace lighters, are some of the more dangerous items. Matches may lose a flaming head when struck, and burning tobacco may drop on a dress, but research has shown that it is unlikely that most garment fabrics will be ignited by a smoldering cigarette.[15] One of the more common and dangerous sources of ignition is a heating appliance against which a flaring skirt may be pressed without the wearer's knowledge, or a burning candle that may readily ignite the loose sleeve at the dinner table. Again, the main item that predisposes to ready ignition is the design and the weight of the fabric, rather than the type of cloth, so long as the cloth itself is reasonably ignitable.

Regulation of Flammable Fabrics

The flammability of flammable fabrics is regulated in the United States by the Consumer Product Safety Commission (CPSC) on the federal level and also by states. Many states and regulatory and control agencies have separate and broader-ranging regulations. This is especially true with respect to carpets, wall coverings, and furnishings to be used in hotels, theaters, hospitals, and other public buildings. Some local building codes establish strict requirements for flammability of products in such high-risk occupancies. Of the state regulations, those promulgated by California tend to be the most rigorous. Because manufacturers who sell in California also sell in other states, the California regulations function as de facto national regulations.

The federal and state regulations are based on specific test methods. Descriptions of the relevant test methods are described in the Flammability Testing section.

REGULATION OF FLAMMABLE FABRICS
Overview of Federal Jurisdiction
The Flammable Fabrics Act (FFA) was passed in 1953 to regulate the manufacture of all highly flammable wearing apparel and fabrics when such materials were to be offered for sale on an interstate basis. Under the FFA, regulation rested with the U.S. Department of Commerce (DOC).

The Consumer Product Safety Act of 1972 transferred all enforcement of the FFA to the CPSC, which has broad jurisdiction over household products and their safety and now promulgates standards and enforces them. Since then, a number of important regulations have been enacted with regard to flammable fabric standards. Since 1972 the Department of Health and Human Services (previously the Department of Health, Education, and Welfare) has set standards for interior finishes and upholstery fabrics used in hospitals and nursing homes. The fabrics and cushioning materials used in the interiors of motor vehicles have had to pass certain tests prescribed by the Department of

Transportation (DOT) since 1972 (U.S. Motor Vehicle Safety Standard No. 302). (See the discussion in Chapter 9 on MVSS 302.) DOT regulations on the flammability of aircraft interior finishes were first passed in 1968 as FAA Title 14. The current tests for aircraft upholstery deal with their resistance to radiant heat from a large nearby fire (e.g., a jet fuel spill) and smoke emission.

Clothing Flammability

The first standard published under the FFA was CS-191-53, which measured flame spread on fabric after direct flame ignition. This Standard for the Flammability of Clothing Textiles is now known as 16 CFR 1610. It should be noted that the CPSC General Wearing Apparel Standard 16 CFR 1610 is a very weak requirement (involving the application of a needlelike flame to the surface of the test specimen for one second and observing fire spread.) Because newspaper will pass that test, the standard will exclude only the most dangerous fabrics. Passing this test does not mean the fabric is fire safe! Tests 16 CFR 1615 and 1616 require the application of a larger flame to the exposed edge of the test specimen for 3 seconds. In any such tests, only a fabric specimen is tested. The finished garment may well pose higher ignition risks. In 1967, the act was amended and expanded to include textiles intended for nonclothing uses, such as interior finishes and furnishings. Enforcement responsibilities were divided between DOC and the Federal Trade Commission (FTC).

Flammability of Carpets and Rugs

The first standard created under the expanded FFA regulated the flammability of large carpets and rugs. This standard, which is still mandatory for carpets and rugs sold in the United States, measures horizontal flame spread upon ignition by methenamine tablet (a reproducible ignition source). This Standard for the Surface Flammability of Carpets and Rugs (DOC FF 1-70) is currently known as 16 CFR 1630. The flammability of small area carpets and rugs was addressed in 1970 in DOC FF 2-70 (now known as 16 CFR 1631).

Flammability of Children's Sleepwear

Children's sleepwear was the next concern. In 1971, the smaller sizes (0 to 6X) of sleepwear sold in the United States came under the control of DOC FF 3-71 (now known as 16 CFR 1615). The larger sizes (7 to 14) were incorporated into the regulations under DOC FF 5-74 (now 16 CFR 1616). Both standards are fairly stringent, dealing with ignition of vertical fabric surfaces by open flames.

Mattress Flammability

Because of the frequency and severity of fires started in furniture and mattresses, the most important and most controversial standards have to do with the flammability of mattresses and upholstered furniture sold in the United States. Since 1973 there has been a national standard for the cigarette ignition resistance of all mattresses, futons, and mattress pads sold in the United States (DOC FF 4-72, now 16 CFR 1632, revised in 1984). This standard mandates that mattresses and mattress pads must not be ignitable by ordinary cigarettes placed directly on the mattress or between sheets covering that mattress. Prior to 1973, the standard mattress consisted of coil springs covered with a layer of insulating felt covered by thick batting, protected by a fabric covering or ticking. Although the cotton batting was susceptible to smoldering ignition, it was difficult to ignite by open flame. In compliance with the 1973 law, most manufacturers began using a non-fire-retardant polyurethane pad as a supplement or a complete replacement for the cotton batting. Although the mattresses made between 1973 and 2007 are, indeed, resistant to ignition from cigarettes, they are much more likely to ignite from open flames and to continue to burn than are their predecessors. Some manufacturers have produced mattresses made with a fire-retardant-treated cotton or flame-resistant polyurethane pad to optimize all forms of fire resistance. These materials are marginally more expensive than the unsafe products, however, and cost-conscious manufacturers have been slow to adopt such materials.

FIGURE 11-1 Foam mattress topper burned readily when ignited by a match flame at its edge. *Courtesy of John Jerome and Jim Albers.*

Mattresses made for hospitals and prisons generally must meet more stringent standards (discussed in a following section). Since 1977 the state of California has required all mattresses that do not meet both smoldering and open-flame standards to bear a consumer warning label stating that the mattress is susceptible to open-flame ignition.

In August 2001 the California legislature enacted a law (Assembly Bill 603) requiring all mattresses, futons, and mattress sets manufactured on or after January 1, 2005, and sold in California (including those for residential use) to be resistant to a large open flame. The goal of the legislation was to reduce the incidence of fire deaths, injuries, and property losses associated with bedroom fires.

California *Technical Bulletin (TB) 603: Standard for Mattresses, Box Springs and Futons* required that all mattresses, box springs, and futons comply with new flame-initiated testing procedures. The test was designed to simulate real-world flaming ignition sources and measure the heat released from such items as they burn.[16] This test has been supplanted by CPSC test 16 CFR 1633, which is described later. Compliance with such standards is almost exclusively achieved by the use of a variety of fire-blocking layers or barriers rather than chemical retardants. These layers involve PVC cloth, metal films, and other passive blocking systems.

Unfortunately, flammable bedding materials such as aftermarket foam bed-toppers, comforters, pillows, and duvets are not covered under any current fire safety specification. Such materials are readily ignited by small flames (as shown in Figures 11-1 and 5-20) and continue to be a threat to life safety despite strenuous efforts to include them in fire standards. The California Bureau of Home Furnishings in 2003 published a draft test standard for such items (denoted as TB 604). The CPSC in 2007 published a notice of proposed regulation for bedclothes flammability (16 CFR 1634), but it has not progressed further. It would have mandated that such products pass a flame test such as that in Figure 11-2. It has recently been shelved.

Upholstered Furniture

Upholstered furniture is subject, to a lesser degree, to the same sorts of regulations. Although there are national industry standards regarding the flammability of upholstered furniture, these are voluntary and aimed at smoldering ignition sources only. Since 1975

FIGURE 11-2 Comforter built to meet California's proposed safety standard (TB 604) resists flaming ignition, and fire spread is dramatically reduced. Compare with typical comforter shown in Figure 5-20. *Courtesy of Bureau of Electronics & Appliance Repair, Home Furnishings.*

the state of California has required all upholstered furniture (and its component stuffing materials) sold in the state to pass a standard test for flame and smoldering resistance (California TB 116 and 117). The only exception is furniture whose intended use is solely outdoors or recreational equipment whose use would preclude exposure to fire hazards (exercise benches and the like). Furniture must bear consumer warning labels as to which standards it has met.

TB 116: *Flame Retardance of Upholstered Funiture* is a voluntary standard that requires that finished articles of furniture be resistant to cigarette ignition. TB 117 is a mandatory requirement for furniture sold in California that specifies that all furniture components must comply with a series of small-scale fire tests.[17] These tests were designed to measure the resistance of the furniture's cushioning or padding to both flaming and smoldering ignition sources.

In recent years California has developed rigorous standards and flammability test procedures for mattresses and upholstered furniture for use in public buildings such as jails, hospitals, nursing homes, hotels and motels, college dormitories, public auditoriums, and so on.[18] In 1980, California TB 121: *Flammability Test Procedure for Mattresses for Use in High Risk Occupancies*, developed by the California Bureau of Home Furnishings, was adopted by the California Board of Corrections as a mandatory requirement for all California detention facilities.[19]

In 1984 the flammability test procedure for seating furniture for use in public occupancies was first developed by the California Bureau of Home Furnishings and Thermal Insulation.[20] It was then published in 1988 as the *Flammability Test Procedure for Seating Furniture for Office and Public Occupancies*, which is now known as California Technical Bulletin 133 (TB 133).[21] TB 133 became a mandatory law in California in 1992 and has also been adopted by several other states and a number of local jurisdictions, including Illinois, Ohio, Minnesota, Massachusetts, the city of Boston, and the New York–New Jersey Port Authority. A generic version has been adopted as *ASTM E 1537: Standard Test Method for Fire Testing of Upholstered Furniture*.[22] Finally, TB 129: *Flammability Testing Procedures for Mattresses for Use in Public Buildings* was published in 1992. TB 129 has also been widely adopted, and a generic version has been published as ASTM E 1590.[23]

The California Department of Consumer Affairs maintains a nationally recognized testing laboratory as part of its Bureau of Home Furnishings and Thermal Insulation. This laboratory is responsible for testing fabrics, apparel, and home furnishings to ensure they comply with the stringent state standards. This laboratory reports that there is no way to predict the time frame of smoldering versus flaming combustion in modern upholstered furniture. In its tests, ignition by cigarette may lead to flaming ignition in times ranging from less than 1 hour to 9 or more hours.[24] This variability has been confirmed in a number of full-scale furniture tests conducted by the authors.

The earliest flame-retardant urethane foams suffered reduced flame retardance as the foam aged, especially at high temperatures. The additives being used today for the most part are stable under normal living conditions, but a portion of the California law requires that foams meet flammability tests before and after aging. The additives used in some foams to make them more resistant to flaming ignition actually made them more susceptible to smoldering ignition. It appears that manufacturers must be very careful in their choice of flame retardants, adhesives, and even mold release agents to ensure the best fire retardance properties in the finished product. Several flexible foams have recently been made that have excellent fire resistance. Combustion-modified, high-resiliency (CMHR) foams supplied by several domestic polyurethane foam manufacturers have exceptional fire resistance but may be less flexible and of much higher density than standard foam. There is also a so-called low-smoke neoprene foam that has excellent flame resistance. The CMHR foam has been useful in high-risk facilities such as jails and other special applications, but until the advent of melamine urethanes (manufactured by BASF) its physical limitation made it less desirable for regular use. The newest foams promise excellent fire resistance and long life along with light weight, flexibility, and a cost advantage over other fire-retardant foams.[25] A variety of highly fire resistant foams have been introduced, but except for specialty applications, they have not proven viable in the consumer market.

Furniture Testing

Laboratory testing of furnishings can provide some general guidance to investigators as to the ignition and flame-spread characteristics involved, but as Krasny et al. describe, there are many approaches to testing and regulations.[26] Ignition sources vary from radiant sources such as cigarettes or glowing electric heating elements to flaming sources from matches and methenamine pills to crumpled paper to gas burners of various heat release rates and configurations, depending on the severity of the test. Information can be gathered from rate of heat release (calorimetry), total heat released over a fixed time (typically 180 seconds), rate of mass loss, total mass loss (absolute or percentage), visible charring, smoke production, rate of flame spread, or time to transition to flame. (Pass-fail criteria can be based on any one or more of those factors.) The test samples may be individual components, small-scale mock-ups, full-scale mock-ups, prototypes, or real furniture sampled from stores. The small-scale mock-ups have to be tested to demonstrate their correlation to full-scale performance before they can be used in regulatory testing. During the early 1990s, the European Union (EU) sponsored the largest-ever research project on the fire behavior of upholstered furniture [published as Combustion Behavior of Upholstered Furniture (CBUF)]. Some of the better-known or more widely used tests for upholstered furniture are summarized in Table 11-1. Of particular interest are the tests that include flame ignition, such as BS 5852 and BS 6807 (UK), TB 133 and TB 129, and some of the ASTM tests. The International Organization for Standardization (ISO) 9705 corner room configuration has also been studied for the information it can provide to modeling real-world performance.[27] In the United States, the Upholstered Furniture Advisory Council (UFAC) has acted as an adviser to the industry on flammability standards.

TABLE 11-1	Summary of Important Furniture Item Flammability Tests				
TEST	**CIGARETTE IGNITION**	**FLAME IGNITION**	**HRR**	**SMOKE**	**TOXICITY**
ISO 8191 (Nordtest 039)	X	X			
BS 5852	X	X			
ASTM E 1352 (Mock-ups)	X				
ASTM E 1353 (Components)	X				
NFPA 261 (Mock-ups)	X				
NFPA 260 (Components)	X				
California TB 116, 117	X				
Upholstered Furniture Action Council (UFAC)	X				
16 CFR 1633 (TB 603)[a]		X	X		
Business and Institutional Furniture Manufacturers Association (BIFMA)	X	X			
Code of Federal Regulations Part 1632[a]	X				
BS 6807[a]	X	X			
California TB 133		X	X	X	X
California TB 129[a]		X	X	X	X
California TB 121[a]		X		X	X
ASTM E 1537		X		X	X
ASTM E 1590[a]		X		X	X
ASTM E 162[b]		X			
ASTM D 3675[b]		X			
ISO 5660			X	X	X
ASTM E 1474			X	X	X
ASTM E 1354			X	X	X
NFPA 264-A			X	X	X
ASTM F 1550 M[c]			X		
ISO TR 5924				X	
ASTM E 662				X	
ASTM E 906			X	X	
ISO TR 9122					X
ISO TR 6543					X
ASTM E 1678					X

Reprinted from FIRE BEHAVIOR OF UPHOLSTERED FURNITURE AND MATTRESSES by John F. Krasny, William J. Peter, and Vytenis Babrauskas, Chapter 3, "Test Methods, Standards and Regulations," pp. 85–86, © 2001, with permission of Elsevier.

[a]Mattress test.
[b]Surface flammability test.
[c]Vandalized specimen.

FIGURE 11-3 Furniture Calorimeter. Combustion products are drawn into the exhaust hood, where the flow rate, concentration of oxygen, carbon dioxide, and carbon monoxide are measured. The heat produced is then calculated from the rate of oxygen consumption based on the value 13 MJ of heat per 1 kg of oxygen. *Source: Reprinted with permission from the* Society of Fire Protection Engineers, *2nd edition, National Fire Protection Association, Quincy, MA, 1996. This reprinted material is not the complete and official position of the National Fire Protection Association on the referenced subject, which is represented only by the book in its entirety.*

Labels in figure: Smoke transmission path; Exhaust fan; Gas sampling ring probe; Velocity probe; Thermocouple; Water-cooled skirt; Waste basket simulator burner; Gardon gauge; Weighing platform

furniture calorimeter ■ An instrumental system for measuring the amount of heat produced by the combustion of a moderate-sized fuel package, and the rate at which the heat is produced.

The heat release rate of furniture can be measured by bench-scale testing of models or mock-ups of the furniture; however, more accurate results are generally produced by testing in room calorimeters (such as those used in the California Technical Bulletin methods) or an oxygen-depletion **furniture calorimeter** of the type shown in Figure 11-3. The maximum heat release rates measured in such devices for upholstered chairs ranged from 370 to 2,500 kW; for sofas, 2,500 to 3,000 kW; for curtains, 150 to 600 kW; and for pillows from 16 kW (feathers) to 35 to 43 kW (urethane foam) to 117 kW (latex foam).[28] As we saw in Chapter 3, the heat release rates of common household furnishings are critical to evaluating the potential for development of the fire. It should be remembered that the rates as measured in a furniture calorimeter are not directly transferable to predicting their performance in a room fire. The maximum heat release rate of any such fuel is dependent to some extent on the radiant heat that fuel is receiving from the fire environment. Heat feedback in a room fire may well increase the maximum heat release rate produced over what is recorded in a calorimeter test. At the same time, the furniture calorimeter allows maximum ventilation, while a fire in a room may well reach ventilation limits that reduce the maximum heat release rate.

Flammability Testing

The testing of upholstered furniture and apparel fabrics is fairly complex and will not be treated in great depth here. The American Society for Testing and Materials (ASTM International) and various regulatory agencies have developed tests that address particular fabrics under reproducible laboratory conditions. Needless to say, variations in construction, use, and the nature of fire exposure have a great influence on the flammability of a particular fabric item. These variations cannot be duplicated completely in laboratory tests, and so, in real life, the fire hazards of such items may not be fully realized from evaluating these tests. Some generalities can be offered, however, as in the case of open-weave versus tight-weave fabrics. Apparel fabrics, because of their thinness, do not retain heat very well and are therefore less susceptible to ignition by a glowing heat source such as a cigarette than are heavy upholstery fabrics and stuffings, which retain heat until it builds up to flaming ignition. The standardized tests reflect these general tendencies, and some typical examples are outlined here.

FLAMMABILITY TESTS FOR FEDERAL REGULATIONS

Flammability of Clothing Textiles (Title 16 CFR 1610—U.S.)

Standard 16 CFR 1610 requires that a 10-in. (25-cm) piece of fabric that is placed in a holder at a 45° angle (see Figure 11-4) and whose surface is contacted by a small flame

FIGURE 11-4 16 CFR 1610 fabric test: 10-in. (25-cm) strip of fabric is held in a metal frame at 45°. Flame is applied to fabric at bottom for 1 second. *Courtesy of Salvatore Messina, CEO, The Govmark Organization, Inc.*

for 1 second not ignite and spread flame up the length of the sample in less than 3.5 seconds for smooth fabrics or 4.0 seconds for napped fabrics.[29]

Flammability of Vinyl Plastic Film (Title 16 CFR 1611—U.S.)
Standard 16 CFR 1611 requires that vinyl plastic film (for wearing apparel) placed in a holder at a 45° angle and ignited not burn at a rate exceeding 1.2 in. (3.0 cm) per second.

Flammability of Carpets and Rugs [Title 16 CFR 1630 (Large Carpets) and CFR 1631 (Small Carpets)]
Standards 16 CFR 1630 and 16 CFR 1631 require that a burning methenamine tablet placed in the center of a specimen not cause a char more than 3 in. (7.6 cm) in any direction. (The burning tablet or pill is a reproducible equivalent of a burning match, that is, ~50W for 1 minute, as shown in Figure 11-5.)

Flammability of Mattresses and Pads (16 CFR 1632—U.S.)
A minimum of nine regular tobacco cigarettes are burned at various locations on the bare mattress—quilted and smooth portions, tape edge, tufted pockets, and so forth. The char length of the mattress surface must not be more than 2 in. (3 cm) in any direction from any cigarette. The test is repeated with the cigarettes placed between two sheets covering the mattress.

Flammability of Mattresses and Box Springs (16 CFR 1633—U.S.)
The top and side of the mattress or mattress set are exposed to two 18-kW gas burners for a period of time (as shown in Figure 11-6). Failure of the test can be either (1) a peak

FIGURE 11-5 Methenamine pill on carpet test (16 CFR 1630). *Courtesy of John D. DeHaan.*

heat release rate of 200 kW or greater within 30 minutes of ignition or (2) a total heat release of 15 MJ or greater within 10 minutes of ignition. As of July 1, 2007, all mattresses sold for residential use in the United States must have passed the 16 CFR 1633 test.

Flammability of Children's Sleepwear (16 CFR 1615 and 1616—U.S.)

Each of five 3.5 × 10-in. (8 × 25-cm) specimens is suspended vertically in holders in a cabinet and exposed to a small gas flame along its bottom edge for 3 seconds (as shown in

FIGURE 11-6 Mattress and box springs built to meet California's safety standard (TB 603) or 16 CFR 1633 are subjected to test burner flames for 70 seconds but meet stringent limits on heat release rate and total heat for 30 minutes. Compare with fire in a typical mattress in Figure 5-20. *Courtesy of Bureau of Electronics & Appliance Repair, Home Furnishings.*

FIGURE 11-7 Test apparatus for 16 CFR 1615 (children's sleepwear): 10 × 3.5-in. (25 × 8.0-cm) strip hung vertically. Flame is applied to bottom edge.

Fabric strip

Steel cabinet

Glass door

Bunsen burner

Figure 11-7). The specimens cannot have an average char length of more than 7 in. (18 cm); no single specimen can have a char length of 10 in. (25 cm) (full burn); and no single sample can have flaming material on the bottom of the cabinet 10 seconds after the ignition source is removed. This test is required for finished items (as produced or after one washing and drying) and after the items have been washed and dried 50 times.

FLAMMABILITY TESTS FOR CALIFORNIA REGULATIONS

Flame Retardance of Upholstered Furniture
(State of California TB 116 and 117)

For testing under TB 116, the article to be tested must be the finished product ready for sale to the consumer or a prototype mock-up of actual components that duplicate the design and structure of the finished product. Three regular tobacco cigarettes are ignited and placed along each of the following structural features: smooth surface, welting, quilted and tufted locations, crevices, and tops of arms and backs (as illustrated in Figure 11-8).

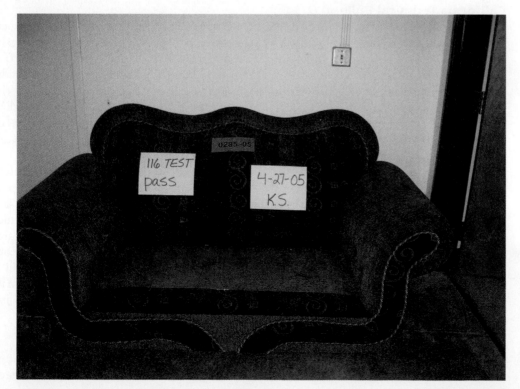

FIGURE 11-8 TB 116 9 (cigarette ignition) test: Cigarettes are placed at various locations on the backrest, arms, and cushions. *Courtesy of Bureau of Electronics & Appliance Repair, Home Furnishings.*

All cigarettes are covered with a single layer of cotton sheeting. Furniture fails if obvious flaming combustion occurs or if a char develops more than 2 in. (5 cm) in any direction from a test cigarette.[30] Filling materials are tested separately (as per TB 117) by inserting a vertical specimen into the flame of a gas burner for 12 seconds and observing the length of time for which flame and glowing combustion are maintained after the ignition flame is extinguished and the length of the resulting char pattern.[31] TB 117 requires both open-flame and cigarette tests to be performed on all furniture fillings.

Flame Retardance of Mattresses and Seating Furniture for Public Buildings (State of California TB 121 and 129)

Prison mattresses are tested using TB 121, a severe fire test using 10 double sheets of newspaper in a waste container ignited directly beneath the penal institution mattress or bed pad. The test is conducted in a specially designed full-scale test facility. The performance of the mattress is measured using temperature, smoke opacity, carbon monoxide production, and heat release rate.[32]

Mattresses intended for health care facilities, convalescent homes, college dormitories, and other multiple occupancies are tested using the protocol described in TB 129. It is a full-scale test in which complete mattresses or other bedding components are tested in a specially designed test facility. Ignition is by direct application of a T-shaped propane burner (producing ~17 kW) to the bottom edge of the mattress for 180 seconds. Test criteria are based on heat release rates achieved.[33]

Seating furniture for use in public buildings is tested under the protocol described in TB 133. The test is carried out in a specially designed room calorimeter (like TB 121 and TB 129). Ignition is induced by applying a large, calibrated square propane burner to the seat surface of the furniture for 80 seconds. Failure is dependent on the maximum heat release rate and total heat release observed.[34] TB 133 tests can be performed by about a dozen fire test labs in the United States. The concept of TB 133 testing requirements has been promoted by the National Association of Fire Marshals, the International Association of Fire Fighters, and several furniture manufacturers' associations.[35]

California Technical Bulletin 603

The TB 603 test protocol, performed in a full-scale fire test facility capable of calorimetric measurements, applies two large gas-jet flame sources (18 kW), ignited simultaneously, to the bed set (mattress with a box spring or foundation if normally used with one), one on the top surface for 70 seconds and a second flame on the lower edge for 50 seconds. Sample burning is observed for 30 minutes, unless flashover appears inevitable or the sample self-extinguishes. If the peak heat release rate in the test room exceeds 200 kW during this period or the total heat release exceeds 25 MJ in the first 10 minutes, the product fails the standard.[36] TB 603 has now been supplanted by the CPSC national standard, 16 CFR 1633, which became a national requirement on July 1, 2007.

Mattresses meeting TB 603/16 CFR 1633 should exhibit a slower level of fire growth when subjected to open-flame fires than those currently on the market and should thus provide a widened window of escape time for occupants before untenable conditions are reached. As can be seen in Figure 11-6, there is a dramatic improvement in fire performance in TB 603/16 CFR 1633–compliant mattresses. In addition, the federal mattress standard, 16 CFR 1632, requiring mattresses, futons, and mattress pads to be cigarette resistant, continues as law in all 50 states. Compare comforter fire tests in Figures 5-20 and Figures 11-2 to see what role such bedding materials can play in fires. As shown in Figure 11-1, aftermarket mattress toppers are easily ignited with a match and can produce an intense, sustained fire.

GENERAL OBSERVATIONS

As part of the extensive testing conducted since 1975 by the California Bureau of Home Furnishings and Thermal Insulation (BHFTI), some general observations have been made regarding cigarette ignition of furniture and mattresses. The key factor in predicting the smoldering propensity of upholstered furniture is the type of fibers used in the exterior cover fabric. Furniture covered in fabrics made predominantly of cellulosic fibers is much more likely to smolder than that covered in thermoplastic-fiber fabrics. The tendency to smolder increases with increasing weight of cellulosic fabric. Furniture covered with thermoplastic fabrics, particularly with heavier fabrics, tends to be very resistant to ignition by cigarette, even if the stuffing is cotton batting.[37] In the BHFTI testing, cigarette-induced furniture ignition required from 40 minutes to a number of hours before flaming combustion begins. In some cases, open flaming never began. In others, flaming combustion proceeded so quickly that a typical article of furniture, once flaming, could create flashover conditions in a normal-size room in as little as 2 to 3 minutes.[38] Extensive testing by the Center for Fire Research at the National Institute of Standards and Technology (NIST) has revealed that the transition to flame from a smoldering cigarette fire is quite variable. In one series of tests on non-flame-retardant cotton fabric over cotton and urethane foam padding, 22 chairs were tested; 10 never went to flame (self-extinguished or smoldered to completion). The remaining 12 identical chairs took from 29 to 63 minutes to transition to flame (with an average of 44 minutes). In various NIST tests, smolder-to-flame in upholstered chairs required as little as 22 minutes to as much as 306 minutes.[39] This range tallies with the authors' experience, where the shortest time for flames to be observed was 22 minutes when the cigarette was placed between intact upholstered surfaces of the armrest and seat cushion. When the outer covering is torn and the cigarette is placed directly in contact with non-flame-retardant cotton padding, ignition has been observed in 18 minutes. On other occasions, flaming ignition has taken up to 3 hours. Cigarette ignition of kapok stuffing and latex foam rubber has also been created (with ignition about 1 hour after insertion of the cigarette).

CONSIDERATIONS FOR FIRE INVESTIGATORS

Many variables contribute to susceptibility and time factors. These include fuel moisture, ambient temperature, position of the cigarette, and nature of the outer fabric covering(s) (thin or thick, cellulosic or synthetic, presence of a barrier layer, etc.). The presence of external air currents can also affect flame transition, because a draft can increase the combustion rate. As mentioned in the TB 116 test, a single layer of cloth over a smoldering cigarette can reduce losses by convection and radiation and make ignition more likely. Holleyhead published an extensive review of the furniture fire test literature and explored a number of the furniture-related variables.[40] The investigator needs to be aware of the variables related to both the furniture and the scene, and what effects they can have on ignitability. It should be noted that tests are conducted with full cigarettes, and most accidentally dropped cigarettes are likely to be only partial ones. Ignition requires sufficient contact, which may not occur with partial cigarettes or dropped embers. The bottom line is that cigarettes and open flames are competent ignition sources for furniture depending on the nature of the fabrics involved. Open flames will ignite modern fabrics almost on contact (3–5 seconds), whereas a smoldering cigarette will require 20 minutes or more, and then only if the other factors are favorable. Ignition from open flames is a more substantial risk today than just 10 years ago due to the popularity of candle usage (see Chapter 6). Home decorating schemes and "aromatherapy" uses mean candles are in use for prolonged periods around furnishings and bedding. These sources have to be considered in addition to "traditional" risks from matches and lighters.

Thermoplastic fabrics tend to melt and shrink when exposed to the radiant heat of a growing fire, which can expose the underlying padding to flames and radiant heat, producing very fast fire growth. Polypropylene (polyolefin) and polyvinyl fabrics pose particular hazards in this regard. The inclusion of fire-blocking layers (fiberglass, PBI, coated fabrics, and even neoprene foam bonded to a textile fabric) can dramatically reduce the maximum heat release rate demonstrated by these furnishings.[41] One recent observation by NIST researchers is that as polyurethane foams burn they decompose and "melt" into a semiliquid state that pools under the chair, sofa, or mattress to support a very intense pool fire. Using inert fillers to inhibit the flow of such products resulted in a significantly reduced maximum heat release rate.[42] When carpets are involved, the types of face yarn and backing are major factors. Because of its very low melting point, polypropylene fiber carpet melts very readily to expose combustible padding beneath.

Fire investigators today are realizing that the types of fibers, fabrics, and padding involved in furniture, mattresses, clothing, and carpet all can affect ignition, fire spread, fire intensity, and generation of toxic gases. As a result, these items need to be sampled and preserved for possible identification. In the future, fire investigators will have to determine whether the furnishings have met appropriate fire standards. Meeting standards can be accomplished either by design (such as barrier layers), chemical structure, or treatment with fire retardants. The mere presence of a tag or label does not prove the product was compliant with those standards. CPSC and the California Deptartment of Consumer Affairs report numerous failures of consumer products, some complete with false or forged tags.

Summary

In general, it should be remembered that tests applied to fabrics and finished household goods reflect only the ignitability of the product—its susceptibility to being accidentally ignited. These tests do not determine whether a fabric will burn and contribute to a room's fuel load in a given room fire. All fabrics will burn if the temperature, ventilation, and duration of the fire are sufficient; even highly fire-resistant fabrics of Nomex or PBI will burn under the right conditions. These comments on fabrics and their fire tests are offered to assist the investigator in assessing the logical consequences of ignition from small flames or glowing embers that come in contact with apparel or upholstered furniture.

One lesson from the extensive tests conducted by the Fire Research Station in the United Kingdom as part of the inquiry into the Stardust Disco disaster (1981) was that even when individual materials and products pass appropriate flammability tests, they may interact in complex ways to create extraordinary fire conditions. In the Stardust Disco fire, the combination of furnishings and wall coverings produced radiative feedback, drop down of flaming residues, and piloted ignition of smoldering upholstery that spread a fire throughout an open seating area much more quickly than anyone could have predicted. It was only through extensive full-scale simulations that the fire could be reenacted. The fire investigator must keep those lessons in mind when evaluating the contributions of individual fuels in any fire.[43]

Review Questions

1. What is the chemical difference between natural fibers and synthetic fibers? Name three examples of each.
2. What fire behavior is peculiar to cellulosic fibers? Why does it not occur with most synthetics?
3. Name three of the U.S. government regulations on flammability of fabrics and describe each application.
4. What is a typical time frame for cigarette ignition of upholstered furniture?
5. What does an oxygen-depletion calorimeter do?
6. What is the California flame resistance standard (TB 603) and why is it so different from previous standards?
7. Describe three of the fabric flammability tests used today.
8. What is the methenamine pill test and what is it used for?
9. Why is polypropylene carpet such a special consideration in fire investigations?
10. What is the USMV SS 302 and what are its limitations?

References

1. D. Miller, R. Chowdhury, and M. Green, "2004–2006 Residential Fire Loss Estimates," (Washington, DC: U.S. Consumer Product Safety Commission, October 2009), 9–10.
2. M. Ahrens, "Home Fires that Began with Upholstered Furniture" (Quincy, MA: Fire Analysis and Research Division, National Fire Protection Association, 2008).
3. Ibid.
4. Miller et al., "2004–2006 Residential Fire Loss Estimates."
5. J. R. Hall, "The U.S. Smoking Material Fire Problem" (Quincy, MA: Fire Analysis and Research Division, National Fire Protection Association, 2008).
6. K. D. Rohr, "Selections for Products First Ignited in U.S. Home Fires: Soft Good and Wearing Apparel" (Quincy, MA: Fire Analysis and Research Division, National Fire Protection Association, 2005.
7. J. F. Hoebel, G. H. Damant, S. M. Spivak, and G. N. Berlin, "Clothing Related Burn Casualties: An Overlooked Problem," *Fire Technology* 46(3), July 2010, 629–649.
8. J. R. Hall, Jr. "Targeting Upholstered Furniture Fires," *NFPA Journal* (March–April 2001): 57–60; J. F. Krasny, W. J. Parker, and V. Babrauskas, *Fire Behavior of Upholstered Furniture and Mattresses* (Norwich NY: William Andrew, 2000), 7.
9. G. H. Damant, "The Flexible Polyurethane Foam Industry's Response to Tough Fire Safety Regulations," *Journal of Fire Sciences* 4 (1986).

Revised by G. H. Damant, personal communication, January 1996.

10. R. P. Foster and J. B. Zicherman, "Is There a Time Bomb in the Sofa?" *Trial* (November 2005): 58–63.

11. A. K. Mehta and F. Wong, "Hazards of Burn Injuries from Apparel Fabrics," NTIS COM-73–10960, Fuels Research Laboratory, MIT, February 1983.

12. J. Holmes, "Some Clothes May Play with Fire," *Insight* (June 9, 1986): 46.

13. V. M. Bhatnagar, ed., *Advances in Fire Retardant Textiles* (Westport, CT: Technomic, 1975); S. Backer, G. C. Tesoro, T. Y. Yoong, and N. A. Moussa, *Textile Fabric Flammability* (Cambridge, MA: MIT Press, 1976).

14. R. McRae, Calgary Police, personal communication, June 2000; J. W. Munday, personal communication, March 2001.

15. Mehta and Wong, "Hazards of Burn Injuries from Apparel Fabrics"; J. L. Sanderson, "Cigarette Fires in Fabrics," *Fire Findings* 15, no. 3 (Summer 2007): 1–3.

16. *Technical Bulletin 603: Requirements and Test Procedures for Resistance of a Mattress/Box Spring Set to a Large Open Flame* (North Highlands, CA: California Department of Consumer Affairs, January 2002).

17. *Technical Bulletin 117: Requirements, Test Procedure and Apparatus for Testing the Flame Retardance of Resilient Filling Materials Used in Upholstered Furniture* (North Highlands, CA: California Department of Consumer Affairs, January 1980).

18. G. H. Damant, "Recent United States Developments in Tests and Materials for the Flammability of Furnishings," *Journal of the Textile Institute* 85, no. 4 (1994a).

19. *Technical Bulletin 121: Flammability Test Procedure for Mattresses for Use in High Risk Occupancies* (North Highlands, CA: California Department of Consumer Affairs, April 1980).

20. P. S. Wortman, S. S. Williams, and G. H. Damant, "Development of a Fire Test for Furniture for High Risk and Public Occupancies" in *Proceedings of the International Conference on Fire Safety* (Sunnyvale CA: Product Safety Commission, 1984), 55–67.

21. *Technical Bulletin 133: Flammability Test Procedure for Seating Furniture for Use in Public Occupancies* (North Highlands, CA: California Department of Consumer Affairs, 1988).

22. Damant, "Recent United States Developments in Tests and Materials."

23. *Technical Bulletin 129: Flammability Testing Procedures for Mattresses for Use in Public Buildings* (North Highlands, CA: California Department of Consumer Affairs, 1992).

24. V. Babrauskas and J. F. Krasny, "Upholstered Furniture Transition from Smoldering to Flaming," *Journal of Forensic Sciences* 42 (1997): 1029–31; J. A. McCormack, G. H. Damant, S. S. Williams, and J. F. Mikami, "Flaming Combustion of Upholstered Furniture Ignited by Smoldering Cigarettes" (North Highlands, CA: California Department of Consumer Affairs), International Conference on Fire Safety, San Francisco, January 1986.

25. Damant, "The Flexible Polyurethane Foam Industry's Response"; Damant, "Recent United States Developments in Tests and Materials."

26. Krasny et al., *Fire Behavior of Upholstered Furniture and Mattresses.*

27. B. Sundstrom, ed., *Fire Safety of Upholstered Furniture: The Full Report of the European Commission Research Programme (CBUF)* (London: Interscience Communications, 1995).

28. Krasny et al., *Fire Behavior of Upholstered Furniture and Mattresses.*

29. *Guide to Fabric Flammability,* U.S. Consumer Product Safety Commission, June 1976, 1–5.

30. *Technical Bulletin 116: Requirements, Test Procedure and Apparatus for Testing the Flame Retardance of Upholstered Furniture* (North Highlands, CA: California Department of Consumer Affairs, January 1980).

31. California Department of Consumer Affairs, *Technical Bulletin 117.*

32. California Department of Consumer Affairs, *Technical Bulletin 121.*

33. California Department of Consumer Affairs, *Technical Bulletin 129.*

34. G. H. Damant et al., "The California TB 133 Test: Some Background and Experience," *Fire Safety Journal,* 16 (1990): 1–12.

35. Krasny et al., *Fire Behavior of Upholstered Furniture and Mattresses.*

36. California Department of Consumer Affairs, *Technical Bulletin 603.*

37. G. H. Damant, "Cigarette Ignition of Upholstered Furniture," Inter-City Testing and Consulting, Sacramento, CA, 1994; also in *Journal of Fire Sciences* (September–October) 1995.

38. Krasny et al., *Fire Behavior of Upholstered Furniture and Mattresses.*

39. Ibid.

40. R. Holleyhead, "Ignition of Solid Materials and Furniture by Lighted Cigarettes: A Review," *Science and Justice* 39, no. 2 (1999): 75–102.

41. Damant, "Recent United States Developments in Tests and Materials."

42. bfrl.nist.gov.

43. P. T. Pigott, "The Fire at the Stardust, Dublin: The Public Inquiry and Its Findings," *Fire Safety Journal* 7 (1984): 207–12; W. A. Morris, "Stardust Disco Investigation: Some Observations on the Full-Scale Fire Tests," *Fire Safety Journal* 7 (1984): 255–65.

12

Explosions and Explosive Combustion

Courtesy of Det. Richard Edwards (retired), Los Angeles County Sheriff's Department, Whittier, CA.

KEY TERMS

backdraft, *p. 502*

brisance, *p. 511*

deflagration, *p. 489*

detonation, *p. 489*

explosive, *p. 489*

fragmentation, *p. 522*

high explosive, *p. 507*

low explosive, *p. 507*

Molotov cocktail, *p. 519*

seat of explosion, *p. 520*

stoichiometric mixture, *p. 495*

OBJECTIVES

After reading this chapter, you should be able to:

- Explain the basic concepts and definitions of explosions.
- Identify the common causes of diffuse-phase explosions by gases, vapors, and dust suspensions.
- Identify the common types of dense-phase explosives, their uses, and whether they are high or low explosives.
- List the proper methodology for conducting an explosion scene investigation.
- Describe several techniques used during scene examination, evidence recovery, and laboratory analysis.
- Locate the latest explosives statistics published by the U.S. Bomb Data Center.

Fires and explosions so frequently accompany each other that no treatise on fire and its investigation would be complete without some special consideration of the nature and role of explosions. Explosions range in violence from the diffused type of rolling, progressive flame resulting from the combustion of a rich mixture of flammable gases or vapors in air, to the violent, almost instantaneous detonation of condensed-phase explosives. The destructive effects of these explosions vary widely and may involve broken windows and dislodged bricks from a diffuse gas explosion or the shattering blow of a high-explosive detonation. In the most general terms, an explosion may be defined as the sudden conversion of potential chemical or mechanical energy to kinetic physical energy in the form of pressure, heat, and sometimes light as gases are generated very rapidly or released. There are four basic types of explosions that the fire investigator may encounter: chemical, mechanical, electrical, and nuclear. By far the most common will be chemical or mechanical, and chemical explosions will be the focus of most of this chapter. Mechanical explosions involve the sudden release of a gas or liquid under extreme pressure as its container, vessel, or pipe bursts. Electrical explosions occur when large currents suddenly flow through air, oil (in transformers or switch gear), or undersized conductors. The extremely rapid heating causes the expansion of surrounding gases, the boiling, and expansion of the oil, or evaporation of the conductor. This expansion occurs so rapidly (as in the "crack" of a nearby lightning strike) that explosive effects can be created, especially if the event is confined. Such events can also trigger deflagrating chemical explosions. Nuclear explosions cause massive blast effects by extremely rapid heating of surrounding gases and evaporating solids. (It is most unlikely that the reader will be involved in such a scene!) All the foregoing processes involve the sudden release or production of gases that exert force and inflict damage to nearby materials.

Chemical Explosions

As you will see, the conversion of chemical energy to kinetic energy can take on a variety of forms, and the energy released can exhibit a range of effects. The processes involved and the speed with which the reaction progresses depend not only on the chemistry of the materials involved but on their physical form and environment (i.e., containment). For the purposes of this book, explosive materials can be thought of as falling into two general categories—diffuse explosive mixtures and condensed-phase or concentrated explosives—and each will be examined in turn.

While diffuse explosive mixtures are most often combinations of a combustible gas with air, the condensed-phase explosion is not dependent on oxygen from the air but results from a combustive oxidation of the explosive. An **explosive**, then, can be defined as any material capable of undergoing an almost instantaneous conversion from one form to another, producing gases and heat as it does so. An explosive can be a material that contains a mixture of both a fuel for the combustion and an internal oxidizer, or one that is unstable and capable of rearranging its chemical structure when stimulated, with highly exothermic effects. Black powder is an example of the first "mixture" type, while dynamites, TNT, nitroglycerine, PETN, and lead azide are examples of the second (single compound). All contain in themselves all the essential components for almost instantaneous conversion to gases with great kinetic energy (heat). In this chapter, the nature of explosive combustion will be explored and the processes, effects, and diagnostic signs of various mechanisms will be discussed.

explosive ■ Any material that can undergo a sudden conversion of physical form to a gas with a release of energy.

Key Terms and Concepts

Oxidation reactions can proceed at various rates, from the processes of rusting or decay, which may progress slowly and go unnoticed, to the extremely fast detonation of some explosives, which proceed so quickly that they cannot be directly observed. It is when oxidation proceeds at a sufficiently rapid rate to generate conveniently measurable amounts of heat and light that it is called *combustion*. One can actually accelerate the rusting process of steel by heating steel wool in an oxygen-rich environment; it is then oxidized with a flash of flame, undergoing true combustion.

Because some compounds and elements can undergo an oxidation reaction without oxygen itself being present, the term *oxidation* must be reduced to its most basic meaning—the loss of electrons from a reactant (to be gained by another reactant). Therefore, oxidation can be defined as the combination of one compound or element with another that involves the loss of electrons from its atomic structure.

Turner and McCreery offer some excellent definitions of the various classes of oxidation reactions (chemical explosions):[1]

Combustion	A rapid oxidation with the evolution of heat and light.
Deflagration	A very rapid oxidation with the evolution of heat and light as well as the generation of a very low intensity pressure wave of moving gases. Note this is a low-energy pressure wave and not a shock wave.
Explosion	A very rapid oxidation with the evolution of considerable heat accompanied by a disruptive effect (the disruptive effect being a result of pressure and confinement). All reaction products are gaseous.
Detonation	An extremely rapid, almost instantaneous oxidation reaction (oxidation here meaning the loss of electrons, not necessarily the combination with elemental oxygen) with the evolution of considerable heat accompanied by a very violent disruptive effect and intense, high-speed shock wave. The detonation shock wave is supersonic (faster than the speed of sound) and requires no confinement for its propagation. All reaction products are gaseous.

deflagration ■ A very rapid oxidation with the evolution of heat and light and the generation of a low-energy pressure wave that can accomplish damage. The reaction proceeds between fuel elements at subsonic speeds.

detonation ■ An extremely rapid reaction that generates very high temperatures and an intense pressure/shock wave that produces violently disruptive effects. It propagates through the material at supersonic speeds.

Note that these classes of reactions are not separate and discrete categories but portions of a continuum of reaction rates—from the very slow to the ultrafast. Because the term *explosion* is used to describe the entire continuum from rapid combustion to detonation, it has been suggested that to minimize confusion, the categories be redefined as combustion, deflagration, and detonation, as follows:

Combustion	Rapid oxidation that generates heat and light but does not generate sufficient gases to produce a pressure wave.
Deflagration	A very rapid oxidation producing heat, light, and a pressure wave that can have a disruptive effect on the surroundings. This is an oxidation process that takes place at the surface of the reacting fuel (whether vapor or solid). This process can proceed at speeds up to 1,000 m/s (3,280 ft/s).
Detonation	An extremely rapid reaction that generates very high temperatures and an intense pressure/shock wave that produces violently disruptive effects. The process occurs through the reacting material at speeds in excess of 1,000 m/s (3,280 ft/s)* as the shock wave is transmitted at supersonic speeds.[2]

*Sometimes rounded to 3,000 ft/s. The speed of sound in air at 20°C (68°F) and sea level is 334 m/s (1,100 ft/s), but in denser materials it is much higher. Because the "average" density of high explosives is about 1.5 g/cc, the "average" speed of sound in those materials is about 1,000 m/s (3,280 ft/s).

These are, in fact, the definitions used by the Institute of Makers of Explosives and the Bureau of Alcohol, Tobacco, Firearms and Explosives. Because there is less confusion with this scheme and its transition criteria (pressure and speed) are fairly clearly defined, it will be used as the basis for the discussions in this chapter. One can see from these definitions that each category produces a set of physical effects that will leave distinctive diagnostic signs of pressure and speed in a structure or vehicle. A careful evaluation of these diagnostic signs at a fire-explosion scene will allow the analytical investigator to estimate the nature of the explosion that occurred.

Virtually all chemical explosions create gases and heat by their functioning. The postcombustion gases usually occupy a larger volume than the prereaction components (a great deal larger in the case of solid explosives), and the creation of great volumes of gaseous products accounts for some gaseous phase effects. The heat created, however, causes the majority of the expansion and thereby the majority of the blast effects.

Explosions are characterized by both the pressures produced and the rates at which those pressures rise. Harris demonstrated that for hydrocarbon fuels (methane, ethane, propane, etc.), the number of moles of precombustion gases is about the same as the number of moles of postcombustion gases.[3] This means that the expansion of gases is caused primarily by the thermal expansion caused by the heat produced. Under ideal conditions, this heating [from 27°C (80°F; 300 K) to 2,027°C (3,680°F; 2,300 K), for example] could cause overpressures of up to 8.3 bar, 827 kPa, or 120 psi. The pressures produced by real-world deflagrations of hydrocarbon fuels are usually limited by the failure pressure of the confining vessel or structure. Let us first examine the most common type of explosive combustion in structure fires—that of a diffuse-phase gas, vapor, or dust explosion.

Diffuse-Phase Explosions

Diffuse-phase explosions are the rapid combustion of a fuel (gas, vapor, mist, or dust) premixed with air or oxygen and then ignited. The reaction proceeds as a flame front extends from the ignition source into the premixed fuel/air mixture and then grows larger in volume and surface area with time. Its effects are dependent on its type of fuel, concentration, ignition mechanism, and degree of confinement. Each will be discussed in turn.

GASES

Perhaps the most common example of an accidental diffuse-phase explosion is that of natural (or LP gas) escaping in quantity into a confined space and mixing with air and then being ignited. Depending on the amount of gas escaping, the amount of air in the space, and the amount of leakage and mixing that takes place, the reaction mixture can be within its explosive range (see Chapter 4), too rich, or too lean. As we saw in Chapter 4, when a volume of natural gas is uniformly mixed with about 10 volumes of air, a mixture close to its ideal explosive ratio, ~9 percent, results. If this mixture fills a room and is ignited, the combustion is a deflagration that progresses radially in all directions away from the ignition source and through the room at a rate of up to 3 m/s (10 ft/s). (The rate is pressure and temperature dependent and will increase as the ignition proceeds and the flame front expands, so it can become considerably higher.) The combustion produces a great amount of heat very suddenly, and the combustion gases (and remaining atmospheric gases—nitrogen, etc.) are quickly heated and expand violently. This expansion is, of course, resisted by the walls, floor, and ceiling of the room. The resulting overpressure, which is typically a few pounds per square inch (psi), several kilopascals (kPa), or fraction of a bar[*] pushes the walls, floor, and ceiling, or breakable portions like windows and doors, outward. If enough gas is premixed with room air, the destruction can be extensive, as shown in Figures 12-1a–m.

ON SCENE

A natural gas explosion destroyed this four-unit apartment building when the gas shutoff valve was removed from the 19-mm (3/4-in.) delivery pipe serving the gas range in a vacant unit. Over an estimated 45-minute interval, gas filled the apartment until it was ignited in the vicinity of the wall furnace in the living room. The resulting deflagration lifted the roof of the entire building and scattered large and small pieces in all directions. The blast took out two of the walls of the living room and inflicted serious damage to the living room of the adjoining apartment, injuring its occupant. The wall where the wall heater was located was left intact, as was the longitudinal wall that separated the living room and the kitchen. Spot fires were ignited in the furniture by burning debris (cellulose ceiling insulation) from the attic. The plume of gas released from the open gas line charred the back of the stove and the side of the adjoining cabinet unit as it burned post-blast (see Figures 12-1a–m).

FIGURE 12-1A This overhead photo shows the debris field extending more than 100 ft in all directions. Note the car in the street struck by a large portion of the front wall of the structure. Wall debris crushed the rear of the convertible top, sparing the mother and child inside any injury. *Courtesy of Det. Richard Edwards (retired) Los Angeles County Sheriff's Department, Whittier, CA.*

FIGURE 12-1B Overhead, looking down onto the top floor units. The source of gas was in the kitchen (upper left). The source of the ignition was in/near the wall heater in the living room (upper center of photo). *Courtesy of Det. Richard Edwards (retired) Los Angeles County Sheriff's Department, Whittier, CA.*

FIGURE 12-1C A view showing the direction and intensity of the damage. The living room of the vacant unit had the highest concentration of gas. The blast propagated into an occupied apartment (left) and dissipated its energy as it moved out. Fire damage to the ceiling joists resulted from post-blast fires in the furniture. The blast propagated into the small hallway and then into the bedrooms, propelling the walls out toward the street. *Courtesy of Det. Richard Edwards (retired) Los Angeles County Sheriff's Department, Whittier, CA.*

[*]1 atmosphere = 14.7 psi = 1.013 bar = 101.32 kPa, so 1 psi = 6.89 kPa = 0.07 bar = 70 mbar = 27.67 in. w.c.

FIGURE 12-1D A view into the kitchen showing the water heater (center). If the water heater had been the ignition source, there would have been more damage to the rooms to the left of the main dividing wall. *Courtesy of Det. Richard Edwards (retired) Los Angeles County Sheriff's Department, Whittier, CA.*

FIGURE 12-1E A view from behind the stove showing the plume damage to the side of the cabinet and the flex line with no shutoff valve. The open gas line is in the wall on the right. *Courtesy of Det. Richard Edwards (retired) Los Angeles County Sheriff's Department, Whittier, CA.*

FIGURE 12-1F Another view of the kitchen. The stove had been moved away from the wall and the burners and top moved before the fire (in an attempt to remove the pressure regulator from the inside). The unvented hood and adjacent cabinets aimed the flow of venting gas (from behind the stove) toward the living room door (on right). *Courtesy of Det. Richard Edwards (retired) Los Angeles County Sheriff's Department, Whittier, CA.*

FIGURE 12-1G Damage to the ceiling of the ground-floor apartment (below the origin) shows the downward deflection of joists. *Courtesy of Det. Richard Edwards (retired) Los Angeles County Sheriff's Department, Whittier, CA.*

FIGURE 12-1H Localized fire damage from burning ceiling insulation. *Courtesy of Det. Richard Edwards (retired) Los Angeles County Sheriff's Department, Whittier, CA.*

FIGURE 12-1I Localized fire damage to the occupied apartment from a sofa ignited by burning debris from the attic. *Courtesy of Det. Richard Edwards (retired) Los Angeles County Sheriff's Department, Whittier, CA.*

FIGURE 12-1J Fresh tool marks on the stove pressure regulator showed the unsuccessful attempt at removal. *Courtesy of Det. Richard Edwards (retired) Los Angeles County Sheriff's Department, Whittier, CA.*

FIGURE 12-1K Documentation included status of gas service to building. *Courtesy of Det. Richard Edwards (retired) Los Angeles County Sheriff's Department, Whittier, CA.*

FIGURE 12-1L Undamaged gas line and burner for wall furnace (most probable ignition source). Note the undamaged louvers (rear). *Courtesy of Det. Richard Edwards (retired) Los Angeles County Sheriff's Department, Whittier, CA.*

FIGURE 12-1M The floor plan showing the directions of blast propagation through the top floor.

FIGURE 12-2A Broken gas pipe filled this cabin from below the floor where it was ignited. *Courtesy of Lamont "Monty" McGill, Fire/Explosion Investigator.*

FIGURE 12-2B TV remote embedded in ceiling when floor and TV were lifted by explosion. *Courtesy of Lamont "Monty" McGill, Fire/Explosion Investigator.*

This explosion process meets our definition of a deflagration because of the speed of propagation and moderate pressure that develop. With the heat generated in such a short time, items that are most readily ignited, charred, or damaged will be affected. Eyebrows and scalp hair may be singed, but flesh will not be deeply burned, because a longer time is needed for heat to penetrate and damage skin than hair. Clothing and thin draperies may be scorched or set on fire, but furnishings will be little affected because finer, less massive fabrics ignite more readily than upholstery and wood. The pressures produced may move walls outward, break doors and windows, and even lift the roof or floors (see Figures 12-2a and b). However, a person in the room (especially one close to the source of ignition) may not be seriously injured by the blast itself. The compressive pressure wave near the origin is usually slight enough not to cause permanent injury and is moving away from the ignition source, and the flash fire may be transitory enough that an occupant may not be seriously burned, and combustibles may not be ignited. The pressures are not such that a localized "seat" of shattered or pulverized material is produced.

The combustion rate grows as the surface area of the flame front expands. As a result, anything that increases the surface area by turbulent mixing can increase the combustion rate. This mixing can be caused by furniture or by passage of the flame front through doorways or across barriers. Expansion of gases in the room of origin may also pressurize and heat the mixture in adjoining rooms. Since higher pressures and temperatures increase the intensity of the combustion and its propagation rate, the reaction accelerates. While a gas/air explosion originates at its ignition source, the greatest destruction will as a result occur at distances from the ignition source (while still within the premixed volume).

Ideal Mixture

Every fuel has a ratio with its oxidizer at which its production of heat and reaction velocity are at their maximum. This is called the ideal or **stoichiometric mixture**. It is often close to the midpoint between the lean limit (LEL) and rich limit (UEL). The total damage done by a gas or vapor/air mixture will be maximized, since the amount of gaseous product and heat produced and the rate will all be at their respective maxima.

stoichiometric mixture ▪ A reaction mixture in which the reactants and products are chemically balanced.

Rich Mixture

Consider now the "rich" mixture, where the escaping natural gas has mixed with the room air to form a mixture of 15 percent gas in air, or consider that the mixture of gas and room air is not uniform and there are pockets in which the gas concentration is much higher than in other parts of the room. When the mixture is ignited, there is an explosive "thump," but there is insufficient air (oxygen) available for immediate and total combustion. Much of the natural gas is not burned, and the speed and force of the reaction are diminished. There is fuel left over and it has been heated. The combustion products and remaining fuel expand when heated but cool immediately afterward, and air is drawn into the partial vacuum that results. This mixing provides more oxygen to combine with the remaining hot gas, and a rolling fire results as the turbulence caused by the initial combustion mixes the remaining fuel and air. Note that because the reaction takes place over a longer period of time than in the case of an ideal mixture, a significant pressure wave may not develop, and the reaction may more accurately be classified as a combustion than as a deflagration. A victim may generally describe the event as, "I heard a 'whoosh' and was on fire." Interestingly, a concussive "thump" will be more apparent to a listener outside the room than to someone within. Destruction by rich-mixture ignitions may far exceed that caused by theoretically perfect mixtures of gas and air, because such explosions are followed by fire, and the fire may consume a building that otherwise would have been subjected only to a sudden expansion that might move its walls but not destroy its contents.

It should be noted that the preceding discussion centered around natural gas (methane/ethane) or LP gas. Other gases behave quite differently. Hydrogen and acetylene, for instance, burn at much higher speeds than methane and exert much higher pressures. Acetylene can also detonate under some conditions. The chemical properties of any potential fuel gas must be carefully considered when assessing a blast scene. (Detonations of gaseous fuels will be discussed later.)

Lean Mixture

The "lean" mixture combusts differently than the others in an explosion. Because the propagation rate for a gas/air mixture is greatest when the mixture is near its ideal or stoichiometric ratio, the lean-mixture explosion may be heard or felt as sharper than in the rich-mixture case. The force of a lean-mixture deflagration may be considerable, and its noise is very startling. However, with insufficient fuel to develop maximum speed, heat, and pressure during the blast or lingering, rolling post-blast fire, the damage from a lean-mixture deflagration may be minor compared with that of the other mixtures. No following fire is expected from a lean-mixture deflagration except for filmy materials like sheer curtains that may be ignited (and then go on to set other fuels alight). (See Figures 12-3a–g for an example.) Generally, although not in every instance, an explosion of this type produces purely physical forces and not an established fire.

FIGURE 12-3A Wood-frame residence looks undamaged at first sight. Note the dead shrub and dying tree (despite drip irrigation on both) in foreground. This was an indication of possible underground migration of natural gas. *Courtesy of John D. DeHaan.*

FIGURE 12-3B Bottom of wall is bowed outward by more than 12 cm (5 in.) at base. Note that windows (small, double-glazed, aluminum-framed) are still intact. *Courtesy of John D. DeHaan.*

FIGURE 12-3C Interior view of same wall shows the sill plate split at the bolt, and drywall and molding damaged. *Courtesy of John D. DeHaan.*

FIGURE 12-3D Wall of bedroom opposite main explosion shows outward deflection of nearly 7 cm (3 in.) at its center. No thermal effects were noted in this bedroom. *Courtesy of John D. DeHaan.*

FIGURE 12-3E Closet doors were blown off their tracks. *Courtesy of John D. DeHaan.*

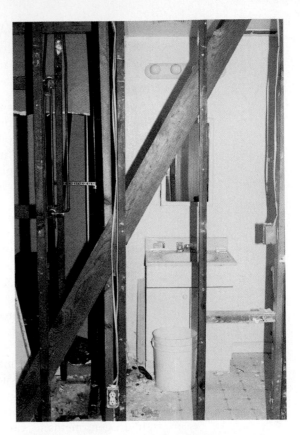

FIGURE 12-3F This interior wall was blown apart when the light switch (far right) was thrown, as were other two walls surrounding the shower stall pan (at far left). The occupant was standing in the bathroom doorway (extreme right) when she turned off the switch, and was unhurt. *Courtesy of John D. DeHaan.*

FIGURE 12-3G Gas had migrated under concrete slab (apparently following an abandoned sewer connection), percolated under the pan, and filled all three adjacent walls. Ignition propagated through all three walls, blowing drywall off both sides. Note loose soil under shower pan and scorching on nearest studs. *Courtesy of John D. DeHaan.*

VAPORS AND VAPOR DENSITY

Not all explosions involve fuel gases. Many involve the combustion of vapors generated by the evaporation of flammable liquids.

Vapor Density

As we discussed in Chapter 4, explosions that result from the ignition of vapors from liquid fuels are much more likely to produce post-explosion fires than will those of gases. The explosion itself produces heat to volatilize remaining liquid fuel and provides an ignition source for the "new" vapors being generated. Gaseous fuels will be largely consumed or physically displaced, leaving little for continuous flaming fire afterward (unless there is a continuous source of new gaseous fuel, such as a broken supply pipe) (as seen in Figure 12-4).

The vapor density of organic liquid vapors must also be considered in connection with special circumstances of the environment. For example, machines are sometimes degreased with hydrocarbon solvents. Not only is there possibly a high concentration of flammable vapors in the area, but drains and service pits provide lower regions into which the vapors can sink. If such vapors accumulate in such a depression, it is there that the hazardous condition is first established. Any arc or flame in this vicinity can cause an explosion, even though it may be separated from the working area by some distance. Drains, sewers, heating ducts, elevator shafts, tunnels, and utility shafts are all possible areas for such remote ignition. The density of flammable liquid vapors can produce rapid flame spread through the layer where the premixed fuel/air mixture is in the flammable range.

The previous discussion centered on natural gas deflagrations because natural gas is lighter than air and mixes freely with it. Thus, it is a good example of the effects of various concentrations of fuel/air mixtures. Natural gas, however, is one of a few gases (hydrogen, methane, carbon monoxide) that are lighter than air. It is important to note

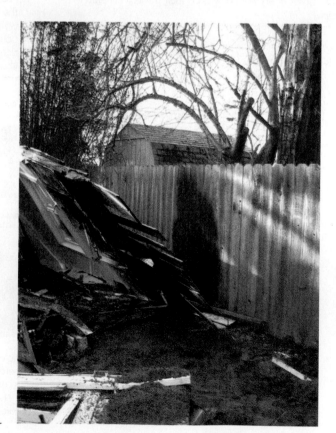

FIGURE 12-4 Flame plume from broken gas line charred the interior of the displaced wall and nearby fence after a massive natural gas explosion. *Courtesy of John D. DeHaan.*

that all vapors generated by evaporation of ignitable liquids at ordinary temperatures are *heavier* than air (at the same temperature). Pentane vapors, for example, are about 2.5 times as heavy as air. Most petroleum distillates are less volatile than pentane, and their vapors are even heavier. Fresh automotive gasoline contains up to 8 percent *n*-butane and its related isomers, as well as pentane, so its initial evaporation and ignition properties are dominated by those of butane and pentane.[4]

Solvents

Most solvents (lacquer thinners, petroleum ether, acetone, and the like) have vapors like those of gasoline that are heavier than air; their mixing properties in air must be thoroughly understood by the fire investigator. Dense vapors like these can be visualized as having fluid properties like maple syrup. When released, the vapors form a pool a few centimeters deep and then spread in a thick layer, seeping under doors, flowing down stairs, and collecting in pools in low spots of the room or structure (see Figure 4-7). These vapors are heavy enough to concentrate at the bottom of such low spots and there form a very rich mixture. The vapors then mix slowly into overlying still air by diffusion, creating a gradient of concentration from the too rich to the too lean. The combustion of such a gradient will usually progress quickly, but a rolling fire may be sustained as the rich layers are mixed by the turbulent combustion of the overlying layers. Ignition of gasoline vapors in the house shown in Figures 12-5a–e produced enough force to blow window assemblies from the building, but the following fire was limited to one room due to an excess of fuel vapors and insufficient ventilation elsewhere (producing a "too rich" mixture).

Mechanical Mixing

If sufficient mixing of gases or vapors with air occurs before ignition, there can be a massive deflagration, as shown in Figure 12-2. An idealized layer-cake condition will exist only, of course, in a space where there is no mechanical mixing of the vapor/air layers. This mixing can be caused by open windows, doors, chimneys, machinery, or even people walking about. Such mechanical mixing will complicate the interpretation of the resulting fire. Propane releases are very sensitive to thermal mixing. Solar heating onto or into a compartment can cause enough circulation in the room to mix a propane/air mixture throughout. This may serve to dilute small amounts to concentrations below the LEL or create ignitable mixtures throughout a room, despite the vapor density of pure propane.

In one case examined by one of the authors, the reconstruction of an explosion and fire aboard a fishing vessel showed that the LP gases, which had leaked from a stove and

FIGURE 12-5A A gasoline vapor explosion severely damaged this abandoned house. *Courtesy of Jamie Novak, Novak Investigations and St. Paul Fire Dept.*

FIGURE 12-5B Only the room where windows had blown out suffered fire damage. *Courtesy of Jamie Novak, Novak Investigations and St. Paul Fire Dept.*

FIGURE 12-5C Gasoline had been poured throughout the house, but there was inadequate oxygen to allow full propagation. *Courtesy of Jamie Novak, Novak Investigations and St. Paul Fire Dept.*

FIGURE 12-5D Furnace ductwork was badly damaged by over-pressure within. *Courtesy of Jamie Novak, Novak Investigations and St. Paul Fire Dept.*

FIGURE 12-5E In the basement, the mixture was so rich that raw gasoline was still detected. Gasoline had pooled around the furnace and water heater (possibly ignited by the furnace). *Courtesy of Jamie Novak, Novak Investigations and St. Paul Fire Dept.*

accumulated in the bottom of the boat, were ignited only when the boat's owner walked about in them, stirring them up. Residues of the gas were detectable in the victim's clothing (the pants of which were torn and scorched only below the knees) and in tissue samples from the lungs.

Draft Conditions and Pilot Lights

Draft conditions must be considered whenever remote points of ignition play a part in the ignition of flammable vapors. The propagation of a layer of vapor along a horizontal surface from a container or spill of flammable liquid can be predicted to occur at a particular rate [on the order of 2–5 cm/s (0.8–2 in./s) for pentane at 20°C (68°F)] in still air depending on the volatility of the liquid and the temperature of the surroundings.[5] Even small drafts caused by fans, mechanical equipment, or the pilot light of a stove or heater can cause a positive flow of vapors toward an ignition source at a much faster rate than expected. A pilot light creates small volumes of hot gas, which rise because of the reduced density of hot gases; cool air then flows into the base of the flame, giving rise to a slow but definite flow of air toward the flame. If heavy vapors are being released, even at some distance, this slow air flow will move the vapors toward the pilot and may result in ignition. As discussed in Chapter 3, the layer of vapors lies close to the floor, and if the combustion chamber is elevated, the vapors will not accumulate to an ignitable concentration. Pilot lights are particularly treacherous to the arsonist pouring flammable liquids. They not only provide an open-flame ignition source but also draw the explosive vapors toward it. Ignition of the main burner (or start-up of the fan) causes the flow to occur much more quickly.

DEFLAGRATIONS

Although fires involving the vapors of flammable liquids can fall into the general descriptive class of combustion—with rolling flame fronts and minimal pressures produced—depending on circumstances of fuel, mixing, concentration, and confinement, they can readily achieve the next, more violent class of oxidation, deflagration. For a pressure effect to be produced, however, the deflagrating gas or vapor/air mixture must be confined to some extent. This confinement normally takes the form of a rigid container or even a room or structure. If very large quantities of gaseous materials are involved, sufficient confinement may be produced by the inertia of the gas cloud itself. Although small quantities of a vapor/air mixture may deflagrate when confined in a rigid container, very large masses of the mixture must be present to provide a confinement effect in open air to produce what is sometimes called an *unconfined vapor cloud explosion* (UVCE). This open-air confinement effect is noticeable only in large clouds of airborne vapor above very large spills of volatile liquids. This is a more common occurrence in refinery or industrial premises where a large quantity of volatile liquid fuel is released quickly (often at high temperature and pressure). This quickly produces a very large vapor cloud that can then be ignited by any appropriate source in contact with it. This was the mechanism, for instance, for the huge UVCE at Flixborough, UK (1974) when a large quantity of cyclohexane was released from a failed pipe joint. The explosion destroyed the large plant and killed 28 employees.[6]

One of the means of discriminating a deflagration from other violent forms of combustion is based on the time required for it to occur. A deflagration may take from several milliseconds to as much as a second to develop.[7] As we shall see, the time scales for detonations are much, much shorter. Similarly, the peak pressures developed by deflagrations are many times lower than those developed by detonations. Naturally, these lower pressures and slower rates of pressure rise have a different effect on their surroundings than those from detonations. We shall treat these differences later on when the diagnostic features of both mechanisms are discussed.

Gaseous fuels are susceptible to transitioning from combustion to deflagration to detonation. This is especially true if they are heated and pressurized in contact with enough oxygen to support combustion. Heat and pressurization extend the explosive range, so what is not an explosive mixture at room temperature may be explosive under the conditions of actual use. As gases expand during combustion there is often a "pressure-piling" action that raises pressures and temperatures ahead of the actual reaction front. The geometry of the confining vessel is also critical to deflagration/detonation transitions. If the mixture is confined in a pipe (or even long corridor) whose length (l) is significantly greater than its width or diameter (D) (typically $l/D > 100$; for some gases, it is as low as $l/D > 10$), the flame front stretches out due to frictional drag at its sides. This increases the surface area of the flame front and increases the combustion rate. When this acceleration reaches a certain point, the reaction transitions to a detonation.[8] The detonation velocities of some gas mixtures at room temperature and atmospheric pressure are shown in Table 12-1. Note that all of them are above the 1,000 m/s (3,280 ft/s) threshold described previously, even though their densities are close to that of air.

Interestingly, the detonation limits of such mixtures are not the same as the flammability/explosive limits discussed earlier, and the maximum velocities are not necessarily achieved at a "central" or ideal concentration. Detonation velocities ranging from 1,500 m/s to 2,000 m/s (between 4.2 and 50 percent) (4,920 to 6,560 ft/s) for acetylene in air were

TABLE 12-1	Detonation Velocities of Gas/Vapor Mixtures
GAS/VAPOR MIXTURE	**DETONATION VELOCITY**
Acetylene/air	2,414 m/s (7,920 ft/s)
Acetylene/oxygen	2,716 m/s (8,911 ft/s)
Hydrogen/oxygen	2,821 m/s (9,255 ft/s)
Methane/oxygen	2,146 m/s (7,041 ft/s)
Methane/air	1,880 m/s (6,168 ft/s)
Pentane/air	1,680 m/s (5,512 ft/s)

Pure acetylene (98%) at 5 kg/cm^2 pressure had a reported detonation velocity of 1,050 m/s (3,445 ft/s)

Source: B. Lewis and G. Von Elbe, *Combustion, Flames and Explosions of Gases,* 2nd ed. (New York: Academic Press, 1961), 534.

measured by Breton, depending on concentration.[9] Pure (98 percent) acetylene under pressure (5 kg/cm^2) detonates at 1,050 m/s (3,445 ft/s). Hydrogen detonation limits in air are reported to be 18.3 to 59 percent; carbon monoxide and hydrogen in air: 19 to 59 percent; and acetylene/air: 4.2 to 50 percent.[10]

The strength of the initiating source also can play a role in whether the reaction will be a deflagration or detonation. The stronger the source, the more likely a detonation will take place in a susceptible explosive or mixture.

Backdraft

backdraft ▪ A deflagrative explosion of gases and smoke from an established fire that has depleted the oxygen content of a room or structure, most often initiated by introducing oxygen through ventilation or structure failure.

One special case of deflagrations that many investigators do not always consider is that of **backdraft** or smoke explosions. The gases, vapors, and solids produced by a smoky fire constitute a vapor-state fuel. If supplied with enough oxygen and a source of ignition, accumulations of smoke can deflagrate with sufficient energy to break windows, move light walls, and injure firefighters, producing pressures of 0.03 to 0.1 bar (0.5 to 1.5 psi).[11] The mechanisms responsible for smoke explosions are better understood today thanks to recent research.[12] See Figures 12-6a and b for an illustration of the explosive potential of

(a)

(b)

FIGURES 12-6A AND B Front windows and side wall of garage blown out by smoke explosion observed by arriving firefighters. Fire ignited against wood siding outside front door. *Courtesy of Jamie Novak, Novak Investigations and St. Paul Fire Dept.*

a backdraft. Depending on the combustion rate of the accumulated smoke, there may not be sufficient overpressure to inflict physical damage. The expansion of the burning smoke layer, however, may fill portions of the building that were clear moments before and force smoke out door and window openings.

Dust Suspensions

The special case of a solid fuel in a diffuse-phase explosion must be mentioned—that is, the suspension of dust or powdered fuel in air. Coal dust in mines, flour dust in granaries, metallic dust in machine shops, sawdust in lumber mills, or any fuel in powdered form can be dispersed in a cloud to form an explosive fuel/air mixture. Although we do not normally think of a solid fuel deflagrating in a diffuse-phase mixture, it can happen. As in gas- or vapor-based mixtures, there are ranges of concentration (typically measured in grams of fuel per cubic meter of air) for each fuel that will support combustion. The behavior of the ignited solid/air mixture parallels that described for rich, lean, and ideal mixtures of gases and vapors. Unlike gases, however, there are restrictions on the particle size for each fuel that will sustain combustion. Generally, the finer the dust, the more likely it is to both (1) remain airborne in suitable concentrations in air and (2) combust readily to contribute to a self-sustaining reaction. Organic dusts such as cornstarch, soy flour, sugar and wheat starch are capable of creating maximum pressures of 8.5–9.9 bar (850–990 kPa or 123–144 psi) under ideal conditions (fine particles, suitable concentrations, strong ignition source).[13] One major difference between gas/air deflagrations and dust explosions is that accumulations of dust on surfaces are agitated by the initial deflagration, and this additional fuel will propagate the deflagration much longer, often gaining considerable energy in later stages.[14] More energetic ignition sources are frequently required for dust suspensions than for gas/air mixtures.

It is thought that airborne dusts of pyrotechnic mixtures such as aluminum powder and potassium perchlorate flash powders are responsible for the massive destruction produced when illicit "fireworks" factories explode.

Turbulence caused by the initial combustion of fireworks can aerate such mixtures, allowing almost instantaneous combustion throughout (because flame will then come into contact with the fuel/oxidizer particles throughout the dust cloud). Such unstable materials are capable (like acetylene and ethylene oxide) of transition to a detonation even when not airborne dusts.

IGNITION

The hazard of open flames with regard to flammable fuel/air mixtures has been touched on, but naturally other sources of heat can initiate such mixtures. It is important to remember that no matter what other concentrations of vapors are present in the area, the fuel mixture *must* be present at combustible concentrations in the immediate area of the heat source to be ignited. With a fire, continued heating of a pyrolyzable fuel

 CHEMICAL SAFETY BOARD (CSB)

The U.S. Chemical Safety and Hazard Investigation Board (called the CSB) is a nonregulatory, independent federal agency charged with investigating industrial chemical accidents. Headquartered in Washington, DC, the agency's board members are appointed by the president and confirmed by the Senate.

The CSB conducts origin and cause investigations of chemical accidents, such as dust, flammable vapor, gas, fire, and chemical explosions at fixed industrial facilities. The CSB does not issue fines or citations but makes recommendations as to prevention measures to plants, regulatory agencies such as the Occupational Safety and Health Administration (OSHA), the Environmental Protection Agency (EPA), industry organizations, and labor groups.[15]

ILLICIT EXPLOSIVE DEVICES

The term *fireworks* is a serious misnomer when referring to "M-80" type devices (as shown in Figure 12-7). With their explosive filler—1 to 20 g of highly reactive flash powder—and cardboard containers, they constitute a significant risk from thermal and blast effects. The solid end plugs produce missile and fragmentation effects capable of causing injury and even death. They are dangerous and are controlled as illicit explosive devices under federal law in the United States.

FIGURE 12-7 Illicit explosive devices such as these "M-80s" are commonly available. *Courtesy of John D. DeHaan.*

will create the conditions for the fire but not for the explosion. Any hot spot or region, regardless of its size, can initiate an explosion of vapors if it meets the following criteria:

- There has to be contact between the ignition source and the fuel.
- At some point, the temperature of the source is well above the autoignition temperature of the flammable material involved (i.e., it has enough energy).
- At the ignition point (in time and space), the fuel/air mixture is within its explosive range.
- These conditions of contact last long enough for enough heat to be transferred to the fuel to trigger a self-sustaining combustion.

These criteria apply to all flames; electric arcs, including those created by static electric discharge; and hot solids such as heated metal and incandescent sparks. Interestingly, smoldering combustion, as on the tip of a lighted cigarette, is generally incapable of initiating an explosion in a fuel vapor/air mixture, as discussed in Chapter 6.

Unlike an open flame that produces a continuous source of high temperature, an electric arc may be only of short duration. Either the vapors in contact with the arc are in their explosive range and explode, or they are not, in which case nothing happens as a result of the arc. A similar effect is produced when a small charge of high explosive is used to disperse gasoline from a plastic or glass container. The high explosive generates gases at very high temperatures (thousands of degrees), but they are of short duration, and the explosion pushes the gasoline (fuel) and oxygen mixture away from the ignition source. More often than not, no ignition occurs. If a metal container is used for the gasoline, however, the metal fragments (shrapnel) are superheated by the initiating blast and frictional tearing and will often act as point sources of ignition as they retain their heat longer than do the gases of the reaction.

Condensed-Phase Explosions

Condensed-phase explosions involve solid (or sometimes liquid or gel) explosives in which either the fuel and oxidizer are mechanically mixed or the monomolecular material itself is capable of undergoing extremely rapid conversion. Explosions of higher energy than the deflagrations described previously usually involve a condensed-phase or dense-phase chemical, because detonation of a diffuse-phase system is very rare. (Such detonations may occur in high-pressure or high-temperature gas systems or in systems involving especially sensitive compounds such as acetylene or ethylene oxide.)

Oxidation reactions that fall into the detonation category differ greatly from combustion and deflagration reactions in a number of ways, as graphically represented in Figure 12-8. There are major differences, of course, in the reaction time and in the reaction volume of the explosive reaction.

Compared with the milliseconds or seconds required for a deflagration, detonations occur throughout the charge within microseconds. As will become apparent, the forces generated and their effects on the surroundings will be different more because of this time relation than because of the total force generated. The pre-blast reaction volume of a stick of dynamite in a room is the small volume occupied by that stick, virtually a point source from which all the explosive forces are generated as the post-blast gases occupy some 1,500 times the pre-blast volume. If the same room is filled with an explosive mixture of vapors and air, the explosive volume is that of the entire room, which, upon combustion, tries to occupy up to 8 times its original volume. In physical terms, this is the chief distinction between diffuse- and condensed-phase explosions. (In chemical terms, of course, there are many other differences, discussed later.) Even though the total forces produced (i.e., total work done) may be similar, the specific effects of the two types of explosions will be quite different. While the energy released by the complete deflagration of a quantity of gasoline can be compared to that released by a detonation quantity of high explosives, the damage potential of the two events is vastly different due to the time frames and dynamics involved. The energy released by TNT detonating is on the order of 4.56 MJ/kg, while the heat of combustion of methane is 55 MJ/kg and that of gasoline is 46 MJ/kg.[16]

Most condensed-phase high explosives require an initiating mechanical shock that triggers a cascade effect of reactions throughout the reaction volume. It is thought that

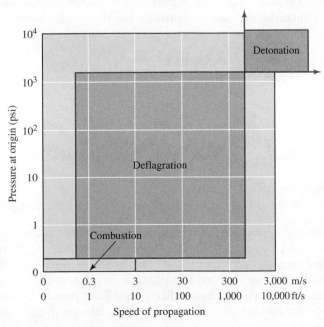

FIGURE 12-8 Phase diagram showing ranges of pressure and speed of propagation for general states of explosive combustion, deflagration, and detonation.

materials that detonate have chemical bonds sensitive to energy applied in the form of pressure and are disrupted by the transmittal of pressure or shock throughout the mass of the explosive charge.

The characteristic feature of detonation is the shock wave that propagates the reaction moving through the explosive at a speed greater than the speed of sound in that matrix. Unlike the high-speed oxidations of gaseous explosive mixtures, detonations of a condensed-phase explosive do not require confinement except when small quantities of material are involved. Many sensitive materials will still detonate when unconfined, but not very efficiently. The pressures produced will be low, and a considerable amount of the material will remain undetonated in what is called a *low-yield* or *nonideal* explosion. Some materials will detonate when heated to a critical temperature (which is often pressure dependent).

All explosives have a critical diameter, however. If a mass of explosive is too thin or too small in cross section there will be no propagation. Such critical diameters are taken into account when commercial preparations such as dynamite cartridges, grenades, or high-explosive boosters are being designed but are ignored in many improvised criminal devices.

CHEMICAL AND PHYSICAL PROPERTIES

As you might expect, materials that explode or detonate have a different chemistry from those that require the addition of an external source of oxygen to enable the reaction. The molecules of materials that can detonate have the capability of shattering into fragments when properly stimulated. As those molecular fragments recombine, they release large quantities of heat in a fraction of the time it takes for combustion to take place.

Nitrogen compounds are probably the most common types of chemicals used in the formulation of explosive materials. This is because many nitrogen compounds are basically unstable from the standpoint of their compositional (potential) chemical energy. Nitrogen is an element that resists chemical combination with other elements. This means that the compounds of nitrogen tend suddenly to rearrange themselves to yield elemental nitrogen. In doing so, they release their potential chemical energy as heat and light in an explosively sudden manner. Oxygen is often combined with nitrogen in various carbon-containing organic compounds in common explosives. In the explosion, nitrogen is released from its chemical structure along with a great deal of energy, and the oxygen combines with carbon compounds during the reaction to produce even more energy.

Materials that undergo explosive decomposition almost always contain one or more characteristic chemical structures that are responsible for the sudden releases of great energy. These structures include the following groups:

- —NO_2 and —NO_3 in organic and inorganic nitrates
- —N_3 in inorganic and organic azides
- —NX_2, where X is a halogen
- —$ON\equiv C$ in fulminates
- —$OClO_2$ and —$OClO_3$ in inorganic and organic chlorates and perchlorates, respectively
- —O—O— and —O—O—O— in inorganic and organic peroxides and ozonides, respectively
- —$C\equiv C$— in acetylene and metal acetylides
- [M—C], a metal atom bonded with carbon in some organometallic compounds[17]

In addition to compounds that contain the preceding groupings, there is a wide variety of mixtures containing both fuel and oxidizer that are subject to deflagration. These produce generally low-energy deflagrating explosions but can produce severe damage when encountered in improvised explosive devices. For instance, black powder is a mixture of carbon, sulfur, and potassium nitrate (as the oxidizer), and flash powder is commonly a mixture of aluminum powder and sulfur (as fuels) and a perchlorate as the oxidizer.

In addition to their similar chemistries, materials that explode or detonate have several performance properties in common—they all produce large amounts of heat and when properly initiated convert nearly completely to gaseous products (with the rare exception of the metal acetylides that produce carbon and metal and no gaseous products). These gases (and atmospheric gases nearby) in turn expand rapidly because of the heat generated. Most explosive materials are sensitive to heat in some degree. Raising the temperature will cause explosion or detonation, but unlike combustible vapors, explosive materials, due to their chemistry, do not have precise ignition temperatures.[18] The ignition temperature measured depends on the rate of rise of the temperature. The faster the temperature rises, the lower the autoignition temperature appears to be.

Types and Characteristics of Explosives

In addition to being sensitive to heat, most explosives are also sensitive, to a greater or lesser degree, to initiation by mechanical shock. Thus, the sensitivity of an explosive material to both such effects is of interest to the fire and safety investigator. Based on the sensitivities and oxidation processes of these materials, explosives are often classified as propellants or **low explosives** or as **high explosives**. Explosions may also be characterized as high order or low order, as will be discussed.

Common explosives and their compositions are summarized in Table 12-2. A full treatment of the formulation, use, and hazards of these high explosives is beyond the scope of this book. On the rare occasions when the fire investigator encounters such materials, several texts in this area can be recommended.[19]

low explosive ▪ Any material designed to function by deflagration.

high explosive ▪ Any material designed to function by, and capable of, detonation.

PROPELLANTS OR LOW EXPLOSIVES

Propellants or low explosives are combustible materials that carry both fuel and the necessary oxygen for their combustion. Although both single-base smokeless powder (nitrocellulose) and black powder fall into this category, their reactions to heat and confinement are somewhat different. Nitrocellulose propellants ignite at approximately 190°C (374°F) and combust when burned in small quantities at normal atmospheric pressures.[20] This oxidation rate is pressure (confinement) dependent. Modest confining pressures elevate it to a deflagration, which becomes extremely fast at the pressure developed inside a cartridge when confined by a bullet. For example, the burning rate of Bullseye smokeless powder at normal atmospheric pressure is 0.23 mm/s (0.009 in./s), at 34.5 bar (3.45 MPa; 500 psi) its burning rate is 6 mm/s (0.24 in./s), while at 2,410 bar (241 MPa; 35,000 psi) (a typical breech pressure inside a weapon), its burning rate is 300 mm/s (11.8 in./s).[21] Double-base smokeless powders contain nitroglycerin in addition to nitrocellulose and can be sensitive to sympathetic (pressure) initiation and detonation if nitroglycerin is present in high enough concentration. They do not detonate under most conditions. Black powder, on the other hand, deflagrates even when burned in small quantities at atmospheric pressure and can produce high-intensity blast effects when confined by only a moderate pressure (such as the inertia of the surrounding charge itself). Black powder has a burning rate of 17 mm/s (0.67 in./s) at atmospheric pressure, 30 mm/s (1.2 in./s) at 34.5 bar (3.45 MPa; 500 psi), and only 62 mm/s (2.44 in./s) at 2,410 bar (241 MPa; 35,000 psi).[22] It is not subject to sympathetic (pressure) initiation and does not detonate at any time.

Both smokeless powder and black powder function as small-arms propellants when they are ignited as thin flakes or small granules (as illustrated in Figure 12-9). The finer the size, the faster they will burn to completion. Deflagrations can produce pressures of up to 2,760 bar (276 MPa; 40,000 psi) (solid explosives) with reaction velocities of up to 1,000 m/s (3,280 ft/s) (depending on chemistry, confinement, and manner of initiation). Depending on their means of ignition, geometry, and confinement, some materials, like double-base smokeless powder and potassium perchlorate/aluminum ($KClO_4$/Al) flash powders, can detonate. Under some conditions, then, they can be considered to be high explosives.

TABLE 12-2   Common Explosives

TYPE	COMPOSITION	CHARACTERISTICS
High Explosives *Primary/Main Charge* Straight dynamite	Nitroglycerin (NG) Ethylene glycol dinitrate (EGDN) Sodium nitrate Wood pulp	Water resistant High strength
Gelatin dynamite	NG/EGDN Nitrocellulose (gelling agent) Wood pulp Sulfur (in some cases)	Water resistant High strength
Ammonia dynamite	NG/EGDN Ammonium nitrate (NH_4NO_3) Sodium nitrate ($NaNO_3$) Wood pulp Sulfur	Less expensive Reduced water resistance
Nitrostarch dynamite	Nitrostarch Aluminum flakes	Similar to nitrocellulose No NG headaches Does not separate or decompose
Military dynamite	RDX Trinitrotoluene (TNT) Cornstarch Oil	Insensitive to shock, friction Stable in storage
TNT	Trinitrotoluene	Military uses High brisance
Initiators PETN	Pentaerythritol tetranitrate	High brisance Also used in detonating cord and Semtex
RDX	Cyclotrimethylenetrinitramine	High brisance Also used in detonating cord and Semtex
Blasting Agents Slurries	Used in bulk or prepackaged form Ammonium nitrate or sodium nitrate, often with fuels such as carbon or aluminum	Main charge only Require booster Can detonate
Water gels	Ammonium nitrate Sodium nitrate Sensitized with organic nitrates (e.g., monomethylamine nitrate) In emulsion form	Require booster Can detonate
Low Explosives	Commonly used in improvised explosive devices. Also used in explosive hardware and tool applications	Deflagrate, but can produce detonation-like effects if adequately confined
Black powder	Potassium nitrate Sulfur Charcoal	Easily ignitable Filler in safety fuse Firearm propellant
Single-base smokeless powder	Nitrocellulose	Firearm propellant
Double-base smokeless powder	Nitrocellulose Nitroglycerin	Firearm propellant

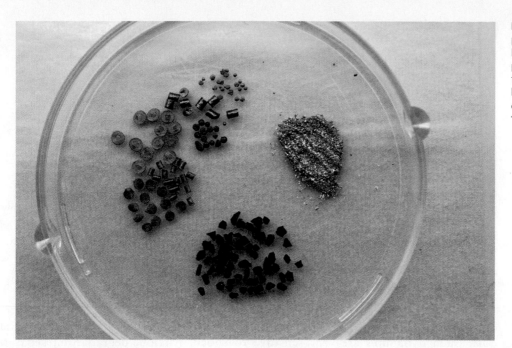

FIGURE 12-9 Common pipe bomb fillers. Upper left: various smokeless powders. Upper right: flash powder. Bottom: black powder. *Courtesy of Wayne Moorehead.*

HIGH EXPLOSIVES

High explosive materials are intended to function by detonation and will detonate when properly initiated by heat or shock. It is thought that the localized heating produced by mechanical shock is sufficient to rupture some of their chemical bonds and initiate a chain reaction in which heat from one bond disruption is quickly transmitted to surrounding bonds in a cascade, which results in the explosion of the entire mass within a few microseconds. As shown in Table 12-3, high explosive detonations have velocities of 1,000 to 9,000 m/s (3,280 to 29,500 ft/s) with maximum pressures (at the surface of the explosive charge) of more than 68.9 bar (6,890 MPa; 1,000,000 psi). Such localized high pressures result in a "seat" of most intense damage—often shattering or pulverizing nearby surfaces, as shown in Figure 12-10.

Mohanty reported the densities and detonation velocities of common high explosives to be as listed in Table 12-3.

Properties of Straight Dynamite

At one time, probably the most common high explosive was ordinary nitroglycerin or *straight* dynamite. Straight dynamite is compounded with nitroglycerin (NG) and

TABLE 12-3	Densities and Detonation Velocities	
EXPLOSIVE	**DENSITY (g/cm³)**	**DETONATION VELOCITY**
ANFO	0.84	4,600 m/s (15,090 ft/s)
Emulsion	1.20	5,200 m/s (17,060 ft/s)
Dynamite	1.45	5,400 m/s (17,720 ft/s)
TNT	1.64	6,930 m/s (22,740 ft/s)
PETN	1.67	8,000 m/s (26,250 ft/s)
RDX	1.77	8,700 m/s (28,540 ft/s)

Source: A. Beveridge, ed., *Forensic Investigation of Explosions* (London: Taylor & Francis, 1998), 22.

ethylene glycol dinitrate (EGDN) soaked into a solid absorbent or filler such as sawdust. Such dynamites are very sensitive to shock, friction, degradation, separation (weeping), and thermal ignition and are very rarely used today except for "ditching" applications, where water resistance and sympathetic initiation are desirable features. The strength of a straight dynamite is based on the actual percentage of NG present (by weight).

Properties of Gelatin Dynamite

The next form of dynamite is *gelatin* dynamite, which uses a combination of NG dissolved in nitrocellulose. These formulations are water resistant and very energetic but are expensive to make and prone to separation, "weeping" liquid nitroglycerin. These expensive components (NG and nitrocellulose) are sometimes replaced (all or in part) with ammonium nitrate or sodium nitrate. Such combinations are still effective high explosives but are sensitive to water, which can render them unusable. When ammonium nitrate is used, the product is called an *ammonia* or *semigelatin* dynamite. These ammonia dynamites are rated in strength *equivalent* to a particular percentage of NG (i.e., 40 percent ammonia dynamite does not contain 40 percent NG but contains the energy equal to 40 percent–strength straight dynamite).

Alternative High Explosives

Due to the expense and danger of making and storing nitroglycerin dynamites, they have been largely replaced by water-gel dynamites that use organic nitrates as sensitizing agents and ammonium nitrate as the main explosive ingredient. Today, nitroglycerin dynamites are made only in small quantities for special applications. Some dynamites are made with nitrostarch as the explosive agent. This dynamite contains no nitroglycerin, and the attendant risks are absent. The water-gel explosives are based on formulations of ammonium nitrate and organic nitrates such as monomethylamine nitrate or monoethanolamine in a water-based gel or guar gum or similar agent. There are also slurry-type blasting agents that contain ammonium or sodium nitrates and aluminum or coal tar powders as fuel. Such materials are more stable than nitroglycerin but less sensitive, requiring a substantial detonation to trigger them. These explosives degrade to inert compounds rather than to more dangerous forms (e.g., straight dynamite "weeping" nitroglycerin). There are limitations, however. Due to their physical form as an emulsion, suspension, or slurry containing water, they are less sensitive at cold temperatures, are rendered unusable by freezing, and cannot be used in wet applications.

HIGH EXPLOSIVE CATEGORIES

High explosives may be further categorized as *primary* or *initiating* explosives and *main charge* explosives. Initiators like mercury fulminate, lead azide, picric acid salts, and others are highly sensitive to mechanical shock, friction, heat, and electricity and are commonly used in blasting caps and primers to initiate a larger explosion. Main charge explosives include dynamite, TNT, PETN, RDX, and the like. When dispersed, main charge explosives often simply burn, but in any mass at all (more than a few grams) these materials can detonate. Their actual performance may depend on the size and shape of the charge and the manner of initiation. High explosives are rated by factors related to their intended application. Generally, commercial explosives are used in "pushing" or "heaving" applications, while military explosives are usually needed for their "shattering" power. The effectiveness of such explosives is rated by their **brisance** or shattering power. Commercial high explosives are formulated in dilutions or mixtures of components to control not only the energy released but this shatter strength or brisance, because different applications require different explosive properties.

brisance ■ The amount of shattering effect that can be produced by a high explosive.

Blasting Agents

Blasting agents are explosive materials that generally require a high-explosive booster to initiate a detonation. Ammonium nitrate/fuel oil (ANFO) and its variants are the most common blasting agents. They are often sold in bulk rather than as individual units. See Appendix G for a list of the materials considered by the U.S. Treasury Department; U.S. Department of Justice; and the Bureau of Alcohol, Tobacco, Firearms and Explosives to be explosives and blasting agents under federal law.

Binaries

There are a small number of high explosives called *binaries* made up of two components that by themselves are not explosives. When mixed together binaries form a cap-sensitive high explosive. Mixtures of nitromethane and ammonium nitrate or nitromethane and organic amines have been marketed as binary explosives.[23] Tannerite, a filler for exploding firearms targets, is sold as a binary today. It consists of ammonium nitrate and ammonium perchlorate mixture as one component and fine aluminum powder as the second. Mixed together, they constitute a powerful, impact-sensitive high explosive.

Improvised Explosives

A number of high explosives can be synthesized by fairly simple chemical processes to be used in improvised explosive devices (IEDs). These include nitrogen triiodide (NI_3), triacetone triperoxide (TATP), hexamethylene triperoxide diamine (HMTD), hydroxyl ammonium nitrate (HAN), ammonium dinitro amide (ADN), and metallic azides, fulminates, and acetylides. These are powerful and sensitive primary explosives (initiated by heat, friction, shock, or even static electricity) that would leave minimal detectable residues in post-blast debris if properly made and functioned (high order).[24] Other high explosive blasting agents that have been easily synthesized and used in criminal bombings include ammonium nitrate/sugar and urea nitrate. These mixtures were estimated to have detonation velocities of 3,730 m/s for urea nitrate and 3,560 m/s for ammonium nitrate/sugar.[25] Silver acetylide [$Ag-C \equiv C-Ag$] is an example of a rare category of explosive that produces only solid residues (carbon and metallic silver) and no gas during its detonation. The enormous heat produced, however, heats the surrounding air to several thousand degrees, and the resulting expansion of gases exerts explosive force. Despite its unstable and sensitive nature, TATP has become the most widely used improvised explosive. It can be formulated using acetone, hydrogen peroxide, and acid from easily obtainable consumer products (e.g., nail polish remover, hair bleaches, and batteries). The processes are simple and can be carried out with a few jars, filters, and ice.

COMPONENTS

High explosives usually require the use of initiators, time fuse, and boosters.

Blasting Caps

Since high-explosive main charges often require a detonation to initiate them, a small detonation source can be readily contained in a blasting cap, as shown in Figure 12-11. It consists of an igniter compound, time delay, primary explosive and, typically, a high explosive. An electric blasting cap has two leg wires and a bridge wire embedded in the ignitor. Applying electric current causes the bridge wire to overheat quickly, igniting the deflagrating material. A delay is often built in to increase the time between application of current and initiation of the primary explosive (typically from 10 to 200 milliseconds). The primary charge detonates, triggering the high explosive filler, which in turn, provides the shock needed to cause the main charge to detonate. All this is contained within a metal (often aluminum) sleeve or tube, about 5 mm (3/16 in.) in diameter and 40–65 mm (1-1/2–3 in.) in length. Nonelectric blasting caps have a hollow end that is crimped over the open end of a traditional safety fuse. There are today a variety of "shock tube" caps that can be triggered only by the detonation of a fine coating of high explosive on the inside of a small-diameter plastic tube. Such caps are safer to use than traditional caps, since they cannot be triggered by stray electric currents or flame. The propagation of the detonating coating is confined to the interior of the plastic tube.

Boosters

Boosters are small, specially formed high explosive charges (which may be "ice cube," finger-sized sleeves, or cylindrical in shape) made of TNT, PETN, or RDX. They are used as intermediates between blasting caps and main charge explosives. High explosive detonating cord is sometimes threaded through boosters or even used as substitute booster by knotting it around or threading it through a main charge high explosive. Detonating cord is filled with PETN or RDX high explosive and itself requires a blasting cap to initiate it.

Time Fuse

A variety of fuse types may be encountered by fire and explosion investigators. A common type is pyrotechnic or hobby fuse, in which a core of pyrotechnic powder is wrapped with windings of fine string and coated with nitrocellulose lacquer. When ignited, it burns with a visible exposed flame. Although intended to burn at rates of 12–25 mm/s (1/2–1 in./s), it can burn faster when crimped or crushed.

Safety fuse consists of a core of finely ground black powder protected by wrappings of string, tar, and plastic. Such a fuse is ignited at one end, and a small spurt of flame is emitted at the other end. There is no exposed flame anywhere along its length. It burns at 80–100 cm/min (30–40 in./min), and due to its waterproofing layers, it can burn underwater.

DuPont No. 6 Commercial Nonelectric Blasting Cap

FIGURE 12-11 Typical blasting caps. Top: non-electric. Bottom: electric

DuPont No. 6 Commercial Electric Blasting Cap

Igniter cord consists of a perchlorate-based incendiary filler with a multiple spiral wrap of fine nickel-chrome-iron wire. It burns very quickly and is rarely seen in improvised devices.

Improvised fuses can include string or paper impregnated with black powder or pyrotechnic powders. Fuse materials removed from commercial fireworks such as quick match (black powder core wrapped in layers of paper) are also encountered.

Firing Trains

The combination of initial heat source, fuse, blasting cap, booster, and main charge is often called the *firing train* or "fuzing" system. High explosives generally require a small detonation to initiate them but will not be initiated by a simple flame or electric arc. Time delays, blasting caps, and boosters may not be present, depending on the explosive involved and the intended use. It is often useful for the investigator to consider what evidence is present (or should be sought) to fulfill the necessary roles: ignition source, delay, ignition, detonation. The accidental explosion of a munitions ship in December 1917 in the harbor of Halifax (Nova Scotia) is a good example of an accident with a complete firing train. Collision with another ship caused the release and ignition of drums of benzene (as deck cargo). The fire triggered the detonation of picric acid explosive, which in turn caused the detonation of the main cargo of TNT. A total of 2,925 tons of high explosive disintegrated the ship and flattened the center of Halifax, causing some 2,000 deaths and injuring 9,000.[26]

HIGH-ORDER/LOW-ORDER EXPLOSIONS

When one is describing the mechanism and effectiveness of an explosive process, there is considerable confusion regarding the terms *high order* and *low order*. Only high explosives can achieve *detonation*, the term referring to the propagation of a supersonic reaction wave within the explosive charge. This shock wave actually causes the rupture of the chemical bonds holding the explosive together. The entire mass of explosive converts to gas and heat at supersonic speeds, causing extremely high temperatures and pressures as well as shock, all of which bring about the shattering effect commonly associated with such explosives. All other explosive mechanisms are deflagrations (by our definition), since they propagate at subsonic speeds. The energy and heat produced in the initial charge are transferred within the charge at lower speeds (often by particle-to-particle transfer of thermal energy), and combustion of the individual fuel elements then proceeds at the surface of each particle. Heat and gases are produced at a slower rate, and lower pressures are produced. The external effects are less shattering than with a detonation, but they can be substantial in terms of the total kinetic energy produced.

Confusion often arises when someone concludes that a high explosive used means that a high-order (or high-yield) explosion took place. High explosives need not initiate and propagate perfectly each time. This is particularly true with improvised explosive devices. Commercial explosive charges and military munitions are carefully designed to convert as completely to kinetic energy as possible so that the explosive effects are predictable. Such complete conversion is referred to as *ideal*, *high-yield*, or *high-order detonation*. Improvised devices contain shapes and sizes of charges that cannot detonate completely, even when high explosives are used. (Explosives have characteristic minimum diameters below which detonation is incomplete.) If the initiating charge or device is inadequate in size (energy) or poorly placed in the main charge, detonation may not take place at all. For this reason, even when traditional high explosives (TNT, nitroglycerine, PETN, dynamite, RDX) are used, they may not detonate completely or at all. This is referred to as *low-yield*, *nonideal*, or *low-order detonation*. Premixed gas or vapor fuel/air mixtures at their ideal or stoichiometric mixture can be said to sometimes propagate high order because they function at maximum efficiency with no residual starting material.

Low explosives (black powder, nitrocellulose) do not detonate even when confined and initiated under ideal circumstances and, because of the inefficiency or ineffectiveness of the thermal processes involved, do not explode "high order." Double-base smokeless powder, because it contains both a low explosive (nitrocellulose) and a high explosive (nitroglycerin), can act as either low or high explosive, depending on its initiation and confinement. When a powder such as Alliant Red Dot or Bullseye with a high nitroglycerin content is used and is initiated with a suitable blasting cap, it can detonate, especially if an improvised container of high bursting strength (such as a gas cylinder) is used, but will rarely reach ideal or high-order conditions. In most improvised devices, the containment is weak or the initiation is inadequate, and the resulting blast, while terribly destructive, is not ideal (high order) and there will be intact, unreacted explosive scattered about. So when asked, "Did such-and-such a material explode high order?" the correct answer is usually, "It depends." That determination often depends on careful assessment of thermal and mechanical effects (indicators) at both macro and microscopic scale.

Mechanical Explosions

In a mechanical explosion, a container, vessel, or pipe bursts when internal gas or liquid pressures exceed the tensile strength of the container. The resulting catastrophic release of high-pressure gas meets our general definition of explosion and can produce blast effects (pressure, shock, and fragmentation), which can damage or destroy vehicles and buildings. Because an exothermic chemical reaction is not involved, heat is not released (except in the tearing or shattering of metals). If liquid is released at temperatures or pressures far above ambient conditions, it vaporizes nearly instantaneously, and the rapidly expanding vapors exert their destructive forces on nearby surfaces. An example of a massive mechanical explosion involving steam is shown in Figure 12-12. Such mechanical explosions may occur by themselves or before, during, or after fires. When associated with a fire, they may be the cause of the fire, or simply add to its destruction. Once the general nature of the explosion has been deduced, the search for pieces of the explosive device or the accidental mechanism responsible, or for residues of the chemicals involved (if any), can be undertaken more efficiently.

Mechanical explosions can produce blast effects that may be difficult to distinguish from those of chemical explosions. In either case, fluids or gases at very high pressures

FIGURE 12-12 A massive Northern-class steam locomotive lost its boiler assembly due to a mechanical (steam) explosion, probably caused by flooding cold water into a "dry" hot boiler. The remains of the boiler are on the far side of the track. *Courtesy of John D. DeHaan.*

strike surfaces, distorting and sometimes shattering them to produce energetic secondary missiles, or shrapnel. Cylinders of compressed gases, aerosol cans, storage tanks, hydraulic lines, pipelines, and similar systems can be overpressurized due to mechanical failures of compressors, relief valves, or regulators. Such containers may also be exposed to heat with a resulting increase in internal pressure. When the pressures developed within the line or vessel exceed its bursting strength, it will fail, and the pressurized fluid will be released, potentially with catastrophic results. The escaping gases inflict shock or pressure effects on nearby surfaces and may propel resulting fragments at high velocities. Mechanical explosions can sometimes produce a localized seat of damage, similar to that produced by condensed-phase explosives. Note that they are usually gaseous phenomena, but that changes in chemical or physical state are not involved, and that heat is not produced (except heat created by the rapid tearing of a metal container). The absence of fire or other thermal damage is often a reliable indicator that a mechanical explosion occurred. When external heating has been involved, it will be seen that heated metals have lower tensile strengths than do those same metals at ordinary temperatures. An externally heated vessel, then, is most likely to fail where it is heated. See Figure 12-13 for an example of a mechanical explosion involving a medical oxygen cylinder that failed (where it was exposed to flames) with tremendous concussive force, and see Figure 12-14 for an exploded fire extinguisher. Propane cylinders often fail with a typical "fish-mouth" hole where exposed to fire that produces little explosive effect. The location and orientation of that failure will direct the "flow" of released gases, sometimes causing the container itself to become a projectile.

ACID, GAS, OR BOTTLE BOMBS

A variety of improvised explosive devices function in general by creating a gas, usually non-flammable, by chemical reaction that causes the mechanical explosion of a confining

FIGURE 12-13 Medical oxygen tank (aluminum alloy) failed where it was heated in an enveloping fire. *Courtesy of Detective Mark Allen, Tempe Police Department, Tempe, AZ.*

FIGURE 12-14 Fire extinguisher casing after mechanical explosion caused by exposure to sustained fire.
Courtesy of Vic Massenkoff, Contra Costa County Fire Protection District.

vessel, most commonly a plastic soft-drink bottle. The reactions may produce CO_2 with little evolution of heat or may be strongly exothermic and produce steam as well as gas. Such devices usually involve common household products, so they are a favorite of adolescents. They can produce high pressures and toxic or caustic gases capable of inflicting serious injuries as well as mechanical destruction.[27] Some of the more common combinations are shown in Table 12-4.

The mixtures that produce flammable gas (hydrogen or acetylene) can have a double explosive effect—the mechanical failure of the confining container followed by the deflagration of the turbulently mixed fuel/air mixture. This deflagration may be nearly

TABLE 12-4	Common Ingredients Found in Acid, Gas, and Bottle Bombs
COMBINATION	**PRODUCT OF REACTION**
Dry ice and water	Carbon dioxide
Alka-Seltzer and vinegar	Carbon dioxide
Calcium carbide and water	Acetylene gas
Sugar, dry bleach, and water	Steam, bleach
Aluminum foil and Drano	Hydrogen
Metal foil and hydrochloric acid	Hydrogen
Toilet, pool, or spa treatment and alcohol/ammonia or hydrogen peroxide (Newest products contain chlorinated or brominated hydantoins or chlorinated isocyanuric acids)	Chlorine gas or bromine gas[a] [a]Also nitrogen trichloride or nitrogen tribromide explosives

FIGURE 12-15 Melted and heat-distorted 2-L soda bottle is the remains of an acid–metal bottle bomb. Residues of unreacted hydrochloric acid and metal are present. *Courtesy of Wayne Moorehead, Forensic Consultant.*

instantaneous if an open-flame ignition source (such as a lighted candle) is placed nearby, or delayed if the mixture has to spread to find an ignition source.

The pressures produced by the failure of a soft-drink bottle can be on the order of 200 psi. Chemical hazards from residual acid, base, or bleach can cause injuries to persons nearby. Due to the hazards of explosive, incendiary, and chemical injuries, states now consider these devices to be explosive devices for prosecution.

The post-blast solid residues may be very limited in amount and subject to degradation from exposure to water. As seen in Figure 12-15, the aluminum (or other metal used) and unreacted acid or base are often found. The bottle used will have characteristic deformation due to the heat produced and internal pressure. Dry-ice devices fail with extreme cold fractures and no visible residues. Due to the unpredictable time intervals between mixing and explosion, the maker of the device is sometimes among the injured.

BLEVEs

If the original contents are liquid rather than gas, heating causes the liquid to reach a temperature far above its normal boiling point, since the liquid would ordinarily shed its excess heat as vapors to maintain its boiling point temperature. A failure of the vessel results in a BLEVE (boiling-liquid, expanding-vapor explosion). The gases initially involved are supplemented by the superheated liquid that turns to vapor immediately on contact with normal atmospheric pressure. Such explosions can be extremely destructive because of the energy contained in the liquid, even when the liquid and its vapor are noncombustible. (BLEVEs involving 40-gallon water heaters have been known to demolish houses.) BLEVEs involving superheated water or steam explosions are notable for their absence of fire/thermal effects. When the contents are flammable, the ensuing fires can be awesome (Big Loud Explosion, Very Exciting). Every explosion investigator must be aware of the possibility of mechanical explosions and examine the scene for remnants of gas cylinders, boilers, containers, or pipelines that could be the origin. When a liquid storage vessel is exposed to an external fire, the liquid inside in contact with the heated vessel wall absorbs the heat by conduction, while the wall in contact with only gas or vapor is not cooled. This may result in a tide mark on the exterior of the vessel that can tell the investigator what the level of liquid was in the vessel during the fire or the position of the vessel during fire exposure. This also means a vessel engulfed in fire will usually fail first where it is not in contact with liquid contents, because that portion will reach its tensile failure temperature before the portion that is in contact with liquid.[28]

Electrical Explosions

The passage of high currents through air, insulating oil, or undersized conductors can cause very rapid heating, which, in turn, causes extremely rapid expansion of surrounding gases or liquids. The ultimate example of the explosive effects of arc-heated air is the lightning bolt, which can involve current flows of over 100,000 A. In the case of oil (as in oil-filled transformers or switchgear) the oil boils and turns to a vapor upon failure of the vessel. This vapor may be ignited by the continuing electric arc to produce an extensive fireball. Conductors can vaporize, producing a gaseous vapor moving at high speeds. Such failures can cause flash burns as well as blast and pressure damage, as illustrated in Figures 12-16a–c.

FIGURE 12-16A Electrical explosion caused when service person shorted out 10 kV at switch gear. Blast pattern on wall. *Courtesy of Paul Spencer, London Fire Brigade.*

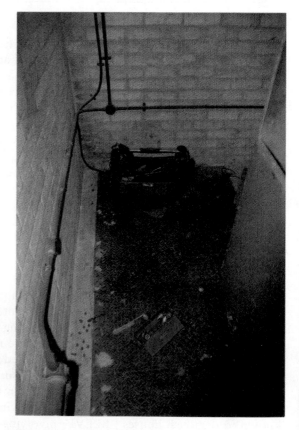

FIGURE 12-16B Fire-damaged tools; door of cabinet on floor around corner. *Courtesy of Paul Spencer, London Fire Brigade.*

FIGURE 12-16C Flash pattern on insulators. Serviceman fatally injured by arc blast. *Courtesy of Paul Spencer, London Fire Brigade.*

Investigation of Explosions

Explosions can occur as the result of accidental mechanisms in materials in industrial applications or processes, household mishaps, or the deliberate criminal act of setting off a bomb. In rare instances, a bomb may be military ordnance that is being misused, but much more often, a bombing involves an improvised explosive device (IED). The ATF reported that there were an average of 576 bombings and attempted bombings reported in the United States in 2004–2007,[29] with some 59 percent being pipe, tubing, or gas cylinder containers (pipe bombs). Smokeless powder, black powder, and flash powder were the dominant explosive fillers used.[30] (From 1993 to 1997, some 34 percent of the 38 devices used in explosive incidents were pipe bombs with various types of low explosive fillers.)[31] This trend does not appear to have changed in the years since, because such materials are so readily available. A vast majority of the remaining devices were flammable liquid fillers in bottles (**Molotov cocktails**).

It is the responsibility of the investigator *not* to presume that every event is an accident (or a bomb) but to treat each scene as a *potential* crime scene from the standpoint of scene security, preservation, and documentation. Evidence that an explosion is the result of a particular accidental cause can be just as important as evidence that proves it was a criminal act, and requires the same standard of care.

The examination of an explosion scene is generally carried out in the same manner as a thorough and proper fire investigation and, in fact, is most often carried out as part of the examination of the fire of which the explosion was a part. However, unlike at a fire scene, it is most important that evidence or structural members not be moved prior to careful examination, because the relationship between the various pieces may be critical to reconstructing the cause of the explosion, much more so than in fire evidence.

Molotov cocktail ■
A breakable container filled with flammable liquid, usually thrown. It may be ignited by a flaming wick or by chemical means.

Witness interviews are especially important in explosion investigations to reveal what someone was doing or where they were when the event occurred and what was observed. Surveillance cameras are becoming more commonplace, and they have captured critical events just before and during explosions.

THE SCENE SEARCH

A preliminary search of the scene will reveal the extent and direction of the spread of fragments (blast pattern) of the building, vehicle, device, or victim, as well as an apparent **seat of explosion**. Scene overview (and photographic documentation) can best be accomplished from a fire department aerial platform. Figures 12-17a–d show how much more information can be gained from an aerial view of a structure explosion compared with a ground-level view of the same scene. The distance from the apparent seat to the farthest discovered fragment should be multiplied by at least 1.5 (150 percent) to establish a preliminary perimeter. This perimeter may have to be expanded as the search progresses and reveals a specific directionality to the spread (as with pipe bombs, whose end caps may be launched hundreds of meters away) but will suffice to protect most of the evidence.

seat of explosion ■ The area of most intense physical damage caused by high explosive pressures and shock waves in the vicinity of a solid or liquid explosive.

FIGURE 12-17A A view of initial propagation of a structural propane explosion from ground level. *Courtesy of Jamie Novak, Novak Investigations and St. Paul Fire Dept.*

FIGURE 12-17B Massive propagation (fireball) outside structure. *Courtesy of Jamie Novak, Novak Investigations and St. Paul Fire Dept.*

FIGURE 12-17C High-energy shattering of large structure illustrates the power of a deflagration at near-ideal concentration. *Courtesy of Jamie Novak, Novak Investigations and St. Paul Fire Dept.*

FIGURE 12-17D Aerial view of same scene records the distribution of fragments (note cruciform distribution perpendicular to the four main walls). *Courtesy of Jamie Novak, Novak Investigations and St. Paul Fire Dept.*

(a)

(b)

FIGURES 12-18A AND B Survey photos document debris field of this bombing. The rear doors of the ambulance were found nearly 65 m (200 ft) from the vehicle (at the wire fence in a). *Courtesy of John D. DeHaan.*

Security of an explosion scene is paramount. No one but the searchers should be allowed access during the early stages. Figures 12-18a and b show an example of a large debris field from a vehicle bombing. Unlike evidence at fire scenes or ordinary crime scenes, where the evidence is readily visible, explosion evidence may be tiny fragments of paper, wire, metal, or wood that are easily overlooked and may stick, unnoticed, to the shoes of anyone in the scene and thus be carried off and lost forever.

The perimeter should be clearly marked with banner tape, rope, or barricades and closely monitored to ensure that only authorized personnel essential to the search are admitted. (A scene entry log such as that in Appendix I may be useful in controlling the number of people wandering through a scene.) The perimeter can be expanded if searching and evaluation indicates there is some directionality to the distribution of fragments. Deflagrations, especially of diffuse-phase mixtures, tend to produce omnidirectional distribution, but devices like pipe bombs can produce directional effects, as pipe caps are often expelled great distances along the axis of the device.

Documentation includes thorough photographic coverage of the scene as well as diagrams (plan views) showing locations and distances of debris. This was done in the past with tape measures and protractors, and more recently with Total Station surveying systems. The advent of laser scanning systems that can accurately document locations up to 500 m (1,600 ft) in day or night conditions is likely to set new standards of accuracy and ease of use.[32] See Figures 12-19a and b for an example of the usefulness of such scanning technology.

The size of the search sectors will be determined by the size of the scene and the number of searchers. They should be small [less than 0.4 m^2 (4 ft^2)] near the seat, becoming larger

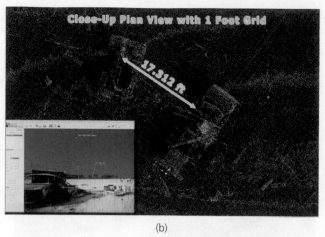

(a)   (b)

FIGURES 12-19A AND B Documentation of an explosion scene using a Leica C10 laser scanner allows production of aerial and eye-level views. *Courtesy of Leica Geosystems and Precision Simulations.*

[up to 10 m^2 (100 ft^2)] toward the margins of the search area where the debris is lightly scattered. The sectors can be marked with chalk, string, or rope with stakes, or with tape stuck on the ground and must be numbered for later reference.

Once the scene is secured and the sectors marked, the search can begin, with a perimeter search and photographic documentation, just as in a fire scene. The size, shape, and conditions of the scene and the number of *qualified* searchers available will determine the organization of the search. Grid, sector, and spiral search patterns have all been used with success. The search should always start from the outside perimeter and work inward. This allows better preservation of the critical explosion seat and improved protection against souvenir hunters. The primary goal of the scene search and documentation is the identification of displaced debris or fragments and measurement of their displacement (distance from starting position).

The Four Rs

The scene search can be thought of as the first step in the investigative process, which may be remembered as the "Four Rs": Recognition, Recovery, Reassembly, and Reconstruction.

Recognition at the scene includes assessment of the blast damage (location, intensity, direction) as well as recognition of pieces of physical evidence to be used in the later stages. Physical evidence of an accidental situation or mechanism that brought about an explosion can be just as important as the evidence of a deliberate act of bombing. The training and experience of examiners or searchers is critical to the investigation, because important evidence can be small and appear inconsequential. Such evidence is easily over-looked by untrained searchers.

Recovery involves not only the physical process of picking up evidence but its documentation and preservation as well. This process must take into account the fact that explosions produce blast, pressure, **fragmentation**, and thermal effects. *Reassembly* of the device or mechanism may take place within the laboratory or at the scene but is a necessary test to see whether the entire chain of circumstances is accounted for. Finally, the scenario must be *reconstructed*. What was present? What was the initiator? How did it happen? Who was present?

The Four Cs

Just establishing that there was an explosion of a particular kind of material (gas, liquid, or solid) is usually not adequate. One way to remember all the necessary elements is the

fragmentation ■ The fast-moving solid pieces created by an explosion. Primary fragmentation is that of the explosive container itself; secondary fragmentation is that of the target shattered by an explosion.

"Four Cs": Container, Concealment, Content, and Connections. While these are most applicable to criminal bombings, they are useful even for assessing accidental explosions.

Container   What evidence is there for the means of containment of gases, vapors, or low explosives sufficient for the blast effects we see? This may be fragments of pipe, furnace, wall, or vehicle. Fragments of the device are called *primary fragmentation*; fragments of everything else shattered by the blast are called *secondary fragmentation*. High explosives, of course, require no container to develop their maximum explosive effect. The absence of any container, then, may be construed to imply that a high explosive was used. Blast effects must corroborate this hypothesis.

Concealment   Bombs are often concealed in some manner to prevent their premature discovery. Are there fragments of wrapping, container, or other mode of concealment? Is the location of the blast seat such that an explosive device could have been concealed there? Is there an intended target from whom the device had to be concealed? The concealing container does not form the actual containment of the explosive charge and can be anything from a paper bag or suitcase to a vehicle. The question is: Did this item belong at the scene naturally or was it brought here?

Content   What was the actual explosive material? In improvised devices or accidental explosions, the explosive is never totally consumed. Residues may be visible to the naked eye and be easily found or they may require sophisticated laboratory analysis, but they can be found if the investigator thinks to recover debris that was close to the initiating explosive charge. In accidental explosions, gases, vapors, or solids can accumulate from manufacturing processes or poor housekeeping, or inadequate safety measures may bring about destruction of normal stock. Interviews and review of fire inspections or previous accidents may reveal the content of potential explosive products.

Connections   Finally, connections must be examined—not just electrical wires, but any physical arrangement or sequence of events that could initiate an explosion. How did it start? In bombings, such considerations include where the device was placed (vehicle, dwelling, business), how it was triggered (movement, timer, on-command, etc.), and what the sequence was from power source to initiator to main charge (sometimes called the firing train). The answers to such questions help reconstruct the entire event, including the intent of the bomber—to destroy, to intimidate, or to kill? In accidental situations, the same questions help establish cause and effect so that proper liability can be determined, and more important, similar accidents can be avoided in the future. In accidental electrical explosions, an appropriate source of very high amperage electrical power must be identified, along with a conductive path. It must be remembered that some portions of the path may be obliterated by movement, fire damage, or the passage of current itself.

SPEED AND FORCE OF REACTION

When it comes to interpreting explosion damage at a fire scene, it is important to remember the differences between deflagrations and detonations as they have been described here. The differences in speed of reaction and in the forces and temperatures generated are responsible for characteristic effects on a structure or vehicle. In determining whether a combustion, deflagration, or a detonation took place, one guideline to keep in mind is *the smaller the pieces and the farther they are thrown, the higher the energy of explosion—combustion, deflagration, detonation*. The time required for a deflagrative explosion is on the order of hundredths of a second as compared with that of a detonation, which is on the order of thousandths or millionths of a second. The relative slowness of a deflagration is important in distinguishing its effects on targets at close range, because they will be subjected to a more or less progressive pressure pulse that is subsonic with respect to the target material. This gives the material a chance to spread the pressure out, resulting in less shattering and more pushing of larger pieces. The abrupt pressure or shock wave from a detonation is supersonic

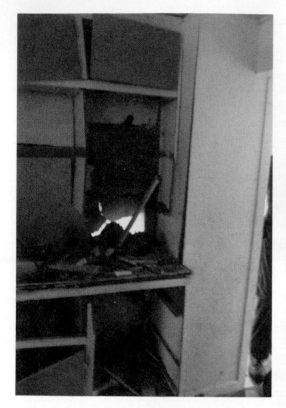

FIGURE 12-20A Localized shattering of wooden cabinets by 230 g (0.5) lb of military dynamite. *Courtesy of Lamont "Monty" McGill, Fire/Explosion Investigator.*

FIGURE 12-20B Results of deflagration of 1 gallon of gasoline (vaporized by a small explosive charge). The roof was lifted 1.2 m (4 ft) into the air, and unsupported walls were flattened. Note the surface scorching of the painted wall on right. *Courtesy of Lamont "Monty" McGill, Fire/Explosion Investigator.*

at close range; the material cannot respond and is shattered rather than pushed. This intense shattering effect gives rise to the localized damage called the "seat." As energy dissipates, the damage becomes less severe. The shock wave from a detonation compresses air into a very thin layer that does a great deal of the shattering. It is followed by a high-pressure front that does a great deal of the physical movement. The effects of an explosion may be categorized as blast wave, overpressure, missile, and thermal.

Figures 12-20a and b show the difference in effect between a deflagration of gasoline vapor and the detonation of military dynamite in a wood-frame structure. The pressures produced by a deflagration are very low in the vicinity of the ignition source and grow as the flame front expands through the premixed fuel/air mixture. The lower, more uniform pressures of a deflagration usually do not produce a localized seat of more intense damage. Published tests by DeHaan have shown that pressures produced by a deflagration of a shallow hexane vapor layer (0.1–0.2 m deep) (4–8 in.) on the floor of a compartment (3.6 × 2.4 × 2.4 m in size) (10 × 8 × 8 ft) equilibrate at all surfaces of the compartment within a few milliseconds of one another. This pressure was seen to grow over a period of about 100 milliseconds until the failure pressure of the vent panel (5 to 6 kPa; 0.75 psi) was reached.[33] Such equalized pressure would cause the failure of the weakest part of the compartment (in this case, the vent panel). In a real-world building this weakest part may be the roof, the windows, the bottoms of walls weakened by decay, or other components. The stronger the structure, the greater the confinement, and the higher the pressures that can be developed by a deflagration, from less than 7 kPa (1 psi) to 700 kPa (100 psi) or more, depending on the chemistry and quantity of fuel gas/air mixture present.[34]

(a)   (b)

FIGURES 12-21A AND B Natural gas explosion (see also Figure 4-19) in an apartment tower block (second floor from top). The blast wave reflected between facing tower blocks. Notice the broken windows on alternate floors of both buildings. *Courtesy of Fire Investigation Unit, Metropolitan Police, London, England.*

Figure 12-3 showed the result of a natural gas explosion in a wood-frame residence where the toeplates were split and walls deflected out at their bases 4 to 12 cm (2 to 5 in.), but the small double-glazed, aluminum-framed windows remained intact. In a detonation, the confinement of the structure plays no significant role in the pressures developed. Any pressure pulses can be reflected from large flat surfaces and bounce back to induce damage around corners or over or under barriers. The more energetic the explosion, the more pronounced such effects can be. Reflected pressure pulses can add to the originating pulse and create even more damage in the room of origin. Figure 12-21 is an example of a pressure pulse reflected back and forth between two high-rise buildings that broke windows on alternate floors.

Note that when the pressure or shock wave is transmitted through air, it drops off in a predictably rapid fashion, as reflected in Figure 12-22. Finally, the temperatures produced in a deflagration would not be expected to exceed 1,500°C (2,700°F) but can reach several thousand degrees in a detonation. As one would expect, these differences in speed, pressure, and temperature produce different effects on the target materials. As we saw in Figure 12-17, a near-stoichiometric mixture of propane ignited below grade vented upward, completely shattering the building above. An analyst familiar with the differences can accurately estimate the type of explosion, and often the location of the ignition, that took place by examining fragments of the target and their distribution. Beveridge and others have published excellent descriptions of the characteristic features.[35]

Explosion Damage

The pressures produced by a deflagration are usually fairly low and do not *generally* shatter or pulverize, but they can cause serious structural damage to a building. Exceptions are shown in Figures 12-17c and 12-23. The pressure wave moved at subsonic speed and

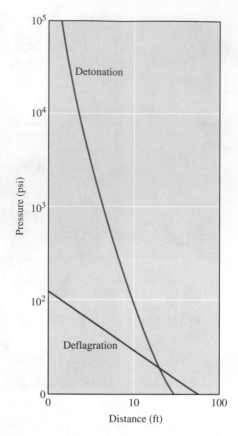

FIGURE 12-22 Pressures as a function of distance for the detonation of 100 lb of explosive and a typical large deflagration. *Courtesy of Dr. J. H. Burgoyne and Partners, London, England.*

FIGURE 12-23 Massive deflagration of near-ideal mixture of LP gas and air produced extensive fragmentation and high velocities. *Courtesy of Denise DeMars, Streich DeMars Inc., and Jamie Novak, Novak Investigations and St. Paul Fire Dept.*

retained its destructive force longer as it moved away from the source, where an ideal mixture of LP gas and air was achieved. Because the rate of reaction of a deflagrating fuel/air mixture increases with the expanding surface area of the expanding flame front, the force exerted increases as the distance from the ignition source increases. For this reason, the area of most destruction of a fuel/air deflagration will usually not be in the vicinity of the ignition source but rather some distance away from it. This pattern makes establishing the actual ignition source very difficult. Once the limit of the premixed volume is reached, the energy falls off with distance. The mixing and turbulence caused by the passage of the flame front around furniture and through doorways can dramatically increase the surface area of the flame front. Pre-ignition heating and pressurization of the fuel/air mixture in adjoining rooms can also result in more energetic deflagration than in the room of ignition.[36] This in turn increases the damage done by the extensions of the blast into areas away from the ignition source. The deflagration pressure wave can be vented by the loss of doors or windows in a structure, minimizing damage to the walls. The normal appearance of such a scene, then, is substantial destruction over a wide area with large fragments and projectiles. (See Table 12-5 for typical blast effects.)

In contrast, a detonation, with its peak pressures in the tens of thousands of pounds per square inch, shatters and pulverizes materials in its immediate vicinity and creates a shock wave with a massive positive-pressure front with a negative-pressure wave or zone right behind it. The negative-pressure wave creates a partial vacuum that is filled by air rushing back in toward the explosion seat. This current can be strong enough to move light debris or even topple weakened masonry walls in a direction opposite the original blast. Windows can be "sucked out" of a wall as a blast wave moves parallel to it. The negative-pressure pulse that follows the positive-pressure wave from a confined deflagration is low in energy. The hexane vapor tests described earlier revealed a negative-pressure pulse about one third as strong as the positive-pressure peak.[37] This may be enough to move draperies back through broken windows, pull pictures off walls, pull unsupported doors into a room, or cause further collapse of structural members weakened by the positive pressure.

The detonation pressure wave initially moves at supersonic speeds but very quickly dissipates, making its damage highly localized. Because of its supersonic speed, the pressure wave is not subject to venting and can destroy a nearby wall whether it has windows and doors or not. The normal appearance of a detonation scene, then, is highly localized destruction with pulverization, shattering, and a large number of very small pieces, some of which will be projected large distances away. Keep in mind that all explosive detonations produce what becomes a low-energy pressure wave at large distances from the source.

The distance that window glass fragments are thrown from their origin is directly proportional to the pressure on the glass pane at the time of failure. Plain glass is thrown 50–60 m (180–200 ft) by a pressure of 27.6–34.5 kPa (4–5 psi), 30–35 m (100–120 ft) by pressures on the order of 13.8 kPa (2 psi), and 12–15 m (40–50 ft) by overpressures of 4.83–6.89 kPa (0.7–1 psi).[38] Wire-reinforced or plastic-film-laminated glass, which has higher failure pressures, will be thrown shorter distances by the same pressures. High explosives have been used to vaporize flammable liquids in combination devices used as antipersonnel weapons. Such scenes bear post-fire evidence of both types of explosions. In the immediate vicinity of the device, the shattering effects of the high explosive predominate, as shown in Figure 12-23. Elsewhere in the structure the effects of the deflagrating gasoline vapor are the most apparent, as was seen in Figure 7-46. The pressures produced by detonations often pulverize nearby surfaces, creating a crater or seat, whose depth and dimensions reflect the size and sometimes the shape of the explosive charge. Such seats are not produced by diffuse-phase explosions.[39]

Pre-fire and Trans-fire Explosions

Because explosions can occur before a fire (as the means of ignition) or during a fire, the presence or absence of pyrolysis products on debris ejected from a building can reveal

TABLE 12-5 | Structural Damage Produced by Blast

PRESSURE (psi)	DAMAGE
0.02	Annoying noise (137 dB), of low frequency (10–15 cps)
0.03	Occasional breaking of large glass windows already under strain
0.04	Loud noise (143 dB); sonic-boom glass failure
0.1	Breakage of windows (small) under strain
0.15	Typical pressure for glass failure
0.3	"Safe distance" (no serious damage below this value with a probability of .05) Missile limit Some damage to house ceilings; 10 percent of window glass broken
0.4	Limited minor structural damage
0.5–1.0	Large and small windows usually shattered; occasional damage to window frames
0.7	Minor damage to house structures
1.0	Partial demolition of houses; made uninhabitable
1–2	Corrugated asbestos shattered Corrugated steel or aluminum panels and fastenings fail, followed by buckling Wood panels (standard housing) and fastenings fail; panels blown in Persons knocked down
1.3	Steel frame of clad building slightly distorted
2	Partial collapse of walls and roofs of houses
2–3	Concrete or cinderblock walls, not reinforced, shattered
2.3	Lower limit of serious structural damage
2.5	50 percent destruction of brickwork of house
3	Heavy machines (weight: 1,400 kg; 3,000 lb) in industrial buildings suffer little damage Steel-frame buildings distorted and pulled away from foundations
3–4	Frameless, self-framing steel panel buildings demolished Rupture of oil storage tanks
4	Cladding of light industrial buildings ruptured
5	Wooden utility poles (telegraph, etc.) snapped Tall hydraulic press (weight: 18,000 kg; 40,000 lb) in building slightly damaged
5–7	Nearly complete destruction of houses
5.1–14.5	Eardrum rupture
7	Loaded railroad cars overturned
5.8–8.7	Reinforced concrete structures severely damaged
7–8	Brick panels, 20–30 cm (8–12 in.) thick, not reinforced, fail by shearing or flexure
9	Loaded railroad boxcars completely demolished
10	Probable total destruction of buildings Heavy (3,200 kg; 7,000 lb) machine tools moved and badly damaged Very heavy (5,500 kg; 12,000 lb) machine tools survive
29–72.5	Lung damage
102–218	Lethality
300	Limit of crater lip

Sources: V. J. Clancey, "Explosion Damage," paper presented at Sixth International Meeting of Forensic Sciences, Edinburgh, 1972; G. Kinney and K. Graham, *Explosive Shocks in Air* (New York: Springer-Verlag, 1985).

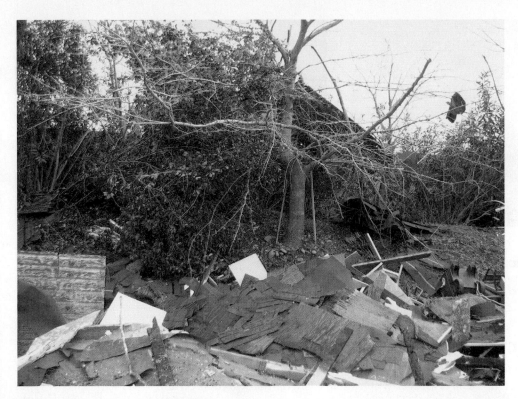

FIGURE 12-24 The underside of plywood roofing panels (foreground) bear "flame-washed" effect from fuel-rich natural gas explosion in the attic of the structure. *Courtesy of John D. DeHaan.*

whether the fire was underway before the explosion. Smoke explosions or BLEVEs involving ignitable liquids occurring *during* a fire will often produce ejected debris that show thermal effects or combustion residues, while debris from deflagrations or detonations that initiate a fire usually have little or no such residues. If it is energetic enough to disrupt the ceiling or roof of the confining structure, a natural gas explosion will cause the combustion and release of the accumulated gas (as was seen in Figure 12-2). This results in minimal post-blast burning except in the vicinity of the source of the gas. Although the pressures of a deflagration inside a structure equilibrate at the speed of sound and, as a result, usually produce no localized blast effects, the pre-explosion distribution of fuel gas or vapor may produce localized *thermal* effects as the overrich layers or pockets of fuel burn off. Scorching or "flame-washed" effects may be caused by fuel/air deflagrations in wood structures (see Figure 12-24.) The distribution of melting or scorching of materials (particularly thin, low-mass materials like paper or plastic) may reveal whether the pre-blast accumulation layer of gas or vapor was lighter or heavier than air (see Figure 4-9). As debris is documented and collected, its appearance should be noted.

When flammable liquid vapors in a closed room are ignited, the flame propagates only through the vapor layer near floor level, as observed numerous times during tests by DeHaan.[40] Ide demonstrated extension of flames out the lower portion of a door opening when gasoline vapors were ignited in the adjoining room. These flames charred the lower legs of trousers worn by a mannequin placed in the doorway.[41] Observation and documentation of the distribution of thermal effects are critical to evaluation of the pre-blast distribution of vapors or gases. See Figures 12-25a and b for an example of thermal effects from a vapor explosion where liquid LPG was released into a room.

(a)

(b)

FIGURES 12-25A AND B Ignition of commercial LP gas released into this structure produced floor-level charring of vinyl tile and walls. Notice pronounced irregular charred areas and limited extension to baseboards and some walls. *Courtesy of Senior Special Agent Steven Bauer, ATF (retired).*

SCENE EVALUATION AND HYPOTHESIS FORMATION

Once the preliminary evaluation of the explosion debris and thermal patterns is completed, the investigator will be able to direct the search in certain areas with specific types of evidence in mind. The search of an explosion scene can be inward spiral, sector search, strip search, or grid search. Searches of larger scenes will be aided by laying out grid lines with string or rope and processing grid sectors in sequence. This is especially useful if search personnel are limited in number.

A detailed sketch of the scene is most useful at this stage of the investigation. Because items of evidence can be described on the sketch, along with measurements of their distance from a reference point and a reconstruction of their line of flight, such a sketch can make relationships between those items apparent. If properly done, the sketch will show how the paths tend to intersect at one area. This indicates that the seat of the explosion (or the point of ignition of a deflagrating fuel/air mixture) was probably in that area and allows the investigator to focus the search for pieces of the device (or ignition source) itself in that room or area. The distribution, distance, and trajectory of window glass fragments are especially important, since windows are usually broken in an explosion (see Figure 12-26). A sample of the glass should be preserved or at least documented as to thickness and type (safety, plate, plain, wired, etc.). If the distance, thickness, and type are known, the pressure produced inside the structure can be estimated.

EVIDENCE RECOVERY

In spite of the ultrahigh temperatures and pressures developed in a high-order detonation, some pieces of the device, however small, will remain in an identifiable form. Safety fuse is particularly resistant to blast effects and will survive, sometimes even with a remainder of the blasting cap still attached. Due to the frequency with which pipe bombs are used, the examination of a scene should include an examination of the debris for any broken bottles, pressure cylinders, or fragments of pipe or pipe cap found in the debris. The fragmentation of pipe, when used as a pipe bomb, follows a very predictable pattern. The larger the total number of fragments and the smaller the fragments, the higher the energy of the explosive filler.[42] This information can be useful in quickly establishing the nature of the device and explosive filler and the type of other evidentiary materials that should be searched for.

Small fragments of metal or plastic pipe, wire, or cardboard may easily be overlooked by inexperienced searchers or compromised by being stepped on. Explosion scenes should be searched grid-by-grid (preferably by two different searchers), and any item or debris that looks out of place photographed and recovered for later evaluation. Some examiners prefer to have each grid swept up after the visual search and all materials bagged for later examination under controlled conditions. Sifting screens can be used to sift for fragments, but unlike fire scene evidence, the debris must be sifted dry, since many of the post-blast residues of low explosives are water soluble and will be lost if the debris is wet sifted. The clothing and the bodies of any victims (living or deceased) must be carefully examined for the presence of fragments (either primary or secondary). The patterns of thermal or impact injury to the clothing and skin may also be critical in reconstructing the event (whether the explosion was an accident or a bombing) and must be documented by photograph and diagram.[43] Medical treatment can quickly alter the damage patterns of injuries, so documentation and recovery of evidence from victims must take place as soon as possible. X-rays are irreplaceable in assisting the search for fragments, but it should be remembered that PVC and other plastics have the same physical texture and X-ray density as cartilage or bone in the body and can very easily be overlooked during X-ray or visual examination of the wounds. (Whitaker et al. reported velocities of PVC pipe nipple fragments to be as high as 290 km/h (465 mi/h; 80 m/s) and capable of inducing injuries despite their low mass.[44]

Low explosives like black powder or smokeless powder are not completely consumed when used in improvised devices (unless there has been a significant post-blast fire), and traces will often adhere to larger pieces of the containers (such as the nipple of a pipe bomb) or to large, cool surfaces near the exploding device or buried into nearby wood surfaces. Organic explosives can be absorbed into plastic surfaces such as wire insulation, vinyl upholstery or luggage, or plastic bottles. It is much better to recover any item suspected of bearing explosive resides for laboratory examination and extraction than to attempt to swab the contaminated surface and submit just the swabs. For this reason, all loose debris from the seat or crater should always be collected and preserved in clean new cans or jars. Polyethylene plastic bags and paper envelopes will allow volatile organic explosives to migrate into and through the container with the possibility of generating cross-contamination concerns and should be avoided for the blast seat debris or bulk organic explosives, if possible. Polyethylene (Ziploc) bags are suitable for small components and small items where high concentrations of organic explosive residues are not expected. Special plastic bags such as nylon or Kapak (by Ampac) are suitable for explosive evidence, but ragged, sharp-edged debris can puncture them, so careful handling is required.

Large immobile objects such as buildings will require swabbing with clean cotton gauze or cotton swabs wetted with methyl or isopropyl alcohol, distilled water, or acetone.[45] The swabs then need to be sealed in an appropriate clean glass or plastic vial. The hands and fingernails of persons suspected of handling explosives can be sampled with swabs, and fingernail scrapings recovered with clean flat wooden toothpicks. As with fire debris evidence, maintaining the collected evidence in cool, dark surroundings until submitted to the lab will minimize degradation or evaporation. Any comparison materials or reference specimens (bulk explosives) should, of course, be kept isolated from scene debris.

LABORATORY ANALYSIS

Although high explosives like nitroglycerin or TNT can detonate with nearly 100 percent efficiency, traces of the unexploded material can usually be detected on portions of the device, its container, or nearby surfaces, especially with the sensitivity of today's laboratory techniques. These tests require solvent extraction of selected fragments of debris and analysis of the solution by very sensitive laboratory methods including thin-layer **chromatography**, capillary electrophoresis, gas chromatography, high-pressure liquid chromatography (HPLC), ion chromatography, and mass spectrometry. Forensic scientists who have extensive experience in explosives casework [such as those at the U.S. Department of Justice—Bureau of Alcohol, Tobacco, Firearms and Explosives, and the Defense Explosives Research Agency (DERA) in the United Kingdom] have developed effective analytical schemes for the detection and identification of traces of high explosives and the more commonly used low explosives. The methods used in such tests are beyond the scope of this text, and the reader is referred to the specialist literature.[46]

Military high explosives in their original packaging and many slurry- and water-gel blasting agents often leave virtually no detectable chemical residues if properly initiated. Highly sensitive analytical methods such as ion mobility spectrometry and liquid chromatography/mass spectrometry have been shown to be useful in identifying such traces.[47] Even residues of peroxide-based explosives have been identified post-blast.[48] Portions of the packaging such as the plastic wrappers, the heavy metal wire staples crimped on the ends of some cartridges, and even the metal end panels from TNT charges have been known to survive. Metallurgical tests on fragments may reveal the speed of reactions to which a container was subjected, even when the explosive itself cannot be identified.[49] Such tests are of particular value in examining debris from aircraft incidents, in which the mechanical impact can shatter metal components but not at the velocities created by detonating high explosives.

Portions of the timing or triggering mechanism of IEDs will survive if they are not in immediate contact with the explosive charge. Fragments of clockwork or electronic timers and the batteries used to power timers and electric blasting caps are identifiable by experienced laboratory analysts in many instances. Much of the laboratory reconstruction and identification of bomb components is visual, so the more complete the photographic documentation of the scene (with close-ups of the evidence in place before recovery) and the more thorough the search and recovery process, the higher the chances of success will be.

Recently, more techniques have been developed for examining melted components such as electrical or duct tape, and characterizing unburned smokeless powder.[50] DNA has been identified on post-blast bomb fragments. New analytical methods are being introduced that make it possible to identify "batches" or sources of explosives on the basis of ion ratios.[52]

The identification of explosive residues and the reconstruction of explosive devices are specialized talents not found in all forensic laboratories. In addition to the extensive training required of personnel, the correct and comprehensive analysis of explosive-related evidence requires specialized equipment and techniques and extensive reference collections of such esoterica as pipe fittings, clocks, batteries, blasting caps, and fuses. The investigator is encouraged to contact nearby laboratories to determine what services are available before submitting evidence to them.

Gas chromatography/mass spectrometry using vapor concentration recovery methods (described in Chapter 14) has been successfully applied to identify gaseous fuels (propane, butane, pentane) trapped in clothing of victims (also in lung tissue and skin of decedents).[53] Clothing, including shoes of explosion victims (living or deceased), should be collected as soon as possible and stored separately in airtight metal cans or glass jars for analysis. Porous materials such as upholstery *if* exposed to high concentrations of gases or vapors and not exposed to extensive fire afterward may contain identifiable residue.

CASE STUDY

In a recent case, the primary victim of an explosion and fire in a parts painting operation died without regaining consciousness. A coworker reported seeing the victim turning away from a 209-L (55-gal) steel drum where he was in the habit of discharging cleaning solvent (xylene) from his electrostatic paint sprayer. The witness saw a small flame at the bung hole on the top of the drum, but before he could warn his coworker, the flame progressed into the drum, causing it to explode from within. The drum separated at its bottom seam and was propelled into the roof structure, where it deformed a steel truss, spewing many gallons of waste xylene and paint onto the work area and the victim. Under most conditions there would not be propagation because of xylene's high flash point (ca. 30°C; 85°F). This incident occurred in midsummer in a non-air-conditioned factory where the air temperature was reported to be in excess of 32°C (90°F) at the time. The aerosol discharge of the xylene from the spray gun created an ignitable vapor at the mouth of the drum. Manufacturer data indicated that the high-voltage discharge tip of the spray gun operated at a high voltage (>15 kV) when the trigger was pulled. Bringing the tip close to the metal drum during discharge apparently allowed an arc discharge to occur to cause ignition of the xylene vapors.[54]

INCIDENT ANALYSIS

The analysis of explosion incidents follows the same scientific method described in Chapter 1—gathering data, formulating and testing hypotheses, and reaching a testable, defensible conclusion. Recent studies have focused on the damage and deformation of road signs, power poles, and adjacent vehicles as a means of reconstructing the amounts of blasting agent in large car bombs.[55] Some of the analytical tools useful in fire reconstruction (such as computer models) do not work on diffuse combustion explosion events due to the very different dynamics of gas movement, pressure, and heat production. See Figure 12-23 for an example of a very destructive LP gas deflagration. It is easier to analyze a condensed-phase explosion because it is a more predictable single-pressure event. Combustion explosions can be very complex, with multiple propagations dependent on the geometry of the building, distribution of fuel, failure mechanisms, and the location of the ignition. (In tests observed by one of the authors, ignition of a uniform natural gas/air mixture produced minimal flame extension out the vent/window and modest pressures when the mixture was ignited from the center of the room. When a similar mixture was ignited from the rear wall, a mass of unburned fuel was pushed through the window to create a very large, energetic fireball outside the compartment. The pressure from this *external* deflagration stripped drywall lining off the walls of the test compartment. Mniszewski and Campbell have offered some examples of analytical tools that can be of use, including some CFD computer models of explosions.[56] Models of fuel concentration and distribution are suggested as ways of testing hypotheses about fuel sources and ignition mechanisms. Recently, the CFD software FLACS has been shown to be a very powerful analytical tool for modeling gas or vapor dispersion and deflagration dynamics.[57]

Bomb Arson Tracking System (BATS)

Since 2003 the U.S. Bomb Data Center (USBDC) and ATF have developed a remarkable Web-based data system for tracking fire, explosion, and explosives-related incidents of all types occurring across the United States. The Bomb Arson Tracking System (BATS), which is provided free of charge, allows any police or fire agency to enter incident information, including location, description, area of origin, scene details, names of victims, witnesses, or suspects for inclusion in a database searchable by any other law enforcement or fire investigation agency. Each agency is responsible for administering its own BATS account.

BATS is part of the USBDC's mission to collect, analyze, and disseminate timely and relevant information for use by U.S. federal, state, local, and tribal agencies. BATS has been designated by the U.S. Attorney General as the sole repository for information pertaining to explosives- and arson-related incidents. It combines data on IEDs and suspected bomb events, stolen and recovered explosives, and fire investigations (structural, vehicle, and wildland) so that trends and possible links can be identified. The system automatically encodes all locations into GIS coordinates and offers mapping options. (See Appendix J for a sample of BATS entry screens.)

BATS is capable of generating and printing out standardized reports for consistency throughout the fire and explosion investigation community. Printed BATS reports are customized with each participating agency's header and seals. It can function as a case management tool as well. BATS has drop-down menus for fire descriptors and scene details, as well as provisions for narratives and paste-ins of photos, diagrams, log sheets, and supplemental reports. It has provisions for "read-only" or "read/write" access designated by the user to collaborate with other agencies on joint investigations. BATS can function as a bomb/arson unit management system as well, because time spent on cases, court, training, travel, and maintenance can be tracked individually and as a unit.

Information can be designated as "Restricted" or "Unrestricted." If a case is designated as "Restricted," only contact information for the investigator logging in the case is provided. The USBDC provides two options to public safety agencies for access to BATS: Law Enforcement and Non-Law Enforcement. Users from an agency with an NCIC-issued ORI number have full access from any computer with Internet access. (Fire department investigators gain full access through cooperative agreements with local police agencies or after receiving an ORI from their state CJIS coordinator.) As of summer 2009, they can also choose the "Non-Law Enforcement" role, which limits their access to their own and other non-LE agency information. Data are released only to agencies authorized specifically by the user.

Because BATS has been designated as the sole reporting system by Congress and the attorney general, regular reporting provides investigators with a powerful tool for access to critical incident information. Arson investigators can compare local case specifics in real time against a jurisdictional database without the limitations that NFIRS reporting includes. (NFIRS is intended for documenting fire department responses to fire incidents, not for documenting investigations. Its information is subject to unrestricted access by all fire department personnel. According to the U.S. Fire Administration, NFIRS is not intended as a case management system).

In September 2009 the House passed a bill (H.R. 1727: Managing Arson Through Criminal History; the MATCH Act) establishing a *National Arsonist & Bomber Registry*. The legislation would authorize the attorney general to establish and maintain a database within ATF that would be built into BATS. The registry would require convicted arsonists and bombers to report where they live, work, and go to school and would include descriptions of vehicles they own, an up-to-date photograph, and a photocopied I.D. In addition to that information, the database would include finger and palm prints. This legislation will provide an important tool to law enforcement by enabling investigators to track arsonists and bombers, regardless of where they live, across jurisdictions. By querying a zip code to locate convicted arsonists living or working in the vicinity of a fire, or by querying the components database for previous incendiary or bomb components that match those used in an unsolved arson or bombing investigation, the process of narrowing down suspects would be expedited. The bill, H.R. 1727, passed the House on September 30, 2009, and on October 2009, was read in the Senate and referred to the Committee on the Judiciary. The bill was allowed to die without action.

BATS is being used by police and fire agencies across the United States, and it promises to dramatically improve the quality of information sharing and comparisons for the fire and explosion investigation community. For more information, public investigators are encouraged to contact the U.S. Bomb Data Center (usbdc@atf.gov) or BATS (www.BATS.gov).

CHAPTER **REVIEW**

Summary

Explosions may occur as part of the progress of a normal fire, as the manner of initiation of a fire, or by themselves with no accompanying fire. Like fires, they may be the result of accident or criminal intent. Although the destructive power of an explosion is released in a much shorter time than that of a fire, it should now be apparent that many similarities exist between the two processes. These similarities should make it easier for the careful investigator to reconstruct the sequence of events for an explosion in the same manner as for a normal fire. The investigator should be able to demonstrate or prove (1) the nature and identity of the explosive material, (2) a competent source of ignition for that explosive, and (3) the manner or sequence of events that led to the initiation. The same attention to small details and methodical approach are required for both types of scenes. The reader is referred to the excellent outline of scene investigation offered in the NIJ *Guide for Explosion and Bombing Scene Investigation*[58] and Thurman's *Practical Bomb Scene Investigation*.[59] In spite of the awesome destruction of some explosions, there always remains some evidence for the skillful investigator to recover and interpret.

Review Questions

1. Name the four categories of "explosion" and describe the features of each.
2. What is a mechanical explosion?
3. Name three major differences between a condensed- or solid-phase explosion and a diffuse-(gas/vapor) phase explosion.
4. What is a blast propagation diagram and how is it useful in an explosion investigation?
5. Name five examples of high explosives.
6. What are primary high explosives?
7. How do low explosives function?
8. What are the four Rs of an explosion scene investigation?
9. Why is it important to map the distance and direction of travel of debris at an explosion scene?
10. What are the four Cs of an explosion device reconstruction?

References

1. C. F. Turner and J. W. McCreery, *The Chemistry of Fire and Hazardous Materials* (Boston: Allyn and Bacon, 1981), 92.
2. DuPont de Nemours, Inc., *The Blasters Handbook*, 175th anniversary ed. (Wilmington, DE: DuPont de Nemours Inc., 1978), 33.
3. R. J. Harris, *The Investigation and Control of Gas Explosions in Buildings and Heating Plant* (New York: E & FN Spon, 1983), 126.
4. J. D. DeHaan, "The Reconstruction of Fires Involving Highly Flammable Hydrocarbon Fuels," Ph.D. dissertation, Strathclyde University, 1995. (Also, University Microfilms, Ann Arbor, Michigan, 1996.)
5. Ibid.
6. J. E. S. Venart, "Flixbourgh (*sic*): A Final Resolution" in *Proceedings Interflam 1999* (London: Interscience Communications), 257–72.
7. R. Zalosh, "Explosions," chap. 8 in *Fire Protection Handbook*, 20th ed. (Quincy, MA: National Fire Protection Association, 2008), vol. 1, chap. 8, sec. 2; J. D. DeHaan et al., "Deflagrations Involving Stratified Heavier-than-Air Vapor/Air Mixtures," *Fire Safety Journal* 36 (2001): 693–710.
8. B. Lewis and G. Von Elbe, *Combustion, Flames and Explosions of Gases*, 2nd ed. (New York: Academic Press, 1961).
9. Ibid., 533.
10. Ibid., 533.
11. Harris, *The Investigation and Control of Gas Explosions*.
12. C. M. Fleischmann, "Research on Backdraft Phenomena," *Fire Engineers Journal*, 57, no. 186 (January 1997): 9–13; R. W. Bukowski, "Modelling a Backdraft Incident," *NFPA Journal* (November–December 1995): 85–89; see also *Fire Engineers Journal* 56, no. 185 (November 1996): 14–17; R. Chitty, "Management Summary of Survey of Backdraught," *Fire Engineers Journal* 56, no. 18 (November 1996): 12–13.

13. R. G. Zalosh, "Explosion Protection," Chaps. 3–15 in *SFPE Handbook of Fire Protection Engineering*, 4th ed. (Quincy, MA: NFPA, 2008).
14. A. B. Spencer, "Dust: When a Nuisance Becomes Deadly," NFPA *Journal* (November–December 2008): 57–61.
15. U.S. Chemical Safety and Hazard Investigation Board, 2175 K Street NW, Suite 400, Washington, DC 20037, http://www.chemsafety.gov; S. J. Selk, "Explosion Investigation and Reconstruction Using Multidisciplinary Methods," paper presented at American Academy of Forensic Sciences, February 2008.
16. Kirk-Othmer, *Encyclopedia of Chemical Technology*, vols. 1– 27, 4th ed. (New York: Wiley, 1998).
17. T. Urbanski, *Chemistry and Technology of Explosives*, vol. 1 (New York: Pergamon, 1964).
18. H. Henkin and R. McHill, "Rates of Explosive Decomposition of Explosives," *Industrial Engineering Chemistry* 44 (1952): 1391.
19. Urbanski, *Chemistry and Technology of Explosives*; DuPont de Nemours Inc., *The Blasters Handbook*; J. J. Manon, "Explosives: Their Classification and Characteristics," *Engineering and Mining Journal* (October 1976); "The Chemistry and Physics of Explosives," *Engineering and Mining Journal* (December 1976); "Explosives: The Mechanics of Detonation," *Engineering and Mining Journal* (January 1977); "Commercial Applications of Explosives," *Engineering and Mining Journal* (February 1977).
20. T. L. Davis, *The Chemistry of Powder and Explosives* (Hollywood, CA: Angriff Press, 1943).
21. P. W. Cooper and S. R. Kurowski, *Introduction to the Technology of Explosives* (New York: Wiley-VCH, 1996), 43.
22. Ibid., 43.
23. Ibid., 43.
24. D. Muller et al., "Improved Method for the Detection of TATP after Explosion," *Journal of Forensic Sciences* 49, no. 5 (September 2004): 935–40.
25. S. A. Phillips et al., "Physical and Chemical Evidence Remaining after the Explosion of Large Improvised Bombs, *Journal of Forensic Science* 45, no. 2 (2000): 324–32.
26. D. B. Flemming, "Explosion in Halifax Harbor" (Halifax, NS: Formac, 2004).
27. K. McDonald and M. Shaw, "Identification of Household Chemicals Used in Small Bombs via Analysis of Residual Materials," paper presented at American Academy of Forensic Sciences, February 2009.
28. P. K. Raj, "Exposure of a Liquefied Gas Container to an External Fire," *J. Hazardous Materials* A122 (2005): 37–49.
29. U.S. Bomb Data Center, ATF, Washington DC, 2009.
30. U.S. Department of Justice, *2003 Annual Statistical Report* (Washington, DC: U.S. Bomb Data Center, ATF, March 2004).
31. U.S. Treasury, *1997 Arson and Explosive Incidents Report* (Washington, DC: BATF, 1999).
32. T. G. Gersbeck, "Advancing the Process of Post-Blast Investigation," paper presented at AAFS, Washington DC, February 2008; see also M. G. Haag and T. Grissim, "Technical Overview and Application of 3-D Laser Scanning for Shooting Reconstruction and Crime Scene Reconstruction," paper presented at AAFS, Washington DC, February 2008.
33. DeHaan et al., "Deflagrations Involving Stratified Heavier-than-Air Vapor/Air Mixtures."
34. Harris, *The Investigation and Control of Gas Explosions.*
35. A. Beveridge, ed., *Forensic Investigation of Explosions* (London: Taylor & Francis, 1998); H. J. Yallop, *Explosion Investigation* (Harrogate, England: Forensic Science Society Press, 1980).
36. R. Pape, K. R. Mniszewski, and A. Longinow, "Explosion Phenomena and Effects of Explosions on Structures," Practice Periodical on Structural Design and Construction, American Society of Civil Engineers, 2009 (in press).
37. DeHaan et al., "Deflagrations Involving Stratified Heavier-than-Air Vapor/Air Mixtures."
38. Harris, *The Investigation and Control of Gas Explosions.*
39. G. Kinney and K. Grahm, *"Explosive Shocks in Air"* (New York: Springer-Verlag, 1985).
40. DeHaan et al., "Deflagrations Involving Stratified Heavier-than-Air Vapor/Air Mixtures."
41. R. H. Ide, "Petrol Vapour Explosion: A Reconstruction" in *Proceedings Interflam 1999*, (London: Interscience Communications), 757–63.
42. J. C. Oxley et al., "Improvised Explosive Devices: Pipe Bombs," *Journal of Forensic Sciences* 46, no. 3 (2001): 510–34; K. M. Whitaker et al, "Coming Apart at the Seams: The Anatomy of a Pipe Bomb Explosion," paper presented at AAFS, Denver, CO, February 2009.
43. J. D. DeHaan and W. R. Dietz, "Bomb and Explosion Investigations," National Association of Medical Examiners, September 1987.
44. Whitaker, "Coming Apart at the Seams."
45. Phillips, "Physical and Chemical Evidence Remaining."
46. C. R. Midkiff and R. E. Tontarski, "Detection and Characterization of Explosives and Explosive Residue: A Review," paper presented at 11th International INTERPOL Forensic Science Symposium, Lyon, France, November 1995; Beveridge, *Forensic Investigation of Explosions*; The Working Group on Fires and Explosions, *Guidelines for Explosives Analysis* (University of Central Florida, Orlando, FL: TWGFEX, November 2004); J. Yinon and S. Zitrin, *Modern Methods and Applications in Analysis of Explosives* (Chichester, UK: Wiley, 1993).
47. J. Cochran et al., "Optimization of Explosives Analysis by Gas Chromatography and Liquid Chromatography," paper presented at AAFS, Denver, CO, February 2009;

M. N. Bottegal, B. R. McCord, and L. Adams, "Analysis of Post-Blast Residues of Black Powder Substitutes by Chromatography and Capillary Electrophoresis," paper presented at AAFS, San Antonio, TX, February 2007; A. Gapeev, J. Yinon, and M. E. Sigman, "LC/MS of Explosives: RDX Characterization through Impurity Profiles," paper presented at AAFS, Dallas, TX, February 2004); M. E. Sigman et al, "Matrix Effects on Explosives Recovery and Detection," paper presented at AAFS, Dallas, TX, February 2004. C. H. Wu and R. A. Yost, "Evaluation of Different Atmospheric Pressure Ionization (API) Sources for Use in Explosives Detection," paper presented at AAFS, Denver, CO, February 2009; M. E. Sigman, K. L. Steele, and C. D. Clark, "Discrimination of C-4 Plastic Explosives by GC/MS Analysis of Impurities Associated with the Manufacturing Process of RDX," paper presented at AAFS, San Antonio, TX, February 2007.

48. M. N. Bottegal et al., "Analysis of Trace Hydrogen Peroxides by HPLC-ED and HPLC-FD," paper presented at AAFS, San Antonio, TX, February 2007; M. E. Sigman and D. Clark, "Optimized Analysis of Triperoxide by GC-MS," paper presented at AAFS, San Antonio, TX, February 2007.

49. L. S. Chumbley and F. C. Lacks, "Analysis of Explosive Damage in Metals Using Orientation Imaging Microscopy," *Journal of Forensic Sciences* 50, no. 1 (January 2005): 104–10.

50. E. M. Briley and J. V. Goodpaster, "Practical Applications of Pattern Recognition to the Post-Blast Analysis of Black Electrical Tape," paper presented at AAFS San Antonio, TX, February 2007; A. L. Hobbs et al., "Use of a Database for Significance Assessment and Sourcing of Duct Tapes," paper presented at AAFS Seattle, WA, February 2006; J. G. Rankin, J. A. Meyers, and J. T. Harris, "GCIR as a Tool for Analysis of Smokeless Powder Residues from IEDs,"

paper presented at AAFS, Denver, CO, February 2009; I. Corbin, M. Blas, and B. R. McCord, "Analysis of Smokeless Powder Components by Capillary Chromatography–TOF Mass Spectrometry," paper presented at AAFS, Denver, CO, February 2009; W. Moorehead, A. Tibbetts, and A. Basconcillo, "The Characterization of Reloading Smokeless Powders toward Brand Identification," paper presented at AAFS, Seattle, WA, February 2006.

51. K. L. Sewell et al., "Analysis of Nuclear DNA from Exploded Bomb Fragments and Spent Cartridge Casings," paper presented at AAFS, San Antonio, TX, February 2007.

52. J. Ehleringer and M. Lott, "Stable Isotopes of Explosives Provide Useful Forensic Information," paper presented at AAFS, San Antonio, TX, February 2007.

53. K. Kimura et al., "Gasoline and Kerosene Components in Blood: A Forensic Analysis," *Human Toxicology* 7 (1988); J. Schuberth, "Post-Mortem Test for Low Boiling Arson Residues of Gasoline by GC/MS," *Journal of Chromatography* 662 (1994).

54. DeHaan, J., unpublished case report, 2005.

55. Phillips, "Physical and Chemical Evidence Remaining."

56. K. R. Mniszewski and J. A. Campbell, "Analytical Tools for Gas Explosion Investigation" in *Proceedings Interflam 1998* (London: Interscience Communications).

57. S. G. Davis et al., "Advanced Methods for Determining the Origin of Vapor Cloud Explosions. Case Study: 2006 Danvers Explosion Investigation," *Proceedings ISFI 2010* (Sarasota, FL: International Symposium on Fire Investigation Science and Technology, 2010), 197–207.

58. NIJ Technical Working Group for Bombing Scene Investigation, *A Guide for Explosion and Bombing Scene Investigation* (Washington, DC: National Institute of Justice, June 2000).

59. J. T. Thurman, *Practical Bomb Scene Investigation* (Boca Raton, FL: CRC/Taylor & Francis, 2006).

13
Chemical Fires and Hazardous Materials

Courtesy of U.S. Department of Justice, Drug Enforcement Administration.

KEY TERMS

nonflammable, *p. 557*

OBJECTIVES

After reading this chapter, you should be able to:

- Identify the common causes of fires involving chemicals and hazardous materials.
- List the typical flammable gases found in residential and commercial structures.
- List the typical types of hazardous liquids found in residential and commercial structures.
- List the typical types of solid fuels found in residential and commercial structures that can form incendiary mixtures.
- Describe the chemical compounds associated with the production of improvised high explosives.
- Explain the explosion and fire hazards associated with clandestine drug laboratories and marijuana cultivation.
- Understand the concepts of the warning systems used to label and transport hazardous materials.

Not every building or vehicle contains the same predictable fuels of wood, paper, textiles, and petroleum products. If they did, the fire investigator's task would be fairly simple—the physical properties, flammability, and fire behavior would all fall into fairly narrow categories for easy analysis and reconstruction. Unfortunately, this is not the case; many fires involve and, in fact, begin because of hazardous materials of various types. Not only can these materials make the ignition of a fire much more likely than with "normal" fuels, they can make the fire much more intense, dramatically increase the rate of fire spread, and make the suppression of the fire more hazardous by producing toxic gases. It is beyond the scope of this volume to discuss all possible hazardous materials—solids, liquids, and gases. The reader is referred to specialized texts in this area.[1] Instead, the characteristics of the materials most commonly encountered will be discussed, along with the effects they may be expected to have on a fire scene.

Gases

Hazardous gases are most commonly found in commercial structures like welding and automotive shops, hospitals, laboratories, and fertilizer-manufacturing plants. They may be considered hazardous because of their inherent toxic properties or flammability hazards or simply because even inert gases, when compressed into cylinders for storage, can produce explosive pressures when exposed to an enveloping fire. For the purposes of this book, the toxic gases will not be considered. The reader is referred to other references in that area. (See *NPFA 497, 2008 ed.: Recommended Practice for the Classification of Flammable Liquids, Gases, or Vapors and of Hazardous (Classified) Locations for Electrical Installations in Chemical Process Areas.*)

Compressed gases are usually packaged in metal bottles at pressures up to 20,000 kPa (3,000 psi) for shipment and storage. The containers are normally checked periodically to pressures at several times that level, but when heated, the contents can exceed those pressures until the container fails. The resulting mechanical explosion causes damage to nearby structures similar to that produced by a low-order explosion (see Chapter 12). The damage may often be unidirectional because the container will generally fail in only one direction—along a longitudinal seam or at a welded top or valve, for instance. Even if the gas is inert, the rapidly expanding volume can do serious mechanical damage. If it is flammable, the fuel made available to the fire may have a dramatic effect on the extent of the fire and its rate of spread. A somewhat different problem is encountered when the gas is subject to explosive or detonating decomposition when heated under pressure. Under those conditions, the result is equivalent to the detonation of a large quantity of a high explosive. When such a detonation occurs, massive destruction of the structure surrounding the exploding tank is usually the result along with the creation of an explosion "seat." Fortunately, gases that undergo such explosive decomposition are fairly rare. The chemical properties of some of the more commonly encountered flammable and explosive gases were included in Table 4-1 but will be discussed in more detail here.

HYDROCARBONS

Hydrocarbon gases include methane, ethane, propane, butane, ethylene, and acetylene. They are discussed next, in order of chemical structure from the simplest to the more complex.

Methane (CH$_4$)

Methane is a colorless, odorless gas also known as swamp gas, marsh gas, or sewer gas because it is produced by the decomposition of organic materials in swamps, marshes, sewers, and landfill dumps. Methane is the major constituent of natural gas for heating and power generation. It burns readily in air and can explode if pressurized when heated. It is much lighter than air, having a vapor density of 0.55. Methane has a fairly narrow explosive range, from 5 to 14 percent by volume in air. Methane was once made by heating coal or coke to make town gas or producer gas for illumination and cooking. This process resulted in high concentrations of carbon monoxide (CO) along with the methane. Today it is most often recovered during oil drilling, but methane is also being made as a commercial fuel by the controlled fermentation of agricultural waste products in biogas conversion facilities. The product is odorized with sulfur or mercaptan compounds to ensure leaks are detected; however, under some circumstances, the odorant may be "scrubbed" out of the gas, resulting in an undetected or odorless leak. Methane from sewers or swamps has a strong odor due to the sulfur compounds (chiefly hydrogen sulfide) that accompany it.

Ethane (C$_2$H$_6$)

This colorless, odorless gas is the other major constituent of natural gas fuel. It has a vapor density of 1.03, so it mixes readily with air to form an explosive mixture at concentrations of 3 to 12.5 percent by volume.

Propane (C$_3$H$_8$)

Propane is one of the most common hydrocarbon fuels, since it is a very flammable fuel that is readily liquefied for storage in a variety of containers for vehicular, heating, cooking, and industrial applications. Propane has a boiling point of −42°C (−44°F) and a vapor density of 1.52. Its flammable mixture range is between 2.2 and 9.6 percent in air. It is considered to pose a severe explosive hazard when exposed to flame in suitable concentration.

Butane (C$_4$H$_{10}$)

Butane is a colorless gas that is most often encountered as a component of liquefied petroleum gas used as fuel for heating, cooking, or industrial uses. It is very flammable and will ignite at concentrations between 1.9 and 8.5 percent in air. The gas itself has a negligible odor but is usually (but not always) odorized with methyl mercaptan or sulfur compounds to allow detection of leakage. The gas has a vapor density of about 2.0 and tends to form a low-lying cloud along the ground. Under some circumstances it may be encountered as a liquid, since it has a boiling point of −0.5°C (30°F). Butane from hand-torch cylinders or lighter-refill containers is often used as an extraction solvent in clandestine drug labs or in the production of marijuana "oil" or "hash oil."

Acetylene (C$_2$H$_2$)

Acetylene is a flammable gas used extensively in welding operations. It is particularly hazardous in an enveloping fire because it has a very wide explosive range and a vapor density of 0.9, allowing it to mix readily with air. Acetylene forms an explosive mixture in concentrations between 2.5 percent and 81 percent by volume, but pure acetylene can detonate under certain conditions of heat and pressure. Even when simply burning in air, acetylene can create considerable damage because of the extremely high temperatures developed. In most commercial applications, the acetylene is dissolved in acetone in a cylinder containing a porous cement-like filler. If the contents are released while the cylinder is on its side, liquid acetone will be released along with the gas. Heating of acetylene cylinders can trigger decomposition that can cause explosions some time later, even after the cylinders are cooled.

Ethylene (C_2H_4)

Ethylene is a colorless, pleasant-smelling gas that burns or explodes very readily. It has a vapor density of 0.96, so it mixes readily with air. Ethylene forms an explosive mixture in air at concentrations between 3 percent and 30 percent. It can readily react with oxidizing materials and has been known to explode its compressed gas cylinder when exposed to heat and mechanical shock.

OTHER GASES

Nonhydrocarbons of particular interest to the fire invetsigator include hydrogen, oxygen, ammonia, and ethylene oxide.

Hydrogen (H_2)

Hydrogen is a colorless, odorless gas that is used in a variety of chemical, medical, and industrial processes. It is formed as a by-product of many processes including battery charging and electroplating. Hydrogen is the lightest gas, with a vapor density of 0.07, and dissipates very quickly in open air. It has a wide range of explosive concentrations—from 4 to 74 percent. Because of its flammability and explosive limits, hydrogen is considered by some to be the most hazardous of the compressed gases. Hydrogen gas may also be liberated by the dissociation of water as certain metals, such as aluminum, magnesium, or titanium, burn. For this reason, firefighters should exercise extreme caution when considering the application of water to burning metal fires.

Oxygen (O_2)

While it is not a flammable fuel or toxic gas, oxygen in pressurized tanks can have considerable effect on a fire. Oxygen may be present in industrial applications as a compressed gas or as a supercold liquid (boiling point −183°C; −297°F). When present as a gas, oxygen can dramatically increase the ignitability of ordinary combustibles, the temperatures developed by burning materials, and their rates of combustion. Organic greases in compressed oxygen atmospheres become so readily ignitable (by friction or electrical discharge) that they are forbidden. Many ordinary combustibles when exposed to liquid oxygen become so readily oxidizable that they become virtually explosive. Even materials such as synthetic fabrics and polymers, which are not considered flammable, become hazardous when exposed to liquid oxygen or saturated with gaseous oxygen. Sacks of carbon black saturated with liquid oxygen were used as a cap-sensitive blasting agent for many years.

Ammonia (NH_3)

Ammonia is a colorless, sharp-smelling gas that is used in very large quantities as a fertilizer in agriculture and as a refrigerant in industrial plants. Anhydrous ammonia is the liquefied form of the gas. While not often thought of as an explosive gas, ammonia can explode when mixed with air. It has a vapor density of 0.6, so it behaves like natural gas (methane) when released. Its explosive range is 16 to 25 percent by volume. Ammonia can be corrosive to metal valves, especially brass. It is also extremely toxic, especially because ammonia readily dissolves in water, forming the caustic ammonium hydroxide, which attacks the mucous membranes of the eyes, mouth, and respiratory tract.

Ethylene Oxide (CH_2OCH_2)

Technically a flammable liquid, but with a boiling point of 11°C (51°F), ethylene oxide is usually encountered in gaseous form. It is water soluble, and even dilute (5 percent) solutions in water may be flammable. Like acetylene, ethylene oxide is ignitable over a wide flammability range (3.6 to 100 percent) and can detonate in pure form. It has a vapor density (1.52) similar to that of propane. Ethylene oxide can ignite spontaneously on contact with pure oxides of iron or aluminum, or with alkali metal hydroxides (sodium, calcium, potassium).

Liquids

There are many more hazardous liquids of concern to fire investigators than there are gases, because they are found in more places and because many liquids volatilize to form explosive or combustible vapors at temperatures developed in structure fires. Of course, virtually all the petroleum distillates of concern to fire investigators fall into the category of hazardous liquids because of their flammability. The general properties of these fuels were discussed in Chapter 4 and will not be repeated here. Because of extreme fire hazards and the frequency of occurrence of some particular liquids in industrial fires and clandestine drug laboratories, they will be described in detail here. With very few exceptions, the liquids of interest are clear and colorless (to light amber in color) with various viscosities. Some have characteristic odors, and it is suggested that, because of the somewhat subjective nature of smell, the fire investigator acquire samples of some of the more commonly encountered hazardous liquids and lightly sniff them firsthand to learn to recognize particular fluids by their distinctive smells. The general chemical and physical properties of some commonly encountered liquids and their vapors were described in Chapter 4 but are summarized in Table 13-1.

SOLVENTS

Solvents of particular interest to the fire investigator, grouped by chemical structure, include methyl alcohol, ethyl alcohol, acetone, ethyl ether, methyl ethyl ketone, rubbing alcohol, carbon disulfide, aromatics, bromobenzene, cyclohexanone, nitroethane and nitromethane, and petroleum products.

Methyl Alcohol
Methyl alcohol is a clear, colorless liquid used in many chemical manufacturing processes including drug, plastic, paint, varnish, and glue manufacture. It is also used as a fuel for high-performance auto and boat engines and is also known as methanol, wood alcohol, or wood spirits. Methyl alcohol's low flash point, wide flammability limits, and a vapor density near that of air make it a very dangerous liquid. Because methyl alcohol burns leaving no carbon residues, its flame is nearly transparent and blue in color. It can be burning fiercely and not be visible until the flames impinge on another combustible that produces normal luminous flames. Methyl alcohol's flames have a very high average temperature. It burns to form carbon dioxide and water vapor, but if combustion is incomplete, it produces formaldehyde, a highly toxic material.

Ethyl Alcohol
Although it is best known as a beverage, ethyl alcohol (ethanol or grain alcohol) is not encountered in its pure form except as a laboratory reagent. It is normally stored in water solutions of various concentrations. Its flammability properties depend on its concentration, sometimes referred to (especially in beverages) as "proof" (50 proof = 25 percent alcohol, 100 proof = 50 percent alcohol, and so on). Grain alcohol intended for industrial purposes is most often at 95 percent concentration (190 proof) and is denatured by the addition of a small quantity of a poisonous substance to make it undrinkable. This denaturant may be methyl alcohol, sulfuric acid, benzene, gasoline, or other toxic materials.

Although ethyl alcohol will burn at higher concentrations, it does not produce as much heat per unit mass as do similar volatile petroleum products. It, too, burns with a clear blue flame with little smoke. Because of the relatively low emissivity of burning ethyl

TABLE 13-1 | Properties of Hazardous Liquids

HAZARDOUS LIQUIDS	FLASH POINT (°C)	(°F)	FLAMMABLE LIMITS (%) LOWER	UPPER	VAPOR DENSITY	AUTOIGNITION TEMPERATURE (°C)	(°F)
Methyl alcohol (methanol)	11	52	6.7	36	1.1	385	725
Ethyl alcohol (ethanol)	13	55	3.3	19	1.6	365	689
Acetone	−18	0	2.6	12.8	2.0	465	869
Ethyl ether (diethyl ether) (ether)	−45	−49	1.9	36	2.6	160	320
Methyl ethyl ketone (MEK) (butanone)	−6	21	1.8	10	2.5	515	960
Isopropyl alcohol (rubbing alcohol)	12	53	2.0	12.7	2.1	399	750
Carbon disulfide (carbon bisulfide)	−30	−22	1.3	50	2.6	90	194
Benzene (benzol)	−11	12	1.3	7.1	2.8	560	1,040
Toluene (toluol)	4	40	1.2	7.1	3.1	480	896
p-Xylene	27	81	1.1	7.0	3.7	530	986
o-Xylene	32	90	1.0	6.0	3.7	465	869
m-Xylene	29	84	1.1	7.0	3.7	530	986
Gasoline (100 octane)	−36	−38	1.4	7.6	NA	456	853
Styrene (monomer) (vinyl benzene)	32	90	1.1	6.1	3.6	490	914
Camp stove fuel (white gas, naphtha)	−28	−18	1.5	7.4	NA	335	635
Paint thinner (various)	~40	~100	~0.8	~6	NA	~210	~410
Lacquer thinner (see toluene)							
Petroleum ether (benzine)	<−18	<0	1.1	5.9	2.5	288	550
Cyclohexanone	44	111	1.1 @ 212°C	—	3.4	420	788
Nitroethane	28	82	3.4	—	2.6	414	778
Nitromethane	35	95	7.3	—	2.1	418	785
Diethylamine	<−18	<0	1.8	10.1	2.5	312	594
Gum turpentine	35	95	0.8	UNK	—	253	488

Source: NFPA, *Fire Protection Guide to Hazardous Materials,* 2001 ed. (Quincy, MA: NFPA, 2001).

NA = not applicable; UNK = unknown.

alcohol, it is not a very good radiator of heat, and fires ignited with it grow more slowly than those involving petroleum distillates.

Acetone
A very common solvent in the manufacture of plastics, fabric dyes, lacquers, paints, oils, waxes, and other products, acetone may be encountered in clandestine drug and explosive laboratories. It is a colorless, watery liquid with a mintlike odor, and burns with a yellow flame.

Ethyl Ether
Also known as diethyl ether or ether, ethyl ether is used as a solvent in pharmaceutical formulations and is very common in clandestine drug laboratories. It has been used for many years as an inhalant anesthetic in medicine, but ethyl ether also has wide use as a solvent for

fats, greases, and resins. Because of its extreme volatility, wide range of flammable concentrations, and low ignition temperature, ethyl ether is extremely dangerous as a fresh liquid. Upon aging, it can form unstable peroxides, especially when exposed to air. These peroxides are sensitive to initiation by heat or shock and can cause a detonation. Burning vapors of ethyl ether may be difficult to detect, as the flame produced is nearly transparent.

Methyl Ethyl Ketone (MEK)
Also called butanone, MEK is used in paints, varnishes, lacquers, glues, and general solvents. It may be found as an intermediate solvent in clandestine drug laboratories. It is similar in its physical and combustion properties to acetone.

Cyclohexanone
Cyclohexanone is another ketone like acetone and MEK, with similar physical and chemical properties. Cyclohexanone is a moderate fire hazard that is used in clandestine drug laboratories in conjunction with other flammable solvents.

Rubbing Alcohol (Isopropyl Alcohol)
Usually sold in water solutions (70 percent), isopropyl alcohol is a flammable liquid readily available to consumers. It burns with a yellow, sootless flame.

Carbon Disulfide
A very common solvent in industrial processes, carbon disulfide is highly dangerous in the presence of heat, flame, sparks, or even friction. At normal temperatures it is a toxic material that affects the central nervous system like a narcotic or anesthetic. When heated, carbon disulfide gives off highly toxic sulfur oxides as it decomposes. Because of its volatility and extremely low autoignition temperature (<100°C; 212°F), carbon disulfide is considered one of the most hazardous materials known. It can explode readily at vapor concentrations of around 5 percent but has a wide flammability range of 1.3 to 50 percent.

Aromatics
The aromatics—benzene, toluene, and the xylenes—are often found in combinations of one another because of their similar chemical properties. All of them produce toxic vapors, and all have been linked to some extent with causing cancer in animals chronically exposed to aromatics. Benzene has the lowest flash point and greatest volatility; the properties of an aromatic mixture will largely be determined by the benzene concentration. Benzene is common in the plastic-, resin-, wax-, and rubber-manufacturing industries. Toluene and the xylenes may also be found as solvents for paints, lacquers, dyes, and petroleum products. They are used in making explosives and high-performance aviation fuel. All aromatics burn with a yellow, very sooty flame.

Bromobenzene
Bromobenzene is a low fire and explosion hazard because of its relatively high boiling point (156.2°C; 313°F) and low vapor pressure. It is common in clandestine drug laboratories and can generate toxic bromine-compound fumes when heated to decomposition.

Nitroethane and Nitromethane
Like other nitrated materials, these colorless, slightly oily liquids pose a potential fire and, particularly, explosion hazard. Both have relatively low flash points and ignition temperatures and will generate poisonous nitrogen oxides when heated. When they are contaminated with virtually any organic compound, particularly organic amines, they become potent shock- and heat-sensitive explosives.

Nitroethane and nitromethane pose a potential fire and, particularly, explosion hazard. Although nitromethane is used in large quantities as a lacquer and varnish solvent, when sensitized it becomes an explosive with 95 percent the strength of TNT. Nitromethane is often found in smaller quantities as a fuel for model cars and airplanes and as a fuel additive for gasoline engines.

Petroleum Products

Because virtually all petroleum products of significance in accidental or incendiary fires are a mixture of many separate chemical compounds that constitute a boiling point range or "cut," it should be kept in mind that the flammability hazard of a simple mixture of chemicals is largely determined by the properties of the lightest, most volatile component. For example, although automotive gasoline contains more than 100 compounds, many of which have flash points above 50°C (122°F), the flash point (FP) of gasoline is close to that of its most volatile major component, *n*-pentane (FP: −49°C; −56°F). Petroleum ether constitutes a narrow range of hydrocarbons that distill from petroleum at temperatures between 70°C and 90°C (158°F–194°F). Petroleum ether contains *n*-pentane, *n*-hexane, and *n*-heptane, and hydrocarbons of similar properties. It may be called benzine, benzin, or even naphtha depending on its components and its intended use. It may be considered to have the same hazardous properties as gasoline. Because of its excellent solvent properties, high volatility, and negligible residue, petroleum ether is widely used in pharmaceuticals and is often encountered in clandestine drug operations.

MISCELLANEOUS LIQUIDS

Other liquids of particular interest to fire investigators include organic amines, turpentine, and glycerin.

Organic Amines

Diethylamine and dimethylamine, like most organic amines, have a characteristic fishy odor. While dimethylamine is a gas at room temperature (boiling point: 7°C; 44°F) it is most commonly encountered as a water solution that is considered to be a flammable liquid. Dimethylamine has an ignition temperature of 400°C (725°F). Diethylamine is a liquid at room temperature (boiling point: 57°C; 134°F) but has a very low flash point (−23°C; −9°F) and an ignition temperature of only 312°C (594°F). Both diethylamine and dimethylamine are extremely toxic and are important precursors in the illicit manufacturing of LSD and methadone and may be encountered in fires involving clandestine laboratories.

Turpentine

Gum turpentine, not to be confused with mineral spirits or mineral turps, is an oil obtained by steam distillation of the resins of various pine and fir woods. It consists largely of the terpenes and other oleoresins found in the wood of those trees.[2] Because it is a natural product, turpentine will vary depending on the variety of woods from which it is made and on the gums or oils added to it for particular properties. Turpentine is considered a flammable liquid, although its high flash point (35°C; 95°F) makes it difficult to ignite at room temperature in the absence of a wick. Turpentines of various types are widely used in the paint, rubber, and plastics industries.

Glycerin

Glycerin (also glycerine and glycol) is a combustible liquid with a very high flash point that is completely soluble in water. Because it is safe for foods, it is widely used in food wrappings and cosmetics as a plasticizer. It is also used in paints, cigarettes, and the manufacture of nitroglycerin explosives. It is becoming widely used as an antifreeze additive in domestic fire sprinkler systems. Unlike ethylene glycol, glycerin is compatible with plastic sprinkler piping and cements.

Solids

Most fuels in a structure fire are solids—wood, cloth, carpet, paper, and the like—converted to combustible vapors by pyrolytic action of the flames. There are some solids that, by virtue of their chemical properties, pose special hazards in terms of their flammability or speed of combustion. These solids will most often be encountered as part of the fuel load of commercial or manufacturing structures when they burn. As such, they can contribute significantly to the speed of and intensity of the fire. If the fire investigator is not aware of the potential damage that can be caused by these compounds, he or she is liable to misinterpret the diagnostic signs left by their combustion. In addition, it is precisely because of the ease of ignition or the intensity of reaction that certain materials are used to initiate incendiary fires in all types of targets. It is because of their unusual fire properties that certain hazardous solids will be examined here.

INCENDIARY MIXTURES

The most common incendiary solids are simple mechanical mixtures of various components—usually a fuel and an oxidizer, sometimes with initiators, catalysts, and the like, added to increase the efficiency of the reaction. Some of these mixtures include black powder, flash powder, matches, safety flares or fusees, thermite, Cadweld, and aluminium powder/plaster.

Black Powder

An explosive used in blasting, early firearms, and fireworks, black powder, or gunpowder, is a mixture of potassium nitrate, charcoal, and sulfur, usually in ratios of about 75:15:10. (The ratio is changed to affect the rate of burning.) When dry, black powder can be ignited by a small spark, flame, or friction. It has an approximate AIT of 350°C (662°F). It can be rendered inert by absorbing water from the atmosphere. Temperatures on the order of 2,000°C (3,600°F) are produced during its combustion. When unconfined, black powder deflagrates and can be used as a trailer to transmit flame from one point to another. (The chemistry of explosive combustion of materials was treated in more depth in Chapter 12.) It forms the central flammable core in safety fuse. Black powder is now made in the United States at only one plant in Pennsylvania for military, antique firearms, and fireworks users.[3]

Flash Powder

Flash powder, which is a mixture of a readily oxidizable metal such as powdered aluminum or magnesium with a strong oxidizer such as potassium perchlorate or barium nitrate, will produce an intensely hot flame in an almost instantaneous deflagration. (Sulfur is usually present to promote uniform burning.) These mixtures require minimal energy in the form of spark, heat, or friction to ignite and can produce temperatures in excess of 3,000°C (5,400°F). They are most often encountered in small explosive devices (including those sold as cherry bombs or M-80s) and their larger, more destructive variants—military artillery simulators. Because of their sensitivity and destructive potential, devices using them should be opened or dismantled only by qualified EOD personnel, even for evidentiary analysis. As discussed previously, the illicit factories set up to make devices using these fillers pose serious explosion hazards because of airborne dusts, thin films on working surfaces, and product stored in bulk. Some powders include a sulfur and potassium chlorate mixture. The self-ignition propensity of this mixture has been well known since the 1890s. It has been banned from commercial fireworks by most Western countries but is occasionally found in devices made in China or Mexico.[4]

Matches

Although innocuous enough in single use, the ignitable heads of common matches can be combined into substantial incendiary devices. Safety matches contain sulfur (fuel), potassium chlorate (oxidizer), and a variety of glues, binders, and inorganic fillers. When ignited by flame or spark, this mixture produces considerable heat and evolves sufficient gases to permit match heads to be used as propellants in improvised firearms. Such matches normally require frictional contact with a striker containing red phosphorus and powdered glass in a binder. By contrast, "strike-anywhere" matches usually consist of two layers of material with different flammability characteristics. Both contain potassium chlorate, phosphorus sesquisulfide (P_4S_3), rosin, and powdered glass, but the upper "igniter" button contains a higher percentage of P_4S_3 and glass, while the lower contains paraffin and sulfur.[5] This combination allows the tip to ignite when struck against any hard surface, while the lower part has good flame characteristics but poor frictional sensitivity. Match strikers in bulk are extracted with a volatile solvent to provide phosphorus for methamphetamine production.

Safety Flares or Fusees

Safety flares and fusees contain a "fuel" of sulfur, wax, and sawdust and an oxidizer of strontium nitrate (for burning with a red flame) and potassium perchlorate. These devices burn with a very hot (in excess of 2,000°C; 3,600°F) flame and leave a white or gray lumpy residue consisting almost entirely of oxides of strontium. The mixture requires a high-heat source for ignition, and an igniter button is built into the end of each flare to provide it (see Figures 8-23 and 8-24).

Thermite (or Thermit)

A mixture of powdered aluminum and iron oxide (Fe_2O_3), thermite, once ignited, will burn with extreme violence, spattering molten metal and producing temperatures in excess of 2,500°C (4,500°F)—that is, sufficient heat to melt iron, steel, and even concrete. Fortunately, the reaction requires considerable heat input to begin. Usually, a magnesium strip is ignited as an initiator. Once ignited, a thermite fire is very difficult to extinguish because it supplies its own oxygen.

Cadweld

A mixture of powdered aluminum–copper alloy and iron powder that functions in the same way as thermite, Cadweld produces a pool of molten copper as a residue and is used to weld large copper buss bars in heavy-duty electrical equipment. The temperature produced is on the order of 2,000°C (3,600°F). The mixture requires a high-temperature igniter, but the commercially prepackaged containers include a small quantity of magnesium powder so that the whole sample can be ignited with a match.

Aluminum Powder/Plaster

A mixture of fine aluminum powder, mixed with plaster of Paris and cast into ice cube–size blocks with a magnesium ribbon igniter was used as an improvised incendiary device in World War II and has much more recently been suspected as an ignition source in an arson case. It burns similarly to thermite, with a very hot, intense flame.

OXIDIZING SALTS

Although they will not be treated individually here, all compounds containing chlorate or perchlorate salts must be considered potentially hazardous materials because of their potency as oxidizing agents. These materials in combination with virtually any oxidizable substance can produce deflagrations with potential for serious structural damage. The same holds true for nitrate salts of inorganics, or nitrate esters of organics (like cellulose). Virtually all nitrogen-containing compounds are unstable to some extent, and nitrates are generally even more unstable as very active oxidizers. Some inorganic nitrates such as sodium nitrate are used as a substitute for more expensive explosives in dynamites and blasting agents. Although ammonium nitrate in its pure form will explode when heated, when it is contaminated with any organic material (such as charred paper or oil) to act as a fuel, it explodes with less stimulation and considerably more force.[6]

This phenomenon is responsible for the frequent use of ammonium nitrate–fuel oil (ANFO) as a blasting agent for both legitimate and criminal purposes. For many years ammonium nitrate fertilizer was thought to be so stable that dynamite was used to break up masses of it in damp warehouses. It was when these same piles of fertilizer were contaminated with oil that they detonated in sympathy with the smaller dynamite charges. Exposure to fire can also cause detonation. This mechanism was the cause of the Texas City,

Texas, explosion of 1947.[7] Sodium chlorate recovered from home-hobbyist welding torches or from weed killers has been used as an oxidizer in improvised explosive and incendiary devices.

REACTIVE METALS

Reactive metals include sodium metal, potassium, phosphorus, and magnesium.

Sodium Metal

Sodium metal is used in drug synthesis and related organic chemistry reactions. It is a soft metal that looks like silvery white putty. It is normally stored in a kerosene or oil bath because it reacts strongly with any water available, even that in the air or on human skin. When exposed to dry heat, sodium metal gives off very toxic oxides and in a general fire will flame spontaneously in air. It has an ignition temperature of about 115°C (240°F). In water, it generates large quantities of explosive hydrogen gas and caustic sodium hydroxide (lye) in a violently exothermic reaction, which spews flame and spatters hot lye and molten sodium.

Potassium

Potassium is used less frequently in chemical synthesis than sodium metal but presents the same serious hazards as sodium. In its pure form it is a silvery metallic crystal. Potassium generates hydrogen gas, potassium hydroxide, and a great deal of heat on contact with water. In addition, potassium can react with atmospheric oxygen to form explosive peroxides and a superoxide that form a yellowish layer on the metal. These oxides can detonate when initiated by friction. In moist air, potassium can develop sufficient heat to ignite spontaneously.

Phosphorus

Phosphorus occurs in two forms of waxlike metal: one white (or yellow) and the other red. The two forms have quite different properties. White phosphorus is one of the few known materials that will burst into flames when exposed to air, so it is stored under water or a petroleum distillate. It has an ignition temperature of 30°C (86°F) and will react with most oxidizers, producing very toxic phosphorus oxides. In spite of its flammability, white phosphorus is not often encountered in accidental or incendiary fires.[8] It has been used in the form of a dispersion of finely powdered white phosphorus in a volatile solvent, called "Fenian fire." When this substance is thrown onto a target, the solvent quickly evaporates, and the phosphorus bursts into flame.

If exposed to sunlight, white phosphorus can crystallize into the more stable red phosphorus, which does not burn when exposed to air and has an ignition temperature of 360°C (500°F). It will burn, however, and react violently, even explode, on contact with strong oxidizers. It is used as an igniter in frictional ignition devices, including those on matches and safety flares. It has found increasing favor in illicit methamphetamine manufacturing, where a mixture of red phosphorus and other precursors is often allowed to react unmonitored.

Magnesium

When people imagine the hottest flame possible, they often think of burning magnesium. Although not readily ignitable in large masses with small surface areas, magnesium shavings, powder, or ribbon can readily be lit with a common match. Magnesium fires can reach temperatures of 3,000°C (5,400°F) and produce toxic oxides, most often a dense, white, fluffy powder. When heated, magnesium, can react with water or steam and produce flammable hydrogen gas as well as heat. See Figure 5-26 for an example of a magnesium fire.

Clandestine Laboratories

As used here, the term *clandestine laboratories* refers to clandestine drug labs, marijuana cultivation, and clandestine explosives laboratories.

CLANDESTINE DRUG LABORATORIES

A significant number of structure fires are caused every year by accidental ignition of the solvents or vapors associated with the illicit manufacturing of drugs. The Drug Enforcement Administration (DEA) reported that seizures of clandestine drug labs and dumpsites have decreased dramatically since 2005, when 12,619 were seized, down to 5,910 in 2007, and 6,783 in 2008. The highest numbers are no longer recorded in California (286–470 in that period); they are now reported in larger numbers in Missouri, Indiana, Illinois, and Tennessee.[9] Some 97 percent of the labs seized were reportedly producing methamphetamine, but amphetamine, phencyclidine (PCP), methadone, methaqualone (very rare), and hallucinogens such as lysergic acid diethylamide (LSD), and methylene dioxyamphetamine (MDA) or methylene dioxymethamphetamine (MDMA) are also found.[10]

The most common method of large-scale (10 lb or more) illicit manufacture of methamphetamine uses hydriodic acid (HI) and red phosphorus. Hydriodic acid is extremely corrosive, and red phosphorus can be converted to the more reactive white (or yellow) forms by the application of heat. Large glass flasks containing these reactants readily overheat and boil over, starting fires in all manner of buildings. Small batches of methamphetamine can be made in a bathroom sink with just a few glass jars (as illustrated in Figures 13-1a and b). The reaction uses anhydrous ammonia, metallic (elemental) lithium (or sodium), and iodine. The lithium may be purchased or extracted from lithium batteries; the iodine may be purchased in solid form or extracted from medical iodine solutions (tinctures), and the phosphorus extracted from the striker strips of matchbooks. The starting materials themselves are often extracted from large quantities of over-the-counter cold tablets. Anhydrous ammonia, used extensively as an agricultural fertilizer, has been stolen in quantities from gallons to tractor-trailer loads. Sometimes the anhydrous ammonia is transferred to empty propane tanks. It is corrosive to the tank fittings and will cause them to leak badly, as shown in Figure 13-2.

Until recently, Freon was used as the extraction solvent, but with the increasing scarcity of fluorocarbons, flammable solvents are used. If bulk anhydrous ammonia is not available, ammonium nitrate fertilizer can be converted to anhydrous ammonia using household lye. Empty lye containers or fertilizer bags would be significant evidence of the use of this process.

(a)

(b)

FIGURES 13-1A AND B Before-and-after (fire) photos of a mock-up of a common small-batch methamphetamine lab. Jars, cans, and tubing could easily be overlooked as evidence if the fire damage were more extensive. *Courtesy of John D. DeHaan.*

FIGURE 13-2 LP bottle used to transport anhydrous ammonia with cross-threaded "adapter." Green/blue corrosion on brass valve will eventually cause leakage of valve. Note red phosphorus reaction in glass jar behind bottle. *Courtesy of the U.S. Department of Justice, Drug Enforcement Administration.*

All these extraction and manufacturing processes involve the use of large quantities of flammable liquids—methyl alcohol, butane, acetone, lacquer thinner, ethyl ether, toluene, petroleum ether, camping fuel, and even gasoline—as solvents (which are then often evaporated by heating on hotplates or cookstoves). Many of the intermediate reactions take place in such solutions, and almost invariably the final product is "salted out" of a solution using hydrogen chloride (HCl) gas or even table salt. The resulting mixture must then be allowed to dry, and the room or building then fills with the vapors of evaporating solvent—often to explosive concentrations. The explosions are initiated by the electric arc of a switch or motor, the overheating of reactant materials, the flames of a cookstove, or the match of an anxious observer. The ensuing fire is fueled by the large volumes of available flammable liquids on hand. Because of the potential profits, the large-scale laboratories can be surprisingly sophisticated, well equipped, and capable of turning out large quantities of fairly pure dangerous drugs. (See Figure 13-3, for example. An example of a typical "backyard" meth lab is shown in Figure 13-4.)

The newest twist in clandestine methamphetamine manufacture is called the "one pot" method. Here, pseudoephedrine from cold medications is mixed with ammonium nitrate, lithium metal (from lithium batteries), lye (sodium hydroxide), water, and a flammable solvent such as ethyl ether or white gas in a single container such as a 2-L soft drink bottle, or similar jar or metal can. The reaction produces heat and a great deal of hydrogen gas. If the container is not vented repeatedly, the pressure can cause a mechanical explosion of the container. This explosion releases hydrogen, hot flammable liquid, hot lithium metal, and caustic hydroxides. In about 20 minutes, the chemical reaction produces methamphetamine base (oil) in the solvent, which then must be "salted out" with HCl gas. This gas is often generated in a separate jar with table salt and battery acid, or metal strips with hydrochloric acid. The final result, once filtered and washed, is a relatively pure meth with no telltale odors of bulk chemicals.[11]

FIGURE 13-3 Clandestine drug laboratories can include reactive and toxic materials in elaborate and expensive glassware and other preparation equipment. Investigators should be cautious so as to avoid disturbing any reactions in progress. *Courtesy of the U.S. Department of Justice, Drug Enforcement Administration.*

To avoid detection, these laboratories are usually located in remote houses, garages, shacks, barns, or even trailers—anywhere electric power can be made available. More frequently, clandestine labs are appearing in motorhomes and camping trailers as isolated, self-contained plants with their own power generators or batteries and closed water-cooling systems. Clandestine drug labs have also been found in motels, rented houses, or apartments in urban areas, sometimes close to police and fire facilities.

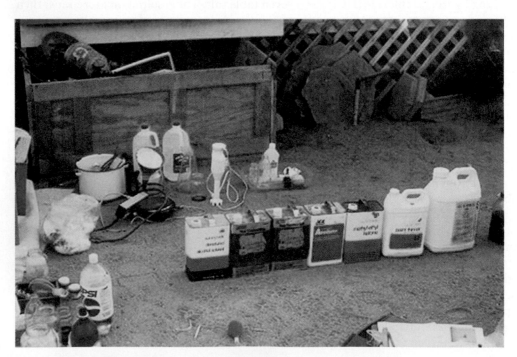

FIGURE 13-4 Typical "backyard" meth lab with hot plate, mixers, extraction solvents and extraction/reaction vessels (soda bottles). *Courtesy of the U.S. Department of Justice, Drug Enforcement Administration.*

FIGURE 13-5 Pseudo-ephedrine extraction of powdered diet tablets using automotive fuel additive. Lithium will be extracted from the batteries. *Courtesy of the U.S. Department of Justice, Drug Enforcement Administration.*

The finding of one or more of the following at a fire scene should warn the fire investigator that he or she is dealing with a potential clandestine drug lab:

- Large numbers of empty containers for the following:
  - Cold/allergy medicines or diet aids with pseudoephedrine (see Figure 13-5)
  - Lithium batteries (e.g., for cameras, watches, or calculators)
  - Butane (lighter refills)
  - Lye (drain cleaner)
  - Camping fuel or automotive fuel additive (see Figure 13-5)
  - Muriatic acid (hydrochloric acid)
  - Iodine solution
  - Dimethyl sulfone (MSM) (used as a diluent or cutting agent)
- Cylinders or tanks of the following:
  - Anhydrous ammonia
  - HCl (hydrogen chloride)
  - H_2 (hydrogen)
  - LP gases (modified or corroded)
- Matchbooks without striker strips (in very large quantities)
- Multiple hotplates, electric skillets, or camp stoves
- Large quantities of alcohol (methyl, ethyl, or isopropyl), acetone, or MEK
- Ice-water baths in sinks, pans, or beer coolers with miscellaneous jars or tubing
- Residues dumped outside that look like oatmeal

Because of the complexity of the chemistries involved and the nature of the hazardous materials present—flammable solvents; organic amines; methylamine; strong acids [sulfuric (H_2SO_4), hydrochloric (HCl), and especially hydriodic (HI)]; caustic bases; dangerous gases such as H_2, HI, and HCl; and extremely reactive materials such as anhydrides, phosphorus, metallic lithium, sodium, and potassium (not to mention the toxic intermediate products including iodine vapors and phosphine gas)—these scenes are very hazardous to investigate. At the first appearance of stored chemicals, glassware, laboratory equipment, pumps, ovens, and related apparatus, the assistance of a qualified forensic drug chemist is essential, and the

Flammable, caustic, and highly toxic chemicals in gaseous, liquid and solid forms may be encountered in such labs. Proper PPE is essential.

agency should seal off the scene and notify the appropriate narcotics enforcement team or DEA. In the interim, personnel should not turn on equipment that is off, should not turn off equipment that is on, and should not remove bottles, jars, or flasks from ice-water baths, if present. Many of the synthesis reactions are strongly exothermic and are being cooled by immersion or circulation to prevent a runaway buildup of heat that can cause the reactant mass to explode. All such scenes should be approached with extreme caution because of the hazards of the solids, liquids, and even gaseous materials present.

MARIJUANA CULTIVATION

"Grow-ops" for the concealed cultivation of marijuana have been found in every type of structure in nearly every locale in North America—private residences, commercial premises, apartments, storage rental units, underground mines, even motorhomes and trailers. Except for the flammable solvents used for extraction of "hashish" oils, their main fire risk is due to the massive power needs involved in the high-intensity lamps used to promote rapid growth. Power connections (often illicit bypasses) can be haphazard or very sophisticated. Electrical fires caused by these connections are the most common means by which these "grow-ops" are discovered (as in Figures 13-6a–c). Fans, pumps (for water or hydroponic solutions), timers, and the lamps all provide ignition sources.[12] The high-intensity

FIGURE 13-6 Marijuana "grow operations" are often detected when illicit power connections (meter bypasses) fail. *Courtesy of Vic Massenkoff, Contra Costa County Fire Dept.*

lamps used (often metal halide) can also ignite structural materials or growing plants by direct contact, or ignite nearby wood by radiant or conducted heat.

CLANDESTINE EXPLOSIVES LABORATORIES

As discussed in Chapter 12, improvised explosive mixtures are becoming more common. These typically involve hydrogen peroxide solutions and acetone (from consumer products). The intermediate and final products are all hazards due to their extreme sensitivity and explosive power. The volatile solvents pose the usual fire risks. EOD units should handle all suspected explosive evidence and decontamination.

Warnings

There are two different warning systems in common use in the United States to warn those handling or transporting hazardous materials of the types and severities of dangers posed if a material is exposed to heat or fire. These systems are the *NFPA 704* system and the Department of Transportation Hazardous Materials Transportation System.

NFPA 704 SYSTEM

NFPA 704, Standard System for the Identification of the Hazards of Materials for Emergency Response, is used on storage tanks, exteriors of buildings, entry doors, pipelines, and the like. This system warns of the hazards relative to health, flammability, and chemical reactivity with a numbering system in which 0 represents no significant hazard, and 4 represents the highest degree of that hazard. The NFPA label is in the form of a diamond-shaped sign, as shown in Figure 13-7.

The topmost quadrant (often color-coded red) is the fire hazard. The left-hand quadrant is the health hazard (often color-coded blue), and the right-hand quadrant is the chemical reactivity hazard (susceptible to release of energy) and is usually color-coded yellow. The bottom quadrant is for special warnings when appropriate: the three-bladed propeller for radiation hazard; the letter W with a horizontal line through it to designate materials that are reactive to water (and therefore water should not be applied during fire emergencies). The letters OXY in the lower quadrant signify that the material is an oxidizer, while the letter P designates materials that can undergo polymerization (an exothermic reaction). The numeric designations are graded as follows:

A. Health Hazard
4: Materials that can cause death or major injury upon very short exposure even if medical aid is given immediately
3: Materials that can cause serious temporary or residual injury on short exposure
2: Materials that could cause temporary incapacitation upon intense or repeated exposure unless medical aid is offered immediately

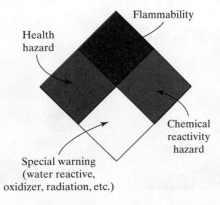

Health hazard

Flammability

Chemical reactivity hazard

Special warning (water reactive, oxidizer, radiation, etc.)

FIGURE 13-7 *NFPA 704* system for labeling hazardous material containers. *Courtesy of Meyer, E. E., Chemistry of Hazardous Materials, 5th ed. Upper Saddle River, NJ: Pearson/Brady, 2010.*

1: Materials that can cause irritation only if no medical aid given

0: Materials that offer no hazard beyond that of any combustible material

B. Flammability

4: Materials that will evaporate or vaporize completely on exposure to air at normal temperature and pressure, will mix with air, and will burn readily

3: Liquids and solids that can be ignited under almost all ambient conditions

2: Materials that will ignite if heated moderately or exposed to high ambient temperatures

1: Materials that must be preheated before they will ignite

0: Materials that will not burn

C. Reactivity

4: Materials that are capable of detonation or explosive decomposition at normal temperatures

3: Materials that are capable of detonation or explosive reaction but require a strong initiator, or heating under confinement, or water reactivity

2: Materials that become unstable at elevated temperatures and pressures, may react violently with water, or form potentially explosive mixtures with water

1: Materials that can become unstable at elevated temperatures or pressures or react with water

0: Materials that are stable even under fire exposure conditions, and are not reactive with water

FEDERAL HAZARDOUS MATERIALS TRANSPORTATION SYSTEM

There are also designations for hazardous materials under the U.S. Dept. of Transportation under the Federal Hazardous Materials Transportation Law (49 C.F.R. Parts 171–179). Under this system, each material classified as hazardous has a three- or four-digit number preceded by the prefix UN for United Nations (international shipping) or NA for North America (United States Canada, and Mexico). DOT labels are diamond-shaped and color-coded to a specific hazard class and usually include a warning pictograph (Figure 13-8). They are affixed to the outside of shipping containers (boxes, drums,

FIGURE 13-8 DOT placard on fuel truck (Biodiesel). *Courtesy of John D. DeHaan.*

etc.) and in some cases to be on placards on a vehicle transporting them. The classifications are as follows:

Class 1: Explosives, with subdivisions reflecting sensitivity: 1.1 for detonating explosives, to 1.6 for materials with a negligible probability of accidental initiation

Class 2: Gases (flammable, **nonflammable**, poisonous)

Class 3: Flammable liquids (materials with a flash point of less than 60.5°C (141°F)

Class 4: Flammable solids, spontaneously combustible materials, or materials that are dangerous when wet

Class 5: Oxidizers and organic peroxides

Class 6: Poisonous materials and infectious materials

Class 7: Radioactive materials

Class 8: Corrosive materials

Class 9: Miscellaneous hazardous materials

nonflammable ■ Material that will not burn under most conditions.

The reader is encouraged to review the emergency response literature for more specific information. Note that under this classification system, liquids with a flash point of less than 60°C (141°F) are classified as flammable liquids (as opposed to the 38°C (100°F) classification used by NFPA.

Summary

Hazardous materials of the types discussed in this chapter are one of the problem areas facing the fire investigator. They can be the cause of the fire or explosion (as the first fuel ignited or source of heat) or involved in the fire as an additional fuel. Because of their reactivity or intensity of reaction, their presence can significantly change the behavior of a fire. If a fire appears to violate the rules, the investigator must consider these materials whenever the circumstances indicate. Like all other fuels, however, they follow basic principles of combustion.

Review Questions

1. Name three of the most hazardous flammable hydrocarbon gases.
2. Name three flammable nonhydrocarbon gases.
3. Why are ethyl ether and carbon disulfide so dangerous from a flammability standpoint?
4. What is thermite (or thermit) and what residues does it leave behind?
5. Name three metal–oxidizer mixtures that can be used as incendiary materials.
6. Name four of the chemicals used in illicitly manufacturing methamphetamine.
7. What solvents have been found in clandestine drug labs?
8. List five of the waste materials or discards that indicate a clandestine meth lab is (or was) present.
9. Why are reactive metals so dangerous?
10. Why do nitromethane and nitroethane pose a special safety risk?

References

1. C. F. Turner and J. W. McCreery, *Chemistry of Fire and Hazardous Materials* (Boston: Allyn and Bacon, 1980); *DOT Hazardous Materials Emergency Response Guidebook* (Washington, DC: U.S. Government Printing Office, 2008) (see http://www.phmsa.dot.gov/hazmat/erg for English and Spanish editions); NFPA, *Fire Protection Guide to Hazardous Materials,* 2001 ed. (Quincy, MA: National Fire Protection Association, 2001); N. I. Sax, *Dangerous Properties of Industrial Materials,* 10th ed. (New York: Van Nostrand Reinhold, 2000).
2. M. A. Trimpe, "What the Arson Investigator Should Know about Turpentine," *Fire and Arson Investigator* 44, no. 1 (September 1993): 53–55.
3. J. A. Conkling, "Chemistry of Fireworks," *Chemical and Engineering News,* June 29, 1981, 24–32.
4. H. G. Tanner, "Instability of Sulfur–Potassium Chlorate Mixtures," *Journal of Chemical Education* 36, no. 2 (February 1959): 58.
5. H. Ellern, *Modern Pyrotechnics* (New York: Chemical Publishing, 1961).
6. USFA, "Six Firefighter Fatalities in Construction Site Explosion, Kansas City, Missouri (November 29, 1988)," FEMA, USFA National Fire Data Center.
7. H. W. Stephens, *The Texas City Disaster—1947* (Austin, TX: University of Texas Press, 1997).
8. E. Meyer, *Chemistry of Hazardous Materials,* 5th ed. (Upper Saddle River, NJ: Pearson/Brady, 2009).
9. U.S. Department of Justice, Drug Enforcement Administration, www.justice.gov/dea/index.htm, March 2009.
10. U.S. Department of Justice, Drug Enforcement Administration, EPIC, September 2000.
11. Forensic Applications Consulting Technologies Inc., Bailey, CO, personal communications November 24, 2009; D. Hunt, S. Kuck, and L. Truitt, *Methamphetamine Use: Lessons Learned,* final report to the National Institute of Justice (NCJ 209730), available at www.ncjrs.gov/pdffiles1/nij/grants/209730.pdf (Cambridge, MA: Abt Associates, January 31, 2006).
12. C. Reed and A. Reed, "Fire Investigation of Clandestine Marijuana Grow Operations" in *Proceedings Fire and Materials 2005* (London: Interscience Communications), 365–78.

14
Laboratory Services

Courtesy of John D. DeHaan.

adsorption, *p. 570*

chromatography, *p. 562*

fingerprints, *p. 596*

mass spectrometry, *p. 566*

spoliation, *p. 595*

OBJECTIVES

After reading this chapter, you should be able to:

■ Understand the role of laboratory services in fire and explosion investigations.

■ Describe the various types of services and tests available from forensic and fire testing laboratories.

■ List and identify the common types of evidence found at fire scenes.

■ Identify examples of how failure analysis by forensic engineers can be used to help determine fire causation.

■ Explain the basic concepts of how volatile accelerants are identified using gas chromatography and mass spectrometry.

■ List the various types of techniques used to isolate and identify volatile residues.

■ Identify the common types of materials found in chemical incendiaries.

■ Describe investigative methodology for identification and preservation of non-fire-related criminal evidence.

■ Search the Internet for the appropriate and up-to-date guidance on standards used by public and private forensic laboratories.

Even an experienced fire investigator often requires an analysis or interpretation of the physical evidence relating to fire causation or spread that entails the services of a laboratory specialist. In fact, it is the wise and experienced investigators who recognize the limitations of their own knowledge, beyond which they must take advantage of other experts to ensure that the investigation is complete and accurate. The less experienced may try to overextend their expertise and offer erroneous or misleading interpretations of evidence, which can embarrass them and their clients.

Today, courts expect fire investigators to rely on properly conducted tests for critical data. It is well worth the time and effort for the investigator to anticipate the need for such experts and determine their availability before they are needed in a crucial case.*

Availability of Laboratory Services

FORENSIC LABORATORIES

Most states in the United States have at least one state criminalistics laboratory that can provide most of the analytical services a fire investigator will need. In fact, many states have a system of regional laboratories to minimize travel time and expense. These laboratories, operating under the auspices of the state's department of justice, public safety, state prosecutor, state police, or similar law enforcement authority, work in cooperation with the city or county laboratories found in many areas to provide a wide range of services. On a national level, the laboratories of the Federal Bureau of Investigation and Bureau of Alcohol, Tobacco, Firearms and Explosives (ATF) are well known. In Canada, fire debris analysis is provided by the Centre of Forensic Sciences in Toronto and Sault St. Marie, the *Laboratoire de sciences judiciaires et de médecine légale* in Montreal, and through the Forensic Science and Investigation Service of the Royal Canadian Mounted Police, which operates laboratories in six cities across the country. With some exceptions, the services from these public laboratories are available at no cost to personnel from public fire or police agencies and to investigators from some semiprivate agencies such as transit authorities, public utilities, and railroads. Due to limited laboratory resources, there may be restrictions on which agencies they are permitted to serve. In addition, there may be a case priority scheme in operation that relegates fire-related evidence to low-priority status. Because the priority is often set on the basis of whether there were deaths or injuries or on a court date's having been set, that information should be provided to the

---

*Many such experts may be described as *forensic scientists*, forensic chemists, forensic engineers, and the like. The term *forensic* means having to do with a court of law, so such terms refer to the application of a scientific discipline to problems having to do with the law.

lab when the evidence is brought in to ensure fastest possible service time. The investigator should contact the nearest laboratory before its services are needed, so that time is not wasted just in finding a laboratory.

Private investigators, whether retained by public agencies such as public defenders or by private or corporate clients in civil or criminal cases, have a more difficult and costly problem. Although a few public laboratories will service public defenders and even private-sector investigators on a contractual basis, for the most part, private-sector investigators must turn to private laboratories. Many large insurance and engineering firms, in addition to providing laboratory services to their own investigators, offer such services to outside clients on a fee basis. There are also small, private laboratories that can perform appropriate tests on a per-case, per-analysis, or hourly fee basis and provide expert testimony as to their results. Finally, there are qualified consultants who will evaluate the evidence and provide interpretation and expert testimony while contracting with others for the actual laboratory work. There are many more qualified, private-sector facilities available now than just a few years ago. Many also do work (on a fee basis) for public-sector agencies whose own labs cannot provide timely services.

FIRE TESTING LABORATORIES

There are also many tests that forensic laboratories cannot provide. These include heat release rate testing, flame spread, ignitability, and room (even multiroom) fire tests. Public agencies in the United States may qualify for assistance from the Fire Research Laboratory operated by ATF in Annandale, Maryland. Private laboratories such as Omega Point (Elmendorf, Texas), Southwestern Research Institute (San Antonio, TX), MDE (Seattle, WA), Underwriters Laboratories (Northbrook, IL), Chilworth Pacific Fire Testing, (Kelso, WA), and Western Fire Center (Longview, WA) have provided qualified testing and analysis services. Some of the ASTM tests such labs can provide have been described in previous chapters.

EXPERT QUALIFICATIONS

The most important consideration is not whether a lab has the necessary equipment but whether it has the experience and specialist training necessary to provide accurate interpretation. The effects of fire, evaporation, exposure, and even sample collection and storage on physical evidence from fires must be well understood by the forensic laboratory analyst if critical errors in interpretation are to be avoided. Simply having a chart or numerical reading does not do the investigator any good. Only a *qualified* analyst will be able to correctly interpret that data and guide the investigator in assaying the contribution of the evidence to the case. Unfortunately, judging someone's qualifications is much more difficult than dialing a phone number, and there have been instances of malfeasance and even fraud on the part of some laboratories.

The International Association of Arson Investigators (IAAI) first addressed the issue of laboratory qualifications by establishing a Forensic Science Committee and developing consensus guidelines on the analysis and interpretation of fire-related evidence.[1] The American Society for Testing and Materials (ASTM) used that IAAI document as the foundation for what is now a series of test methods, guides, and practices covering many of the techniques used in fire debris analysis, including *ASTM E1618: Standard Test Method for Ignitable Liquid Residues in Extracts from Fire Debris Samples by Gas Chromatography-Mass Spectrometry*[2] and *ASTM E1492: Practice for Receiving, Documenting, Storing, and Retrieving Evidence in a Forensic Science Laboratory.*[3] (Additional ASTM methods will be described in following sections.) These methods were developed by consensus of the scientists involved and they can be used by any analyst. Adherence to them indicates a professional attitude on the part of the analyst. The American Society of Crime Laboratory Directors (ASCLD) has established a protocol for accreditation of forensic laboratories that calls for use of such standardized methods for analysis and interpretation, as well as record keeping. Finally, the American Board of Criminalistics (ABC) has established a certification program for individuals involved in fire debris analysis. Scientists who pass a written specialty knowledge exam and who participate in a quality assurance testing program may be designated to be a Fellow of the ABC (F-ABC) in that specialty.

In addition, active membership in one of the regional or national forensic science organizations is a hallmark of a professional forensic scientist. Fire testing laboratory personnel will often have membership in ASTM International (formerly known as the American Society for Testing and Materials), the IAAI, the National Fire Protection Association (NFPA), or the Society of Fire Protection Engineers (SFPE) as part of their professional qualifications. These qualifications should aid the fire investigator in finding a suitable laboratory and a qualified analyst.

The selection of a qualified lab is not easy and there are too many variables of expertise, qualifications, analysis time, and costs to provide specific guidelines here. It is clear that courts expect some sort of scientific data to be offered in support of the investigator's conclusions and that the data be demonstrated to be accurate and reliable. With the information offered here on types of evidence, possible analyses, and strengths and weaknesses of evidence and techniques, the investigator should be able to confer knowledgeably with various experts and determine their suitability for the particular problem at hand.

Identification of Volatile Accelerants

By far the predominant service provided by laboratories to fire investigators is the analysis of fire debris for suspected volatile accelerants. In one laboratory study, ignitable liquids (flammable and combustible liquids) were found in 49 percent of all arson cases submitted to a major crime laboratory over a 3-year period.[4] Data from a more recent survey verify the same percentages.[5] Although fire databases do not include information on the frequency of ignitable liquid use in setting fires, it is the consensus of fire and arson investigators in the United States that a large percentage (more than 50 percent) of detected arsons involve their use. The use of ignitable liquids, most commonly petroleum products, has been common throughout the history of twentieth-century incendiarism. Also, ignitable liquids are the first fuel ignited in an estimated 30 percent of all accidental building fires.[6] Many ignitable liquid accelerants are used in excess, and their residues remain for ready detection by instrumental analysis. Unfortunately, they can be used in moderation or on combustible substrates like paper or plastic, and the resulting fire can consume or evaporate all traces of the accelerant. One of the authors participated in a number of structure fire experiments in which a petroleum product such as gasoline had been used to ignite a fire, and the debris, even though collected directly after the fire, packaged properly, and analyzed by the best available methods, failed to reveal any accelerant. A negative laboratory finding is not proof that an accelerant was *not* used; it merely fails to establish its presence after the fire. On the other hand, petroleum products are detectable in very low concentrations in a wide variety of consumer products and building materials, so the interpretation of "trace" levels in debris must take such background volatiles into consideration (to be discussed further). In addition, petroleum products change in their physical and chemical properties as they burn or evaporate during (and after) a fire, and their specific identification can be a challenge even to experienced laboratory analysts.

chromatography ■ Chemical procedure that allows the separation of compounds based on differences in their chemical affinities for two materials in different physical states, e.g., gas/liquid, liquid/solid.

GAS CHROMATOGRAPHY

Gas **chromatography** (GC) as an accepted laboratory procedure has been in use for the last 50 years or so. Gas chromatography is used both for the screening of samples to determine which contain adequate volatile ignitable liquids for identification and for the identification itself once the volatile has been isolated. Although gas chromatographs (called GCs) vary considerably in their complexity, their basic principles of operation are quite simple. All types of chromatography are a means of separating mixtures of materials on the basis of slight differences in their physical or chemical properties.

As represented in Figure 14-1, gas chromatography uses a stream of gas (nitrogen or helium) as a carrier to move a mixture of gaseous materials along a long column or tube filled or coated with a separating compound. The components in the mixture interact with the separating compound by alternately dissolving in it and then volatilizing to be swept farther along the column by the carrier. To ensure that all the compounds in the unknown mixture remain in the gaseous state, the whole column is kept in an oven to maintain it at a precisely controlled temperature. The mixture is injected onto the column at one end, and at the other there is a detector, usually a small flame whose electrical properties, monitored by electronic circuits, change whenever a compound comes off the column [called flame ionization detection (FID)].

The whole process is analogous to a stream of vehicle traffic routed through a forest. The motorcycles speed right through, zigzagging around the trees and come out the other side of the forest first to encounter a radar speed trap. The small cars have more trouble finding a path and come out shortly afterward to encounter the same radar. The big cars take a long time to find spaces big enough to allow passage, and they exit after a longer delay, followed by the trucks that take the longest of all.

In the case of volatile hydrocarbon analysis, the "traffic" would be the unknown hydrocarbon mixture, and the "forest" would be the GC column. The analyst can adjust the spacing of the "trees" by choosing the correct chemicals for the separating compound and carefully controlling the temperature and carrier flow so that each type of "traffic"

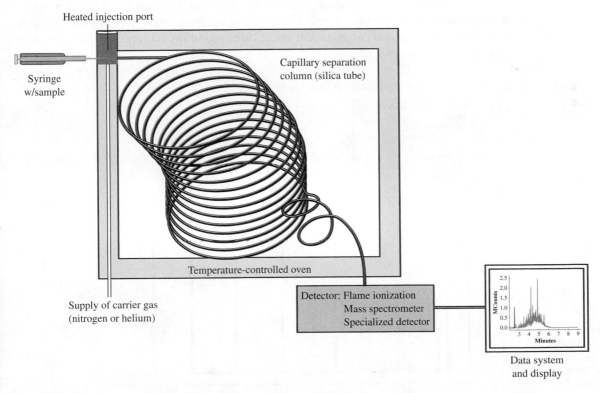

FIGURE 14-1 Schematic of gas chromatograph used for fire debris analysis.

comes out in a neat, readily identifiable group. To extend the "forest-traffic" analogy further, the proper choice of GC column and conditions for the hydrocarbons found in fire debris allow the lab to identify the type of "traffic" introduced by analysis of the groups of "vehicles" that result. From simple, insulated box ovens that kept the column at a set temperature, ovens have evolved to models that are programmed to provide steady or increasing temperatures at preset time intervals to maximize the separation of similar compounds. The technology of detectors has progressed from simple thermal detectors to flame detectors that can be made sensitive only to compounds containing specific elements, or to mass spectrometers.

Although the column can be selected to separate a mixture of compounds on the basis of their chemical properties, most separations in volatile accelerant analysis are made on the basis of differences in the boiling points of the various components. Nearly all the components of petroleum distillates of interest are either aliphatic (straight-chain) or aromatic (ring) in structure and thus will have very similar chemical properties within those two groups. A typical separation of a series of aliphatic hydrocarbons is shown in Figure 14-2. In this typical chromatogram, one can see that the heavier a molecule is, the less volatile it is, and the later it will appear on the chromatogram. Columns for GC analysis of suspected arson residues are usually set up to offer the best separation of components having volatilities between those of pentane (C_5H_{12}) and triacontane ($C_{30}H_{62}$), since that range includes all the commonly encountered petroleum products. Although the mainstay of columns was for many years a packed tubular column, averaging 3 mm (1/8 in. in diameter and 3 m (10 ft) in length, it has now been supplanted by a new generation of very narrow [0.1 to 0.25 mm (0.004–0.01 in.) in diameter] capillary columns of various lengths.[7] Early capillary columns required very long lengths [50 m (167 ft) or more] to achieve good separations, but these extreme lengths usually required long analysis times. Such columns were used primarily for research or development, in which short analysis times could be sacrificed for sensitivity and selectivity. The newest columns provide extremely high performance on short columns (10 to 25 m, or roughly 33 to 82 ft) using high temperature-programming rates. The result is analysis times that are shorter than those possible with packed columns and faster turnaround times. Capillary columns also

Scan Range: 1–1760 Time Range: 0.00–14.99 min.

FIGURE 14-2 Capillary GC performance standard, including C_8 to C_{20} even *n*-alkanes: (A) toluene; (B) ethylbenzene; (C) *p*-xylene, *m*-xylene; (D) *o*-xylene; (E) 3-ethyltoluene; (F) 2-ethyltoluene; (G) 1,2,4-trimethylbenzene. *Courtesy of David Brien, EFI Global, Rocklin, CA.*

Scan Range: 1–1760 Time Range: 0.00–14.98 min.

RIC all paintthinner.sms

MCounts

Minutes

FIGURE 14-3 Gas chromatogram of paint thinner. *Courtesy of David Brien, EFI Global, Rocklin, CA.*

use smaller sample sizes than were ever possible with packed columns, an advantage of much merit in forensic work, where the sample size is often critically small. Sensitivity is now in the nanoliter range (10^{-9} liter).[8] Examples of capillary gas chromatograms are seen in Figures 14-3–14-5. Since gas chromatography is a separation technique, the result of such GC-FID analysis is not a specific identification of each compound but a pattern of peaks that the analyst attempts to match to a reference chromatogram to make the identification. This approach is described in *ASTM E1387*: *Standard Test Method for Ignitable Liquid Residues in Extracts of Samples of Fire Debris by Gas Chromatograph.*[9] Due to the limitations of the GC-FID method, *ASTM E1387* was withdrawn in 2008 and

Scan Range: 1–1760 Time Range: 0.00–14.98 min.

RIC all kerosene.sms

MCounts

Minutes

FIGURE 14-4 Gas chromatogram of kerosene. *Courtesy of David Brien, EFI Global, Rocklin, CA.*

FIGURE 14-5 Gas chromatogram of gasoline (fresh unevaporated) (TIC). *Courtesy of David Brien, EFI Global, Rocklin, CA.*

Scan Range: 1–1760 Time Range: 0.00–14.98 min.

RIC all gasoline001.sms

MCounts

1.5

1.0

0.5

0.0

3   4   5   6   7   8   9

Minutes

was replaced by the GC-mass spectrometry (GC/MS) method *ASTM E1618*, described in the next section. GC-FID is still used for special applications such as samples of very low molecular weight.

In addition, progress in computer technology has dramatically improved the data collection and handling in gas chromatography. Runs (and irreplaceable samples) are no longer lost when the recording conditions are not set just right, since the data system collects all the data and then displays whatever portion the analyst specifies. The displayed data can expand or condense portions of the chromatogram to make identification easier. Collected chromatograms are also filed on disks and can be compared against a library of stored chromatograms of known products. Such libraries can be swapped between instruments and even between labs, permitting ready comparison.

GAS CHROMATOGRAPHY/MASS SPECTROMETRY (GC/MS)

mass spectrometry ■ Analysis of organic molecules by fragmenting them and separating them by size.

Flame ionization detectors (FIDs) are sensitive and stable detectors, but they yield little chemical information about the compounds that pass through to produce a signal. **Mass spectrometry** allows the analyst to break apart each compound into small submolecular pieces and, by counting those pieces, establish the chemical structure of the original molecule. Mass spectrometry alone is an analytical technique that needs to be fed one compound at a time, while gas chromatography is a separation technique that is good at separating things into groups. Because most of the volatile residues in fire debris involve complex mixtures, coupling the two together gives the best tool. While GC/MS has been available for more than 30 years, only recently has it been made small, inexpensive, and user-friendly enough to make it convenient for routine fire debris analysis. It has become the "baseline" technique for fire debris analysis because it can yield so much more information than GC-FID.[10] In addition to displaying a mass spectrum for each peak, it can be asked to scan for selected ions, those that are characteristic for a particular chemical species of interest, such as the ion having a "weight" of 91, which is characteristic for toluene. This means that the GC/MS can display a chromatogram just for aromatics, for instance, ignoring all the peaks that do not yield that specific ion. This can be accomplished by displaying one ion at a time or by "summing" three or four ions that are characteristic for the species of interest.[11] A typical set of ion profiles for gasoline is shown in

Figure 14-6, where the ions for alkanes (ion "weights" of 57, 71, 85, and 99), cycloparaffins/alkenes (ions 55, 69, 83), simple aromatics (ions 91, 105, 119), and naphthalenes (ions 128, 142, 156) are shown separately. Note that the species responsible for the major peaks in the gasoline chromatogram (Figure 14-5) can now be identified. The pattern of peaks, their proportions, and the relative concentrations (higher concentrations of

FIGURE 14-6 Typical combined ion profile analyses of gasoline showing alkanes, cycloparaffins, aromatics, and naphthalenes as separate chromatograms. *Courtesy of David Brien, EFI Global, Rocklin, CA.*

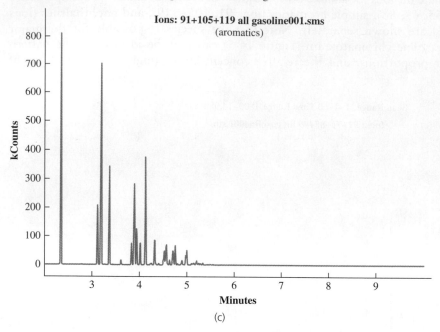

Scan Range: 1–1760 Time Range: 0.00–14.98 min.

Ions: 91+105+119 all gasoline001.sms
(aromatics)

(c)

Scan Range: 1–1760 Time Range: 0.00-14.98 min.

Ions: 128+142+156 all gasoline001.sms
(naphthenics)

(d)

FIGURE 14-6 (Continued)

aromatics than other species) yields very discriminatory information. This can make it possible to pick out characteristic patterns even amid complex chromatograms with high backgrounds. The "counts" of separate ions also can be compiled into a single chromatogram called a *total ion chromatogram* or TIC, which captures the same information as a GC-FID analysis and displays it as a single chromatogram for pattern recognition. Some particularly difficult samples may require *target compound analysis* (TCA), in

568   **Chapter 14** Laboratory Services

which the detection of a "set" of particular compounds permits identification of a petroleum product among complex background peaks.[12] TCA has also been used to compare gasolines with suspected sources. Some particularly difficult samples may require GC/MS/MS, or tandem mass spectrometry. This technique takes the molecular species created by the ionization process of MS and breaks them into even smaller pieces, yielding more specific information. Higher sensitivity is also claimed, but the method has found more application in explosives analysis than in routine fire debris analysis.[13]

Because of its sensitivity and analytical power (resolution), GC/MS has become the standard forensic technique for fire debris analysis.[14]

Currently, most forensic GC analyses for suspected accelerants are carried out using a capillary column of moderate length (10–25 m; 31–76 ft), using dimethyl silicone, phenyl silicone, or an equivalent nonpolar bonded phase) in a programmed temperature range of 50°C to 250°C (122°F to 482°F), and employing a mass spectrometer.

Because the MS "detector" has to be turned off for the time period when the extraction solvent used is passing through, volatiles such as methanol, ethanol, hexane, acetone, or toluene will not be detected. If such a product is suspected, the lab should be advised so that a GC-FID analysis can be conducted or an appropriate alternative isolation protocol followed [such as heated head space or solid-phase microextraction (SPME)], which uses no solvent.

SAMPLE HANDLING AND ISOLATION OF VOLATILE RESIDUES

As described in Chapter 7, the best containers for debris suspected of containing volatile ignitable liquids are clean, lined metal paint cans, sealable glass jars, or bags of nylon or a suitable copolymer (e.g., nylon/acrylonitrile/methacrylate made by Ampac). The first step in the normal examination protocol of such debris is to briefly open the container (in the case of bags or jars, this will usually not be necessary) and to inspect the contents. The nature of the contents (wood, carpet, soil, etc.) may provide guidance in selecting an analytical technique or predicting what interferences are likely to be encountered. The contents should be checked against the labeled or indicated contents to ensure there have been no errors in packaging or labeling that could compromise the value of the evidence. The examination must be brief to minimize losses of volatiles that may be present only at trace levels. Very strong odors of a volatile detected during this examination may indicate that a particular test protocol should be used. The major problem with fire debris analysis is having a small quantity of product mixed with a great deal of solids or liquids that may provide contamination. Each of the techniques described here has advantages and disadvantages that must be evaluated in light of the character of the sample.

Heated Headspace

Since any volatiles present will reach vapor equilibrium in a sealed container, advantage can be taken of the vapors in the air cavity (headspace) of the container. Heating the sample container to 50°C to 60°C (120°F to 140°F) will ensure at least a partial vaporization of heavier hydrocarbons prior to sampling. Some analysts prefer heating the container above 100°C (212°F) to force even higher boiling materials (such as fuel oil) into the vapor stage. It is the experience of most other analysts, however, that all but the heavy ends of kerosene will be detectable in most samples heated to moderate temperatures (60°C; 140°F) and, more important, that excessive temperatures (>80°C; 176°F) encourage the degradation of synthetics (carpet, upholstery, etc.) in the debris. These degradation products add to the complexity of the chromatogram, possibly masking low levels of real accelerants while not contributing significantly to the sensitivity of the method. If a heavy product is suspected, isolation using solvent extraction or dynamic headspace will result in more complete recovery and analysis.

Once the sample container has equilibrated at a moderate temperature, it is punctured and a small sample (0.1 to 3 mL) of the headspace vapor is drawn off using a gas-tight syringe. This vapor is then injected directly into the gas chromatograph. For many mixtures, positive results of this injection may be adequate to permit characterization of the volatiles present (as described in *ASTM E1388: Standard Practice for Sampling of Headspace Vapors from Fire Debris Samples*).[15] Positive results also signal the presence of very volatile residues that may be lost by careless handling, or alcohols or ketones that require special handling. Because this technique is not as sensitive as others described, a negative result here merely indicates that no high concentration of volatile exists and that another isolation technique is suggested. The heated headspace technique is fast, convenient, and requires virtually no sample handling, while providing some useful information.

Passive Headspace Diffusion (Charcoal Sampling)

adsorption ■ Trapping of gaseous materials on the surface of a solid substrate.

Another technique that capitalizes on the vapor equilibrium in a sealed container while concentrating some of the vapors in a more condensed form is passive **adsorption**. This method includes several similar techniques that rely on the adsorptivity of activated charcoal. Originally developed by forensic scientists in England using a charcoal-coated wire, this technique requires no manipulation of the sample or its original container.[16] As originally described, the method called for a coated wire to be inserted into the airspace of the sample container (as shown in Figure 14-7) and allowed to equilibrate for at least 2 hours. The wire was then removed from the container and inserted into the pyrolysis unit of a GC, where it was raised instantly to a very high temperature, driving the adsorbed hydrocarbons off into the column. This technique has been adapted in a variety of methods described in *ASTM E1412: Practice for Separation and Concentration of Ignitable Liquid Residues from Fire Debris Samples by Passive Headspace Concentration.*[17] These adaptations include the insertion of a plastic or glass bead coated with activated charcoal, or a sampling tube or bag containing charcoal into the airspace of the container.[18] The container is resealed and allowed to equilibrate, usually at room temperature, for 12 to 15 hours or at 60°C to 80°C (140°F to 176°F) for 2 hours. The sampling device is removed, and adsorbed volatiles are extracted from the charcoal with a small quantity of solvent such as pentane, diethyl ether, or carbon disulfide for injection into the GC. Today a carbon strip is used. These techniques are very sensitive to all types of volatiles, require no heating or manipulation of the samples, and can be repeated again and again because they take up only a small fraction of the volatiles each time. They are suitable for all kinds of debris and are especially valuable for debris that will be examined for trace evidence or latent fingerprints.

A variant of this technique has become the single most widely used sampling and concentration method for fire debris. In this form, a small strip of activated carbon fiber mat (sometimes called a *C-strip*) is suspended in the container, at room temperature or at moderately elevated temperatures (60°C–80°C). It is then extracted with carbon disulfide or diethyl ether. The technique has excellent sensitivity [0.1 microliter (μL) under some conditions] and no interferences. It is described in full in *ASTM E1412*.

Solid-Phase Microextraction

A variant of passive activated-carbon techniques, called *solid phase microextraction* (SPME), uses a solid-phase adsorbent bonded to a fiber that is inserted into the can of fire debris through a hollow needle, as shown in Figure 14-8. After the fiber has been exposed to the heated contents of the can for 5 to 15 minutes, it is withdrawn and then inserted directly into the injection port of a GC. The heat of the injection port then thermally desorbs the volatiles directly into the GC column. The technique offers extreme simplicity (with no manipulation of the sample or any extract) and low cost, since it involves no solvents and no modifications to existing gas chromatographs.[19] SPME been found suitable for fire debris analysis, and *ASTM E2154: Standard Practice for Separation and Concentration of Ignitable Liquid Residues from Fire Debris Samples by Passive Headspace Concentration with Solid Phase Microextraction (SPME),* applies.[20] This technique appears to be particularly well suited for isolating petroleum product residues from water and skin tissue.[21] The small volume of adsorbent can be overwhelmed by high concentrations of petroleum distillates, causing distortion of the peak pattern in the GC

FIGURE 14-7 Passive charcoal strip sampling of fire debris.

FIGURE 14-8 Solid-phase microextraction uses a thin fiber of adsorbent exposed to the headspace of a debris container. The fiber is then placed directly into the injection port of a GC for desorption.

analysis. SPME has the advantage of using no extraction solvent, so all volatiles are readily detectable in a single run. Its limited sample capacity means it is easily overloaded by extended exposure or high concentration, which can lead to distortion of its GC profile.[22]

Dynamic Headspace (Swept Headspace)

This technique, sometimes called the *charcoal trap method*, employs a charcoal or polymeric matrix through which the air from the evidence container is drawn. It was adapted from methods developed for trapping ultralow concentrations of hydrocarbons for environmental monitoring.[23] The sample container is fitted with a modified lid or a septum that allows for the introduction of a heated carrier gas (purified room air or nitrogen). The container is warmed while vapors are drawn off by a low vacuum through a cartridge filter containing activated charcoal or a molecular trapping agent such as Tenax (see Figure 14-9). Once extraction is complete (20 to 45 minutes), the charcoal can be washed with carbon disulfide or a like solvent, and the resulting solution is ready for GC analysis. Tenax or charcoal may be thermally desorbed to avoid the handling of toxic solvents, contamination, or any further dilution. The technique requires the use of traps (commercially prepared or laboratory prepared) and specially built apparatus, described in *ASTM E1413: Standard Practice for Separation and Concentration of Ignitable Liquid Residues from Fire Debris Samples by Dynamic Headspace Concentration.*[24] It is very useful because of its sensitivity (less than 1 μL) and its applicability to all ignitable liquid residues (from gasoline to fuel oil) including alcohols and ketones. The GC analysis of Tenax traps, which are thermally desorbed, has been automated, making it more suitable for laboratories with a large caseload. The technique does have the drawback that all the volatiles in a sample can be swept out (and pulled through the trap) if the time and temperature conditions are not monitored closely and are too severe for the sample size. Excessive extraction causes loss of the lighter fractions, and the process can generally be carried out only once on a sample.

Steam Distillation

The oldest separation technique, originating from classical chemistry, requires only modest but specialized distillation equipment. The specimen is removed from its container and boiled with water in a glass flask (as illustrated in Figure 14-10). The steam, carrying the volatiles with it, is condensed and trapped. It is suitable for volatiles that form an *azeotropic* (nonmixing) layer with water; therefore, it cannot be used to recover alcohol or acetone. The technique is suitable for petroleum distillates as heavy as paint thinner; heavier products (e.g., fuel oil) require the use of ethylene glycol, which boils at a higher temperature.[25] Light hydrocarbons may be lost in recovery. Detection limits are on the

FIGURE 14-9 Swept headspace technique is the most efficient means of collecting volatile hydrocarbons but requires a source of carrier gas, a heat source, and a vapor trap of charcoal or other adsorbent. It is not considered a nondestructive isolation technique.

FIGURE 14-10 Steam distillation of arson debris requires only water, a boiling flask, a heating mantle, a condenser, and some form of receiver to collect distillates.

order of 50 μL (for gasoline), so it is not as sensitive as charcoal-trapping methods, but it has the advantage of providing a neat liquid sample with no extraneous solvents.[26] The method is described in *ASTM E1385: Standard Practice for Separation and Concentration of Ignitable Liquid Residues from Fire Debris Samples by Steam Distillation*.[27]

Solvent Extraction

This very direct and simple method for extracting volatiles has been in use for many years. Debris is extracted or washed with a small quantity of *n*-pentane, *n*-hexane, dichloromethane, or carbon disulfide that has been checked to ensure its purity. The liquid extract is filtered and then concentrated by evaporation in a stream of warm air until a small quantity remains for testing, as shown in Figure 14-11. Solvent extraction is especially suitable for

FIGURE 14-11 Solvent extraction is not very sensitive but may be used to extract trace volatiles from nonabsorbent materials like metal, glass, or stone.

treating small, nonabsorbent specimens (glass, rock, metal) or washing the interior of containers used to transport ignitable liquid accelerants. It provides good recovery for many petroleum products, especially the heavier ends of kerosene and diesel fuel, but it is not suitable for isolating the lighter petroleum distillates due to evaporative losses. Its overall sensitivity is about the same as for steam distillation with the main drawback that it will extract residues of partially pyrolyzed carpet or foam rubber, which often interfere with the detection of low levels of hydrocarbons of evidentiary significance.[28] This technique requires only simple glassware and high-purity solvents. It is described in detail in *ASTM E1386: Standard Practice for Separation and Concentration of Ignitable Liquid Residues from Fire Debris Samples by Solvent Extraction.*[29]

IDENTIFICATION OF VOLATILE RESIDUES

Once the suspected ignitable liquid residues have been isolated, attention can be turned toward identifying them as precisely as possible. Because recoveries are typically on the order of several microliters, gas chromatography/mass spectrometry will be the technique of first choice because of the quantity of information it yields while consuming 1 µL or less of the sample. The identification of the volatiles (usually a petroleum distillate or product) present is then usually based on a comparison of the pattern of peaks generated in the chromatogram of the unknown against a library of chromatograms of known materials made under the same conditions. A standardized approach to the classification and characterization of volatiles is described in *ASTM E1618.* Classification of the volatile into of the basic volatility categories—light, medium, or heavy—is based on the presence of a particular range or sequence of hydrocarbons. The products are further categorized as to their chemical nature—gasoline, true distillates, de-aromatized distillates, isoparaffins, aromatics, naphthenic-paraffinics, *n*-alkanes, oxygenates, and other miscellaneous volatiles—based on the MS data collected. This classification scheme is intended to make description and comparisons easier for the laboratory analyst, but because consumer products may represent different classes of products, and a single product may have more than one consumer designation, the system is not always well understood by the investigator. For this reason, it is recommended that when an analyst describes an ignitable liquid residue as a particular classification, examples of some common consumer products should be included, as listed in Table 14-1, which is based on *ASTM E1618.*

A growing number of ignitable liquids are not petroleum distillates and therefore do not lend themselves to a simple distillate classification scheme based on peak pattern alone; for instance, there are aromatic specialty solvents used in insecticides, enamel reducers, and pavement sealers, isoparaffinic mixtures used in consumer products such as "odorless" paint thinners, and naphthenic–paraffinic petroleum products used in lamp oils (and sold as "kerosene" for portable heaters). The GC-FID patterns of these products are not sufficient to permit their identification, and GC/MS is required.

Table 14-1 is considerably revised from previous versions because of the influx of petroleum products in various household materials such as naphthenic-paraffinic products that are not traditional petroleum distillates. Their "peak patterns" in some cases are not readily differentiated from the patterns generated by some pyrolysis products. Thus, the use of GC/MS is necessary in a larger percentage of cases, because the GC/MS can yield information about what specific chemical species are present. For instance, isoparaffinic hydrocarbon products are being used in many consumer products, from copier toners to insect sprays. Because they are not true petroleum distillates, they should not be classified as such, although in some instances they display a boiling point range that fits light or medium distillates. When such products are identified, a list of the consumer products that may contain them should be offered in the report to give the investigator a clue as to what product may have been used to start the fire or that a product normally at the scene may have become involved accidentally in the fire. Because the contents of many products are not always described accurately on the labels, comparison samples

TABLE 14-1 Classification of Ignitable Liquids in Fire Debris

CLASS	LIGHT (C$_4$–C$_9$)	MEDIUM (C$_8$–C$_{13}$)	HEAVY (C$_8$–C$_{20+}$)
Gasoline (auto)	Fresh gasoline in range C$_4$ to C$_{12}$		
Petroleum distillate	Petroleum ether Some camp fuels	Some charcoal starters Some paint thinners	Kerosene Diesel fuel
De-aromatized distillates	Some camp fuels	Some charcoal starters Some paint thinners	Some charcoal starters Odorless kerosenes
Isoparaffinics	Aviation gas Special solvents	Some copier toners Some paint thinners	Specialty solvents Some copier toners
Aromatic products	Xylene/toluene Some parts cleaners	Some insecticide solvents Parts cleaners Fuel additives	Cleaning solvents
Naphthenics/paraffinics	Solvents	Some charcoal starters Lamp oils	Some lamp oils Some insecticide solvents Industrial solvents
n-Alkane products	Solvents Pentane/hexane	Candle oils Copier toners	Copier toners NCR papers (carbonless forms) Candle oils
Oxygenated solvents	Alcohols/ketones Surfacers Some lacquer thinners Fuel additives	Some lacquer thinners Industrial solvents Turpentines (gum)	
Other: Miscellaneous	Single-component products, blended products		

Adapted from ASTM Standard E1618, Standard Test Method for Ignitable Liquid Residues in Extracts from Fire Debris Samples by Gas Chromatography-Mass Spectrometry, copyright ASTM International, 100 Barr Harbor Drive, West Conshohocken, PA 19428. A copy of the complete active standard may be obtained from ASTM, www.astm.org.

(preferably in their original containers) of any product in the vicinity of the fire's origin should be collected and submitted for analysis if there is a chance they could have been involved in the fire. Because the analyst is expected to match the GC/MS data from the unknown to that of a known (reference) material, having immediate access to a comparison sample from the scene is very important. This is especially true when the material is an unusual proprietary or specialty solvent.

This classification scheme reflects changing consumer uses and products and a much-changed petroleum refining process, as well as improved sensitivity and discrimination of the lab methods used. There is also a desire to provide the investigator with the most accurate and meaningful (useful) information. The investigator must strive to understand what these product designations mean and use that information to evaluate various accidental and intentional sources of such residues in fire debris.

Special detectors may also be used to gain more information about a volatile. Each requires injection on a separate GC column with a specific detector in place of the FID, but some GCs have the ability to split a single injection onto two separate columns, each with its own detector. Photoionization detection (PID), for example, can be used in parallel with FID because it can be made specific to aromatic, olefinic, or aliphatic hydrocarbons. Another detector, originally intended for the detection of nitrogen- and phosphorus-containing compounds, has been modified to specifically detect the oxygenated volatiles (ketones, alcohols, and ethers) that have become widely used as fuel additives to minimize air pollution.[30]

When distillation is used to isolate a neat volatile, infrared (IR) spectroscopy can yield considerable information about its chemical structure. About 20 μL of sample is needed for conventional liquid-cell spectrophotometry, but a specialized FTIR microcell has been designed for use as a detector on a GC column when smaller quantities are recovered. While the sensitivity of the method is limited, it can aid in the identification of unusual liquids, complex pyrolysis products, and some natural products (terpenes).[31]

If sufficient quantities of a suspected ignitable liquid are recovered, physical properties like flash point can be measured. The criminal charges against a suspected arsonist may depend on the flammability of the accelerant in his or her possession, and a flash point determination may be required. A precise determination requires a considerable sample and specialized equipment, but a rough estimate can be made by placing several drops of the liquid on a watch glass and placing it in a freezer. On removal from the freezer, the glass will warm up gradually, and a small ignition source passed over the specimen will cause a flash when the flash point is exceeded. If the liquid has a flash point above ambient (usually 25°C; 77°F), the glass can be placed on a hot plate and the sample tested in the same manner. In rare circumstances, steam distillation has been able to recover enough volatile accelerant to permit flash point determination by Setaflash tester (which requires 2 ml of sample).[32] Ultraviolet (UV)/visible fluorescence techniques permit extensive characterization of petroleum distillates, and these techniques, along with high-pressure liquid chromatography (HPLC), may be useful in special circumstances because they provide information that supplements gas chromatography (especially for heavy petroleum products).[33]

Evaporation and Degradation

Petroleum products behave in a fairly complex but predictable manner when burned or simply exposed to air to evaporate. Due to their higher vapor pressures, the most volatile components evaporate more quickly than those less volatile, and this changes the peak profile of the product when it is subjected to gas chromatography. As we can see from Figure 14-12, a gasoline profile changes dramatically as it evaporates, complicating its identification. When the gasoline is burned, this same process occurs, only much more rapidly. It is not possible to determine whether a petroleum product burned or merely

FIGURE 14-12 GC/MS comparison of fresh, 50 percent evaporated and 90 percent evaporated gasoline. Note decreased amounts of "light" products and relative increases in heavier components as the gasoline evaporates. *Courtesy of David Brien, EFI Global, Rocklin, CA.*

evaporated at room temperature by examination of its chromatogram.[34] In the case of gasoline, characteristic aromatic compounds are found in residues of gasoline that has evaporated down to less than 3 percent of its original volume. Therefore, evaporated or "weathered" gasoline can still be identified if those key or target compounds are identified in the debris, as illustrated in Figures 14-13 and 14-14. Unfortunately, this is not true

Scan Range: 1–1760 Time Range: 0.00–14.99 min.

Ions: 57+71+85+99 all 50% wxgasoline.sms

(a)

Scan Range: 1–1760 Time Range: 0.00–14.99 min.

Ions: 55+69+83 all 50% wxgasoline.sms

(b)

FIGURES 14-13A–D Combined ion chromatograms of 50 percent evaporated gasoline: (a) alkanes, (b) alkenes and cycloparaffins, (c) aromatics, and (d) naphthalenes. *Courtesy of David Brien, EFI Global, Rocklin, CA.*

Scan Range: 1–1760 Time Range: 0.00–14.99 min.

Ions: 91+105+119 all 50% wxgasoline.sms

(c)

Scan Range: 1–1760 Time Range: 0.00–14.99 min.

Ions: 128+142+156 all 50% wxgasoline.sms

(d)

FIGURE 14-13 (Continued)

for simple petroleum distillates like mineral spirits (paint thinners). As these substances evaporate, their residues assume more and more of the characteristics of the next heavier class of petroleum distillate. Thus, it is not always possible to conclusively establish the exact nature of the original product.

In addition, petroleum products exposed to moist garden soil can undergo microbiological degradation. This degradation involves the consumption of both aliphatic and aromatic compounds, so the peak profile of gasoline, for instance, can be significantly changed.[35] This degradation is minimized if the samples are kept frozen until analysis.

Evaporation and degradation effects combined with contamination from pyrolysis products in the debris can make the identification of a specific petroleum product from post-fire debris difficult if not impossible. The increased use of non-distillate petroleum

Scan Range: 1–1760 Time Range: 0.00–14.99 min.

Ions: 57+71+85+99 all 90% wxgasoline.sms

(a)

Scan Range: 1–1760 Time Range: 0.00–14.99 min.

Ions: 55+69+83 all 90% wxgasoline.sms

(b)

FIGURES 14-14A–D Combined ion chromatograms of heavily (90 percent) evaporated gasoline: (a) alkanes, (b) alkenes and cycloparaffins, (c) aromatics, and (d) naphthalenes. *Courtesy of David Brien, EFI Global, Rocklin, CA.*

products (see Chapter 4) has resulted in GC chromatograms that are not so easily classified or identified. Examples in Figure 14-15 show the complex nature of these products, making GC/MS analysis, and comparison standards essential. In most cases, the laboratory can be expected to characterize the product or its general family—for example, light (petroleum ether), medium (paint thinners), or heavy (kerosene/diesel) petroleum distillates; lacquer thinners; and so on—from residues left in the debris unless the

Scan Range: 1–1760 Time Range: 0.00–14.99 min.

Ions: 91+105+119 all 90% wxgasoline.sms

(c)

Scan Range: 1–1760 Time Range: 0.00–14.99 min.

Ions: 128+142+156 all 90% wxgasoline.sms

(d)

FIGURE 14-14 (Continued)

(a)

(b)

FIGURES 14-15A–C
Total ion chromatograms of petroleum products:
(a) lamp oil, (b) gum turpentine, and (c) pyrolyzed polyethylene.

Scan Range: 1–1760 Time Range: 0.00–14.99 min.

RIC all polyethylene.sms

(c)

FIGURE 14-15 (Continued)

liquid fuel has been nearly or totally consumed. Any petroleum distillate, when used as an accelerant, can be burned away completely, leaving no readily detectable traces if the fire is hot enough and long enough, and the accelerant is exposed directly to its effects. Fortunately, in most incendiary-caused fires, accelerants are used in excess, and the resulting fires do not burn with sufficient intensity to completely destroy them.

Laboratories are often asked to determine the presence of petroleum distillate accelerants from the soot accumulated on window glass taken from the scene. As we discussed in Chapter 7, the appearance of the soot is most directly linked to the state of combustion in the room and the amount of ventilation. Fuel-rich fires produce heavy soot; lean or well-ventilated fires produce little or none. The synthetic fabrics and rubbers that constitute a large percentage of the fuel load in modern structures produce an oily soot very similar chemically to that produced by petroleum distillate accelerants. It has been suggested that there should be chemical species in the soot produced by gasoline that are characteristic for that fuel and are not produced by any "normal" fuel. It was once thought that the combination of bromine and lead in leaded automotive gasoline would produce soot with a unique elemental profile. Unfortunately, some carpets contain both elements, so there is a potential for background interference.[36] More important, lead and its related additives have been phased out of modern fuels, making them very unlikely to be detected. There are several organic molecules, chiefly polynuclear aromatic hydrocarbons that appear to be produced by burning gasoline but are not produced by burning synthetic materials.[37] The most extensive work done in this area is that of Pinorini et al., who used GC combined with elemental analysis, X-ray diffraction, and scanning electron microscopy (SEM) to characterize soot from various ignitable liquids and from synthetic fibers and materials.[38] They discovered that under controlled conditions, soot from the liquids could be distinguished from that produced by solid fuels using multiple techniques. Unfortunately, the presence of water from suppression produced nonreproducible condensation of the soot and limited the discrimination value of even the combination of techniques. Some analysts have reported that freshly made carbon soot will adsorb petroleum distillate fumes from the room air just as activated charcoal adsorbs them in laboratory extractions. If a fire occurs and a too-rich mixture of fuel is present, the soot formed during the early stages may retain enough of the accelerant to permit identification upon extraction of the soot. Occasional analysis of soot on glass from accelerated vehicle fires has been observed, but no systematic study has been reported.

Despite partial evaporation, most common petroleum distillates have a distinctive GC/MS pattern that can be compared with the patterns of known compounds. The experienced analyst will know what the effects of fire and evaporation can be and will take them into account when comparing chromatograms. All petroleum distillates are overlapping "cuts" or fractions of a continuous spectrum of compounds. Therefore, it is possible for a complex hydrocarbon mixture to be sufficiently affected by fire that its chromatogram cannot readily be identified. It could, for instance, represent a heavily evaporated product such as kerosene, or a relatively undamaged product like a heavier solvent. Today, the introduction of biofuels or synthetic fuels further complicates the interpretation issue. Some of these have unusual gas chromatographic profiles that do not readily fall into the existing classification schemes.[39]

Source Identification

As to the identification of a specific source for a product, the range of techniques available now enables the analyst to rule out many potential sources and thereby shorten the list of "possibles." The sophistication of GC/MS and today's data handling systems make it possible to recognize delicate differences in the patterns of GC peaks between products that were thought to be indistinguishable. Determining the significance of these differences is entirely another matter, requiring extensive field sampling and validation of the reproducibility of patterns detected. As with many mass-produced materials that start from different feedstocks in different manufacturing facilities, changes can be produced by many factors. To complicate matters, when gasoline, for instance, is distributed, it mixes with gasoline from other sources already in storage and transportation facilities, and its composition changes. Under some circumstances, such as having a liquid gasoline sample that has not undergone significant burning or evaporation, and a small number of potential sources, it is possible to state with some confidence that certain sources can be completely excluded and that another represents a source that is indistinguishable by any of its individual features.[40] Rather than give the investigator a false lead with a possible identification, the laboratory will usually suggest several alternative identifications in its report. The investigator can evaluate the suggestions and determine which, if any, fit the circumstances best. If the investigator does not clearly understand the identification or characterization, the problem should be discussed with the laboratory before proceeding.

INTERPRETATION OF GC RESULTS

It should be noted that most of the newer isolation techniques, especially when combined with capillary column GC/MS, are more sensitive than the techniques used less than a decade ago. In many cases they are more sensitive than the human nose, often relied upon to screen for the best debris samples at the scene. This means that sample collection may depend not only on the detection of unusual odors but also on the presence of suspicious burn patterns and on aids such as canine detection. Comparison samples of floor coverings or other fuels involved are much more desirable now than ever before because of the contributions made by the pyrolysis of synthetic materials, binders, and adhesives, which can interfere with the analyis.[41] Recent work has been published by Stauffer exploring the chemical mechanisms of polymer pyrolysis and the contributions each type can make to the GC profile.[42] This work has improved the discrimination of pyrolysis products and petroleum products substantially (see Figures 14-15a–c).

Sensitivity

As the minimum detection level of GC/MS has decreased, analysts have begun to see just what a petroleum-based world we live in. From the volatile products in roofing paper, copy paper, carbonless forms, magazines, and newspaper print, the solvents in glues, and adhesives used in floor coverings and footwear, to residues of dry cleaning solvents, insecticides, and even cleaning agents (like the limonene used in "green" grease removers), petroleum products can be detected in all manner of consumer goods today.[43] Without specific identification of the product and comparison against a reference sample, the mere presence of traces of a volatile petroleum product in fire debris may well be meaningless. Much of the sensitivity issue came to light when canine detection was being debated. A properly trained canine nose can detect petroleum products at very low levels with a high degree of accuracy, sometimes at concentrations below even what a qualified lab can properly identify, so there may be samples in which a canine correctly detects traces of gasoline that the lab cannot confirm if it adheres to the accepted standard practices. But the issue is not simply sensitivity; there are far too many other possible (innocent) sources of petroleum products. It is the accuracy of the identification of the actual product present that is essential. A canine can signal only yes or no; it cannot identify whether it is a trace of the insecticide (with its aromatic blend solvent) the owner used the week before or gasoline that an arsonist used to set the fire. Only the laboratory with its verified and accepted methods and criteria can distinguish those products. The canine cannot discriminate between the residues of carpet glue used to repair a section of carpet from the residues of isoparaffinic solvent used as an accelerant. Despite the advice of the IAAI Forensic Science Committee and the NFPA Technical Committee on Fire Investigations (*NFPA 921*), a number of courts have admitted evidence of canine alerts as indications of the presence of ignitable liquid accelerants in the absence of laboratory identifications.[44] Studies have shown that even the best-trained canine teams cannot discriminate all possible pyrolysis products and background contaminants from accelerants and will alert to a percentage of targets that really do not contain ignitable liquid accelerants.[45] Without a specific verifiable identification of exactly what is present, the investigator cannot decide the significance of such positive alerts. It is wrong to assert the absolute reliability of a canine alert.

Another issue of extreme sensitivity is erroneous reconstruction. If someone steps in a puddle of gasoline leaking from an electrical generator or spilled from a fuel can outside the scene and then walks through the scene, traces *theoretically* might be found in several adjacent locations. A study by Armstrong et al. demonstrated that detectable gasoline residues could be tracked onto clean substrates for a step or two, but not beyond.[46] Similar results have been recently reported by Kravtsova and Bagby.[47] Transfers by absorption onto clothing of gasoline vapors in a room have also been shown to be remotely possible but very weak.[48] It has also been suggested that exhaust gases from gasoline-powered air fans at fire scenes could contaminate exposed charred material. Tests have revealed that although such contamination at trace levels is possible, it is very unlikely to affect waterlogged debris.[49] A blind reliance on the mere presence of trace residues could lead to the erroneous conclusion that gasoline was used in several places throughout the scene.

The investigator must ask: Why was this found in these concentrations, and how else might it have gotten there? Fire investigators and forensic analysts need to ask: What is the meaning of this "positive" at the concentration found here? With today's sensitive techniques, the soil gathered from the side of a busy highway will yield identifiable gasoline residues at parts-per-billion levels, just from the traces accumulated from leaks and spills from passing vehicles. Is that finding of any significance? Does that mean that someone poured gasoline along the side of the road and set fire to it to commit arson? Lentini demonstrated that floor-coating solvents (stain, varnish, and oil) were readily detectable in interior woods for more than 2 years after application.[50] Hetzel and Moss detected waterproofing solvent after more than 2 weeks of severe exterior exposure.[51] A recent case showed that the gasoline used as a solvent for interior floor varnish some 15 years before the fire was still identifiable when the wood was heated and the volatiles trapped and analyzed by GC/MS. Scientists and investigators of 25 years ago only dreamed about the sensitivity that is available today, but intelligence must be applied to interpret findings accurately. The mere presence of a volatile is no longer sufficient; rather, it is the concentration at which it becomes significant that is critical to its interpretation.[52]

Comparison Samples

With the sensitivity of today's detection methods (including canines) and the complexity of today's world of petroleum products, it is essential the investigator recognize the critical importance of comparison samples. If debris is collected at a scene for lab analysis, unless there is a broken Molotov cocktail or a can of gasoline lying next to the burned area, comparison samples should be collected (using clean gloves and cleaned tools). These samples should include the floor covering, any padding or underlay, and the floor beneath, not only near the point of origin but also at another location in the same room (preferably one protected under furniture). If an ignitable liquid is suspected, comparison samples should also be collected (in the original containers, if possible) of any solvents, cleaners, or insecticides that may have been stored or used in the vicinity. This may involve interviewing the owner, tenant, or maintenance staff to ascertain what might have been used in the course of business prior to the fire.

If any material is used to absorb suspected liquid residues (such as cotton, flour, Fuller's earth, diatomaceous earth, or kitty litter), a comparison sample of the absorbent must be submitted in a separate, clean, sealed container. Any item or product used for absorbing suspected ignitable liquids must be tested and shown not to contribute any volatiles that would interfere with GC or GC/MS identification. One commercial product, ILA, was introduced in 2003 but was later found to degrade upon extended storage and contribute volatile breakdown products to the GC analysis.[53] Contamination is the major concern regarding use of activated charcoal or hazmat absorption mats. They are such aggressive absorption media that they are easily contaminated by exposure to vehicle exhaust or other atmospheric sources of hydrocarbons. If foams, wetting agents, or other additives are used during suppression, they can affect the chromatographic data obtained from the sample but have not been reported to prevent accurate GC/MS detection and identification.[54] The fire investigator should note the use of any additives in suppression or post-fire hazard control and at least inform the lab. Getting a comparison sample is strongly recommended.

Detection on Hands

Various methods have been suggested over the years for the recovery of volatile products from human skin, particularly from the hands of suspects. Dry swabbing and solvent swabbing have been tried with no success. Volatiles on living skin are rapidly driven off by evaporation or soak into the epidermal and dermal tissues, preventing "surface" residues. There have been reports of successful detection of ignitable liquid residues by enclosing the hand in a PVC plastic or vinyl glove or polymeric bag with a charcoal strip or SPME collection device and warming the hand gently with a radiant heat lamp (or simply enclosing the hand in a PVC glove and using the glove as the absorbent).[55] For this technique to be successful, there has to have been a considerable quantity of fuel on the hand originally, and the collection has to be attempted within 1 to 2 hours of the contact. This short time frame precludes useful testing of suspects taken into custody several hours after contact. Similar testing has been more successful on tissues excised from a body during postmortem and tested in small vials using C-strips. Gasoline and similar volatiles can be detected on shoes and clothing for up to 6 hours if the garments are worn, and longer if they are removed and piled together.[56]

The very nature of fire, fire debris, collection, and analysis means that no one can actually quantify the results of fire debris analysis, but from experience, experiments, and test fires the analyst can establish a feel for the levels detected. The question to be answered then becomes: Is this material significant at this concentration? The sensitivity and discrimination power of today's fire debris methods is a great tool for the investigator, but it is one that must be used carefully and judiciously if the investigations are to be accurate and just.

Chemical Incendiaries

Volatile accelerants are not the only arson evidence that is amenable to laboratory testing. A small but significant percentage of all detected arsons use a chemical incendiary as an initiator or primary accelerant.[57] Fortunately, most chemical arson sets leave residues with distinctive chemical properties, although physically the residues may be nondescript and easily overlooked.

Safety flares or fusees are the most commonly used chemical incendiary, presumably because they are effective, predictable, and readily available. Red-burning flares contain strontium nitrate, potassium perchlorate, sulfur, wax, and sawdust.[58] Their post-fire residue consists of a lumpy inert mass, white, gray, or greenish-white in color (see Figure 8.23.). It is not soluble in water, but exposure to water turns it to a pasty mush.

The residue contains oxides of strontium and various sulfides and sulfates. Although strontium is found in all natural formations of calcium, only flare residues contain strontium salts in nearly pure form. Elemental analysis by emission spectroscopy, atomic absorption, or X-ray analysis will readily reveal these high concentrations of strontium in suspected debris. The igniter-striker button on the cap and the wooden plug in the base of the stick are resistant to burning and may be found in the fire debris and their manufacturer may be identified by their physical features.[59]

IMPROVISED MIXTURES

The solid chlorine tablets or powder used for swimming pool purification will react with organic liquids such as the glycols in automotive brake fluid or some hair dressings to produce a very hot flame that lasts for several seconds. This reaction occurs after a time delay that varies with the concentration of the chlorine-based components and the degree of mixing between the solid and liquid phases. The reaction has been studied and a mechanism for the production of ethylene, acetaldehyde, and formaldehyde by the action of the hypochlorite on the ethylene glycol has been suggested and verified.[60] The production of a cloud of readily ignitable vapors would certainly account for the dramatic yellow-orange fireball of flames that occurs in these reactions, as seen in Figure 16.11. The starting solid compounds are usually not completely consumed and are not immediately water soluble and may be visibly detected after the fire. The presence of hypochlorite salts can be detected by infrared (IR) spectrometry or chemical spot tests to confirm the identity of the oxidizer. Debris containing residues of such chlorine-based products may give off the odor of household bleach when wetted.

Potassium permanganate is used in a variety of chemical incendiary mixtures, including a hypergolic (self-igniting) mixture with glycerin. It is soluble in water and can be washed away. Even a weak water solution of it, however, will have a pronounced red, green, or brown color, depending on the oxidation state of the manganese ion. Such chemical ions are readily identified by chemical test or by IR spectrometry if even a minute chip of the solid chemical is recovered.

Flash powders contain a fine metallic powder, usually aluminum, and an oxidizer such as potassium perchlorate or barium nitrate. Elemental analysis of the powdery residue left after ignition will reveal the aluminum, chlorine, potassium, or barium. The perchlorate or nitrate residues are detectable by micro chemical test if any trace of the mixture remains unreacted.[61] Mixtures of oxidizers (chlorates, nitrates, perchlorates, or sulfates) have been shown to produce extremely hot fires with metallic aluminum or magnesium powders as fuels. With a suitable binder to control combustion rates and prevent aerosolization, these materials will burn progressively and have been used for decades as solid rocket fuels (propellants). In most rocket fuel applications the binder is a rubber, but improvised mixtures have used diesel fuel as a binder. These mixtures have sometimes been referred to as "high-temperature accelerants." Tests have demonstrated that such mixtures produce extremely large volumes of hot gases, very high heat fluxes, and extremely high temperatures (of both flame and gases). Even large compartments have been brought to flashover conditions in 2 to 3 minutes by the combustion of about 110 kg (250 lb) of an improvised mix of aluminum and magnesium with potassium nitrate and diesel fuel.[62] Sulfates (of calcium or barium) will also act as oxidizers.

Potassium chlorate is easily identified by chemical or instrumental tests if it can be recovered in its unreacted state. When it reacts in an incendiary mixture, however, its residues are almost entirely chloride salts, which are innocuous and which if recovered from debris, could not be readily identified as having come from a device.

Sugar reacts with sulfuric acid (H_2SO_4) to leave a puffy brown or black ash of elemental carbon. It may have a faint smell of burning marshmallows, but for the most part it is innocuous in appearance and is easily overlooked. The acid present may char wood, paper, or cloth in its concentrated form. It is very soluble in water, but mineral acids like H_2SO_4 do not evaporate, and they remain active and corrosive for some time. A burning sensation in the skin on contact with debris indicates that the debris should be checked for acids. The pH (acidity) can easily be measured, and chemical tests for sulfate or nitrate ions carried out. Reactive metals such as potassium or sodium can be used as an ignition device because they generate great amounts of heat and gaseous hydrogen on contact with water. The residues will contain potassium hydroxide or sodium hydroxide (lye)—both very strong caustics that can cause skin irritation and burns. The pH of water containing such residues will be very alkaline (basic), and chemical or spectrophotometric tests will reveal the metal present. It should be noted that the residues from these and similar devices are very corrosive to metal. Debris suspected of containing such corrosives should be placed in glass jars with plastic or phenolic lids, or in nylon or polyester/polyolefin bags—not in metal cans, paper bags, or polyethylene bags.

Because white phosphorus ignites on contact with air, it has been used as an incendiary device in the past. (Fenian fire is a suspension of crushed white phosphorus in an organic solvent. When the container is smashed, the solvent evaporates, exposing the phosphorus to air with immediate ignition.) Elemental analysis of the debris may reveal an elevated concentration of phosphorus, and there are chemical spot tests, although it tends to be a difficult element to detect and confirm. A common source of phosphorus in incendiary mixtures is the household match. Characterization of the components in match heads can be carried out by elemental analysis alone or combined with microscopic analysis using SEM.[63]

As described in Chapter 12, a variety of heat- and gas-producing chemical reactions can occur by accidental or intentional combination of household products. Many of these produce only enough heat to evaporate water to create a steam explosion, with no fire or flame. Typical of these are lye (drain cleaner), aluminum foil and water, dry bleach, sugar and water, or acid and metal. Still others create flammable gases like acetylene (calcium carbide and water) or hydrogen (mineral acid and zinc) that if properly ignited, can cause fires as well as explosions. Residues include unreacted starting materials, metallic oxides, chlorides, or sulfates. The heavy soot of acetylene/air combustion may be visible on adjacent target surfaces. Many so-called bottle bombs or acid bombs involve such materials. They are capable of producing high temperatures and high-pressure mechanical explosions. They can cause fragmentation injuries, chemical and thermal burns, as well as property damage. Many states and the federal authorities include such chemical mixtures in their explosive device criminal codes today. Fingerprints may survive on the typical 2-L plastic bottles used (Figure 14-16).

LABORATORY METHODS

Recent advances in high-pressure liquid chromatography (HPLC) and the related technique of ion chromatography have made it possible to identify numerous inorganic ions in a water extract of post-fire (or post-blast) debris. As with GC, the characterization of an ion of interest is based on its detection by a suitable detector at a particular retention time as it flows through a separatory column. Single-column analyses are now capable of

FIGURE 14-16 Bottle bomb remains may bear fingerprints and DNA. *Courtesy of Wayne Moorehead, Forensic Consultant.*

separating and detecting all the significant anions: Cl^-, ClO_3^-, ClO_4^-, NO_2^-, NO_3^-, SO_4^{2-}, etc. Others can separate the cations Na^+, K^+, NH_4^+, and even sugars, commonly found in explosive and incendiary materials.[64]

Many improvised chemical incendiaries are described in the literature (too many to describe here).[65] Whenever such an incendiary is suspected, samples of nearby materials should be collected for comparison. Sometimes, it is not the presence of a particular element or ion but an elevated concentration with respect to that found in the background that betrays an incendiary. Residues of most chemical incendiaries (sulfates, nitrates, chlorides) are widely dispersed and innocuous in appearance, so selection of "negative control" areas is not as straightforward as selection of such areas for liquid accelerants. Random selection may be necessary, with chemical analysis revealing areas where no residues are found. Many residues are water soluble, so exposure to rain or suppression water may compromise their detection. Aluminum metal–fueled mixtures have been seen to produce microscopic spheres of aluminum and aluminum oxide, which are visually unlike the residues of unaccelerated combustion (oxidation) of aluminum. SEM may assist identification in such cases.[66]

General Fire Evidence

Considerations related to laboratory analysis of fire evidence include the following:

- Identification of charred or burned materials
- Burned documents
- Failure analysis by forensic engineers
- Evaluation of appliances and wiring
- Miscellaneous laboratory tests
- Spoliation

IDENTIFICATION OF CHARRED OR BURNED MATERIALS

One of the most common questions a fire investigator asks is, What is (or was) this stuff—and does it belong at the scene? In earlier chapters we discussed the importance of looking for things that appear to be out of place. In a fire, many items, both trivial and

FIGURE 14-17 A suspected incendiary device can be reassembled to help establish the nature of the container and aid in search for fingerprint and trace evidence. *Courtesy of the State of California, Department of Justice, Bureau of Forensic Services.*

significant, can be sufficiently damaged that a casual examination will not result in their identification. Plastic, rubber, or metal objects having a relatively low melting point are typical problems. The identification may be based on the shape of the remaining portion, X-rays, portions of labels or details cast into the object that survive, or a determination of the type of material present. Even though the plastic or metal can be chemically identified, if only minute amounts remain unburned, this information may not be of use in establishing the specific item. Some materials, like polyethylene, polystyrene, or zinc diecast, seem to be universal in their applications. Some plastic bottles may be identified by their shape or casting marks, but the knowledge that a puddle of debris is melted polyethylene is of little use in figuring out which of thousands of polyethylene objects it might represent. Glass objects, whether broken by mechanical or thermal shock, can be pieced back together (as shown in Figure 14-17) if there has not been excessive melting of the pieces.

Laboratory identification may consist of visual examination, reassembly, cleaning, or microscopic, chemical, or instrumental tests of considerable complexity, depending on the nature of the material, the quantity recovered, and the extent of damage. By verifying the composition of even partial remains, these services may be crucial in detecting the substitution of less valuable items for heavily insured objects like jewelry, art, or clothes in cases of insurance fraud, or even in confirming their complete removal in cases of fraud or theft. If an ignition or time-delay device is suspected, the laboratory may be able to reconstruct the device. The nature or specific details of the device may then be of use in identifying the perpetrator, linking him or her with the scene, or in interrogating suspects. Figure 14-18 is an example of a common household object that careful lab examination revealed to have been used as a time-delay device (similar to Figure 16-8).

BURNED DOCUMENTS

The reconstruction and identification services described earlier can extend to fire-damaged documents. Writing paper has an ignition temperature of approximately 250°C (480°F), and single sheets of paper exposed to fire in air will readily char and burn almost completely.

FIGURE 14-18 Plastic milk bottle: Was it a device or innocent "victim"? (Compare with Figure 16-8.) *Courtesy of Jamie Novak, Novak Investigations and St. Paul Fire Dept.*

However, most documents in files, stacks, or books will not burn completely but will char and if intact, may be identified. Identification in the laboratory may employ visual examination under visible, UV, or IR light; photographic techniques; or treatment with glycerin, mineral oil, or organic solvents to improve the contrast between the paper and the inks. Some clay-coated papers and writing papers with a high percentage of rag content are more resistant to fire and are more easily restored than cheaper wood-pulp papers like newsprint.

The most critical problem is the fragility of the charred papers. Even fragmentary remains can sometimes be identified, but the more intact the document, the better the chances of identification. When charred paper is to be recovered, the investigator should take every precaution against physical destruction of the remains. Documents should be disturbed as little as possible and carefully eased, using a piece of stiff paper or thin cardboard, onto a cushion of loosely fluffed cotton wool. This cotton should be placed into a rigid box of suitable size and a layer of fluffed cotton wool placed over the documents to keep them in place. (Rigid plastic containers like sandwich boxes or those used for cottage cheese are excellent for small items like matchbooks.) The box should then be hand-carried to the laboratory.

Reconstruction of burned documents is best done under controlled laboratory conditions by experienced personnel with proper photographic support in case characters are visible only fleetingly. For these reasons, the investigator should not attempt to perform *any* tests in the field. Virtually anything done to a document in the way of coating, spraying, or treating it with solvent will interfere with laboratory tests. Many general crime laboratories and qualified document examiners have the expertise and equipment needed to perform such tests properly.

FAILURE ANALYSIS BY FORENSIC ENGINEERS

If a mechanical system failed and caused a fire, it may be necessary to determine the causation of the original mechanical fault. A registered mechanical or materials engineer is often able to tell whether a driveshaft or bearing failed as a result of misuse, overloading, poor maintenance, or design error. The failure of hydraulic systems may produce leaks of combustible hydraulic fluid that can contribute to the initiation or spread of a fire, and the experience of a hydraulic engineer may be required. Because the civil liability of the manufacturer or user of a mechanical system may be in the millions of dollars, it is very important to have an accurate estimation of the condition of that system. Such determinations are beyond the capabilities of most crime laboratory personnel, because they require specialized knowledge of the materials or processes involved in the product's manufacture. A registry of qualified professional engineers is offered in many states to assist investigators.

EVALUATION OF APPLIANCES AND WIRING

The guidelines offered in Chapter 10 will assist the investigator in making an accurate assessment of the condition of appliances or wiring and the contribution they may have made to a fire. Many diagnostic signs, however, require microscopic examinations, metallurgical tests, or continuity and conductance tests that require laboratory examination and interpretation by a specialist. For instance, whether a switch was on or off at the time of the fire requires careful dismantling and examination of the contacts, housing, and switch mechanism. Due to fire damage, such dissection may easily result in movement of the components and destruction of their precise relationships. X-ray analysis may be used if the contacts and housing are arranged to permit cross-sectional viewing. Twibell and Lomas described a novel resin-casting technique that allows cross-sectioning of the switch unit while ensuring that the components do not move.[67] Soft X-rays (of the energies used for medical procedures, 75 to 85 kV) have also been demonstrated to be of value in examining thermal cutoffs (TCOs) used to protect coffeemakers and other similar appliances from overheating, and cartridge-type fuses. Such X-rays can be used to determine their continuity and often to establish the mechanism of their operation or cause of failure. The causes for electrical product failures are sometimes revealed by X-ray examination (as seen in Figure 14-19b). In the case of fuses, those that failed due to moderate overcurrents were often differentiable from those that failed in a short circuit.[68] X-rays are also useful in establishing the operability of fire-damaged smoke detectors, as shown in Figure 10-55b.

Under some conditions, whether a copper wire was melted by overheating due to excessive current or by the heat of the fire may be determined by the degree of crystallization within the conductor, because the entire cross section of a conductor will be heated by an overcurrent, while the surface of a conductor is likely to be affected to a greater degree than the core if the heat is applied from the exterior.

Some years ago, it was suggested that elemental analysis of the surface of some fire-damaged copper conductors could yield information to determine whether the melting was

FIGURE 14-19A Fire-damaged switch: on/off or defective?
Courtesy of Helmut Brosz and Peter Brosz, Brosz and Associates.

FIGURE 14-19B X-ray of switch reveals internal connections.
Courtesy of Helmut Brosz and Peter Brosz, Brosz and Associates.

the cause or result of the fire. Other researchers, however, suggested that the absorption of gases from the ambient atmosphere while the metal was molten do not necessarily reflect the composition of fire gases in the vicinity.[69] Others contended that the composition of fire gases, whether they are absorbed or not, do not discriminate between arcing that occurs to cause the fire and arcs that occur as a result of external charring.[70] Because most post-fire indicators are the summation of the fire's effects, testing would have to examine a wide variety of fire exposures. One landmark paper discussing double-blind tests revealed that diagnosis of "fire cause" versus "fire effect" using the Auger emission spectroscopy (AES) elemental profile method was no better than random chance.[71] Currently, there is no 100 percent reliable means of establishing whether melting of copper conductors was the result of fire exposure or electric overcurrent unless there are very small globules created (as shown in Figure 10-47).[72] As a general rule, the finer the droplet size, the higher the energy (and temperature) of the molten copper. Unless there was electric arcing, the conditions needed to produce aerosol-type beads do not exist in normally fueled fires. Even then, proof of electrical activity does not prove the fire was started by an electrical fault (only that the conductor was energized).

Metallurgists can conduct crystallographic tests, but, unfortunately, elemental analysis requires expertise and equipment available only in fairly sophisticated materials- or surface-science laboratories. Elemental analysis by scanning electron microscopy/ energy dispersive X-ray spectrometry (SEM/EDS) may be of value in establishing contact between conductors of different metals (copper wire against a zinc-plated steel receptacle box, for instance) (as shown in Figure 10-51). This finding may help confirm a particular accidental ignition mechanism.[73] The presence of metallic contaminants on insulators in switches and relays has been linked to arc tracking and electrical equipment fires. SEM/EDS can help identify the nature and origin of such contamination. Recent work by Carey and Howitt on arc faults between copper conductors in fire environments has demonstrated the potential value of microscopic and elemental features.[74] Failures such as stress cracking, contamination, microorganism-induced corrosion (MIC), or erosion require analysis by a combination of microscopic, metallurgical, and chemical techniques. These, too, require specialized resources and knowledge not found in public forensic labs.

It should be kept in mind that although the laboratory specialist can perform tests far beyond the capability of the investigator, there are practical limits. It is possible for a fire to be sufficiently hot and prolonged to destroy enough of the diagnostic signs that may have once been present that the required determination cannot be made with any degree of certainty. Every laboratory has its quota of miracles that it can perform!

MISCELLANEOUS LABORATORY TESTS

There are a number of other determinations useful in the assessment of a fire that the laboratory is well suited to perform. Usually, very high localized temperatures produced during the fire may give an indication of the use of an incendiary ignition device or an unusually high fuel load (if post-flashover conditions are considered). These temperatures may be indicated by the melting (or melted state) of various materials in the vicinity. Because small variations in composition may result in considerable differences in melting points, the objects of interest should be identified in the laboratory and their melting points measured directly if at all possible. SEM/EDS has been shown to be of value in establishing the nature of interactions between aluminum and copper in forming eutectic alloys during a fire.[75] SEM can show differences in crystal structure, and when teamed with elemental analysis (SEM/EDS), it can demonstrate mixing of various elements (as was seen in Figure 10-51). This finding may be important when melting (of switch or thermostat contacts) or alloying is found. Such melting or alloying may be the result of normal fire exposure or the failure mechanism that caused the fire.

Flash Point

The flash points of materials found at the fire scene, either as part of the scene itself or as part of an incendiary device, should also be determined in a laboratory. Flash point (FP) determinations are made using one of several types of apparatus:

- Tagliabue (Tag) Closed Cup (FP below 175°F) (80°C)
- Pensky-Martens Closed Cup (FP above 175°F) (80°C)
- Tag Open Cup
- Cleveland Open Cup

The ASTM has established standard methods for the use of these devices (previously described), and the reader is referred there for more details.[76] Unfortunately, many laboratories cannot measure the flash point of a flammable liquid if that point is below ambient (room) temperature. In addition, the flash point apparatus most commonly found in analytical labs requires at least 50 mL (2 fl oz) for each determination. This is far more than is normally recovered as residue from an incendiary device and precludes such analysis. However, a flash point apparatus called the Setaflash Tester has found acceptance in forensic laboratories because it will accept samples as small as 2 mL (0.08 fl oz) as well as permit subambient measurements. It is an approved ASTM test method.[77]

Melting or Softening Points

The temperatures at which plastics melt, soften, decompose, or char may be of great value in scene evaluation (see Figure 14-20). Lab methods [including thermogravimetric analysis (TGA) or differential scanning calorimetry (DSC)] are available in many material testing or engineering labs. The melting points of even very small samples can be tested using a Mettler Thermometric Hot Stage Microscope.

Mechanical Condition

In reconstructing the scene, visual examination is often adequate to ascertain whether a door lock mechanism was locked or unlocked at the time of the fire (see Figure 14-21).

FIGURE 14-20 The softening temperature of the plastic face of this wall clock, if determined by lab analysis, could reveal the temperature of the smoke layer or amount of radiant heat to which it was exposed. *Courtesy of Jamie Novak, Novak Investigations and St. Paul Fire Dept.*

FIGURE 14-21 Door lock assembly. *Courtesy of Det. Mike Dalton, Knox County Sheriff's Dept., Knoxville, TN.*

The lock mechanisms are usually of brass or zinc die cast, both of which have relatively low melting points. As a result, they will often seize at temperatures produced in normal structure fires. When recovered at a scene, they should be photographed in place and then worked to clearly indicate their orientation and photographed in the second position. LP-fueled microtorches have been used to melt the internal components of door locks to permit forcible entry while leaving the external housing intact. Internal (lab) examination of critical door locks should be considered before assuming that the lock was normally inoperative.

If a door was not exposed to too much direct fire, it is possible to ascertain whether it was open or closed at the time of the fire by examining its hinges. In a closed door, the plates of a typical butt hinge are protected from fire by the door jamb and the edge of the door. The spine of the hinge will be exposed to the fire and will therefore be more heavily damaged than the more protected plates, as shown in Figures 14-22a and b. In contrast, the spine and plates of the hinges of an open door will be exposed to roughly the same amount of heat, and damage will be uniform on both. If the fire was not too extreme, discoloration of the metal and remnants of paint on the hinge may indicate the relative positions of the two plates. It is possible for the fire damage to be severe enough to erase the diagnostic signs for this determination. Because such fire damage is more likely near the ceiling of a structure, if it is important to know the position of a door, all (usually three) hinges from a door should be recovered and labeled as top, bottom, and middle. Photos of the hinges in place may also be helpful (along with photos of any protected

(a)   (b)

FIGURES 14-22A–B As this door hinge is moved, the differential in damage proves it was in the closed position when exposed to the fire. *Courtesy of Greg Lampkin, Knox County, TN, Fire Investigation Unit.*

area on the floor or carpet). These determinations, like the others in this section, are within the capabilities of many experienced laboratory analysts.

Techniques for evaluating soot deposits on the components of smoke detectors have recently been published.[78] The mechanical vibration of an operating detector horn causes characteristic patterns of soot deposits. These can be used to establish whether the alarm was sounding during the fire. Soot patterns can also reveal if the battery inside was connected or missing.[79]

Fire and Smoke Hazards

In accidental fires, the specialist fire laboratory can be of assistance in assessing the contribution made by various materials to the fuel load. Carpets, particularly modern ones with polypropylene-blend face yarns, can represent a hazard because of their flammability (flame spread) properties. They have been shown to be ignitable by small localized fires and then to spread fire across an entire room over a period of hours. Although furnishings sold in some states must pass flammability tests before they can be sold, this is not the case nationwide. Furnishings with either synthetic or cotton coverings and cellulosic, latex, or polyurethane fillings can therefore present a possible starting point for an accidental fire, depending on the nature of the ignition source. Furniture with synthetic fabrics and padding can spread fire very quickly, and so their identification may be an issue in determining why a fire grew so fast that escape was impaired. If any of the coverings or fillings remain unburned, it may be possible to identify them, duplicate them, and test the combination for susceptibility to accidental ignition (see Chapter 11). Such testing, although it can be done informally by most laboratories, requires considerable experience if the results are to be used in civil litigation. Some states have agencies, like California's Department of Consumer Affairs, Bureau of Home Furnishings and Thermal Insulation, that will test such hazards on a contract basis; otherwise, a private consumer or materials laboratory will have to be contacted. The simple identification of the fiber constituents of carpets or upholstery, or the rubber (elastomer) in padding may be useful in assessing the contributions those products might make to ignition or flame spread. Such identifications can be accomplished by many forensic labs using pyrolysis gas chromatography (pyro GC) or Fourier transform infrared spectrometry (FTIR).

As part of a reconstruction of the cause and manner of death in fatal fires it may be necessary to determine the origin of toxic gases that caused death, or of heavy smoke that prevented the escape of still-conscious victims from a small fire. Although the production of toxic gases and dense smoke is largely a variable of the temperatures and ventilation during the actual fire, some laboratory tests can be carried out to evaluate potential hazards. Work of this type has been done to test upholstery materials for closed environments such as airplane interiors, but it is costly and time consuming and may not be available to most fire investigators. The NFPA has developed standard tests for evaluating the fire hazards of interior finishes, and the reader is referred to its publications for specific guidance.[80]

Self-Heating Mechanisms

Self-heating of vegetable oils, fish oils, and many polymers as they dry has been known to lead to ignition of the materials involved (as discussed in Chapter 6). The identification of vegetable oils from fire debris was a major challenge for laboratory analysts due to the thermal destruction that often occurs in the vicinity of the ignition source. In addition, the vegetable oils involved do not have sufficient vapor pressure to be recovered by heating isolation methods. The chemistry of the materials and processes involved have recently been explained by Stauffer.[81] Sensitive new debris analysis techniques have been developed in which the suspected self-heating oils are extracted with a solvent, concentrated, and chemically derivatized. This process causes their fatty acid components to be rendered susceptible to GC/MS characterization by the same methods used for ignitable liquid residues described previously. These techniques have been published by Stauffer, Coulombe, Reardon, and others and are now used in many forensic labs.[82] This same technique may aid the detection of vegetable oils from potato chips if they are used as an accelerant.

Dryer fire investigations often must take into consideration the possibility of a spontaneous fire caused by the self-heating of kitchen oils in poorly laundered towels. Corn, safflower, and soy oils can all self-heat to the point of open flame when the process is initiated by the temperatures of a dry cycle. Analysis of the rinse water of the washing machine that is paired with the dryer is recommended when the dryer contents are suspected of self-heating. While the dryer drum contents may be nothing more than a charred mass, the washing machine rinse water most likely survived the fire intact, preserving traces of the vegetable oil from the original laundered load. Residues in clothes washer pumps and drains reportedly have also been identified.[83]

As described in Chapter 6, differential scanning calorimetry or critical temperature testing may reveal self-heating propensities of particular combinations of materials. These tests, however, will require bulk quantities for testing rather than residual amounts.

SPOLIATION

Before any testing is conducted, however, the investigator needs to consider what effect a test will have on the condition of the evidence. The term **spoliation** refers to any test or

spoliation ■ The destruction or material alteration of, or failure to save, evidence that could have been used by another in future litigation.

examination that will change the condition of the evidence such that other tests cannot be performed by other concerned parties. In criminal cases, the concept of nondestructive testing has been well established. An examiner is obligated to retain a sample of evidence if testing is going to be destructive, document the condition of the evidence completely beforehand, or ask that a representative of the opposition be present to observe the test. This principle extends to testing of evidence in civil cases as well. When testing of evidence is likely to change that evidence such that the opposing party (or other interested party) cannot test it properly, testing should be delayed until all interested parties are notified and have an opportunity to nominate a representative to take part in the testing. Failure to notify interested parties may expose the investigator and the examiner to civil penalties, including denial of admissibility of the results of the testing. Spoliation issues will be examined in Chapter 17.

Non-Fire-Related Physical Evidence

The most frequent mistake made by fire investigators is to have a single-minded focus on the fire itself and its origin and ignore other types of physical evidence. It is up to the fire investigator to construct a complete scenario of a fire setter's activities at the scene. These activities are not necessarily limited to the setting of the fire itself. Because arson may be used to destroy evidence of other criminal acts, the careful investigator must always be alert to the existence of evidence of "normal" criminal activity.

In particular, the fire investigator needs a basic understanding of fingerprints, blood evidence, impression evidence, physical matches, and trace evidence. Note that some of these types of evidence—particularly impression evidence and physical matches—can play a role in determining origin and cause for accidental as well as incendiary fires.

FINGERPRINTS

fingerprints ■ Unique friction ridge patterns on the palmar surfaces of the hands and fingers (or their impressions).

Many investigators assume, as do arsonists, that a fire destroys all **fingerprints**. That is not true. While it is true that many fingerprints will be lost during a fire, many will survive, even on ignition sources, as illustrated in Figure 14-23. Plastic (three-dimensional) impressions in window putty or *patent* (visible) impressions in paint or blood may remain even after direct exposure to fire. In fact, they may be permanently fixed in place by the heat. Plastic containers used to carry or hold gasoline in an arson fire have been seen to be softened sufficiently by the gasoline to allow the plastic to take on the fingerprint impressions of the person holding the container. If such a container is not exposed directly to the flames, these plastic impressions may survive. The plastic window attacked with a butane mini-torch shown in Figure 14-24 bore both plastic impressions in the softened plastic and patent impressions in the soot.

Latent fingerprints, those requiring some sort of physical or chemical treatment to make them visible, consist basically of five components: water, skin oils, proteins, salts, and contaminants. The water evaporates rapidly or soaks into absorbent surfaces and is usually of little value. Skin oils are the residues normally detected by dusting with a soft brush and fingerprint powder. These oils will remain on hard surfaces for some time or will soak into absorbent materials like paper. They can be evaporated by elevated temperatures, and such "dried out" prints may not respond to dusting. The soot produced by a smoky fire (such as a gasoline pool) may condense on a glass or metal surface to protect the latent prints, however, needing only to be gently brushed or washed in a stream of tap water to reveal the impressions. Exposure to heat may fuse the soot to the ridge details of the print so the print becomes visible when the excess soot is washed away.

In the last 25 years, dozens of techniques have been developed for the detection of latent prints that may aid the inspection of fire-related exhibits. Examination using lasers and high-intensity tunable light sources (including UV and IR wavelengths) makes it possible to penetrate contaminants and make prints visible against a variety of backgrounds.[84]

FIGURE 14-23 Palm print on soda can "developed" by exposure to fire. *Courtesy of John D. DeHaan.*

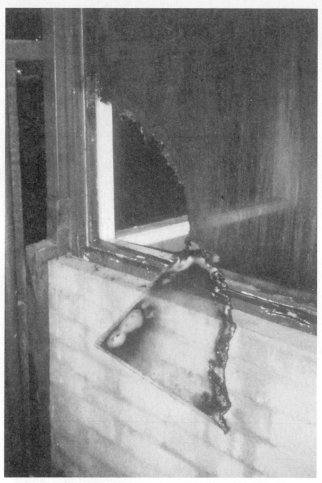

FIGURE 14-24 A plastic window in a school was attacked with a butane mini-torch to permit access. Plastic fingerprint impressions in the softened plastic and patent prints (in soot) were discovered. *Courtesy of Joe Konefal, Arson/Bomb Investigator, Sacramento, CA.*

Cyanoacrylate ester fuming and a variety of fluorescent dye stains and powders further expand the possibilities of recovery of latent prints.[85] All these methods permit the development of latent fingerprints on many surfaces thought previously to be incapable of retaining such impressions. These include surfaces such as leather, wood, vinyl plastics, smooth fabrics, and even charred paper and metal.

Proteins degrade to amino acids that are detected by reaction with the chemical ninhydrin. They are easiest to detect when the latent impressions are on clean, light-colored, porous surfaces like paper or cardboard. The proteins can be denatured by high temperatures so that they no longer react, but if the paper has not been charred by the fire, it may be worth testing. However, exposure to water (from suppression, from condensation of water vapor from the fire, or from wet weather) will dissolve the proteins and amino acids and blur the fingerprints, but the fatty deposits in the latents (skin oils and cosmetics) will not be affected by the water. They can be detected by the physical developer method. Nonporous surfaces that are or have been wet can be examined using small particle reagent (SPR). The salts are the most fire resistant of all latent fingerprint residues, resisting very high temperatures. Like amino acids, however, they are susceptible to water damage. Salts may react with the inks or coatings on paper or with the metal surfaces of cans. In doing so, they may leave an identifiable patent (visible) print in corrosion. This effect may be enhanced by the heat and moisture of the fire, leaving a visible print requiring no further treatment. Other contaminants that can produce patent prints include cosmetics, food residues, grease, motor oil, and blood.

Fingerprint expert Jack Deans in cooperation with Gardiner Associates (UK) recently conducted a series of fingerprint tests on objects subjected to observed room fires (most of which had gone to full room involvement) Deans has successfully recovered latent prints on plastic bottles (by dusting and by cyanoacrylate fuming), on newspaper (used as a torch) and matchboxes by physical developer, on the paper used as a wick in a Molotov cocktail by ninhydrin, and on other plastic surfaces by vacuum metal deposition and forensic light techniques, and has developed latent prints on plastic and glass bottles with SPR. He has also reported success in enhancing patent prints in blood and dirt by amido black treatment and forensic light techniques on a variety of surfaces.[86] Air temperatures in each of these fire tests were measured at 200°C to 500°C (400°F to 1,000°F) at floor level. Suppression of each fire was by normal water hose stream. Any surface that has been protected from direct, prolonged flame contact should be considered as a possible bearer of identifiable prints.[87] See Figures 14-25 and 14-26a and b for examples. In addition, Deans recently reported that the powder from dry chemical fire extinguishers will develop latents on smooth surfaces.[88]

Latent prints have even been recovered from human skin. Recovering such prints requires appropriate skin conditions (smooth, dry, free of hair), and the contact has to

FIGURE 14-25
Enlargement of palm print detail developed on plastic trash bag (postfire) using cyanoacrylate. *Courtesy of Jack Deans, FFS, MFSSoc, New Scotland Yard (Ret.), Fingerprint Consultant, Gardiner Associates, Ltd.*

FIGURE 14-26A Cardboard matchbox amid burned debris on floor. *Courtesy of Jack Deans, FFS, MFSSoc, New Scotland Yard (Ret.), Fingerprint Consultant, Gardiner Associates, Ltd.*

FIGURE 14-26B Fingerprints developed by physical developer on inner tray of the same box. *Courtesy of Jack Deans, FFS, MFSSoc, New Scotland Yard (Ret.), Fingerprint Consultant, Gardiner Associates. Ltd.*

have been of adequate duration. Such prints lose detail after a few hours even if the victim is deceased. If a fire victim has not been dead more than a few hours and the skin was not damaged by the fire, a search for such latents should be considered.[89] There is no more conclusive proof of identity than a fingerprint, making it worth the time and effort to search for it. Because of the rapidity with which these new techniques are being put into service, the investigator should never conclude that latents are impossible to recover before consulting with a knowledgeable specialist in the field.

BLOOD

Bloodstains left at a crime scene have much greater evidentiary value now than ever before because of great strides made in analysis of DNA in body fluids, hairs, and tissue. Above 200°C (392°F), the stain degrades and probably cannot be identified even as blood. If the proteins in a blood or semen stain (or tissue fragment) have not degraded, the evidence can be subjected to DNA typing. DNA analysis has become the primary standard for analysis of blood, tissue, hair, saliva, and other body fluids. Techniques have been developed for the analysis of increasingly smaller quantities, and databases have been assembled for convicted felons, arrested suspects, and cold cases. Mitochondrial DNA allows testing of hairs, bone, and tissue of great antiquity. This testing has become a major factor in the solution of crimes, as it can dramatically increase the identifiability of fire victims and fragmentary remains (proven at the Waco and World Trade Center scenes), as well as the link between evidence at a scene or on a victim and a suspect (even when no blood is shed).[90] Advances in extraction and polymerase chain reaction (PCR) and short tandem repeat (STR) DNA analysis methods have led to significant improvement in the DNA analysis of even burned bone.[91] It is always worth submitting samples to the lab if there is any question of identity or source. Most forensic labs are equipped or prepared to conduct such analyses, and databases are rapidly improving for comparing a scene stain against data from large populations. DNA-typable material is much more resistant to thermal and biological degradation than are the enzymes and protein factors once used. DNA-typable material (from saliva) was recently reportedly recovered from the cloth wick from a glass soft-drink bottle subsequently used for a Molotov cocktail.[92] DNA does not always result in a unique identification, but it does offer a means of associating a stain or tissue fragment with an individual with considerable certainty. If there is any doubt about a suspected stain, it should be collected, kept dry and as cool as possible, and submitted to the laboratory as quickly as possible.

Even in the absence of analysis, bloodstains can be of importance in the reconstruction of crimes (such as murder or assault) and other fire-related activity (such as escape attempts). When dried bloodstains are coated with soot and smoke in an ensuing fire, they are easy to overlook and can be very difficult to interpret even if detected. A number of chemicals have been used to enhance the visibility of faint or tiny bloodstains at crime scenes. Luminol, for instance, requires complete darkness and continual respraying to make bloodstains visible as glowing images (which are very difficult to photograph). One chemical more recently introduced, called leuco crystal violet (LCV), reacts with dried blood to form a deep blue-purple stain that is permanent and easily photographed in ordinary room light. The colorless reagent is applied as a wash or a spray (using appropriate safety equipment). The reaction is nearly instantaneous, and the solvent action washes much of the adhering soot away, enhancing the visibility of the stain pattern. Blood spatters, shoeprints, and even fingerprints in blood can be recovered by such enhancement methods. In one case, LCV was used and revealed blood spatters on the walls of a building more than a year after a murder had taken place and after two subsequent fires in the structure.

IMPRESSION EVIDENCE

Often overlooked, impressions of tools on windows or doors can confirm the fact of forced entry (see Figures 14-27 and 14-28), identify points of entry (or attempted entry), and if tools are recovered, permit identification of the tools responsible. Such tool marks may also be found on desk drawers, filing cabinets, locks, or chains at a scene, confirming that a burglary took place before the fire. The investigator should learn to recognize signs of

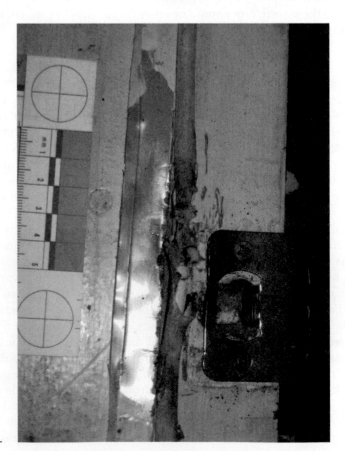

FIGURE 14-27 Tool impressions on the edge of this "jimmied" door include both impressed and striated marks, making it easier to identify the tool responsible. *Courtesy of John D. DeHaan.*

FIGURE 14-28 Doors should also be carefully examined for signs of forcible entry not consistent with the means of entry used by firefighters on the scene, such as this pried and shattered door frame. *Courtesy of Greg Lampkin, Knox County TN Fire Investigation Unit.*

forcible entry and be able to account for all such damage at a fire scene. The best kind of tool-mark evidence is the impression itself. After it has been sketched and photographed, the object (or suitable portion) bearing the impression should be collected. If its size makes recovery impractical, a replica may be cast in silicone rubber (Mikrosil) or dental impression material. Photos of a striated impression are of no use in a laboratory comparison other than to establish the location and orientation of the mark. The submitted tools are tested on a variety of surfaces to duplicate the manner of use and the qualities of the mark-bearing material as closely as possible without risking damage to the tool. The evidence and test marks are examined under 40× to 400× magnification to try to match the striations or contours present. See Figure 14-29 for a typical comparison. Tool impressions can

FIGURE 14-29 Tool-mark comparisons provide the potential of linking a suspect and the tools directly with a crime scene. *Courtesy of the State of California, Department of Justice, Bureau of Forensic Services.*

FIGURE 14-30A This sooted gas can looks normal. *Courtesy of John D. DeHaan.*

FIGURE 14-30B The underside reveals numerous penetrations in the bottom made using a knife. These holes made it possible to use the can as an effective means of spreading the gasoline inside. Tool marks on the cuts may bear striated features to identify the knife used. *Courtesy of John D. DeHaan.*

reveal the nature of the tool used to create or modify incendiary devices, and the manner of their use, which can be used to prove intent, as illustrated in Figures 14-30a and b.

Impressions of footwear and vehicle tires may be found on the periphery of the fire scene. Shoe prints have also been linked to forcible entry, as shown in Figure 14-31. The visibility of shoe prints left as a transfer of dust or light-colored soil may be enhanced by

FIGURE 14-31 A shoe print on the door reveals forced entry despite the modest scorching of wood. *Courtesy of Greg Lampkin, Knox County, TN, Fire Investigation Unit.*

FIGURE 14-32 Good-quality photograph of a shoe impression at a crime scene includes a scale; low-angle, oblique illumination; and carefully focused close-up photography. *Courtesy of Greg Lampkin, Knox County, TN, Fire Investigation Unit.*

the scorching or charring of walls or doors. Although prints impressed in soil cannot generally be collected themselves, they are worth photographing in place and possibly casting with plaster of Paris. Photographs must be taken using low-angle (oblique) lighting to highlight the three-dimensional details; a scale or ruler must be included in the photo, near (not on!) and preferably parallel to the impression; and the film plane of the camera must be parallel to the ground to prevent distortion. See Figure 14-32 for a good example. Although particular tires can sometimes be matched to their impressions, it is more common to establish only a correspondence of size and tread pattern. Footwear, due to its characteristic details and slower rate of change, is more amenable to identification and offers a means of placing the subject at the fire scene.

PHYSICAL MATCHES

One of the few conclusive identifications possible with typical evidence is the physical comparison of torn, cut, or broken edges or surfaces with each other to establish a "jigsaw" fit between them. Pieces of glass from shoes or garments have been matched back to windows, vehicle lamps, or bottles to connect a victim or suspect with a scene. Wrappings from incendiary or explosive devices have been matched to source materials in the possession of the suspect, as have pieces of tape, wire, wood, and rope. The evidentiary value of such positive identifications is very high, and the investigator must be aware of the possibilities when searching a scene for anything that looks out of place or when searching a suspect's property or vehicle. Close cooperation and communication between the lab and the investigator has produced some excellent evidence (see Figure 14-33 for an example).

TRACE EVIDENCE

Trace or transfer evidence such as paint, fibers, soil, or glass can be used to link a suspect with a crime scene even in arson cases. Paint can be transferred to clothing or tools when

FIGURE 14-33 A physical match showing a jigsaw fit that conclusively links these two pieces of vinyl electrical tape as having once been a single piece. One piece is from an incendiary device; the other is from a roll of tape in the suspect's possession. *Courtesy of John D. DeHaan.*

a forcible entry is attempted. Flakes as small as 1 mm (0.04 in.) square are adequate for color or layer sequence comparisons or complete analysis by IR spectroscopy, spectrography, or X-ray analysis. Such properties are generally class characteristics; that is, even a complete correspondence between the chemical properties of known and questioned paints does not imply that the two paints must have come from the same object. Instead, it confirms that they came from the same class of objects painted with the same paint. Multilayered paint chips, however, may be linked to a particular source if they have enough layers to demonstrate a unique origin.

Glass is found in nearly all structures and can reveal a good deal about the events of the crime. It can be broken by either mechanical force or thermally induced stress. Curved fracture ridges on the edges of glass fragments reveal that the breaking force was mechanical. The shape of these conchoidal fracture lines can help establish from which side of the window the force came. The absence of these lines usually means the fracture was caused by thermal stress. The presence or absence of soot, char, or fire debris on the broken edges can establish whether breakage occurred before or after the fire (see examples in Chapter 7). When a glass window is broken, minute chips of glass are scattered as much as 3 m (10 ft) away, especially back in the direction of the breaking force. If a subject is standing nearby, minute glass chips may be found in his or her hair, clothes, hat, pockets, or cuffs. Even very small (less than 1 mm; 0.04 in.) fragments can be compared with a suspected origin, such as a bottle or window. Once again, this usually results in a class characteristics comparison. Glass is made in large batches that vary little in their physical or chemical properties. But such transfer evidence does not require direct or prolonged contact with the source to make it valuable to the investigator.

Fibers can be transferred from the furnishings of the fire scene to the subject or from the subject's clothes to the scene, particularly at the point of entry. Such two-way transfers can be very useful in linking suspect and scene, even though fibers are usually another type of class characteristic evidence. Fibers are compared by microscopic, chemical, elemental, or spectrophotometric tests. Even the most sophisticated analyses must deal with the fact that synthetic fibers are produced in such large quantities that they are used mainly as corroborative or circumstantial evidence.

In a similar way, soils from shoes or clothing can be compared with soil from a fire scene. This may be especially helpful in a wildland fire where a distinctive or unusual type of natural soil is found in the vicinity of the fire's origin or along an access road. In urban environments, artificial contaminants such as slag, cinders, metal filings, or paint chips may contribute to the uniqueness of a soil. Contact with the soil near a workshop or industrial site may result in the transfer of soils that are unique and very much different from the natural soils nearby. Normally, soil comparisons are of marginal evidential value, but in instances such as those presented here—individual soils, especially artificial ones—do occur and should not be overlooked.

CHAPTER REVIEW

Summary

The forensic scientists or criminalists who perform the examinations discussed in this chapter are a breed apart from scientists in research or industrial laboratories. Although they may share the same science or engineering backgrounds as the others, forensic scientists relate their work directly to the real world and its criminal activities. They are uniquely capable of applying scientific tests to evidence, interpreting the results, and reconstructing a sequence of events that led to the production of that evidence. The forensic scientist provides an independent evaluation of the evidence that should challenge the interpretation of the evidence as presented by the fire investigator. It is in the role of suggesting alternative hypotheses and providing data to support or refute those alternatives that the laboratory can be of maximum value. Forensic scientists cannot work in a vacuum, however. To do their job properly, they must interact with the fire investigators, exchanging information and testing ideas and hypotheses. Investigators should encourage this kind of communication; only in this way can better evidence be obtained through fuller use of all that the scientific expert witness has to offer.

Review Questions

1. Name four sources of forensic laboratory assistance available to fire investigators.
2. What can elemental analysis offer to a fire investigator?
3. What does the gas chromatograph do?
4. What is the difference between GC/FID and GC/mass spectrometry?
5. Which isolation technique is the most widely used for preparing extracts of fire debris for volatile analysis in forensic labs?
6. Name three of the complications that can arise when a lab attempts an identification of a volatile product in fire debris.
7. Name three different kinds of chemical incendiaries.
8. Under what conditions can latent prints be developed on objects that have been in a fire?
9. What are three ways DNA analysis can assist in an arson investigation?
10. Name three kinds of impression evidence and describe how each can assist in an arson investigation.

References

1. IAAI Forensic Science and Engineering Committee, "Guidelines for Laboratories Performing Chemical and Instrumental Analysis of Fire Debris Samples," *Fire and Arson Investigator* 38, no. 4 (June 1988).
2. *ASTM E 1618: Standard Test Method for Ignitable Liquid Residues in Extracts from Samples of Fire Debris by Gas Chromatography–Mass Spectrometry* (Conshohocken, PA: ASTM, 2006).
3. *ASTM E 1492: Practice for Receiving, Documenting, Storing, and Retrieving Evidence in a Forensic Science Laboratory* (Conshohocken, PA: ASTM, 2005).
4. J. D. DeHaan, "Laboratory Aspects of Arson: Accelerants, Devices, and Targets," *Fire and Arson Investigator* 29 (January–March 1979): 39–46.
5. V. Babrauskas, *Ignition Handbook* (Issaquah, WA: Fire Science Publishers, 2003).
6. California State Fire Marshal, CFIRS Report, Sacramento, CA, March 1995.
7. A. T. Armstrong and R. S. Wittkower, "Identification of Accelerants in Fire Residues by Capillary Column Gas Chromatography," *Journal of Forensic Sciences* 23 (October 1978); D. Willson, "A Unified Scheme for the Analysis of Light Petroleum Products Used as Fire Accelerants," *Forensic Science* 10 (1977): 243–52.
8. W. Jennings, *Gas Chromatography with Glass Capillary Columns* (New York: Academic Press, 1980).

9. *ASTM E1387: Standard Test Method for Ignitable Liquid Residues in Extracts of Samples of Fire Debris by Gas Chromatography* (West Conshohocken, PA: ASTM) (withdrawn 2008).

10. J. Nowicki, "An Accelerant Classification Scheme Based on Analysis by Gas Chromatography/Mass Spectrometry," *Journal of Forensic Sciences* 35, no. 5 (September 1990); *ASTM E 1618*.

11. M. W. Gilbert, "The Use of Individual Extracted Ion Profiles versus Summed Extracted Ion Profiles in Fire Debris Analysis," *Journal of Forensic Sciences* 43 (1998): 871–76.

12. P. Sandercock and E. D. Pasquier, "Chemical Fingerprinting of Unevaporated Gasoline Samples." *Forensic Science International,* 134 (2003): 1–10; J. G. Rankin and M. Fletcher, "Target Compound Analysis for the Individualization of Gasolines," paper presented at AAFS, Washington DC, February 2007; M. R. Williams, M. E. Sigman and R. G. Ivy, "Individualization of Gasoline by GC/MS and Covariance Mapping," paper presented at AAFS, Washington DC, February 2007.

13. D. A. Sutherland and K. C. Penderell, "GC/MS/MS: An Important Development in Fire Debris Analysis," *Fire and Arson Investigator* (October 2000): 21–26; B. J. de Vos et al., "Detection of Petrol (Gasoline) in Fire Debris by Gas Chromatography/Mass Spectrometry/Mass Spectrometry (GC/MS/MS)," *Journal of Forensic Sciences* 47, no. 4 (July 2002): 736–42.

14. E. Stauffer, J. Dolan, and R. Newman, *Fire Debris Analysis* (New York: Elsevier/Academic Press, 2008).

15. *ASTM E1388: Standard Practice for Sampling of Headspace Vapors from Fire Debris Samples* (West Conshohocken, PA: ASTM, 2008).

16. J. D. Twibell and J. M. Home, "A Novel Method for the Selective Adsorption of Hydrocarbons from the Headspace of Arson Residues," *Nature* 268, 1977.

17. *ASTM E1412: Standard Practice for Separation and Concentration of Ignitable Liquid Residues from Fire Debris Samples by Passive Headspace Concentration with Activated Charcoal* (West Conshohocken, PA: ASTM, 2007).

18. J. Juhala, "A Method of Absorption of Flammable Vapors by Direct Insertion of Activated Charcoal into Debris" (Bridgeport, MI: Bridgeport Crime Lab); reprinted in *Arson Accelerant Detection Course Manual* (Rockville, MD: U.S. Treasury Department, 1980); J. D. Twibell, J. M. Home, and K. W. Smalldon, "A Comparison of Relative Sensitivities of the Adsorption Wire and Other Methods for the Detection of Accelerant Residues in Fire Debris," *Journal of the Forensic Science Society* 22, no. 2 (April 1982): 155–59.

19. K. G. Furton, J. R. Almirall, and J. C. Bruna, "A Novel Method for the Analysis of Gasoline from Fire Debris Using Headspace Solid-Phase Microextraction," *Journal of Forensic Sciences* 41, no. 1 (January 1996): 12–22.

20. *ASTM E 2154: Standard Practice for Separation and Concentration of Ignitable Liquid Residues from Fire Debris Samples by Passive Headspace Concentration with Solid Phase Microextraction* (SPME) (West Conshohocken, PA: ASTM, 2008).

21. J. R. Almirall, J. Bruna, and K. G. Furton, "The Recovery of Accelerants in Aqueous Samples from Fire Debris Using Solid-Phase Microextraction (SPME)," *Science and Justice* 36, no. 4 (1996): 283–87; J. R. Almirall, J. Wang, K. Lothridge, and K. G. Furton, "The Detection and Analysis of Ignitable Liquid Residues Extracted from Human Skin Using SPME/GC," *Journal of Forensic Sciences* 45, no. 2 (2000): 453–61.

22. H. Yoshida, T. Kaneko, and S. Suzuki, "A Solid-Phase Microextraction Method for the Detection of Ignitable Liquids in Fire Debris," *Journal of Forensic Sciences* 53, no. 3 (May 2008): 668–76; J. A. Lloyd and P. L. Edmiston, "Preferential Extraction of Hydrocarbons from Fire Debris Samples by Solid Phase Microextraction," *Journal of Forensic Sciences* 48, no.1 (January 2003): 130–36.

23. J. E. Chrostowski and R. N. Holmes, "Collection and Determination of Accelerant Vapors from Arson Debris," *Arson Analysis Newsletter* 3, no. 5 (1979): 1–17.

24. *ASTM E1413: Standard Practice for Separation and Concentration of Ignitable Liquid Residues from Fire Debris Samples by Dynamic Headspace Concentration* (West Conshohocken, PA: ASTM, 2007.

25. J. Brackett, "Separation of Flammable Material of Petroleum Origin from Evidence Submitted in Cases Involving Fires and Suspected Arson," *Journal of Criminal Law, Criminology and Police Science* 46 (1955): 554–58.

26. S. Woycheshin and J. D. DeHaan, "An Evaluation of Some Arson Distillation Techniques," *Arson Analysis Newsletter* 2 (September 1978).

27. *ASTM E1385: Standard Practice for Separation and Concentration of Ignitable Liquid Residues from Fire Debris Samples by Steam Distillation* (West Conshohocken, PA: ASTM, 2009).

28. Woycheshin and DeHaan, "An Evaluation of Some Arson Distillation Techniques."

29. *ASTM E1386: Standard Practice for Separation and Concentration of Ignitable Liquid Residues from Fire Debris Samples by Solvent Extraction* (West Conshohocken, PA: ASTM, 2005).

30. P. L. Patterson, "Oxygenate Fingerprints of Gasolines." *DET Report No. 18*, October 1990, 8–12.

31. S. E. Hipes et al., "Evaluation of the GC-FTIR for the Analysis of Accelerants in the Presence of Background Matrix Materials," *MAFS Newsletter* 20, no. 1 (1995): 48–76.

32. *ASTM D3278: Standard Test Methods for Flash Point of Flammable Liquids by Setaflash Closed-Cup Tester* (West Conshohocken, PA: ASTM, 1996).

33. J. Alexander et al., "Fluorescence of Petroleum Products II: Three-Dimensional Fluorescence Plots of Gasolines," *Journal of Forensic Sciences* 32 (January 1987): 72–86.

34. D. C. Mann, "Comparison of Automotive Gasolines Using Capillary Gas Chromatography (I and II)," *Journal of Forensic Sciences* 32 (May 1987): 616–28.

35. D. C. Mann and W. R. Gresham, "Microbial Degradation of Gasoline in Soil," *Journal of Forensic Sciences* 35, no. 4 (July 1990): 913–23; K. P. Kirkbride, et al., "Microbial Degradation of Petroleum Hydrocarbons: Implications for Arson Residue Analysis," *Journal of Forensic Sciences* 37, no. 6 (November 1992, 1585–99; D. A. Turner and J. V. Goodpaster, "A Comprehensive Study of Degradation of Ignitable Liquids," paper presented at AAFS, Denver, CO, February 2009; *Journal of Forensic Sciences* (in press).

36. J. Andrasko, et al., "Practical Experiences with Scanning Electron Microscopy in a Forensic Science Laboratory," Linkoping University, NCJRS—Microfiche Program, Report No. 2, 1979.

37. J. Andrasko, "Analysis of Polycyclic Aromatic Hydrocarbons in Soils and Its Application to Forensic Science," Linkoping University, NCJRS—Michofiche Program, Report No. 4, 1979.

38. M. T. Pinorini et al., "Soot as an Indicator in Fire Investigations: Physical and Chemical Analysis," *Journal of Forensic Sciences* 39, no. 4 (July 1994): 33–73.

39. R. J. Kuk and M. V. Spagnola, "Alternate Fuels and their Impact on Fire Debris Analysis," *Journal of Forensic Sciences,* 53, no. 5 (September 2008): 1123–29.

40. J. G. Rankin, J. Wintz, and R. Everette, "Progress in the Individualization of Gasoline Residues," paper presented at AAFS, New Orleans, February 2005; N. D. K. Petraco, "Statistical Discrimination of Liquid Gasoline Samples from Casework," paper presented at AAFS, Washington DC, February 2008; A. T. Barnes et al., "Comparison of Gasolines Using Gas Chromatography/Mass Spectrometry on Target Ion Response," *Journal of Forensic Sciences* 49, no. 5 (September 2004): 1018–27.

41. J. D. DeHaan and K. Bonarius, "Pyrolysis Products of Structure Fires," *Journal of the Forensic Science Society* 28, no. 5 (September 1988): 299–309.

42. E. Stauffer, "Concept of Pyrolysis for Fire Debris Analysts," *Science and Justice* 43 (2003): 29–40.

43. J. J. Lentini, J. A. Dolan, and C. Cherry, "The Petroleum-Laced Background," *Journal of Forensic Sciences* 45, no. 5 (2000): 968–89; S. B. Wells, "The Identification of Isopar H in Vinyl Flooring," *Journal of Forensic Sciences* 61, no. 4 (July 2005): 865–72.

44. IAAI Forensic Science Committee, "Position on Canine Accelerant Detection," *Fire and Arson Investigator* 45, no. 1 (September 1994): 22–23; *NFPA 921: Guide for Fire and Explosion Investigations* (Quincy, MA: National Fire Protection Association, 2008), 138.

45. S. R. Katz and C. R. Midkiff, "Unconfirmed Canine Accelerant Detection: A Reliability Issue in Court," *Journal of Forensic Sciences* 43, no. 2 (1998): 329–33; R. Everett et al., "False Positive and False Negative Rates for Canine Detection of Ignitable Liquids," paper presented at AAFS, Seattle, WA, February 2006; J. D. DeHaan, "Canine Accelerant Detection Validation," *CAC News,* Spring 1994.

46. A. Armstrong et al., "The Evaluation of the Extent of Transporting or 'Tracking' an Identifiable Ignitable Liquid (Gasoline) throughout Fire Scenes during the Investigative Process," *Journal of Forensic Sciences* 49, no. 4 (July 2004): 741–46.

47. Y. Kravtsova and D. Bagby, "Transfer of Gasoline from Footwear to Flooring Materials: Can This Occur at a Fire Scene?" paper presented at AAFS, Washington DC, February 2007.

48. R. J. Kuk and E. C. Diamond, "Transference and Adherence of Gasoline Vapors onto Clothing in an Enclosed Room," paper presented at AAFS, New Orleans, LA, February 2005.

49. T. L. Lang et al., *Detection of Exhaust Components from Ventilation Fans* (Toronto: Centre of Forensic Sciences, 2002).

50. J. J. Lentini, "Persistence of Floor Coating Solvents." *Journal of Forensic Sciences* 46, 6, December 2001, 1470–73.

51. S. S. Hetzel and R. D. Moss, "How Long After Waterproofing a Deck Can You Still Isolate an Ignitable Liquid?" *Journal of Forensic Sciences* 50, no. 2 (March 2005): 369–72.

52. J. D. DeHaan, "Our Changing World: Part 3— Detection Limits: Is More Sensitive Necessarily More Better?" *Fire and Arson Investigator* 52, no. 4 (July 2002).

53. D. E. Byron, "An Introduction to the New Ignitable Liquid Absorbent (ILA)," *Fire and Arson Investigator* (January 2004): 31–32; D. Mann and N. Putaansuu, "Alternative Sampling Methods to Collect Ignitable Liquid Residues from Non-porous Areas Such as Concrete," *Fire and Arson Investigator* 57, no 1 (July 2006): 43–46; M. Nowlan et al., "Use of a Solid Absorbent and an Accelerant Detection Canine for the Detection of Ignitable Liquids Burned in a Structure Fire," *Journal of Forensic Sciences* 52, no. 3 (May 2007): 643–48.

54. E. McGee and T. L. Lang, "A Study of the Effects of a Micelle Encapsulator Fire Suppression Agent on Dynamic Headspace Analysis of Fire Debris Samples," *Journal of Forensic Sciences* 47, no. 2 (March 2002): 267–74; S. A. Coulson, R. K. Morgan-Smith, and D. Noble, "The Effect of Compressed Air Foam on the Detection of Hydrocarbon Fuels in Fire Debris Samples," *Science and Justice* 40, no. 4 (2000): 257–60; B. S. Geraci, G. Hine, and W. Shaw, "CAFS

and Its Impact in Fire Scene Investigations," retrieved from www.firehouse.com, 2008.

55. J. R. Almirall et al., "The Detection and Analysis of Ignitable Liquid Residues Extracted from Human Skin Using SPME/GC," *Journal of Forensic Sciences* 45, no. 2 (March 2000): 453–61.; I. Montani, S. Comment, and O. Delemont, "The Sampling of Ignitable Liquids on Suspects' Hands," *Forensic Science International* 194 (2010): 15–124; M. Darrer, J. Jacquemet-Popiloud, and O. Delemont, "Gasoline on Hands: Preliminary Study on Collection and Persistence," *Forensic Science International,* 175 (2008): 171–78.

56. P. J. Loscalzo, P. R. DeForest, and J. M. Chao, "A Study to Determine the Limit of Detectability of Gasoline Vapor from Simulated Arson Residues," *Journal of Forensic Sciences* 25, no. 1 (January 1980): 162–67; Kuk and Diamond, "Transference and Adherence of Gasoline Vapors."

57. DeHaan, "Laboratory Aspects of Arson."

58. H. Ellern, *Modern Pyrotechnics* (New York: Chemical Publishing, 1961).

59. W. L. Dean, "Examination of Fire Debris for Flare (Fusee) Residues by Energy Dispersive X-Ray Spectrometry," *Arson Analysis Newsletter* 7, 1984.

60. K. P. Kirkbride and H. J. Kobus, "The Explosive Reaction between Swimming Pool Chlorine and Brake Fluid," *Journal of Forensic Sciences* 36, no. 3 (May 1991): 902–7.

61. R. E. Meyers, "A Systematic Approach to the Forensic Examination of Flash Powders," *Journal of Forensic Sciences* 23 (October 1978): 66–73.

62. N. R. Keltner et al., "Investigation of High Temperature Accelerant Fires" in *Proceedings Interflam 1993* (London: Interscience Communications), 607–20.

63. J. Andrasko, "Identification of Burnt Matches by Scanning Electron Microscopy," *Journal of Forensic Sciences* 23, no. 4 (October 1978): 637–42.

64. J. M. Doyle, et al., "A Multicomponent Mobile Phase for Ion Chromatography Applied to Separation of Anions from the Residues of Low Explosives," *Analytical Chemistry* 72, no. 10 (2000): 2302–7.

65. U.S. Army, *Unconventional Warfare Devices and Techniques,* TM 3–201–1, May 1966.

66. D. Mann, unpublished test data, 2005.

67. J. D. Twibell and S. C. Lomas, "The Examination of Fire-Damaged Electrical Switches," *Science and Justice* 35, no. 2 (1995): 113–16.

68. J. D. Twibell and C. C. Christie, "The Forensic Examination of Fuses," *Science and Justice* 35, no. 2 (1995): 141–49.

69. D. G. Howitt, "The Chemical Composition of Copper Arc Beads: A Red Herring for the Fire Investigator," *Fire and Arson Investigator* 48, no. 3 (March 1998): 34–39.

70. R. Henderson, C. Manning, and S. Barnhill, "Questions Concerning the Use of Carbon Content to Identify 'Cause' vs. 'Result' Beads in Fire Investigations," *Fire and Arson Investigator* 48, no. 3 (July 1998): 26–27.

71. B. Béland, "Examination of Arc Beads by Auger Spectroscopy," *Fire and Arson Investigator* 44, no. 4 (June 2004): 20–22.

72. V. Babrauskas, "Fires Due to Electric Arcing: Can 'Cause' Beads Be Distinguished from 'Victim' Beads by Physical or Chemical Testing?" in *Proceedings Fire and Materials 2003* (London: Interscience Communications, 2003).

73. M. J. McVicar, "The Perforation of a Copper Pipe during a Fire," *The/Le Journal, Canadian Association of Fire Investigators* (June 1991): 14–16.

74. N. Carey, PhD dissertation, Strathclyde University, Glasgow, November 2009; D. G. Howitt and S. Pugh, "Direct Observation of Arcing through Char in Copper Wires," paper presented at AAFS, Washington DC, February 2007.

75. E. C. Buc, D. J. Hoffman, and J. Finch, "Failure Analysis of Brass Connectors Exposed to Fire" in *Proceedings Interflam 2004* (London: Interscience Communications), 731–39.

76. O. M. Slye Jr., "Flammable and Combustible Liquids," chap. 5 in *Fire Protection Handbook,* 20th ed. (Quincy, MA: NFPA, 2008), sec. 3.

77. *ASTM D 3278.*

78. C. Worrell et al., "Enhanced Deposition, Acoustic Agglomeration and Chladni Figures in Smoke Detectors," *Fire Technology* 37, no. 4 (2001): 343–62; K. C. Kennedy, et al., "Fire Analysis Tool: Revisited Acoustic Soot Agglomeration in Residential Smoke Alarms" in *Proceedings Interflam 2004* (London: Interscience Communications), 719–24.

79. J. Colwell and A. Reza, "Use of Soot Patterns to Evaluate Smoke Detector Operability," *Fire and Arson Investigator* (July 2003): 42–45.

80. N. R. DeHaan, "Interior Finish," chap. 4-1 in *Fire Protection Handbook,* 16th ed. (Quincy, MA: NFPA, 1991), sec. 6, pp. 6-37–6-47; R. G. Gann and N. P. Bryner, "Combustion Products and Their Effects on Life Safety," vol. 1, chap. 2 in *Fire Protection Handbook,* 20th ed. (Quincy, MA: NFPA), sec. 6.

81. E. Stauffer, "A Review of the Analysis of Vegetable Oil Residues from Fire Debris Samples: Spontaneous Ignition, Vegetable Oils, and the Forensic Approach," *Journal of Forensic Sciences* 50, no. 5 (September 2005): 1091–1100.

82. E. Stauffer, "A Review of the Analysis of Vegetable Oil Residues from Fire Debris Samples: Analytical Scheme, Interpretation of Results and Future Needs," *Journal of Forensic Sciences,* 51, no. 5 (September 2006): 1016–32; R. Coulombe, "Chemical Analysis of Vegetable Oils Following Spontaneous Ignition," *Journal of Forensic Sciences* 47, no. 1 (January 2002): 195–201; A. K. Gambrel and M. R. Reardon, "Extraction, Derivatization and Analysis of Vegetable Oils from Fire Debris," *Journal of Forensic Sciences,* 5,

no. 6 (November 2008): 1372–80; L. M. Schewenk and M. R. Reardon, "Practical Aspects of Analyzing Vegetable Oils in Fire Debris," *Journal of Forensic Sciences* 54, no. 4 (July 2009): 874–80.

83. D. Mann, "Washing Machine Effluent," *Fire Findings* 7, no. 4 (1999): 4.

84. P. Margot and C. J. Lennard, *Techniques for Latent Print Development* (Lausanne, Switzerland: University of Lausanne, 1991).

85. H. C. Lee and R. E. Gaensslen, "Cyanoacrylate 'Super Glue' Fuming for Latent Fingerprints," *The Identification Officer,* Spring 1985; E. R. Menzel and J. M. Duff, "Laser Detection of Latent Fingerprints: Treatment with Fluorescers," *Journal of Forensic Sciences* 24 (1979): 96–100.

86. J. Deans, "Recovery of Fingerprints from Fire Scenes and Associated Evidence," *Science and Justice* 46, no. 3 (July 2006): 153–68.

87. S. M. Bleay, G. Bradshaw, and J. E. Moore, "Fingerprinting Development and Imaging Newsletter; Special Edition: Fire Scenes," Publ. 26/06, Home Office Scientific Development Branch, UK, April 2006.

88. J. Deans, personal communication, September 2009.

89. D. Graham, "Some Technical Aspects of the Demonstration and Visualization of Fingerprints on Human Skin," *Journal of Forensic Sciences* 14 (1969): 1–12; M. Trapecar, "Lifting Techniques for Finger Marks on Human Skin: Previous Enhancement by Swedish Black Powder," *Science and Justice* 49, no. 4 (December 2009): 292–95.

90. J. Di Zinno et al., "The Waco, Texas, Incident: The Use of DNA to Identify Human Remains," *Fifth International Symposium on Human Identification,* London, England, 1995; A. Z. Mundorff, E. J. Bartelink, and E. Mar-Casslh, "DNA Preservation in Skeletal Elements from the World Trade Center Disaster," *Journal of Forensic Sciences* 54, no. 4 (July 2009): 739–45.

91. J. Ye et al., "A Simple and Efficient Method for Extracting DNA from Old and Burned Bone," *Journal of Forensic Sciences* 49, no. 4 (July 2004): 754–60.

92. "Testing Molotov Cocktails for DNA," *Fire Findings* 12, no. 4 (Fall 2004): 14; J. Mann, N. Nic Daéid, and A. Linacre, "An Investigation into the Persistence of DNA on Petrol Bombs" in *Proceedings European Academy of Forensic Sciences,* Istanbul, 2003.

15

Fire-Related Deaths and Injuries

Courtesy of Jamie Novak, Novak Investigations and St. Paul Fire Dept.

OBJECTIVES

After reading this chapter, you should be able to:

- Identify the common causes (or contributory factors) for fire-related deaths and injuries.
- Explain the basic concepts of the team approach used when investigating a fire death.
- List the six common questions to be asked when remains (human or suspected human) are discovered at fire scenes.
- Explain several techniques involved in pathological and toxicological examinations in fire deaths.
- Explain the fire conditions and effects associated with the destruction of a body.
- Understand the role of carbon monoxide in causing deaths in fires.
- List several examples of other toxic gases produced by fires.
- Search the Internet for the autopsy protocol published by the U.S. Fire Administration that is used in firefighter line-of-duty deaths.

PEARSON

myfirekit™

For additional review materials, appendices, and suggested readings, visit **www.bradybooks.com** and follow the **MyBradyKit** link to register for book-specific resources.

Register for **MyFireKit** by following directions on the **MyFireKit** student access card provided with this text. If there is no card, go to **www.bradybooks.com** and follow the **MyFireKit** link to Buy Access from there.

Deaths and injuries frequently accompany fires. Structure fires, whether accidental or incendiary, have a high potential for causing human fatalities. Fire is a leading cause of accidental death in the United States, estimated to have claimed more than 3,400 civilian lives in 2007 and 2008 each (with 16,700 injured in 2008). Approximately 83 percent of all fire deaths occur in the home (representing a fairly constant percentage in recent years), with approximately 4 percent occurring in commercial or industrial buildings and 10 to 11 percent in vehicles.[1] Arson is used as both an agent of murder and to destroy evidence of a previously committed crime.

Every fire investigator must be prepared to deal with fire scenes where there has been a loss of human life. Figures 15-1 through 15-3 illustrate a fire that was first reported as an accidental (smoking) fire but turned out to be set to conceal a natural death (for insurance fraud) that progressed to cause the deaths of a family of four in the apartment above.

FIGURE 15-1 Exterior of apartment building. Family of four died in the upstairs apartment. *Courtesy of Randy Crim, Fire Marshal, Lake Jackson, TX.*

FIGURE 15-2 Interior of ground-floor apartment identified as room of origin. *Courtesy of Randy Crim, Fire Marshal, Lake Jackson, TX.*

FIGURE 15-3 Close-up of recliner with body of occupant. Area of origin of fire, ignited by dropped smoking materials. Suspected accidental fire was later determined to have been set by the wife of the decedent (who had died of natural causes) for insurance fraud. *Courtesy of Randy Crim, Fire Marshal, Lake Jackson, TX.*

Most fatalities in fires are not directly the result of the effects of the flames but rather of the asphyxiation caused by replacement of breathable air by toxic gases. In fact, about three times as many victims die from asphyxiation as from the thermal or physical impact of fire or explosion-caused injuries.[2] Subsequent exposure to fire then causes destruction of the body. No matter what the actual cause of death, the finding of a victim amid the ashes initiates a new series of investigative steps. Deaths from fires are not always instantaneous; they may occur hours, days, or weeks after the fire. For this reason, every fire that produces a serious injury to an occupant or emergency responder (serious enough to warrant hospitalization) should be considered a potential fatal fire and should be treated accordingly.

The Team Effort

Every fire investigation involving a death becomes a team effort, and properly so. The fire investigator, pathologist, toxicologist, radiologist, and odontologist all play vital roles in the investigation, each to a greater or lesser degree depending on the exact circumstances. Without their combined expertise, the cause and manner of death cannot be determined, and the failure of a fire investigator to appreciate the value of forensic medical contributions will seriously compromise the success of the entire investigation.

Whenever human, or suspected human, remains are discovered at a fire scene, the investigator must consider six questions:

1. Are the remains human?
2. Who is the victim?
3. What was the cause of death?
4. What was the manner of death?
5. Was the person alive and conscious at the time of the fire? If so, why did he or she not escape?
6. Was the death due to the fire or only associated with it?

We shall examine each of these points in order.

SPECIES OF REMAINS

Unless the remains can be conclusively identified as those of an animal, they should be presumed to be human, and appropriate personnel (coroner, medical examiner, or sheriff) should be contacted before anything more is disturbed. It is better to err on the side of caution. One investigator who used a pocket knife to slice into what was thought to be a pig carcass in a farm building fire was horrified to learn later that the remains were, indeed, human. The reverse is not unusual. Seriously burned remains of pigs, deer, even those of large dogs, have been mistaken for human remains. In many areas, human remains are the sole province of the coroner or medical examiner by statute, and it is a violation of the law to disturb them. Crime laboratory personnel can be called upon to make a preliminary assessment, but a physical or forensic anthropologist from a local college or university may be required to identify badly charred or fragmentary remains. Even fragmentary remains can be identified as to species if any unburned tissue is present to allow serological tests in the crime laboratory. DNA has been found to be quite resistant to degradation. Even badly burned bones and teeth have yielded some DNA of use in establishing species. The investigator should also consider having the remains of any pets found in the fire examined. The presence of injuries or bullets or the presence of carbon monoxide (CO) or drugs in the body of the pet has sometimes been an important clue in the reconstruction of equivocal fire deaths.

IDENTITY OF THE VICTIM

If the remains are human, it is important to determine the identity of the victim. Because the identification will usually have great bearing on the motive or opportunity in an incendiary fire, such an identification is essential to any thorough fire investigation. It is also possible that remains have been substituted for the purposes of insurance fraud or homicide. In instances of homicide, there must be a positive identification in order to prosecute the responsible party. The identification offered by personal effects such as jewelry or documents such as a driver's license or credit cards should not be relied on except under special circumstances (such as airplane crashes where the identities of a fixed number of possible victims are already suspected) and then *only* if there are *no* disparities between known facts and the remains. Although in some rare cases, visual personal identification can correctly be made by friends or relatives of the decedent, the effects of fire and putrefaction usually make such "identifications" of very limited reliability. Edema can bloat faces, and heat shrinks skin, tightening facial features and sometimes taking years off the estimated age. Heat and soot can also dramatically change the appearance of hair color and skin color, making misidentifications of even basic factors like age, race, or hair color more likely. The emotional trauma of dealing with a relative killed in a fire makes visual personal identification even more unreliable.

For the purpose of identification then, the services of a qualified forensic pathologist, forensic odontologist, and radiologist may be required. Identifying characteristics may include the pathological signs left by illness or previously suffered injuries, unique tattoos, surgical scars, or reconstruction of height and weight by measurement of the long bones of the limbs.

The best form of conclusive identification is often provided by comparison of the teeth of the remains with dental charts and X-rays. The teeth, fillings, replacements (prosthetics), bridges, and similar features can survive even extensive incineration and offer conclusive identification in most cases if there are adequate antemortem records (such as X-rays, impressions, and photographs).[3] Porcelain crowns can resist temperatures up to 1,100°C (2,000°F), and dental amalgam resists temperatures up to 870°C (1,600°F). Some of the newer generations of restorative dental materials degrade at temperatures of 300°C to 500°C (575°F to 930°F).[4] Composite resins have been identified as to manufacturer by scanning electron microscopy/energy dispersive X-ray spectrometry (SEM/EDS) even after thermal damage.[5] Resins and metal fillings can be damaged by direct fire-exposure heat, but the teeth themselves will withstand more severe fire exposure (eventually shattering from failure of the dentin).

Other evidence revealed by the X-rays may lead to an identification, since pacemakers, orthopedic plates, screws, pins, artificial joints, implants, and heart valves may be custom-made for a patient and may be traceable. False teeth may be identified by a numbered tag, microchip, or other feature imbedded in the resin when the plate is made. Previous fractures or unusual skeletal features revealed by radiography have been used to substantiate identifications of seriously burned bodies. The internal structural features of sinus cavities, roots of teeth, and other bones have been used to make identifications if the jaws and exposed teeth are missing and if antemortem X-rays are available.[6]

In addition to providing information of value in the identification of the victim, the radiologist is essential to a fire death investigation, for he or she is able to see beyond the char and destruction and detect both injuries and weapons. Fires may be set to conceal evidence of foul play, so a careful X-ray examination for the presence of injuries, bullets, broken bones, or even knife blades is essential. More than one investigator has been convinced that he or she was dealing with an accidental death until moving the body onto the stretcher revealed a knife still protruding from its fatal wound, or X-rays revealed shotgun pellets in the torso.

Fingerprints should not be discounted as a means of identification if any of the skin tissue of the fingers remains. If the identity of the victim is suspected, even a small portion of fingertip skin may bear sufficient detail for comparison. The epidermis of finger skin can be released by heat-caused skin slippage and may be found in the debris adjacent to the body. The analysis of remaining tissue for its DNA profile is becoming a practical reality for identification, particularly when immediate family members are available to provide blood or tissue samples for comparison with the unknown body. DNA analysis has been shown to provide accurate identifications even after extreme fire exposure and post-fire decomposition.[7] DNA identifications were made even from fragmentary remains after the 2001 World Trade Center attack.[8]

CAUSE OF DEATH

The establishment of cause and manner of death is almost always the province of the medical examiner, coroner, or equivalent official.[9] The cause of death is the actual event that brought death upon the victim. It may be further defined as the event, injury, or disease responsible for initiating the sequence of events that ultimately produced the fatal result. Aside from the entire range of homicidal and accidental causes of death, causes can be quite varied in a fire and include the following:

- Asphyxiation by carbon monoxide or other toxic gases (the most common)
- Burns (incineration)
- Anoxia due to oxygen depletion
- Internal edema caused by the inhalation of hot gases
- Hyperthermia from exposure to high temperatures alone
- Falls
- Electrocution
- Trauma from being struck by falling structural members of the building

Because of the importance of asphyxia and burns, these topics will be treated in greater depth later in this chapter.

In general terms, fire-related causes of death can be linked to the time elapsed between the fire and time of death. If death ensues very rapidly (seconds to minutes to a few hours), the cause is often related to heat, smoke, or carbon monoxide. Deaths in the first day or so are more often associated with shock, fluid loss, or electrolyte imbalance, while those occurring many days or weeks after the fire are usually precipitated by infections or organ failures. It is in this last set of cases that the link between the death and the actual cause may be lost. Here the "cause" of death may be listed as an infection, kidney failure, respiratory failure, toxemia, or even cardiac arrest, but these are mechanisms of death, not causes. The mechanism of death is the immediate medical event that brought about the cessation of life. The erroneous listing of one of these mechanisms as the cause of death obscures the involvement of fire as the *actual* causal event.

MANNER OF DEATH

The manner of death may be defined as the way or circumstances in which the cause of death occurred. This is determined by pathological and toxicological findings supported by a reconstruction of the fire scene and the victim's activities there. This reconstruction requires the fullest cooperation between the fire investigator and the pathologist (and sometimes the criminalist). It is under these circumstances that teamwork is of the highest necessity. There are five generally accepted manners of death in the United States: natural, homicidal, suicidal, accidental, and undetermined. (In some jurisdictions, a finding of death by misadventure or medical intervention is accepted.) Although the largest percentage of deaths at fire scenes will be accidental, investigators should never prejudge the manner of death any more than they should prejudge the cause of fire itself.

VICTIM STATUS AT TIME OF DEATH

Carbon monoxide levels, indications of asphyxiation and the degree of inhalation of flaming gases, and the presence of toxic combustion products in the blood or tissues and soot in the airways all play important parts in establishing the manner of death, because they can help reconstruct the circumstances at the time of the fire as well as help answer the question: Was the victim alive and conscious during the fire? If the victim was conscious, what circumstances prevented escape? Barred windows, inadequately marked exits, mental confusion due to drugs, alcohol, age, or partial asphyxiation; incapacitation by irritant smoke; obscuration of exits by heavy smoke; physical restraints; or physical disability—all are possibilities that must be considered by the investigator. The clothing worn by the victim and the position and location of the body may also be clues to his or her activities at the time of the fire. Items found near the body—a fire extinguisher, flashlight, telephone, purse, car keys, and the like—may provide a clue as to what the victim was doing or attempting to do when overcome.

DEATH DUE TO FIRE VERSUS DEATH ASSOCIATED WITH FIRE

Only a comprehensive investigation including a full pathological and toxicological examination can determine if the death was due to the fire, as we will see shortly. But fire can be part of a ritual or retaliative crime, in which case the victim is already dead due to other causes, or it may be the result of a completely accidental sequence of events. The cause of the fire and the cause of death may or may not be related. The relationship between the victim and the fire may be determined by assessing both in terms of the spread of fire and its heat and smoke. Accidental fires may occur whether the death is accidental, suicidal, natural, or homicidal. Similarly, an arson fire may trigger an accidental or natural death or be part of a homicide. Factors such as smoke detector activation (or, usually, lack thereof) may play a major role in the reconstruction.[10]

Pathological and Toxicological Examination

GENERAL CONSIDERATIONS

Although even partial destruction of a body by fire complicates the work of the pathologist, it is a very rare fire that precludes the discovery of at least some information useful to the investigator in answering the questions outlined earlier. In short, *every* body recovered from a fire, no matter how badly burned it is, should be X-rayed and given a proper forensic postmortem examination. Precautions should be taken at the scene to ensure that all remaining relevant evidence is preserved. The body should be thoroughly photographed and sketched in position before any attempt is made to move it. Once it has been ascertained that the victim is indeed dead, there is no reason to rush removal (unless continuing fire or collapse would further compromise the body). If the normal protocol for on-scene medicolegal processing is to take the internal temperature of the decedent, that should be done with a fire victim. It has recently been found that torso temperatures of an adult are very slow to rise, requiring fire exposure of 30 to 60 minutes.[11] While the internal temperature cannot be used in the traditional death interval formula, a very *low* temperature can reveal that the body was dead for a considerable period of time prior to the fire.

Skin tissue, clothing, and even bones are turned delicately brittle by fire, so extreme care must be exercised in moving the body and its associated debris; the skin, clothing, and related documents in wallets or pockets may all have vital information to offer. It is often useful to wrap the head of the victim with plastic cling-wrap before it is moved. This helps retain the smallest fragments, such as those of teeth and bone embrittled by heat, in their relative positions until they can be documented and examined in the autopsy.

When a fire death is being investigated, the team should include both a fire investigator (for whom the body and its clothing represents a source of fire patterns, often an element of the fuel load, and a potential cause or even point of origin) and a homicide investigator to whom the elements of violent death scenes are more familiar ground. The primary investigators should together conduct the search of debris in the immediate area of the body by careful sectoring and layering (examining small areas in careful sequence from top to bottom). This is where much of the most critical evidence is found and should not be delegated except under extreme circumstances. A fire death scene that has been gridded off with string is shown in Figure 15-4. The examination should include wet sifting of the debris through screens of various sizes to ensure maximum recovery. Pope warns, however, that heavily calcined bone turns the consistency of wet chalk when wetted, so dry sifting by hand is suggested when searching the area where small bone fragments (fingers, teeth, skull fragments) might be found.[12] Sustained exposure to direct fire can vitrify calcined bones to glassy "fossils" that are not affected by water. Debris adhering to the body should be sent with the body where it can be examined under the best possible conditions. The pathologist should examine the remnants of clothing and other fire debris collected with the body for evidence that may help establish identity, cause, or manner of death. If not, the investigator should personally carry out such visual examination.

FIGURE 15-4 A fire death scene gridded off with string to allow comprehensive search and reconstruction.
Courtesy of Lamont "Monty" McGill, Fire/Explosion Investigator.

FIGURE 15-5 Sieving of the debris on the bed under the victim yielded jewelry, bullets, and hardware from furniture that had been piled atop him to add fuel. *Courtesy of Lamont "Monty" McGill, Fire/Explosion Investigator.*

Once the body and associated debris have been removed, the remaining ashes and fire debris in the area should be carefully examined, preferably again by sifting through large-mesh screens and using a metal detector to ensure that all jewelry and other evidence are recovered. Bullet casings and other types of evidence of homicide may also be uncovered in this fashion, as shown in Figure 15-5. The debris within 1 m (3 ft) of the body should be carefully searched for physical evidence to help reconstruct the victim's last actions. Are the remains of a flashlight, car or house keys, dog leash, or keepsakes found nearby?

Clothing is not often completely destroyed by a fire. What the victim was wearing may indicate his or her activity just prior to the fire. Consideration must be given here to the time of the fire relative to the state of dress. Was the victim normally fully clothed at 3 A.M. or in pajamas until noon? If the fire occurred at 3 P.M. and the victim was dressed for bed, was he or she ill or incapacitated prior to the fire? More traditional time of death markers such as rigor mortis and body temperature are seriously skewed by fire exposure, and environmental indicators take on additional importance.

The postmortem examination should be carried out as soon as practical. The body must not be embalmed in the interim, because any such treatment will interfere with toxicological tests vital to the investigation. The postmortem examination is nearly always preceded by a full set of X-rays of the entire body. X-rays are critical because they can reveal bullet fragments, knife blades, broken bones, or other evidence of assault or murder that are not visible from an external examination as well as assist in the identification by the methods described earlier. The sex, approximate age, and height are determined from examination of the skull, pelvic bones, and major limb bones, while skin, hair, or other soft tissues remaining may be used to determine race and complexion. [White hair is turned yellow by temperatures on the order of 140°C (290°F), but other colors of hair remain unchanged up to the temperature at which the hair itself begins to char (260°C; 500°F) according to one author.][13] Observations by one of the authors indicate that dark hair may lighten and turn reddish with some fire exposure. Artificial dyes can react differently than natural pigments. (This may be due to ventilation effects, with an oxidizing atmosphere having one effect, while a fuel-rich reducing atmosphere has quite another.)

In severely charred bodies, computed tomography (CT) and magnetic resonance imaging (MRI) have been used successfully to document injuries caused by burns, as well as vital reactions in soft tissues caused by mechanical trauma.[14]

DESTRUCTION OF THE BODY

One problem the authors have encountered in fire death investigations is a presumption that the condition of the body as recovered from the fire is the same condition it was in at the time of death. This is obviously illogical. The body represents a fuel package and will undergo changes postmortem if it continues to be exposed to fire. It is very rare to encounter heat of sufficient intensity and duration that it destroys all anatomical features. Complete legal incineration (cremation) of an adult human body requires temperatures on the order of 950°C to 1,100°C (1,800°F to 2,100°F) for destruction to be completed within 1 to 2 hours.[15] Even then, identifiable portions of the skull, pelvic bones, and dentition usually survive. Legal cremation occurs in a gas-fired furnace in which the body is fully exposed to intense, well-ventilated flames and radiant heat from the firebrick oven lining. Lower temperatures and less complete exposure would be expected to produce less complete destruction or require longer times. The few recorded experiments on burning bodies and observations of open-air cremations in India indicate that a well-ventilated wood or coal fire with sufficient fuel can destroy most of an adult body in a few hours. A fully involved, wood-frame structure fire lasting one hour resulted in reduction of an adult male body to large bone fragments.[16] Once again, identifiable portions of the skeletal structures still remained. Case and experimental studies indicate that repeated and sustained burnings of an adult human body with mechanical crushing can result in remains that may be unidentifiable, as shown in Figure 15-6.[17] The skeletal remains of children, however, are much more readily destroyed due to their smaller size, lower mass, and less complete calcification. Such remains may be completely consumed in accidental fires.[18] Under most circumstances, the flame temperatures and durations reached in a normally fueled and ventilated fire in a frame dwelling will not be sufficient to destroy the skeletal remains of an adult or even all the soft tissues of the torso because of the significant percentage of water in the tissues and large mass. A body that is supported, even partially,

FIGURE 15-6 Sieved bone fragments from a cadaver burned in a wood-fueled "burn barrel" for 8 hours and then crushed with a wood 2 × 4. *Courtesy of Jamie Novak, Novak Investigations and St. Paul Fire Dept.*

on metal furniture or seat springs provides for better exposure to surrounding flames and may be consumed to a much larger degree than might be expected. This may account for the extensive consumption of bodies in vehicle fires, even of fairly short duration. Here the body may be well exposed to a high-temperature, energetic fire fueled by the upholstery materials and plastic components. In one case in DeHaan's experience, the body of an adult was burned to less than half its weight with one leg, one arm, and half the torso reduced to bone in an accidental fire (of some 30 minutes' estimated duration) in a camping van.[19]

Under some conditions of fuel and ventilation, however, there can be substantial incineration of the body, even without destruction of the dwelling or even other fuels in the vicinity. There have been a number of fire deaths described in the literature where the victims were found with most of the torso consumed, and the fire had not extended beyond fuels in immediate contact with the body.[20] A common scenario is a "cremated" body found in a room whose furnishings are largely undamaged by fire but that bear smoke and soot deposits and often signs of radiant heating. Such cases, fancifully labeled "spontaneous human combustion," are most often described in popular literature with only marginal scientific assessments of the incidents. Nickell and Fischer have published reviews of the historical cases and noted that most of the reports date from the eighteenth and nineteenth centuries, when much less was known about the chemistry and pathology involved.[21]

A careful examination of the well-documented cases reveals that there are certain elements common to them. First, the victims are usually elderly, with indications of partial incapacitation due to alcohol, medication, or infirmity. Many of the victims (but not all) are overweight. There is almost always a room heater, fireplace, or stove present as a source of ignition and steady heat input. Smoking materials are often present, sometimes with a history of previous accidental ignition of bedding or garments (and often with physical evidence of cigarette burns in clothing and furniture to support that history). There is usually a long time (5–12 hours) between the time the victim was last seen and discovery. Most significantly, there are almost always furnishings, bedding, or carpets involved. Such materials would provide not only a continuous source of fuel for an external fire but also a supply of char to act as a wick. With this combination of features, the investigator can appreciate the basics—fuel, in the form of clothing or bedding as the first fuel ignited and then furnishings as well as the body fat to feed later stages; an ignition source—smoking materials or cooking or heating appliances; and finally, the dynamics of heat, fuel, and ventilation to promote a slow, steady flaming fire that may generate only small open flames and insufficient radiant heat to encourage fire growth. In such circumstances, the fat rendered from a burning body will act in the same manner as the fuel in an oil lamp or candle. If the body is positioned so that oils rendered from it can seep into a porous wicking material or drip or drain onto an ignition source, it will continue to fuel the flames. This effect is dramatically enhanced if there are combustible fuels—carpet padding, bedding, upholstery stuffing—that form a rigid, carbonaceous char that can absorb the rendered fat and act as a wick. These observations are supported by experiments reported by Ettling,[22] Gee,[23] and DeHaan et al.[24]

This phenomenon, termed the *candle effect* or *wick effect*, fits with what was discussed in earlier chapters on fuels and fire dynamics. Clothing or bedding is ignited accidentally by an appliance or smoking material; the victim is quickly overcome by smoke or carbon monoxide (alone or in combination with drugs or alcohol) and cannot escape. (Note that the human body is chemically incapable of spontaneous combustion. The ignition source and the initial fuel are always external to the body.) The fire progresses through the normal fuel as a slow, low heat release fire that burns the skin, causes it to split and shrink after some minutes, and begins to render the body fat into fuel. This fuel is essentially animal fat, which has a total heat of combustion on the order of 39 MJ/kg.[25] Figures 15-7 and 15-8 illustrate two such cases. This fat does not burn cleanly, however, and has a water content of some 10 to 20 percent by weight, so its effective heat of

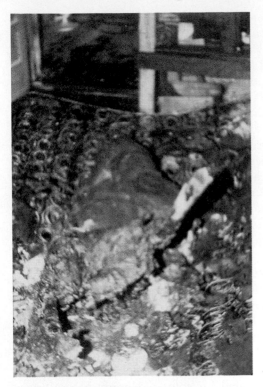

FIGURE 15-7A Lower torso and legs of elderly homicide victim are almost completely destroyed, along with the mattress stuffing. The blanket wrapped around the upper body conceals the remains of the shoulders and head. The victim was not discovered for at least 12 hours after his death, which was caused by bludgeoning and asphyxiation. *Courtesy of St. Martin Parish Sheriff's Department, St. Martinsville, LA.*

FIGURE 15-7B Unburned combustible boxes, clothing, and wallpaper in the same room as the victim are only smoke stained. The corner of the bed, where the victim was found, is in the left foreground. Fire had self-extinguished prior to discovery. *Courtesy of St. Martin Parish Sheriff's Department, St. Martinsville, LA.*

FIGURE 15-8 The body of a female murder victim was set afire in an isolated woodland. Over a period of some 5 hours, the body was consumed between the mid-chest and knees. The head (not visible), shoulders, left arm, right hand, and lower legs remained. Charred leaves beneath the body provided the needed "wick." *Courtesy of the Oregon State Police, Medford, OR.*

combustion is reduced, and the unburned pyrolysis products condense on nearby surfaces as a brown or black oily soot (often noted at such scenes). The heat of combustion is certainly high enough to sustain a modest flame with localized temperatures high enough to destroy tissue and degrade bone over a period of hours. One case in which the body was found while still burning yielded graphic proof of the mechanism. Photographs of the burning body showed low flames, less than 0.4 m (15 in.) high issuing from the torso where body fat was sustaining a hot but localized fire. This flame height correlates to a heat output of less than 40 kW.[26] The extensive destruction of the body in this case is shown in Figure 15-8. Clearly, such a small fire would be enough to raise the temperature of a closed room but not enough to bring the temperature of the room up to ignition temperature, nor even produce enough radiant heat to ignite close-by combustibles.

Note that such situations are difficult to create intentionally because they require a subtle balance of the rate at which fuel is made available and the heat release rate of the resulting flames. DeHaan conducted a test in which a 95-kg (210-lb) pig carcass was used to simulate the fuel load of a normal human body. The carcass was wrapped in a cotton blanket on a carpeted wood floor. The process was initiated by igniting 1 L (1 qt) of gasoline poured across the shoulders of the carcass. The gasoline fire burned itself out in less than 3 minutes. The synthetic carpet surrounding the carcass burned for an additional hour and 20 minutes. At the conclusion of the gasoline-fed fire, there appeared to be only surface scorching of the skin. After about 10 to 15 minutes, as the blanket and carpet continued to burn, the skin shrank and split (with the expansion of the fat and tissue beneath), exposing the subcutaneous body fat, which melted, to be absorbed by the carbonaceous char of the blanket and carpet. It could then support a flaming fire. This rendering process was sustained for some 6.5 hours before extinguishment. During that time about half the body of the pig had been burned to ashes, and the exposed bones had turned light gray and white in color, as seen in Figure 15-9.

FIGURE 15-9 The remains of a 210-lb pig carcass after self-supporting combustion for 6.5 hours. About half the mass of the body was consumed by a fire of some 60 kW. The bones are friable and light gray to white in color. *Courtesy of John D. DeHaan.*

The exposed bones were very friable and tended to shatter to ash and small fragments when handled. Flame temperatures during this process were measured at 812°C to 913°C (1,500°F to 1,675°F), while the heat release rate was a steady 50 to 60 kW during the body combustion. Such a small fire requires only a modest supply of oxygen, so the air leaking into the average room could keep it burning, even with doors and windows closed.[27] The small, smoky but prolonged fire produced limited effects in its surroundings, as seen in Figure 15-10. The results of this test were supported by testing in a cone calorimeter, which established the effective heat of combustion of both human and pork fat to be about the same, ~32 kJ/g. Note that this was *not* a smoldering fire. To support combustion, the body fat has to be presented on a porous material that can act as a wick to support a continuous flame. The range of fire sizes that might be expected from this process range from 20 to 130 kW depending on the surface area of the wick involved.[28] The amount of fuel (body fat) available determines the length of time the process will continue before it self-extinguishes for lack of fuel. Recent tests with dressed human cadavers (ignited only with direct flame) confirmed that the torso will sustain the longest-burning flames due to its larger supply of subcutaneous fat. This can cause more damage to the torso in sustained fires (6 to 7 hours) where the rendered fat can burn under the body.[29] See Figure 15-11 for an example of a body burned nearly to completion (including the torso).

The self-fueling process can be initiated only by a fire external to the body that is sustained long enough to cause the skin to shrink and split. When very badly damaged bodies are found, the immediate thought is that an accelerant had to be present. If an accelerant is used, the initial flames are usually intense enough to ignite other fuels, the room may reach a very high temperature, and some exposed fuels burn much more quickly. A pool of flammable liquid burns only for such a short period of time, however, that the fire is sustained only long enough to burn the clothing and some of the

FIGURE 15-11 Cadaver burned on a box spring with only cotton clothing and a cotton blanket. Ignited with a match, this fire burned for over 6 hours with extensive destruction of the torso. *Courtesy of John D. DeHaan.*

exposed skin and surface tissue away, but not nearly long enough to destroy the body itself or even initiate a self-sustaining fire. Figures 15-12a and b illustrates a test where gasoline was burned on the back of a cadaver with minimal effect to the skin. Experiments by one of the authors revealed that a pour of 1.9 L (0.5 gal) of a flammable camping fuel on a carpeted surface produced flame plume temperatures on the order of 775°C to 950°C (1,430°F to 1,740°F), but only for a period of 2 minutes when ignited in a medium-sized room, at which time the flames self-extinguished as the camping fuel was consumed.[30]

FIGURE 15-12A Gasoline pool burning on the back of a cadaver. Flames lasted less than 1 minute. *Courtesy of Jamie Novak, Novak Investigations and St. Paul Fire Dept.*

FIGURE 15-12B Flames nearly self-extinguished. Some scorching of skin visible, but no deep damage. *Courtesy of Jamie Novak, Novak Investigations and St. Paul Fire Dept.*

This is not nearly long enough to begin the process just described. Most of the cases involving badly burned bodies are ignited with open flame onto clothing or furnishings. In the absence of furnishings to provide the extra fuel needed to maintain the combustion process, it is possible that incidents may be attributed to the flammable nature of some clothing fabrics, with an accidental ignition by contact with a match or candle flame. (See Chapter 11 for a discussion of clothing fire injuries.) The fuel load offered by cellulose acetate, nylon, or other synthetic fibers (in cotton-blend fabrics) in the form of a complete garment could produce severe charring of the body of an unfortunate wearer. Such garments do not, however, provide enough fuel to support extensive, long-term combustion or enough rigid, porous char needed to sustain a fat-fueled fire. The investigator must carefully assess the fuel load present (including the body fat), the nature of any possible "wick" material under the body, potential sources of ignition, the time elements involved, and the extent of damage (or lack thereof) to the surroundings and balance the fire dynamic equation. Such considerations would certainly explain most, if not all, of the incidents described in the literature as "spontaneous human combustion." In a recent case, DeHaan examined a scene where the body of a homicide victim clothed in a wool overcoat and other winter garments burned almost to skeletal remains on a concrete floor. The fire plume never reached the low wooden ceiling overhead. The fire had self-extinguished before discovery.

In summary, the rate at which a body is consumed by fire is dependent on the size of the fire and the extent of exposure of the body to the fire. If the fire is a small, localized fire supported only by the combustion of the subcutaneous body fat and other susceptible tissues, in which the heat release rate of such fires has been measured to range from 20 to 130 kW, the mass loss rate of such fires has been measured to be 3.6 to 14 kg/hr (8 to 30 lb/hr).[31] (In contrast, a fully involved post-flashover room fire of 10 MW produces flame temperatures in excess of 1,000°C (1,850°F) and radiant heat fluxes of 120 to 150 kW/m^2 with commensurately rapid combustion of the body if it is exposed.) This fuel consumption rate has been used to estimate duration of the time the body was burning if the conditions of fire exposure can be documented (i.e., nature of external fuel, maximum fire size, etc.).

With regard to the destruction of the corpse, it must be kept in mind that quite different conditions apply in industrial fires. Fires involving fuels such as magnesium, thermite, or similar incendiary materials can involve temperatures of thousands of degrees, resulting in complete reduction of all the bones to calcium and silicon salts if the fire is of sufficient duration. Accidental, homicidal, or even suicidal incidents involving steel smelting, cast metal foundries, or cement making have resulted in total destruction of bodies.

EFFECTS OF FIRE

In most structure and wildlands fires, however, the pathologist has at least the skull and torso of the body with which to work, and the effects of the fire are much more manageable. Fire (heat) causes the muscles and tendons to contract, flexing the joints, particularly of the limbs. The resulting "pugilistic pose" (see Figure 15-13) or fighter's stance has been mistaken by some investigators for a last-minute self-defense posture on the part of the victim. This is not the case, as it is a commonly found artifact that has nothing to do with the physical activity of the victim just before death. In fact, the flexor muscles of the human body are usually stronger than the opposing extensor muscles. When heated, the shrinkage causes flexion at all affected joints. The final position of the body may have changed from its pre-fire position, as shown in Figures 15-14 and 15-15. Physical restraint of the limbs (binding or position) can prevent heat-induced flexion. Flexion at major joints can cause the burning body to shift position. This shift can cause bodies to fall off chairs or beds. Heating by fire exposure requires considerable time before the temperature of deep internal organs is affected. Even so, a *low* liver temperature of a fire victim may reveal an extended time interval between death and fire exposure.

FIGURE 15-13 Pugilistic posture (flexing of joints from heat exposure) of a fire victim. *Courtesy of Greg Lampkin, Knox County, TN, Fire Investigation Unit.*

Skin shrinkage and combustion exposes underlying bone to direct flame contact. This exposure occurs first where there is minimal thickness of overlying tissue—skull, elbows, knees, and wrists.[32] Fire exposure burns away the organic components of bone and then calcines the mineral portion, which causes the color of the bone to change from tan to charred black to gray to white. Extensive exposure can cause shrinkage and fracturing of calcined bones, which can be mistaken for antemortem injuries. The appearance of fracture edges in the vicinity allows the distinction to be made between postmortem and antemortem fractures by an experienced pathologist or anthropologist. Fire can also cause fractures of the skull, with separations sometimes occurring along the suture lines. Exposure to a sustained fire can cause the skull to calcine and shatter in a variety of fracture

FIGURE 15-14A Pre-fire position of cadaver: on back, arms extended by sides, legs on bed. *Courtesy of Jamie Novak, Novak Investigations and St. Paul Fire Dept.*

FIGURE 15-14B Post-fire position due to heat-induced posturing and partial collapse of mattress: right arm and right leg off bed, left leg flexed. *Courtesy of Jamie Novak Investigations and St. Paul Fire Dept.*

FIGURE 15-15A Near start of the fire: arm extended at side. Fingers are starting to flex toward the fire in blanket. *Courtesy of Jamie Novak, Novak Investigations and St. Paul Fire Dept.*

FIGURE 15-15B During the fire (mattress and body well alight): arm raised, hand flexed down. Note pool fire under bed as rendered body fats burns on noncombustible floor. *Courtesy of Jamie Novak , Novak Investigations and St. Paul Fire Dept.*

FIGURE 15-15C Late in the fire: arm has collapsed away from body. Note rendered fat on floor under wrist and hand. *Courtesy of Jamie Novak, Novak Investigations and St. Paul Fire Dept.*

FIGURE 15-16 Skull exposed to prolonged fire exhibits charring, heat induced fractures, calcination, delamination (exposing the diploe between the inner and outer tables of the skull). *Courtesy of Jamie Novak, Novak Investigations and St. Paul Fire Dept.*

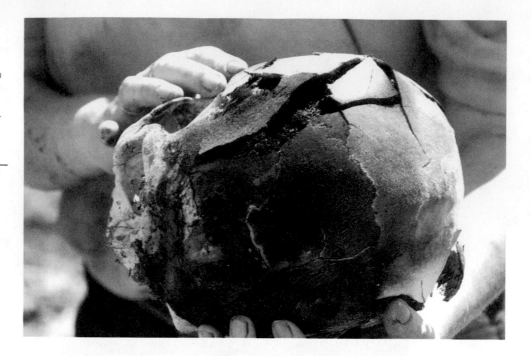

patterns as shown in Figure 15-16. The expansion of tissues, steam, or gases created by the action of the fire within the skull then may cause the fractures to separate and the tissue to exude from the skull. In controlled tests by Pope, no head was seen to "explode" as a result of fire exposure. Fire exposure caused the skull to become very friable and easily broken by post-fire movement or impact by falling debris.[33] It had been suggested that preexisting trauma (such as bullet holes or fractures) prevented such "exploding," so its absence would prove the absence of such trauma. Pope's experiments showed preexisting trauma made no difference. Heat fracturing is usually distinguishable from the depressed skull fractures caused by blunt-force trauma or beveled edges caused by bullet entry or exit. Microscopic analysis of the margins of the fractured bones may determine whether the bone was broken by mechanical or thermal insult.[34]

Bone consists of collagen for tensile strength and hydroxyapatite (calcium phosphate) mineral crystals that provide compressive strength and hardness.[35] Extreme heat exposure can cause the shrinkage of some bones, reportedly up to 25 percent in large bones at temperatures of more than 700°C (1,300°F).[36] Such changes may cause errors in the estimation of height. Bone can crack and split during severe heating. This cracking varies with the pre-fire state of the bones, with fresh bones with moisture still present cracking differently than those that have desiccated due to long-term exposure (such as archeological remains). There are indications that spalling of the fresh bone surfaces (similar to that on concrete) occurs as a result of expansion of trapped moisture. Most flames are also intermittent and turbulent, which cause the surface temperatures to rise and fall quickly, inducing mechanical stresses.[37] Under some sustained fire conditions the calcined bones can vitrify or turn to a glassy green-gray material that is not water soluble, as shown in Figure 15-17.

Tests by Christensen have revealed that bones of the elderly suffering from osteoporosis are much more easily fragmented by fire exposure than are bones of normal density.[38] This may explain why some victims of body-fueled fires have such extensive destruction of the exposed bones (not because the fire was particularly hot or intense).

Pope and Smith have described several means by which mechanical and thermal trauma to skulls can be differentiated.[39] Heating can cause partial delamination of the skull at a fracture, simulating the beveling effect of bullet exits. They recommend reassembly

FIGURE 15-17 Bone fragments recovered from a cadaver "burn barrel" test show charring calcination and vitrification as a result of sustained fire exposure. *Courtesy of Jamie Novak, Novak Investigations and St. Paul Fire Dept.*

of fractured pieces and evaluating fracture patterns and differential heat effects on adjoining pieces to distinguish between fractures caused by thermal stress from those of mechanical origin.

Conversely, damage by firearms projectiles or blast debris may be mistaken for fire damage by the inexperienced. See Figure 15-18a for an example. Fire damage is very rarely localized, so severe tissue and bone disruption in a single, isolated area must be considered in light of damage to surrounding tissues and the circumstances of the fire before it is dismissed as fire-caused. The skin's vital response to the application of heat, to redden and to blister, ceases near the time of death, but fluid-filled blisters can be caused by postmortem heating. Blisters and first- and even second-degree burns can also be caused by contact with gasoline and caustic chemicals, while reddening can be simulated by the desiccation of underlying muscle even after death.[40] The reddening of tissue caused by carbon monoxide inhalation can be obscured by sooting, burning, and other fire effects. Wounds from small-caliber weapons can be overlooked if fire damage is severe, so X-rays are very important (see Figures 15-18b–d).

Small amounts of blood may be forced from the internal tissues by heat effects, which will seep from the nose, ears, or vagina of a burned body. Significant external bleeding is *not* caused by post-fire seepage, so the presence of pooled liquid blood under

FIGURE 15-18A This X-ray of the upper arm of a victim shows shattered bone and missing pieces as a result of a gunshot injury prior to the fire. *Courtesy of Lamont "Monty" McGill, Fire/Explosion Investigator.*

FIGURE 15-18B Bomb squad portable X-ray in use at fatal fire with badly incinerated body. *Courtesy of Jamie Novak, Novak Investigations and St. Paul Fire Dept.*

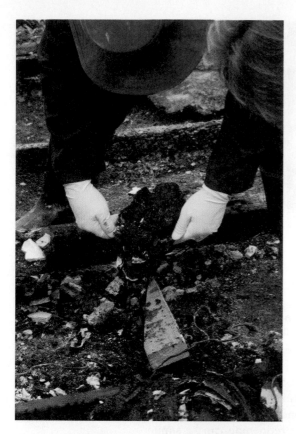

FIGURE 15-18C Positioning the head with minimal disturbance. *Courtesy of Jamie Novak, Novak Investigations and St. Paul Fire Dept.*

FIGURE 15-18D X-ray showing bullet fragments in head (small-caliber entry wound concealed by fire damage). *Courtesy of Jamie Novak, Novak Investigations and St. Paul Fire Dept.*

a body is a strong indication of antemortem violence. Fire exposure sears or cauterizes the tissue as it is exposed, thus minimizing blood loss. Internal bleeding into the extradural or epidural space of the skull (between the *dura mater* and the skull), however, can be caused by fire exposure. An accumulation of blood between the dura and the brain (subdural hematoma), however, should be evaluated as possible evidence of physical trauma. An examination of the burn injuries alone may not reveal whether they were caused immediately before or after death, and the success of the inquiry depends on the extent of cooperation and interaction between the pathologist, homicide investigator, and fire investigator.

Short-term exposure to high temperatures can cause slippage or "gloving" of the exposed skin (epidermis). The underlying dermis will tend to be red and moist-looking. Human skin consists of three major layers—the epidermis or exterior layer of hardened "dead" skin cells, the dermis, and subcutaneous tissue. The epidermis overlays the dermal layer, which includes the germinative layer and nerve endings, glands, and blood supply. Under that is a layer of connective tissue and subcutaneous fat (see Figure 15-19). First-degree burns are reddened areas of the epidermis, as seen in Figure 15-20. Second-degree or partial-thickness burns that expose the dermal layer can be very painful because the underlying pain sensors in the dermis are nearly exposed. Deeper burns destroy the dermis and the nerves (and the capacity for the skin to regenerate). Thermal exposure of skin well after death tends to produce a wrinkled, dry, yellow or brown skin surface, and gas-filled blisters can be caused by fire-induced pyrolysis long after death. Fire-induced shrinkage of the skin and underlying tissues may result in splitting or tearing, which allows the

FIGURE 15-19 Cross section of human skin: thin (0.5 mm; 0.02 in.) epidermal layer, overlying a thicker (1.5 mm; 0.06 in.) dermal layer. The epidermis is joined to the dermis by the dermal papillae (pegs) that bear the growing skin cells that replace those shed from the exterior.

Epidermis

Dermal papillae with germinative layer

Dermis (blood vessels, nerve endings, glands)

Connective tissue/ subcutaneous fat

Nerve endings

Muscle

Vein

protrusion of internal organs from the abdomen or the tongue from the mouth. The margins of such torn tissues will be irregular, and various layers will tear differently as opposed to an incised wound, where the edges are sliced uniformly. Such splitting or tearing must not be confused with antemortem cutting or blunt-force trauma injuries. If the victim lived under hospital care for some time prior to death, normal surgical procedures

FIGURE 15-20 Protected area shows where shirt was in place when heat injury (first-degree burn) occurred. *Courtesy of Greg Lampkin, Knox County, TN, Fire Investigation Unit.*

such as cutting the burned tissue (eschar) to allow circulation, expansion and motion, debridement, or even harvesting of skin from unburned areas may appear to be ante-mortem injuries. Burn injuries are discussed in a later section of this chapter.

OTHER PATHOLOGICAL FINDINGS

There are other pathological findings that may be of significance to the fire investigator. Particles of solid soot and carbon from a smoky fire will be inhaled and will cling to the moist surfaces in the throat and lungs. The finding of such contaminants in the lungs, windpipe, or larynx during the postmortem is a strong indication that the victim was alive at the time of the fire. Soot found adhering just to the inside of the mouth or nose may be the result of soot settling on exposed surfaces (onto the face of a victim lying face up in the room, for instance). It may not be an indication that the victim was alive and breathing. The position of the victim must be taken into account. A body lying face down in a lightly smoke-charged room may not show significant soot in the airways even if the victim was breathing.

Inhalation of extremely hot gases can cause a reflexive constriction of the pharynx (sometimes referred to as *laryngeal spasm*) and cessation of breathing sometimes called *vagal inhibition*. In such cases, the other pathological signs will be consistent with those of strangulation, but some pathologists will not offer much of an opinion because the spasm (if it occurs) relaxes on death, so there is no positive pathological finding. Inhalation of flame or very hot gases can also produce a rapidly developing edema (swelling) of the tongue and pharynx that produces the same symptoms as asphyxiation. In both such causes of death, the carbon monoxide level in the victim's blood may be quite low, especially if the victim was exposed to a sudden flash of flame early in the stage of the fire. A victim who suddenly opens a door from a cool, smoke-free room into a hall-way completely enveloped in flame is typical of such fatalities. Very hot dry gases are rapidly cooled by the evaporation of moisture from the mouth and throat, and fire damage (desiccation or edema) will not normally extend below the larynx. If superheated steam or moisture-laden fire gases are inhaled, there is no evaporative cooling, and thermal damage can penetrate all the way to the lungs. Such damage may occur in flash fires of hydrocarbon fuels, in which the combustion products are carbon dioxide and water vapor at very high temperatures. When such a premixed fuel/air mixture deflagrates in a room, the atmosphere becomes deficient in oxygen (nearing zero) and rich in carbon dioxide. Both these factors contribute to the collapse and death of a person in the room, and yet neither oxygen saturation nor carbon dioxide saturation can be measured in a post-mortem sample because the complexes they form with the blood are comparatively unstable and begin to change as soon as death occurs.[41]

Although instances of it are quite rare, the possibility of suicide by fire should not be discounted by the investigator. Such suicides are often (but not always) accomplished by pouring flammable liquid over oneself and then igniting it with a match. In these cases, low carbon monoxide levels are typical, because death is often caused by shock or asphyxiation as a result of inhaling the flames or superheated gases. The remnants of the clothing found on the body and nearby floor coverings and furnishings may bear residues of the ignitable liquid used and should be put into separate clean metal cans, Kapak (by Ampac) bags, or sealable glass jars to preserve them for gas chromatographic identification of the liquid.

In the absence of ignitable liquids, consideration should be given to the flame-spread characteristics of the clothing worn by the victim (as described in Chapter 11) (in both accidental and intentional fire deaths). Figures 15-21a and b show the burns to clothing and skin to an elderly man who ignited the chair in which he was sitting. He collapsed after rising from the chair. His wife died in the adjacent chair of smoke inhalation. There have been several cases in which bodies were recovered from the closed trunk of a burned car, and other evidence (non-fire) indicated the fire was set deliberately in the interior of

FIGURE 15.21A The chair and pant leg of this elderly fire victim were ignited by discarded smoking materials. The victim arose from the burning chair to collapse nearby. *Courtesy of Chiefs Mike Oakes and Tony Hudson, Clallam County Fire Investigation Team, Port Angeles, WA.*

FIGURE 15.21B Outline of body on carpet produced by soot (generated by the burning chair). *Courtesy of Chiefs Mike Oakes and Tony Hudson, Clallam County Fire Investigation Team, Port Angeles, WA.*

the car with suicidal intent.[42] Suicide cases have also been recorded in which the decedent crawled or jumped into an operating incinerator or industrial furnace, or into an established fire in a structure or vehicle.

Samples of blood and undamaged organs are mandatory. Even in badly burned bodies there may remain some liquid blood and undamaged internal organs (kidney, liver, gallbladder, heart). The blood will be used for toxicological tests to determine the presence of carbon monoxide, cyanide, alcohol, drugs, or poisons, as well as serological or DNA tests to establish the identity of the victim. Even if no liquid blood is present, tissue can be typed as long as it is still intact and not carbonized. (See Chapter 14 for a discussion of

TABLE 15-1 | Summary of Postmortem Tests Desirable in Fire Death Cases

TEST OR ACTIVITY	SUBJECT	TESTED FOR
Blood tests	Major blood vessels	Carbon monoxide
		HCN
	Heart chamber	Blood alcohol level
	Not from body cavity	Drugs (therapeutic and abuse)
Tissue tests	Brain	Drugs
	Kidney	Poisons
	Liver	Volatile hydrocarbons
	Lung	
	Skins near burns	Vital chemical or cellular response to burns
Chemical assay	Stomach contents	Activities prior to death
		Possible time of death
Airway tests	Full longitudinal transection of airways from mouth to lungs	Edema
		Scorching/dehydration
		Soot
		See figures
Internal body temperature	Liver or rectal temperature	May be low due to pre-fire death
		May be elevated due to hyper-thermia antemortem
X-rays (not fluoroscope)	Full body	Unusual features such as fractures, surgical appliances, implants, prostheses
	Debris in body bag	Details of teeth
Clothing[*]		Volatile accelerants
Photographs (in color, with scale)	Major burn damage	
	Any visible trauma	
Attend postmortem[†]		

[*]Collect and air dry clothing, package separately. If volatile accelerants are suspected, seal immediately in clean metal cans or nylon debris bags

[†]It is very useful for the fire investigator to be present when the postmortem is conducted, not only to ensure that appropriate observations are made but also to be on hand to answer any questions that arise during the examination. Few pathologists have extensive knowledge of fire chemistry or fire dynamics, so the investigator is in a good position to advise the pathologist as to fire conditions in the vicinity of the body.

forensic blood analysis.) The organs should be examined for signs of pathology due to disease both as an aid to identification and in determining the cause of death. Ignitable liquid accelerants may be absorbed by the tissues and may be detected in the blood, subcutaneous fat, or internal organs. To ensure the adequacy and accuracy of the investigation, it is very important that organ and tissue specimens be tested for toxic materials and drugs. The presence and level of drugs, particularly alcohol, will often have great bearing on the determination of cause and manner of death. Alcohol is often associated with fire deaths, particularly in accidental fires where the victim, incapacitated by alcohol, succumbs to burns or asphyxiates as a result of even minor fires. In larger fires, the influence of drugs or alcohol may be increased by smoke and toxins, producing confusion in the victim, thereby preventing escape from the fire. See Table 15-1 for a list of postmortem tests recommended. Failure to obtain a postmortem and toxicology exam precludes reliable conclusions as to the real causes of both the death and the fire.

Asphyxiation by carbon monoxide or other toxic combustion gases causes most fire deaths. Direct fire exposure is rarely the significant agent. In the Dupont Plaza Hotel fire (San Juan, Puerto Rico, 1986) and in the Stardust Disco fire (Dublin, Ireland, 1981), a large percentage of the victims had low CO and cyanide levels inconsistent with smoke inhalation deaths.[43] Victims caught in flashover situations in those fires were seen to catch fire by radiant heat as they tried to escape. In the Dupont Plaza fire, the fatalities in the hotel away from the fire had high COHb saturations.[44]

Because only carbon-based fuels are involved in virtually all structure fires, one would expect carbon dioxide (CO_2) to be the predominant gaseous product if combustion is complete. Fires rarely produce flames in which combustion is complete, and the ventilation (amount of oxygen available) and flame temperatures determine the production ratios of CO, CO_2, and other species.[45] In nearly all fires, the inefficiencies of combustion, especially of carbon-based solid fuels, will mean that CO will be present and will play a part in fire deaths. Depending on the composition of the building's contents and furnishings, other toxic gases will be produced that are also of great significance to the fire investigator. We shall discuss those separately.

Carbon Monoxide Asphyxiation

Carbon monoxide (CO) is probably the single most important cause of death from fires. Carbon monoxide, like oxygen in the air, combines with the hemoglobin of the blood. The blood hemoglobin transports all the oxygen to the tissues of the body and is therefore a vital link in the chain that supports oxidative metabolism and life itself. Unfortunately, CO combines with hemoglobin to form a complex called *carboxyhemoglobin* (COHb) with about 200 times the avidity or stability of the oxygen-hemoglobin complex (OHb, called *oxyhemoglobin*).[46] As more and more of the hemoglobin is tied up with CO, there is less available to carry oxygen to the brain and muscles. The amount of CO present in whole blood is normally expressed as a percentage saturation (COHb satn), that is, the percentage of the hemoglobin that is occupied by CO.

The intrusion of a material that displaces the oxygen from oxyhemoglobin and produces a more stable and useless hemoglobin combination serves to starve the cells of needed oxygen, and an internal suffocation results. In addition, CO interferes with the basic energy production process in the body's cells [the adenosine triphosphate (ATP) cycle], so it poisons cells and reduces their viability. CO also bonds with the myoglobin in red muscles (such as the heart) and affects their functioning.[47] In addition to replacing oxygen (asphyxiant), CO also seems to affect the transfer of signals in the nerve cells and can have permanent or delayed neural effects.[48]

These effects are fortunately reversible in a victim given oxygen or fresh air. The O_2 will slowly displace the CO from the carboxyhemoglobin, and the CO content of the latter will diminish. Thus, if a person who was exposed to a dangerous concentration of CO is placed in clean air or, better, given oxygen, he or she will generally recover as a result of this displacement.[49] The COHb level in a living person will begin to decrease as soon as the CO exposure is stopped and fresh air or oxygen is made available. The COHb level will decrease by 25 percent of the initial level after 1¾ to 2 hours in fresh air and after about 40 minutes breathing 100 percent oxygen. It will decrease by 50 percent after about 4–5 hours in air and about 1½ hours of 100 percent oxygen.[50] Elimination will be even faster if the victim is treated in a hyperbaric chamber, which increases the pressure of O_2 and hastens elimination of CO. If a victim received medical treatment before death, such effects must be taken into account when evaluating the postmortem COHb level.

Based on the relative stability of COHb and OHb, it is relatively simple to calculate the carbon monoxide content of the air that may be dangerous. If the concentration of oxygen in air was just 200 times as high as the carbon monoxide, extended breathing of this mixture would result in a mixture of COHb and OHb that would be about equal, and half the effective oxygen-carrying capacity of the blood would be lost. Because air contains about 20 percent oxygen, 1/200 of this value would be 0.1 percent. Thus, breathing a concentration of carbon monoxide of 0.1 percent [1,000 parts per million (ppm)] would eventually result in a potentially fatal 50 percent COHb level. Experience shows that

TABLE 15-2	Accumulation of Carbon Monoxide in Blood	
CO CONCENTRATION (%)	TIME OF EXPOSURE	PERCENT BLOOD SATURATION
0.02–0.03	5–6 hours	23–30
0.04–0.06	4–5 hours	36–44
0.07–0.10	3–4 hours	47–53
0.11–0.15	1½–3 hours	55–60
0.16–0.20	1–1½ hours	61–64
0.20–0.30	1/2–3/4 hours	64–68
0.30–0.50	20–30 minutes	68–73
0.50–1.0	2–15 minutes	73–76

Reprinted, with permission, from the *Journal of Forensic Sciences,* Vol. 7, No. 4, copyright ASTM International, 100 Barr Harbor Drive, West Conshohocken, PA 19428.

the dangerous limits are actually around 0.05 percent, and that a concentration of carbon monoxide of 0.06 percent can be breathed only for a period of 1½ hours without becoming critically dangerous. See Tables 15-2 and 15-3 for concentration and exposure times.

The Stewart equation relates the CO concentration in air, the breathing rate of the victim, and the COHb saturation in the blood.[51] It is usually given as

$$\% \text{ COHb} = (3.317 \times 10^{-5})(\text{ppm CO})^{1.036}(\text{RMV})(t)$$

Where:

$\%$COHb = percent saturation
ppm CO = concentration of CO in the air being inhaled
t = time of exposure
RMV = respiratory minute volume (liters per minute breathing rate)

RMV is dependent on age, sex, and physical activity[52] (see Table 15-4).

Substituting into the equation the values COHb = 50%, RMV = 20 L/min, ppm CO = 100,000 ppm (10%), we obtain

$$t = (3.015 \times 10^4)\,(50)/(1 \times 10^5)(20) = 0.75 \text{ min}$$

TABLE 15-3	Time to Reach 30 Percent Concentration of Carbon Monoxide in Hemoglobin[a]
PERCENT OF CO IN BREATHED AIR	TIME
0.01 (100 ppm)	—
0.025 (250 ppm)	—
0.05 (500 ppm)	2–3 hours
0.1 (1000 ppm)	15–30 minutes
0.2 (2000 ppm)	5–10 minutes
1.0 (10,000 ppm)	1–5 minutes

Reprinted, with permission, from the *Journal of Forensic Sciences,* Vol. 7, No. 4, copyright ASTM International, 100 Barr Harbor Drive, West Conshohocken, PA 19428.

[a]This concentration produces confusion and collapse in some healthy adult individuals.

TABLE 15-4	Respiratory Minute Volume (L/min)			
	ADULT MALE	ADULT FEMALE	CHILD	INFANT
Resting	7.5	6	4.8	1.5
Light activity	20	19	13	4.2

Source: Health Canada, "Investigating Human Exposure to Contaminants in the Environment" (Ottawa, Ontario: Ministry of National Health and Welfare, 1995); www.hc-sc.gc.ca.

This means that in moderate activity (or stress), an adult male can inhale a fatal CO amount in less than 1 minute or in a few breaths if the CO concentration is very high in the air he is breathing (as from a smoldering fire). Note that heavy activity can accelerate the RMV to 50 to 60 L/min for an adult male. This can mean even faster accumulation of CO to fatal concentrations. The "general" effects of CO are shown in Table 15-5.

THE CARBON MONOXIDE HAZARD

COHb is normally found at low concentrations in human blood, as it is a product of the normal breakdown of hemoglobin. Healthy individuals have a COHb saturation of 0.5 to 1 percent. Some diseases can cause COHb levels as high as 3 percent.[53] People exposed to CO in their environments—auto mechanics or anyone using gasoline- or propane-powered equipment inside a building—can have levels of 5 to 15 percent. Smokers will have elevated levels depending on how frequently they smoke. Heavy smokers can have levels as high as 10 percent, and even a nonsmoker sitting for any time in a smoke-filled room can have a COHb level of 3 to 5 percent. For this reason, many laboratories will not report COHb levels below 10 percent. In some instances even low levels would be of evidentiary significance. In a case in which a young, healthy woman, nonsmoker, who lived in a clean, semirural environment was found dead in a burned-out vehicle, a 3 percent COHb saturation was considered as significant indication of brief exposure to the fire before dying.

TABLE 15-5	Physiological Effects of Carbon Monoxide (for Healthy Adults)
PERCENT SATURATION IN BLOOD	PHYSICAL EFFECTS
0–10	Slight loss of mental acuity
10–20	Slight headache, dilation of skin vessels
20–30	Severe headache, throbbing
30–40	Severe headache, weakness, dizziness, confusion, nausea, vomiting, collapse
40–50	Fainting, rapid heartbeat, collapse, death in some individuals
50–60	Fainting, rapid breathing, possible coma, convulsions, respiratory irregularity
60–70	Convulsions, depressed heart action, death
70–80 and higher	Weak pulse, respiratory failure, death

Source: P. E. Besant-Matthews, "Investigation of Fire-Related Deaths," in Arson Investigation, vvi–8 (Sacramento, CA: California District Attorneys Association, 1980).

Regardless of the exposure of the ordinary citizen to carbon monoxide in quantities that appear distinctly in the blood, such exposure is not expected to result in fatalities. Although a 5 percent saturation level of COHb may be found in many ordinary citizens, it usually requires about 20 percent saturation to produce serious symptoms such as headache, nausea, disorientation, and dizziness in a healthy subject (see Table 15-5). At about 40 percent saturation, unconsciousness is expected, although the results may be more or less extreme. Victims determined to have been killed by carbon monoxide asphyxiation have been found to have levels below 30 percent; however, this is unusual. When the individual has preexisting heart or respiratory conditions that compromise the body's oxygen-delivery system or heart muscle, the loss of even 20 to 30 percent of the capacity may be fatal (when combined with CO's effects on heart muscle and nerves). Other toxic gases such as HCN may also play a role, as it is impossible to separate CO from other toxic gases in fire exposure. A low COHb level in a fire victim may indicate death resulted early in a fire caused by massive incineration, inhalation of flame-hot gases, or mechanical trauma (crushing, burial in debris, explosion trauma). Hyperthermia or anoxia can also kill even in areas not directly affected by products of combustion. A negligible COHb saturation does *not* prove the victim died before the fire. Carbon monoxide is very stable in postmortem samples and tissues. Unlike ethanol, it is not produced endogenously in the body. Carbon monoxide is *not* absorbed from a fire environment postmortem; it is absorbed only by inhalation during life.

Cyanide is present in many combustions and is considered to be the most toxic gas in fire environments. It can kill at concentrations as low as 200 ppm in inhaled air, producing 2 to 3 mg/L in blood (normal blood CN levels are on the order of 0.25 mg/L).[54] Unlike COHb, CN concentrations can change with decomposition and storage conditions. It is always found in combination with CO, but its effects seem to be simply additive.

For example, in the Happy Land Social Club (New York, 1990), where the fire was limited to a gasoline-fueled fire in the ground-level entry, the COHb concentrations in victims ranged from 37 to 93 percent with a mean of 76.5 percent. Cyanide blood concentrations ranged from 0 to 5.5 mg/L; nine had zero cyanide detected. CN concentrations did not correlate with amount of smoke inhalation, so its role in causing deaths was doubtful. Burn injuries were relatively minor and apparently played no role in causing the deaths.[55] It is generally agreed that a level of 50 percent saturation or greater is incompatible with life, despite the finding in dead bodies of both lower and higher degrees of saturation. Data in Table 15-6 from Gettler and Freimuth indicate the COHb percentage in a total of 65 cases of fatal asphyxiation.[56] It will be noted that they found some deaths

TABLE 15-6	Fatal Cases of Carbon Monoxide Asphyxiation Classified according to Carbon Monoxide Saturation of the Blood
TOTAL CASES (%)[a]	CO (SATURATION OF BLOOD, %)
4.5	30–40
4.5	40–50
14.5	50–60
28.0	60–70
48.5	70–80

Source: A. O. Gettler and H. C. Freimuth, "Carbon Monoxide in Blood: A Simple and Rapid Estimation," *American Journal of Clinical and Pathological Technology*, Supplement 7, no. 79 (1943).
[a]Total of 65 cases.

below 40 percent and none over 90 percent. One of the authors has encountered laboratory values of over 90 percent in a very few instances (usually when the victim was most likely asleep during the entire exposure).

EFFECT OF RATE OF ABSORPTION

It is well recognized that all individuals show different tolerances to the influence of toxic gases and other environmental influences. Certainly, such variations must exist with carbon monoxide as well. It remains to be explained how the death of one person from carbon monoxide at 40 percent saturation can be consistent with that of another who had accumulated a 90 percent saturation. The relevant factors appear to be as follows:

- Rate of inhalation of the gas
- Activity or its lack, which changes the requirements for oxygen
- Individual variations in susceptibility

The first two factors are firmly linked. The more physical activity, the faster the breathing will be, and the faster a lethal level will be reached. Physical activity increases the demand for oxygen from the voluntary muscles, and oxygen starvation of those issues will tend to incapacitate a person and reduce his or her chances of escape. Case histories reveal that persons at rest (asleep or unconscious) tend to have the highest COHb saturations because they make no demands on their heart. The role of CO interaction with heart muscle may be the strongest factor in human variability. If a person is relatively inactive, the need for oxygen is diminished, the breathing is slower and more shallow, and carbon monoxide in the air will accumulate more slowly than if breathing is faster and deeper. Carbon monoxide in the air will rapidly be accumulated if breathing is deep and at a high rate, as would be required in performing physical work. Thus, it can be stated that if a person suspects that carbon monoxide is present in significant amounts in the ambient air, he or she is better off resting quietly, breathing as little and as shallowly as possible, and avoiding all activity that will increase the breathing rate. Such considerations are often important for firefighters who must exert themselves in fighting a fire, rescuing trapped persons and the like, because the high degree of activity will inevitably increase the hazard considerably. This is one of the reasons firefighters use SCBA during interior firefighting efforts. Physical activity also increases the body's demand for oxygen, and a sudden increase in demand when 40 or 50 percent of the blood is already saturated may bring about collapse.[57] Variations in individual tolerance lie in the area of physiological differences, concentrations of hemoglobin in the blood, and related factors that are generally not under the control of the individual. Much of the toxicological studies cited previously dealt with healthy adult males as test subjects. Effects on the elderly and on small children have not been studied, but it would be expected that CO would have a negative effect on these victims at lower levels than for healthy adults. Elderly victims, whose heart and lung functions may already be impaired as a result of age or illness, are most probably more likely to die when less than 40 percent of their blood is bound up with CO. Infants have a faster respiration rate, and their uptake of CO may be faster than that evidenced in adults. There are also indications that the effects of CO are combined with or even enhanced by the effects of other toxic gases (such as CN or formaldehyde) from the fire, and by the CO_2 and low O_2 levels present.[58]

Fires are complex events, and their environments are not uniform, so the exposure of a person is not very predictable, being highly dependent on what the victim was doing during the fire. Death in a fire may be brought about by a combination of negative factors—heat, flames, oxygen deficiency (hypoxia plus enriched carbon dioxide concentrations), carbon monoxide, and other toxic gases. Each of these factors has what is called a *fractional incapacitation dose.* If a victim is exposed to half of a fatal (incapacitating) dose of CO (developing a level of 30 percent, for instance), oxygen of 12 percent (just able to sustain consciousness), and some heat, the total incapacitation dose may exceed 100 percent, and the person collapses to die from structural collapse or flame contact (incineration). No one of the factors was enough to cause death, but together they created a fatality. It is important to remember how many things can kill someone in a fire beyond CO alone. The fire environment and the nature of the victim's interaction with it must be considered in evaluating the toxicology findings.

SOURCES OF CARBON MONOXIDE

It was stated earlier that virtually all carbon monoxide is the product of incomplete combustion of carbon compounds. Almost every instance of asphyxiation from this gas results from one of the following:

- Fires involving structures, clothing, or furnishings
- Defective heating equipment (including charcoal grills used indoors), the primary cause of nonfire fatalities involving CO
- Automobile or other internal combustion motor exhaust
- Industrial processes in which CO is intentionally generated as a fuel or as part of a manufacturing process or in which it is produced as an unintentional product by incomplete combustion of some fuel

Structural Fires

In every structure fire, carbon monoxide is generated and is the cause of most fire deaths. The rate at which a fire produces CO is controlled by the size of the fire, the efficiency of its combustion, and the availability of oxygen. As we saw previously, smoldering, underventilated, or post-flashover fires produce CO concentrations of 1 to 10 percent in its byproducts, while well-ventilated, free-burning fires produce less than 200 ppm (0.02 percent). It is also produced in forest and wildlands fires, but the ventilation and exposure conditions of such fires make CO fatalities very rare. Fires in large buildings such as hotels produce vast quantities of combustion gases that may "channel" through halls, elevator shafts, and ventilation ducts and collect at areas remote from the actual fire. Deaths may occur as a result of the fire even though no fire is in the vicinity. Flash fires involving premixed fuel/air mixtures (which produce very little CO until the predominant fuel becomes the pool of residual liquid or solid fuels ignited by the flash fire) or conflagrations in enclosed structures can consume enough of the oxygen in the room, replacing it with CO and CO_2, that the little remaining oxygen is incapable of supporting life. Even if the victim avoids inhaling the hot gases, there may not be enough oxygen to maintain consciousness (hypoxia), and the victim collapses.[59] Oxygen concentrations between 15 and 21 percent have no effect on healthy individuals; atmospheric oxygen concentrations below 15 percent can cause dizziness, confusion, and loss of motor skills. Concentrations below 10 percent will lead to unconsciousness and death. In a fire, the carbon dioxide produced dilutes the available oxygen and increases the breathing rate, making the situation even more critical.

Heating Equipment

Of the 480 deaths each year in the United States from CO from nonfire sources, the Consumer Product Safety Commission (CPSC) estimates that about 33 percent are from CO emissions from consumer products and that some 47 percent of those incidents involve heating appliances (gas, oil, or wood).[60] Heating equipment is always suspect when an asphyxiated person is found in a residence with no fire damage and no automobile. Some general principles of the operation of nonelectric heating systems are important to understand. For a dangerous CO level to be created in an environment, two factors are essential:

- The fire (combustion) being maintained in the equipment is generating carbon monoxide in dangerous quantities.
- The gas being generated is escaping into the room air instead of being vented.

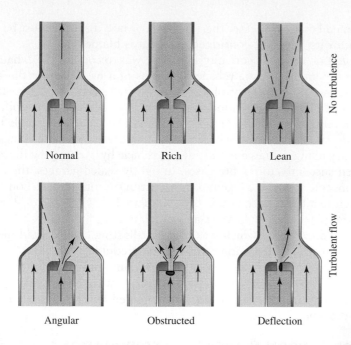

FIGURE 15.22 Typical venturi operation for gas burner under various configurations.

No turbulence

Normal   Rich   Lean

Turbulent flow

Angular   Obstructed   Deflection

These two factors are so important that they must never be overlooked if errors are to be avoided. Most gas appliance burners generate some CO, and the color of the flame is one clue to its efficiency, since an orange or yellow flame is probably producing significant CO in addition to the carbon that is making the flame yellow. A blue flame of natural gas properly adjusted, for example, reflects almost complete conversion of the fuel to CO_2 and H_2O and generates very little CO. Such a flame may not be a hazard even if all its combustion gases are escaping directly into the room (except for producing CO_2 and replacing the oxygen in the room). Venting of gas appliances would not be necessary if the flames were perfectly adjusted at all times and adequate replacement air was provided (see Figure 15-22). Because flames do not always stay perfectly adjusted and there is no way to ensure replacement air, there is a strong and valid argument for the adequate venting that is required by the codes of most jurisdictions. (See Chapter 6 for an example of a malfunctioning water heater.)

Once the cause of generation of excessive amounts of carbon monoxide is established, it is then necessary to locate the reason that the venting of the appliance did not harmlessly remove the gas but rather delivered it into the air that was breathed by the victim. This reason always has to do with the system by which combustion gases are vented or kept from entering the environment of the persons utilizing the appliance. Examination of water heaters, gas stoves, fireplaces, and their vents is beyond the scope of this text (except for the related causes of fire) and will not be dealt with here. However, visual examination of the vents may reveal readily identifiable blockages such as birds nests, trash, or accumulated debris.

"Heating equipment" sometimes includes fireplaces, wood stoves, and even charcoal barbecue grills. All these produce CO (especially the charcoal), so if their products are not properly vented to the outside, a fatal concentration of CO can quickly develop in a room. Charcoal grills being used for indoor cooking as well as heating are responsible for a large number of non-fire CO deaths. Portable LP gas heaters and salamanders can also produce CO if used in confined spaces or if improperly adjusted. Gasoline-powered electric generators and extractor fans can produce lethal CO concentrations if used in confined spaces for a suitable length of time.

Automobile Exhaust/Industrial Processes

Deaths involving CO from non-fire sources such as motor vehicles or industrial processes will not often concern the fire investigator and will not be dealt with in depth here. The investigator should remember that whether there is a fire or not, a source must be identified and a mechanism for its transport confirmed before it is blamed.

For many years, a certain means of committing suicide was to connect the exhaust pipe of an automobile with the interior of the vehicle by means of a hose. With today's emission-controlled engines, the CO content of the exhaust is on the order of 0.1 to 1.0 percent at the tailpipe (in warm engines). Dilution of this low concentration by introducing it into a car produces a level that may be too low to be fatal. The would-be suicide may be waiting for hours for death while the car, whose catalytic converter was not designed for long periods of stationary idling, ignites a fire inside the vehicle by overheating the exhaust system. The victim then succumbs to the fire gases. In tightly sealed garages, the running modern engine may consume so much oxygen that the victim succumbs to carbon dioxide or oxygen depletion before a lethal dose of CO is accumulated.[61]

For many years, the piped gas for heating and appliances in most cities was "town gas," "coal gas," or "producer gas," which was made by heating coal to yield methane, hydrogen, and carbon monoxide. Inhalation of this product could be fatal in a very short time. Today nearly all such systems use natural gas, which contains no carbon monoxide. Would-be kitchen suicides find themselves inhaling large quantities of gas with no fatal results, unless the flammable mixture thus produced is ignited by open flame or electric arc, with spectacular, if tragic, deflagrations resulting.

INVESTIGATION OF CARBON MONOXIDE ASPHYXIATIONS

Make Sure the Victim Actually Died from Carbon Monoxide

Carbon monoxide often produces a bright pink or red color in the victim's skin if present in concentrations greater than 30 percent COHb.[62] This coloration should be used only as a preliminary indicator because there are other conditions that can produce such coloration—for example, death by hypothermia and cyanide poisoning. The pink coloration may not be detectable in all people (except in the gums, nailbeds, or internal organs). It may also be masked by decomposition or fire effects (smoke and thermal effects). The postmortem report must, in such cases, include a COHb value (preferably a numerical value). If there was fatal CO asphyxiation, the value will generally be over 40 percent but may occasionally be lower. If much lower, there will be real doubt as to the cause of death, and the investigation may take a different direction. (The multiple contributions fires can make toward death were discussed earlier in this chapter.) The measurement of CO level in blood is a relatively straightforward test that can be accomplished by chemical assay or gas chromatographic analysis.[63] Levels below 10 percent may be reported by the laboratory as a negative because of the expected "background" levels in smokers. Request an exact percentage. The percentage of smokers and nonsmokers exposed to cigarette smoke is considerably lower today than 20 years ago. In living subjects, Suchard advises that standard pulse oximeters used in emergency rooms do not detect or accurately measure COHb and OHb levels, because COHb maintains a bright red color that other devices interpret as regular OHb.[64]

Determine a Probable Origin

The circumstances of the asphyxiation will normally throw heavy suspicion on a particular source. In a house that did not burn, gas- or oil-fired appliances are first-choice offenders. If a fire occurred in the house, the fire is the most probable source. Such probable sources are generally apparent to the first person on the scene, and they require little effort to identify. CO deaths have occasionally been documented in open air, when a vehicle trapped in mud continued to run, and when swimmers in the water behind a running motorboat or sitting on the shelf behind the transom of a ski boat were killed by CO.[65] Such cases might erroneously be dismissed as simple natural deaths or drownings.

Structural Fires

Because deaths can occur from CO both in the vicinity of the fire and remote from it, a careful and complete inquiry must be made into every death at a fire scene. A complete postmortem is required to ensure that the true cause of death is established. Contributions to the manner of death from CO, smoke, toxic gases, drugs, alcohol, and illness must be taken into account. As in every type of investigation, all the factors must balance out. All the pieces must fit. The investigator must be careful not to jump to a convenient conclusion. Computer modeling techniques (such as NIST's Fire Dynamics Simulator—FDS) allow prediction of species concentration in smoke at any position in the developing fire. Such predictions have been used to predict times to incapacitation of occupants or firefighters.[66]

Carbon monoxide detectors are becoming more common in residences in the United States. They have been credited with reducing deaths from all three major sources of CO—undetected smoldering fires, malfunctioning gas appliances, or exhaust from running vehicles seeping into attached houses. It must be remembered that carbon monoxide has a very wide flammable range (12 to 74 percent) and a vapor density of nearly 1.0. This means that if it is at the same temperature as the air, it will not settle or rise in an enclosure and will diffuse very readily throughout the space to form a deadly mixture from both the life safety and fire standpoints.

Other Toxic Gases

As discussed in previous chapters, many compounds found in building products, furnishings, and household products produce significant quantities of toxic gases as they burn. The actual chemistries of their production are very complex. The final products are dependent not only on the fuel but also on the temperatures involved and the amount of ventilation available (pyrolysis versus oxidation). Considerable work is underway in fire toxicology to enable one to predict the toxic effects of various products.[67]

HYDROGEN CYANIDE AND OTHER TOXIC GASES

Aside from CO, the predominant toxic gas hazard is probably the hydrogen cyanide (HCN) produced by many common polymers that contain nitrogen. Urethanes are the most common because they are present in many coatings, paints, and varnishes; flexible polyurethane foam rubbers are found in nearly all upholstered furniture; and rigid urethane foams are used in ceiling and wall insulation. Other materials that contain nitrogen in their molecular structure will produce HCN as well as nitrogen dioxide (NO_2) as a combustion by-product—hair, wool, nylon, silk, and urea melamine and acrylonitrile polymers. Because these products are nearly universal in modern structures, they must be considered a threat to life in all fires. In deaths where the carbon monoxide level of the blood is found to be low and there appears to be no other cause of death, an assay of the blood or tissues should be carried out for HCN. Exposure to levels of 0.3 percent in air is rapidly fatal, so firefighters and police personnel involved in the rescue operations must exercise due caution.[68] Hydrogen cyanide remains in water-soaked debris for long periods of time, and fire investigators who rely on their sense of smell to detect flammable liquids should be aware that low-level exposure to HCN can reduce the sensitivity of the human nose, so that lethal concentrations cannot be smelled.

Other toxic gases such as nitrous oxide (N_2O) and nitric oxide (NO) are produced by the same nitrogen-containing polymers described earlier. Polymers containing chlorine, predominately polyvinyl chloride (PVC), can produce large amounts of hydrogen chloride (HCl), a very toxic gas that turns into corrosive hydrochloric acid on contact with water, as well as elementary chlorine. PVC pipe is widely used in water and sewer connections in homes, mobile homes, and some light industrial structures and poses a widespread risk both to firefighters and to fire investigators. If Freon refrigerants or related halogen-based fire extinguisher gases are released into a hot fire, hydrochloric acid (HCl) and hydrofluoric acid (HF) may be formed, further adding to inhalant dangers. These products (and the acrolein produced by wood combustion) are strong irritants that cause watering eyes and convulsive coughing that can interfere with escape.

TOXIC GASES FROM SULFUR-CONTAINING POLYMERS

Sulfur-containing polymers—polyether sulfone and vulcanized rubber among them—produce sulfur dioxide (SO_2), hydrogen sulfide (H_2S), and carbonyl sulfide (COS). Carbonyl sulfide is considered to be more toxic than carbon monoxide.[69] Polystyrene, most often found in packaging materials, household products, and light fixture diffusers, produces styrene monomer on combustion, another toxic compound. All polymers and petroleum products produce formaldehyde, acrolein, and similar volatile toxic materials when burning unconfined. In addition to their direct toxic effects, HCl and acrolein are very active irritants. In even modest concentrations, they cause irritation of the mucous membranes of the esophagus and lungs, which can interfere with breathing. It has also been suggested that chemical compounds called *free radicals* produced in flames, if inhaled, interfere directly with the transport of oxygen from the air to the bloodstream. This causes anoxia, which can lead to unconsciousness in 10 to 15 seconds.[70] Such an effect could bring about death in spite of other nonfatal indicators such as low CO levels, minimal toxins in the body, or even burns or soot in the throat and lungs of the deceased. Finally, a method has been reported in which gasoline and kerosene-type hydrocarbons have been detected in the blood of subjects exposed to petroleum distillate vapors. The levels detected by gas chromatography/mass spectrometry (GC/MS) could be produced by breathing an explosive vapor atmosphere.[71] In addition, very sensitive GC/MS methods have been reported that are capable of detecting other volatile traces in blood or tissue samples, which can help in the reconstruction of fire fatality incidents. For instance, the detection of normal alkanes and methyl *t*-butyl ether (MTBE) can reveal the presence of automotive gasoline, while the detection of acetone (found in some paint thinners) can confirm the involvement of such a solvent.[72] Volatiles have reportedly been extracted from human skin and tissues using solid-phase microextraction (SPME) and activated carbon strip.[73]

Other Mechanisms

The human body is able to adapt to hot environments by using blood circulation and evaporative cooling (from the skin and from the mucosal tissues of the mouth, throat, and lungs) to maintain a constant body temperature. If the body core temperature exceeds 43°C (109°F), the brain stops functioning and the person loses consciousness and will die. If exposed to dry air at high temperatures 120°C (250°F) or higher for more than a few minutes, the body loses the ability to maintain its core temperature, with heatstroke occurring at a core body temperature of 40°C (105°F), leading to a lethal cascade of thermoregulatory and physiological events. Heatstroke can occur sooner or at lower temperatures if the air is moisture-laden (which prevents evaporative cooling).[74]

Blunt trauma can also kill people during a fire. Such trauma can result from structural collapse, explosion fragmentation, from falls, or from crashing into door frames or walls while attempting to escape the smoke. The nature of the impact and the location in which the victim was found may suggest the origin of such injuries. Bloodstains on walls and floors may reveal contact with hands, fingers, shoes, and help reconstruct the impact area and the attempted path of escape.

While the gaseous components of smoke are normally considered to be the dangerous components of the smoke from a fire, the soot itself may also play a role. Soot is carbon condensed into solid particles of various sizes. Hot particles of soot are not readily cooled by evaporative cooling when they are inhaled, so they can retain their heat until they lodge in the tissue itself, causing burns and edema. The carbon can also carry chemically active and toxic compounds from the combustion (being activated carbon). When inhaled or ingested, these toxins will be transferred directly to the mucosal tissues and absorbed into the bloodstream. Soot can also be so thick in the smoke that it completely obscures light and any view of an escape route. In extreme cases, the soot is inhaled in such quantities that it physically blocks the airways and causes asphyxiation.

BURN INJURIES

Even when they do not result in death, fires are responsible for an enormous toll in human injuries. As such they may come to the attention of the fire investigator. It is a sad observation, but it is generally true in the United States that fire fatalities are much better documented than fires in which the victims are only injured, yet the civil and criminal involvement can be more substantial in injury fires than in fatal fires. The burn patterns on flesh and clothing are just as important in injury cases as in death cases, and yet, unfortunately, the clothing of a burn victim is likely to be discarded in the emergency room and there will be no attempt to document the injuries until after treatment is well underway, sometimes only after weeks of healing. The investigator must realize that serious burns may result in death after weeks of hospitalization, and that a simple household fire may become a fire death case. That makes it mandatory to take the same steps to preserve physical evidence in injury cases as in immediate-death fires to ensure that clothing, shoes, and other items associated with the victim are preserved for presence of accelerants as well as reconstruction of burn patterns and to see that injuries are documented accurately by means of photographs as well as by notes and diagrams. The notes made by physicians on admission are often sketchy, as the physician is naturally more interested in beginning treatment rather than documenting exactly where the burns are located. Reports of burn injuries often include two values that are sometimes misunderstood by the investigator. First, the depth of injury to the tissue is often reported in degrees:

- First degree: superficial, reddened skin (like ordinary sunburn)
- Second degree: deeper damage with blisters formed and sloughing of epidermis; partial-thickness burns that will heal by cell growth from surviving underlying tissue (dermis)
- Third degree: severe damage to full thickness of skin to underlying tissue; full-thickness burns that can heal only from the edges and generally require skin grafts
- Fourth degree: very severe including underlying tissue with charring
- Fifth degree: underlying muscle and bone involved

Second, the predominant means of estimating the amount of body (skin) surface area that is involved is the "Rule of Nines," whereby major body areas are assigned various values (for adults) as follows:

- 9 percent for the head (front, back, and sides)
- 9 percent for each arm
- 9 percent for the back of each leg
- 9 percent for the front of each leg
- 18 percent for the back of the torso
- 18 percent for the front of the torso
- 1 percent for the genitalia

Different percentages are used for infants and children because of their different body proportions.

Burns, particularly first, second, and third degree, can be created by exposure to radiant or convected heat, or heat conducted from hot surfaces or liquids, or by chemical action. Skin can blister after about 11 seconds of exposure to radiant heat flux of 10 kW/m^2 or after about 5 seconds at 20 kW/m^2.[75] A full-thickness burn can be produced by immersion in hot water for 30 seconds at 54°C (130°F) or 1 second at 70°C (158°F) in adults. A child's skin is more fragile, sustaining full-thickness burns after immersion in hot water for 10 seconds at 54°C (130°F) or 0.5 second at 65°C (149°F).[76] Because pain acts as a deterrent for someone seeking an escape route in a fire, a threshold heat flux of 2.5 kW/m^2 or sustained exposure to hot gases in excess of 120°C (250°F) is considered to be the "incapacitation" level.[77]

FIGURE 15.23 The wearer of these denim pants suffered only localized second-degree burns (in the vicinity of the tear in the upper left leg) despite the fact that gasoline had been splashed across her legs. The tight denim pants over knit thermal underwear conducted enough heat to produce a burn under the tear. *Courtesy of John D. DeHaan.*

The flammability properties of fabrics were treated in Chapter 11. While injuries are usually matchable to burned or melted portions of the garments worn at the time of the fire, burn injuries may not tally with the clothing damage. Clothing of natural fibers can conduct heat sufficiently so that second-degree burns can be produced through cloth even if the cloth remains intact. Because many synthetic cloth fibers are capable of transmitting infrared radiation, burns can be induced by radiant heat through single layers of synthetic cloth. One interesting observation is that some blue denim fabric turns red upon exposure to temperatures in excess of about 60°C (130°F), whereas scorching of the cotton fiber or even melting of the acrylic fiber requires much higher temperatures. All these observations mean that fire injuries to the body can be correlated in many cases with damage to the clothing as long as the physical and chemical properties of the fabrics are appreciated and fire behavior is well understood (see Figure 15-23). A sustained fire supported by heavy clothing can cause third-degree burns and skin splitting, as seen in Figure 15-24.

The thermal inertia of extraskeletal human tissue is not unlike that of polyethylene plastic or even wood, so the length of time the body is exposed will establish how deep the damage will be. A flash fire involving flammable liquid vapors (or any bright radiant heat exposure) is of such short duration that only the epidermis may be affected. The epidermis may start to separate from the underlying dermis, raising blisters (similar to those formed when wallpaper or paint is heated on a wall). The blisters have no tissue in contact beneath and are quickly heated and burned away, sometimes leaving a spotted pattern on the skin of blackened skin surrounding oval patches of freshly exposed dermis (on living or postmortem skin), as seen in Figure 15-25. If an ignitable liquid is burned on a horizontal skin surface, the damage may resemble the halo pattern described earlier for carpet and floor. As seen in Figure 15-26, a ring or halo of blister was raised around the central portion where the evaporating liquid protected the skin. The hair was singed off in both areas. If the skin surface is vertical, the burning liquid may pour off so quickly that the skin is only reddened, unless retained by garments. Due to the mass of the torso, it takes a long time for normal external fire exposure to have an effect on the deep tissues

FIGURE 15.24 Sustained fire from gasoline-dampened clothing induced skin splits and charring. Compare with Figure 15-12B. *Courtesy of Jamie Novak, Novak Investigations and St. Paul Fire Dept.*

and organs. Tests monitoring internal body temperature demonstrate that 60 to 120 minutes of exposure to a generalized low-intensity fire is necessary to raise the core body temperature of a cadaver.[78]

Documentation (by diagram, as in Figure 15-27, and by photography) should be carried out on all burn victims (living and deceased) as soon after the fire as possible. This ensures accurate assessment before medical attention and healing can distort the appearance of this transitory evidence. The burn patterns can then be compared with the burn patterns of clothing and other surfaces at the scene.

FIGURE 15.25 Radiant heat caused postmortem "leopard spot" blisters on this cadaver. *Courtesy of Jamie Novak, Novak Investigations and St. Paul Fire Dept.*

FIGURE 15.26 Cigarette lighter fuel poured on the back of the hand (horizontal surface) and ignited produced a halo pattern blister around the pool. The liquid pool protected the skin under the liquid. *Courtesy of Dr. Roger Berrett, Weybridge, Surrey, UK.*

Body diagram

Front

Back

FIGURE 15.27 Simple body diagram useful for recording extent and depth of burn injuries.

Name_____

Examined by_____ Date _____

MANNER OF DEATH

The determination of manner of death (accidental, suicidal, homicidal, etc.) is usually made by medicolegal authorities in consultation with the fire investigators, since it is the set of circumstances under which the fire injuries (cause of death) were suffered that establishes manner. For the reasons described in the early pages of this chapter, it is clear that the fire death investigation is a three-pronged affair—cause of death, cause of fire, and conditions under which both occurred. The investigator should not presume an accidental death occurred any more than he or she should presume the fire was accidental. A homicidal death or a natural death may result in an accidental fire. An intentionally set fire may result in unintended death. A suicide death may result in an accidental fire or may be the result of an intentionally set fire. Suicides by fire are rare (estimates are on the order of 1 percent of all identified suicides). Most that are identified are the result (sometimes observed) of pouring and igniting an ignitable liquid over oneself, but there are variations. One of the authors knows of one case in which the individual set fire to a pile of wood using kerosene and sat amid the burning material for some time. The decedent apparently changed her mind, since the body was ultimately found several feet from the fire site after the fire had self-extinguished. In another incident that was filmed by a passerby, a man ignited the interior of his parked car, waited until it was fully involved, and then jumped through the passenger window into the flames just as firefighters arrived. Other suicides are less direct, involving setting fire to a structure or a vehicle and then waiting in the smoke-filled enclosure (away from the fire). One case was published by investigators from Montreal (structure fire).[79] DeHaan investigated a structure fire in a residence that resulted in smoke inhalation deaths of four children and the near death of the firesetter and an older child, who survived in the smoke-filled house for several hours.[80] There was also a case in which the individual, despondent over a pending jail term, set fire to his vehicle and then crawled into the trunk of the vehicle, where he was found. He broke a rear window before the fire to ensure adequate ventilation for the fire. A similar case was described by Adair.[81] The full circumstances of events preceding the fire, physical evidence, postmortem findings, and fire engineering analysis of heat and toxic gas (CO) production and routes of incapacitation are needed to reach a reliable conclusion in such challenging cases.

CHAPTER REVIEW

Summary

There is a temptation, especially for the pathologist or death investigator, to regard a fire as a single event (like a gunshot) and assess the physical and toxicological effects seen on the body as the result of exposure to that event. Those familiar with fire know that fire is not uniform in its intensity or the distribution of its toxic gases and that humans can be exposed to it in a tremendous variety of ways over a period of time. The hot and toxic gases in many room fires will be almost completely concentrated in the hot smoke layer, so a person reclining on a bed or the floor of a room may not be exposed to any of it until he or she sits or stands up. Fire can kill by any number of means, and people can react to it in a wide variety of ways—from ignoring it, to going to investigate, to fleeing, to calling for assistance, or escaping and then returning to rescue people, pets, or sentimental items. When one or more persons die in a fire and others escape there is always the question: Why did someone survive when others did not?

In summary, the following are the primary issues facing the fire investigator in dealing with a fire-related death:

1. Assistance of specialized personnel should be obtained as soon as possible.
2. These personnel should meet (preferably ahead of time) to set a protocol for who does what at these scenes.
3. Access to the scene and in particular the area around the body must be tightly controlled.
4. The body should be fully documented with photos and diagrams before it is moved if circumstances permit.
5. Extreme care must be used in the removal of the body and associated debris. The debris around/ under the body should be sifted and checked with a metal detector and X-rays. Gridding and layering may be necessary.
6. The area of the body should be photographed before, during, and after debris removal (by layering).
7. The body should be completely X-rayed prior to examination.
8. A complete forensic autopsy should be carried out in all fire deaths to determine the cause of death.
9. All burns and physical trauma on the body should be documented by photos, diagrams, and notes.
10. The carbon monoxide level in the blood should be determined in all fire fatalities and, if possible, in living victims rescued from the fire.
11. The presence of alcohol, drugs, and toxic materials should be determined in all cases where it is deemed warranted.
12. The body represents an important part of the fuel load of many fires and often bears burn patterns that must be interpreted as part of the burn patterns in the rest of the room. The fire dynamics and ignition analyses may include the body and its clothing.
13. Clothing bears important burn patterns in both fatality and injury cases and must be preserved. Burn injuries suffered by surviving victims must be preserved by photographs as soon as possible after the fire. Residues of ignitable liquids may be detected both in clothing and skin.
14. Fires and people's interactions with them are often complex and prolonged. The cause of death is rarely a single event or condition. Combinations of effects such as shock, CO, hypoxia, hyperthermia, and toxic gases may result in death when no single factor would be fatal.
15. The cause of the fire and the cause of death may or may not be related. The relationship between the victim and the fire may be determined by assessing both in terms of the spread of fire and its heat and smoke. Accidental fires may occur whether the death is accidental, suicidal, natural, or homicidal. Similarly, an arson fire may trigger an accidental or natural death or be part of a homicide. This consideration should extend to vehicle fires. Both accidental and incendiary fires have been linked to numerous deaths.[82]

The fire investigator, homicide investigator, and pathologist must be prepared to exchange all information fully and to work in close cooperation with one another to expedite the identification of the victim and determination of the cause and manner of death. Such cooperation will maximize the information available so as to accurately reconstruct the sequence of events at a fire death scene.

Review Questions

1. List the six major questions that must be answered when burned remains are discovered in a fire scene.
2. Who are the members of the ideal fire death investigation team?
3. What is the "wick effect" and how does it contribute to the destruction of a body?
4. How hot is a commercial crematorium and how long does it take to cremate a normal adult human body?
5. Name five effects heat and fire exposure can have on a human body.
6. What is the relationship between CO concentration, breathing rate, time, and COHb saturation?
7. What is the range of potentially fatal COHb saturations and why is it so large?
8. Describe the "degree" system of burn injuries and relate it to human skin.
9. Name three other gases that can contribute to the death of a fire victim.
10. What are the three elements of a fire death investigation?

References

1. M. J. Karter Jr., "Fire Loss in the United States 2008" (Quincy, MA: National Fire Protection Association, Fire Analysis and Research Division, August 2009); see also M. J. Karter Jr., "U.S. Fire Loss for 2008," *NFPA Journal* (September–October 2009).
2. D. A. Purser, "Toxicity Assessment of Combustion Products," Chap. 2–6 in *SFPE Handbook of Fire Protection Engineering*, 3rd ed. (Bethesda, MD: Society of Fire Protection Engineers, 2002), 2–83, 84.
3. V. F. Delattre, "Burned Beyond Recognition: Systematic Approach to the Dental Identification of Charred Human Remains," *Journal of Forensic Sciences* 45, no. 3 (July 2000): 589–96.
4. F. G. Robinson, F. A. Rueggeberg, and P. E. Lockwood, "Thermal Stability of Direct Dental Esthetic Restorative Materials at Elevated Temperatures," *Journal of Forensic Sciences* 43, no. 6 (November 1998): 1163–67.
5. M. A. Bush, P. J. Bush, and R. G. Miller, "Detection of Classification of Composite Resins in Incinerated Teeth for Forensic Purposes," *Journal of Forensic Sciences* 51, no. 3 (May 2006): 636–42.
6. N. J. Kirk et al., "Skeletal Identification Using the Frontal Sinus Region," *Journal of Forensic Sciences* 47, no. 2 (March 2002): 318–23; see also C. W. Schmidt, "The Recovery and Study of Burned Human Teeth" in *The Analysis of Burned Human Remains*, ed. C. W. Schmidt and S. A. Symes, 55–74 (New York: Academic Press, 2008).
7. J. di Zinno et al., "The Waco Texas Incident: The Use of DNA Analysis in the Identification of Human Remains," paper presented at *Fifth International Symposium on Human Identification*, London, England, 1994; see also D. W. Owsley et al., "The Role of Forensic Anthropology in the Recovery and Analysis of Branch Davidian Compound Victims: Techniques of Analysis," *Journal of Forensic Sciences* 40, no. 3 (May 1995): 341–48.
8. A. Z. Mundorff, E. J. Bartelink, and E. Mar-Cash, "DNA Preservation in Skeletal Remains from the World Trade Center Disaster," *Journal of Forensic Sciences* 54, no. 4 (July 2009): 739–45; National Institute of Justice, *Mass Fatality Incidents: A Guide for Human Forensic Identification* (Washington DC: NIJ, 2005).
9. M. J. Hickman et al., "Medical Examiners' and Coroners' Offices, 2004" (Washington, DC: U.S. Department of Justice, Bureau of Justice Statistics, June 2007).
10. J. M. Fleming, "Smoke Detector Technology and the Investigation of Fatal Fires," *Fire and Arson Investigator* (April 2000): 35–40.
11. J. D. DeHaan and E. J. Pope, "Sustained Combustion of Bodies: Some Observations," American Association for Forensic Science, Denver, February 2009 (in press).
12. E. J. Pope, O. C. Smith, and T. G. Huff, "Exploding Skulls and Other Myths about How the Human Body Burns," *Fire and Arson Investigator* (April 2004): 23–28.
13. T. Watanabe, *Atlas of Legal Medicine* (Philadelphia: J. B. Lippincott, 1968), 136.
14. M. J. Thali et al., "Charred Body: Virtual Autopsy with Multi-Slice Computed Tomography and Magnetic Resonance Imaging," *Journal of Forensic Sciences* 47, no. 1 (March 2002): 1326–32.
15. W. G. Eckert, S. James, and S. Katchis, "Investigation of Cremations and Severely Burned Bodies," *American Journal of Forensic Medicine and Pathology* 9, no. 3 (1988): 188–200; M. Bohnert, T. Rost, and S. Pollak, "The Degree of Destruction of Human Bodies in Relation to Duration of the Fire," *Forensic Science International* 95, no. 1 (July 1998): 11–21.
16. E. J. Pope, "Burning Bodies," paper presented at *American Academy of Forensic Sciences*, New Orleans, LA, February 2005.

17. DeHaan and Pope "Sustained Combustion of Bodies."
18. C. S. Petty, "Fire Death Investigation and Pathology," *Fire and Arson Investigator* (April–June 1968).
19. J. D. DeHaan and F. Fisher, "Reconstruction of a Vehicle Fire Fatality," paper presented at the First Fire Investigation and Research Conference, Fire Research Station, Garston, UK, October 1995; see also "Reconstruction of a Fatal Fire in a Parked Motor Vehicle," *Fire and Arson Investigator* (January 2003): 42–46.
20. C. Palmiere et al., "Ignition of a Human Body by a Modest External Source: A Case Report," *Forensic Science International* 188, no. 1 (July 2009): 1–3 (e17–e19).
21. J. Nickell and J. F. Fischer, "Spontaneous Human Combustion," *Fire and Arson Investigator* 34 (March 1984): 4–11; J. Nickell and J. F. Fischer, "Incredible Cremations: Investigating Spontaneous Combustion Deaths," *The Skeptical Inquirer* 11 (Summer 1987): 352–57.
22. B. V. Ettling, "Consumption of an Animal Carcass in a Fire," *Fire and Arson Investigator* (April–June 1968).
23. D. J. Gee, "A Case of Spontaneous Combustion," *Medicine, Science and Law* 5 (January 1965): 37–38.
24. J. D. DeHaan, S. Nurbakhsh, and S. J. Campbell, "The Combustion of Animal Remains and Its Implications for the Consumption of Human Bodies in Fire," *Science and Justice* 39, no. 1 (January 1999): 27–38.
25. NFPA. *Fire Protection Handbook,* 20th ed. (Quincy, MA: National Fire Protection Association, 2008), app. A.
26. J. D. DeHaan and M. Scanlan, "A Case of Not-So-Spontaneous Human Combustion," *CAC News,* 1996; also in *Fire and Arson Investigator* (June 1997): 14–16.
27. J. D. DeHaan and S. Nurbakhsh, "The Sustained Combustion of an Animal Carcass and Its Implications for the Consumption of Human Bodies in Fires," *Journal of Forensic Sciences* 46, no. 5 (September 2001): 1076–81.
28. DeHaan et al., "The Combustion of Animal Remains."
29. DeHaan and Pope, "Sustained Combustion of Bodies."
30. J. D. DeHaan, "The Dynamics of Flash Fires Involving Flammable Hydrocarbon Liquids," *American Journal of Forensic Medicine and Pathology* 17, no. 1 (March 1996): 24–31.
31. DeHaan and Nurbakhsh, "The Sustained Combustion of an Animal Carcass."
32. S. A. Symes et al., "Patterned Thermal Destruction of Human Remains in a Forensic Setting" in *The Analysis of Burned Human Remains,* ed. C. W. Schmidt and S. A. Symes (New YorK: Academic Press, 2008), 15–54.
33. Pope et al., "Exploding Skulls and Other Myths."
34. N. P. Herrmann and J. L. Bennett, "The Differentiation of Traumatic and Heat-Related Fractures in Burned Bone," *Journal of Forensic Sciences* 44, no. 3 (May 1999): 461–69; E. J. Pope and O. C. Smith, "Identification of Traumatic Injury in Burned Cranial Bone: An Experimental Approach," *Journal of Forensic Sciences* 49, no. 3 (May 2004): 431–37; Symes et al., "Patterned Thermal Destruction of Human Remains."
35. Hermann and Bennett, "The Differentiation of Traumatic and Heat-Related Fractures."
36. Eckert, James, and Katchis, "Investigation of Cremations"; T. J. O. Thompson, "Heat Induced Dimensional Changes in Bone and their Consequences for Forensic Anthropology," *Journal of Forensic Sciences* 50, no. 5 (September 2005): 1008–15.
37. Symes et al., "Patterned Thermal Destruction of Human Remains."
38. A. M. Christensen, "Experiments in the Combustibility of the Human Body," *Journal of Forensic Sciences* 47, no. 3 (May 2002): 466–70.
39. Pope and Smith, "Identification of Traumatic Injury in Burned Cranial Bone."
40. L. A. Simpson and C. W. Cruse, "Gasoline Immersion Injury," *Plastic and Reconstructive Surgery"* 671 (January 1981): 54–57; J. F. Hansbrough et al., "Hydrocarbon Contact Injuries," *Journal of Trauma* 25, no. 3 (1985): 250–52.
41. DeHaan, "The Dynamics of Flash Fires."
42. T. W. Adair et al., "Suicide by Fire in a Car Trunk: A Case with Potential Pitfalls," *Journal of Forensic Sciences* 48, no. 3 (September 2003): 1113–17; J. D. DeHaan, unpublished case report.
43. National Fire Protection Association, *Report on the Fire at the Dupont Plaza Hotel and Casino, December 31, 1986* (Quincy, MA: NFPA, 1987); W. D. Wooley et al., "The Stardust Disco Fire, Dublin 1981: Studies of Combustion Product during Simulation Experiments," *Fire Safety Journal* 7 (1984): 267–83.
44. B. C. Levin et al., "Analysis of Carboxyhemoglobin and Cyanide in Blood from Victims of the Dupont Plaza Hotal Fire in Puerto Rico," *Journal of Forensic Sciences* 35, no. 1 (1990): 151–68.
45. D. G. Penney, ed., *Carbon Monoxide Toxicity* (Boca Raton, FL: CRC Press, 2000).
46. P. B. Hawk, B. L. Oser, and W. H. Summerson, *Practical Physiological Chemistry,* 12th ed. (Philadelphia: Blakiston, 1947).
47. J. M. Feld, "The Physiology and Biochemistry of Combustion Toxicology" in *Proceedings of Fire Risk and Hazard Assessment: Research Applications Symposium* (Quincy, MA: Fire Protection Research Foundation, 2002); H. J. Vreman, R. J. Wong, and D. K. Stevenson, "Carbon Monoxide in Breath, Blood, and Other Tissues," Chap. 2 in *Carbon Monoxide*

Toxicity, ed. D. G. Penney (Boca Raton, FL: CRC Press, 2000).

48. A. C. Berent et al., "Carbon Monoxide Toxicity: A Case Series," *Journal of Veterinary Emergency and Critical Care* 15, no. 2 (2005): 128–35.

49. A. Stolman and C. P. Stewart, *Toxicology, Mechanisms, and Analytical Method,* vol. 1 (New York: Academic Press, 1960).

50. Vreman et al., "Carbon Monoxide in Breath, Blood and Other Tissues."

51. D. J. Icove and J. D. DeHaan, *Forensic Fire Scene Reconstruction*, 2nd ed. (Upper Saddle River, NJ: Brady/Prentice Hall, 2009), 311–12; see also D. A. Purser, "Interactions among Carbon Monoxide, Hydrogen Cyanide, Low Oxygen Hypoxia, Carbon Dioxide, and Inhaled Irritant Gases," Chap. 7 in *Carbon Monoxide Toxicity*, ed. D. G. Penney, 164 (Boca Raton, FL: CRC Press, 2000).

52. Health Canada, "Investigating Human Exposure to Contaminants in the Environment" (Ottawa, Ontario: Ministry of National Health and Welfare, 1995); www.hc-sc.gc.ca.

53. Vreman et al., "Carbon Monoxide in Breath, Blood, and Other Tissues."

54. D. A. Purser, "Assessment of Hazards to Occupants from Smoke, Toxic Gases, and Heat" in *SFPE Handbook of Fire Protection Engineering*, 3rd ed. (Quincy, MA: National Fire Protection Association, 1995), 2–96; see also 172.

55. J. R. Gill, L. Goldfeder, and M. Stajic, "The Happyland Homicides: 87 Deaths Due to Smoke Inhalation," *Journal of Forensic Sciences* 48, no. 1 (January 2003): 161–66.

56. A. O. Gettler and H. C. Freimuth, "Carbon Monoxide in Blood: A Simple and Rapid Estimation," *American Journal of Clinical and Pathological Technology Supplement* 7, no. 79 (1943).

57. M. J. Hazucha, "Effect of Carbon Monoxide on Work and Exercise Capacity in Homes," Chap. 5 in *Carbon Monoxide Toxicity*, ed. D. G. Penney, 102–7 (Boca Raton, FL: CRC Press, 2000).

58. D. A. Purser, "Interactions among Carbon Monoxide, Hydrogen Cyanide, Low Oxygen Hypoxia, Carbon Dioxide, and Inhaled Irritant Gases."

59. DeHaan, "The Dynamics of Flash Fires."

60. J. Flynn, "CO Deaths," *NFPA Journal* (January–February 2008): 33–35; see also J. C. Mah, "Non-Fire Carbon Monoxide Deaths and Injuries Associated with the Use of Consumer Products," (Bethesda, MD: CPSC, October 2000), 57.

61. S. J. deRoux, "Suicidal Asphyxiation of Automobile Emission without Carbon Monoxide Poisoning," *Journal of Forensic Sciences* 51, no. 5 (September 2006): 1158–59; J. R. Suchard, "Motor-Vehicle Related Toxicology," *Topics in Emergency Medicine* 28, no. 1 (2006): 76–84.

62. D. Risser, A. Bonsch, and B. Schneider, "Should Coroners Be Able to Recognize Unintentional Carbon Monoxide Deaths Immediately at the Death Scene?" *Journal of Forensic Sciences* 40, no. 4 (1995): 596–98.

63. Vreman, "Carbon Monoxide in Breath, Blood, and Other Tissues," 35–42.

64. Suchard, "Motor-Vehicle Related Toxicology," 77.

65. M. I. Jumbelic, "Open Air Carbon Monoxide Poisoning," *Journal of Forensic Sciences* 43, no. 1 (1998): 228–30.

66. A. M. Christensen and D. J. Icove, "The Application of NIST's Fire Dynamics Simulator to the Investigation of Carbon Monoxide Exposure in the Deaths of Three Pittsburgh Fire Fighters," *Journal of Forensic Sciences* 49, no. 1 (January 2004): 104–10.

67. Purser, "Interactions among Carbon Monoxide"; W. D. Woolley and P. J. Fardell, "Basic Aspects of Combustion Toxicology," *Fire Safety Journal* 5 (1982): 29–48; K. J. Vorhees and R. Tsao, "Smoke Aerosol Analysis by Pyrolysis—Mass Spectrometry/Pattern Recognition for Assessment of Fuels Involved in Flaming Combustion," *Analytical Chemistry* 57, no. 8 (1985): 1630–36; J. Rio et al., "An Investigation of the Toxic Effects of Combustion Products—Analysis of Smoke Components," *Journal of Analytical Toxicology* 12, no. 5 (September–October 1988): 274–78.

68. NFPA. *Fire Protection Handbook*, 20th ed. (Quincy, MA: National Fire Protection Association, 2008), Chaps. 3–5.

69. C. J. Hilado and C. M. Machado, "Effect of Char Yield and Specific Toxicants on Toxicity of Pyrolysis Gases from Synthetic Polymers," *Fire Technology* 16, no. 59 (February 1979).

70. S. Hattangadi, "A Chemical Clue to Mystery Deaths," *Industrial Chemist*, October 1987, 12–15; W. T. Lowry et al.," Free Radical Production from Controlled Low-Energy Fires: Toxicity Considerations," *Journal of Forensic Sciences* 30, no. 1 (1985): 73–85.

71. K. Kimura et al., "Gasoline and Kerosene Components in Blood—A Forensic Analysis," *Human Toxicology* 7 (1988).

72. J. Schuberth, "Post-Mortem Test for Low-Boiling Arson Residues of Gasoline by Gas Chromatography—Ion Trap Mass Spectrometry," *Journal of Chromatography B* 662, no. 1 (December 1994): 113–17.

73. J. R. Almirall et al., "The Detection and Analysis of Ignitable Liquid Residues Extracted from Human Skin Using SPME/GC, *Journal of Forensic Sciences* 45, no. 2 (2000): 453–61; I. Montani, S. Comment, and O. Delemont, "The Sampling of Ignitable Liquids on Suspects Hands," *Forensic Science International* 194 (2010): 115–24.

74. J. L. DeJong and T. Adams, "Entrapment in Small Enclosed Spaces: A Case Report and Points to Consider Regarding the Mechanism of Death," *Journal of Forensic Sciences* 46, no. 3 (2001): 708–13.

75. Purser, "Assessment of Hazards to Occupants," 2–115.
76. P. E. Besant-Matthews, "Deaths Associated with Burns, Fire, and Arson," Dallas, TX, 1993.
77. Icove and DeHaan, *Forensic Fire Scene Reconstruction*.
78. DeHaan and Pope, "Sustained Combustion of Bodies."
79. A. Savageou, A. Racette, and R. Yesovitch, "Suicide by Inhalation of Carbon Monoxide in a Residential Fire,"
Journal of Forensic Sciences 50, no. 4 (July 2005): 937–38.
80. DeHaan, unpublished case.
81. Adair et al., "Suicide by Fire in a Car Trunk."
82. Karter, "Fire Loss in the United States 2008."

KEY TERMS

arson set, *p. 671*
corpus delicti, *p. 658*

fraud, *p. 662*
motive, *p. 661*

pyromania, *p. 670*
torch, *p. 662*

OBJECTIVES

After reading this chapter, you should be able to:

- Explain the basic concepts of arson as a crime and list its three elements.
- Identify and list the motive classifications for arson as cited by the FBI's National Center for the Analysis of Violent Crime (NCAVC).
- List and describe the typical findings for the location, arrangement, types of fuel, and methods of initiation used by arsonists in four of the classifications.
- List the investigative methodology and deductions for the interpretation of evidence.
- Search the Internet for actual case examples detailing arson cases and motives.

Arson, in its most general terms, may be defined as the willful and malicious burning of a person's property. It must be regarded as a possible, if not probable, cause of every fire investigated. It was estimated at one time that arson fires accounted for up to 31 percent of all reported structure fires.[1] Informal surveys of experienced U.S. fire investigators support that up to 40 percent of all urban structure fires are incendiary (intentionally set). Arson is a serious threat to public safety and well-being in urban, suburban, and even rural areas. Based on data reported by U.S. fire departments in its annual survey, the NFPA estimated there were 30,500 intentionally set structure fires in 2008. This represents a decrease of 6.2 percent from 2007 totals. These estimates (about 6 percent of all structure fires) are the lowest in the 27 years studied. The authors note that recent National Fire Protection Association (NFPA) estimates, however, are based on United States Fire Administration/National Fire Incident Reporting System (USFA/NFIRS) 5.0 data, whose criteria were changed in 2002 to exclude so-called suspicious fires and unknown cause fires. Based on the methodology used in making these estimates and changes in reporting criteria, comparisons with data from previous years are not valid. The reported "intentionally set" fires resulted in an estimated 315 civilian deaths and about $866 million in direct property damage.[2] In 2008, there were an estimated 17,500 intentional fires in vehicles, with property damages of around $139 million. By comparison, in 2002, there were an estimated 68,000 intentional fires in structures, costing $1.9 billion in direct property damage, 630 civilian deaths, and over 2,000 civilians injured.[3] There were also an estimated 30,500 vehicle arsons. Because many fire departments in the United States release many "suspected arson" cases to police authorities, their arson statistics may be inaccurate, and some state agencies do not forward fire statistics to NFIRS at all. As a result, NFIRS incendiary estimates are very conservative.

The U.S. Department of Justice, Federal Bureau of Investigation (FBI), operates and maintains the national collection of statistics on arson crimes known as the Uniform Crime Reporting (UCR) Program. In 2008, based upon reporting from 13,980 law enforcement agencies (representing an estimated population of 250,243,947 persons), there were 56,972 documented arsons in the United States. Of these incidents, 24,750 (43.4 percent) were structures, 16,454 (28.9 percent) were mobile, and 15,768 (27.7 percent) other properties were targets of arsonists. The average monetary loss per property was $16,015, with 18 percent of the cases cleared by an arrest or other exceptional means. Juvenile offenders under

TABLE 16-1

TABLE 16-1 Arson by Type of Property, 2008 (13,980 agencies; 2008 estimated population 250,243,947)

PROPERTY CLASSIFICATION	NUMBER OF ARSON OFFENSES	PERCENT DISTRIBUTION[a]	PERCENT NOT IN USE	AVERAGE DAMAGE	TOTAL CLEARANCES	PERCENT OF ARSONS CLEARED[b]	PERCENT OF CLEARANCES UNDER 18
Total	56,972	100.0		$16,015	10,277	18.0	37.4
Total structure:	24,750	43.4	19.0	29,701	5,520	22.3	35.2
Single occupancy residential	11,435	20.1	21.4	28,788	2,389	20.9	25.7
Other residential	4,062	7.1	13.6	29,343	983	24.2	25.0
Storage	1,629	2.9	20.7	18,021	298	18.3	48.7
Industrial/ manufacturing	251	0.4	22.3	212,388	48	19.1	41.7
Other commercial	2,331	4.1	15.7	55,531	450	19.3	28.0
Community/ public	2,667	4.7	16.2	15,799	857	32.1	65.3
Other structure	2,375	4.2	21.4	13,668	495	20.8	47.1
Total mobile:	16,454	28.9		8,766	1,461	8.9	20.3
Motor vehicles	15,572	27.3		8,186	1,314	8.4	18.7
Other mobile	882	1.5		18,991	147	16.7	34.7
Other	15,768	27.7		2,099	3,296	20.9	48.7

Source: Uniform Crime Report 2008 (Washington DC: U.S. Department of Justice, FBI, 2009); www.fbi.gov.ucr.

[a]Because of rounding, the percentages may not add to 100.0.

[b]Includes arsons cleared by arrest or exceptional means.

18 years of age were responsible in 37.4 percent of the case clearances. Table 16-1 summarizes the FBI's UCR reported statistics for arson during 2008.[4]

The Department on Communities and Local Government in the United Kingdom (UK) estimated that 17.8 percent (9,400) of 52,700 reported dwelling fires in 2007 were deliberate, while 38.7 percent of 31,000 "other building" fires were incendiary.[5] The Fire Protection Association reported that in 2007 to 2008 in the United Kingdom, some 30 percent of all serious structure fires [those involving over £250,000 ($415,000) loss each] were arson. (Over the period 2000 to 2003, arson fires constituted 30 to 35.7 percent of major fires) From December 2007 to November 2008, of 120 reported serious fires 36 (30 percent) were deliberate, causing over £26 million ($43,000,000) in direct damages.[6] The Office of the Deputy Prime Minister estimated that the number of arson structure fires increased about 17 percent over the decade 1994–2003. Including the costs of investigation, prosecution, community losses, and preventive measures, the total cost of arson fires in the United Kingdom is estimated to be £2.8 billion per year.[7]

Arson is a difficult crime to investigate and prosecute, for the following four reasons. First, the scene must be carefully investigated before it can be determined whether a crime actually took place. This is unlike finding a dead body and a bloody knife, which are strong indications that a murder may have taken place. Such indicators gain the immediate attention of investigators and warn them to proceed carefully so as not to compromise any evidence present, unlike a common dwelling fire that may be investigated cursorily, if at all.

Therefore, it is vitally important that *every* fire scene be treated as a potential arson scene (from the standpoint of security, preservation, and evidence) until clear proof of natural or accidental cause is discovered.

Second, the crime itself, if successful, destroys the physical evidence at its origin. In some cases, proving the crime of arson may be likened to a homicide investigation where the victim's body has turned to dust. Evidence of its existence is still there, but it requires careful and methodical analysis. Arson is the one crime that destroys evidence rather than creates it as it progresses.

Third, the official investigation of arson seems to fall into a gap between the area of responsibility of the fire department and that of the police. Fire department personnel may consider their primary responsibility to be public safety through extinguishment of the fire. They may consider the warrants, interviews, and fieldwork necessary for a criminal investigation to be outside their normal authority and more appropriate for the police. However, police investigators realize that the physical evidence of the crime—that is, fire behavior, diagnostic signs, and charred debris—is beyond their training and may decline to deal with it. As a result of this schism, valuable evidence of one type or another is often overlooked.

Fourth, statistical data on arson are likely underreported. In addition to underreporting in the NFIRS program, law enforcement data and trends are also spotty. It has long been known that unless fire officials in concert with police submit proven arson cases to the FBI's Uniform Crime Reporting (UCR) arson system, case clearance and arrests do not appear in national tracking of law enforcement trends.[8]

The Crime of Arson

The crime of arson comprises the following three elements:

1. *There has been a burning of property.* This must be shown to the court to be actual destruction, at least in part, not just scorching or sooting (although some states include any physical or visible impairment of any surface). As used here, "burning" includes destruction by explosion.
2. *The burning is incendiary in origin.* Proof of the existence of an effective incendiary device, no matter how simple it may be, is adequate. Proof must be accomplished by showing specifically how all possible natural or accidental causes have been considered and ruled out.
3. *The burning is shown to be started with malice, that is, with the intent of destroying property (i.e., starts a fire or causes an explosion with the purpose of destroying a building of another or one's own with fire).*[9] The "degree" of the arson charge in most jurisdictions has to do with the occupancy—first degree corresponding to an occupied structure, second to an unoccupied structure, and third to other property.

corpus delicti ■
Literally, the body of the crime. The fundamental facts necessary to prove the commission of a crime.

The investigator must keep these three elements of proof, the **corpus delicti** of the crime, in mind as the scene is examined. To prove a criminal case, of course, there has to be proof that the accused did in fact set the fire in question (or arranged to have it set). Although the mechanics of the examination are carried out in the same fashion as any other thorough fire investigation, evidence to support any and all of these elements must be constantly sought at a possible arson scene. The detection of abnormal fire behavior or unusual fire conditions (as revealed by the diagnostic signs discussed earlier), incendiary devices, fuels, or simply things that are out of place for some reason, should sound an alarm in the examiner's brain that warns of criminal activity. Because the circumstances of arsons are so varied, they cannot all be treated here. We shall touch on characteristics of fire behavior and means of ignition, as well as the motivation of fire setters as they relate to the detection of arson. The motives, devices, intents, and targets for arson are all inter-related to some degree. These relationships will be discussed as well, because the discovery of one factor can be used to suggest or deduce others. *Intent* can be described as the purpose or deliberateness of someone's action (or omission). Evidence linking the suspect to the arson can include proof of access and opportunity, physical evidence such as fingerprints, video or documentary records, or even conduct consistent with his/her involvement.

The crime of arson is not a specific-intent crime. That means the prosecution need prove only that the arsonist meant to cause damage or bodily harm by setting the fire. This absence of specific intent may have an impact on a search for a motive. Arson-related charges may include considerations of intent to cause the specific damage incurred, intent to cause some damage, or even the foreseeability of damage from the act (setting a fire under conditions that would likely lead to severe damage).

Arson Law

English common law, on which most U.S. law is based, defined arson as the willful and malicious burning of the dwelling house of another. This made arson a crime against the security of habitation (because loss of habitation could well cost the lives of its dwellers by exposure to weather or enemies). It required certain elements to be present—the structure had to be a dwelling, it had to belong to another, and it had to be burned as a deliberate or intentional act. As the common law concept of arson became inadequate, statutory law (laws passed by governmental bodies) expanded the definition to include other buildings and property. Omitting the dwelling or occupancy requirement means that other property such as shops, factories, prisons, public buildings, forests, fields, boats, and cars are now included in most arson statutes. By the 1950s most states had adopted an equivalent to the Model Arson Law first proposed in 1940 or had statutes incorporating significant portions of it (see Appendix E for the text of the Model Arson Law). This law brought about some measure of uniformity and recognized the seriousness of the offense in terms of several degrees of arson crime. The first degree held to the common law concept that the burning of a dwelling or associated building is the most serious offense. Lesser degrees depend on the nature of the property, extent of involvement of the accused in a conspiracy, or the nature of unsuccessful attempts to set fire to property. Because the statutes defining the property susceptible to arson and the liability of persons involved vary from state to state, the reader is urged to consult local legal sources.[10]

For many years, arson was considered in the United States to be a minor crime involving only property. In 1979, arson was elevated to Part I crime category status by the FBI, and incidents are now reported under the Uniform Crime Reports. The Arson Control Act of 1982 made this change permanent and focused public and governmental attention on arson as a violent crime directed against people as much as property. This increased awareness has made funding and investigative resources more available in recent years, but they are still woefully inadequate.

Elements of Proof

The finding that a fire has been caused by incendiary (deliberate) means is the starting point of a sometimes frustrating legal prosecution to try and punish the person(s) responsible. Not all incendiary fires are arson, for the crime of arson carries with it certain requisites of proof. The elements of proof that must be established to prove any crime are called the *corpus delicti*, the body of the crime. For arson, there are three elements that must be proven: (1) there has been burning of the property in question, (2) the burning was the result of an intentional act, and (3) it was done with intent, that is, a malicious (willful) criminal act.

"Burning" does not require that the building, vehicle, or timber be completely consumed. It is adequate that there be some destruction of the actual fiber of the material, however small. Material that is merely discolored or scorched does not qualify in most states; that the material has been charred or similarly changed is adequate. Proof of such an element may be carried out by photographic exhibits or the testimony of witnesses. Failure to provide some sort of documentation of this damage will result in dismissal of charges.

The remaining elements of proof (malicious intent) require that two conditions in fact be proven: (1) that the fire was intentionally set and (2) that it was set with malice, that is, with the purpose of destroying property for criminal gain. Legally, there is always a presumption of innocence, even as to the cause of the fire. A court will assume that a fire was caused by accidental means or by natural means, that is, an act of God, unless it can be proven beyond a reasonable doubt that no accidental or natural cause was present.[11] The discovery of an actual incendiary device may be of secondary importance because (in many jurisdictions) arson may be provable in the absence of an ignition device (sometimes referred to as *negative corpus*) as long as every logically conceivable accidental and natural cause has been considered and ruled out. Even when a device is found, the investigator should be careful to rule out other means of ignition as part of the investigation. This proof will defuse any attempt by the opposition counsel to claim that no accidental or natural source was considered, thereby introducing doubt as to the reliability of the expert's conclusion.

Proof of intent is sometimes the most difficult proof to obtain. Note that motive itself is not required, but the discovery of a valid motive may often lead to discovery of intent; for example, a failing business needs money to sustain its operations (motive), but to burn the warehouse to allow an insurance claim would provide needed capital (intent). Intent is most often proven by means of circumstantial evidence—a concept deserving some discussion. The natural and probable consequences of a person's acts are presumed to be the intended consequences. The word *willful* as used in the statutes means intentional or with knowledge and purpose.[C1*] When a person unintentionally sets fire to a building in the course of committing another felony, the felonious intent is also attributed to the arson.[C2]

Although intent is an element of the crime of arson, motive is not. However, motive may be relevant to helping identify the accused as the responsible party but it is not evidence of the crime itself.[C3]

DIRECT AND CIRCUMSTANTIAL EVIDENCE

Evidence in legal proceedings is any means used to help establish or disprove an alleged fact. *Direct* evidence may be in the form of testimony from eyewitnesses who saw someone, for instance, actually touch a match to fuel at the scene. *Black's Law Dictionary* (7th ed.) defines direct evidence as that which conclusively establishes a fact without inference or presumption, such as a description of events or an identification of a perpetrator by an eyewitness to the crime. Because fires in general and arson in particular are rarely witnessed firsthand, direct evidence is rare in fire-related cases. *Circumstantial* evidence is the means by which alleged facts are proven by inference or deduction from other facts that can be testified to firsthand. Such circumstantial evidence may be considered by fiction writers always to be imperfect if not erroneous, but a carefully constructed case with properly interpreted circumstantial evidence is every bit as valid as any case based on eyewitnesses. In fact, circumstantial evidence such as a crime scene reconstruction employing a crime laboratory evaluation of the evidence is usually much more reliable than an eyewitness interpretation of what was seen. Confessions are considered to be a form of

---

*In this chapter, citation numbers that appear preceded by a superscript C refer to legal references that are listed at the end of the chapter under "Court Citations."

direct evidence. Many people believe them to be infallible, self-sufficient proof of guilt, but a confession that is unsubstantiated by independent corroborating evidence is not proof by any means. One need only review the dozens of confessions recorded in each of the heinous crimes of the past century to realize their lack of reliability.

Arson may be proved by direct or circumstantial evidence. However, it is not often that there is an eyewitness to the crime. Most often the evidence is largely circumstantial. This means that the crime and guilt of the accused must be inferred or deduced from other facts that can be established. There are limitations to the conclusions that are based on inferences. That is, the incriminating facts must be consistent with guilt of the accused and inconsistent with any other rational conclusion.[C4,C5] Usually, incendiary origin of fire is established by circumstantial evidence showing that a foreign flammable liquid, its container, or a contrived ignition source was present, or by showing the improbability of an accidental fire—for example, by proving three separate origins.[C6]

Motive

Motive is some inner drive or impulse that causes a person to do something or act in a certain way. Basically it is the cause, reason, or incentive that induces or prompts specific behavior. In a legal context, motive explains why the offender committed the illegal act. Though motive, unlike intent, is not an essential element in criminal prosecution, it often lends support to it.[12] The investigator must be careful not to overemphasize the importance of a possible motive revealed during the investigation. Having a "motive" does not prove criminal action. This caution regarding the distinction between intent and motive is emphasized in *NFPA 921: Guide for Fire and Explosion Investions*.[13]

motive ■ Inner drive or impulse that causes a person to do something or act in a certain way.

The investigator must, however, give consideration to possible motives for every fire, because they may provide clues predicting certain behavior patterns that can lead to identifying and intercepting the culprit, particularly in cases of multiple arson.

Multiple offenses may be categorized in three ways:

- *Serial arson*   several fires set at different locations with an emotional "cooling-off" period between fires (the cooling-off period being from several hours to several weeks)
- *Spree*   several fires set in different locations close to one another in time (i.e., with no cooling-off time)
- *Mass arson*   multiple fires set in the same place at one time[14]

Arson classification by motive is an intuitive attempt to make logical sense of the crime of arson and to accurately categorize firesetters. Combining information about the offender characteristics and/or offense features had developed over the years as an accepted approach. Inciardi in 1970 was one of the first to focus primarily on motive for fire setting, and he classified adult arsonists into six categories: excitement, revenge, insurance fraud, vandalism, crime concealment, and institutionalized firesetters.[15]

In 1987, Icove and Estepp conducted research on arson from a law enforcement point of view when they considered the motives of adult and juvenile arson offenders. In their study, over a 4-year period fire investigators filled out protocols after each arrest. Consistent with previous work by Inciardi, their analysis of these cases again reported the main classification categories as vandalism, excitement, revenge, crime concealment, and fraud. They also included a category of "other" to address instances in which motive could not readily be determined.[16]

The FBI's National Center for the Analysis of Violent Crime (NCAVC) has identified six major motive classifications that apply (either singly or in combination) to the vast majority of arsonists. These can apply to either a single instance of fire setting or multiple offenses. These classifications are also referenced in Chapter 22 of *NFPA 921* (2008 and 2011). The FBI studies on arson motives are *offender-based*; that is, they look at the relationship between the behavioral and the crime scene characteristics of the offender as it relates to motive. A motive-based method of analysis can be used to identify personal traits and characteristics exhibited by an unknown offender.

The NCAVC has classified these motives, which are also cited in the 2008 and 2011 editions of *NFPA 921*, as the following:

- Profit (*NFPA 921*, Sec. 22.4.9.3.6)
- Vandalism (*NFPA 921*, Sec. 22.4.9.3.2)
- Excitement (*NFPA 921*, Sec. 22.4.9.3.3)
- Revenge (*NFPA 921*, Sec. 22.4.9.3.4)
- Crime concealment (*NFPA 921*, Sec. 22.4.9.3.5)
- Extremism (*NFPA 921*, Sec. 22.4.9.3.7)

These motives are discussed in turn, as well as juvenile fire setting, irrational fire setting, and mixed motives.

PROFIT

Arsonists in this category set fires with the expectation of profit, which may be direct monetary gain or indirect gain from a goal other than financial. The direct gains could be from insurance fraud, eliminating or intimidating business competition, extortion, improving property values by removal of unwanted structures (either on the property or nearby to eliminate an eyesore or even to improve a scenic view), escaping financial obligations, or even gaining employment. The profit motive may be extended to include murder by fire of family members with the intent of gaining insurance proceeds or inheritances.

One common, well-recognized motive for insurance fraud is poor financial condition at and before the time of the loss and/or financial distress. That is, however, by no means the only reason people intentionally set fires. While the motive sometimes is "financial need" driven, sometimes the motivation can more aptly be described as "greed" driven. Indeed, there are many, many well-documented cases in which the person responsible for an intentionally set fire was of very substantial means but was motivated by greed, or other motives. For example, a prominent example is the case of Constantine Pappadopoulos in California—in which Mr. Pappadopoulos, a multimillionaire businessman, was caught on an FBI wiretap plotting the fire of his luxurious mansion home with the **torch** he hired to burn it down.

torch ■ A professional fire setter.

In many communities the single most common motive is profit or fraud. It is a major factor in residential and commercial fires. Economic hardships fall on private property owners and businesspeople alike, and they all may seek relief from their problems by burning them. Fraud may be defined as a "deception deliberately practiced in order to secure unfair or unlawful gain." Under common law, burning one's own property was not considered a crime even if one gained by it. This aspect has been remediated by statutory laws in most states that prohibit the destruction of one's own property with the intent of defrauding another party.

Fraud is principally of two types: commercial and residential.

Commercial Fraud

fraud ■ Deception deliberately practiced in order to secure unfair or unlawful gain.

Commercial **fraud** fires may be set or arranged by the owner to destroy old or outmoded stock or equipment, destroy records to avoid taxes or audits, clear an unsalable building from marketable real estate, or gain insurance money to rebuild a better building at a better location. Fires may also be set by a competitor to gain a market advantage or by agents of organized crime for purposes of extortion, insurance fraud, protection rackets, or intimidation. Labor problems are a frequent cause, particularly in the building trades. During construction, buildings are especially susceptible to fires set by pro-union or anti-union agitators. When property values escalate, a business may be sold repeatedly among members of an arson "ring" to inflate its value, with huge insurance settlements on the business when it finally burns. Some rings operate by buying small-to-midsize businesses that may be marginal or even successful, draining the assets, selling off stock, replacing furniture and fittings, and virtually stripping the business physically as well as financially. The remains are then burned, and the insurance due on a "thriving business" is paid to the members of the ring.

Signs of a commercial fraud fire can include the following:

- Substitution of cheap, old furniture, equipment, or stock for new
- Open drawers in file cabinets or desks to encourage destruction of records
- Remodeling underway with minimal or poor-quality materials
- Disabling of fire detection or sprinkler systems

The background investigation should determine recent changes in ownership, value, or tenancy; the state of the business—debts, back orders, financing, and mortgages; the nature of any inventory—contents new or old, seasonal, outmoded, obsolete, or unsalable because of changes in governmental regulations; and the extent of competition. Contact with the insurance agent and adjuster will reveal the amount of insurance, multiple policies, or recent changes in valuation or coverage. All these factors may reveal the motive and intent of a would-be arsonist. Extensive documentation of business conditions may permit a qualified forensic accountant to identify failing businesses, even to the point of calculating the money being lost every month. This can be telling evidence to dispute a claim for generous business interruption benefits.

Residential Fraud

Residential fires may be set by owners with the intent of defrauding the insurance carrier or by tenants wishing to defraud the owner or a welfare agency. The burning of residential properties has been particularly popular since the implementation of Fair Access to Insurance Requirements (FAIR) plans in many states. These government-backed plans guarantee substantial insurance coverage to property owners, in many cases with minimal investigation of the building's value or the owner's prior history. These plans have come under considerable abuse by both individual landlords and organized rings who buy housing units, insure them heavily, and then profit from convenient fires. Many states have insurance statutes that preclude coverage for intentional acts.

Increasing tax rates, physical deterioration of structures (accompanied by legally mandated repairs at great cost), pressure from condominium developers, and statutory rent-control measures all combine to "force" multiple-dwelling landlords to seek relief by destroying the now burdensome structure.

Welfare fraud plays a significant part in residential fires in many urban areas. When large numbers of tenants were being displaced from their housing by arsons, it was decided that welfare agencies should reimburse them for their unplanned costs of temporary housing and new furniture, appliances, and clothes. It was not long before such benefits were being abused. If a move was already being contemplated, who could refuse a free "grubstake" to buy new clothes or furniture? The usable belongings could be moved out or sold off, and up to $3,000 would be provided when the apartment was gutted by a convenient fire.

In addition to the substitutions described previously with respect to commercial fires, there are some diagnostic signs of arson fires peculiar to dwellings. Items having sentimental value (photographs, antiques, family Bibles) or high monetary value (cash, stocks, deeds, insurance policies, stamp or coin collections) are frequently removed before the fire is set. Clothing or furnishings may be removed or substituted with cheap goods. The scene should be inventoried by the investigator for his or her own protection as well as that of the owner. The victim should be asked to provide a complete list of the contents of the destroyed areas. Substitutions may be detected by checking for noncombustible remains like hangers or buckles from clothes and metal frames or springs from furniture. As noted in Chapter 14, a laboratory may be able to identify the remains of suspected valuables and confirm that they have not been replaced with junk. Removal of portable valuables may not entail removal from the property, and a check of the outbuildings may reveal their location. The investigator should determine whether family pets (dogs, cats, birds, horses) survived and if so, whether they were moved or released prior to the fire or escaped during it. Convenient "vacations" may take the responsibles themselves out of the area. In the Pappadopoulos case mentioned previously, the wife was convicted of conspiring with her husband to have their 929 m^2 (10,000 ft^2) home burned while they were vacationing overseas. Mr. Pappadopoulos remains a fugitive in Greece.

Insurance policies should be thoroughly checked with the carrier agent and adjuster for any overinsurance, recent increases in coverage, multiple coverages, or the nature of any particularly valuable items inventoried separately for coverage. The investigation of the victim should include the status of finances, job changes (move or termination), stability of marriage, illnesses, injuries, or lawsuits pending. The history of the victim should be checked for previous fires on or near the property [see Chapter 17 for information on NICB (National Insurance Crime Bureau) resources]. Once a formula is found successful, such fire setters often cannot resist the greedy temptation to cash in again.

Other Profit Motives

In addition to direct insurance or welfare/public assistance fraud, other possibilities for financial gain need to be checked and eliminated. Fires have been set to secure employment in a variety of situations. Firefighters, either seasonal, part-time, or paid volunteers (or full-time firefighters who are paid overtime for off-duty responses), have been found responsible for setting fires to increase their income. Unemployed security guards have been known to set fires to prove their value to the threatened business, with the intent of promoting employment. Wildland fires have been set to improve game-hunting conditions (within a historical context). Construction workers have burned partially completed buildings to ensure continued employment, and heavy-equipment agents have set wildland fires to provide a market for the rental of their equipment. Fires have been set to gain freedom from underpaid contractual and even military obligations.[17] Tragically, there have been cases in which family members (particularly young children) have been killed by set fires, sometimes to gain directly from insurance but also to gain freedom from family obligations or to improve the conditions for another relationship.[18]

VANDALISM

Malicious or mischievous fire setting that damages property, seemingly at random and with no identifiable purpose, is often considered vandalism. It is usually a motivation related to juveniles and adolescents, and targets are often schools or school properties. Abandoned buildings, vehicles, and wildland (vegetation) are also frequently targets. Fires are set when the opportunity arises, often after school or work or on weekends.

Boredom and frustration among youths sometimes lead to a peer-group challenge to create some excitement. It may be either a group effort of the gang or a solo trial as an initiation rite for a new member. This is a group Dr. Dian Williams calls "delinquent" fire setters, since they often have records for other forms of lawbreaking.[19] According to Dr. Williams, such fires may be set as part of (or to conceal) common vandalism (breaking windows, graffiti) or burglary, or as a demonstration of "gang" behavior. This gang behavior may be part of an initiation or to impress others in the gang to gain entry or status. Common vandalism, petty theft, or burglary may accompany such fire setting. Devices, rarely used, will be very simple, with paper trailers or Molotov cocktails being the most complex. Fuels or accelerants may be brought in, but fuels incidental to the scene are usually employed, since the attacks are usually not planned. These fires may be set as a convenient weapon against someone (e.g., revenge motive). Such fire setters can be dangerous because they generally lack any regard for the safety of others. Apprehension is difficult because the fire setter's behavior is unpredictable and may follow no set pattern. Vandalism by setting fires is a very different motive from "fire play" fire setting by youngsters, which will be considered in a later section.

JUVENILE FIRE SETTING

Fire setting by juveniles can extend far beyond vandalism, since it can occur with children too young to form intent or understand the consequences of losing control of a fire or flame (often as a result of the curiosity phase of fire experimentation, typically with boys 3 to 10 years old). Juveniles may also set fires as a solitary activity to express anger or seek attention. In such cases, the fires generally will be small—in closets, beds, waste cans, or garages—but often get out of control and destroy much property and even cost lives. Such fires may also be set as a cry for help because of neglect or an abusive (physical, sexual, or psychological) family situation or in deliberate response to a threatened departure (as in divorce) or unwanted arrival (new step-parent, child, or lover). Such fire setting is often foreshadowed by emotional upheavals in the family (death, divorce, unemployment, alcoholism, even moving), an absent or inattentive parent, or even school or social problems. In one recent case, numerous small household items were set on fire in kitchen drawers and the refrigerator by a troubled 13-year-old girl. Fortunately, the fires self-extinguished and were discovered only later by family members or investigators.

Juvenile fire setting has been seriously underestimated in years past, most probably because many incidents were not reported by families or social welfare workers. It is now clear that the arsons cleared by arrests each year in the United States involve a large percentage of suspects under the age of 18. The investigator should be careful not to assume that young juveniles are not capable of forming the intent for arson. It is estimated that nearly 7 percent of those arrested for arson in the United States in 1994 were under the age of 10.[20]

Traditionally, about half of all arson arrests each year in the United States are juveniles (probably because they are the easier cases in which a suspect can be identified). FBI Uniform Crime Reporting (UCR) statistics show that for the most recent 10-year period, the percentage of persons arrested for arson who were under the age of 18 dropped 13 percent, from 55 percent in 1999 to 48 percent in 2008.[21] Of the 10,784 suspects arrested in 2008, 36.8 percent were under 18 years, 18.7 percent were under 15 years, and 1 percent were under 10 years of age. In the past, arson was considered a crime traditionally committed by males; however, FBI UCR statistics show that of the 10,784 total suspects arrested in 2008, 84.1 percent were male and 15.9 percent were female.[22]

Bob Wood of the United Kingdom has explored some of the complex issues that distinguish between juvenile "fire play" and deliberate fire setting.[23] True curiosity or "fire play" fire setting is considered by many to be a natural part of learning about one's environment. Such fire setters are boys between ages 3 and 10, there is no expressed intent to do damage, and there is rarely repeat behavior (beyond two fires). Such fire setting appears to be opportunistic and spontaneous when matches or a lighter are available and there is no adult supervision.[24] It is clear that both caregivers and fire investigators have to learn to recognize the critical patterns and features so that the problem can be addressed. Through local counseling programs (through police, and public health and community aid agencies) great progress has been made in addressing this national problem.[25] Investigators must appreciate the delicate legal ground involved in interviewing juveniles as suspects or witnesses and should remember that the parents will lie to protect their offspring, even more than the subject will.

EXCITEMENT AND THRILL SEEKING

Fires may be set by those seeking excitement, attention, recognition, and, very rarely, sexual stimulation. Excitement or thrill seeking as a motive for fire setting is closely related to the human psychological need for attention, with the difference largely one of intent—that is, did the fire setter realize that he or she was doing it out of a specific search for excitement, just to gain the attention of someone, or out of anger or desire for revenge? Such anger/revenge motives will be discussed later. Fires have been set as attention-getting devices by persons to call attention to themselves or a crisis situation.

Adults and children as young as 10 have been identified as setting fires to make others aware of physical, emotional, or sexual abuse conditions. These fires may be very small, set in readily ignitable household materials—books, towels, drawer contents, even clothing. Fires set for excitement can range from small trash and dumpster fires to fires involving occupied buildings. The typical excitement-seeking fire setter will set many fires during his or her career, often with the severity (amount of damage and risk to life) increasing with time.[26] The fires are usually set in a neighborhood familiar to him or her (from school, work, recreation, etc.). The arsonist will sometimes stay in the area to watch, possibly even photographing or videotaping the action. Devices are very rarely used; matches or lighters and available combustibles are typical. Although the excitement usually relates to the fire itself, some may be more focused on the fire apparatus, sirens, and lights of the suppression efforts. In these cases, the arsonist begins to set fires just to participate, actively or vicariously, in the suppression. (This has occurred with firefighters seeking the "adrenaline rush" of emergency responses.) This brings us to consider a subclass of the excitement-seeking fire setter—the hero/vanity arsonist.

The excitement inherent in a large fire and its suppression efforts offers many possible motives for would-be arsonists. People seeking attention or respect may set fires to gain that attention, either by being in attendance (part of the crowd), taking part in rescue and suppression efforts, or even by taking responsibility for the fire. Such fire setters are often described as "hero/vanity" arsonists. Firefighters may wish to impress their families or their superiors with their performance and set fires to give themselves more opportunity to do so. Those who get extra pay for responding to calls (as volunteers or as bonus-paid, full-time firefighters) have been known to set nuisance, structure, or wildland fires for the income.[27] Security guards may set fires simply to relieve boredom, force the company to hire additional guards, or prove their value to the firm by discovering the fire and assisting in its extinguishment. Their increased value may improve their chances of a pay raise or forestall a threatened layoff. According to the 2003 USFA report, firefighters who set fires often do so for the chance to be a hero. This heroic role can be in the form of first detecting or reporting fire, rescuing occupants, or acting prominently in its suppression. The ultimate such case could well be that of John Orr, a captain with the Glendale Fire Department. Over a period of years, he is suspected of having set scores of fires (both wildland and structures) where he could intercede as the chief investigator to "solve" many difficult fires.[28] He is thought to have been primarily a vanity fire setter who enjoyed a considerable thrill from working alongside the people looking for the arsonist. His motivation would be difficult to classify.[29] He was convicted in 1992 and 1998 of numerous arsons in commercial premises and wildlands.

The hero/vanity fire setter often has an overriding self-image as a superhero or "big shot," which is mostly bravado. These people often carry a weapon because it is an extension of their power. Attacks are generally unplanned, using materials and circumstances as they are found. He sets fires to provide opportunities to show his courage or dedication or to gain public attention or acclaim.

The investigation of this type of fire setter should include a background check of current personnel, especially those involved with discovering or extinguishing the fire. Because attention seeking may be related to frustrated ambitions to be a police officer or firefighter, recently rejected applicants for these positions should be considered potential suspects.

REVENGE, RETALIATION, OR SPITE

For adult fire setters, revenge or spite is the most common motive. Fires may be set in retribution for some real or imagined insult or hostile behavior. Because of the complex psychological factors, revenge may be part of other motive categories. Targets of revenge fires may be a person, an institution, the values or symbols of society in general, or a group.

Persons

An attack on a person is often in retaliation for a personal grievance. Fires stemming from hatred or spite between individuals can be a spontaneously conceived, one-time occurrence or multiple events with deliberately and carefully preplanned fire setting. The target may be the person's personal possessions, vehicle, or home, and the fires are often accompanied by vandalism or physical destruction prior to the fire. Such revenge-motivated fires can often turn deadly when the "target" or innocent persons are trapped by the fire. The location of the fire is often a significant indicator, because burning the bedroom or the clothes closet is considered to be worse punishment of the victim than just burning the living room or den. Bedding and clothes may be cut or torn up, or paint, ink, or glue may be spread on them and the furnishings before the fire is set. Photographs or memorabilia are sometimes smashed. The attack may be directed at the target of the hate or a substitute. Because the attack is not rationally prepared, it is typically disorganized and unsophisticated. The arsonist is usually reacting to an extreme emotional state. Fuels or accelerants are usually whatever is conveniently at hand at the scene. Time-delay devices are rare, because the perpetrator is not rationally considering escape and an alibi at the time.

One case in the authors' experience defied both of the last two such guidelines, however. An estranged wife used gasoline to soak clothing and furnishings of the husband's vacation home and then attempted to set a candle-and-paper delay device. Unfortunately, she had been alternating her destruction with heavy drinking, and the resulting delay allowed an explosive mixture of gasoline vapor to permeate the house. When she attempted to light the candle, the house exploded in flames, seriously burning her as she fled. In such premeditated cases, fuels, trailers, and delay devices may be found.

All types of interpersonal relationships can result in spite fires. Check marital relationships, friendships, or job-related situations (in revenge for loss of a job or promotion). Such spite fires may be encountered in either the gay or straight community. The tensions released when a heterosexual or homosexual love relationship fails can result in disastrous and tragic consequences. Such fires may be set against either the new lover or the unfaithful former partner. The psychological complexities of gay relationships and the strained or even hostile relationships between the gay and the straight communities make an investigation of such a fire even more challenging.

Institutions

A person who feels he or she has been wronged by an institution such as government, medicine, courts, military services, or even by employers, banks, or other representatives of authority may attack (usually in serial attacks) the buildings housing these agencies. The executives of those agencies may also be selected as targets. The fires can range in severity from nuisance fires in trash containers to planned fires of great destructiveness, although they tend to be mass arsons (multiple sets in the same building). The same target may be attacked repeatedly over a long period.[30] Because the grievance against the institution may be real or imagined, suspects would not necessarily be limited to former employees or disgruntled customers. The Oklahoma City Murrah Federal Building bombing is an example of an attack on an institution. To Timothy McVeigh, the Federal Court House represented the government that, among other things, had attacked the religious compound at Waco, Texas, two years previously. Ted Kaczynski, the Unabomber, had a similar motive, but his targets covered the spectrum of imaginary grievances—against airlines, computer technology, psychologists, and forestry.

Society

Retaliation against society is the motive of someone who, feeling abused, persecuted, and powerless, strikes out against that society in a series of often-random attacks. Such fires are set in a variety of targets, typically using available materials and matches or a lighter for ignition, although they are often premeditated. Interestingly, the societal retaliation fire setter rarely contacts the media to claim responsibility, apparently satisfying him- or herself that something has been done to even the score.[31] Interestingly, Dr. Dian Williams links many terrorism acts with fire or explosions to revenge motivation somewhere in the perpetrator's background.[32]

Groups

These fires are often called *hate crimes* because they are the result of hate or anger that the subject has against a group rather than a person. The targets for these fires may be religious, racial, or fraternal and may be a building or other symbol of the group rather than an individual. Churches or synagogues are common targets, but the meeting places of clubs, organizations, and (recently) rival urban gangs are also struck. In some cases, an emblem (cross, star, logo) may be attacked, regardless of its location.[33] Graffiti and other vandalism may accompany the fire. Premeditation and planning are encountered, and devices (particularly Molotov cocktails) are often used.

Church arsons across the southeastern United States were the impetus for a national task force. One series of arson attacks was carried out by a lone transient with a grudge against religion. Others have been racially motivated or have targeted Jewish houses of worship. The National Church Arson Task Force (NCATF)—which included the FBI, the Bureau of Alcohol, Tobacco and Firearms personnel, U.S. Attorneys, the U.S. Marshals Service, and state and local fire and police personnel—was formed in 1996 in response to a rash of fires at houses of worship, mostly in the South. The task force made significant progress in the fight against such arsons. In a report issued in September 2000, the task force said it opened 945 investigations into arson and attempted arson at houses of worship between January 1995 and August 2000 and arrested 431 suspects in 342 of the cases. This translates into a 36.2 percent arrest rate, which is more than double the 16 percent rate of all arsons as reported by the FBI.[34] As reflected by the convictions from the NCATF efforts, the arsons—at both African American and other houses of worship—were motivated by multiple factors, including racism and religious hatred. Arsonists have burned churches for other reasons, including opportunistic and random vandalism, pyromania, mental health disturbances, feuding with ministers, retribution against religious authorities, parking or neighborhood disputes, cover-up of burglaries, and financial profit. In some cases, the arsonists claimed they believed the church to be an abandoned building.

CONCEALMENT OF ANOTHER CRIME

Because of the destruction brought on by a major structure fire, many people believe that any fire destroys all physical evidence of any criminal activity. Although it is true that even a moderate fire will obscure and render partially useless some types of evidence, blood and physiological fluids will retain much of their evidential value, fingerprints can survive under a variety of conditions, and weapons like knives, guns, or bullets require a great deal of fire exposure before they are rendered completely worthless. See Chapter 14 for further information on physical evidence.

While fire can obliterate fingerprints and facial features on which normal identifications are based, it requires significant combustion over a period of hours before the body can no longer tell its tale of cause and manner of death. (The reader is referred to Chapter 15 for more information on the evidence related to fire deaths.) Nevertheless, fires may be set to impair identification of a victim, destroy evidence of foul play, or simulate an accidental death in the hope that authorities will not examine it too closely.

It must be remembered that fire is one of the human's oldest weapons against fellow humans and, as such, may be used to kill or assault another. In a number of recent cases a flammable liquid was poured or thrown on someone and then ignited intentionally. Such fires have been used as a means of murder of spouses and other family members, recalcitrant lovers, and even of robbery. Such cases usually involve premeditation (to procure the flammable liquid), but fires may occur in the midst of a heated verbal argument or some time afterward. Sadly, some of these incidents are written off as household accidents.

Although fires that occur during or after the commission of another crime are usually thought to be intended to destroy evidence of the crime, it has been suggested that they may be part of an anger-retaliation crime. In murder cases, the subject may want not only to kill but also to obliterate or destroy the victim and will set fire to the body (or clothing or piled furniture) to achieve that end. Because such behavior may increase the thrill or satisfaction of such crimes, when they are part of a series they may include careful planning; complex, even ritualized preparation; and specialized tools. The cunning and organization extend to security arrangements (remote or private locations). According to psychologist Richard Walter, such fire setting tends to be episodic (serial) but follows no set time pattern. It is so much an instrument of aggression that the fire setter will often keep a souvenir from each scene as a trophy of the triumph.[35]

The use of fire as crime concealment is not limited to crimes of personal assault. Fire may be set to obliterate signs of theft, pilferage, burglary, or even embezzlement and fraud. In such cases the manner of preparation and ignition are often afterthoughts to the main criminal activity. Sophisticated devices are rare and accelerants used, if any, are usually those incidental to the scene. Lack of complete ignition, absence of proper fuel or ventilation, or a general appearance of things "out of place" are diagnostic signs associated with this kind of fire. When fuels are moved about to ensure destruction, they are usually the ones involved in the theft—racks of clothing, boxes of stock, or drawers of financial records are positioned with the intent of promoting the highest destruction in their immediate vicinity. Administrative or correctional records may be targeted in educational, military, or court office to prevent detection or punishment.

EXTREMISM (SOCIAL PROTEST AND TERRORISM)

Fire has been used as a weapon of social, political, or religious protest since revolutions first began. If one could not attack an oppressor directly, then one could destroy the oppressor's castle, counting house, or crops. Fire is considered by some political minorities as a legitimate expression of dissent. During the 1960s and 1970s, wholesale arson, often during general riot conditions, was employed against the white "establishment" to express the hostility of racial minorities and to force urban renewal into the then-destroyed areas.[36]

These fires may be started by individuals or by groups and are usually accompanied by ransacking, vandalism, and looting. Even though a fire may have occurred during a period of unrest, it should not be a foregone conclusion that it was started by rioters. After the major urban riots of the 1960s and 1990s, it was discovered that a number of store owners burned their own property for fraudulent purposes, using the riots as cover. (A similar pattern develops whenever there is a serial arsonist operating in an area, with property owners hoping that every fire will be written off as the work of the fire setter.) Devices in riot or unrest cases are usually limited to what is at hand or the ubiquitous Molotov cocktail. Urban street gangs have recently taken up Molotov cocktails as a weapon of intimidation, thus accounting in part for a substantial increase in the number of such weapons reported by police agencies.

Many of the thousands of political movements in the world have spawned waves of fire and explosive incidents as statements of political protest or outright terrorism. The differences between protest and terrorism may be only in the degree of organized planning.

Although an organized terrorist campaign often involves random targets of innocent persons to maintain an aura of anarchy, terrorists, like protesters, often select targets of political or economic significance. Political targets include government offices, newspapers, universities, or political party headquarters. Economic targets were very popular for terrorists in the 1970s and included the offices of big business (auto, oil, computers) and the offices or distribution facilities of utilities, banks, or companies that were thought to have adverse effects on the environment. In the 1980s, political terrorism by fire and explosion in the United States drew such diverse targets as animal welfare (protesting use of animals in lab research), neighborhood protection (stores or industries not deemed suitable for the area), and abortion clinics (which were hit by an extensive series of explosive and incendiary devices in many states). Today, so-called eco-terrorists are targeting lumber companies, automobile dealers, commercial livestock facilities (poultry and fur), property developers, ski lodges, and even university facilities engaged in agricultural, animal, or stem cell (reproductive) research with incendiary device attacks. Terrorists, because of their planning and financing, make more frequent use of sophisticated timing and ignition devices: tamper-proof electronic timers, chemical incendiary mixtures, radio-controlled detonators, as well as large-scale explosive charges. The World Trade Center (New York City, 1993) and the Murrah Federal Building in Oklahoma City were both targets of terrorist vehicle bombs, each estimated to involve over 500 kg (1,100 lb) of blasting agent. The September 2001 attack on the World Trade Center accomplished complete destruction by causing structural collapse by means of massive fires from fuel-laden airliners.[37] Retired ATF Special Agent John Malooly outlined many of the targets in the United States that terrorists could effectively target using simple fire sets. Targets could include major freeways and bridges, commercial or financial centers, places of public assembly, schools, and public transport.[38]

The "Provisional" wing of the Irish Republican Army must be considered an excellent example of a well-financed, experienced terrorist force. It selected a mixture of military and civilian police emplacements and personnel for its main targets (using both explosive and incendiary devices) and added an assortment of random civilian targets to maximize the confusion and fear of the general populace. One device used with devastating effectiveness on occupied civilian targets was a quantity of gasoline dispersed and ignited by a small explosive charge. One or two 20-L (5-gal) jerry cans were placed in close contact with a charge of 0.4 kg (1 lb) of military or commercial high explosive. The device, placed in a window or doorway, produced an instantaneous fireball that filled the room, killing nearly all occupants. The resulting flash fire sometimes caused only superficial damage to the building, because the gasoline was completely consumed while in the gaseous phase. Afterward, the room bore no smell of gasoline, and the obvious diagnostic signs pointed toward a gas explosion rather than an incendiary device.

MIXED MOTIVES

Often, a set of mixed motives may be found in an arsonist. For example, an element of revenge may be found in many vandalism fires. Many arsonists who set fires for excitement may also profit from them. It may also be argued that many extremists have an underlying revenge motivation in setting their fires. The point is that in dealing with human behavior it is necessary to examine each case to determine motive, mixed or otherwise.

IRRATIONAL FIRE SETTING

As we have seen, the vast majority of incendiary fires fits into one of the rational motive categories described here, where the fire setter "sees" some specific benefit is accruing to him or her as a result of setting the fire (even if an outsider cannot perceive any benefit). Irrational or "motiveless" fire setting is sometimes called **pyromania**. Recent evaluation of this term has shown that there is no consensus even among mental health professionals and

pyromania ■ Uncontrollable psychological impulse to start fires.

behavioral scientists as to what constitutes pyromania, and, indeed, whether there really is such a disorder.[39] Some analysts believe that what investigators consider irrational motives for setting fires are perfectly rational to the perpetrator suffering from some mental disorder. Moreover, the motives in the mind of the fire setter fit into the major categories described previously. Subjects may claim that the fires are set merely for the release of tension or anxiety, and they simply "feel better" once the flames are started. The current version of the American Psychiatric Association's *Diagnostic and Statistical Manual of Mental Disorders-IV* (DSM-IV-TR) describes "pyromania" as an impulse-control disorder characterized by "a fascination with, interest in, curiosity about, or attraction to fire and its situational context."[40] The key element is lack of impulse control. If the fire setting were truly uncontrollable, the subject would be setting fires in front of witnesses, even police. If the impulse can be controlled such that the fire starting is away from known observers, it is not "pyromania." Extensive research at the NCAVC challenges the concept of the motiveless fire setter. Analysts there are persuaded from extensive interviews with arsonists that they all have a motive, but some lack the introspection or communication skills to adequately describe their motivation.[41]

Although the preceding is the condition to which most people refer when they use the term *firebug*, it is clear that only a small percentage of all arsonists fit the description of motiveless fire setters. Even David Berkowitz, New York's "Son of Sam" killer, who claimed credit for setting some 2,000 fires in New York over a 3-year period and who is considered by many to have been a pyromaniac, may have better fit the description of a societal retaliation arsonist because he felt inadequate and lonely, and struck out with fire (and ultimately with a gun) in response to those feelings.[42]

Such behavior usually starts with the setting of small trash fires (or grass fires in rural areas), often escalating in target selection to fires in empty buildings, abandoned vehicles, and, in extreme cases, to occupied apartment houses, sometimes with devastating and fatal effects. Serial dumpster fires may be the first signs of the presence of a long-term problem fire setter, who may be a simple excitement-motivated fire setter. While the fires set by such subjects are often planned, they rarely, if ever, employ accelerants, trailers, or time-delay devices. They are most often started in piles of paper or trash in out-of-the-way corners, hallways, or alleys, which may delay their detection. The fires are almost always set under cover of darkness, and the setter is nearly always a low-profile loner. Although it was once a criterion, IQ cannot be used to classify such offenders, since surveys have shown no reliable relationship. Most fire setters are in the normal range of IQ. Fire setting has been the subject of research in the past; however, it is clear that although concepts of fire setting and pyromania have evolved with new knowledge in recent years, research interest is lacking.[43] (Lewis and Yarnell's "classic" study, for instance, has been found to be based on very limited data of questionable reliability.) Dr. Dian Williams has published an excellent study of arsonists—both adult and juvenile—and their "management" in investigations.[44] Because diagnosing someone as a pyromaniac is a clinical diagnosis to be made only by a mental health professional, fire investigators should be very careful not to label the subject of their investigations as such.

The Arson Set

The **arson set** may be defined as the contrivance device or deliberate assembly of first fuel and heat source intended to ignite the fire. The professional and, to a lesser extent, the amateur arsonist want to produce a maximum of destruction no matter how small the first ignition. To this end, most rational arsonists will be sure the fundamental requirements for a large, spreading fire are present: ready availability of fuel, proper ventilation, and reliable ignition. Although much attention has been given to the ignition devices used, the most important considerations to the success of the fire are sufficient fuel and access of air. Only when delayed ignition is essential is there any need for unusual methods of igniting

arson set ■ Device, assembly, or contrivance used to ignite an incendiary fire.

the fire. Many more arsonists choose the time and circumstances of the fire to minimize their chances of being seen than rely on any unusual (and often undependable) igniting device. Perhaps the preoccupation of some investigators with such devices results primarily from their novelty, which adds much interest to their discovery. In one study by DeHaan, only 12.3 percent of the arson cases examined involved the use of any type of ignition (timing) device. By far the most common "set" was the direct pouring of an ignitable liquid with ignition by open flame (match or lighter), which was found in 61.2 percent of the cases.[45] This trend was found in cases examined in the author's laboratory during the 1980s and continues today. The extent and nature of preparations, such as building or assembling a device, preparing the scene (providing kindling of a particular type or arranged in a certain fashion), providing accelerants, ensuring fire spread, and the like, are all indications of the organization characteristics of the fire setter. Such characteristics (organized versus disorganized) can be used to focus on a possible motive and even suggest the characteristics and traits of the arsonist (profile) for use in the investigation.[46]

ARRANGING THE FIRE—LOCATION

Once the target is selected, the arsonist must make several choices that are critical to the success of the fire. If destruction is a specific goal, a point of origin must be selected so that flames will rise into fresh fuel. Although flames will spread fastest across the underside of a ceiling or overhang, it is usually impractical to arrange a point of ignition high up on a wall. The floor adjacent to a combustible wall is the likeliest point for initiating the fire. Walls covered with gypsum sheetrock do not offer much fuel, and once the paint and surface paper burn off, the fire will die back rapidly. Unless holes have been broken or cut through the sheetrock to expose the combustible wall studs or ceiling joists before the fire, such covering will seriously inhibit the spread of the fire. Wooden paneling products vary considerably in their fire resistance but eventually they will all add greatly to the heat release rate (fuel load) of a room. Furniture may be used if present in the correct type and sufficient quantity. Older furniture stuffed with horsehair and covered with wool fabrics is difficult to ignite and does not burn readily once ignited. Nearly all modern furniture contains large quantities of polyurethane foam padding that can be readily ignited with a match and burns with great speed and intensity. Lighter upholstery fabrics of cotton and synthetics can be readily ignitable, adding further to the ease of ignition and the fuel load. In a commercial or industrial building, stocks of finished goods, packing materials, or raw materials may be moved about to provide a large fuel load in a readily combustible form. High rack storage provides ideal fuel distribution and concealment for the fire setter. Unless in-rack sprinklers are fitted, a fire can become very large before detection. It must be remembered that knowledge of fire behavior, detection, and suppression is not always familiar to the fire setter and so he or she may pick a location unaware that the fire will not grow or will be readily detected or extinguished by a sprinkler head immediately above it.

Within the structure, the point selected for the origin may be limited by what areas of the building are readily accessible, and the "ideal" spot may be protected by a locked door, gate, or elevator. The arsonist may then have to select a less perfect location. Basements are very suitable because the walls and ceilings are often not covered, the entire structure above the fire providing maximum fuel. A basement location is remote, providing the best concealment, and the utilities and trash often found in basements provide many possibilities for an "accidental" fire. Any basement fire should be investigated with special care, because of all possible locations, this one is generally most suited to the arsonist's purposes.

Attics, although they provide convenient fuel in exposed structural members, trash accumulations, and points of possible "accidental" ignition from exposed wiring, are generally not selected as a point of origin. The availability of fuel is limited to the contents and the roof and adjacent structure. Insurance fraud fires are sometimes started in the attic, however, for the very reason that the ceiling and roof are destroyed, the building is rendered unusable, and business is interrupted. When the insurance for these claims is collected, repairs are begun, and the building, while being rebuilt, suffers another more complete fire; the insurance on an upgraded building is then collected. One interesting variant seen recently is to set fire to adjoining businesses knowing that the fire setter's property is likely to incur some roof or smoke damage from a large fire. The insurance proceeds are then available while the investigation focuses on the adjacent property. Common or adjacent roofs also provide a convenient route of concealed entry and exit. A history of previous fires in adjacent properties may offer a link to the perpetrator.

Closets provide both concealment and some fuel in the form of stored clothing. Other than nuisance, spite, or juvenile-set (fire play) fires, arson fires in closets are rare. A knowledgeable arsonist realizes that the poor flammability of some clothing, the limited air, and the difficulty of encouraging fire spread to other rooms make a closet a poor choice for an area of origin. Clothes closets and beds, when confirmed as points of origin, may strongly indicate a revenge/spite motive rather than fraud. However, when smoke damage is the desired end (for insurance fraud—to gain remodeling expenses, stock replacement, etc.) rather than a lot of fire, a closet may be the best place to ignite the fire because ventilation and spread can usually be controlled.

In an attempt to simulate accidental ignition, some arsonists ignite the fire in the immediate vicinity of a fireplace, heater, dryer, water heater, or similar heat-producing appliance. The arsonist hopes that a cursory inspection of the scene will place the point of origin in the vicinity of such a "normal" cause, and the fire will be written off as an accidental malfunction of the appliance. All such fires should be examined carefully, and the investigator must not prejudge such fires and fall into the arsonist's trap. With very few exceptions, accidental fires associated with such appliances will start above floor level and travel upward. A fire set with a flammable liquid will usually burn from the floor upward. If the diagnostic signs of fire behavior indicate a very low burn, the entire area should be examined carefully, as described in Chapter 7. Fire starting accidentally in concealed floor spaces or post-flashover room fires, of course, can produce floor-level burns and penetrations.

FUELS

Arson investigators find themselves investigating a fire set by a truly professional "torch" only on rare occasions, probably because of two guidelines used by such professionals: (1) the simpler the better, and (2) use fuels available at the scene. Both guidelines minimize the materials that must be brought to the scene by the arsonist, and the fewer the number of times the individual is seen in the area (especially carrying containers or odd packages) the fewer the chances of being spotted by a witness. Simpler sets minimize the chances that elaborate devices will fail to ignite while they maximize the chances that the entire set will be destroyed by the fire or even normal suppression activities.

The use of accumulated trash as an ignition point obeys both guidelines. Trash piles provide an inherent risk of ignition by accidental causes, and many of them are written off as accidental by investigators without careful examination. In the experience of the authors, a small quantity of ignitable liquid on a pile of loose paper and plastic cartons will ensure complete ignition and yet will be completely undetectable in the post-fire debris. Trash accumulations are frequent in basements, attics, storage rooms, and around back doors (inside and out). A trash-filled utility room is ideal for producing a large destructive fire. These can be difficult fires to establish as arson-caused because there is nearly always some possible means of accidental ignition, such as defective electrical wiring, furnace, water heater, or appliance; and loose paper or empty cardboard or plastic foam boxes make excellent first fuels. Accidental fires involving these must be eliminated

FIGURE 16-1A A gasoline trailer ignited only the bed (left front corner) leaving minimal fire scorch marks, and a clear trailer pattern of carpet protected by the gasoline. *Courtesy of Lamont "Monty" McGill, Fire/Explosion Investigator.*

FIGURE 16-1B This gasoline trailer ignited only the adjacent carpet. *Courtesy of Lamont "Monty" McGill, Fire/Explosion Investigator.*

as causes to prove the crime of arson if the scientific method is being followed. Outside trash piles are often used by the arsonist who does not have access to the inside of the structure. Because such fire setters usually do not intentionally provide a path for the fire to spread to the structure, the fires do not always spread to the interior of the nearby building.

Ignitable Liquids

Many arsonists, especially amateur ones, do not rely on solid fuels available at the scene and will heavily supplement them with ignitable liquid accelerants to ensure complete ignition. As defined previously, an accelerant is as any fuel deliberately added to increase the intensity or speed of spread of an intentional fire. This is often an ignitable liquid. Although such fuels have the ability to produce a great deal of fire in any given location, there is no automatic assurance that the flames so rapidly produced will find other fuel in their path so that they can build into a conflagration (as shown in Figure 16-1). The larger the quantity of ignitable liquid used, the larger the initial fire may be, but if a volatile fuel is involved, the vapor/air mixture in an enclosed room is more likely to be above the upper explosive limit (meaning no ignition). A very large flammable liquid fire may also deplete the oxygen in the room or structure and result in self-extinguishment before other fuels in the room can be ignited. It is difficult (and often ineffective) to transport and disperse more fuel than can readily be carried by hand. Except in rare instances, less than a few liters (gallons) will be used. In instances where large drums of volatile hydrocarbons are readily available (such as in an industrial situation) or where the fuel tanks or vehicles can be used as sources, massive quantities of flammable liquids can be employed. Such unusually high fuel loads can be used to ignite a secondary source of fire such as a large accumulation of automobile tires, which burn ferociously and are all but impossible to extinguish, or the wood framing of the structure itself.

The most frequently used amount of an accelerant is that amount a person can carry conveniently without attracting undue attention. A person can easily carry an amount sufficient for igniting a very large fire, so this is not necessarily a serious limitation. The amateur nearly always uses much more of a volatile accelerant than is necessary. By using it judiciously on top of loosely piled combustibles, a person can start a very satisfactory fire in a building using only a very small quantity of accelerant. If one were trying to burn a large building in an urban area where detection and suppression would quickly follow ignition, it might be desirable to employ a large reservoir of liquid fuel over a period of time. A metal drum of a suitable fuel such as gasoline, paint thinner, or other petroleum

FIGURE 16-2 A large LPG cylinder placed inside this car and set alight caused massive explosive damage. *Courtesy of Det. Sgt. Ken Legat, New Zealand Police, Christchurch.*

product might be placed at the desired point of origin with enough fuel under and around it to start the fire and to heat the drum to the point of rupture. A well-filled drum is not difficult to burst by heating, and the ruptured drum will then spray fuel into the fire for a long time with virtually no possibility of being extinguished by conventional water or fog sprays. There is always the possibility of an explosion from such a device when flammable vapors are suddenly mixed into a large fire, but maximum damage is usually desired by the arsonist. Such a drum placed at the origin of the fire kindled with additional fuel may not be suspected as an instrument of arson because there are many legitimate reasons for the storage of liquid flammables on most industrial or commercial premises. Cylinders of compressed gases such as LPG cylinders may also be used, as shown in Figure 16-2.

If the fire is of very limited duration and has not spread horizontally or progressed to flashover, the approximate amount of flammable liquid used may be estimated from the surface area of the "pool" of fuel. Due to its low viscosity, gasoline will flow freely on a smooth, nonporous floor to a depth of 0.5–1 mm (0.04 in.), so a 1–L (1–qt) quantity will cover a maximum area of about 1–2 m^2 (21 ft^2). If it is *poured* on a semiporous surface (not just scattered or sprayed), such as smooth unpainted wood or concrete, its effective depth will be about 2 to 3 mm (0.08–0.12 in.), so a liter (1 qt) of gasoline will saturate an area of about 0.3 to 0.5 m^2 (3.2–5.4 ft^2). On carpet, the saturation depth is the thickness of the carpet (plus that of an absorbent pad beneath), 10 to 20 mm (0.4 to 0.8 in.). The surface area of a pool can be very small, with 1 L (1 qt) saturating 0.05 to 0.1 m^2 (0.5–1 ft^2).[47]

The amateur arsonist does not always place the liquid accelerant where it will provide the most effective ignition. Instead, most commonly, he or she pours accelerant over carpeting, furniture, and open floor, a pile of paper or trash, or a combination of these. Pouring a flammable liquid onto an absorbent material, such as soil, newspapers, or carpet, allows some fuel a chance to escape combustion. For example, a stack of newspapers soaked in gasoline allows the fire to burn only on the outside of the stack, with the absorbent papers feeding additional fuel to the fire in the same manner as the wick of a kerosene lamp. Fuel near the center of the stack is not exposed to air or even to extreme temperatures and remains for detection days later, especially if the debris has been water-soaked to minimize post-fire evaporation. A pool of paint thinner in a carpet may burn a distinct ring, with the carpet pile in the center of the ring barely scorched and a substantial portion of the fuel still present where it was not exposed to the flame–air interface. Fuel soaked into the soil likewise is poorly consumed and then only at the surface. Loscalzo, DeForest, and Chao poured gasoline on plywood, carpet, and soil, ignited it, and tested detectability of the residue at various time intervals. Depending on the length of time the gasoline was originally allowed to burn, gasoline residues were identifiable by gas chromatography as long as 24 hours after burning on carpet and as long as 162 hours after on soil.[48] More recent tests showed that small quantities of kerosene (unburned) were detectable by standard forensic methods on denim jeans for only 24 hours and on shoes for only 72 hours, stored indoors at 22°C (68°F). Unburned gasoline was detectable for less than 24 hours on shoes and denim jeans under the same conditions but was still detectable on carpet after 7 days of storage outside at 5°C to 15°C (40°F to 50°F).[49] Recent tests using the most modern, more sensitive techniques but with smaller amounts of gasoline showed detectable quantities only up to 4 hours on clothing.[50]

The splattering of liquid fuel on a floor or furnishings will often produce detectable residues on shoes and pants cuffs, even more so if the arsonist steps into the liquid. The gas chromatographic profile of such liquid transfers will readily be distinguishable from the light volatiles transferred by vapor contact alone. The clothing items worn by a suspect in an "accelerant" fire should always be collected as soon as possible and packaged *separately* in appropriate cans or Kapak (made by Ampac) bags pending lab analysis. Because of the relatively short "lifetime" of such traces, a positive lab finding is a useful indicator of recent contact.

The sensitivity of today's methods and the detection of background petroleum products in innocent clothing (especially athletic shoes) make it important to package items separately and collect comparison samples (of liquids and clothing). A common explanation for gasoline traces on hands is that gasoline was spilled while fueling a vehicle. Tests have shown that this process does not produce detectable amounts on the hands if properly conducted. Even when a living subject's hands are deliberately doused with 1 ml of gasoline, gasoline is detectable by the most sensitive method for only 2 hours after contact.[51] (See Chapter 14.)

The pouring of a flammable liquid directly on the floor is probably the most commonly detected form of arson. It is readily detected because a considerable portion of the fuel remains unburned, and what does burn can leave characteristic burn patterns or diagnostic signs. Poor ignition is due in part to the amateur arsonist's misconceptions about the properties of such fuels. On a smooth, nonabsorbent, tight floor, little or no damage may result if the burning of the fuel did not raise the floor covering to its ignition point. (As seen in Chapter 4, the liquid fuel tends to protect the floor covering by evaporation rather than damaging it.) On a floor with cracks, holes, or joints into which the fuel can penetrate, holes and lines will burn into the floor in a characteristic manner (see Figure 16-3). Remember that if the room subsequently goes to flashover, the radiant heat produced may obliterate many of the more subtle indicators on the floor. Sustained post-flashover burning will quickly burn wood floors to complete penetration. (Many of the misinterpretations of burn patterns as indicators of arson were discussed in Chapter 7.) On ground-level floors, sometimes enough flammable liquid is poured out that it seeps completely through the floor and into the soil beneath. In such cases, the wooden subflooring, joists, and soil should be checked for the presence of volatile fuels. Another common mistake for the amateur is to dump the fuel in the center of the room without ensuring that the flames will reach furniture, walls, or other fuels to keep the fire going (as in Figure 16-1).

Not all sets or trailers involve ignitable liquids. As seen in Figure 16-4, crumpled newspaper or wax paper can be used. Such materials may burn quickly and completely to leave very little residue.

FIGURE 16-3 Gasoline pool fire on wood plank floor after 1 minute. Gasoline flashed off flat areas and continued to burn only in joints. *Courtesy of Jamie Novak, Novak Investigations and St. Paul Fire Dept.*

FIGURE 16-4A Trailer of crumpled newspapers. *Courtesy of Lamont "Monty" McGill, Fire/Explosion Investigator.*

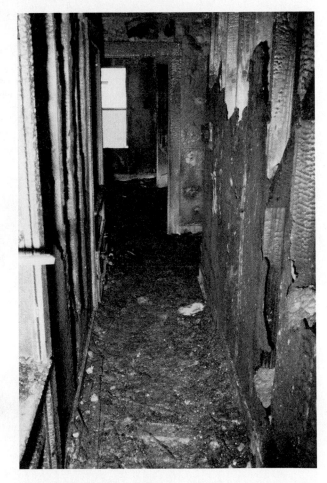

FIGURE 16-4B After fire in hall: minimal residues. *Courtesy of Lamont "Monty" McGill, Fire/Explosion Investigator.*

Multiple Ignition Points

Multiple ignition points are common in arsons. Not satisfied to gamble on a single point of ignition, the arsonist thinks that "more is better," with the intent of producing a larger fire that spreads more quickly and widely than that from one point. These multiple points are usually connected by a *trailer* of flammable liquid, smokeless gunpowder, rags, twisted ropes of newspaper, waxed paper, or wood packing material (excelsior). In a few cases, trailers of flammable liquid were poured from electrical outlet to outlet along a baseboard, apparently in hopes of simulating massive electrical fires. Although multiple sets can provide a more complete destruction of the target, they present many problems that the arsonist would just as soon avoid. They require additional materials and fuel, and there is also the chance that one of the elements will fail, leaving intact sets to be discovered. It requires more time to distribute fuel in several sites than one, and in the process of distribution, vapors from the earlier sets may be forming an explosive mixture that may deflagrate, involving the arsonist. The arsonist's shoes and clothing are often splattered with the fuel and may ignite. The hazards from these sources are likely to outweigh any advantage that might be gained from multiple points of ignition, especially if the fuel has a high volatility, like gasoline or lacquer thinner. The time required also prolongs the chances that the earlier sets will be detected and the perpetrator will be spotted while still at work. A low-volatility fuel such as kerosene may be suitable for such operations, but it requires a "wick" to provide an ignitable vapor, since it cannot be ignited as a pool on a nonporous surface.

FALSE MULTIPLE ORIGINS

It should be noted that multiple points of ignition are not necessarily proof of arson. Flames from a large ordinary combustible fire can melt and ignite the polystyrene light diffusers used in ceiling lighting fixtures, draperies, or the asphalt from a collapsing roof. These flaming "liquids" can produce multiple, apparently unconnected, points of origin throughout a large area. Such multiple points have been mistaken in the past for arson sets. The investigator should be careful in documenting the scene to establish where such light fixtures, draperies, or roof vents were prior to the fire. (See Figures 16-5a–c for an example

FIGURE 16-5A Fire started with a candle: initial vertical spread. *Courtesy of Jamie Novak, Novak Investigations and St. Paul Fire Dept.*

FIGURE 16-5B Burning draperies fall. *Courtesy of Jamie Novak, Novak Investigations and St. Paul Fire Dept.*

FIGURE 16-5C Chair in corner ignites to produce a false second origin. *Courtesy of Jamie Novak, Novak Investigations and St. Paul Fire Dept.*

of a candle fire spread by draperies to look like separate origins.) A sudden surge on the line power or a lightning strike can induce a surge in line voltage that can cause ignition of multiple appliances (or anything plugged in). That hypothesis has to be ruled out, but since most power line transformer– or lightning-caused surges will affect multiple dwellings, neighboring properties should be checked.

Type of Accelerant

As we have seen previously, the type of flammable liquid used as an accelerant is determined in great part by the experience, motivation, and intent of the arsonist. In a premeditated structure fire, some thought will usually be given to selecting an appropriate fuel. Gasoline is by far the most commonly detected accelerant because it is an efficient and easily ignited flammable liquid and because it can be purchased and transported even in large quantities without arousing the suspicions of the seller or authorities.

Paint thinners and kerosenes are the next most common fuels.[52] They are effective fuels and can be purchased in moderate quantities without suspicion because they have legitimate uses even in the private home. Turpentines and lacquer thinners are almost never encountered as premeditated fuels.

The circumstances or living conditions may dictate what fuel is most convenient— kerosene or even camping fuel may be closer at hand than gasoline. The authors know of one fire set that used gum turpentine because it was readily available as part of a remodeling project. Gum turpentine is very costly and not often used in bulk by private individuals. The chemistries of new paints make the use of solvents more of an exception than the rule it once was, since most paints sold today for household use are the water-based latex type, which do not require ignitable liquids as thinning or cleaning agents.

Although most arsonists seem to give little thought to the accelerant's volatility, it is a property that has great bearing on the fuel's suitability to the task. Extremely volatile fuels such as gasoline, lacquer thinner, or alcohols may evaporate excessively at high temperatures, interfering with the intended ignition and spread of the fire. The vapors generated can travel considerable distances and produce explosive mixtures in the vicinity of distant "accidental" sources of ignition, most typically the pilot light on the water heater or furnace. This makes it necessary to set the accelerant very quickly and to ignite it from

Chapter 16 Arson as a Crime   **679**

a safe distance, thereby reducing the arsonist's control of the set. Kerosenes, paint thinners, and proprietary lantern fuels are safer to use because of their lower volatility. The varieties of lantern and campstove fuels can vary considerably in their volatilities and flash points. Because they may contain heavier hydrocarbons with commensurately higher boiling points, however, they are more resistant to evaporation and will remain in the debris longer as residues than will, say, lacquer thinners. Some fire setters use a 50/50 mix of diesel fuel and gasoline, thinking the diesel fuel will make it "safer" to use. Unfortunately, the vapor pressure of the gasoline dominates the ignitability of the vapor mixture. Until there is less than 5 percent gasoline in the mix, the liquid will ignite just the same as gasoline.

At the opposite end of the petroleum distillate scale is fuel oil. It is very difficult to burn except when it is dispersed into a mist and well mixed with air. When applied to paper, cloth, or trash, it can burn reasonably well from the wick and thereby add to the fuel load, but it will not be ignited as a pool.

Alcohols, both methyl (wood) and ethyl (grain), are as volatile as gasoline and carry the same hazards. Because of their flammability and their reduced chances of detection (due to their miscibility with water), they would appear to be good choices as flammable accelerants. They are not, however, efficient fuels, and it takes a larger quantity of burning alcohol to generate sufficient heat to bring other nearby combustibles to their ignition points. Alcohols will still burn when mixed with some water, but with a lower risk as well as effectiveness. Potable liquors of 90 to 100 proof (45 to 50 percent alcohol) can be used as accelerants, though not very effective ones. Alcohols may also be encountered in chafing dish heaters and campfire starting fuels (Figure 16-6). They produce very little damage to the surface beneath and leave little if any residues post-fire.

FIGURE 16-6 Campfire starters may be either wax/sawdust sticks or viscous ignitable liquids such as the one here. *Courtesy of John D. DeHaan.*

The foregoing discussions deal with premeditated fire sets involving accelerants. Depending on the motivation and opportunity of the arsonist, such fuels may not be acquired beforehand. When it comes to fires set using materials already at the scene, ingenuity knows no bounds. Glues, dye solvents, rubber cements and their solvents, mop cleaners, uncured polyester resin, paints, brush cleaners, and the like are common in homes, schools, and businesses, and all have been used as effective arson accelerants. Under these circumstances an inventory of the scene and a thorough interview with owners, tenants, and maintenance people may be vital to the success of the inquiry. Although it is commonly known that the duplicating fluid used in the old-style "spirit" duplicators or mimeograph machines is highly flammable methyl alcohol, it is not commonly known that the copier toner used in some older office "dry" copiers contains a petroleum distillate or isoparaffinic blend similar to paint thinners. Many offices have cases of this fluid stored where it is easily accessible to burglars intent on erasing signs of their presence. The same holds true for commercial insecticides that are often dissolved in a paint thinner–like petroleum distillate or aromatic solvents. When such unusual fuels are suspected by the investigator, these suspicions should be relayed to the laboratory examiner along with control samples of all possible fluids observed at the scene.

METHOD OF INITIATION

The methods chosen by the arsonist to initiate the fire will be closely related to the type of fuels used and sometimes to the manner selected to best avoid detection. If liquid accelerants have been used, an almost immediate ignition is desired because the vaporizing fuel may be lost or form an undesirable explosive hazard. With less volatile materials, the time of delay may still be considerable, but it is impossible to leave open-flame devices such as a lighted candle. These will not delay the fire for long. If the primary fuel is not volatile, a delayed-action igniter may be suitable, but in general, the development of the fire will be less certain. The choice of primary ignition devices include flames, fuses, smoldering materials, cigarettes, electric arcs, glowing wire, chemical ignition, black powder and flashpowder, and improvised chemical delays.

Flames

Simplest of all ignition devices to produce, the flame of a match or lighter is easy to obtain, to apply, and to eliminate as evidence. The flame is a certain means of starting the fire because it is a fire already and requires only propagation. Whatever is left at the scene, for example, a burnt match, is likely to be destroyed to the point of escaping detection.

One exception is the common candle, which often is not sufficiently destroyed to remain undetected. Melted wax will persist in a fire if it accumulates on a noncombustible surface even though the form of the candle itself will be lost. Remnants of the wick and metal wick anchor tab will often survive a generalized fire.

Another source of a flame, often overlooked as being an accidental malfunction, is an altered gas pilot light assembly or a broken gas line that is set to throw a blowtorch-like flame on nearby combustibles. A special case is that of a flamethrower, weed burner, or similar device that generates and maintains a large flame for a protracted time. The use of such a device is not common but is highly effective even when a refractory fuel like fuel oil is employed. It is possible to ignite even large timbers and to spread a fire widely through a variety of combustibles with such a flame source. In one sawmill fire, it was determined that a weed burner was the igniter because of extensive deep burns on the bases of heavy timbers that were lying on soil where no other source of such intense flame was possible. The fire was started in this fashion in a number of places so that overall development of the fire was very fast and destruction extensive. The weed burner was found nearby.

Molotov Cocktails and Delay Devices

Of all the remote-ignition devices known, the Molotov cocktail is the most common. It consists of a breakable glass container of a flammable liquid that depends on a lighted wick or a chemical incendiary device to provide ignition. Its effects are maximized when it can be thrown against a hard surface (wood is often not hard enough) to break the bottle and mechanically atomize the gasoline into an explosive mist. The ignition then produces a fireball that can spread at very high speeds (up to 10 m/s; 40 ft/s). Although a Molotov is usually ignited by its own flaming wick, hypergolic mixtures have been used to avoid the use of an open flame. A packet containing sugar and potassium chlorate is taped or tied to the outside of the bottle, and sulfuric acid is added to the gasoline inside. Upon breaking of the container, the acid reacts with the sugar and chlorate to initiate a very hot flame that, in turn, ignites the gasoline. The burning time of the gasoline from a thrown quart-size Molotov (as shown in Figure 16-7) is on the order of approximately 10 seconds, so the target has to be ignited in that time, or only scorch marks will result. Although the production of a thickened "napalm" has been recommended in terrorist literature, the minimally enhanced burning effects are generally not worth the extra time and effort required to mix the gasoline with soap flakes, polystyrene foam, or paraffin wax.

Although the use of such a device is convenient from the standpoint of the arsonist, its easy detection at a fire scene makes it less than ideal (for him or her). A broken bottle, often a broken window, and perhaps eyewitnesses accompany its use. Tracing it to the arsonist may be difficult but often is not. The broken bottle is usually found at the bottom of the fire, where it is exposed to the minimum amount of heat, so fire damage to the fragments is minimal. The portions of the bottle bearing the most information about its origin—the neck with its labeling and the base with its cast-in production data (date and place of production and product code)—are the pieces most resistant to mechanical and fire damage and usually survive intact. The investigator should endeavor to collect all or nearly all the pieces to allow reconstruction of the device for identification or even a search for fingerprints. (As noted in Chapter 14, the oily components of latent prints are somewhat soluble in gasoline and may be washed off by the accelerant as it seeps from a broken container, but new techniques now promise recovery of prints even in the presence of gasoline.) As described previously, DNA has been identified on the mouth and wick of Molotov cocktails that have burned.

(a)   (b)

FIGURE 16-8 (a) Delay device—plastic container of gasoline with sponge wedged in handle. Sponge set alight, flames melt top of bottle and ignite gasoline within. (b) Post-fire. *Courtesy of Jamie Novak, Novak Investigations and St. Paul Fire Dept.*

Plastic bottles or containers filled with fuel have been encountered in a number of fires (see Figures 16-8a and b). These may be placed over room heaters to soften and melt but are more commonly surrounded by loose paper and trash that is set afire to provide a means of softening them and igniting the spilled contents. (There also may be a timing delay such as a cigarette-and-match device.) If properly supported by combustibles, these containers can burn sufficiently to make identification impossible.

One variation on this is the use of a toy balloon or plastic bag containing gasoline that is suspended by a string and set swinging over a candle. Once the oscillations grow smaller, the balloon stops over the candle, which then melts a hole in the bag; the fuel released quickly spreads the fire to the surrounding trash. Although effective if small quantities of accelerant are used, the amateur will often use a large trash bag with lots of fuel that is more likely to extinguish the candle as it rushes out than be ignited by it.

Fuse

A fuse must be considered as a means of timing delay, a trailer connecting several points of origin, and a means of directing a flame to a remote location. There are three basic types of fuse: (1) pyrotechnic, which employs a train of flammable powder packed around a string core; (2) safety, which contains a train of black powder within layers of string and

FIGURE 16-9 Pyrotechnic or rocket fuse burning on plywood produces an external flame that scorches the surface beneath it. *Courtesy of John D. DeHaan.*

asphalt wrappings and sometimes a continuous tube of water-resistant plastic; and (3) igniter cord. All three types of fuses should be considered essentially a remote-control match, a means of transmitting a live flame from one point to another at a predictable rate.

The pyrotechnic fuse has an exposed flame at all points along its length and will scorch and even ignite combustibles upon which it rests (as shown in Figure 16-9). Pyrotechnic fuse, also known as hobby or rocket fuse, is widely available in hobby stores, while safety fuse is almost always sold through suppliers of commercial blasting supplies. As time-delay devices, pyrotechnic fuses can vary considerably in their burning rates, from 30 s/ft (2 ft/min; 0.6 m/min) to as quickly as 10 s/ft (6 ft/min; 1.8 m/min). Safety fuse has an exposed flame only at its ends. Although the asphalt coating will melt and smolder, the wrappings will remain in place, and damage to surroundings will be minimal (as shown in Figure 16-10). Safety fuse is made to exacting standards to burn predictably at 30 to 40 s/ft (0.5–0.6 in./s; 1.2–1.5 cm/s) at sea level.

Igniter cord has a core of pyrotechnic powder wrapped in wire and is used only in special blasting operations where its high temperatures and quick burning times are desirable. After burning, a multiple spiral of fine wire is left behind.

Smoldering Materials

Although mattresses, sawdust piles, and other fuel materials with limited access to air may smolder and ultimately develop into a large fire, they are not common ignition sources for

FIGURE 16-10 Burned safety fuse leaves a trail of melted wax on the surface but scorches the wood only at the end where the flame exits (lower left corner). *Courtesy of John D. DeHaan.*

arson sets. Their major use is in delaying the outbreak of the fire, allowing time for escape. Unfortunately, such fires are unreliable in their behavior, and the heavy smoke they produce increases the chances of detection before the "real" fire can take place. Slow-burning materials like a cattail, incense, or punk have been used as delay devices when wrapped with matches. The matches in turn are set to ignite a black-powder or pyrotechnic initiator (as seen in Figure 16-6). Smoldering ignition sources alone will not ignite liquid accelerants, which require open flame.

Cigarettes

Cigarettes are rather uncommon as the source of ignition for a successful structure fire. If they are set to ignite liquid flammables, they will almost certainly fail. If they are used to start a smoldering fire (typically in a couch, chair, or bed), the results will be highly uncertain. Prank fires started with a cigarette are more likely than a planned and deliberate arson, because the prankster may throw a lighted cigarette into appropriate fuel as a whim or experiment. The lighted cigarette in the wastebasket is a source of accidental fires from carelessness, but a fire can also result from a prank. When thrown into crumpled paper, a lighted cigarette tends to char its way through the papers it encounters without providing a sufficient source of heat to initiate self-sustaining combustion.

Although the cigarette alone may be of little value in igniting fires, there are ways of combining it with other flame sources to enhance its value. It is commonly found in combination with matches. If the lighted cigarette is placed between two layers of paper matches in a matchbook, when it burns down to the heads it will ignite them, and the entire book will burst into flame. Another variation is to tie a number of matches around the outside of a cigarette, which is then allowed to burn down to them. Such a device is fairly resistant to burning to completion (especially the matchbook), and its remains can be found by careful examination of the point of origin (see Figure 8-22). These are the most commonly detected delay devices in grass and wildland fires. Although they rarely burn so completely as to avoid detection, they are usually charred sufficiently so that fingerprints, physical comparisons, and saliva (DNA) testing may not be possible. Improvements in DNA techniques and more extensive DNA databases are making it more likely to identify suspects from cigarette-butt testing, however.

Electric Arcs

As we saw in Chapter 10, electric arcs create a great deal of energy but usually for only a short period of time. As a tool of the arsonist, the arc can be used to ignite only certain types of fuels. Most such fuels are explosive in nature, either explosive fuel/air mixtures or sensitive low explosives like black powder or flashpowder. A fire ignited by an arc, then, will be characterized by an initial explosion. The only alternative to the explosive fuel as an initial fuel is to convert the arc into a flame. This may be done by a powder trail or a readily ignitable flashpowder. The use of solid fuels in conjunction with arcs is uncertain and not an efficient method of kindling fires.

Because arcs can be generated by electrical means, delay devices using electric arcs would seem to offer many advantages to the arsonist. However, without elaborate arrangements of ignitable vapors, explosives, or similar active fuels, and corresponding elaborate electrical devices that will, in part, survive the fire, it is difficult to ignite a fire by such means. Any custom-made device that survives the flames increases the chances of identifying the fire as incendiary in nature and provides possible clues with which to track down the perpetrator. Such devices are therefore usually avoided by professional arsonists. Some ingenious devices have been employed that used the arc generated when a light switch is turned on or a motor is started. For example, a short circuit can be set up between two wires of a multiwire power cord (even by the simple expedient of a wire staple through the insulation). The short-circuited portion is then put in contact with readily ignited fuel. Because the short circuit is minor it is difficult to trace, but the heat generated by even a brief arc can start a fire if nearby fuel is susceptible. Even the most ingenious devices are unreliable, however, and are not often encountered.

Solar

There have been reports of focusing mirrors (see Chapter 6) or convex lenses set up with ignitable fuel at their focal point. Only when the sun would be in the right position relative to the set device would there be enough energy to cause ignition. The resulting residue—broken glass and a charred framework—could easily be overlooked during the scene examination.

Candles

Candles have been used alone (burning down into combustibles) or in combination with other devices for many years. Wax candles are useful timing delays (their combustion rate is determined by their design and diameter—the smaller the diameter, the "faster" they will burn down). They are, however, easily extinguished by drafts, and some can fall over easily, making them less reliable as ignition sources. Residues of wax or unburned wick may be impossible to find on heavily burned substrates like upholstery.

Glowing Wire

Although they are rare, devices using an electric current to heat a wire to incandescence to ignite a combustible material in contact with it have been reported by some investigators. Such a device would allow the fire to be ignited by any electric circuit, which could by completed by a timer, by remote radio control, or even by an unwitting bystander. The elements used could come from a heater, toaster, lightbulb, or flashbulb. (The latter, of course, includes its own limited supply of flammable fuel.) One device reported in the underground literature involves drilling a hole in the base of an electric lightbulb, partially filling it with black powder or incendiary mixture, and screwing it back in place. When the light is turned on, the glowing filament ignites the powder, and the heat from the combustion shatters the lamp, allowing the burning powder to fall onto flammable fuels below. A more commonly encountered glowing-wire device is an electric heater element connected to a power source through a timer. The heater is wrapped or packed with paper or cloth to provide fuel for ignition. This fuel is often supplemented with gasoline in plastic bags or bottles set on the cloth or paper.

Chemical Ignition

A wide variety of chemical igniter systems have been devised by arsonists. Fortunately, they are sufficiently sophisticated and unpredictable that they are employed only rarely. They can be as simple as metallic sodium or potassium brought into contact with water. Small chunks of such materials are dropped into water that reacts with the metal, producing a flash of flame that can ignite volatile fuel/air mixtures under the correct conditions.

Incendiary mixtures can be more complex, such as thermite—a mixture of aluminum oxide and iron filings that when ignited by magnesium "starter" produces an intense fire. This fire can reach a temperature of thousands of degrees and burn through steel and concrete. It leaves a significant amount of residue that is chemically distinctive but may be innocuous in appearance and easily overlooked. The magnesium igniter may be difficult to obtain and initiate on its own, so the use of thermite devices is fortunately rare.

Black Powder and Flash Powder

Mixtures of aluminum powder and potassium perchlorate or potassium nitrate must be considered effective chemical initiators. Black powder or flashpowder from small "firecrackers" can ignite exposed fuels or larger powder charges. The flashpowder in an M-80-type explosive device (as in Figure 12-7) can shatter a glass bottle of gasoline and ignite a sizable fireball in the manner of a well-thrown Molotov cocktail.

Improvised Chemical Delays

Dry chlorine of the type used for swimming pool purification can spontaneously ignite a mixture with glycol-based fluids, such as brake fluid or hair dressing (see Figure 16-11).

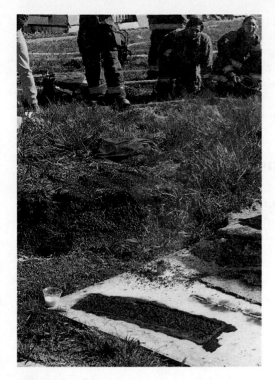

FIGURE 16-11 Clear flame plume from glycerin–pool chlorine mixture is very hot but low in radiant heat and lasts only a few seconds. *Courtesy of Keith Parker, Marin County Fire Department, Woodacre, CA.*

FIGURE 16-12 Post-fire residue of device in Figure 16-11 is not visually distinctive; however, residues of the foam cup and some unreacted pool chlorine are still present. *Courtesy of Keith Parker, Marin County, Fire Department, Woodacre, CA.*

Residues from such mixtures or from acid–chlorate mixtures previously described are not remarkable in their post-fire appearance and are also easily mistaken for plaster debris (as in Figure 16-12). A wide variety of other chemical systems are susceptible to spontaneous ignition under proper circumstances, and the means are known for controlling the reaction time.[53] These means include physical ones where a physical barrier separating the reactants is allowed to fail or where a chemical solvent, which keeps the reaction from proceeding, is allowed to evaporate. Other timing factors include particle size, amount of mixing, and the temperature of the environment or of the reactants. The influence of these controlling factors serves to make the performance of such mixtures unpredictable. The residues are variable depending on the original mixture and the completeness of the reaction. They may be overlooked or misinterpreted by even the best scene investigators. If there is reason to believe that a fire is arson and that such an "exotic" means of ignition was used, then the investigator is wise to enlist the aid of a competent chemist or criminalist. Such improvised chemical mixtures can also be used as the main accelerant. Mixtures of metallic aluminum or magnesium powder and strong oxidizers (chlorates, perchlorates, nitrates, or even sulfates) have been demonstrated to be capable of generating very high temperatures and extremely high heat release rates. These incendiary mixtures are similar to flash powder formulations but contain a binder (wax, rubber, or diesel fuel) to prevent aerosolization and explosion while acting as an additional fuel.

Deductions from the Interpretation of Evidence

All the preceding considerations are important to the arsonist deciding which procedure to use in starting the fire. They are of equal consequence to the fire investigator who must correlate the observed facts about the fire with the possibility that any fire may be deliberately set. They also serve as a guide to the thinking and experience of the perpetrator. Inept or inexperienced arsonists may set themselves on fire, cause unnecessary or premature explosions, or kindle a fire that is not self-sustaining or capable of developing into a large conflagration. Such occurrences as well as the availability of fuels at the scene and the means of procuring other fuels or devices that can be traced all serve as indicators of the possible identity of the arsonist. Because arsonists are sometimes very ingenious, the investigator must develop a comparable ingenuity in investigating cases as well as a highly developed observational ability.

CRIMINAL INVESTIGATIVE ANALYSIS OR PROFILING

One of the most interesting developments of recent years has been the improvement of offender profiles, especially those dealing with serial crimes. These profiles offer investigative leads of the inductive type—predicting the age, race, sex, occupation, socioeconomic status, and mental state of an offender—based on the data extracted from an analysis of the crimes. Several textbooks address criminal investigative analysis and offender profiling.[54]

Creating a profile involves an in-depth study of the crime scene and the victim as well as psychological and sociological data. The FBI's approach used at the NCAVC for creating a profile mirrors the scientific method in that it follows a series of five overlapping stages that resemble a medical diagnosis. These stages consist of:

1. Collecting and assessing data
2. Reconstructing the situation
3. Formulating a hypothesis
4. Testing the hypothesis against known information
5. Reporting the results[55]

Determination of an apparent motive is the primary step in developing an offender profile that is designed to focus on two fundamental questions: Who committed the crime? and Where can they be apprehended? Data evaluated on incidents include police and fire reports (including origin and cause determination), date, time, and description of target. The areas involved need to be mapped, showing the locations of all fires (not just those identified as arson-caused), and demographic data must be available. All incidents are analyzed for trends or patterns in time or place of occurrence, sequence, target, or means of ignition. (One of the possible values of the new BATS system described in Chapter 12 is trend analysis linked to geographic mapping.) Predictions of date, time, place, or target can be tested against newly reported incidents as a pattern of behavior is elucidated. These models are sufficiently sophisticated to allow the prediction of the subject's state of mind and behavior after the fire as well—immediately after and long term, and even his or her reaction to police contact.[56] Some studies have been based on carefully structured interviews with hundreds of convicted arsonists, both male and female.[57] They have greatly expanded the knowledge base of fire setters and improved the accuracy and precision of analytical and predictive methods.[58]

Geographic Profiling

Offender-based classification of arson extends beyond merely identifying the motive. Investigators should also closely examine the *geographic locations* selected by the arsonist. An analysis of these sites may reveal much about the intended target, add insight into the arsonist's motive, and provide a potential surveillance schedule for attempting to identify and apprehend the arsonist. The results of a joint research effort on the geography of serial violent crime illustrate the patterns found in serial arson cases. Results indicate those criminal offenders who repeatedly set fires exhibit temporal, target-specific, and spatial patterns that are often associated with their *modus operandi* (method of operation or "m.o.") and motive.[59]

With geographic profiling, the analysis of times and locations of events has shown great promise in evaluating criminal trends, allocating preventive or response resources, or assisting investigators. Because the locations are linked to fire details, similarities in modus operandi can be spotted, aiding in the identification of serial arsonists. Depending on the motive classification described earlier, some serial fire setters will target selected areas by convenience, accessibility, travel time, or target type. Geographic profiling has been very useful in establishing such patterns, estimating the residence or base location of the perpetrator, or predicting future targets.[60]

The two primary references in the discipline of geographical profiling are by two pioneers—Kim Rossmo in Canada and David Canter in the United Kingdom[61] followed by extensive work in Australia by Richard N. Kocsis.[62] A Web-based commercial computer program, *CrimePoint Fire and Arson,* allows any investigator to exploit this valuable investigation tool.[63]

Work by Katarina Fritzon in the United Kingdom examined and modeled the spatial relationship between the distance traveled and behavioral motivational aspects of arsonists. This research confirmed that arsonists whose behavior contained a strong emotional component tended to travel much shorter distances, on average, than other arsonists. Wherever an individual was targeted in a revenge-motivated attack, the arsonist was prepared to travel a greater distance. Also, arsonists who had recently separated from a partner were more likely to travel greater distances, due to the target-specific nature of the crime. Trends were also seen between the age of the arsonist and the distance traveled.[64]

Tracking Serial Offenders

Work by Brett Martinez lays out the strategy for the identification and tracking of serial arsonists. Using basic crime analysis techniques, his textbook assists an experienced fire investigator in detecting and identifying temporal and geographic patterns of arson fires as well as understanding the motivation behind these crimes. Martinez provides examples of how to present information in a final report for dissemination to all personnel involved in the investigation.[65]

ANALYTICAL REASONING

Although the physical evidence of an arson set and its laboratory analysis are often of great value to the investigator in reconstructing the crime and identifying the perpetrator, successful prosecution of the case is dependent almost entirely on the expertise of the investigator and his or her skill in reliably interpreting the diagnostic signs of fire behavior at the fire scene. Many cases have been brought to successful completion even in the absence of conclusive laboratory results with respect to the presence of flammable or incendiary materials. Analytical reasoning skills found useful to successful FBI profilers consisted of experience in the investigation of crime, understanding human behavior and motivations, and a capacity for logical and objective reasoning, and intuition.[66]

When an investigator considers motive, it must be remembered that in virtually any in-depth inquiry, facts will be uncovered that may be construed as "motive." Everyone who has a mortgage and an insecure job has a potential motive for burning the house. Anyone faced with expensive repair estimates on a vehicle is tempted to "sell it to the insurance company." Means and opportunity are, likewise, available to any person who lives in the house or drives the car. It is not enough (in fact, it is malfeasance) to simply establish a possible motive and use that to bootstrap a fire investigation that on its own merit would be an "undetermined" fire into an "arson" and then pursue it through legal channels. The determination of whether a fire is arson or accidental has to be established on the strength of the information from the scene investigation, supported by interview data regarding the fire event. If that call cannot be made at the conclusion of the active investigation, then the cause of the fire should be considered "undetermined."

The investigator must be prepared to analyze each fire carefully for the basic elements described previously in this volume. Is the identified area of origin specific enough to permit reliable assessment of ignition sources and first fuels in the area? Is there enough fuel normally to account for the destruction seen, or was additional fuel needed? Is there a reasonable accidental mechanism for igniting the normal fuels present? Could the suspected deliberate ignition device accomplish the fire spread observed? Does the arrangement of fuels and contents appear normal or contrived? Are the residues of ignitable liquid detected the result of normal room contents or proof of an intentionally added accelerant? Was the direction and rate of fire spread normal for the kind of fuel and amount of ventilation present? As we have seen, fires in today's furnishings can grow very quickly, achieve flashover, and create extremely high temperatures throughout a compartment without the presence of any accelerant or even in the absence of a deliberate ignition.

The extensive data collected by fire researchers regarding ignition and fire spread are readily available today. The development of computer models of fire growth and spread continues every year. Both fairly simple and very sophisticated computer programs exist today that allow the investigator to enter data on the fuel, room dimensions, ventilation, and insulation and predict the time and intensity of the developing fire, as well as predict the onset of flashover. Some aspects of these models as investigative tools will be described in Chapter 17.

ELIMINATION OF ACCIDENTAL OR NATURAL CAUSES

Because one of the elements of proof of the crime of arson is the process of elimination of all reasonably possible accidental or natural sources of ignition, the investigator should not overlook such causes even in the most obvious fires. The process of elimination of such possible causes should be carefully documented as part of the thorough investigation, because they will be of great concern in later judicial proceedings. This is where the scientific method offers the best means for evaluating the conclusion reached. The more alternative hypotheses that have been formed, tested, and rejected, the more reliable the final hypothesis or conclusion. The process also prepares the expert for cross-examination. Practices such as *ASTM E620: Standard Practice for Reporting Opinions of Technical Experts* recommend that an expert report elucidate what other explanations for the facts have been formulated and tested.[67] It should be noted that this practice includes reporting of data that helped rule out alternative hypotheses considered.

Careless smoking or waste disposal methods are still common causes of structure fires. The presence of people in the wilderness with their campfires, matches, cigarettes, heaters, trucks, and motorcycles presents an awesome risk to wildlands. Accidental fires annually burn far more acreage than any natural cause. Sparks from machinery and hot exhaust systems from vehicles must be considered. Overloaded electrical systems and heating or electrical systems in violation of building codes must also be evaluated. All appliances or equipment that uses electricity or gas must be considered as possible sources of ignition. Decorations, particularly Christmas decorations, may involve candles, temporary wiring, and electric lamps. Cooking utensils and exhaust systems (particularly those in restaurant kitchens) combine both a source of heat and readily combustible fuels (foods and greases). It seems that common sense cannot be legislated, and people constantly defy logic and common sense in their storage and handling of hazardous materials. Gasoline is used to clean car parts or remove tar from clothes without regard for its fire hazard and so may be found in kitchens and laundries. Aerosol products are found in profusion in homes and commercial locations, and every can holds LP propellant gases (and may also have an ignitable liquid solvent). They are easily punctured or overheated and can cause significant fire danger. Even what appear to be deliberate sets may be found to be accidental: electrical insulation destroyed by rodents or insects, matches gathered by pack rats, or even sources of ignition knocked over into combustible household furnishings by pets, with disastrous consequences. Natural means of ignition such as lightning, overheating by sunlight (more prevalent in these energy-conscious times), accumulations of static electricity, disasters such as earthquakes, and wind also occasionally cause serious fires. In spite of appearances to the contrary, man's destructive hand does not play a part in all fires. The use of a fuel–ignition matrix such as described in Chapter 6 can help ensure that all likely ignition sources have been evaluated in the context of the scene at hand.[68]

While a comprehensive fire investigation may be able to eliminate all possible accidental causes on the basis of physical data alone, other information (if reliable) can make it feasible. Witnesses (or surveillance cameras) that saw the start of the fire and saw a perpetrator in the act of producing a flame, for instance, would be difficult to overturn. Witnesses who reported no fuel (or insufficient) fuel or no competent accidental ignition sources should be considered. A proper fire engineering analysis (based on a complete and reliable reconstruction of pre-fire conditions) can test various hypotheses regarding the development of the fire and production of fire patterns. This analysis can reveal which ignition scenarios can be reliably eliminated. *NFPA 921,* in discussing this analysis, suggests that it is more often more practical to eliminate all accidental sources of ignition except for the use of an open flame.[69] Sources of open flame (match or lighter) can readily be removed from the scene, leaving no physical evidence behind. The hypothesis being tested is really: Can the "first fuel" be ignited by any source other than open flame? In the absence of conclusive physical evidence of the deliberate cause of the fire—residues of a foreign ignitable liquid, the remains of an incendiary device, and a clearly defined area of origin uncompromised by post-flashover fire or structural collapse, the investigator must be very cautious in concluding that arson took place.

CHAPTER REVIEW

Summary

Arson is a crime of violence that inflicts terrible losses of life, property, and prosperity on everyone in a community. Three elements must be proven to substantiate a charge of arson: a burning (destruction) of property; the burning was intentional, not accidental in cause; and the fire was set with malice. Motives for the crime include vandalism, excitement, revenge/spite, profit, crime concealment, and extremism (terrorism). The selection of the target, as well as the preparations and manner of setting the fire, may yield useful clues as to the motive, intent, and identity of the fire setter. Incendiary devices can range from the simple flame of a match or lighter to the use of accelerants to elaborate time delay or remote ignition devices. A variety of heat-producing sources can be employed: electric appliances, chemical reactions, cigarettes, fuses, even solar energy. Criminal behavior analysis has been found useful in solving serial arson crimes, as has geographic mapping or "profiling."

Arson is the most difficult crime to solve and prosecute (as national statistics prove) because of the destruction of evidence and the problems of linking a specific person to the crime even after it has been established that a crime was committed. The investigator must be cautious that establishing a motive alone (or even motive, means, and opportunity together) does not constitute proof of guilt of the crime of arson. The court will treat a fire (no matter how serious) as an accident until the elements of the crime are *positively* demonstrated by the investigator and prosecutor. One of the major elements of such proof is the exclusion of any reasonably accidental mechanism for the fire. This means that the investigator must carefully apply the scientific method to all the information gathered and create and evaluate as many reasonable alternative (accidental) hypotheses before concluding the fire was deliberately set.

Review Questions

1. Define the three elements of the crime of arson.
2. Name the six general classifications of motive for fire setting as defined by the NCAVC.
3. Briefly describe three ways in which fraud for financial gain can be exhibited in an arson.
4. Into what general categories do targets of revenge or retaliation fall?
5. What is the current status of irrational (motive-less) fire setting?
6. Name three different types of locations for fire sets and the possible intent behind each.
7. Name four different sources of heat used to ignite incendiary fires.
8. What is the difference between arson and an incendiary fire?
9. Describe geographic profiling and its application to the investigation of serial arsonists.
10. What steps must a good investigator take to ensure that a sound case has been assembled to prove arson?

Court Citations

C1. *People v. Andrews,* 234 Cal. App. 2d 69; 44, Cal. Rptr. 94 (1965).
C2. *People v. Fanshawe,* 137 NY 68; 32 N.E. 1102 (1893).
C3. *Pointer v. United States,* 151, U.S. 396; 14 S. Ct. 410 (1894).
C4. *People v. Lepkojes,* 48 Cal. App. 654; 192 P. 160 (1920).
C5. *People v. Kessler,* 62 Cal. App. 2d 817; 145 P. 2d 656 (1944).
C6. *People v. Saunders,* 13 Cal. App. 743: 110 P. 825 (1910).

References

1. J. F. Boudreau et al., "Arson and Arson Investigation: Survey and Assessment" (Washington, DC: National Institute of Law Enforcement and Criminal Justice (NILECJ), October 1977).

2. M. J. Karter Jr., "Fire Loss in the United Stated during 2008" (Quincy, MA: National Fire Protection Association, Fire Analysis & Research Division, August 2009); see also M. J. Karter Jr., "U.S. Fire Loss for 2008," *NFPA Journal* (September–October, 2009).

3. M. J. Karter Jr., "Fire Loss in the United States during 2003," *NFPA Journal* (October 2004).

4. FBI Uniform Crime Reporting Program, Table 2, Arson by Type of Property in the United States, 2008, Washington, DC; retrieved from http://www.fbi.gov/ucr/ucr.htm, 2009.

5. Communities and Local Government Fire Statistics (UK): 2007 (London); retrieved August 2009 from www.communities.gov.uk.

6. "Twelve-Month Analysis of £250,000 plus Fires in the UK," *Fire Risk Management Journal* (November 2009): 6.

7. Office of Deputy Prime Minister, *Economic Cost of Fire in England and Wales* (London: ODPM, 2003).

8. P. G. Jackson, "Assessing the Validity of Official Data on Arson," *Criminology* 26, no. 1 (1988): 181–95.

9. J. F. Decker and B. L. Ottley, Arson *Law and Prosecution* (Durham, NC: Carolina Academic Press, 2009), 337–41.

10. Anon. "Model Arson Law," *Encyclopedia of Crime and Justice*, vol. 1 (2002); http://law.jrank.org.

11. For a detailed discussion on this topic, see J. Q. Whitman, *The Origins of Reasonable Doubt: Theological Roots of the Criminal Trial* (New Haven, CT: Yale University Press, 2007).

12. A. O. Rider, "The Firesetter: A Psychological Profile," *FBI Law Enforcement Bulletin* 49 (June–July 1980).

13. *NFPA 921: Guide for Fire and Explosion Investigations* (Quincy, MA: National Fire Protection Association, 2008, 2011), sec. 22.4.9.2.1.1, 188.

14. A. D. Sapp et al., *A Motive-Based Offender Analysis of Serial Arsonists* (Washington DC: Department of Justice, 1994); see also D. J. Icove and M. H. Estepp, "Motive-Based Offender Profiles of Arson and Fire-Related Crimes," *FBI Law Enforcement Bulletin* 56, no. 4 (April 1987): 17–23.

15. J. A. Inciardi, "The Adult Firesetter: A Typology," *Criminology* (August 1970): 145–55.

16. D. J. Icove and M. H. Estepp, "Motive-Based Offender Profiles of Arson and Fire-Related Crimes," *FBI Law Enforcement Bulletin* 56, no. 4 (April 1987): 17–23.

17. A. D. Sapp et al., "Arsons aboard Naval Ships" (Washington, DC: National Center for the Analysis of Violent Crime, FBI, 1993).

18. T. G. Huff, "Filicide by Fire: The Worst Crime," *Fire and Arson Investigator* (June 1993).

19. D. L. Williams, *Understanding the Arsonist: From Assessment to Confession* (Tucson, AZ: Lawyers & Judges Publishing, 2006).

20. R. Estrin, "Juveniles: Nation's Top Firebugs," Oakland *Tribune,* January 20, 1996, A-4.

21. FBI Uniform Crime Reporting Program, Table 32, Ten-Year Arrest Trends, 1999–2008, Washington, DC (http://www.fbi.gov/ucr/ucr.htm), 2009.

22. FBI Uniform Crime Reporting Program, Tables 39, 40 and 41, Arrests for Arson by Age and Sex, 2008, Washington, DC (http://www.fbi.gov/ucr/ucr.htm), 2009.

23. B. Wood, "Children's Fire Setting Behaviour," *Fire Engineers Journal* (November 1995): 31–33.

24. Williams, *Understanding the Arsonist.*

25. Factory Mutual Engineering Corp., *Arson: Not Only for Profit* (Norwood, MA: Factory Mutual, 1985); Williams, *Understanding the Arsonist.*

26. Sapp et al., *A Motive-Based Offender Analysis.*

27. USFA, "Special Report: Firefighter Arson," USFA-TR-141 (Washington, DC: U.S. Fire Administration, January 2003).

28. J. Wambaugh, *Fire Lover* (New York: HarperCollins, 2002).

29. Williams, *Understanding the Arsonist.*

30. Sapp et al., *A Motive-Based Offender Analysis.*

31. Ibid.

32. Williams, *Understanding the Arsonist.*

33. Sapp et al., *A Motive-Based Offender Analysis.*

34. http://www.justice.gov/crt/church_arson/arson98.php, 2005; see also Independent Insurance Institute, iii.org, 2005).

35. R. D. Walter, "Arson: Who Burns and Destroys with Design?" paper presented at the Tenth Australian International Forensic Sciences Meeting, Brisbane, Australia, May 1988.

36. P. C. Hoffer, *Seven Fires* (New York: Public Affairs, 2006), chap. 5.

37. K. McGrattan and C. Bouldin, "Simulating the Fires in the World Trade Center" in *Proceedings Interflam 2004* (London: Interscience Communications), 999–1008; J. Quintiere, "A Predicted Timeline of Failure in the WTC Towers" in *Proceedings Interflam 2004* (London: Interscience Communications), 1009–22.

38. J. E. Malooly, "Fire as the Terrorists' Weapon" (Chicago, IL: American Academy of Forensic Scientists, February 2003).

39. L. H. Gold, "Psychiatric Profile of the Firesetter," *Journal of Forensic Sciences* 77 (October 1962); Sapp et al., *A Motive-Based Offender Analysis;* T. G. Huff, G. P. Gary, and D. J. Icove, *The Myth of Pyromania* (Quantico, VA: National Center for the Analysis of Violent Crime, 1997); M. Gardiner, *Arson and the Arsonist: A Need for Further Research,* Project Report (London: Polytechnic of Central London, June 1992).

40. American Psychiatric Association, *Diagnostic and Statistical Manual of Mental Disorders (DSM-IV-TR),* 4th ed. (Arlington, VA: American Psychiatric Publishing, June 2000).

41. Huff et al., *The Myth of Pyromania.*

42. A. O. Rider, "The Firesetter: Part II," *FBI Law Enforcement Bulletin* 49, July 1980.

43. N. D. C. Lewis and H. Yarnell, *Pathological Firesetting (Pyromania)*, Nervous and Mental Disease Monograph No. 82, (New York: Coolidge Foundation, 1951); J. James, "Psychological Motives for Arson," *Popular Government* March 1965; R. G. Vreeland and M. B. Wilter, *Psychology of Firesetting: Review and Appraisal* (Washington, DC: U.S. Government Printing Office, January 1979).

44. Williams, *Understanding the Arsonist*.

45. J. D. DeHaan, "Laboratory Aspects of Arson," *Fire and Arson Investigator* 29, no. 26 (March 1979).

46. J. E. Douglas et al., *Crime Classification Manual* (New York: Lexington Books, 1991).

47. J. D. DeHaan, "The Reconstruction of Fires Involving Highly Flammable Hydrocarbon Liquids," PhD dissertation, Strathclyde University, Glasgow, 1995.

48. P. J. Loscalzo, P. R. DeForest, and J. M. Chao, "A Study to Determine the Limit of Detectability of Gasoline Vapor from Simulated Arson Residues," *Journal of Forensic Sciences* 25, no. 1 (January 1980): 162–67.

49. T. E. Folkman et al., "Evaporation Rate of Gasoline from Shoes, Clothing Wood, and Carpet Materials and Kerosene from Shoes and Clothing," *Canadian Society of Forensic Sciences Journal* 23, nos. 2–3 (June–September 1990): 49–59.

50. N. Sester et al., "Evaluation of Gasoline Traces Transferred on Hands during Car Tank Filling" in *Proceedings European Academy of Forensic Science*, Cracow, Poland, 2000.

51. M. Darrer, J. Jacquenet-Papilloud, and O. Delemont, "Gasoline on Hands: Preliminary Study on Collection and Persistence," *Forensic Science International* 175 (2008): 171–78.

52. DeHaan, "Laboratory Aspects of Arson"; V. Babrauskas, *Ignition Handbook* (Issaquah, WA: Fire Science, 2003), 683.

53. U.S. Army, *Unconventional Warfare Devices and Techniques*, TM31-201-1, May 1986. Reprinted as *Improvised Incendiary Mixtures* by various publishers.

54. B. Turvey, *Criminal Profiling: An Introduction to Behavioral Evidence Analysis*, 3rd ed. (New York: Academic Press, 2008. R. Girod, *Profiling the Criminal Mind: Behavioral Science and Criminal Investigative Analysis* (Bloomington, IN: iUniverse, 2004); J. H. Campbell and D. Denevi, *Profilers: Leading Investigators Take You Inside the Criminal Mind* (Amherst, NY: Prometheus Press, 2004); J. Esherick, *Criminal Psychology and Personality Profiling* (Broomall, PA: Mason Crest, 2005); R. N. Kocsis, *Criminal Profiling: Principles and Practice* (Totowa, NJ: Humana Press, 2006).

55. J. Douglas and M. Olshaker, *The Anatomy of Motive* (New York: Scribner, 1999).

56. P. Horbert, "Profiling the Arsonist," paper presented at the Kansas City Arson Task Force Seminar, Kansas City, Missouri, May 1988; D. J. Icove and M. O. Soliman, *Arson Information Management System: User's Guide and Documentation* (St. Louis, MO: International Association of Arson Investigators, 1983); J. E. Douglas and A. E. Burgess, "Criminal Profiling: A Viable Investigative Tool against Violent Crime," *FBI Law Enforcement Bulletin,* December 1986; D. J. Icove and M. H. Estepp, "Motive-Based Offender Profiles of Arson and Fire-Related Crimes," *FBI Law Enforcement Bulletin,* April 1987; D. J. Icove, "Automated Crime Profiling," *FBI Law Enforcement Bulletin,* December 1986.

57. A. D. Sapp et al., *A Report of Essential Findings from a Study of Serial Arsonists* (Quantico, VA: National Center for the Analysis of Violent Crime (NCAVC), 1993).

58. Sapp, *A Motive-Based Offender Analysis*.

59. D. J. Icove, E. C. Escowitz, and T. J. Huff, "The Geography of Violent Crime: Serial Arsonists," *Agenda and Abstracts, 8th Annual Geographical Resources Analysis Support System, (GRASS) Users Conference,* Reston, VA, March 1983; K. Fritzon, "An Examination of the Relationship between Distance Traveled and Motivational Aspects of Firesetting Behavior," *Journal of Environmental Psychology* 21 (2001): 45–60.

60. D. J. Icove and J. D. DeHaan, *Forensic Fire Scene Reconstruction,* 2nd ed. (Upper Saddle River, NJ: Pearson/Brady, 2009), 255–59; 262–70.

61. D. K. Rossmo, *Geographic Profiling* (Boca Raton, FL: CRC Press, 2000); D. V. Canter and P. Larkin, "The Environmental Range of Serial Rapists," *Journal of Environmental Psychology* 13 (1994): 63–69; D. Canter, *Mapping Murder* (London: Virgin Books, 2003).

62. R. N. Kocsis, "Psychological Profiling of Serial Arson Offenses: An Assessment of Skills and Accuracy," *Criminal Justice and Behavior* 31, no. 3 (June 2004): 341–61; R. N. Kocsis, "Offense Location Patterns: Geographic Profiles," Chap. 10 in *Criminal Profiling: Principles and Practice* (Totowa, NJ: Humana, 2006).

63. Crime Point Fire and Arson, Crime Prevention Analysis Lab, Burnaby, BC, Canada.

64. K. Fritzon, "An Examination of the Relationship between Distance Traveled and Motivational Aspects of Firesetting Behavior," *Journal of Environmental Psychology* 21 (2001): 45–60.

65. B. Martinez, *Multiple Fire Setters: The Process of Tracking and Identification* (Tulsa, OK: PennWell Books, 2002).

66. R. R. Hazelwood et al., "Criminal Investigative Analysis: An Overview," in *Practical Aspects of Rape Investigation: A Multidisciplinary Approach,* 2nd ed., ed. R. R. Hazelwood and A. W. Burgess, 115–26 (Boca Raton, FL: CRC Press, 1995).

67. *ASTM E620: Standard Practice for Reporting Opinions of Technical Experts* (West Conshohocken, PA: ASTM, 2004).

68. L. Bilancia, "The Ignition Matrix," paper presented at California Conference of Arson Investigators Conference, March 2009.

69. *NFPA 921: Process of Elimination* (Quincy, MA: National Fire Protection Association, 2008), sec. 18.2, 156; and (2011), sec. 18.6.5, 174.

CHAPTER

17
Other Investigative Topics

*Used with permission of Retired Major
J. Ron McCardle, Florida State Fire
Marshal's Office.*

KEY TERMS

seat of fire, *p. 713*

OBJECTIVES

After reading this chapter, you should be able to:

- Identify specialized tools that are relevant and useful in fire investigation.
- List common safety hazards encountered at fire scenes.
- Discuss the use of fire modeling and the questions to be asked when selecting and using a model.
- Select a particular fire model and discuss how it meets the various ASTM standards.
- Discuss the elements of proof necessary in investigating and prosecuting arson case investigations.
- Inventory various sources of information available to fire investigators.
- Explain the concept of spoliation as it applies to fire investigation.
- Identify the elements necessary to include in a comprehensive fire investigation report.
- Discuss the requirements for expert witness testimony, scientific method, and exhibits.
- Search the Internet for expert witness requirements in your jurisdiction in both state and federal court.

Safety and Health

It is important that the fire scene investigation be carried out thoroughly, professionally, and safely. There are many dangers to a fire scene—structural collapse, electricity, toxic fumes and dusts, water-filled holes, fire-weakened floors and stairs, falling glass, and other penetration hazards are just a few typical ones. The preliminary scene assessment should include a survey of the hazards the scene presents, since some of them may have to be controlled before the detailed scene examination can even begin. This is a requirement cited by *NFPA 1033: Standard for Professional Qualifications for Fire Investigator.*[1] Steps may have to be taken, for instance, to shore up or even demolish unsupported walls (particularly masonry), roofs, or ceiling structures, remove large glass panels from upper floors, and pump out standing water. The assessment can be carried out with the assistance of the fire department safety officer, a county or city building engineer, a chemist with hazardous materials experience, or even with experienced demolition contractors. The scene in Figure 17-1 illustrates multiple hazards typical of fires in older communities: a multistory masonry building that has been reduced to a facade and partially collapsed interior walls (all a collapse risk), severe ice and cold weather conditions, a multiuse building with undetermined contents, and the usual risks of multiple electrical systems and raw sewage.

The preliminary scene assessment should include a survey of the hazards the scene presents.

FIGURE 17-1 Ice-covered remains of fire-damaged multistory building. Hazards include structural collapse, ice, cold, multiple electrical systems, unknown building contents, and open sewer lines. *Courtesy of Don Perkins, Fire Cause Analysis.*

The investigator should come prepared to deal with the simplest safety problems by having suitable protective clothing (simple overalls or synthetic fabric suits for hazardous material exposure), gloves (leather for penetration protection, heavy rubber for chemical exposure, latex for debris handling), safety helmet with face shield, boots with safety toes and sole plates, and masks for both dusts and toxic fumes. Dusts (charcoal dust, asbestos, plaster, etc.) are present at every dry scene and pose significant risks of inhalation. Simple, disposable nuisance-dust masks should be used at all dry scenes or whenever there is a risk of exposure to airborne biological pathogens (blood, dried tissue). Mask-type respirators with appropriate filter cartridges can be used at scenes with modest levels of some toxic fumes, but some scenes may require full self-contained breathing apparatus (SCBA) such as shown in Figure 17-2.

Carbon-fiber composite materials, once solely found in aircraft components are now found in a wide variety of automotive applications to reduce weight while increasing strength. Combustion of such materials produces gaseous and aerosols of toxic pyrolysis products from the polymers involved as well as airborne microscopic fibers that are small enough to be inhaled or enter the skin. While the long-term health risks of inhaled fibers are not clearly defined, fire investigators need to be aware of inhaling any potentially toxic aerosol chemical and take appropriate steps to minimize their exposure.[2]

Specialized training is needed to properly deal with biological (sewage, infectious waste) and chemical (toxics, corrosives) hazards, and the investigator should seek assistance from qualified personnel before risking injury. Some fire departments will insist that the carbon monoxide (CO) levels be checked in a building before an investigator is allowed in. Clip-on personal carbon monoxide monitoring devices are useful at "fresh" scenes when CO may still be accumulating from residual smoldering.[3]

Electrocution is a constant hazard in both buildings and even outdoors. In older buildings with multiple occupancies, conversions, floating neutrals, and even "back fed" power, it is very dangerous to assume that because the overhead service outside the front door has been disconnected, there is no danger inside. The investigator is encouraged to assume electrical lines are still "hot" until proven otherwise and to use tools such as voltage probes (Figure 17-3) to check individual outlets and connections.

Stay Safe

The investigator is encouraged to assume electrical lines are still "hot" until proven otherwise.

FIGURE 17-2 Asbestos was one of the hazards identified in this large scene. Full "hazmat" precautions were initiated, which required all investigative agencies and support services to wear personal protection equipment. Fire investigators and fire engineers from the Office of the Fire Marshal were on scene for five days. *Courtesy of Pierre Yelle, Office of the Fire Marshal, Ontario, Canada.*

FIGURE 17-3 Voltage probes help investigators establish whether wires or equipment are energized. On left: Non-contact type, battery powered. On right: Contact-type AC or DC up to 600 V, no batteries needed. *Courtesy of John D. DeHaan.*

Large industrial or commercial sites may have large battery arrays to serve as uninterruptible power supplies (UPSs) in the event of utility failures. A new danger today is the increasing installation of solar cell power arrays. Such sources are often back-fed into the line supply via a separate shutoff and can deliver significant electrical power to the system of either a home, business, or industrial plant. Because that power may not be disabled when the main breakers are thrown, there is a significant electrocution risk.

Scene examinations should not be carried out alone, and some means of emergency communications should be at hand (cell phone, radio, or telephone). Severe weather, such as ice or high winds, even darkness, makes scenes unduly dangerous as well as obscuring critical evidence and is a justifiable reason for postponing a search. No scene is worth risking death or serious injury. A comprehensive guide to safety and health issues for fire investigators has been published by IFSTA.[4] The Fire Protection Association (UK) has also published an excellent guide to safety at fire scenes, and the reader is referred there for more complete information.[5] Safety considerations are also outlined in *NFPA 921: Guide for Fire and Explosion Investigation*.[6] It should be noted that if any governmental body declares a fire scene to be a hazardous material site, all responders must follow Occupational Safety and Health Administration (OSHA) regulations. OSHA has authority to enforce stringent safety regulations at a fire scene if employees are involved (including mandatory annual training),[7] and local building authorities in many jurisdictions have the authority to prohibit examination until a scene can be made safe. The investigator is encouraged to seek counsel on these matters before launching the examination.

No scene investigation is worth risking death or serious injury.

Fire Modeling

Fire modeling can be in the form of bench-scale (miniature) physical models into which are introduced dyed liquids, smoke, or hot gases to re-create the flow of smoke and heat or in the form of mathematical analyses. Because the use of physical scale models is rare in fire investigations, this text focuses on mathematical fire models.

MATHEMATICAL FIRE MODELING

Mathematical fire models help us understand complex fire processes, such as the relationship of heat release rate to other factors, including ventilation and heat of vaporization. Such methods also can help us relate post-fire indicators to fire events and behaviors and can help us isolate and analyze unknown factors in fire, such as for example, the effects of ventilation, ignition location, and fuel type.

Examining these variables with a fire model is much easier and more cost-effective than constructing and burning life-size models. Further, such modeling is entirely consistent with, and indeed at the heart of the scientific method—creating (or discerning) alternative hypothesis and testing them.

Fire modeling also helps satisfy the court's demand that fire investigation, analysis, and conclusions follow and utilize the scientific method. Mathematical fire modeling is at least one way the "error rate"—or at least the effect that unknown or unknowable variables may have on estimations of fire processes such as size, rate of growth, likely path of fire travel or spread, the likelihood of flashover, and other variables—can be measured and explored and more effectively explained to the court and/or trier of fact.

Mathematical fire modeling can be as simple as using the algebraic equations for flame height, radiative heat loss, heat transfer, and others (some of which were discussed in this book). These simple calculations can be done on a pocket calculator with scientific functions. Such calculations have also been codified in software packages such as FPETOOL (developed at NIST) and ASKFRS (developed at the Fire Research Station, UK) or Fire Dynamics Tools (NRC). They are PC compatible and available from those agencies for download or purchase.

Since the late 1980s dozens of computer models have been introduced as a way of understanding the fire environment. A survey by Factory Mutual Research in 1991 revealed that there were 31 computer programs available to study fires in compartments and another 31 to predict special behaviors like smoke and fire travel, sprinkler activation, or evacuation.[8] Today, there are many more. (The reader is referred to *www.firemodelsurvey.com* for current models.) Some of these models will run on a personal computer, others require a mainframe or even a supercomputer.

ZONE MODELS

Many fire models in use today are *zone models*; that is, they predict behavior based on movement of gases in the two large zones normally found in a developing compartment fire: a hot gas layer (ceiling) and a normal air layer (floor). ASET-BX (Available Safe Egress Time-Basic), FPETOOL, and CFAST (Consolidated Fire and Smoke Transport) or FASTlite are some of the most commonly used zone models.[9] (A more comprehensive set of programs, HAZARD-I, is also occasionally used.) These models calculate the temperature and depth (height) of the hot smoke layer interface in rooms (sometimes in a series of rooms) and are useful in assessing fire and smoke spread, especially where occupant escape is a factor. The predicted temperature of the hot smoke layer is used to predict when a room is untenable and when it is likely to progress to flashover. CFAST has been recently upgraded by NIST (Version 6.1.1) to become a very powerful and well-validated model. Many programs are available for free download from the Internet.

FIELD MODELS

A few models like JASMINE (Analysis of Smoke Movement in Enclosures) or SOFIE (Simulation of Fire in Enclosures) are field models, which look at the energy and mass transfers occurring in numerous very small volumes of space in a room.[10] The computing power needed to run these field models (sometimes called CFD, or Computational Fluid Dynamics, models) often used to limit them to mainframe computers, but with today's powerful personal computers, they are more widely used. The Center for Fire Research (now the Building and Fire Research Laboratories) at NIST developed a very advanced CFD field model called the Fire Dynamics Simulator (FDS). FDS uses the thermal and fire behavior properties of furnishings, ceiling and wall linings, and floor coverings to calculate fire growth and spread. A second NIST-developed program, called Smokeview, is a visualization program that displays the results of an FDS model as either snapshots or as two- or three-dimensional animations.[11] (See Figures 17-4 and 17-5 for examples.) Some newer zone models have been developed such as BRANZ-FIRE, which predicts and evaluates performance and hazards of complex environments such as combustible room linings, and FRS's CRISP (Computation of Risk Indices by Simulation Procedures), which evaluates multiroom fire scenes using a two-layer zone model but incorporates complex models of human behavior and movement.[12] Many of these programs are valid only up to flashover because the massive turbulence created by full involvement overwhelms the zone models. Some cannot deal with multiple rooms or changes in ventilation, so users must be aware of limitations that may make some programs unsuitable for studying some fires.[13]

Fire modeling has been used by NIST in numerous forensic engineering analyses and reconstructions since the mid-1980s. For a more detailed review of these cases, the reader is referred to Table 6-5 in Chapter 6, "Fire Modeling," in *Forensic Fire Scene Reconstruction*.[14] This textbook also addresses the other uses of fire modeling, such as in detector activation, witness observations, and the determination of the origins and causes of fires by comparing visual fire pattern damage with predicted temperature and heat flux to exposed compartment surfaces.

FIGURE 17-4 FDS prediction of the temperature zones of The Station nightclub 90 seconds after ignition of combustible foam insulation on stage. (Top) Without sprinklers (prediction compared against video recording of actual fire). (Lower) Same scenario with normal sprinklers functioning. *Courtesy of NIST.*

FIGURE 17-5 Smokeview (FDS) prediction of the fire and smoke conditions inside The Station 60 seconds after ignition. (Top) Without sprinklers. (Lower) With sprinklers. *Courtesy of NIST.*

MODELS FOR SPECIALIZED APPLICATIONS

Models for specialized applications include the following:

- *Post-flashover*   COMPF, SFIRE-4 (time–temperature history for energy, mass and species—evaluating structural integrity in fire exposure in post-flashover rooms)
- *Thermal and structural response*   FIRES-T3, HEATING7, TASEF (finite element calculations about structural fire endurance of a building or particular components using known failure conditions such as loss of tensile strength as a function of temperature)
- *Fire protection*   DETACT-QS, DETACT-T2, LAVENT (sprinkler and detector response times for specific fires)
- *Smoke movement*   CONTAM96, Airnet, MFIRE (dispersion of smoke and gaseous species)
- *Egress*   lsafe, buildingEXODUS, EESCAPE, ELVAC, EVACS, EXITT, EXIT 89, EVACNET, WAYOUT, SIMULEX [probabilistic (stochastic) modeling of escape of people from fire using predicted smoke conditions, and occupant and egress variables)]

- *Glass breakage*   BREAK1
- *Wildland*   BEHAVE Plus, Prometheus, FARSITE

ASTM AND CRITICAL MODELING ISSUES

It is particularly important to point out to potential users of fire models that there are existing ASTM guides that deal with critical modeling issues. Those most relevant to our discussion here are the following:

- *ASTM E1355-05a: Standard Guide for Evaluating the Predictive Capability of Deterministic Fire Models*
- *ASTM E1591-07: Standard Guide for Obtaining Data for Deterministic Fire Models*
- *ASTM E1895-07: Standard Guide for Determining Uses and Limitations of Deterministic Fire Models*
- *ASTM E1472-07: Guide for Documenting Computer Software for Fire Models*

All these ASTM standards contain vital information and recommendations for someone contemplating the use of a fire model. As they are published by ASTM International (and developed by ASTM Committee E-05 on Fire Standards), they represent peer-reviewed guides with which all fire model users should be familiar. They outline the documentation (of both the scene and the model) necessary for demonstrating the reliability of models used. They are intended for evaluation of zone models but are applicable to field models as well.

WHAT SHOULD WE ASK ABOUT ANY MODEL WE USE?

Fire investigators should be able to answer the following questions before using any model, by applying the criteria in the ASTM guides and evaluating the case at hand:

- Is it applicable?
- Is it the right tool for the job?
- Does it give accurate results?
- How often does it predict events that do not occur in real fires?
- How sensitive is it to changes in input?
- What is its error rate?
- Has it been used to predict events in real fire tests?
- Has it been validated?
- Where did it come from?
- Where was it published?
- What supporting (or contradictory) data have been published?

If we use such models, we quickly realize that the input data they depend on are usually significantly more extensive than investigators are used to gathering. This paucity of data is more often than not the result of careless or incomplete documentation of the scene. This problem has recently been addressed by the inclusion of recommended data collection forms in Appendix J of this text and in *Forensic Fire Scene Reconstruction*,[15] and NFPA 921. It is a rare scene that is so completely destroyed that basic dimensions, structural and finish materials, and furnishings (type and placement) cannot be established by careful examination. Even in instances where there has been massive destruction, interviews, examination of nearby "exemplar" structures, or recovery of pre-fire photos or videos, can often fill in many of the missing pieces.

NIST gives the following warning to users of fire models, particularly their Fire Dynamics Simulator (FDS):

> Users are warned that FDS is intended for use only by those competent in the fields of fluid dynamics, thermodynamics, heat transfer, combustion, and fire science, and is intended

only to supplement the informed judgment of the qualified user. The software package is a computer model that may or may not have predictive capability when applied to a specific set of factual circumstances. Lack of accurate predictions by the model could lead to erroneous conclusions with regard to fire safety. All results should be evaluated by an informed user.[16]

FIRE ASSESSMENT

As outlined in *ASTM E1895-07,* for instance, the user's first step should be to define the scope of the fire assessment and then determine whether fire modeling is an appropriate tool.[17] Then, the user should determine which models are available and suitable to run on the available computer hardware considering the size and complexity of the problem. For the models being considered, the available documentation should be acquired and evaluated in terms of guidance offered in *ASTM E1472-07: Guide for Documenting Computer Software for Fire Models.*[18] The limitations of the candidate models must be compared with the problem to be solved—one-room versus multiroom, pre-flashover versus post-flashover.

Although it is possible to modify existing models to deal with particular problems, any modifications must be made in cooperation with the original model developer and then subjected to suitable validation as outlined in *ASTM E1355-05a: Standard Guide for Evaluating the Predictive Capability of Deterministic Fire Models.*[19] Other tools, such as small- or large-scale fire tests or mathematical calculations, should be considered as well.

Once a model is selected, the following steps are recommended:

1. Verify the known limitations of the model—room dimensions, fire size, or ventilation.
2. Determine the underlying assumptions (two-layer zone or CFD/field model) and assess their impact on the results.
3. Determine the characteristic variables.
4. Determine what input data is required and where it can be obtained.
5. Determine the rigor of the mathematics involved and check to see whether it will give an answer given the constraints of the problem.
6. Determine the extent of validation to establish its appropriateness for the problem. Validation processes are described in *ASTM E1355.*
7. If validation data are not available, conduct a sensitivity analysis to establish the effect of changing critical variables.
8. Thoroughly document the model "run," including all input data, all assumptions made, and any and all modifications (including validation to support the accuracy of those modifications).

DOCUMENTATION

The documentation for a computer fire model should include a technical guide or user's manual (as described in *E1472*). The source code for the model should also be made available to any potential user. Some well-known programs, such as FPETOOL and FDS, are available for downloading from NIST at no charge. Other programs, such as BRANZ-FIRE, SIMULEX and ASKFRS, must be purchased as a "user license" for a given period of time. The assumptions used, known numerical and physical limitations, and the physical and mathematical treatments used must also be made available to the user. An excellent source of current information is the www.firemodelsurvey.com website, where information about more than 150 models is available. It describes the abilities and limitations of current fire models. A summary of this information was published by Olenick and Carpenter.[20]

A good example of a documentation guide for a fire modeling package is *FPETOOL: Fire Protection Engineering Tools for Hazard Estimation,* by H. E. Nelson.[21] FPETOOL is a collection of analytical tools about fire behavior and properties with simplifications (as assumptions) to make approximations rather than exact predictions using portable or desktop computers. It consists of the following three main elements:

- *Fireform (Fire Formulas)*   A collection of fire safety calculations
- *Makefire*   A series of procedures to produce fire input data files for use with Fire Simulator
- *Fire Sim*   An integrated set of equations (i.e., a model) designed to allow the user to create a fire case study in a Lotus spreadsheet format with specifications of room and vent dimensions; fuel characteristics; and ceiling, wall and floor materials; and input fire to predict layer temperature, flashover, and tenability factors

The FPETOOL guide describes the main elements of the package, its hardware and software requirements, mathematics, underlying assumptions, and comparisons of FPETOOL (Fire Simulator routine) with fire test data. A separate user guide was published as *FPETOOL User's Guide*.[22]

The FIRM-QB zone model developed by Marc Janssens has a technical description, program description, and user's manual all included in his excellent book, *An Introduction to Mathematical Fire Modeling*.[23] The entire program code and supporting documentation are included in a CD-ROM packaged with the book.

MODEL EVALUATION

According to *ASTM E1355,* the evaluation process consists of four steps:

1. Define the scenarios for which the evaluation is to be conducted.
2. Validate the theoretical basis and assumptions used in the model.
3. Verify the mathematical and numerical robustness of the model.
4. Evaluate/quantify the uncertainty and accuracy.

ASTM E1355 offers the following definitions regarding models.

Evaluation   The process of quantifying the accuracy of chosen results from a model when applied for a specific use.

Validation   The process of determining the correctness of the assumptions and governing equations implemented in a model when applied to the entire class of problems addressed by the model.

Verification   The process of determining the correctness of the solution of a system of governing equations in a model. Verification does not imply the solution of the correct set of governing equations, only that the given set of equations is solved correctly.

The steps identified in *ASTM E1355* are not isolated ones. As Janssens points out:

Step 4 is usually based on a comparison between model output and experimental data and provides an indirect method for validation (Step 2) and verification (Step 3) of a model for scenarios of interest (Step 1). It is generally assumed that the model equations are solved correctly and the terms validation and evaluation are therefore often used interchangeably. It is very rare for anyone but the model's developer to spend the time necessary to carry out steps 2 and 3, although an independent reviewer or researcher may. Portions of the mathematics are sometimes compared to other analytical results. Step 4 can be carried out by comparing a model's predictive results to full-scale tests done specifically for that purpose, tests done by others and published in the literature, results of standard room fire tests done in accordance with *ASTM E603-07*, or even against observations or reconstructions of real fires (historical fire data).[24]

A good example of a comparison of a compartment fire test against the predictions of mathematical calculations, zone models, and a field model was published by Spearpoint, Mowrer, and McGrattan in 1999.[25] It showed how accurately calculations and zone models (Fire Simulator, FAST, and FIRST) agreed with test data. The fire model used was ES3D (a predecessor to FDS). Its predictions (including early fire development) were not as accurate as the calculations and indicated further development was needed.

In a more recent study conducted by Rein, Bar-Ilan, Fernandez-Pello, and Alvares in 2006, the researchers compared three models for the simulation of three accidental fire scenarios in a single-family dwelling.[26] The modeling approaches used were an analytical model of fire growth, the CFAST zone model, and the FDS field model. The research found that in spite of the differences in these three modeling approaches, the results were found in relatively good agreement, particularly in the early stages of the fire.

Model uncertainty is based on repeated runs of similar data and sensitivity analyses to identify critical data. For a complex model with many inputs, it is usually prohibitively time-consuming to run repeated runs with different single inputs. Mathematical techniques have been used to reduce the computer time needed for each run (such as linking multiple personal computers to conduct parallel processing).

The accuracy with which FDS predicts temperatures and heat release rates has been validated by large-scale fire tests. Testing has shown that FDS temperature predictions were within 15 percent of the measured temperatures, and heat release rates were within 20 percent of measured values. Results, however, are often presented as ranges to account for some uncertainty. FDS (Version 5) is the primary fire-modeling tool used by NIST today and has been used in major fire investigations involving large losses and death.[27]

When model results are compared with a full-scale test fire, it is usually assumed that the real-fire data are the gold standard. The complexities of the room fire environment and the variables of turbulence make it impossible, however, to get exactly the same measurements when the real fire is repeated. Estimates of the uncertainties inherent in some measurements can be as high as ± 30 percent.[28]

For example, work by Beard evaluating the Dalmarnock Tests in 2006 looked at modeling a test fire started in a sofa in a two-bedroom apartment in a high-rise residential complex.[29] Fire modeling users were given the details of the apartment, arrangement, and materials but not the experimental results. The users were asked to predict the time courses of variables such as the heat release rate, temperatures, and smoke obscuration. Of the ten modelers, two used the CFAST zone model and eight selected the FDS field model. Beard noted that generally the predictions fared poorly in comparison with the actual experimental results.

However, recent work sponsored in part by the NRC examined the verification and validation (V & V) of fire models, including CFAST and FDS.[30] Although the NRC's study centered on fire hazards particular to nuclear power plants, it addressed many of the concerns raised by those who think fire models are not accurate or appropriate for forensic fire scene reconstruction. These concerns include the ability of the models to accurately predict common features of fires, such as upper-layer temperatures and heat fluxes. One of the features of this report was a comparison between actual fire test results and predictions of hand calculations, zone models, and field models. As shown in Figure 17-6, there is generally good agreement among all the models and the variability of real-world fires.

Experienced fire scientists realize the limitation of applying models to real-world fire investigations. These models, in general, are designed to start with the ignition of a fire under preset conditions and predict the time factors and conditions of growth and, sometimes, decay. They are not designed to re-create a particular fire by working backward from a set of final observations to determine what the starting (or even intermediate) conditions were. Models can be used for field investigations, but the process begins with what amounts to an initial guess about the starting conditions (a scenario), and then variants of that scenario (created by changing the input data) are analyzed until the results compare favorably with the post-fire conditions observed. Application of a model results in one set of conditions that could have produced the observed results but not necessarily

FIGURE 17-6 Comparison of hot gas layer temperatures measured in full-scale tests compared with predictions of hand calculations (□), zone models (•), and FDS models (♦). A predicted ±13 percent variability range is included. Note that the hand calculations tended to overpredict layer temperatures, whereas both zone and field model predictions were generally within variability limits. *Adapted from M. H. Salley, J. Dreisbach, K. Hill, R. Kassawara, B. Najafi, F. Joglar, A. Hammins, K. McGrattan, R. Peacock, and B. Gautier. "Verification and Validation: How to Determine the Accuracy of Fire Models." Fire Protection Engineering (Spring 2007): 34–44.*

the actual set of conditions that did exist. The investigator must remember that a model is just that—a model—and not proof of how a particular fire happened. It is essential to understand that any model, whether it is a simple calculation (such as the Fire Dynamics Tools spreadsheets[31]), a simplified model with a few values to input (like FPETOOL or ASKFRS), a predictive model of general fire behavior in a room, or a complex zone model of fire spreading in multiple compartments requiring extensive data, if given faulty data will give faulty results. Note that some of the more advanced models will accommodate a change in ventilation (such as a door opening or a window breaking (note that this means not simply cracking or "failing" but breaking out so as to materially affect the ventilation conditions in the room).

Any of these models can be forced to give misleading results to fit the post-fire appearance by manipulating the input values. This manipulation may not necessarily be intentional but may be due to lack of knowledge and skill on the part of the user. Some of the models make assumptions about the size, shape, ventilation, and other features about the room to give a general prediction. Many will ask for specific data about the room: its dimensions, its ventilation, contents, building materials, and other factors. The investigator must be prepared to provide these data, which are rarely collected during a scene investigation. A data collection form (a sample is shown in Appendix J) will help the investigator gather critical data for later analysis and reconstruction, even if modeling is not carried out. If these data are not supplied, many programs will assign a default value, and the user must then be fully aware of what the default values are, what they mean, and what effect they will have on the prediction if the actual fire scene values are different.

Many of the simpler models cannot accommodate changes during fire development. Because one rarely investigates a fire whose ventilation conditions do not change during the fire (windows failing, doors opening), such models are of limited use in actually predicting the final result. Very few of the computer models will accurately predict the post-flashover behavior of a fire because of the extremely complex behavior of the turbulent gases, and the user must be cautious not to rely on predictions past flashover (even if the computer model is willing to offer them!). Some of the programs have a built-in warning that advises the user that conditions are likely to produce flashover and requests specific acknowledgment to proceed. Few of the models will work on smoldering fires, even though they are sometimes the most dangerous because of the toxic gases generated. Some of the models require the user to estimate the size (HRR) of the initial fuel package. This applies only to open-flame progression, not smoldering initial fires (in furniture, for example). Fortunately a great deal of data are available regarding the heat release curves for a variety of furniture and other first-fuel packages.[32] Some models will not accommodate unusually intense fires (more than 2 or 3 MW), or ultra-fast-developing (accelerated) fires as input. A few will default to a "standard time–temperature curve" model for the input fire. This default is acceptable when one is trying to evaluate the influence of some factor later in the fire, but very few real-room fires behave as a smoothly increasing curve. Instead, they are a series of steps, with each step representing the HRR of another fuel package as it ignites, decaying as that package burns away. As we saw in Chapter 3, if the steps are large enough or occur quickly enough, the fire will grow to full involvement.

TESTING COMPLEX COMPUTER MODELS

One of the major differences between zone and field models is that field models often include routines that calculate the growing fire based on first principles of thermal response, heat flux, and flame spread. Of course, this requires that the initial and secondary fuels be identified and their physical and thermal properties carefully defined as input data.

The limitations of this text are such that detailed descriptions of complex fire models cannot be included here. The reader is referred to the references included here or to the www.firemodelsurvey.com website. The selection of which model to use requires the analyst to ask the following questions about the model:

General

- Appropriateness—is the model's output useful and applicable?
- Limitations (time, ventilation, output)
- Resources—computer speed and capacity needed
- Experience needed
- Input data needed—what default conditions apply if input data are incomplete and how do those presumed defaults affect the results?
- Sensitivity—what happens to output when input data are changed?

Accuracy

- Are the results/outputs realistic? Would events occur that way in a real fire?
- Has the model been used to predict the outcome of a test burn—such as tenability, temperatures, or time to flashover? How accurately did it predict the actual fire results?
- Was the model "fine-tuned" to make its predictions more accurate? (Was the program run multiple times with different data to slightly "tweak" the result?)
- How were the data collected in the test burn? Direct observation, thermocouple measurements, radiometers?
- Where were the measurement/observation point(s)?

As noted previously, the fire environment can be so complex that temperature or radiometric data collected in one location may not be representative of the entire room, leading to possible variations of as much as ±30 percent.

Reproducibility

- If the program is run with the same data by the same person, does it give the same answer?
- If the program is run with the same data by a different person, does it give the same answer?

Robustness

- Is the program applicable to different situations? Has it been tested and demonstrated to give accurate results if starting conditions are very different?
- Has it been shown to give reliable results for:
  - Small fire in a big room versus large fire in a small room?
  - Adequate ventilation versus underventilated?
  - "Ultra-fast" t^2 fire versus "medium" t^2 fire?
- Was the testing done by "game playing" or thorough testing of alternative hypotheses?

Analysis

- What features of fuel, starting conditions, initial fire size are required input data?
- What assumptions have been made by the analyst based on personal judgment or "bias"?
- How do such assumptions affect the prediction?

Generally, if the models are used by knowledgeable users with good input data, they are capable of assessing the probable events for a fire scenario. Although precise predictions are not the intended end product of most fire modeling programs, they can be useful for making general estimates about a given fire and for evaluating the effects of what-if hypotheses.[33] Very often in fire investigations, the time involved is the most critical factor, and yet it is one of the most elusive. Models offer an opportunity to change one variable at a time in a particular fire and see what effect it may have on time of development. Evaluations currently being carried out to compare the results of computer model predictions with real-world fires have demonstrated generally good agreement with full-scale lab tests.

Critical Analysis of Cases

To ensure the appropriate application of computer modeling to fire scenario hypotheses can best be approached using *critical analysis* or *critical thinking*, a field of study popularized by Dr. Richard Paul and Dr. Linda Elder.[34] Critical thinking improves the quality of the thought process by providing structure and intellectual standards, thus enabling the person conducting the fire model to make appropriate and rational decisions in correctly choosing, applying, and interpreting the results.

Paul and Elder suggest that eight elements of reasoning be applied to any scientific thinking process. As with the application of the scientific method's iterative process, each element assists in eliciting inferences that assist in generating or considering new evidence necessary to answer questions or solve problems. The elements are summarized next with illustrations of their application to fire modeling.

1. *Purpose: Scientific reasoning has a purpose to its approach.* State the purpose of the chosen fire model, distinguish the basis of its selection, and periodically check to ensure that it indeed was the best choice. (*Example:* The choice of a field model over a zone model to evaluate smoke distribution and toxicity.)

2. *Questions: Scientific reasoning attempts to either answer questions or solve problems.* Clearly and precisely state the question to be answered by the model, break the question into subquestions, and determine whether the question has one right answer or requires additional reasoning taking into consideration other hypotheses or points of view. (*Example:* Determination of tenability based upon subproblems of smoke, heat, and concentrations of noxious by-products of combustion.)

3. *Information: Scientific reasoning is based upon data, information, and evidence.* State and document what information or data are being used to feed the model, determine whether sufficient data are available, and describe what experience you have had to

validate these data. Make sure you also search for data that oppose your opinion, and develop alternative hypotheses. (*Example:* The selection of heat release rate curves based upon the availability of past historical testing versus data generated through actual test burns using similar materials.)

4. *Inferences/Conclusions: Scientific reasoning uses inferences or interpretations to draw conclusions and thus give new meaning to scientific data.* Clearly describe how and by what assumptions the conclusions were reached, check that the inferences are consistent, and determine if there is another way to interpret the information or arrive at the same conclusion. (*Example:* Comparison of burn patterns based upon visual observations with patterns predicted from field modeling.)

5. *Concepts: Scientific reasoning is expressed and shaped by concepts and theories.* Identify and clearly list the key scientific concepts that underpin your hypothesis and model. Recheck that the scientific precision is bounded, and consider alternative concepts. (*Example:* The knowledge of smoke distribution theory along with ample scientific testing, on-scene observations, and witnessed tests.)

6. *Assumptions: Scientific reasoning is often based upon assumptions.* Clearly and precisely identify your assumptions, and assess whether they are justifiable. Be aware of how your assumptions shape your point of view. (*Example:* The positions of doors, windows, and other paths of leakage out of a compartment in calculations of the minimum heat release rate for flashover.)

7. *Implications/Consequences: Scientific reasoning has both implications and consequences that follow from data and reasoning.* Consider all possible implications and consequences, trace the reasoning paths, and search for both negative and positive implications. (*Example:* The use of three separate fire models to test the hypothesis of the actual activation time of a smoke detector.)

8. *Points of View: Scientific reasoning is done from a chosen point of view.* Consider what your point of view is and whether it has scientific footings. Remember to identify and consider other scientific points of view along with their respective strengths and weaknesses. (*Example:* The determination that additional fuels were needed to cause flashover in a near-vacant compartment and not taking into consideration fuels contained in the walls, flooring, and ceiling.)

Examples of the use of critical thinking are well-founded in other fields.[35] Its use in assessing, choosing, and evaluating fire models and their results can clearly enhance the results of their application.

Search and Seizure

Evidence to be used in fire investigation is subject to laws of search and seizure. Evidence that was unlawfully seized cannot be used in court. By extension, any information developed as a result of an illegal search must also be held inadmissible. This principle is called the *exclusionary rule,* and it controls all evidence to be used in criminal prosecutions. The legality of a search depends on the court's interpretation of how a fundamental principle of the U.S. Constitution, expressed in the Fourth Amendment, applies to the case:

> The right of the people to be secure in their persons, houses, papers, and effects against unreasonable searches and seizures shall not be violated, and no warrants shall issue, but upon probable cause, supported by oath or affirmation, and particularly describing the place to be searched, and the persons or things to be seized.

It is essential that every investigator appreciate the requirements for lawful searches of fire-damaged property and abide by them to ensure that the evidence gathered, analyzed, and presented will be held valid by the court. It was previously mentioned that it is necessary for the fire service to maintain control and custody of the scene until the investigation is completed. In light of recent Supreme Court decisions, there are several avenues to a permissible search:

■ The fire department has a public obligation to investigate all fires (not only to identify arson but to prevent accidental fires). The investigation should begin promptly when the fire is actually extinguished, but it need not be carried out immediately if conditions such as darkness, smoke, or fumes would prevent a full examination. Its start may be delayed for a reasonable amount of time (and be carried out for a reasonable time) after extinguishment *as long as custody and security of the scene are maintained*. The mere custody of the scene by a fire department, however, does not ensure that a court will not find a delayed start search to be in violation of the Fourth Amendment.

■ The investigation can be carried out immediately in the absence of a search warrant if there is a high probability that evidence of its causation will be lost (due to flames, water, or deliberate destruction or removal) if the search is delayed (so-called exigent circumstances).

■ A signed consent can be obtained if the person legally responsible for the property can be located. A prepared consent form such as that shown in Figure 17-7 is suggested.

CFD-OFI CASE FILE #_____

Chicago Fire Department
Office of Fire Investigations

Consent Form For Fire Scene Examination

I, _____ , have been requested to
          (name)

consent to an examination of my property located at_____

_____

_____ ,
         (full description and exact address of property)

This fire scene examination will be conducted in order to determine the origin, cause and circumstances surrounding a fire/explosion which occurred on this property on

_____ , _____ at _____ .
  (day)   (date)   (time)

I am the lawful owner/occupant/agent of this property. I have been advised of my constitutional rights to refuse any further entry, and to require that a search warrant be obtained prior to any examination.

I have been further advised that if I do consent to an examination, any evidence found as a result of such examination can be seized and used . . . . in a court of appropriate jurisdiction, and that I may withdraw my consent at any time prior to the conclusion of the fire/explosion scene examination, and authorize:

_____ _____ _____
(name)   (rank or title)   (badge or star#)

Chicago Fire Department — Office of Fire Investigations

(full name of authorized agency conducting examination)

to conduct a complete fire/explosion examination of the herein described property. Further permission is granted to remove from this property any document, papers, objects or material deemed pertinent to the investigation of this fire/explosion.

_____   _____
(signed)   (date)

Witnesses:

_____   _____
(signed)   (date)

_____   _____
(signed)   (date)

FIGURE 17-7 Consent to Search form. *Courtesy of Chicago Fire Department, Office of Fire Investigations.*

Permission must be given voluntarily; any coercion used may invalidate the search and its evidence. Consent searches can be thought of as limited by the owner's right to refuse, the right to restrict the area being searched, and the right to withdraw permission at any time during the search. Failure to recognize these limitations may make a "consent to search" invalid (or at least restrict the admissibility of evidence seized).

- It is permissible to search a property that has been abandoned by the owner, however, this circumstance may be virtually impossible to prove and is therefore rarely used.

- Evidence may also be seized legally if it is in open view (from a point where the public would have legal access), if it is in plain view of a police officer carrying out his or her normal duties (including hot pursuit of an individual into his or her home), or if it is in an area under the immediate control of, or accessed by, a person being placed under arrest.

- In many jurisdictions, administrative or inspection warrants may be issued by local authorities. Such a warrant is an order, in writing and signed by a judge, directed to a state or local official, ordering that an inspection required or authorized by law or regulations be carried out. An inspection warrant is usually issued only on cause supported by an affidavit, particularly describing the place to be searched and its purpose. It is usually issued for a limited period of time, and the ordered inspection may be prohibited during certain hours or without advance notice. In some cases, such a warrant is considered to be purely administrative in intent, and if clear evidence of a criminal origin is seen in plain view as a result of such a search, that evidence may be seized, but a criminal search warrant must be obtained before the search can proceed any further.

- A full criminal search warrant may be sought. It is an order in writing, signed by a magistrate, that directs a peace officer to search for particular property and bring it before the court. In most jurisdictions, a search warrant requires that cause be shown, supported by an affidavit of facts in hand. The affidavit must describe the person, property, and place to be searched and include a description of the evidence to be sought. A full description of search warrant procedures is offered in the literature.[36]

A search warrant should be obtained if the owner or his or her designated representative has made significant efforts to secure the property (locking gates, erecting fences, boarding windows, etc.), all of which re-establishes a reasonable expectation of privacy. Although insurance investigators have certain rights to examine property on behalf of the insurer, an attempt to use the investigator as an agent of the police or fire authorities will invoke Fourth Amendment protection.

SEARCH AND SEIZURE COURT DECISIONS

A series of landmark cases from the period 1967–2000 affect the admissibility of evidence recovered at fire scenes. A brief commentary on each is offered; full details should be obtained from the case citation.

Camara v. Municipal Court of San Francisco[C1]*

In this case, a housing authority inspector conducting an annual safety inspection of a living area was refused entry by the tenant. The California Supreme Court held that a warrant was required for entry no matter what statutory authority was present. An "inspection warrant" that functions as an administrative search warrant can be issued under such circumstances. The probable cause for the issuance of such a warrant can include elapsed time since last inspection, a preset schedule of such inspections, or conditions visible from the exterior of the dwelling.

See v. City of Seattle[C2]

A fire chief performing a routine fire inspection demanded entry to a commercial warehouse. The chief was denied entry by the owner, who claimed this was an invasion of his

*In this chapter, citation numbers that appear preceded by a superscript C refer to legal references that are listed at the end of the chapter under "Court Citations."

expected rights of privacy. The Supreme Court held that due to considerations of public safety, different standards were expected of commercial areas and that inspections prior to or commensurate with licensing or granting a permit were authorized.

United States v. Biswell[C3]

A sporting goods/firearms dealer gave consent for a search that ultimately revealed the presence of sawed-off illegal weapons. He claimed that the consent search was not legitimate; the U.S. Supreme Court held that the firearms business is closely regulated by the government for the purposes of public safety, and a business conducting firearms trade must expect such inspections, and that consent in this case was submitting to a lawful authority.

Marshall v. Barlow's, Inc.[C4]

An OSHA inspector doing an inspection by statutory authority in Pocatello, Idaho, was denied entry by a plant manager. The manager subsequently obtained an injunction prohibiting reentry of the OSHA inspector on the grounds that he had certain expected standards of privacy. The Supreme Court held that a plumbing and electrical business such as his was not closely regulated by the government and that his expected standards of privacy were higher than those for the firearms or liquor trades.

Michigan v. Tyler[C5]

The landmark 1978 decision that was expected to have the greatest impact on arson cases is *Michigan v. Tyler*. The order of events can be summarized as follows: 2400 hours—fire in progress; 0200 hours—fire chief arrives and is given containers of suspected flammables recovered from the interior of the dwelling by the fire suppression crews; 0400—the fire was extinguished. An attempt was made at 0400 hours to enter the building; however, due to heat, smoke, and darkness, no suitable search could be carried out. At that time, all fire and police personnel were secured from the scene. At 0800 hours the fire chief returned with a fire inspector who examined the building with no results. At 0900 hours an assistant chief and a police investigative team reexamined the building and found evidence of an arson trailer. At that time, the Michigan State Police Arson Investigation Team was requested. That team arrived later that day for examination, photography, and collection of evidence at the scene during several subsequent searches. The U.S. Supreme Court held that the returns to the crime scene at 0800 and 0900 were permissible, since they were essentially a continuation of the examination from 0400 (which had to be discontinued due to poor conditions). The evidence gathered by the Michigan State Police Arson Team, however, was suppressed. The court held that the prosecution's contention that all rights of privacy by the owner were given up by committing a crime, that is, destroying that property, was not substantiable. However, an investigation begun and continued for a *reasonable* amount of time for the benefit of public safety at the scene was permissible. Prior to an extended examination or repeated examinations over a long period of time, a warrant was required to notify the owner of the intended search. The definition of "reasonable period of time" is entirely flexible, depending on the type of building involved, the time and nature of suppression efforts, the complexity of the scene, and the extent of emergency situations. The Court reinforced the statement that it is the duty of fire officials to determine the cause of fires. The circumstances defining "a reasonable length of time" for an investigator to conduct a warrantless search must be considered by the investigator at the time of the incident. If there is any doubt, obtain a warrant.

Michigan v. Clifford[C6]

This 1984 case refined several of the issues not clearly defined in *Tyler*. The private residence involved was heavily damaged by fire during the night while unoccupied. At the conclusion of firefighting, all fire and police personnel left the scene. A team of arson investigators arrived at the scene some 5 hours later. In the interim, the property owners had retained salvage crews to pump water from the basement and board up windows. These efforts were underway when the investigators began the investigation without consent or warrants. A suspected point of origin and evidence of an incendiary device were found in the basement. The investigators then proceeded to search the remainder of the house, finding that much of the furnishings and family possessions had been removed. When charged with arson, the defendants argued that the searches were conducted without consent, warrants, or emergency conditions requiring immediate examination and that the evidence from those searches should be suppressed. The State argued that it should be exempt from all warrants requirements in fire scene searches. The Supreme Court held that the searches were unconstitutional on several grounds, which distinguished the case from *Tyler*. First, there was no continuous custody of the scene. The search was not begun until several hours after police and fire units had left. The search in *Clifford* was neither prompted by emergency conditions (impending loss or destruction of evidence) nor was it a continuation of an earlier search. Further, the structure was a private residence, and the Court felt that there was a higher expectation of privacy in such a dwelling than in a business such as in *Tyler*. That expectation of privacy was expressed in the steps the owners had taken in ordering salvage and security arrangements. The search begun hours after the fire was extinguished required a consent, newly discovered exigency, or an appropriate warrant. The Court also distinguished between the two phases of the search. Even if the basement search had been conducted legally, it did not justify the upstairs search without a criminal search warrant. The basement search revealed that the fire could well have been of incendiary origin. At that point, the findings could have been used as probable cause to justify seeking a criminal warrant. The investigators failed to do this, so the upstairs search could not be undertaken legally without a warrant. It would appear that no matter what the basis for conducting a fire investigation search—consent, emergency, or administrative warrant—once evidence is discovered indicating the fire could have been criminally set, the investigator would be well advised, before proceeding, to obtain a criminal search warrant based on the probable cause revealed.[37]

Mincey v. Arizona[C7]

A drug arrest/raid ended in a shootout with an undercover police officer, several plainclothes police officers, Mincey, his girlfriend, and an associate shot. The scene was searched at that time for additional confederates. A shooting team was called in to search the crime scene because a police officer and several suspects had been shot. The subsequent crime scene search lasted 4 days with considerable destruction of the house and its contents. The court held that evidence in plain view discovered during a search for other confederates or in similar emergency situations is admissible. Once the emergency situation ends, however, the requirement for a warrant prior to further searching resumes. Once again, the determination of "reasonable time" and "end of emergency" rests with the investigative officer and superiors but is open to interpretation by the courts.

Romero v. Superior Court[C8]

In this case, the search of a burning building for a victim revealed weapons and dynamite. A locker suspected of containing a large quantity of ammunition was opened by the fire department. In emergency situations, Section 13875 of the California Health and Safety Code grants peace officer duties to firefighters at the scene. Thereby, this search, which was conducted for safety reasons, was legitimate.

Two additional points should be raised in the matter of crime scene searches. If a fire official has obtained an administrative inspection warrant and is conducting a search pursuant to that warrant, criminal evidence in plain view from where the investigator has a legitimate right to be—that is, where permitted to go under the protection of that inspection warrant—is admissible in criminal proceedings. It is interesting to note that insurance investigators are considered private citizens in the eyes of the law while working on their own, and not subject to the same Fourth Amendment proscriptions. If they are engaged by a public fire or police agent to work as associates in an investigation, they become (legally) deputy agents and are subject to the same laws as police and other public officials.

Swan v. Superior Court[C9]

This decision helped define "reasonable rights" of expected privacy. After an April 4 fire, a search conducted on April 24 of the boarded-up, burned premises was held to be illegal and evidence gained therein was suppressed because the boarding up indicated an expectation of privacy. An earlier search had also revealed evidence of arson, and the court held that subsequent reentry was to gather evidence for the prosecution and not to determine the cause and origin of the fire. In 1973, U.S. v. Green[C10] allowed the submission of 29 counterfeiting plates that were discovered at the point of origin of a fire by an investigating fire marshal. The fire marshal was acting in line with his legislative authority in conducting a search for the seat of the fire that happened to be coincidental with the plates. U.S. v. Delgado[C11] held that a search of a desk drawer was permissible if the evidence present might be susceptible to fire and water damage.

People v. Holloway[C12]

The Illinois Supreme Court suppressed evidence that was found during an origin and cause investigation that was begun 2.5 hours after the fire had been extinguished and firefighters had left the scene. It was held that the investigation should have begun promptly and that the privilege to begin an investigation without a warrant does not survive a departure and return to unprotected property in the absence of an emergency situation.

Passerin v. State of Delaware[C13]

The Delaware Supreme Court held that reentry of a scene 24 hours after the fire had been extinguished and 21 hours after the first origin and cause investigation had ended without a warrant was not permissible because no exigent circumstances were present. On remand, the Court commented that because the first search had resulted in probable cause that a crime had been committed, a proper search warrant was needed for any subsequent entries. Woods and Wallace (1991)[38] and Cabral (1995)[39] have extensively examined Fourth Amendment issues and the relevant court decisions, and the reader is referred there for more details.

Horton v. California[C14]

Once firefighters enter a building, they may seize evidence that is in plain view.

U.S. v. Finnigan[C15]

Discovery of explosives in a building in plain view of responding officers provided probable cause to support a search warrant to explore the remainder of the premises for other explosives.

Sources of Information

Because of the complexity and sophistication of many arson investigations, particularly those of arson-for-profit schemes, the investigator should not hesitate to use all possible sources of information. Many of these have been suggested in the literature.[40] Particularly valuable sources at the federal level include the ATF, FBI, U.S. Fire Administration (Federal Emergency Management Agency, FEMA), and the U.S. Postal Service.[41] The new Bomb Arson Tracking System (BATS) from ATF (described in Chapter 12) promises to be a very valuable aid and resource to fire and explosion investigators.[42] State government sources include the motor vehicle and driver licensing agencies, fire marshal, corrections department, and the prosecutor, attorney general, or state police departments. On a local level, sheriffs, police, probation, and district (or prosecuting) attorney's departments or the offices of recorder, county clerk, or fire prevention may be of value.

Task forces involving city, county, state, and federal agencies have proven their ability to solve complex serial arsons involving dozens of targets. These investigations have successfully identified single arsonists setting dozens of fires across regions or multistate areas, as well as organized arson-for-profit rings. They can bring together experience, knowledge, and resources that no single agency could duplicate, but they do require the mutual trust and understanding of not only the investigators directly involved but of the command ranks of all the agencies represented.[43] A sample agreement setting up a successful interagency task force is included in Appendix L.

Private companies such as utilities (phone, gas, water, cable TV), Better Business Bureaus, and real estate offices can offer some aid in the way of data on the financial well-being of the business or individuals affected by the fire.[44] In a more general sense, the National Insurance Crime Bureau (NICB) provides resources on insurance fraud fires (combining the Insurance Crime Prevention Institute with the National Auto Theft Bureau). The NICB is a nonprofit, private organization set up to facilitate contact between insurance investigators and law enforcement. All insurance companies can submit data on fire, fraud, stolen vehicles, and the like, both nationally and internationally, to NICB and public agencies. Public agency investigators then access this data through the regional representatives for NICB.

By far the richest sources of information regarding the financial motivators of fraud fires are the insurance companies and their adjusters. Until recently, however, insurance information could not be given to police agencies without risking severe civil damages for violating the client's right to confidentiality. In the last few years, all 50 states have enacted some form of what has become known as *arson reporting immunity law.* These laws generally have several important provisions. First, they require insurance companies to notify the state fire marshal or other authorized agency when the insurer suspects that a fire loss on property it insures was caused by incendiary means. Second, these laws grant immunity from criminal prosecution or civil liability to these insurance companies for releasing information about these fires. Third, the insurance company can be fined or denied license to operate if it intentionally withholds this kind of information. While such laws have generally made it easier to successfully detect and prosecute insurance fraud cases, insurance companies are still very cautious about releasing information to police or fire agencies, sometimes preferring to risk possible fines than to risk bad-faith civil judgments. It is best to contact your state's insurance board for specific guidance.

Many major insurance companies in the United States have established Special Investigation Units (mandated by law in some states). These SIU investigators may be retired police officers with extensive arson/fraud investigation experience, staff adjusters given special training, or outside (consultant) employees with specialty experience and knowledge.

The insurance company representative with whom most investigators have contact is the adjuster. There are several types of adjusters, each with some differences in authority. Property adjusters (fire, flood, tornado, etc.) deal with first-party direct claims on damage and determine the amount of monetary loss. Casualty adjusters deal with liability issues on third-party claims. In a fire, the investigation is aimed at determining a cause and origin and then providing an estimate of the repair or replacement of the damaged structure. If warranted, the adjuster may be authorized to pay the insured for the loss. Independent adjusters are employed on behalf of the insurance company and therefore have no vested interest in the company or the insured. Staff adjusters are employees of the insurers and therefore may have an interest in the well-being of the company and the insured. Public adjusters, in contrast, are employed directly by the insureds to represent them in the claim adjustment negotiations. They have no vested interest except in the amount of the claim settlement, of which they normally take a percentage fee.

Spoliation

As defined in Chapter 14, spoliation of evidence is the destruction or material alteration of, or failure to save, evidence that could have been used by another in future litigation (or in litigation that is thought reasonably possible in future). The available remedies for spoliation of evidence continue to evolve but, depending on the jurisdiction, may include criminal penalties, adverse inferences, sanctions, and/or tort liability. For a summary of U.S. court case citations and opinions on the spoliation of evidence, see Table 4-4 in Chapter 4 of *Forensic Fire Scene Reconstruction*.[45]

Criminal investigators and forensic scientists are familiar with the concept of preserving physical evidence or a portion thereof to permit retesting by opposition experts. It is the responsibility, however, of anyone who handles or examines evidence to see that evidence or documents are not lost, destroyed, or changed in such a way that they cannot be analyzed or interpreted by someone else, whether the evidence is in a criminal or civil case.[46]

The investigation of a fire or explosion will invariably involve moving or changing some portions of the evidence even to make an initial evaluation. Because such changes are unavoidable, the state of the evidence must be documented as it was found.

PUBLIC-SECTOR INVESTIGATORS AND SPOLIATION

The public investigator in most jurisdictions has the right to seize evidence of criminal activity but not to remove evidence that clearly has no connection to the proof of arson (unless it is needed to show that certain accidental ignition sources were considered and eliminated). The public-sector investigator should not remove evidence of accidental ignition for personal examination or training purposes. For the most part, public-sector investigators are protected from penalties in spoliation cases as long as they are carrying out their official duties. Willful destruction of evidence or its removal from a scene may expose them to sanctions (criminal or civil). Evidence that is seized for prosecution of an arson case must be properly documented via photographs and diagrams, properly packaged and tagged, and maintained with a proper chain of custody. Failure to protect evidence in a criminal case may compromise the prosecution of the case and open the door for civil spoliation claims against the investigator and his or her agency. Preservation of evidence of fire cause is not limited to the material concerned with the arson set or accidental cause itself. It may (depending on the circumstances) extend to other heat sources in the area that have been examined and eliminated as not being responsible. They should not be discarded or destroyed simply because they have been eliminated.

PRIVATE-SECTOR INVESTIGATORS AND SPOLIATION

The private-sector investigator or forensic expert does not have governmental immunity. Before removing evidence from a scene, the investigator should make reasonable efforts to notify known or possible defendants, plaintiffs, or other interested parties of the loss and the intent to remove evidence. If the identity of possible defendants is not known, documentation by photography (with complete photo log) and sketching is the best alternative. If evidence is retained for testing, known parties (manufacturer, installer, etc.) should be notified in writing before any testing or changes are carried out, so that they can arrange to examine the evidence before testing or arrange for their own experts to be on hand during the examination. The recovery of exemplar or comparison samples of undamaged materials from the scene is also suggested. All investigators must be certain that all the parts of an item are collected (when collecting an appliance, the cord, plugs, and receptacle involved should be collected).

CONSEQUENCES OF SPOLIATION

If an investigator (or someone else handling evidence) destroys, materially alters, or fails to preserve such evidence, a number of adverse consequences may result. Those remedies may include denying permission to introduce any testimony regarding the lost evidence, or instructing the jury that the testing would have resulted in results favorable to the party denied access (adverse inferences), or to civil and even criminal penalties against the person responsible for the loss or destruction. Though quite a few jurisdictions have gone so far as to make spoliation of evidence a criminal offense, such laws are only rarely enforced. (See, e.g., Cal. Penal code §135 [misdemeanor]; 720 Ill.Comp.Stat.5/31-4 [class 4 felony]; N.Y. Penal Law §215.40 [class E felony].)

The most controversial, and as yet unsettled , remedy for spoliation of evidence has been as an independent cause of action for either intentional or negligent spoliation of evidence. Both remedies were first recognized in California. The California Court originated the tort of "intentional spoliation of evidence" in 1984 in *Smith v. Superior Court*.[C22] In *Smith*, the plaintiff was blinded by injuries when, while she was driving a car, the wheel of a passing van flew off and crashed through her windshield. The van was taken for repair to the company that had installed the van's custom wheels. The defendant repair company, however, later destroyed certain automotive parts key to the plaintiff's claim, despite an agreement to maintain those parts. The California Court of Appeal held that the plaintiff could bring an independent lawsuit against the company for intentional spoliation of evidence.

Soon after *Smith* was decided, the state of California adopted a separate cause of action for "negligent spoliation of evidence" in *Velasco v. Commercial Bldng. Maintenance Co.* In *Velasco*, a maintenance worker at a law firm discarded an unmarked paper bag sitting on an attorney's desk.[C17] The bag contained pieces of a glass bottle that were evidence in a product liability claim brought by the attorney's client, Pedro Velasco. Though the court found the maintenance worker not liable (since he would have no way of knowing the unmarked bag of glass was anything other than trash), it found that an action for "negligent spoliation" could be brought under the right circumstances (in this case against the lawyer, for failing adequately to mark or protect this important evidence). *Velasco* was also notable because it involved the actions of a "third party" in spoliation, that is, one who was not a party to the litigation.

Since *Velasco* more than 25 jurisdictions have addressed the issue of whether to recognize an independent tort for the spoliation of evidence. However, the "tort of spoliation of evidence has not been widely adopted in other jurisdictions, nor has much agreement emerged on its contours and limitations."[C18]

In reality, an analysis of the trends in spoliation involves four separate possible torts: (1) intentional spoliation by a party to underlying litigation; (2) negligent spoliation by a party to underlying litigation; (3) intentional spoliation by a third party not a party to underlying litigation; and (4) negligent spoliation by a third party not a party to the underlying litigation.[47]

Significantly, in 1998 and 1999, the California Supreme Court rejected the *Smith* and *Velasco* rulings; and the state that created these causes of action no longer allows intentional or negligent spoliation of evidence as a separate cause of action for first- or third-party conduct.[C19] *Cedars-Sinai Med. Ctr. v. Superior Court* held that there is no tort cause of action in California for first-party intentional spoliation of evidence. *Temple Community Hosp. v. Superior Court* (1999) held that there is no tort cause of action against a third party for intentional spoliation of evidence.[C20] *Farmers Ins. Exch. v. Superior Court* (2000) held that there is no tort cause of action for negligent spoliation of evidence.[C21] *Coprich v. Superior Court* (2000) also held that there is no tort cause of action for negligent spoliation of evidence (but indicated that plaintiff might be able to assert a claim for breach of a contractual duty to preserve evidence).[C22] *Penn v. Prestige Stations, Inc.* (2000) held that the ban on spoliation tort claims applies retroactively, reversing judgments obtained prior to *Cedars-Sinai*.[C23]

Similarly, the trend in the great majority of jurisdictions in recent years has been to curtail the actionability of separate causes of action for spoliation, especially in the context of alleged spoliation by a party to the underlying litigation.[C24] These courts have reasoned that such an action would be superfluous, given the available remedies within the matter at hand.

As of 2010, seven states recognize the tort of intentional spoliation as a tort applicable to first parties, namely, Indiana, Ohio, Connecticut, Louisiana, Montana, New Jersey, and West Virginia. States that appear to recognize the tort of negligent spoliation are the District of Columbia, Kansas, Pennsylvania, Illinois, and New Jersey.[48] Florida recognizes causes of action for both intentional and negligent first-party spoliation. The majority of jurisdictions also do not recognize third-party spoliation as an independent claim. As one court held,

> the benefits of recognizing a tort cause of action, in order to deter third party spoliation of evidence and compensate victims of such misconduct, are outweighed by the burden to litigants, witnesses and the judicial system that would be imposed by potentially endless litigation over a speculative loss, and by the cost to society of promoting onerous records and evidence retention policies.[C25]

Likewise, the majority of jurisdictions also have rejected a cause of action for negligent third-party spoliation. For example, in *MetLife Auto & Home v. Joe Basil Chevrolet, Inc.,* a home insured by MetLife sustained over $330,000 in damage from a fire that began in an aftermarket electronic remote starter system installed in the dashboard of a Chevy Tahoe parked in the garage.[C26] Royal Insurance Company, which insured the dealer who owned the car, took possession of the vehicle after indemnifying its insured, the car dealer. Even though it voluntarily agreed to preserve the evidence for inspection by others, Royal allowed the car to be sent to a junkyard and scrapped. The court held that the societal costs of mandating the preservation of anything that might conceivably be evidence may be enormous and concluded that it did not recognize a cause of action against negligent third-party spoliators.

However, a minority of jurisdictions have recognized third-party spoliation as an independent tort. For example, the West Virginia Supreme Court of Appeals, in *Hannah v. Heeter,* ruled that both intentional and negligent spoliation by third parties are recognized causes of action.[C27] In so ruling, the court concluded that to rule otherwise would be inconsistent with "our policy of providing a remedy for every wrong and compensating victims of tortious conduct." The *Hannah* court further advised that "a duty to preserve evidence for a pending or potential civil action may arise in a third party to a civil action through a contract, agreement, statute, administrative rule, voluntary assumption of duty by the third party, or other special circumstances." The court set out six elements for the bringing of a claim for negligent spoliation of evidence:

1. The existence of a pending or potential civil action;
2. Actual knowledge of the pending or potential civil action;
3. A duty to preserve evidence arising from a contract, agreement, statute, administrative rule, voluntary assumption of duty, or other special circumstances;
4. Spoliation of the evidence;
5. The spoliated evidence was vital to a party's ability to prevail in a pending or potential civil damages action; and
6. Damages.

For intentional spoliation to be proven, the court required seven elements:

1. A pending or potential civil action;
2. Knowledge by the spoliator of the pending or potential civil action;
3. Willful destruction of evidence;
4. The spoliated evidence was vital to a party's ability to prevail in the pending or potential civil action;
5. The intent of the spoliator to defeat a party's ability to prevail in the pending or potential civil action;
6. The party's inability to prevail in the civil action; and
7. Proof of damages.

As it stands now, the law is in a state of flux. Although most jurisdictions have reasoned that sufficient remedies are available without a separate cause of action for first-party spoliation of evidence, eight jurisdictions still retain independent causes of action for intentional and/or negligent spoliation, and this area of the law is by no means settled. Similarly, although the majority of jurisdictions also do not recognize an independent cause of action for negligent spoliation of evidence, some jurisdictions still do permit this separate remedy.[C28]

Firefighters and fire investigators, particularly investigators in the private sector, are now being cautioned that they have a duty to preserve evidence.[49] This duty to preserve may be required by statute, agreement or contract (protocols, engagement letter, or insurance agreement), court order, industry standards, or some other special relationship or circumstance.

Forensic engineers involved with failure analysis investigations are cautioned regarding the spoliation of evidence, and they are advised to become familiar with the requirements of *ASTM E 860: Standard Practice for Examining and Preparing Items That Are or May Become Involved in Criminal or Civil Litigation*, for preservation of evidence and notification of potential interested parties when destructive testing is required. The consequences of not being familiar and complying with *ASTM E 860* could subject the engineer to personal liability for professional malpractice.[50]

Even without an independent cause of action for spoliation, however, several potent remedies are available when material evidence is destroyed, disposed of, altered, or lost. Thus, the proper and safest course of action should be carefully to safeguard evidence until it can be confirmed that any potential claimants no longer require access.

Chain of Evidence

The documentation of the whereabouts of any item of physical evidence is as important as the evidence itself. This documentation, called the *chain of evidence* or *chain of custody*, was discussed in a previous section of evidence collection (Chapter 7). Suffice it to say that the chain of evidence provides a record of every person who had custody of an item. The evidence presented in court must be documented as being the exact item removed from its place of discovery. If the investigator has never let the item out of his or her personal control, the matter is simple. The investigator can testify, "That is the very same item that I recovered, I placed it in this container, and sealed it, and it is in the same condition as when I first picked it up." If that object has been turned over to others for storage, photography, or laboratory examination, it is within the power of the court to call every person involved into court and have him or her testify as to its identity and preservation. The chain of evidence is a record of all such custody. Most often, it is accepted as proof of continuity, but if the evidence appears to have been altered or substituted, the chain provides a record on which to base any inquiries. The documentation begins, not with the physical recovery of the item of evidence, but from its first discovery as something of interest, and includes notes, sketches, and photos (see Appendixes F and I).

It is important to remember that the chain of evidence is to the case what an anchor chain is to a ship. It is only as strong as its weakest link. If even one link breaks, the chain is useless and the case (or the ship) can be lost. Extra care must be exercised by the investigator to keep a chain of custody as short as possible and to keep careful records of its every step.

Report Writing

Thorough fire scene investigation and accurate and insightful reconstruction of the origin and cause of a fire are of no value if the investigator cannot transmit the painstakingly gathered information to others. Although some of this is done through verbal reports and court testimony, by far the most critically important means of communicating is through the written report. The preliminary report is often a brief summary of the what, who, and when of a fire, sometimes even a checklist on a preprinted form. These reports get the basics of the case—names of property owners, dates, property, witnesses—but little else. The investigation report, however, is an official record of the investigator's activities and findings. It must reflect all the important information gathered as to the fire, its circumstances, and the people involved, as well as present the conclusions with regard to origin and cause. The report must be complete so that someone reading it later will have full knowledge of what was found, even if the investigator is not there to interpret it. It will represent the investigator's efforts to police and fire personnel, lawyers, victims, court, insurance companies, judges, and juries. The report must address not only the investigator's final conclusion of the origin and cause of the fire but also the evidence that supports that conclusion. It must also discuss the alternative explanations for the fire and what data were evaluated to eliminate those alternatives. Although report content and format may be constrained by company or agency policies or court rules, many investigators today are following the suggested practices in *ASTM E620-04: Standard Practice for Reporting Opinions of Technical Experts*.[51] This information, in addition to that already listed, includes the name of the person or organization requesting the report, date of preparation, list of items examined, all pertinent facts upon which the opinion was based (evaluated in accordance with *ASTM E678: Standard Practice for Evaluation of Scientific or Technical Data*—the scientific method), expert opinions and the logic and reasoning behind those opinions, and a signature of the reporting expert.[52] One way to ensure the completeness of a report is to think to oneself, "If I were to die tomorrow and this report were the only information on this incident, how completely and accurately would it reflect what I know about this fire?" It is surprising how many gaps one finds in a report when it is regarded as the ultimate document. With completeness and accuracy as the primary considerations, it is only necessary, then, to make it as clear, concise, and readable as possible. Reporting formats and styles will vary from agency to agency, but let us take a simple report and illustrate some major considerations.

REPORT SUMMARY

The most readable reports begin with a short summary of the facts and conclusions to be presented in the body of the report. In this way, the reader is given an idea of where the report is heading without reading the whole document. This summary often includes the following:

- Date, time, and location of fire
- Date, time and circumstances of scene examination
- Nature of fire damage
- Basic conclusions about the origin and cause of the fire.

THE SCENE

Valuable information about the fire cause can be obtained from a history of the building or property involved. Information to be gathered from witnesses or documentary sources and reflected in the report may include a description of the building or property prior to the fire (age, construction, use, tenancy, owner, insurance, financial history, and the occurrence of other fires in area). The names and addresses of witnesses are included along with their statements.

The *absence* of a history of previous problems—flickering lights, power failures, gas smells—may be useful in eliminating particular accidental sources and may be useful in establishing an arson corpus. Such information can be gathered from residents, owners, neighbors, recent visitors, and utility company records.

The fire itself—time and manner of discovery and by whom, time and nature of response, and extent of all firefighting operations—should be documented. Interviews with fire personnel regarding their observations as to fire conditions and extent should reflect unusual flames, smoke, odors, or the presence of vehicles, bystanders, or obvious features of the building such as open doors and windows. Weather conditions at the time of the fire (wind, rain, lightning, temperature) all may play a role in the ignition or spread of a fire and should be included. The pre-fire condition of the building, its contents, and operational systems (electrical, HVAC, heat/smoke detection, suppression, and security) are essential to an accurate and reliable investigation.

THE INVESTIGATION

The observations of the investigator in retracing the progress of the fire must be recorded to substantiate any conclusions reached as to origin and causation. It is not adequate to simply state that the origin was identified as a particular location. The diagnostic signs— char depth, heat horizons, low burns, unusual fuels, and so on—used to reach that conclusion must be listed. These observations lend credence to the investigator's work and allow one to convince a jury of the "reality" of the scene.

Once the origin has been described, the cause can be discussed. What fuel and source of ignition appeared at the origin to start the fire? What signs were there of accidental, natural, or incendiary origin? If it appears to have been accidental, subsequent examinations of the appliance or device responsible may have been carried out. If it appears to have been incendiary, what possible sources of natural or accidental ignition were present, and how were they excluded? If an incendiary origin is suspected, how was it started, and what other signs of criminal activity—broken windows, forced locks, ransacked office, bloodstains, tool marks, tools, or weapons—were found? In short, the investigative report must reflect all evidence found and how that evidence was interpreted.

REPORT CONCLUSIONS

In concluding the investigative report, all follow-up and supplemental inquiries should be cited along with recommended later courses of action. A list of evidence collected and a summary of what is to be done with it is an important part of this supplemental information.

In some cases, a separate report of all investigative activities may be required for judicial processing. This is especially true in some complex arson cases in which documentary evidence from financial, insurance, or governmental sources may be a part of the proof of intent. This report should be complete in its scope and coverage and often will include photos, sketches, diagrams, affidavits, laboratory reports, and other supporting evidence. This report should be organized in such a way that the prosecutor handling an arson case or a civil attorney handling a lawsuit can see all the elements of the case laid out in a clear, well-organized manner. As in other phases of investigation, careful and thorough preparation of this final report will offer great rewards in terms of more successful prosecutions or litigations. Reports from consultants or specialists involved to investigate or develop portions of an investigation, of course, will not be as comprehensive. Those reports should be as complete and comprehensible to any future user.

If a report is to represent the total case to be presented to a prosecutor for consideration, a great deal more material will be included. Such a report should include comprehensive lists of defendants, victims and witnesses, evidence and lab reports, documentary evidence, and interviews. Such reports may also include assessments of motives, m.o. and financial status.[53]

REPORT WRITING BASICS

No matter to what use the report will be put, the following basics should be followed to avoid problems:

■   Write simply and clearly. Others reading this report will not be firefighters; use everyday terms and avoid jargon or code numbers. Use the "active" form of writing rather than the passive.

- Do not report unsupported allegations or suppositions. Your report may be read to a court verbatim, and such suppositions may be used to show bias.
- Because the report is the product of a professional, it should look professional. Errors in spelling or grammar are embarrassing and may give the impression that all the work was sloppy. Keep a dictionary handy. Using simple active sentences will keep grammatical errors to a minimum.
- The report will be relied upon to tell the complete story of the fire by those who refer to it later, so be sure it is complete and all facts reported are accurate.
- Once the report is written, preserve all supporting notes, unless there is a specific departmental protocol that requires notes to be destroyed. Destruction of notes as a personal policy leaves questions in the minds of jurors and offers a line of challenges from the opposition lawyer.[54]

Courtroom Testimony

The final test for a fire investigation is its presentation in court. The fire investigator will often be called on to offer testimony as to what evidence was found at a fire scene and what conclusions were drawn from that evidence.

There are differences with respect to the admissibility of testimony given by eyewitnesses and experts. In general, experts are allowed to express opinions, and eyewitnesses may testify only to facts or events observed. In some cases, testimony as to the cause of a fire by an eyewitness may be acceptable as a statement of fact when the testimony does not extend past the facts observed. Likewise, in some cases, opinions by the expert are excluded when an opinion is given as to the ultimate cause or origin of the fire (holding this to be an issue for the jury).

THE EXPERT WITNESS

Anyone with special knowledge or experience with respect to the subject in question may qualify as an expert through a showing of training or experience to testify as an expert. In the case of arson, the expert giving opinion testimony as to the cause or origin of a fire may be a fire investigator, criminalist, or one of any number of specialists—chemist, engineer, and so on—depending on the nature of the fire. Testimony by such an individual stating that the fire is of "incendiary origin" or "deliberately set" is admissible whenever it is determined that the cause or origin is not within the common knowledge of the layperson. Opinion testimony tending to negate incendiary origin is admissible from experts but usually not from a lay (nonexpert) witness.

Impact of *Frye, Daubert,* and Related Cases

The standards for admitting expert testimony have evolved significantly over the twentieth century. In 1923, the Circuit Court of Appeal for the District of Columbia developed the first test for determining the reliability and thus admissibility of expert testimony. In *Frye v. United States,* the court held that for scientific evidence to be admissible, the party proffering the evidence must establish that the method used to produce it has been generally accepted as reliable in the "scientific community."[C29]

Since 1975, Federal Rule of Civil Procedure 702 has been the standard for admission of expert testimony in federal (as well as many state) courts. Essentially, it required that for a person to qualify as an expert, the court must determine the witness's credentials and methodology used by the expert on two bases: relevance and reliability. In 1993 the U.S. Supreme Court in *Daubert v. Merrell Dow Pharmaceuticals, Inc.* explained that whenever the expert's factual basis, data, principles, method, or their application are challenged, the court (trial judge), acting as a "gatekeeper," must determine whether the expert testimony has a reliable basis in the knowledge and experience of the relevant discipline.[C30] The trial judge also must determine whether the expert is proposing to testify to scientific knowledge that will assist the court in understanding or determining a fact in issue. If an expert is to testify on a scientific theory or technique, then it must be evaluated on four factors: the theory or technique (1) has been or can be tested, (2) has been subject to peer review and publication, (3) has a known or potential rate of error, and (4) has gained general acceptance in the relevant scientific community. Recognizing that those factors would not apply in every case, the Court stated that those four factors do not constitute a definitive checklist or test for all cases and that the trial court must gear its evaluation to the particular facts of each case. The fire investigation community therefore could present other factors for consideration by the court as to the admission of expert testimony in a fire case.

Although the Court explained that this decision was based on the 1975 Federal Rules, many questioned whether the case applied outside of the medical scientific community. *U.S. v. Markum* contended, for instance, that fire investigation for origin and cause determination is based on scientific knowledge, and therefore, a proposed expert should be able to demonstrate the appropriate levels of understanding of the sciences involved.[C31] The expert should also be able to demonstrate the accuracy and reliability of the conclusions reached as part of the determination, as well as the data or information relied upon. This argument was rejected in 1993 in *Markum,* in which it was reinforced that experience alone can qualify in fire causation.

The states accepting *Daubert* standards in state courts include Connecticut, Indiana, Kentucky, Louisiana, Massachusetts, Missouri, New Mexico, Oklahoma, South Dakota, Texas, and West Virginia. *Frye* states include Alaska, Arizona, California, Colorado, Florida, Illinois, Kansas, Maryland, Michigan, Nebraska, New York, Pennsylvania, and Washington. States with their own tests, or typically a *Frye-plus* test, include Arkansas, Delaware, Georgia, Iowa, Minnesota, Montana, North Carolina, Oregon, Utah, Vermont, and Wyoming.[55]

In *Michigan Millers Mut. Ins. Corp. v. Benfield,* it was contended that a fire investigator could offer an opinion that a fire was arson simply as a result of looking at the evidence (with no samples, no tests, and no proof of excluding other possible causes).[C32] It was suggested that fire investigation was not a science and therefore should not be held to the standards in *Daubert.* The U.S. Court of Appeals (11th Circuit) rejected that reasoning and remanded the case for trial, concluding that more substantiation was needed than just the unsupported opinion of a fire expert. It said that there was no reason to admit opinion evidence that was connected to existing data only by the *ipse dixit* of the expert (*ipse dixit*—it is so because I say so—is defined as a bare assertion resting on the authority of an individual).

In 1997, the Supreme Court clarified *Daubert* in *General Electric Company v. Joiner.*[C33] In *Joiner,* the Court reinforced the notion that the trial judge is the "gatekeeper" and that the trial judge is supposed to draw his or her own conclusions about the methodology and data employed by the expert as well as the ultimate conclusions reached. Thus, after *Joiner,* under *Daubert,* the trial judge not only must decide whether a witness is qualified to testify as an expert (as before *Daubert* and *Joiner*) but also must scrutinize the methods employed by the expert, the application of those methods to the facts of the case, and even the ultimate conclusions reached by the expert. All these must meet the approval of the trial judge before they can ever be presented to the jury.

Joiner also clarified that the appropriate standard of appellate review of a district court's decision to exclude expert evidence is the abuse of discretion standard. Thus, unless the trial court can be shown to have abused its (very broad) discretion, the trial judge's decision will not be disturbed on appeal.

In 1999, in *Kumho Tire v. Carmichael,* the Supreme Court, again referring to Rule 702, held that the standard for admission applies not only to scientific testimony but also to any technical testimony or to that based on specialized knowledge.[C34] Fire investigators potentially fall into all three categories. Further, the Court confirmed that just because the expert declares he or she is an expert, the trial court must still examine the expert's credentials and methodology on a case-by-case basis.

In December 2000 Congress expanded the language of Rule 702 regarding experts in an effort to put this issue to rest. The new language states that an expert witness may testify if (1) the testimony is based upon sufficient facts or data, (2) the testimony is the product of reliable principles and methods, and (3) the witness has applied the principles and methods reliably to the facts of the case. There are several important conclusions that might be drawn in applying the new Rule 702, *Daubert, Joiner,* and *Kumho Tire* to the field of fire investigation, First, in all jurisdictions applying the *Daubert* rule, the expert testimony of fire investigators will be subject to challenge on the reliability of the investigator's methods, techniques, and conclusions. Second, the trial judge will determine the reliability of the expert's methods and conclusions, using a framework of criteria that may include some or all of the *Daubert* criteria, or other different criteria appropriate to the particular case. This evaluation will take place before expert is permitted to give any testimony before the jury. Third, it will be the fire expert's burden to show that the methods he or she employed in the fire scene analysis are reliable, and the conclusions drawn also are reliable. The judge must be convinced of the investigator's methods and conclusions before the expert's testimony will be allowed at trial. Fourth, once the trial judge makes his or her decision, that ruling rarely will be overturned on appeal. Fifth, to convince the judge of the reliability of his or her methods and conclusions, the investigator must be prepared to validate and verify every facet of the investigation. The investigator must show *why* a particular method was used and *how* the method has been shown to be reliable. Similarly, the investigator will need to show *why* the conclusions reached are valid, based upon those methods, with documentation of the validity of those conclusions.

Some state jurisdictions still rely on *Frye v. United States* rather than *Daubert* (or FRE 702). Under *Frye,* the court essentially examines whether the technique has been generally accepted within a relevant expert community (the fourth prong of *Daubert*). One way to demonstrate that the methods and techniques employed by the investigator are "generally accepted" is to show the court that they are consistent with the standards, recommended practices, texts, and guides that are presented as a variety of authoritative treatises in the discipline of fire investigation. Among the treatises on which an expert may rely are this textbook, *NFPA 921* (and *1033*), *Forensic Fire Scene Reconstruction* (Icove & DeHaan)), NFPA standards and guides, ASTM practices and methods, and other publications that are recognized by the fire investigation community.

In *Weisgram v. Marley Co.,* a North Dakota court struck the testimony of three fire experts.[C35] In that case, an electrical engineer and a metallurgist were disqualified because the court concluded their opinions were based on speculation and conjecture rather than on provable facts and evidence relevant to the case and on a reliable methodology. The testimony of a fire captain was partially admitted as to origin of the fire but excluded as to cause where he expanded his opinion to electrical and chemical issues for which he had no substantial knowledge, training, or experience.[56]

Normally, the weight of evidence as to the cause of a fire is determined by the same tests applied to other evidence. In other words, the jury is not bound to accept it. Also experts may be allowed to introduce evidence of experiments to sustain a theory of how the fire started if the conditions surrounding the experiment are not too dissimilar from the actual event (i.e., relevance) must be demonstrated.[C36]

In *Pride v. Bic Corp.,* a man mysteriously caught fire while inspecting a pipe behind his house.[C37] In an ensuing product liability action, the widow's theory was that her deceased husband's butane lighter first failed to extinguish, igniting the man's clothing, then exploded, dousing the man with isobutane and fueling a conflagration that ultimately caused his death. The widow offered three experts: (1) a mechanical engineer who had testified in numerous product liability suits, on subjects ranging "from car seat belts to manure spreaders"; (2) a firefighter who had previously testified in Bic lighter cases on causes and origins of fires; and (3) an analytical chemist. The engineer opined, based on his inspection of the lighter, that the exploding-lighter scenario was the most likely cause of the fire and that the mishap was caused by a manufacturing defect and also by failure of the lighter's design to incorporate redundant safety features. The firefighter opined that the lighter was the most likely cause of the fire based on elimination of other plausible causes as well as information suggesting that the fire had started in the victim's breast pocket. The chemist opined, based on information regarding the condition of plastic from the lighter, that the lighter had exploded. After a *Daubert* hearing, the trial court excluded the preferred testimony of all three experts. The appellate court upheld the trial court's exclusions of the evidence on the basis that none of widow's experts conducted replicable laboratory tests showing that explosion of the lighter was consistent with a product defect. The engineer's testimony regarding manufacturing defects was contradicted by the widow's other witnesses and by defense experts' lab tests. The firefighter admitted he was not an engineer, had performed no tests, and was not an expert in lighters. The chemist admitted his lack of expertise in fire investigations and that he did not personally examine the lighter. The chemist also designed a lab experiment to test his hypothesis but said he "chickened out and shut the experiment down."[57]

Impact of *MagneTek*

A 2004 decision of the U.S. Court of Appeals for the Tenth Circuit has been the subject of great discussion for its discussion of both legal issues and fire science. This case underscored the importance of properly evaluating and presenting expert testimony in fire litigation cases. *Truck Insurance Exchange v. MagneTek, Inc.* not only demonstrated the application of the *Daubert* standard for the admissibility of expert testimony but effectively contradicted a long-standing principle of fire science that had been used in a significant number of cases to prove the cause of a fire.[C38]

The *MagneTek* case involved a subrogation action filed by Truck Insurance Exchange against the manufacturer of a fluorescent light ballast that was alleged to have caused a fire that destroyed a restaurant in Lakewood, Colorado. Investigations by both the local fire protection district and a private fire investigation firm hired by the insurer determined that the fire had started in a void space between the basement storage room ceiling and the kitchen floor. In the basement, the investigators found the remains of a fluorescent light fixture that had been mounted to the ceiling of the storage room. They determined the light fixture had been located in the area of origin of the fire and concluded that the fire had been caused by an apparent failure of the ballast in the light fixture.

The investigators theorized that the heat from the ballast had caused "pyrolysis" to occur in the adjacent wood structure of the ceiling, causing "pyrophoric carbon" to form over a prolonged time, which would be capable of ignition at temperatures substantially below the normal ignition range of 204°C (400°F) or more for fresh whole wood. The phenomenon of pyrolysis in the formation of "pyrophoric carbon" has been the subject of a number of studies, reviews, and articles by fire investigators and fire scientists. It has been cited as the cause of a number of fires having no other apparent explanation, often linked to heated pipes in structures within walls, ceilings, and floor areas. As the *MagneTek* court case noted, however, the validity of this phenomenon has been discussed and debated by fire investigators and fire scientists for a number of years.

Pyrolysis and the formation of "pyrophoric carbon" was the foundation of the plaintiff's case against MagneTek. The ballast in the light fixture showed no signs of failure in the thermal protector, which would have limited the heat generated by the ballast to about 111°C (232°F). Even with the exemplar ballast whose thermal protector failed to perform as it had been designed, the temperatures generated did not exceed 171°C (340°F). The investigators admitted the ignition temperature of fresh whole wood is typically at least 204°C (400°F), and the temperatures from the ballast *alone* would not have been sufficient to cause ignition. Their theory that the ballast had caused the fire depended on the concept of pyrolysis to allow ignition to occur at a lower temperature within the range of the temperatures generated by the ballast.

The court noted that under *NFPA 921* an investigator offering the hypothesis of an appliance fire must first determine the ignition temperature of the available fuel in the area and then must determine the ability of the appliance or device to generate temperatures at or above the ignition temperature of the fuel. In that regard, the experts failed on both counts. The ignition temperature of the fuel (wood) exposed to the fixture in this case could not be scientifically proved to be below the 204°C (400°F) threshold for the ignition of most types of wood, and the tests of the ballasts had shown that even with a failed thermal protector, a ballast could not generate temperatures anywhere near that range. Their hypothesis would have to be based on either an unreliable theory (pyrolysis) or unsubstantiated assumptions and speculations about the temperature of the ballast that contradicted their own test results. As such, their testimony could not be admitted.

This decision has significant implications for the litigation of fire cases everywhere. It demonstrates the importance of developing sound and scientifically verifiable theories for proving the cause of a fire as required by the standards of the *Daubert* decision. Moreover, it provides a compelling example of how a theory that has not been validated and generally recognized by others in the scientific community may not withstand a *Daubert* challenge in court.

A scientific analysis and assessment of the *MagneTek* case decision indicates that the court may not have understood several important aspects of fire science, including pyrolysis, which is not newly discovered nor scientifically disputed. In the *MagneTek* case, the Court rejected under *Daubert* the "pyrolysis theory" of long-term, low-temperature ignition as being unreliable. Fire investigators who wish to explain this phenomenon in court must be prepared to make clear and convincing arguments using authoritative treatises and clearly defined definitions.

In the *Ignition Handbook*, pyrolysis is clearly defined as "the chemical degradation of a substance by the action of heat."[58] This definition includes both oxidative and nonoxidative pyrolysis, taking into account the cases where oxygen plays a role. Pyrolysis is a process that has been studied for many decades. Without pyrolysis to break down their molecular structures, most solid fuels will not burn. Unfortunately, the published court decision declares "pyrolysis" to be inadequately studied and documented (presumably intending to refer to "pyrophoric process"). It is unclear what effect this misstatement of the court will have on future deliberations.

Even more problematically, the court effectively denied the existence of any ignition due to self-heating, because it declared that ignitable substances have a "handbook" ignition temperature, and a hot object below that temperature will not be a competent source of ignition. In actual fires, self-heating substances do not have a unique ignition temperature. Furthermore, if a substance (e.g., wood) can ignite owing to external, short-term heating or to a prolonged process involving self-heating, the minimum temperature for the latter has no relation to the published ignition temperatures for the former. In this respect, the citation from *NFPA 921* regarding ignition temperatures is in error. A full discussion of the legal and scientific issues is included in *Forensic Fire Scene Reconstruction*.[59]

Other Considerations for Experts

The responsibilities of the expert witness are much greater than those of a factual or eyewitness. The success of the case will very often depend on how well the fire investigator performs even under the duress of intensive cross-examination. Whether the investigator is there as only one of several witnesses testifying as to case findings or whether he or she is acting as the investigating officer, guiding and assisting the attorney, the conduct and testimony offered must be of the highest caliber. The investigator should be aware of (and it is hoped, be able to meet) the qualifications of knowledge, skills, and abilities for a professional fire investigator as laid out in *NFPA 1033: Standard for Professional Qualifications for Fire Investigator*.[59] It does little for an expert witness's qualifications when the expert seems to know less about the professional requirements for his or her profession than does the opposing counsel.

In a recent series of articles about the *Daubert, Kumho,* and *Benfield* cases, Professor Vincent Brannigan claimed that merely employing the scientific method to test hypotheses about past events does not make it science and that the critical ability of a scientific endeavor is to allow the prediction of future events by creating testable hypotheses.[60] Barring the exceptional cases in which a full-scale re-creation of a fire event is carried out, the hypotheses tested in nearly all fire cases must be based on data from past experiences as to what effects one predicts that certain processes or events will have. This does not necessarily make a fire inquiry nonscientific, any more than not being able to re-create the birth of the universe does not make hypotheses about the process any less scientific, as long as relevant and accurate methods are used to gather data and test pieces of the puzzle. A great deal of reliable scientific data have already been published regarding basic fire processes and fire growth (especially since 1985), and much more about the effects of flammable liquid fires on rooms, furnishings, and people.[61] A great deal of these data have debunked some of the mythology used for so long in fire investigation (see Chapter 7) and confirmed the reliability of other indicators. It is these data that investigators today can use to verify their conclusions about a fire, if the time is taken to do so. Research is continuing in many different venues, in both the public and private sectors.

PRETRIAL PREPARATION

It is essential that the investigator be prepared for the testimony. Notes and reports *must* be reviewed and any questions or problems discussed with counsel. Any errors or discrepancies discovered during this review should be brought to the attention of counsel and resolved, if possible, before testimony. If errors are discovered, it will be better for all concerned if they are freely admitted to upon direct examination rather than during a potentially hostile cross-examination.

An effective expert witness is often in the role of a teacher in the courtroom. Visual aids are very valuable in teaching the jury about fire patterns, physical evidence, and fire behavior. "Before-and-after" illustrations can be very effective. Showing the jury what the building was like before the fire via photos, computer-generated overheads, cut-aways, and "walk-throughs" can be helpful. Some examples of such aids were published recently.[62] Digital projectors and "Elmo" overhead projectors can be of great value, but only if the illustrations are well thought out and prepared beforehand. Throwing a few photos on a projector is not adequate. The "CSI effect" has been to heighten the expectations of judges and juries about the quality and quantity of "scientific" information presented. Figures 17-4 and 17-5 are examples of NIST's FDS/Smokeview modeling showing the dramatic difference in predicted conditions in the Station Nightclub fire (February 2003) with and without sprinklers. The visual and measurement data that can be collected today by laser scanners and scanning digital cameras can streamline the process and vastly improve the quality of the presentation (as described in Chapter 7 and Appendix M).

Charts, diagrams, or other visual aids must be planned to ensure their completeness and accuracy. A pretrial conference is very important, so that both the attorney and the witness know the strengths and weaknesses of the testimony before it is presented. Make a list of your qualifications. Study it and provide a copy to counsel before court. Such preparations are equally important when preparing for depositions in a civil case because, to a large extent, they take the place of the preliminary hearing in a criminal case.

To give yourself the best mental "edge" before testimony, allow extra time for travel to court and try to arrive a few minutes early to provide a breather beforehand. Dress conservatively and in a manner suitable to the venue (a three-piece suit is fine for big cities but may be overdressing for a small rural court). Avoid flamboyant clothes or distracting jewelry. The jury has to be able to concentrate on what is said.

TESTIMONY

Testify in a clear, firm voice. Avoid slang, professional jargon, or profanity (unless quoting what someone else said). Consider *every* question and its answer carefully before responding. Listen carefully to each question. Answering in a rapid, offhand manner can be a risky error. As an expert witness you have a right (subject to review by the court) to answer every question in full and explain your answers. Speak to the jury (or the court if there is no jury), and watch the reactions of your audience. You are there to explain things about which they will have no other information. If the jury cannot hear you or is confused by technical language, your main value as a source of irreplaceable evidence will be lost.

Keep calm and give the same careful consideration to every question no matter which side is asking it. Above all, tell the truth. A skilled cross-examiner can discredit any witness who exaggerates testimony, embellishes it, or fills in unreliable information. Being caught in a web of even such "little" lies will impugn your credibility and may discredit your whole case in the eyes of the jury. Remember that the opposition counsel's job is specifically to test your knowledge, probe the accuracy of your information, and reveal the errors or weaknesses of your conclusions. The professional investigator does not become angry with the opposition or try to engage in an argument. A professional comes thoroughly prepared, testifies to the limits of the information, explains and defends conclusions, and provides a source of reliable information to the trier of facts. It is only in this way that the fire investigator sees that justice is served in the case at hand.

Scientific Method

In the opening chapter of this book, the concept of the "scientific method of fire investigation" was introduced. As you will recall, this means to apply the scientific method of logical inquiry to the investigation. It is an approach most good fire investigators already apply. It simply means the investigation follows a series of logical steps:

1. Observe an event (such as a fire scene).
2. Define the problem (usually to determine how the fire started).
3. Collect data (information about the event—witness interviews, scene examination, fuel load, videos or photos of the fire in progress).
4. Formulate a hypothesis (an estimate of what the first fuel was, what ignition source was present, how long the fire took to grow, etc.).
5. Test the hypothesis against all available data and collect new information (this includes testing alternative hypotheses (possibilities).
6. Revise the main or working hypothesis as needed to fit the available data, possibly collecting more data and doing further analysis (a feedback loop).
7. Reach a final hypothesis and review it against all data and ensure that other possibilities are excluded.
8. Report a final conclusion (along with other possible explanations that cannot be excluded based on the information available).

It is step 7 that is so crucial to the fire investigation process and one that triers of fact will be looking for in the future as a result of *Daubert* and *Kumho*. The scientific method goes beyond just collecting information and interpreting data, it is a way of using those interpretations to predict other future events. It is the accuracy with which future events can be predicted that is the real test of a "final" hypothesis as new data are gathered from these later events. If applied to mechanical systems, the scientific method, yields immediately testable "final" hypotheses: My car breaks down; I apply some basic data gathering and hypothesis testing: is it ignition, is it fuel, is it fuel pump, fuel filter, etc.? I can test most of those hypotheses with a wrench or pliers by the side of the road. I find the fuel filter blocked as part of my testing, flush it out, and reinstall it, with my "final" conclusion being that this is what stopped my car. How do I test it? By getting in the car, starting it, and driving off. That test data affirm what I predicted from previous experience, that unblocking a plugged fuel filter will allow me to go on. But what if that was not the whole (final) solution? The car dies again a few miles later, and more testing reveals mud in the gas tank. To test that I have to drain the tank, flush it, and try to drive it again.

With fire investigation, testing of the "final" hypothesis is more difficult, because the investigator rarely has other data (such as a surveillance video that shows the fire starting in exactly the way he or she concluded). Occasionally, however, off-site surveillance video has proven to be the key or confirming evidence of causation.[63] It is very rare that one has the opportunity or the resources to rebuild a room and set fire to it to see if it really does burn the way one predicts it would. It is nearly impossible to duplicate the fuels and precise conditions that may affect the ignition and growth of some fires. The analytical fire investigator is forced to verify the data (observations of indicators for instance) by comparing them against the physics and chemistry of heat and combustion and the results of controlled tests of similar scenarios. (Such information may be gathered either from personal testing or published test results.) As an expert investigator, one has to be able to demonstrate to the court that the data collected and the method by which they were analyzed are reliable and relevant.

Another important step is implied in step 5, and that is the formulation and testing of alternative scenarios or hypotheses. One way to test the conclusions one has reached is to see if one can create another scenario that will yield exactly the same data. The more alternatives one can create and test (meaning that they have to be provable hypotheses) and show that they do not fit or can be excluded (because of a, b, or c data that do not support them), the more confident one can be that one's final hypothesis is the correct and only answer (at least based on available data). For example, the investigator observes at a fire scene that the floor of a room is very badly burned in an irregular fashion. Rather than just concluding that flammable liquids were spread around generously, a good investigator knows that this irregular burning can happen in several ways, so there is a need to test them all. Was there falldown of combustible ceiling or roof materials? If observations at the scene tell the investigator there were no falling decorations or collapsing ceiling or roof, one can eliminate that possibility. What about flashover? Were there enough ordinary fuels in the room and enough ventilation to allow flashover to occur? If so, that possibility cannot be excluded unless more data are gathered. Did the firefighters see full room involvement? Was there floor-to-ceiling damage? If not, flashover as a cause of these low burns can be eliminated because published tests show what happens in post-flashover room fires. What about a flammable-liquid hypothesis? That can and must be tested, too. Were there residues detected by lab tests after the fire? Were there empty containers strewn about? Were there odors detected? Even if all those data support the mechanism that all the floor damage was the result of flammable liquids, in reaching a "final" conclusion, the good investigator also must test alternatives of cause: Was this deliberate or accidental? Did those containers belong there? Was there a leak in a vehicle or storage tank that introduced flammable liquids? Is there any physical proof that someone deliberately spread the liquids (toolmarks at forced entry points, fingerprints on cans, containers manually opened rather than by fire effects, etc.). Establishing the first fuel ignited is only part of the fire investigation puzzle, and demonstrating intent versus stupidity or bad luck often requires a great deal of additional inquiry and effort.

If the investigator cannot exclude all alternatives based on available data he or she will have to report the fire as undetermined (possibly explaining what the various unexcluded possibilities are). Guesswork is not the same as hypothesis forming. Guesses cannot be verified, whereas good, sound hypotheses can be tested, verified, or excluded. It is

FIGURE 17-8 The 50-year-old building, a single-story block construction with common framed roof constructed of fiberglass shingles. The flooring of the main sanctuary was an above-grade wood frame with tongue-and-groove heart pine wood construction. *Used with permission of Retired Major J. Ron McCardle, Florida State Fire Marshal's Office.*

possible that new data (video, witnesses, weather conditions, etc.) will come along at a later date that will resolve the issue. Some fire scenes provide so little reliable data, however, that several possibilities still exist, and no matter how clever, experienced, or well-trained the investigator may be, those fires will remain undetermined. The case shown in Figures 17-8–17-11 is a good example of testing alternative hypotheses during a thorough scene examination.

FIGURE 17-9 The most heavily damaged area was in the sanctuary along the first row of pews, where the entire length of the pew and the floor area beneath was destroyed. *Used with permission of Retired Major J. Ron McCardle, Florida State Fire Marshal's Office.*

FIGURE 17-10 After many hours of reconstructing the scene, interviewing witnesses including those who reported seeing someone in the area, utilizing K-9s, and obtaining assistance from LP gas experts, a gas line was found running beneath the first row of pews. Inspection of the floor joists revealed fire patterns with travel upward from the ground level. During the sifting and sampling of debris beneath the floor and testing the copper gas lines for any loss of pressure, a rupture was found in the copper gas tubing. *Used with permission of Retired Major J. Ron McCardle, Florida State Fire Marshal's Office.*

FIGURE 17-11 The area was inspected from the bucket truck of the county electric company, and a chipped-out section of a large pine tree was located. This appeared to have been hit by a lightning strike. The lightning strike jumped to the top of an electrical pole at the point where the electrical ground was attached. It then traveled the service drop to the meter box assembly, then eventually to the copper gas line, then onto grounded support, where it breached the line below the first row of pews and ignited the gas. This lightning was confirmed by running a lightning strike report for that location and time. *Used with permission of Retired Major J. Ron McCardle, Florida State Fire Marshal's Office.*

FIGURE 17-12 An analysis of the electrical system of the church revealed that the electrical meter and meter base appeared to have experienced a surge of current that traveled through the grounding connection. *Used with Permission of Retired Major J. Ron McCardle, Florida State Fire Marshal's Office.*

A rural wood-frame church was found on fire at 7:34 P.M. one summer evening with fire venting from two windows and the roof of the sanctuary (Figure 17-8). Interior fire damage included *significant* destruction of the first pew and the wood floor beneath it (Figure 17-9). The primary hypothesis was that someone had forced entry by one of the windows and set one pew on fire in an attempt to destroy the church. However, no residues of ignitable liquid were detected. Excavation discovered a copper gas line running beneath the church floor with a large rupture located beneath the floor under the burned pew. Examination of the electrical system revealed damage to the electric meter, meter base, and grounding rod connected to the meter. When the the ground wire for the meter was traced, the copper gas line was found lying atop the ground wire. Corrosion normal to the copper pipe was cleaned off for several inches around the contact point (Figure 17-10). The electric company responded with a "bucket" service truck. Seen from the bucket, a chipped-out section of a large pine was visible evidence of a recent lightning strike. The lightning jumped from the tree to the power pole, then via the service drop to the meter box to the ground wire. The current then energized the copper gas line, which was breached by arcing to a "grounded" support. That arcing breached the copper gas line and ignited the now-escaping gas (Figure 17-11). The fire plume then burned through the church floor and onto the pew above. The lightning strike hypothesis was confirmed by obtaining a StrikeNet report for that location and time.[64]

Summary

Every fire deserves careful and thorough investigation. Whether it is a small fire started by a faulty light switch, a commercial fire started by an arsonist, or a huge wildlands fire started by a careless camper, the origin and cause should be identified so that, possibly, we can prevent such destruction from being repeated. Incendiary fires inflict terrible damages on our society via direct dollar losses, public costs, and personal damages and suffering. Absent positive proof of an incendiary cause, the only way to be sure a fire was incendiary in origin, is to be certain that all accidental and natural causes are ruled out using the ignition matrix described in Chapter 7 and thorough application of hypothesis formation and testing.

With proper training, fire investigators can apply a methodical, analytical approach to all fires. The fire investigation can be broken down into two parts that are closely interrelated: the scene and its examination, and the field or "legwork" investigation. The scene investigation is quite critical, because the physical evidence present has a finite lifetime, and if the investigation is not begun soon enough, the materials of interest will simply disappear. A matter of hours is usually the minimum delay involved; however, many investigations do not begin until days or weeks after the incident, thus compromising vast quantities of information that could be obtained from a fresh fire scene.

What does the scene investigator look for? One of the major objectives is to locate the seat of the fire, that is, the location where the fire itself began. Numerous fire indicators such as char patterns, heat and smoke horizons, direction of fire spread, localized destruction of structural materials, and the like, are all useful indicators to the trained investigator. These must be thoroughly documented for later reference. There are many new tools available to improve the quality of documentation.

Once at the seat or beginning of the fire, what were the fuels in the vicinity? Which of these was the most likely to be ignited by virtue of its location or its thermal properties? What kind of ignition source would be needed (flaming or glowing, large or small) to ensure ignition in the time frame indicated? What sources were there in the vicinity? Could it have been accidental? Were there appliances? Electrical outlets?

Heat sources? Electrical lamps? Candles? Can the cause of the fire be specifically ruled out as accidental? This "negative corpus" approach is very much dependent on the skill of the investigator in ruling out all possible noncriminal sources of ignition. Was there more than one source of ignition at the scene? Use of an ignition matrix such as that described in Chapter 6 will ensure that all combustions of ignition sources and first fuels have been evaluated. In many cases, the identification of several true points of origin is sufficient indication of an arson-caused fire (as long as accidental causes, such as a power surge, are eliminated). A thorough knowledge of the physics and chemistry of fire, the behavior of materials when exposed to fire, the heat release rates of furnishings and how they affect the fire's development, sources of ignition, and the nature of electric appliances and electricity-related phenomena are all essential to the accurate reconstruction of the fire. Using recognized fire engineering principles, the expert must be able to re-create the fire's ignition and growth, accounting for as many of the observed indicators as possible, in a manner that is defensible from sound scientific and engineering principles.

Laboratory analysis can play an important part in the origin and cause investigation. Comparison samples of floor coverings and other fuels from the scene can be identified by chemical or microscopic tests for evaluation of their ignition properties. Larger control samples (acquired from a manufacturer or retail source but shown to be the same as those found at the scene) can be subjected to various standard tests to demonstrate their response to ignition sources. Such tests will often answer important questions like: What temperatures and heat release rates are produced? What happens when the material is exposed to radiant heat rather than direct flame? Will these materials heat spontaneously or degrade if exposed to particular environmental conditions? What residues will be produced (volatile and nonvolatile char) if such fuels burn? Are such residues found in the fire debris? These are questions that the forensic or fire engineering laboratory can help answer.

The functions of photography and note-taking, as in any other type of crime scene, are, in fact, even more important on a fire scene. The physical relationships of objects and structures in fire scenes are often crucial. These can be recorded only by proper photography and crime scene sketching.

Along with the scene investigation, the other equally important aspect of the investigation is field-work. This is the area where police investigative talents are required. The arson investigator is required to interview witnesses, possible suspects, and insurance investigators and adjusters, and conduct background investigation into the financial and psychological well-being of the individuals involved. Fundamental to the proper conduct of a field investigation is the knowledge of what evidence is needed to substantiate a charge of arson. In many cases this knowledge is lacking, and the investigator cannot provide the prosecutor with the information he or she needs. Communications between the fire investigators and the prosecutorial agencies are the only means by which these gaps can be filled. It is helpful if the prosecutor has visited the scene, preferably during the fire or during the early stages of the investigation. Many arson task forces that include prosecutors have provisions for calling out one of them any time a fire involves major property losses or loss of life. The on-scene exposure improves the quality of understanding among the team members, thereby improving the quality of the cases that may result.

Once the scene and field investigations are completed, the physical evidence collected must be analyzed in the laboratory. There are five aspects of physical evidence: documentation, collection, preservation, analysis, and interpretation. The documentation of evidence before it is collected has to be thorough so as to establish the chain of evidence conclusively. Documentation should be complete, so that another qualified investigator can review it, establish agreement with the conclusions, and offer testimony regarding the scene. Although it is preferable that an investigator visit the scene before it is cleared, it is sometimes not possible. In that case the quality of the documentation will determine the value and reliability of later determinations. In one notable case in the authors' experience, the fire investigator who conducted the "interior" portion of a large scene examination died before the case could be tried. An expert was brought in to testify as to the origin and cause of the fire based upon the documentation of the previous examiner and verification by other investigators at the scene. The court found that fire investigation experts commonly rely on photographs, videotapes, diagrams, sketches, interview reports, and interviews with persons present at the scene to formulate an opinion and

that they can be reasonably relied upon as a basis for one's expert opinion.[C39] The condition of evidence has to be maintained so that opposing experts have an opportunity to examine it.

In the absence of a thorough understanding of fire evidence, many investigators will sample large portions of the scene entirely at random and submit large numbers of samples to the laboratory for screening. This "shotgun" approach is notoriously unsuccessful. In many instances, the material of interest, such as a volatile flammable accelerant, is in hidden and remote corners of the scene and not readily subject to random collection procedures. When canine or electronic hydrocarbon detectors are available to help screen samples for the ones most likely to yield identifiable volatiles, they should be used. The investigator should be able to apply rules of common logic to the collection procedure before proceeding. Once the investigator is thoroughly familiarized with the properties of the materials with which he or she is dealing and the complications that can be brought on by poor preservation, common sense will generally lead to the proper packaging for any kind of evidence.

The analysis of the material submitted is generally the domain of the laboratory. The scientists in most crime laboratories are thoroughly trained and highly experienced in the application of all available scientific methodologies to the solution of analytical problems. Forensic science is the application of scientific procedures to the analysis and interpretation of physical events such as those encountered in fire scenes. Without reliable interpretation of the bits and pieces of information that were obtained from the fire scene, the investigator may still be left with unanswered questions as to how, when, where, and by whom.

The final test of any investigation is the courtroom—whether civil or criminal. Today's investigator needs to have given the analysis and interpretation of his or her facts a great deal of consideration and be prepared to demonstrate to the court how the conclusions were reached, what the evidence was for each conclusion, and how other possible explanations were evaluated and eliminated. The expert may be expected to prove how the methods used to investigate the fire and reach conclusions about it are soundly backed by acceptance within the fire investigation community as well as scientific data, and are not just opinions drawn from thin air (or hazy recollections). Whatever the investigator can do to be prepared and improve communications will be to everyone's benefit. Only an accurately and completely informed court can be expected to render justice.

Review Questions

1. What is the first step in ensuring personal safety during a fire investigation?
2. What are three of the basic types of fire modeling?
3. What are the major differences between zone models and field (CFD) models?
4. What steps does the investigator need to take before using a computer fire model?
5. Name three ways a public investigator can gain legal access to a property to conduct a search for evidence.
6. What was the major impact of *Michigan v. Clifford*?
7. Name 10 sources of information an investigator can use in a fire investigation.
8. Define spoliation as it applies to fire investigators—public and private.
9. What information should be contained in an acceptable fire expert's report?
10. What are three of the major Supreme Court decisions that affect fire investigators today?

Court Citations

C1. *Camara v. Municipal Court of San Francisco*, 387 U.S. 523; 87 S. Ct. 1727 (1967).

C2. *See v. Seattle*, 387 U.S. 541; 87 S. Ct. 1737 (1967).

C3. *United States v. Biswell*, 406 U.S. 311; 92 S. Ct. 1593 (1972).

C4. *Marshall v. Barlow's, Inc.*, 436 U.S. 307; 98 S. Ct. 1816 (1978).

C5. *Michigan v. Tyler*, 436 U.S. 499; 98 S. Ct. 1942 (1978).

C6. *Michigan v. Clifford*, 464 U.S. 287; 104 S. Ct. 641 (1984).

C7. *Mincey v. Arizona*, 437 U.S. 385; 98 S. Ct. 2408 (1978).

C8. *Romero v. Superior Court*, 266 Cal. App. 2d 714; 72 Cal. Rptr. 430 (1968).

C9. *Swan v. Superior Court*, 8 Cal. App. 3d 392; 87 Cal. Rptr. 280 (1970).

C10. *United States v. Green*, 474 F.2d 1385 (1973).

C11. *United States v. Delgado*, 459 F.2d 471 (1973).

C12. *People v. Holloway*, 86 Ill. 2d 78, 426 N.E.2d 871 (1981).

C13. *Passerin v. Delaware*, 419 A.2d 917 (1980).

C14. *Horton v. California*, 496 U.S. 128, 110 S. Ct. 2301 (1981).

C15. *United States v. Finnigin*, 113 F.3d 1182, 1186 (10th Cir. 1997).

C16. *Smith v. Superior Ct.*, 151 Cal. App. 3d 491, 198 Cal. Rptr. 829 (1984).

C17. *Velasco v. Commercial Bldng. Maintenance Co.*, 169 Cal. App. 3d 874 215 Cal. Rptr. 504 (1985).

C18. *Dowdle Butane Gas Co., Inc. v. Moore*, 831 So.2d 1124, 1128 (Miss. 2002).

C19. *Cedars-Sinai Med. Ctr. v. Superior Court* 18 Cal. 4th 1, 74 Cal. Rptr. 2nd 248 (1998).

C20. *Temple Community Hosp. v. Superior Court*, 20 Cal. 4th 464, (1999).

C21. *Farmers Ins. Exch. v. Superior Court*, 79 Cal. App. 4th 1400, (2000).

C22. *Coprich v. Superior Court*, 80 Cal. App. 4th 1081, (2000).

C23. *Penn v. Prestige Stations, Inc.* 83 Cal. App. 4th 336, (2000).

C24. *Trevino v. Ortega*, 969 S.W.2d 950, 961 (Tex. 1998).

C25. *Temple Community Hosp. v. Superior Ct.*, 20 Cal. 4th 464, 470 (1999).

C26. *MetLife Auto & Home v. Joe Basil Chevrolet, Inc.*, 303 A.D.2d 30 (N.Y. App. Div. 2002).

C27. *Hannah v. Heeter*, 584 S.E. 2d 560, 568–70 (W. Va. 2003).

C28. *Smith v. Atkinson*, 771 So.2d 429 (Ala. 2000); *Holmes v. Amerex Rent-A-Car*, 710 A.2d 846 (D.C. 1998); *Boyd v. Travelers Ins. Co.*, 652 N.E.2d 267 (Ill. 1995).

C29. *Frye v. United States*, 293 F. 1013 (D.C. Cir. 1923).

C30. *Daubert v. Merrell Dow Pharmaceuticals, Inc.*, 509 U.S. 579; 113 S. Ct. 2786 (1993).

C31. *United States v. Markum*, 4 F.3d 891 (10th Cir. 1993).

C32. *Michigan Millers Mut. Ins. Corp. v. Benfield*, 140 F.3d 915 (11th Cir. 1998).

C33. *General Electric Company v. Joiner*, 522 U.S. 136, 188 S. Ct. 512 (1997).

C34. *Kumho Tire Co. v. Carmichael*, 526 U.S. 137; 119 S. Ct. 1167 (1999).

C35. *Weisgram v. Marley Co.*, 169 F.3d 514 (9th Cir. 1999).

C36. *People v. Sherman*, 97 Cal. App. 2d 245; 217 P.2d 715 (1950).

C37. *Pride v. Bic Corp.*, 218 F.3d 566 (6th Cir. 2000).

C38. *Truck Insurance Exchange v. Magnetek*, 360 F3d 1206 (2004).

C39. *United States v. Gardner*, 211 F.3d 1049 (2000).

References

1. *NFPA 1033: Standard for Professional Qualifications for Fire Investigator* (Quincy, MA: National Fire Protection Association, 2009), chap. 4, sec. 4.1.3.
2. U.S. DOT, "Health Hazards of Combustion Products from Aircraft Composite Material," DOT/FAA/AR-98/34 (Washington, DC: Office of Aviation Research, September, 1998).
3. KWJ Engineering, Inc., Newark, CA; http:www.kwjengineering.com.
4. M. L. Donohue, *Safety and Health Guidelines for Fire and Explosion Investigators* (Stillwater, OK: Fire Protection Publications, 2002).
5. J. W. Munday, *Safety at Scenes of Fire and Related Incidents* (London: Fire Protection Association, 1994); available online.
6. *NFPA 921: Guide for Fire and Explosion Investigations* (Quincy, MA: National Fire Protection Association, 2008, 2011), chap. 12.
7. D. Icove, V. Wherry, and R. Schroeder, *Combating Arson for Profit*, 2nd ed. (Columbus, OH: Battelle Press, 1998).
8. R. Friedman, *Survey of Computer Models for Fire and Smoke*, 2nd ed. (Norwood, MA: Factory Mutual Research Corp., 1991).
9. B. T. Hume, *Fire Models: A Guide for Fire Prevention Officers*, Report No. 6/93 (London: Home Office Fire Research and Development Group, 1993); H. E. Nelson, *FPETOOL: Fire Protection Engineering Tools for Hazard Estimation* (Gaithersburg, MD: U.S. Department of Commerce, National Institute for Standards and Technology, October 1990).
10. B. Karlsson and J. G. Quintiere, *Enclosure Fire Dynamics* (Boca Raton, FL: CRC Press, 1999), chap. 10.
11. D. Madrzykowski, "The Future of Fire Investigation," *Fire Chief*, October 1, 2000.
12. A. Brown, "A Review of Fire Modeling Computer Software: Branzfire V.99.005," University of Central Lancashire, UK, 2001.
13. J. D. DeHaan, "Reliability of Fire Tests and Computer Modeling in Fire Scene Reconstruction," Part I: *Fire and Arson Investigator* (January 2005): 40–45; Part II: (April 2005): 35–45.
14. D. J. Icove and J. D. DeHaan, *Forensic Fire Scene Reconstruction*, 2nd ed. (Upper Saddle River, NJ: Pearson/Brady, 2009).
15. Ibid.
16. K. McGrattan et al., Fire Dynamics Simulator (Version 5), Technical Reference Guide, vol. 1: Mathematical Model (Gaithersburg, MD: Fire Research Division, Building and Fire Research Laboratory, April 6, 2010).
17. *ASTM E1895-07: Standard Guide for Determining Uses and Limitations of Deterministic Fire Models* (West Conshohocken, PA: ASTM, 2007).
18. *ASTM E1472-07: Guide for Documenting Computer Software for Fire Models* (West Conshohocken, PA: ASTM, 2007).
19. *ASTM E1355-05a: Standard Guide for Evaluating the Predictive Capability of Deterministic Fire Models* (West Conshohocken, PA: ASTM, 2005).
20. S. M. Olenick and D. J. Carpenter, "An Updated International Survey of Computer Models for Fire and Smoke," *Fire Protection Engineering* 13 (May 2003): 87–110.
21. H. E. Nelson, *FPETOOL: Fire Protection Engineering Tools for Hazard Estimation*, NISTIR 4380 (Gaithersburg, MD: National Institute of Standards and Technology, Center for Fire Research, October 1990).
22. H. E. Nelson, *FPETOOL User's Guide*, NISTIR 4439 (Gaithersburg, MD: U.S. Department of Commerce, National Institute of Standards and Technology, October 1990).
23. M. L. Janssens, *An Introduction to Mathematical Fire Modeling* (Lancaster, PA: Technomic, 2000).
24. Ibid., 147–53.
25. M. Spearpoint, F. W. Mowrer, and K. McGrattan, "Simulation of a Compartment Flashover Using Hand Calculations, Zone Models and a Field Model" in *Proceedings ICFRE*, Chicago, IL, 1999, 3–14.
26. G. Rein et al., "A Comparison of Three Models for the Simulation of Accidental Fires," *Journal of Fire Protection Engineering*, 16, no. 3 (2006): 183–209.
27. J. Sutula, "Applications of the Fire Dynamics Simulator in Fire Protection Engineering Consulting," *Journal of Fire Protection Engineering* 14 (Spring 2002): 35–43.
28. DeHaan, "Reliability of Fire Tests."
29. A. N. Beard, "Role Models," *Fire Risk Management Journal* (August 2009): 21–22; www.frmjournal.com.
30. M. H. Salley et al., "Verification and Validation: How to Determine the Accuracy of Fire Models, *Fire Protection Engineering* 34 (Spring 2007): 34–44.
31. N. Iqbal and M. H. Salley, *Fire Dynamics Tools (FDTs)* (Washington, DC: U.S. Nuclear Regulatory Commission, November 2004).
32. S. Sardqvist, *Initial Fires*, Lund University, Lund, Sweden, April 1993; J. F. Krasny, W. Parker, and V. Babrauskas, *Fire Behavior of Upholstered Furniture and Mattresses* (Norwich, NY: William Andrews, 2001).
33. Icove and DeHaan, *Forensic Fire Scene Reconstruction*, chap. 6.
34. R. Paul and L. Elder, "The Guide to Scientific Thinking" (Dillon Bay, CA: The Foundation for Critical Thinking, 2003).
35. D. T. Moore, "Critical Thinking and Intelligence Analysis," Occasional Paper No. 14, (Washington, DC: Joint Military Intelligence College, May 2006), 8–10.

36. *Search Warrants* (Sacramento, CA: California District Attorney's Association, 1978); J. R. Davis, "The Warrant Requirement in Crime Scene Searches," *FBI Law Enforcement Journal* 47 (November–December 1978).

37. S. R. Rubenstein, "*Michigan v. Clifford et al.,*" *Fire and Arson Investigator* 35 (June 1985).

38. E. K. Woods and D. H. Wallace, "Investigating Arson: Coping with Constitutional Constraints," *Security Management,* November 1991; reprinted in *Arizona Arson Investigator,* February 1994.

39. M. J. Cabral, "The Legal Aspects of Arson Investigation: The Tyler/Clifford Legacy" in *Proceedings International Symposium on the Forensic Aspects of Arson Investigation,* Washington, DC, August 1995.

40. J. O. Campane, "Chains, Wheels and the Single Conspiracy," *FBI Law Enforcement Journal* 50 (August 1981): 24–31; LEAA, *Investigating Arson-for-Profit Schemes* (Washington, DC: U.S. Department of Justice; U.S. Government Printing Office, 1978).

41. FEMA, *Arson Resource Directory* (Emmitsburg, MD: U.S. Fire Administration, 1993).

42. www.BATS.gov.

43. E. Strauss, "Paul Kenneth Keller," *National Fire and Arson Report* 11, nos. 4–6 (1993); E. Comeau, "Working Against Arson," *NFPA Journal* (January–February 2001): 65–68.

44. Icove, Wherry, Schroeder, *Combating Arson for Profit.*

45. Icove and DeHaan, *Forensic Fire Scene Reconstruction,* 232.

46. P. A. Lynch, "After the Fire's Out: Spoliation of Evidence and the Line Firefighter," *Fire Engineering,* January 1998.

47. B. S. Wilhoit, "Spoliation of Evidence: The Viability of Four Emerging Torts," *UCLA Law Review* 46 (1998): 631.

48. 70 A.L.R. 4th 984 "Intentional Spoliation of Evidence, Interfering with Prospective Civil Action, as Actionable" (and cases cited therein); 101 A.L.R. 5th 61 "Negligent Spoliation of Evidence, Interfering with Prospective Civil Action, as Actionable" (and cases cited therein).

49. Lynch, "After the Fire's Out"; K. J. Clinkinbeard and G. A. King, "Spoliation: Can the Investigator be Sued for Destruction of Evidence?" paper presented at *ISFI 2008: International Symposium on Fire Investigation Science and Technology,* Cincinnati, OH, 2008.

50. R. A. Dean, "Legal Issues Involved in Failure Analysis," *Journal of Failure Analysis and Prevention* 4, no. 2 (April 2004): 7–16.

51. *ASTM E620-04: Standard Practice for Reporting Opinions of Technical Experts* (West Conshohocken, PA: ASTM, 2004).

52. *ASTM E678: Standard Practice for Evaluation of Scientific or Technical Data* (West Conshohocken, PA: ASTM, 2007).

53. D. J. Icove and M. W. Dalton, "A Comprehensive Prosecution Report Format for Arson Cases" in *Proceedings ISFI International Fire Science,* Cincinnati, OH, 2008; Icove, Wherry and Schroeder, *Combating Arson for Profit.*"

54. M. Cabral, California Conference of Arson Investigators Seminar, November 2009.

55. R. D. Kolar, "Scientific and Other Expert Testimony: Understand It; Keep It Out; Get It In," Appendix 1, Federation of Defense and Corporate Council, *FDCC Quarterly* (Spring 2007): 207–35.

56. E. Stauss and C. Hogan, "Three Experts Excluded: Court Examines Fire Expert Testimony under Reliability Microscope," *Fire and Arson Investigator* (April 2001): 14–18.

57. D. Wood, analysis of *Pride v. Bic,* private communication, 2005.

58. V. Babrauskas, *Ignition Handbook* (Issaquah, WA: Fire Science, 2003).

59. Icove and DeHaan, *Forensic Fire Scene Reconstruction,* 49–53.

59. *NFPA 1033,* 2nd ed., 2009.

60. V. M. Brannigan, "Arson, Scientific Evidence and the Daubert Case," *Fire Chief* (August 1998): 98–109; V. M. Brannigan, "Kumho Tire Verdict: Daubert Applies to All Evidence," *Fire Chief* (July 1999): 98–109; V. M. Brannigan and J. Torero, "The Expert's New Clothes: Arson 'Science' after Kumho Tire," *Fire Chief* (July 1999): 60–65.

61. J. D. DeHaan, "The Reconstruction of Fires Involving Highly Flammable Hydrocarbon Liquids." PhD dissertation, Strathclyde University, Glasgow, UK, 1995; published by UMI, Ann Arbor MI, 1996; A. D. Putorti Jr., *Flammable and Combustible Liquid Spill/Burn Patterns* (Washington, DC: National Institute of Justice, Office of Justice Programs (NIJ/OJP), March 2001; A. D. Putorti Jr., *Full-Scale Room Burn Pattern Study* (Washington, DC: NIJ/OJP), December 1997; J. H. Shanley, *USFA Fire Burn Pattern Tests* (Washington, DC: U.S. Fire Administration, FEMA, July 1997).

62. J. D. DeHaan, "Advanced Tools for Use in Forensic Fire Scene Investigation, Reconstruction and Documentation" in *Proceedings Interflam 2004* (London: Interscience Communications, 2004).

63. R. S. Pepler, "An Atypical Cause of Fire: CCTV Evidence to the Rescue" (Dallas, TX: AAFS, February 2004).

64. R. McCardle, Florida State Fire Marshal (retired), June 2009.

SUGGESTED READING

Chapter 1

Arson—The Neglected Epidemic: Report of the Senate Select Committee of Fire Services. California State Senate, June 29–30, 1977.

Badger, S. G. "U.S. Multiple-Death Fires for 2007." *NFPA Journal* (September–October 2008): 52–56.

Bilancia, L. "The Scientific Method and Fire Investigation." *The Fire Place,* Washington IAAI (Spring 2005).

Boudreau, J. F., G. Denault, and K. Quon. *Arson and Arson Investigation: Survey and Assessment.* Washington, DC: Law Enforcement Assistance Administration, October 1977.

DeHaan, J. D. "Arson: A Problem of Attitudes." *Washington State Chapter IAAI Newsletter* (April–May 1995).

Hazen, M. H., and R. M. Hazen. *Keepers of the Flame: The Role of Fire in American Culture.* Princeton, NJ: Princeton University Press, 1992.

Heys, S., and A. B. Goodwin. *The Winecoff Fire.* Atlanta: Longstreet Press, 1993.

Home Office (UK). *Safer Communities: Towards Effective Arson Control.* London, 1999.

Icove, D. J. "Incendiary Fire Analysis and Investigation." Open Learning Fire Science Program Course. Lexington, MA: Ginn, 1983.

National Fire Academy. "Incendiary Fire Analysis and Investigation, 2nd ed." Emmitsburg, MD: FEMA, 1996.

National Fire Protection Association. *NFPA 921: Guide for Fire and Explosion Investigations.* Quincy, MA: NFPA, 2008 and 2011.

National Institute of Justice. *Fire and Arson Scene Evidence: A Guide for Public Safety Personnel.* Washington, DC: U.S. Department of Justice, 2008 and 2011.

National Fire Protection Association. *NFPA 1033: Standard for Professional Qualifications for Fire Investigator* Quincy, MA: NFPA, 2009.

Riggs, S. *Quick Reference: Critical Tasks in Fire Investigations.* Indianapolis, IN: Public Agency Training Council/Fire Science Training Institute.

Rossotti, H. *Fire: Technology, Symbolism, Ecology, Science, Hazard.* Oxford: Oxford University Press, 1993.

Chapter 2

Babrauskas, V. *Ignition Handbook.* Issaquah, WA: Fire Science Publishers, 2003.

Brenton, J. R., and D. D. Drysdale. "Flames in Fires and Explosions." *Fire Engineers Journal* 57, no. 189 (1997): 27–29.

Drysdale, D. *An Introduction to Fire Dynamics,* 2nd ed. Chichester, UK: Wiley, 1999.

Drysdale, D. "Chemistry and Physics of Fire." Chapter 2-1 in *Fire Protection Handbook,* 19th ed. Quincy, MA: NFPA, 2003.

Faraday, M. 1861. *The Chemical History of a Candle.* Reprint, Atlanta: Cherokee, 1993.

Friedman, R. *Principles of Fire Protection Chemistry and Physics,* 3rd ed. Sudbury, MA: Jones and Bartlett, 2008.

Kunzig, R. "Infernal Combustion." *Discover,* January 2001, 35–36.

Meyer, Eugene. *Chemistry of Hazardous Materials,* 5th ed. Upper Saddle River, NJ: Pearson/Brady, 2009.

Quintiere, J. G. *Principles of Fire Behavior.* Albany, NY: Delmar, 1998.

Stauffer, E. J., A. Dolan, and R. Newman. *Fire Debris Analysis.* Burlington MA: Academic Press, 2008.

Chapter 3

Babrauskas, V. "Estimating Room Flashover Potential." *Fire Technology* 16, no. 2 (May 1980).

Babrauskas, V. "HRR—The Role of Heat Release Rates in Describing Fires." *Fire and Arson Investigator* (June 1997): 54–57.

Babrauskas, V., and D. Holmes. "Heat Release Rate." Part I: *Fire Findings* 6, no. 1 (Winter 1998): 7–11; Part II: *Fire Findings* 6, no. 2 (Spring 1998): 2–11; Part III: *Fire Findings* 6, no. 3 (Summer 1998).

Bales, R. F. *The Great Chicago Fire and the Myth of Mrs. O'Leary's Cow.* Jefferson, NC: McFarland, 2002.

Browne, F. L. *Theories of the Combustion of Wood and Its Controls.* Madison. WI: U.S. Forest Products Laboratory, Report 2136, U.S. Department of Agriculture, 1958. Available at http://www.fpl.fs.fed.us/.

Brannigan, F. L., and G. P. Corbett. *Brannigan's Building Construction for the Fire Service,* 4th ed. Quincy, MA: National Fire Protection Association, 2007.

Crombie, R. *The Great Chicago Fire.* Nashville, TN: Rutledge Hill, 1994.

DeHaan, J. D. "Compartment Test Fires (1996–98)." *CAC News* (Spring 2001).

Drysdale, D. "The Flashover Phenomenon." *Fire Engineers Journal* (1998).

Fristrom, R. M. "The Mechanism of Combustion in Flames." *Chemical and Engineering News* 150 (October 14, 1963).

Gess, D., and W. Lutz. *Firestorm at Peshtigo.* New York: Henry Holt, 2002.

Gorbett, G. E., and J. L. Pharr. *Fire Dynamics,* Upper Saddle River, NJ: Pearson/Brady, 2011.

Hamer, M. "The Night That Luck Ran Out." *New Scientist* 7 (July 1988).

Karlsson, B., and J. G. Quintiere. *Enclosure Fire Dynamics.* Boca Raton, FL: CRC Press, 2000.

Krasny, J. F., W. J. Parker, and V. Babrauskas. *Fire Behavior of Upholstered Furniture and Mattresses.* Norwich, NY: William Andrew, 2000.

National Fire Protection Association. *NFPA 555: Guide on Methods for Evaluating Potential for Room Flashover*. Quincy, MA: NFPA, 2000.

National Fire Protection Association. "Report on the Fire at the Dupont Hotel and Casino—December 31, 1986." Quincy, MA: NFPA, 1987.

Society of Fire Protection Engineers. *Predicting 1st and 2nd Degree Skin Burns from Thermal Radiation*. Bethesda, MD: SFPE, March 2000.

Chapter 4

Eaton, T. E. "Physical Properties of Gasoline." *Journal of Applied Fire Science* 1, no. 2 (1990–91): 175–83.

Ellis, W. H. "Solvent Flash Points, Expected and Unexpected." *Journal of Coatings Technology* 48 (March 1976): 614.

Fried, V., H. F. Hameka, and U. Blukis. *Physical Chemistry*. New York: Macmillan, 1977.

Glassman, I., and F. L. Dryer. "Flame Spreading across Liquid Fuels." *Fire Safety Journal*, 3 (1980–81): 123–38.

Harris, R. J. *The Investigation and Control of Gas Explosions in Buildings and Heating Plant*. New York: E & FN Spon, 1983.

LaPointe, N. R., C. T. Adams, and J. Washington. "Autoignition of Gasoline on Hot Surfaces." *Fire and Arson Investigator* (October 2005): 18–21.

Leffler, W. L. *Petroleum Refining in Nontechnical Language*, 3rd ed. Tulsa, OK: PennWell Corp., 2000.

National Fire Protection Association. *NFPA 54: National Fuel Gas Code*. Quincy, MA: NFPA, 1999.

National Fire Protection Association. *NFPA 58: NFPA Liquefied Petroleum Gas Code*. Quincy, MA: NFPA, 1998.

Chapter 5

Colonna, G. P., and B. M. Cohn. "Plastics and Rubbers." Chap. 8.10 in *Fire Protection Handbook*, 19th ed. Quincy, MA: NFPA, 2003.

Fire, F. L. *Combustibility of Plastics*. New York: Van Nostrand Reinhold, 1991.

Fire, F. L. "Plastics and Fire Investigations." *Fire and Arson Investigator* 36 (December 1985).

Hilado, C. J. *Flammability Handbook for Plastics*, 5th ed. Lancaster, PA: Technomics, 1998.

Marris, P. "Steel Determination." *Fire Engineering Journal and Fire Protection* (May 2005): 31–35.

National Fire Protection Association. *NFPA 705–Field Flame Test for Textiles and Films: Recommended Practice*. Quincy, MA: NFPA, 2009.

Shipp, M., K. Shaw, and P. Morgan. "Fire Behavior of Sandwich Panels Constructions." *Proceedings Interflam 1999*. London: Interscience Communications, 1999.

Stamm, A. J. "Thermal Degradation of Wood and Cellulose." *Industrial and Engineering Chemistry* 48 (1956).

Stauffer, E., J. A. Dolan, and R. Newman. "Chemistry and Physics of Fire and Liquid Fuels." Chapter 4 in *Fire Debris Analysis*. Burlington MA: Elsevier-Academic Press, 2008.

U.S. Department of Agriculture. *The Encyclopedia of Wood*. New York: Skyhorse, 2007.

White, R. H. "Wood-Based Paneling as Thermal Barriers." *Forest Products Laboratory Research Paper*. FPL 408. June 1982.

"Wood and Wood-Based Products." Chapter 8.3 in *Fire Protection Handbook*, 19th ed. Quincy, MA: NFPA, 2003.

Chapter 6

Ackland, P., and A. E. Willey. *Fire Investigation in Commercial Kitchen Systems*. Summerland, B.C., Canada: Philip Ackland Holdings, 2005.

ASTM International. *ASTM E2187-02: Standard Test Method for Measuring Ignition Strength of Cigarettes*. West Conshohocken, PA: ASTM, 2002.

Babrauskas, V. "*Truck Insurance v. Magnetek:* Lessons to Be Learned Concerning Presentation of Scientific Information." *California Fire/Arson Investigator* (October 2004): 6.

Babrauskas, V., B. F. Gray, and M. L. Janssens. "Prudent Practices for the Design and Installation of Heat-Producing Devices near Wood Materials." *Fire and Materials* 31 (2007): 125–35.

Babrauskas, V., and J. Krasny. *Fire Behavior of Upholstered Furniture*. NBS Monograph 173. Gaithersburg, MD: U.S. Department of Commerce, National Bureau of Standards, 1985.

Cuzzillo, B. R. *Pyrophoria*. Ph.D. dissertation, University of California Berkeley, 1997.

Damant, G. H. "Should Polyurethane Foam Be Banned?" *Fire and Arson Investigator* 42, no. 2 (December 1991): 48–52.

DeHaan, J. D. "Our Changing World: Part V—The Candle Threat." *Fire and Arson Investigator* (January 2003): 28–39.

Flynn, B. "Trends in Home Heating and Cooking Fires." *NFPA Journal* (March–April 2001): 63–65.

Henderson, R. W., and G. R. Lightsey. "Safety Device for Preventing Uncontrolled Flareup in Wick-Fed Liquid Fuel Burners." U.S. Patent, 5,338,185, U.S. Patent Office, August 16, 1994.

Holleyhead, R. "Ignition of Solid Materials and Furniture by Lighted Cigarettes: A Review." *Science and Justice* 39, no. 2 (1999): 75–102.

Sanderson, J. L. "Kitchen Fires." *Fire Findings* 5, no. 3 (Summer 1997): 6.

U.S. Fire Administration. *Agricultural Storage Fires*. Topical Fire Research Series, USFA 2, no. 7. Emmitsburg, MD: FEMA, January 2003.

U.S. Fire Administration. *Heating Fires in Residential Structures*. Topical Fire Research Series, USFA 2, no. 5. Emmitsburg, MD: FEMA, January 2001.

Chapter 7

Abrams, M. S. "Performance of Concrete Structures Exposed to Fires." Skokie, IL: Portland Cement Association, 1977.

Abrams, M. S., A. H. Gustfero, and E. A. B. Salse. "Fire Tests of Concrete Joist Floors and Roofs." Skokie, IL: Portland Cement Association, 1971.

ASTM International. *ASTM E84: Standard Test Method for Surface Burning Characteristics of Building Materials.* Philadelphia, PA: ASTM, 2007.

ASTM International. *ASTM E860: Standard Practice for Examining and Testing of Items That Are or May Become Involved in Product Liability Litigation.* Philadelphia, PA: ASTM, 2007.

Better Homes and Gardens. Home Designer Deluxe. Coeur d'Alene, ID: Chief Architect, 2005.

Buchanan, A. H. *Structural Design for Fire Safety.* Chichester, UK: Wiley, 2002.

Chitty, R., and J. Fraser-Mitchell. *Fire Safety Engineering: Reference Guide.* London: IHS BRE Press, 2003.

Cooke, R. A., and R. H. Ide. *Principles of Fire Investigation.* Leicester, England: Institution of Fire Engineers, 1985.

Corry, R. A. *Pocket Guide to Accelerant Evidence Collection,* 2nd ed. Boston: Massachusetts Chapter, 2001.

Cramer, S. M., O. M. Friday, R. H. White, and G. Sriprutkiat. "Mechanical Properties of Gypsum Board at Elevated Temperatures." *Proceedings: Fire and Materials 2003.* London: Interscience Communications, pp. 33–42.

Dietz, W. R. "Physical Evidence of Arson: Its Recognition, Collection, and Packaging." *Fire and Arson Investigator* (June 1991).

Duncan, C. D. *Advanced Crime Scene Photography.* Boca Raton, FL: CRC Press, 2010.

Earls, A. R. "Manufactured Homes Revisited." *NFPA Journal* (March–April 2001): 49–51.

Eaton, T. E. "Underfloor Fires." *Fire and Arson Investigator* 38, no. 1 (September 1987).

"High Rise Office Building Fires." *News about Fire.* Quincy, MA: NFPA. Reprinted in *Fire and Arson Investigator* (September 1993).

Hutchinson, W., ed. *Admission of Photographic Material.* Elmhurst, IL: International Fire Photographers Association, 1981.

Ide, R. H. "Investigative Excavation at Fire Scenes." *Science and Justice* (Spring 1997): 210–11.

Kitchen, A. "Concrete Ideas." *Fire Protection and Fire Engineers Journal* (October 2004): 40–41.

Kocisko, M. J. "Absorption of Ignitable Liquids into Polyethylene/Polyvinylidine Dichloride Bags." *Journal of Forensic Sciences* 46, no. 2 (2001): 356–62.

Lewko, J. "Camera Techniques at Fire Scenes." *Fire and Arson Investigator* (December 1983 and March 1984).

Lie, T. T., T. D. Lin, D. E. Allen, and M. S. Abrams. Fire Resistance of Reinforced Concrete Columns, National Research Council of Canada, Division of Building Research, NRCC 23065, Ottawa, Ontario, 1984.

Lin, T. D., et al. "Fire Resistance of Reinforced Concrete Columns." Skokie, IL: Portland Cement Association, 1992.

Lyons, P. R. *Techniques of Fire Photography,* Quincy, MA: NFPA, 1978.

Malhotra, H. L. "Spalling of Concrete in Fires." Technical Note 118. London: CIRIA, 1984.

NIJ Technical Working Group on Fire/Arson Scene Investigation. *Fire and Arson Scene Evidence: A Guide for Public Safety Personnel.* Washington, DC: National Institute of Justice, June 2000.

Portland Cement Association. *Fundamentals of Concrete.* Skokie, IL: Portland Cement Association, 2001.

Posey, E. P., and J. E. Posey. "Outline for Fire Scene Documentation." *Fire and Arson Investigator* 38 (March 1988).

Putorti, A. D., Jr. *Flammable and Combustible Liquid Spill/Burn Pattern.* NIJ Report 604-00. Rockville, MD: National Criminal Justice Reference Service, National Law Enforcement and Corrections Technology Center, March 2001.

Putorti, A. D., Jr. *Full-Scale Room Burn Pattern Study.* NIJ Report 601-97. Rockville, MD: National Criminal Justice Reference Service, National Law Enforcement and Corrections Technology Center, December 1997.

Sanderson, J. L. "Burn Pattern Testing." *Fire Findings* 9, no. 3 (Summer 2001): 7–11.

Sanderson, J. L. "Floor-Level Burning: What Kinds of Damage Do Burning Accelerants Cause?" *Fire Findings* 9, no. 1 (Winter 2001): 1–4.

Stensaas, J. P. "Determination of Fire Patterns Due to Liquid Accelerants on Floor Covering." *Proceedings Interflam 2001.* London: Interscience Communications, 2001, pp. 1491–96.

Chapter 8

Brown, D. J., *Under a Flaming Sky: The Great Hinckley Firestorm of 1894.* New York: Harper Perennial, 2007.

Cheney, P., and A. Sullivan. *Grassfires: Fuel, Weather and Fire Behavior.* Collingwood, NSW 3066, Australia: CSIRO, 1997.

Evans, D. D., R. G. Rehm, and E. S. Baker. *Physics-Based Modeling for WUI Fire Spread: Simplified Model Algorithm for Ignition of Structures by Burning Vegetation.* NISTIR 7179. Gaithersburg, MD: NIST, October 2004.

Ford, R. T. "Fire Scene Investigation of Grassland and Forest Fires." *Fire and Arson Investigator* 38 (March 1988).

Ford, R. T. *Investigation of Wildfires.* Bend, OR: Maverick Publications, 1995.

Ford, R. T. (attributed to Jones, A. W.) "Investigation of Wildland Fires" in *Arson Investigation.* California District Attorneys Association, 1980.

Maclean, J. N. *The Thirty Mile Fire.* New York: Henry Holt, 2007.

Manzello, S. L., T. G. Cleary, J. R. Shields, and J. C. Yang. "Ignition of Mulch and Grasses by Firebrands in Wildland-Urban Interface Fires." *International J. Wildland Fire* 15, no. 3 (2006): 427–31.

Manzello, S. L., S. H. Park, T. G. Cleary, and J. R. Shields. *Developing Rapid Response Instrumentation Packages to Quantify Structure Ignition Mechanisms in Wildland Urban Interface (WUI) Fires.* In *Proceedings Fire and Materials 2009: 11th International Conference.* London: Interscience Communications, 2009, pp. 215–24.

Maranghides, A., and W. Mell. "A Case Study of a Community Affected by the Witch and Guejito Fires." NIST Tech Note 1635, Gaithersburg, MD: NIST, April 2009.

Mitchell, J. W. "Power Lines and Catastrophic Wildland Fires in Southern California." *Proceedings Fire and Materials 2009.* London: Interscience Communications, 2009, pp. 225–280.

Rehm, R. G., *Effects of Winds from Burning Structures on Ground-Fire Propagation at the Wildland-Urban Interface. Final Report.* NIST GCR 06-892. Gaithersburg, MD: NIST, April 2006.

Steensland, P. "Long-Term Thermal Residency in Woody Debris Piles." Presented at International Association of Arson Investigators, Annual Training Conference, May 2008.

TriData Corporation. "Urban Wildlands Fire." *Fire and Arson Investigator* 38 (March 1988).

Tulloch, M. H. "Photonics Technology Assists Oakland Firestorm Efforts." *Photonics Spectra* (December 1991): 18.

U.S. Forest Service. *Wildfire Cause and Origin Determination Handbook* (FI-210 Course Manual). Boise, ID: NWCG, May 2005.

Vaisala Global Atmospherics, 2705 E. Medina Rd., Suite 111, Tucson AZ 85706, (800) 283–4557, www.thunderstorm.vaisala.com.

Wildfire Cause Determination Handbook: NFES 1874. NWCG Handbook 1, PMS 412-1 (Including FI-110 and FI-210 course curricula), 2005. [National Interagency Fire Center, Attn: Great Basin Cache Supply Office, 3833 S. Development Ave., Boise, ID 83705; Fax (208) 387-5573].

Chapter 9

Bloom, J., and C. Bloom. *Motorhome and Recreational Vehicle Fire Investigation,* 2nd ed. Grants Pass, OR: Bloom Fire Investigation, 2005.

Bourn, G., T. Callahan, L. Dodge, J. Mulik, D. Naegeli, K. Shouse, L. Smith, and K. Whitney. "Development of a Dedicated Ethanol Ultra Low Emission Vehicle (ULEV): Final Report." Golden CO: National Renewable Energy Laboratory (NREL), September 1998.

Brannigan, V., and E. Buc. "Testing the Fire Hazard of Materials: Problem for Regulators." *Proceedings Fire and Materials 2005.* London: Interscience Communications, 2005, pp. 175–86.

Cole, L. S. *A Survey of Vehicle Fire Causes.* Novato, CA: Lee Books, 1988.

Cole, L. S. *Investigation of Motor Vehicle Fires,* 4th ed. San Anselmo, CA: Lee Books, 2001.

Cole, L. S., and R. C. Herold. *The Investigation of Recreational Boat Fires.* Novato, CA: Lee Books, 1983.

Davis, S., S. Kelly, and V. Somandepall. "Hot Surface Ignition of Performance Fuels." *Fire Technology* 46 (April 2010): 363–74.

Dodge, L., et al. "Development of a Dedicated Ethanol Ultra Low Emission Vehicle (ULEV): Final Report." Golden CO: National Renewable Energy Laboratory (NREL), September 1998.

Gardiner, D., M. Bardon, and G. Pucher. "An Experimental and Modeling Study of the Flammability of Fuel Tank Headspace Vapors from High Ethanol Content Fuels." Golden, CO: National Renewable Energy Laboratory (NREL), U.S. Department of Energy, October 2008.

Holleyhead, R. "Thermal Radiation and Ship Fires." *Proceedings Ninth IAFS.* Oxford, September 1984.

Merrall, S., and S. Chenery. "Vehicle Fires: Explaining the Rise in Vehicle Arson." London: Office of Deputy Prime Minister, February 2005.

National Fire Protection Association. *NFPA 556: Guide on Methods for Evaluating Fire Hazard to Occupants of Passenger Road Vehicles.* Quincy, MA: NFPA (proposed for adoption 2010).

National Fire Protection Association. *NFPA 921: Guide for Fire and Explosion Investigations.* Quincy, MA: NFPA, 2008 and 2011, Chap. 25.

Shaw, A., W. Epling, C. McKenns, and B. Wichman. "Evaluation of the Ignition of Diesel Fuels on Hot Surfaces." *Fire Technology* 46 (April 2010): 407–23.

Thomas, G., and M. M. Witts. *Shipwreck: The Strange Fate of the Morro Castle.* New York: Dorset, 1972.

Chapter 10

ASTM E2345-0: Standard Practice for Investigating Electrical Incidents, Vol. 14.02. West Conshohocken, PA: American Society for Testing and Materials, April 2004.

Babrauskas, V. "How Do Electrical Wiring Faults Lead to Structure Ignitions?" In *Proceedings Fire and Materials Conference* (San Francisco, 2001). London: Interscience Communications, 2001, pp. 39–51.

Béland, B. "Electrical Damages—Cause or Consequence?" *Journal of Forensic Sciences* 29 (July 1984): 747–61.

Béland, B. "Electricity as a Cause of Fires." SFPE Technical Report 84–5, May 1989.

Béland, B., and C. Fortier. "Arc Tracking in Relation to Fire Investigation." *Fire and Arson Investigator* (March 1995).

Brannigan, F. L., R. G. Bright, and N. H. Jason, eds. *Fire Investigation Handbook*. NBS 134, Washington, DC: U.S. Department of Commerce. U.S. Government Printing Office, August 1980.

Brown, K. "Resistance Heating of Electrical Terminal Connections on Receptacles." Berkeley, CA: Fire Cause Analysis, 2000.

Ettling, B. V. "Glowing Connections." *Fire and Arson Investigator* 33 (March 1983): 3–5.

Gregory, G. D., and G. W. Scott. "The Arc-Fault Circuit Interrupter: An Emerging Product." *Institute of Electrical and Electronics Engineers, Transactions of Industry Applications* 34, no. 5 (September–October 1998).

Hagimoto, Y., K. Kinoshita, and T. Hagiwara. "Phenomenon of Glow at the Electrical Contacts of Copper Wires." *National Research Institute of Police Science Reports* 41, no. 3 (August 1988). (Also at www.tcforensic.com.au).

The Investigation of Fires of Electrical Origin. Kent County Constabulary, UK, 1992.

Korinek, C. W. "Classification of Electrical Overheating Modes." *Fire and Arson Investigator* (July 2001): 30–37.

Lentini, J. J. "Examination of Televisions in Arson Fraud Investigations." *Fire and Arson Investigator* 33 (June 1983): 15–22.

Liebler, G. E. "Ground versus Earth: There Is a Difference." *Journal of Forensic Sciences* 29, no. 2 (April 1984): 550–56.

Malcolm, J. D. "Documenting the Electrical System." *Fire and Arson Investigator* 36 (December 1985).

Meese, W. J., and R. W. Beausoleil. *Exploratory Study of Glowing Electrical Connections*. Gaithersburg, MD: National Bureau of Standards, October 1977.

National Fire Protection Association. *National Electrical Code*. Quincy, MA: NFPA, 2008.

National Fire Protection Association. *NFPA 921*. Quincy, MA: NFPA, 2008 and 2011, Chap. 8.

Sanderson, J. L. "Exterior Electrical Exam." *Fire Findings* 10, no. 1 (Winter 2002): 1–3.

Sanderson, J., and J. Finneran. "Floating Neutral Problems." *Fire Findings* 3, no. 1 (Winter 1995): 9–11.

Chapter 11

California Department of Consumer Affairs. *Technical Bulletin 117: Requirements, Test Procedures and Apparatus for Testing the Flame Retardance of Resilient Filling Materials Used in Upholstered Furniture*. North Highlands, CA: California Department of Consumer Affairs, January 1980.

California Department of Consumer Affairs. *Technical Bulletin 133: Flammability Test Procedures for Seating Furniture for Use in Public Occupancies*. North Highlands, CA: California Department of Consumer Affairs, 1988.

Cleary, J., T. Ohlemiller, and K. Villa. "Influence of Ignition Source on the Flame Fire Hazard of Upholstered Furniture." NIST Report NISTIR 4847. Gaithersburg, MD: U.S. Department of Commerce, NIST, June 1992.

Damant, G. H., and S. Nurbakhsh, "Heat Release Testing of Furniture and Bedding Products." Fifth Fire Safety Conference and Exhibition, Orlando, FL, February 1993.

McCormack, J. A., G. H. Damant, and C. J. Hilado. "Flammability Studies of 700 Articles of Upholstered Furniture." *Journal of Fire Sciences* (September–October 1988).

O'Malley, B. "Cigarettes and Sofas." *Mother Jones* (July 1979): 56–63.

Sundstrom, B., ed. *Fire Safety of Upholstered Furniture: The Full Report of the European Commission Research Programme (CBUF)*. (London: Interscience Communications, 1995.

Weiger, P. R. "Adapting Tests for Interior Finishes." *NFPA Journal* (March–April) 2001.

Chapter 12

Baker, W. E., et al. *Explosion Hazards and Evaluation*. Amsterdam: Elsevier, 1983.

Béland, B., "An Unusual Explosion." *Fire and Arson Investigator* (October 2006).

Beveridge, A., ed. *Forensic Investigation of Explosions*. London: Taylor & Francis, 1998.

Conkling, J. A. *The Chemistry of Pyrotechnics*. New York: Marcel Dekker, 1985.

Explosives and Propellants from Commonly Available Materials. New York: Gordon, 1988.

FBI Bomb Data Center. *1997 Bombing Incidents*. Washington, DC: U.S. Department of Justice, 1997.

Harris, R. J. *The Investigation and Control of Gas Explosions*. New York: E & FN Spon, 1983.

Harris, R. J., M. R. Marshall, and D. J. Moppett. "The Response of Glass Windows to Explosion Pressures." Institute of Chemical Engineering Symposium, Series No. 49, 1977.

Hough, E., "London: A Whole City Response." *Crisis/ Response* 1, no. 4 (2008): 9–17.

Leslie, I. R. M., and A. M. Birk. "State of the Art Review of Pressure Liquefied Gas Container Failure Modes and Associated Projectile Hazards." *Hazardous Materials* 28, no. 3 (1991): 326–65.

Martin, R. J., A. Reza, and L. W. Anderson. "What Is an Explosive? A Case Study of an Investigation for the Insurance Industry." Journal of *J. Loss Prevention in the Process Industries* 13 (2000): 491–97.

Meyer, R. *Explosives*. New York: Verlag Chimie-Weinheim, 1977.

Midkiff, C. R., and R. E. Tontarski. "Detection and Characterization of Explosives and Explosive Residue: A Review." 11th International INTERPOL Forensic Science Symposium, Lyon, France, November 1995.

Newhouser, R. *Introduction to Explosives*. Gaithersburg, MD: International Association of Chiefs of Police, 1972.

NIJ Technical Working Group for Bombing Scene Investigation. *A Guide for Explosion and Bombing Scene Investigation*. Washington, DC: National Institute of Justice, June 2000.

Quest's Dynamic Fireball Radiation Model from the Quest Quarterly. Issue 3. Norman, OK: Quest Consultants Inc., Fall 1999.

Smith, K. D., et al. "Detection of Smokeless Powder Residue on Pipe Bombs by Micellar Electrokinetic Capillary Electrophoresis." *Journal of Forensic Sciences* 44, no. 4 (1999): 789–94.

Stuart, D. "Shipboard Explosion and Fire." *Fire and Arson Investigator* (June 1994).

Stull, D. R. *Fundamentals of Fire and Explosion*. New York: American Institute of Chemical Engineers, 1977.

Thompson, R. Q., et al. "Aqueous Recovery from Cotton Swabs of Organic Explosives Residues Followed by Solid Phase Extraction." *Journal of Forensic Sciences* 44, no. 4 (1999): 795–804.

Yallop, H. J. *Explosion Investigation*. Harrrogate, North Yorkshire, UK: Forensic Science Society Press, 1980.

Yinon, J., and S. Zitrin. *Modern Methods and Applications on Analysis of Explosives*. New York: Wiley, 1993.

Zalosh, R. G. "Explosion Protection." *SFPE Handbook of Fire Protection Engineering*, 4th ed. Quincy, MA: NFPA, 2008, Chaps. 3–15.

Zalosh, R., "Explosions." *Fire Protection Handbook*, 20th ed. Quincy, MA: National Fire Protection Association, 2008, Vol. 1, Chap. 8, Sec. 2.

Chapter 13

Christian, D. R. *Forensic Investigation of Clandestine Laboratories*. Boca Raton, FL: CRC Press, 2004.

Ellern, H. *Modern Pyrotechnics*. New York: Chemical Publishing, 1961.

Federal Emergency Management Agency. "Six Firefighter Fatalities in Construction Site Explosion, Kansas City, Missouri (November 29, 1988)", FEMA, USFA National Fire Data Center.

Hildebrand, M. S., and G. G. Noll. *Hazardous Materials for Fire and Explosion Investigators*. Chester, MD: Red Hat, 1998.

National Fire Protection Association. *Fire Protection Guide to Hazardous Materials*. Quincy, MA: NFPA, 2001.

National Fire Protection Association. *NFPA 70: Standard System for the Identification of the Hazards of Materials for Emergency Response*. Quincy, MA: NFPA, 2007.

National Fire Protection Association. *NPFA 497: Recommended Practice for the Classification of Flammable Liquids, Gases, or Vapors and of Hazardous (Classified) Locations for Electrical Installations in Chemical Process Areas*. Quincy, MA: NPA, 2008.

Stephens, H. W. *The Texas City Disaster—1947*. Austin, TX: University of Texas Press, 1997.

U.S. Department of Health and Human Services. *NIOSH Pocket Guide to Chemical Hazards*. Washington, DC: U.S. Government Printing Office, June 1990.

Chapter 14

Bertsch, W., Q. W. Zhang, and G. Holzer. "Using the Tools of Chromatography, Mass Spectrometry and Automated Data Processing in the Detection of Arson." *Journal of High Resolution Chromatography* 13, no. 9 (September 1990): 597–605.

Coulombe, R. "Chemical Markers in Weathered Gasoline." *Journal of Forensic Sciences* 40, no. 5 (September 1995): 867–73.

Dolan, J. A., and E. Stauffer. "Aromatic Content in Medium Range Distillate Products; Part I: An Examination of Various Liquids." *Journal of Forensic Sciences* 49, no. 5 (September 2004): 992–1000.

Frontela, L., J. A. Pozas, and L. A. Picabea. "A Comparison of Extraction and Adsorption Methods for the Recovery of Accelerants from Arson Debris." *Forensic Science International* 75 (1995), 11–23.

Frysinger, G. S. and R. B. Gaines. "Forensic Analysis of Ignitable Liquids in Fire Debris by Comprehensive Two-Dimensional Gas Chromatography." *Journal of Forensic Sciences* 47, no. 3 (May 2002): 471–82.

Fultz, M. L., and J. D. DeHaan, "Gas Chromatography in Arson and Explosives Analysis." In *Gas Chromatography in Forensic Science*, edited by I. Tebbett. Chichester, UK: Ellis Horwood, 1992.

Gresham, J. "Investigators, Introducing the Mini Torch." *Fire and Arson Investigator* 47, no. 1 (September 1996): 19–20.

Harris, A. C., and J. F. Wheeler. "GC-MS of Ignitable Liquids Using Solvent-Desorbed SPME for Automated Analysis." *Journal of Forensic Sciences* 48, no. 1 (January 2003): 41–46.

Kuk, R. J., and M. V. Spagnola. "Extraction of Alternative Fuels from Fire Debris Samples." *Journal of Forensic Sciences* 53, no. 5 (September 2008): 1123–29.

Lennard, C. J., V. T. Rochaix, and P. Margot, P. "A GC-MS Database of Target Compound Chromatograms for the Identification of Arson Accelerants." *Science and Justice* 35, no. 1 (January 1995): 19–30.

Levinson, D. W. "Copper Metallurgy as a Diagnostic Tool for Analysis of the Origin of Building Fires." *Fire Technology* 13, no. 5 (August 1977): 211–22.

Mat Desa, N. S., and N. Nic Daeid. "The Application of Multivariate Pattern Recognition for Petroleum-Based Accelerant Identification and Classification." Paper presented at European Academy of Forensic Sciences, Glasgow, September 2009.

McCord, B. R., K. A. Hargadon, K. E. Hall, and S. G. Burmeister. "Forensic Analysis of Explosives Using Ion Chromatographic Methods." *Analytica Chimica Acta* 288, nos. 1–2 (March 1994): 43–56.

McDonald, K., and M. Shaw. "Identification of Household Chemicals Used in Small Bombs via Analysis of Residual Materials." Paper presented at American Academy of Forensic Sciences, New Orleans, February 2005.

Midkiff, C. R. "Arson and Explosive Investigation." In *Forensic Science Handbook*, 2nd ed., edited by R.

Saferstein, 513–15. Upper Saddle River, NJ: Pearson Education, 2002.

Newman, R., M. Gilbert, and K. Lothridge. *GC/MS Guide to Ignitable Liquids*. Boca Raton, FL: CRC Press, 1996.

Phelps, J. L., C. E. Chasteen, and M. M. Render. "Extraction and Analysis of Low Molecular Weight Alcohols and Acetone from Fire Debris Using Passive Headspace Concentration," *Journal of Forensic Sciences* 39, no. 1 (January 1994): 194–206.

Sheff, L. M., and J. A. Siegel. "Fluorescence of Petroleum Products V: Three-Dimensional Fluorescence Spectroscopy and Capillary Gas Chromatography of Neat and Evaporated Gasoline Samples." *Journal of Forensic Sciences* 39, no. 5 (September 1994): 1201–14.

Small, D. "Hot Evidence." *Investigative Practice Journal* (July 24, 2008): 27–29.

Stauffer, E., and D. Byron, D. "Alternative Fuels in Fire Debris Analysis: Biodiesel Basics." *Journal of Forensic Sciences* 52, no. 2 (March 2007): 371–79.

Chapter 15

Artz, C. P., J. A. Moncrieff, and B. A. Pruitt Jr. *Burns—A Team Approach*. Philadelphia: W. B. Saunders, 1979.

Bradtmiller, B., and J. E. Buikstra. "Effects of Burning on Human Bone Microstructure: A Preliminary Study." *Journal of Forensic Sciences* 29, no. 2 (1984): 535–40.

Brogan, R. "The Speed Street Fire—Was It Murder by or Prior to—the Fire?" *Fire and Arson Investigator* (July 2000): 17–21.

Gamlin, L., and B. Price. "Bonfires and Brimstone." *New Scientist,* November 5, 1988.

Hill, I. P. "Immediate Causes of Death in Fires." *Medicine, Science and the Law* 29, no. 4 (1989): 287–92.

Landrock, A. H. *Handbook of Plastics Flammability and Combustion Technology*. Park Ridge, NJ: Noyes, 1983.

Langlois, N. E. I., et al. "Gas Analysis Following the Diversion of Catalytically Converted Exhaust Fumes into the Interior of a Vehicle." *Forensic Science International* 74 (1995): 187–91.

Leitch, D. A. *Guide to Fatal Fire Investigations*. Leicester, UK: Institution of Fire Engineers, 1993.

Nickell, J., and J. F. Fisher. "Spontaneous Human Combustion: The 'Cinder Woman Mystery—Part II.'" *Fire and Arson Investigator* 34 (June 1984).

Penney, D. G., ed., *Carbon Monoxide Toxicity*. Boca Raton, FL: CRC Press, 2000.

Pitts, W. M., E. L. Johnson, and N. P Bryner, "Carbon Monoxide Formation in Fires by High Temperature Anaerobic Wood Pyrolysis." Twenty-Fifth Investigation Symposium on Combustion, The Combustion Institute, 1994, 1455–62.

Purser, D. A., "Assessment of Hazards to Occupants from Smoke, Toxic Gases, and Heat." Chap. 2–6 in *SFPE Handbook of Fire Protection Engineering*, 4th ed., Sec. 2. Quincy, MA: National Fire Protection Association, 2008.

Rho, Yong-Myun, M.D. "The Investigation of Fatal Fires." *FBI Law Enforcement Bulletin* 55, no. 8 (August 1986).

Rogers, D. "The Equivocal Fire Death." *Fire and Arson Investigator* (September 1997): 14–16.

Rossiter, W. D. "The Investigation of Fatal Fires." *Fire and Arson Investigator* (January–March 1968).

Schmidt, C. W., and S. A. Symes. *The Analysis of Burned Human Remains*. Amsterdam: Academic Press, 2008.

Sopher, I. M. "The Role of the Forensic Pathologist in Arson and Related Investigations." *FBI Law Enforcement Bulletin* (September 1972).

Stuerwald, J. E. "Arson Homicide Investigation." *Fire and Arson Investigator* (October–December 1970).

Chapter 16

Comeau, E., and R. F. Tunkel. "A Field Guide to Incident Analysis." *National Fire and Rescue Magazine,* May 2000.

Doley, R. "Making Sense of Arson through Classification." *Psychiatry, Psychology and Law* 10, no. 2 (2003): 346–52.

Geller, J. L., "Firesetting: A Burning Issue." Chapter 9 in *Criminal Profiling: Principles and Practice*. Totowa, NJ: Humana, 2006.

Huff, T. G. *Fire-Setting Fire Fighters: Arsonists in the Fire Department—Identification and Prevention*. Quantico, VA: National Center for the Analysis of Violent Crime, 1994. Published in *On Scene*, IAFC, August 1994.

Icove, D. J. "Principles of Incendiary Fire Analysis." Ph.D. dissertation, College of Engineering, University of Tennessee, Knoxville, 1979.

Icove, D. J., and M. W. Dalton. "A Comprehensive Prosecution Report Format for Arson Cases." Paper presented at ISFI 2008, International Symposium on Fire Investigation Science and Technology, Cincinnati, Ohio, May, 2008.

Icove, D. J., V. B. Wherry, and J. D. Schroeder. *Combating Arson-for-Profit: Advanced Techniques for Investigators*, 2nd ed. Columbus, OH: Battelle, 1988.

Kidd, S. "Arson: The Problem That Won't Go Away." *Fire Engineers Journal* 58, no. 193 (March 1998): 12–17.

Little, C. M. "Recent Decisions Upholding Criminal Convictions for Use of Gasoline or Fire as a Deadly Weapon." *Fire and Arson Investigator* 50, no. 2 (January 2000): 17–19.

Office of Deputy Prime Minister. *The Burning Issue: Research and Strategies for Reducing Arson*. London: ODPM, August 2002.

Rider, A. O. *The Firesetter: A Psychological Profile*, 2nd ed. Edited by T. G. Huff. Washington, DC: U.S. Department of Justice, 1993.

Sapp, A. D., and T. G. Huff. *Arson Homicide: Findings from a National Study*. Washington, DC: National Center for the Analysis of Violent Crime (NCAVC), U.S. Department of Justice, 1994.

Wood, B. "Arson Profiling—A Geographical, Demographic and Motivational Perspective." *Fire Engineers Journal* 60 (September 2000): 208.

Chapter 17

Arson Case Briefs. Washington, DC; U.S. Department of the Treasury, Bureau of Alcohol, Tobacco, and Firearms, November 1985.

Arson Investigation and Prosecution, 2nd ed. Sacramento, CA: California District Attorney's Association, 1995.

Arson Prosecution—Issues and Strategies. Washington, DC: FEMA, U.S. Fire Administration (FA-78), May 1988.

Avato, S. J. "Scientific Methodology." *Fire and Arson Investigator* (July 2000): 35–36.

Bukowski, R. W. "Modeling a Backdraft: The Fire at 62 Watts Street." *NFPA Journal* 89, no. 6 (November–December 1995): 85–89.

Bukowski, R. W., and R. C. Spetzler. "Analysis of the Happyland Social Club Fire with Hazard I." *Fire and Arson Investigator* 42, no. 3 (March 1992): 36–47.

Decker, J. F., and B. L. Ottley. *Arson Law and Prosecution.* Durham, NC: Carolina Academic Press, 2009.

DeHaan, J. D. "Advanced Tools for Use in Forensic Fire Scene Investigation." In *Proceedings Interflam 2004.* London: Interscience Communications, 2004, pp. 1079–1100.

Donahue, M. L. *Safety and Health Guidelines for Fire and Explosion Investigators.* Stillwater, OK: OSU, Fire Protection Publications, 2002.

Drysdale, D. D. "Learning from Experience." *Fire Engineers Journal* (November 2000): 34–39.

Ellington, J. M. "Some Perspectives in Fire Modeling by a Fire Investigator." *Fire and Arson Investigator,* June 1995.

Factory Mutual. *A Pocket Guide to Arson and Fire Investigation.* London, UK, Factory Mutual International and Fire Service College, 1994.

Factory Mutual Data Sheets, "Factory Mutual Global Property Loss Prevention Data Sheets," To obtain see: http://www.fmglobal.com/fmglobalregistration/, Johnston, Rhode Island, last retrieved, March 4, 2011.

Ferrall, R. T. "Arson Information: Who, What, Where?" *FBI Law Enforcement Bulletin* 50 (May 1981): 16–22.

Fire Investigation Handbook. Palos Hills, IL: National Insurance Crime Bureau, 1995.

Giannelli, P. C., and E. J. Imwinkelreid. *Scientific Evidence,* 4th ed. Newark, NJ: Lexis /Matthew Bender, 2008.

Gorbett, G. E., "Computer Fire Models for Fire Investigation and Reconstructon." In *Proceedings International Symposium on Fire Investigation Science and Technology,* Cincinnati, OH, May 2008, 23–34.

Hammett, T. M. "Cracking Down on Arson: Issues for Investigators and Prosecutors." Rockville, MD: National Criminal Justice Reference Service, 1987.

Hewitt, T-D. "On the Hot Seat." *NFPA Journal* (January–February 2000): 43–46.

Icove, D. J., and J. D. DeHaan. *Forensic Fire Scene Reconstruction,* 2nd ed. Upper Saddle River, NJ: Pearson/Brady, 2009.

Iqbal, N., and M. H. Salley. *Fire Dynamics Tools (FDTs).* NUREG 1805. Washington, DC: U.S. Government Printing Office, 2004.

Keierleber, J. A., and T. L. Bohan. "Ten Years After Daubert: The Status of the States." *Journal of Forensic Sciences* 50, no. 5 (September 2005): 1154–63.

Lentini, J. J. "A Calculated Arson." *Fire and Arson Investigator* (April 1999): 20–25.

Little, C. M. "Recent Decisions Upholding Criminal Convictions for Use of Gasoline or Fire as a Deadly Weapon, Dangerous Weapon or Dangerous Instrument." *Fire and Arson Investigator* (January 2000): 17–19.

Lynch, P. A. "Illinois Court Reverses Sanctions for Alleged Spoliation." *California Fire/ Arson Investigator,* July 2005.

McDowell, J. L., and G. C. Burton. "Computer Modeling and the Fire Investigator." *Fire and Arson Investigator* (March 1998): 10–12.

McGrattan, K., et al. *Fire Dynamics Simulator: Technical Reference Guide.* NISTIR 6467. Gaithersburg, MD: NIST, January 2000.

Munday, J. W. *Safety at Scenes of Fire and Related Incidents.* Moreton-in-Marsh, UK: Fire Protection Association, 1996.

Nelson, H. E., and S. Deal. "Comparing Compartment Fires with Compartment Fire Models." In *Fire Safety Science: Proceedings of the Third International Symposium,* ed. G. Cox and B. Langford. London: Elsevier Applied Science, 1991, 719–26.

Peacock, R. D., et al. *CFAST—The Consolidated Model of Fire Growth and Smoke Transport.* NIST TN 1299. Gaithersburg, MD: National Institute of Standards and Technology, February 1993.

Quintiere, J. G. *Principles of Fire Behavior.* Albany, NY: Delmar, 1998.

Urbas, J. "Use of Modern Test Methods in Fire Engineering and Litigation." *Fire and Arson Investigator* (December 1997): 12–15.

GLOSSARY

A

absolute temperature   Temperature above absolute zero (in K or °R).

accelerant   A fuel (usually a flammable liquid) that is used to initiate or increase the intensity or speed of spread of fire.

accident   Unplanned or unintentional event; an event occurring without design or intent.

accidental fire   A fire occurring without design or intent.

adiabatic   Conditions of equilibrium of temperature and pressure.

adsorption   Trapping of gaseous materials on the surface of a solid substrate.

aliphatic   Hydrocarbons with straight-chain structures of the carbon atoms; normal hydrocarbon.

alligatoring   Rectangular patterns of char formed on burned wood.

ambient   Surrounding conditions.

ampacity   Current-carrying capacity of electric conductors (expressed in amperes).

ampere   Quantity of electrical charge passing a point in an electrical circuit per unit of time.

annealing   Loss of temper in metal caused by heating.

appliance   Equipment, usually nonindustrial, that is installed or connected as a unit to perform one or more functions such as clothes washing, air conditioning, food mixing, cooking, etc. Normally built in standardized sizes and types.

arc   Flow of current across an air gap (or another non-conductor) between two conductors.

area of transition   Mixture of directional fire indicators in a wildlands fire.

area of origin   The general locale in which a fire was ignited.

aromatic   Hydrocarbon compound whose structure is based on a benzene ring.

arson   The intentional setting of a fire with intent to damage or defraud.

arson set   Device, assembly, or contrivance used to ignite an incendiary fire.

atom   Smallest unit of an element that still retains its properties.

autoignition   Ignition by surrounding temperature in the absence of an external source of ignition; nonpiloted ignition.

autoignition temperature   The temperature at which a material will ignite in the absence of any external pilot source of heat; spontaneous ignition temperature.

B

backdraft   A deflagrative explosion of gases and smoke from an established fire that has depleted the oxygen content of a room or structure, most often initiated by introducing oxygen through ventilation or structural failure.

backing   Slow extension of a fire downslope or into wind in the opposite direction of its main spread.

BLEVE   Boiling-liquid, expanding-vapor explosion. A mechanical explosion caused by the heating of a liquid in a sealed vessel to a temperature far above its boiling point.

boiling point   The (pressure-dependent) temperature at which a liquid changes to its gas phase.

branch circuit   The circuit conductors between the outlet(s) and the final overcurrent device protecting that circuit.

brisance   The amount of shattering effect that can be produced by a high explosive.

Btu   British thermal unit. A standardized measure of heat, it is the heat energy required to raise the temperature of 1 pound of water 1 degree Fahrenheit.

buoyancy   Tendency or ability to rise or float in air or liquid as a result of a difference in density.

burn patterns   Patterns created when applied heat fluxes are above the critical thresholds to scorch, melt, char or ignite a surface.

C

calcination   Loss of water of crystallization caused by heating.

cellulosic   Plant-based material based on a natural polymer of sugars.

chain of evidence (chain of custody)   Written documentation of possession of items of physical evidence from their recovery to their submission in court.

char   Carbonaceous remains of burned organic materials.

char depth   The measurement of pyrolytic or combustion damage to a wood surface compared with its original surface height.

chromatography   Chemical procedure that allows the separation of compounds based on differences in their chemical affinities for two materials in different physical states, e.g., gas/liquid, liquid/solid.

circuit breaker   A device designed to open a circuit automatically at a predetermined overcurrent without injury to itself when properly applied within its ratings.

clean burn   An area of wall or ceiling where the charred organic residues have been burned away by direct flame contact.

combustible A material that will ignite and burn when sufficient heat is applied and when an appropriate oxidizer is present.

combustible liquid A liquid having a flash point at or above 38°C (100°F).

combustion Oxidation that generates detectable heat and light.

concealed wiring Wiring rendered inaccessible by the structure or finish of the building. Wiring in covered raceways is considered concealed.

conduction Process of transferring heat through a material or between materials by direct physical contact.

conductor Any material capable of permitting the flow of electrons: (1) bare—a conductor having no covering or electrical insulation whatsoever; (2) covered—a conductor encased within a material whose composition or thickness is not considered insulative; (3) insulated—a conductor encased within a material recognized by the electrical code as insulation.

convection Process of transferring heat by movement of a fluid. In convective flow, the warm fluid becomes less dense than the surrounding fluid and rises, inducing a circulation.

corpus delicti Literally, the body of the crime. The fundamental facts necessary to prove the commission of a crime.

crazing Stress cracks in glass as the result of rapid cooling.

crowning Rapid extension of fire through the porous array of leaves, needles, and fine fuels above 2 m.

D

deep-seated Fire that has gained headway and built up sufficient heat in a structure to require great cooling for extinguishment; fire that has burrowed deep into combustible fuels (as opposed to a surface fire); deep charring of structural members.

deflagration A very rapid oxidation with the evolution of heat and light and the generation of a low-energy pressure wave that can accomplish damage. The reaction proceeds between fuel elements at subsonic speeds.

detonation An extremely rapid reaction that generates very high temperatures and an intense pressure/shock wave that produces violently disruptive effects. It propagates through the material at supersonic speeds.

device Any chemical or mechanical contrivance or means used to start a fire or explosion.

diatomic Molecules consisting of two atoms of an element.

direct attack Application of hose streams or other extinguishing agents directly on a fire, rather than attempting extinguishment by generating steam within a structure.

dropdown The collapse of burning material in a room that induces separate, low-level ignition; falldown.

duty Conditions of use in electrical service: (1) continuous duty—operation at substantially constant load for an indefinitely long time; (2) intermittent duty—operation for alternate intervals of (a) load/no load; (b) load/rest; or (c) load/no load/rest; (3) periodic duty—intermittent operation in which the load conditions are regularly recurrent; (4) short-time duty—operation at substantially constant load for a short and definitely specified time; (5) varying duty—operation at loads, and for intervals of time, both of which are subject to wide variation.

E

elastomers Fibers or materials having elastic properties.

endothermic Absorbing heat during a chemical reaction.

entrainment The mixing of two or more fluids as a result of laminar flow or movement.

eutectic An alloy of two materials having special physical or chemical properties, typically having the lowest melting point of any combination of the two.

exothermic Generating or giving off heat during a chemical reaction.

explosion The sudden conversion of potential energy (chemical or mechanical) into kinetic energy with the production of heat, gases, and mechanical pressure.

explosion-proof apparatus Apparatus enclosed in a case that is capable of withstanding an explosion of a specified gas or vapor within it or designed to prevent the ignition of a specific gas or vapor surrounding it by arcs, flashes, or explosion of fuels within, and that operates at such an external temperature that a surrounding flammable atmosphere will not be ignited by it.

explosive Any material that can undergo a sudden conversion of physical form to a gas with a release of energy.

explosive limits (flammability limits) The lower and upper concentrations of an air-gas or air-vapor mixture in which combustion or deflagration will be supported.

exposure Property that may be endangered by radiant heat from a fire in another structure or an outside fire. Generally, property within 40 feet is considered an exposure risk, but larger fires can endanger property much farther away.

F

fingerprints Unique friction ridge patterns on the palmar surfaces of the hands and fingers (or their impressions).

fire Rapid oxidation with the evolution of heat and light; uncontrolled combustion.

fire behavior The manner in which a fuel ignites, flame develops, and fire spreads. Unusual fire behavior may reveal the presence of added fuel or accelerants.

fire load The total amount of fuel that might be involved in a fire, as measured by the amount of heat that would evolve from its combustion (expressed in units of heat).

fire patterns Indicators of damage (or relative lack of damage) created by a combination of smoke, flames, and heat effects.

fire-resistive A structure or assembly of materials built to provide a predetermined degree of fire resistance as defined in building or fire prevention codes (calling for 1-, 2-, or 4-hour fire resistance).

firestorm Overwhelming progression of fire through structures or wildlands caused by a combination of convective and radiative processes.

fire wall A solid wall of masonry or other noncombustible material capable of preventing passage of fire for a prescribed time (usually extends through the roof with parapets).

flame A luminous cloud of burning gases.

flameover The flaming ignition of the hot gas layer in a developing compartment fire.

flame point Temperature at which a flame is sustained by evaporation or pyrolysis of a fuel; fire point.

flame resistant Material or surface that does not maintain or propagate a flame once an outside source of flame has been removed.

flame spread The rate at which flames extend across the surface of a material (usually under specific conditions).

flaming combustion Rapid oxidation of gases and vapors that generates detectable heat and light.

flammable (same as **inflammable**) A combustible material that ignites easily, burns intensely, or has a rapid rate of flame spread.

flammable liquid A liquid having a flash point below 38°C (100°F).

flanking Lateral spread of fire in a direction at right angles to the main direction of fire growth.

flash point Temperature at which an ignitable vapor is first produced by a liquid fuel.

flashback The ignition of a gas or vapor from an ignition source back to a fuel source (often seen with flammable liquids).

flashover The final stage of the process of fire growth; when all exposed surfaces of combustible fuels within a compartment are ignited, the room is said to have undergone flashover.

fragmentation The fast-moving solid pieces created by an explosion. Primary fragmentation is that of the explosive container itself; secondary fragmentation is that of the target shattered by an explosion.

fuel load All combustibles in a fire area, whether part of the structure, finish, or furnishings.

fully involved The entire area of a fire building so involved with heat, smoke, and flame that entry is not possible until some measure of control has been obtained with hose stream attacks.

fraud Deception deliberately practiced in order to secure unfair or unlawful gain.

furniture calorimeter An instrumental system for measuring the amount of heat produced by the combustion of a moderate-sized fuel package, and the rate at which the heat is produced.

G

ghost marks Stained outlines of floor tiles produced by the dissolution and combustion of tile adhesive.

glowing combustion The rapid oxidation of a solid fuel directly with atmospheric oxygen creating light and heat in the absence of flames.

ground A conducting connection, whether intentional or accidental, between an electrical circuit or equipment, and the earth, or to some conducting body that serves in place of the earth.

ground fault An interruption of the normal ground return path of electricity in a structure that leads to unintended current flows.

H

heat horizon The demarcation (usually horizontal) of fire damage revealed by the charring, burning, or discoloration of paint or wall coverings.

heat of combustion The quantity of heat released from a fuel during combustion measured in kilojoules per gram (kJ/g) or Btu per pound (Btu/lb).

heat release rate The rate at which heat is generated by a source, usually measured in watts, joules per second, or Btu per second.

heat transfer Spread of thermal energy by convection, conduction, or radiation.

high explosive Any material designed to function by, and capable of, detonation.

hot-plate ignition (hot-surface ignition) Temperature of a metal test plate at which liquid fuel ignites on contact.

hot set Direct ignition of available fuels with an open flame (match or lighter).

hydrocarbon Chemical compound containing only hydrogen and carbon.

I

ignitable liquid Classification for liquid fuels including both flammable and combustible classes.

ignition energy The quantity of energy that must be transferred into a fuel/oxidizer combination to trigger a self-sustaining combustion.

ignition temperature (same as **autoignition**) The minimum temperature to which a substance must be heated in air to ignite independently of the heating source (i.e., in the absence of any other ignition source; nonpiloted ignition).

incendiary fire A deliberately set fire.

incipient Beginning stages of a fire.

indicators Observable (and usually measurable) changes in appearance caused by heat, flame, or smoke.

indirect application A method of firefighting by applying water fog into heated atmospheres to obtain heat absorption and smothering action by generating steam.

inorganic   Containing elements other than carbon, oxygen, nitrogen, and hydrogen.

J

joule   The SI unit of measurement of heat energy (1 J = 0.238 cal = 0.00095 Btu).

L

ladder fuels   Intermediate-height fuels between the ground litter and the crowns of trees overhead.

low explosive   Any material designed to function by deflagration.

M

mass spectrometry   Analysis of organic molecules by fragmenting them and separating them by size.

Molotov cocktail   A breakable container filled with flammable liquid, usually thrown. It may be ignited by a flaming wick or by chemical means.

motive   Inner drive or impulse that causes a person to do something or act in a certain way.

N

naphthenics   Class of hydrocarbons based on cycloalkanes.

natural fibers   Fibers of animal or vegetable origin without chemical manipulation.

nonflammable   Material that will not burn under most conditions.

normal hydrocarbons (*n*-alkanes)   Hydrocarbons having straight-chain structures with no side branching; aliphatics.

O

odorant   Organic chemical (often a mercaptan) added to natural gas or LP gas prior to its distribution to make it detectable by a person with normal sense of smell at a concentration less than 20% of its LEL.

olefinic   Hydrocarbons containing double carbon-carbon bonds (denoted C═C); nonsaturated hydrocarbons; alkenes.

open neutral   In an American single-phase, 120/240 V residential system, the neutral return leg is not connected to the service ground.

open wiring   Uninsulated conductors or insulated conductors without grounded metallic sheaths or shields, installed above ground but not inside enclosures or appliances; knob-and-tube wiring.

organic   Compounds based on carbon.

outlet   A point on the wiring system at which current is taken to supply equipment or appliances by means of receptacles or direct connections.

overhaul   The firefighting operation of eliminating hidden flames, glowing embers, or sparks that may rekindle the fire, usually accompanied by the removal of structural contents.

oxidation   Combination of an element with oxygen; chemical conversion involving a loss of electrons.

P

paraffinic   Hydrocarbon compounds involving no double or triple C—C bonds; alkanes; saturated aliphatic hydrocarbons.

piloted ignition   Ignition aided by the presence of a separate external ignition source such as a flame or electric arc.

piloted ignition temperature   The minimum temperature at which a fuel will sustain a flame when exposed to a pilot source.

plume   The convective column of hot gases generated by a flame.

point of origin   The specific location at which a fire was ignited.

pyrolysis   The chemical decomposition of substances through the action of heat, in the absence of oxygen.

pyromania   Uncontrollable psychological impulse to start fires.

pyrophoric   Capable of igniting on exposure to atmospheric oxygen at normal temperatures.

R

raceway   Any channel for holding wires, cables, or busbars that is designed expressly for, and is used solely for, this purpose.

radiation   Transfer of heat by electromagnetic waves.

receptacle   A contact device installed as the outlet for the connection of an appliance by means of a plug.

rekindle   Reignition of a fire due to latent heat, sparks, or embers.

resistance   Opposition to the passage of an electric current.

rollout   Exiting of burner flames in an appliance due to massive overfueling.

S

salvage   Procedures to reduce incidental losses from smoke, water, and weather following fires, generally by removal or covering of contents.

saturated   Hydrocarbons that have no double or triple C—C bonds.

seat of explosion   The area of most intense physical damage caused by high explosive pressures and shock waves in the vicinity of a solid or liquid explosive.

self-heating   An exothermic chemical or biological process that can generate enough heat to become an ignition source; spontaneous combustion.

self-ignition   *See* autoignition.

service The conductors and equipment for delivering electricity from the supply system to the equipment of the premises served.

service conductors The supply conductors that extend from the street main or transformer to the equipment of the premises served.

service drop The overhead service conductors from the pole, transformer, or other aerial support to the service entrance equipment on the structure, including any splices.

short circuit Direct contact between a current-carrying conductor and another conductor producing unwanted current flow.

smoke horizon Surface deposits that reveal the height at which smoke and soot stained the walls and windows of a room without thermal damage.

smoldering combustion The direct combination of a solid fuel with atmospheric oxygen to generate heat in the absence of gaseous flames; see glowing combustion.

soot The carbon-based solid residue created by incomplete combustion of carbon-based fuels.

spalling Crumbling or fracturing of a concrete or brick surface as a result of exposure to thermal or mechanical stress.

spark Superheated, incandescent particle.

spoliation The destruction or material alteration of, or failure to save, evidence that could have been used by another in future litigation.

spontaneous ignition Condition in which a chemical or biological process generates sufficient heat to ignite the reacting materials. *See* **self-heating**.

spot fires Fires started by airborne embers some distance away from the main body of the fire.

stoichiometric mixture A reaction mixture in which the reactants and products are chemically balanced.

stoichiometry Balance of chemical reactants and products.

suspicious Fire cause has not been determined, but there are indications that the fire was deliberately set and all accidental fire causes have been eliminated.

synthetic Material that is human-made, usually referring to organic polymers.

synthetic fibers Manmade fibers of chemical origin.

T

thermal protector A device against overheating that is responsive to temperature or current and will protect the equipment against overheating due to overload or failure to start.

thermoplastic Polymer that can undergo reversible melting without appreciable chemical change.

thermosetting (resin) Polymer that decomposes or degrades as it is heated rather than melts.

torch A professional fire setter.

trailers Long trails of fast-burning materials used to spread a fire throughout a structure.

V

vapor The gaseous phase of a liquid or solid that can be returned to that phase by the application of pressure.

vapor density The ratio of the weight of a given volume of gas or vapor to that of an equal volume of air.

vented The fire has extended outside the structure or compartment by destroying the windows or burning an opening in the roof or walls.

ventilation A technique for opening a burning building to allow the escape of heated gases and smoke to prevent explosive concentrations (smoke explosions or backdrafts) and to allow the advancement of hose lines into the structure; air supply to a room fire.

volatile A liquid having a low boiling point; one that is readily evaporated into the vapor state.

volt The basic unit of electromotive force.

voltage, nominal A value assigned to a circuit or system for the purpose of conveniently designating its voltage class (120/240, 480Y/277, 600, etc.).

W

watt Unit of heat release, 1 watt = 1 joule/second; unit of power or work (in electrical circuits, equivalent to voltage multiplied by amperes).

INDEX

Photography, 238–43
  aircraft, 335
  digital images, 239–40
  infrared, 335
  in motor vehicle fire investigation, 376–77
  in wildland fire investigation, 335
Photoionization detection (PID), 574
Piloted ignition, defined, 185
Piloted ignition temperature, defined, 185
Pilot lights:
  in arson, 681
  in vapor explosions, 500–501
Pipes, 206
Plantings, 206
Plasterboard, 77–78
Plastic building materials, 149–50
Plastic fuel containers, 90–93
Plastics, 143–56
  autoignition temperatures for, 144–46
  behavior of, 145–49
  characteristics of, 143–45
  combustion of, 143
  fire investigation considerations, 149–56
  in furnishings, 150–56
  melting point, 295
  in motor vehicle fire, 372
Plugged vent, heater, 184
Plume, defined, 35
Plywood, 79, 138
Polyester, 470
Polyether sulfone, 644
Polyethylene fabrics, 470
Polymers (see Plastics)
Polypropylene:
  carpet, 154–55
  fabrics, 470
Polystyrene, 644
  fabrics, 470
  foams, 146–47
Polyurethane foams, 146–47
Polyvinyl chloride, 145
Portable electric appliances, 186–87
Portable gas heater, 185–86
Portable heaters, 187–88
Portable stoves, 187–88
Positive pressure ventilation (PPV), 62, 251
Post-flashover fire, 61
Post-flashover stage of fire, 58, 61–62
Postmortem examination, 618–19
Potassium, 549
Potassium chlorate, 586

Potassium permanganate, 585
Power lines, in wildland fires, 346–47
Pre-fire conditions, reconstruction of, 249
Premixed flames, 37
Pride v. Bic Corp., 723
Primary arcing, 453
Profiling, geographic, 688–89
Profit motive, in arson, 662–63
Propane, 100, 541
  in LP gas, 113–14
  in motorhome fire, 388
Propellants, 507–8
Propene, 25
Protected areas, 304–10
Pseudoephedrine, 551
Pugilistic pose, 625
PVC, 145
Pyrolysis:
  defined, 23, 28, 125
  of liquid fuels, 112
  of solid fuels, 125–26
  of wood, 28
Pyromania, 670–71
Pyrophoric, defined, 132
Pyrophoric carbon, 132
Pyrotechnic fuse, 512, 684

Q

Q_{10} value, 42

R

Radiant heat, 47–50
  from fireplace, 190
  in floor burns, 276–79
  in floor covering ignition, 270–71
  as ignition source, 177–78
Radiation, 47–50
Rags, in motor vehicle fire, 370
Railroads, 199, 349
Range, electric, 442
Rate of reaction, 42–43
Rats, 223–24
Rayon, 471
Reaction, rate of, 42–43
Reactive materials, 549
Reassembly, in explosion investigation, 522
Receptacles, 411, 424–25, 451
Recognition, in explosion investigation, 522
Reconstruction, in explosion investigation, 522
Recovery, in explosion investigation, 522
Recreational vehicle fire, 387–90

Red phosphorus, 550
Reducing conditions, 35
Rekindling, defined, 345
Remote Automated Weather Stations (RAWS), 355–56
Report writing, 718–21
Residence times, 175
Residential fraud, 663–64
Resins:
  as fuel, 137
  thermosetting, 125
Resistance, defined, 169
Retaliation, in arson, 667–68
Revenge, in arson, 667–68
Rich mixture, gas, 495
Road flares, 352–53, 548, 585
Rocks, in wildland fires, 341
Rodents, 223–24
Rolling fire, 90
Rollout, 184
Romero v. Superior Court, 712–13
Roof fires, 314–15
Room fire sequence, 54–63
Rubber:
  foam, 146–148
  vulcanized, 644
Rubbing alcohol, 544–45
Rugs:
  flammability regulations, 473
  flammability testing, 479
Rule of nines, 645

S

Safety flares, 352–53, 548, 585
Safety fuse, 512
Salts, oxidizing, 548–49
Salvage, 236, 238
Sandwich panel construction, 72
Saturated compounds, 24
Saturated vapor pressure, 87
Scanning electron microscopy/energy dispersive X-ray spectrometry (SEM/EDS), 459, 581, 591
Scene:
  in death investigation, 616–17
  in explosion investigation, 520–23
  in motor vehicle fire investigation, 377–78
  safety and health at, 696–98
  search patterns, 255–57
  security, 236, 238, 521
  in wildland fire investigation, 335–38
Scientific method, 11–15, 722–31
  critical analysis, 707–8
  defined, 11
  steps in, 12–14